MS&A

Volume 19

More information about this series at http://www.springer.com/series/8377

Daniele Antonio Di Pietro • Jérôme Droniou

The Hybrid High-Order Method for Polytopal Meshes

Design, Analysis, and Applications

Springer

Daniele Antonio Di Pietro
Institut Montpelliérain Alexander Grothendieck
Université de Montpellier
Montpellier, France

Jérôme Droniou
School of Mathematics
Monash University
Clayton
VIC, Australia

ISSN 2037-5255 ISSN 2037-5263 (electronic)
MS&A
ISBN 978-3-030-37202-6 ISBN 978-3-030-37203-3 (eBook)
https://doi.org/10.1007/978-3-030-37203-3

Cover illustration by the Authors: Streamlines for the three-dimensional lid-driven cavity flow on a Voronoi mesh

This Springer imprint is published by the registered company Springer Nature Switzerland AG.
The registered company address is: Gewerbestrasse 11, 6330 Cham, Switzerland

Preface

Originally introduced in [146, 158], Hybrid High-Order (HHO) methods provide a framework for the discretisation of models based on Partial Differential Equations (PDEs) with features that set them apart from traditional ones. The construction hinges on discrete unknowns that are broken polynomials on the mesh and on its skeleton, from which two key ingredients are devised:

1. *Local reconstructions* obtained by solving small, embarrassingly parallel problems inside each element, and typically conceived so that their composition with the natural interpolator of sufficiently smooth functions yields a physics- and problem-dependent projector on local polynomial spaces;
2. *Stabilisation terms* that penalise residuals designed at the element level so as to ensure stability while preserving the approximation properties of the reconstruction.

These ingredients are combined to formulate local contributions, which are then assembled as in standard Finite Element Methods. From this construction, several appealing features ensue: the support of polytopal meshes and arbitrary approximation orders in any space dimension, an enhanced compliance with the physics, and a reduced computational cost thanks to the compact stencil along with the possibility of locally eliminating a large portion of the unknowns. This monograph provides an introduction to the design and the mathematical aspects of HHO methods for diffusive problems, along with a panel of applications to advanced models in computational mechanics.

The support of polytopal meshes is perhaps the most defining feature of HHO methods. Mesh generation and adaptation is often the bottleneck of computer-assisted modelling: despite the enormous progress in this field, traditional unstructured meshes suffer from intrinsic drawbacks, and disposing of discretisation methods that deliver high-order approximations on polytopal meshes constitutes a veritable technological jump. Let us consider a few examples. Capturing geometric microstructures in the domain traditionally requires the use of small elements, which can significantly add to the computational burden. With polyhedral meshes, on the other hand, one can incorporate such geometric features into larger agglomerated elements [17, 38], thus achieving a significant cost reduction without compromising the accuracy. In the context of conforming Finite Element Methods, local mesh adaptation requires special strategies to either prevent or treat hanging nodes. When conforming mesh refinement is performed, one typically faces the choice of accepting a degradation of the mesh quality or renouncing the benefit of nested meshes. Polyhedral methods, on the other hand, can usually treat hanging nodes seamlessly and even support innovative strategies such as adaptive coarsening [38]; see, e.g., the a posteriori-based adaptive HHO algorithm devised in [154] for electrostatic models. The seamless support of meshes that are nonconforming in the traditional sense is also crucial for models that feature inner interfaces. In geosciences applications, e.g., accounting for the presence of fractures or faults in the subsoil is paramount to accurately reproduce the flow patterns. In petroleum basin modelling, fractures are typically incorporated into the numerical models by the mutual sliding of two portions of a corner-point grid along the fracture plane, resulting in highly nonconforming meshes [192]. Polytopal methods offer a true advantage in this case, as no special strategy is required to handle this situation; see, e.g., [106, 107], where HHO methods for Darcy flow and passive transport in fractured porous media are devised and analysed, or [55], where fracture networks are simulated using Virtual Element Methods.

Another key feature of HHO methods is compliance with the physics, meaning that they can incorporate fundamental properties of the model into the design. Let us examine a few examples. In the context of diffusive conservation laws in divergence form, e.g., HHO schemes can typically be interpreted as enforcing polynomial moments of local balances with conservative numerical approximations of the fluxes; see, e.g., [129] and [155, Section 4.3.2.5] concerning scalar linear diffusion problems, [68, 147] concerning linear and nonlinear elasticity, [58, 73] on linear and nonlinear poroelasticity, and [70] for the incompressible Navier–Stokes equations. Another example of compliance with the physics is provided by the robustness with respect to the variations of the problem coefficients; see, e.g., [149] concerning anisotropic heterogeneous diffusion problems in mixed formulation. In some cases, robustness can be extended to obtain a seamless treatment of singular limits of PDE models. A first example is provided by the Péclet-robust HHO scheme of [159], which fully supports locally degenerate diffusion with exact solutions that exhibit jumps inside the domain; see also [157] for further insight into this topic. Another example is provided by [69], where the authors propose an HHO scheme for the Brinkman problem that is fully robust in the Darcy limit. In this case, a novel approach to the analysis was also devised in order to identify the local (Darcy- or Stokes-dominated) regime inside each element. A third example of a physics-compliant HHO scheme is provided by [70, 151], where non-dissipative discretisations of the convective term in the incompressible Navier–Stokes equations are proposed.

The construction of HHO methods is conceived so as to enable efficient sequential and parallel implementations in arbitrary space dimension [114]. Both the reconstruction and the stabilisation terms are devised at the element level, and the coupling among neighbouring elements only occurs via the common face unknowns. As a result, element unknowns can be eliminated prior to the assembly step by computing a local Schur complement; see, e.g., [155, Section 4.3.2.4] and [160, Section 3.3.1]. This procedure is often referred to as "static condensation" in the Finite Elements literature, a term reminiscent of its origins in computational mechanics [206, 214]. For the HHO approximation of degree $k \geq 0$ of a scalar three-dimensional diffusion problem, the number of degrees of freedom after static condensation is $\frac{1}{2}(k+2)(k+1)N_{\text{faces}}$, with N_{faces} denoting the number of non-Dirichlet mesh faces. For high polynomial degrees, this is a dramatic improvement over, e.g., vanilla implementations of Discontinuous Galerkin methods, where the number of degrees of freedom is $\frac{1}{6}(k+3)(k+2)(k+1)N_{\text{el}}$, with N_{el} denoting the number of mesh elements. The improvement is even more dramatic when considering more complex models such as those encountered in incompressible fluid mechanics. As noticed in [160] in the context of the linear Stokes problem, it is possible in this case to devise a static condensation strategy that, for any polynomial degree, leads to global systems with only one pressure unknown per element; cf. also [151, Remark 6] for the extension of such a strategy to the fully nonlinear Navier–Stokes problem.

To close this introductory section, we provide a brief historical overview of the development of HHO methods. Hybrid High-Order methods in primal form were originally introduced in [146] in the context of quasi-incompressible linear elasticity models, with the first fully referenced publication [158] dating back to 2014. The early steps in the development of the method were, on the one hand, the identification of equilibrated tractions [147] and, on the other hand, its extension to locally variable coefficients [148] and more general, possibly degenerate scalar second-order models [159]. In parallel, the Mixed High-Order method was introduced in [149], and its link with primal HHO methods first recognised in [9]. Among the first applications to engineering problems, we can cite the a posteriori-driven adaptive algorithm for electrostatics devised in [154]; see also [162]. A keystone in the understanding of the relations between HHO and other hybrid methods is [129], where bridges are built with Hybridisable Discontinuous Galerkin and High-Order Mimetic methods. Further progress in this direction is made in [56], where a unified framework comprising a large number of mixed and hybrid methods is proposed, and in [163], where HHO methods are bridged with Nonconforming Virtual Elements, and a stable gradient reconstruction is proposed which enables the interpretation of both methods as Gradient Discretisation Methods [178]. At the same time, a large effort was undertaken for the application, and corresponding analysis, of HHO methods to complex models, which are more realistic from the point of view of scientific and engineering applications. A first remarkable contribution is the HHO method for the Cahn–Hilliard problem designed and analysed in [105], which constitutes the first application to nonlinear problems. Landmark contributions in this direction are the papers [141, 142], which study the application of the HHO method to fully nonlinear, Leray–Lions type elliptic models. Here, a systematic development of the tools for the design and analysis of HHO methods for nonlinear problems was undertaken, leading to key functional analysis results of broader applicability. A first example of such results is the development of a framework for the study of the approximation properties of projectors on local polynomial spaces, which also resulted in a change of the canonical way of introducing HHO methods [155]. Another valuable set of results are Sobolev embeddings and compactness results for bounded sequences of HHO functions, which form the cornerstone of the convergence analysis by compactness [171, Section 1.2]. The application of HHO methods to more complex models of engineering interest subsequently focused mainly on incompressible fluid mechanics, solid mechanics, and geosciences. Concerning the first applicative field, key contributions include the HHO method for the Stokes problem developed in [9], later hacked in [99, 161] to robustly handle small values of the viscosity and large irrotational body forces, and the HHO methods for the Navier–Stokes problem developed and analysed in [151] and [70], where a conservative formulation based on Temam's device is proposed. A recent contribution inspired by the HHO literature is [98], where Bingham pipe flows are considered. We can also cite [8], where the influence of dominant convection on the order of convergence is evaluated using the Oseen model. Still in the context of fluid dynamics, [69] proposes and analyses an HHO method for the Brinkman model, that is fully robust in both the Stokes and Darcy limits. Key contributions to the application of HHO methods to problems in solid mechanics include [68], concerning nonlinear elastic models valid under the small deformation assumption, [60] on the application of HHO methods to Kirchhoff–Love plate bending problems, and, more recently [4], on finite deformations of hyperelastic materials. Applications of the HHO technology to problems in geosciences include fluid flows in fractured porous media [106, 107], miscible fluid flows in porous media [14], as well as linear and nonlinear poroelasticity [58, 73] possibly including stochastic coefficients [72]. Finally, more recent methodological developments include, in particular, the extension of the HHO method to meshes with curved faces [66], the treatment of unfitted interface problems [89] based on the CutFEM technology [90], and its application to highly oscillatory elliptic models [115]. We also point out here the analysis framework for methods in fully discrete formulation developed in [143], which shows the benefits of this approach, originally adopted in the context of HHO methods, when applied to other recent polytopal technologies.

Other Polytopal Methods

Discretisation methods that support polytopal meshes and, possibly, arbitrary approximation orders have experienced a vigorous development over the last decade. Novel approaches to the design and analysis have been developed or rediscovered borrowing ideas from other branches of mathematics such as topology and geometry, or expanding past their initial limits the original ideas underlying Finite Element or Finite Volume Methods. A brief historical perspective focusing on diffusive problems is sketched in what follows.

Since their introduction, usually attributed to Tihonov and Samarskiĭ [271], Finite Volume Methods have been extensively used for the discretisation of PDE models expressed as linear or nonlinear conservation laws in divergence form. The classical Two-Point Flux Approximation scheme requires, however, strict conditions of mesh-data compliance for consistency; see, e.g., the reference monograph [188]. Several lowest-order polytopal methods have therefore been developed in the Finite Volumes spirit in an attempt to circumvent these conditions. In Multi-Point Flux Approximation methods [1–3, 7, 181], consistent approximations of the flux are obtained through local reconstructions involving elements that share a common node. In Mixed [172] and Hybrid [190] Finite Volume Methods, a similar result is achieved by introducing auxiliary unknowns at faces, which can in some cases be eliminated resorting to local interpolation. In the Discrete Duality Finite Volume Method [137, 166, 208], two simultaneous discrete versions of the conservation law are solved on primal and dual meshes, using both vertex and cell unknowns. All of the above-mentioned methods possess local conservation properties on the primal mesh and enable an explicit identification of continuous approximations of the normal trace of the flux. We refer the reader to [156, 171] for a more comprehensive list and a critical review of Finite Volumes and related methods on generic polytopal meshes. Over the years, polytopal Finite Volume Methods have been applied to a variety of linear and nonlinear PDE models including, e.g., fully nonlinear elliptic problems in divergence form [15, 169], miscible fluid flows in porous media [102, 104], the incompressible Navier–Stokes equations [173, 202, 203], etc.

While classical Finite Element Methods are restricted to certain element geometries, extensions to more general meshes are possible in some cases. Any polytopal mesh can be split, e.g., into a simplicial conforming submesh over which standard Finite Element Methods can be applied, possibly with some modifications. In [221], this idea is applied to construct mixed Finite Element Methods on polytopal meshes by solving local problems; see also [275, Section 7] for a discussion on this and related approaches. Conforming $\mathbb{P}^1$ elements on a simplicial submesh are at the core of the Vertex Approximated Gradient method, see [191] and [178, Section 8.5], which additionally uses barycentric eliminations to prevent the submeshing from introducing additional unknowns, and mass-lumping to mitigate the issues of $\mathbb{P}^1$ elements on coarse meshes in case of heterogeneities. It is also possible to construct Finite Elements on certain types of polytopal elements without resorting to submeshing. Various constructions of H^1-conforming polygonal Finite Elements are explored, e.g., in [265], where the authors compare Laplace [111, 210], Wachspress [276], and mean value shape functions on convex polygons. Further related works include [266, 267]. Similar constructions are also possible for mixed Finite Elements. In [219], $\boldsymbol{H}(\mathrm{div})$-conforming Finite Elements on polygons and certain polyhedra are constructed using barycentric coordinates. A general construction of scalar and vector Finite Elements families on convex polygonal and polyhedral elements is proposed in [198], where basis functions are expressed in terms of barycentric coordinates and their gradients; this approach is inspired by the Finite Elements Exterior Calculus formalism [21, 24, 25]. Similar ideas are developed in [109] to devise $\boldsymbol{H}(\mathrm{curl})$ and $\boldsymbol{H}(\mathrm{div})$ Finite Elements on polytopal meshes with the minimum number of degrees of freedom ensuring the appropriate global continuity.

Constructions extending classical properties of nonconforming and penalised Finite Elements also include the ones at the root of Cell-Centered Galerkin [140] and generalised Crouzeix–Raviart [152] methods. In the former case, a special subspace of piecewise linear polynomials with optimal approximation properties is identified and used within an interior penalty Discontinuous Galerkin formulation. In the latter case, a construction is proposed yielding the continuity at interfaces of the average values of traces on a simplicial submesh which need not be constructed in practical computer implementations.

Lowest-order polytopal methods have also been developed starting from points of view entirely different from those underpinning Finite Elements and Finite Volumes. A particularly fruitful (direct or indirect) source of inspiration has been the classical work of Whitney on geometric integration [279].

Mimetic Finite Differences are derived by mimicking the Stokes theorem to formulate discrete counterparts of differential operators and L^2-products; see, e.g., [222, 235] concerning the first extensions of this approach to polygonal and polyhedral meshes, [82] on the convergence theory for mixed/hybrid versions of the method (with primary unknowns on the mesh faces), and [86] on the nodal version (with unknowns at the mesh vertices). A complete exposition of the mathematical aspects underpinning the Mimetic Finite Difference method can be found in the research monograph [47], where a panel of applications to various models is also discussed. The recent paper [176] showed that the mixed/hybrid flavour of the Mimetic Finite Difference method [82, 83] is algebraically identical to Mixed and Hybrid Finite Volume Methods, and that all these methods can be regarded as three different presentations of a generic family, the Hybrid Mimetic Mixed method. The relation between Hybrid Mimetic Mixed and the lowest-order version of HHO methods has been studied in [158, Section 2.5] for pure diffusion and in [159, Section 5.4] for advection-diffusion-reaction.

In the Discrete Geometric Approach, originally introduced in [131] and extended to polyhedral meshes in [132, 133], as well as in Compatible Discrete Operators [62, 64], formal links with the continuous operators are expressed in terms of Tonti diagrams [272, 273]. The latter enable the identification of analogies among physical theories based on notions borrowed from algebraic topology. Compatible Discrete Operator methods have been applied to a variety of models mainly issued from applications in fluid mechanics, and including steady incompressible creeping flows [63] and advection-diffusion equations [96]. The Discrete Geometric Approach, on the other hand, has a variety of applications to problems in electromagnetism, including electrostatics [260], the Schrödinger equation [261], eddy currents with a cohomology-based approach [165], and the explicit solution of the Maxwell equations [134]. Both the Compatible Discrete Operator method and the Discrete Geometric Approach are strongly related to Hybrid Mimetic Mixed and nodal Mimetic Finite Difference methods [61], and also to the lowest-order version of Mixed High-Order methods [149, Section 3.5].

Several of the methods discussed above have been lately bridged or incorporated into more recent technologies; see, e.g., [56].

High-order discretisations on general meshes that are possibly physics compliant can be obtained by the Discontinuous Galerkin approach. Discontinuous Galerkin methods were originally introduced in [256] for the approximation of first-order PDE models, while their application to the discretisation of second-order PDE models on standard meshes has been considered starting from the late 1970s in [20, 28, 32, 168, 278], building on the original work of Nitsche [246, 247] on the weak enforcement of boundary conditions. A second stage was inaugurated by the pioneering works [120–122, 124, 125] in the late 1980s tackling hyperbolic and parabolic problems, which led to an impetuous development, further boosted by the landmark papers [36, 37] on the application to the full viscous compressible Navier–Stokes model. This second stage culminated in the unified analysis of [23]. A third development stage was thrusted by [38] (see also the Ph.D. thesis [269] from which this work emanates), where Discontinuous Galerkin methods were first applied to polyhedral meshes obtained by element agglomeration, and by [144], where a complete set of analysis tools for polytopal Discontinuous Galerkin methods was first established; we also refer to [16], where bubble stabilisation techniques on polygonal meshes in dimension 2 were developed. Subsequent works that deserve to be mentioned here include [17, 91, 92]. Despite their enormous success, Discontinuous Galerkin methods can have practical limitations in some cases. For problems in incompressible fluid mechanics, inf–sup stability is in general not available for equal-order approximations on general meshes; see, e.g., the discussion in [145, Sections 6.1.2 and 6.1.5]. This typically requires the introduction of non-physical pressure stabilisation terms. For similar reasons, ad hoc strategies are required for quasi-incompressible problems in linear elasticity; see, e.g., [153] and references therein. Additionally, unless special measures are taken, denoting by d the space dimension, k the polynomial degree, and $N_{\rm el}$ the number of mesh elements, the number of discrete unknowns in Discontinuous Galerkin methods for scalar problems grows as $\binom{k+d}{d}N_{\rm el}$, and can therefore become unbearably large, particularly for three-dimensional problems. Notice that remedies are possible, including, e.g., the variation with fewer coupled unknowns very recently pointed out in [236] and valid for general polyhedral meshes or, in the context of the discretisation of conservation laws on standard meshes, the use of nodal bases such as the ones discussed in [209, Chapter 6]. An extensive comparison of Discontinuous Galerkin and HHO methods on a variety of flat and curved two- and three-dimensional meshes, including an assessment of the respective computational cost, is contained in [66], to which we refer for further details.

A very fruitful attempt to overcome the limitations of Discontinuous Galerkin methods was undertaken in [100, 127], leading to Hybridisable Discontinuous Galerkin methods. The key idea here is, starting from problems in mixed formulation, to introduce auxiliary face unknowns enforcing the continuity of numerical fluxes through interfaces. The resulting methods are amenable to hybridisation and static condensation and have a more favourable scaling of the number of discrete unknowns in terms of the polynomial degree when compared to Discontinuous Galerkin methods. While most of the literature focuses on standard meshes, it has been recognised, e.g., in [129] that the canonical versions of Hybridisable Discontinuous Galerkin methods naturally extend to more general polytopal meshes. A different paradigm for the extension to certain polytopal elements is provided by the very recent M-decomposition techniques [118, 119, 130]. A recurrent research topic in the Hybridisable Discontinuous Galerkin literature is the identification of superconvergent variations, motivated by the analogy with classical mixed Finite Elements pointed out in [127]. In this respect, HHO methods have brought two significant conceptual advances (see [129]): first, local reconstructions have been incorporated into the formulation of the method rather than being used for post-processing; second, subtle local stabilisation terms that satisfy richer consistency properties have been identified. As a result, superconvergence of the scalar variable in HHO methods is built-in rather than serendipitous. Achieving similar results for Hybridisable Discontinuous Galerkin methods is possible using enhanced element spaces [228, 249]. Several links between Hybridisable Discontinuous Galerkin and other methods have been pointed out over the years. Specifically, links with the Local Discontinuous Galerkin method of [100, 123] and with the Staggered Discontinuous Galerkin method of [112] are highlighted in [116, Section 6], where it is also shown that the Weak Galerkin methods of [240, 241, 277] enter the Hybridisable Discontinuous Galerkin framework; see also [117] on this subject.

Another important family of arbitrary-order discretisation methods that support general polytopal meshes is that of Virtual Elements. These can be described as Finite Element Methods where explicit expressions for the basis functions are not available at each point; hence the term "virtual" in reference to the function space they span. The degrees of freedom are selected so as to allow the computation of suitable (problem-dependent) projections of virtual functions onto local polynomial spaces, which are used in turn to formulate the Galerkin consistency terms. The polynomial projections are typically nonconforming, so that this procedure results in a variational crime [263]. For this reason, stabilisation terms inspired by Mimetic Finite Differences are required, which can be interpreted as penalisations of the difference between the virtual solution and its polynomial projection. In their original formulation, Virtual Elements were developed based on conforming virtual spaces; see, e.g., [45, 46] for the H^1-conforming case, [87] for the $\boldsymbol{H}(\mathrm{div})$-conforming case, [43] for the H^s-conforming case with $s \geq 1$, and [48] for a more general overview of both the $\boldsymbol{H}(\mathrm{div})$- and $\boldsymbol{H}(\mathrm{curl})$-conforming cases. More recently, a high-order version of the Mimetic Difference Method for a pure diffusion problem was proposed in [234], which was later reinterpreted as a nonconforming Virtual Element Method and analysed in [27]. It was first recognised in [129, Section 2.4] that the resulting method could be interpreted as a variation of the original HHO method with depleted element unknowns. An attempt to provide a unifying perspective on various families of hybrid and mixed methods was made in [56] where, in particular, bridges were built between HHO and Mixed High-Order methods on the one hand, and mixed and nonconforming Virtual Element Methods on the other. Unified analyses of conforming and nonconforming Virtual Element Methods for diffusive problems have been recently proposed in [95], where a standard Finite Elements approach is adopted, and in [143], based on the third

Strang lemma; see Appendix A. For further insight on Virtual Element Methods we refer the reader to Sect. 5.5.

Outline of the Book

This book is subdivided into two parts comprising, respectively, five and four chapters. Part I lays the foundations of HHO methods by introducing the discrete setting and discussing the construction and analysis of HHO schemes for linear models with diffusion. Part II addresses the application of HHO methods to advanced models: nonlinear diffusion operators, linear elasticity, and incompressible flows. The exposition is completed by two appendices: the first one provides the main analysis tools for HHO discretisations of linear models, while the second one covers the principles related to the practical implementation of HHO methods.

For each of the models considered in this book, the exposition follows the same steps: first, we introduce the appropriate local reconstruction operators, use them to build the local contribution to the discrete problem, and show how the latter is assembled; second, we discuss the well-posedness of the discrete problem, highlighting its key stability properties; third, we carry out a convergence analysis. For linear problems, the third step follows the abstract analysis framework of Appendix A. For nonlinear problems, we develop ad hoc analysis strategies while still taking inspiration from the concepts of stability and consistency used in the linear case.

We mention here that, throughout the book, we make the following abuse of language (which is somehow standard in the context of numerical methods): when writing "polynomial of degree k" we actually mean "polynomial of degree k or less". We also often do not make explicit that these polynomials are actually polynomials of several variables (and that the degree always refers to the total degree).

Let us also notice that, the focus of this book being on polytopal methods, the exposition typically concentrates on space dimensions $d \geq 2$. The one-dimensional case entails simple modifications, briefly mentioned in Remarks 1.13 and 2.10.

Foundations

Chapter 1 establishes the setting for the development and analysis of HHO methods. We start by discussing the appropriate notion of polytopal mesh. The main difference with respect to the corresponding notion for Discontinuous Galerkin methods (see, e.g., [145, Chapter 1]) lies in the definition of mesh faces which, for HHO methods, are portions of hyperplanes. Since this book focuses on the so-called h-convergence analysis (with h denoting, as usual, the meshsize), we next introduce the notion of regular mesh sequence, which generalises the classical one encountered in Finite Element Methods; see, e.g., [113, p. 111]. This concept is central to derive the basic geometric and functional inequalities needed for the analysis. Notice that, in this manuscript, we do not address the p- or hp-versions of the HHO method, where convergence is attained by increasing the polynomial degree rather than (or on top of) reducing the meshsize; see Remark 2.30 for references on this subject. The following step consists in introducing some relevant function spaces: Lebesgue spaces, global and broken Sobolev spaces, as well as local and broken polynomial spaces. We also prove some key functional results, namely (direct and) inverse Lebesgue and Sobolev embeddings on local polynomial spaces, as well as continuous and discrete local trace inequalities on mesh faces. The last section of Chap. 1 addresses the cornerstone of HHO methods, namely projectors on local polynomial spaces. We focus on two particularly relevant instances: L^2-orthogonal projectors, obtained by minimising the difference with respect to the projected function in the L^2-norm, and elliptic projectors, where the H^1-seminorm of the difference is minimised instead. In both cases, we study the continuity and approximation properties of these projectors following the approach proposed in [141, 142] and based on the classical results of [179]; see also [77, Chapter 4].

In **Chap. 2** we introduce the basic principles of HHO methods using as a model problem the Poisson problem. The starting point for the local construction is the following key remark: given an integer $k \geq 0$ and a mesh element T, the elliptic projection of degree $(k + 1)$ of a smooth function v can be computed using only the L^2-orthogonal projections of v of degree k on T and on each of its faces. This suggests the construction of a scheme where we take as discrete unknowns polynomials of degree k over T and its faces, without imposing any continuity property between the mesh element unknown and the face unknowns, or among the face unknowns themselves (which can, therefore, exhibit jumps at the element vertices in two space dimensions and at the element edges in three space dimensions). The natural local interpolator consists in L^2-projecting smooth functions onto polynomials of degree k over T, and polynomials of degree k over each face of T. Starting from this set of discrete unknowns, we devise inside the mesh element a potential reconstruction of degree $(k + 1)$, in such a way that its composition with the local interpolator coincides with the elliptic projector. This local reconstruction emulates an integration by parts formula over T with element-based and face-based unknowns playing the role of the function in volume and boundary integrals, respectively. From the reconstruction, we build a local contribution composed of two terms: the first is a consistent Galerkin contribution, while the second is a stabilisation for which a set of abstract design conditions are provided. Having defined a local contribution, the next step consists in formulating and studying the discrete problem. We start by defining a global space of discrete unknowns which incorporates the single-valuedness of unknowns attached to interfaces, as well as the Dirichlet condition on boundary faces. Vectors of discrete unknowns in this space satisfy a discrete counterpart of the Poincaré inequality, which yields the well-posedness of the discrete problem obtained by element by element assembly of local contributions. The discrete problem can be equivalently reformulated in terms of local balances, with numerical normal traces of the flux that are continuous across mesh interfaces. We next estimate the discretisation error applying the abstract results of Appendix A. Specifically, we show that the energy norm of the error converges as h^{k+1} and that, under the usual elliptic regularity assumption, its L^2-norm converges as h^{k+2}. The latter result highlights one of the distinctive features of HHO methods [129], namely the fact that element-based discrete unknowns superconverge to the L^2-orthogonal projection of degree k of the exact solution. To close the chapter, we briefly discuss the extension to more general boundary conditions and provide numerical evidence supporting the theoretical estimates in both two and three space dimensions.

In **Chap. 3** we consider the application of the HHO method to more complex models. We first treat the case of anisotropic and heterogeneous diffusion under the assumption that the diffusion coefficient is piecewise constant on a partition of the domain to which the mesh complies. This kind of model is relevant, e.g., in geosciences applications, where it is used to describe Darcy flows in porous media. The starting point is in this case an oblique version of the elliptic projector which embeds a dependence on the local diffusion coefficient. Having assumed that the latter is constant inside each element, the oblique elliptic projection of degree $(k+1)$ of a smooth function can still be computed from its L^2-orthogonal projections of degree k on the element and each of its faces. Thus, we can build an HHO method from the same set of discrete unknowns as for the Poisson problem by introducing a diffusion-dependent potential reconstruction which, combined with the local interpolator, yields the oblique elliptic projector. A delicate point in this case is the robustness of the method with respect to the variations of the diffusion coefficient. We give a detailed account of this point in the analysis by tracking the dependence of the multiplicative constants on these variations. Specifically, we derive an energy-norm error estimate that is (1) fully robust with respect to the heterogeneity of the diffusion coefficient, meaning that this error is uniform with respect to the jumps of the coefficient across interfaces, and (2) partially robust with respect to the diffusion anisotropy, with a mild dependence on the square root of the local anisotropy ratio. We then extend the model by including first-order transport terms and reaction terms. The discretisation of the former hinges on two contributions devised at the local level: (1) a local reconstruction of the advective derivative which emulates an integration by parts formula (but which, in general, does not return a projector when composed with the local interpolator), and (2) a stabilisation term which introduces some upwinding by penalising the difference between element- and face-based unknowns. A key point consists in this case in ensuring the robustness of the method when the advection term is locally dominant, corresponding to large values of a local Péclet number. We provide a detailed account of this point in the analysis by tracking the dependence of the constants in the energy-norm error estimate. Specifically, we show that, when using discrete unknowns of degree $k \geq 0$, each element T of diameter h_T contributes to the error with a term of order $h_T^{k+\frac{1}{2}}$ in the advection-dominated regime and, coherently with the results for pure diffusion, of order h_T^{k+1} in the diffusion-dominated regime. This estimate also covers all intermediate regimes and carries out to the singular limit corresponding to locally vanishing diffusion, thus making the method fully robust with respect to dominant advection. We also prove error estimates in the L^2-norm under the usual elliptic regularity assumption. Notice that, in this case, we cannot prevent a dependence on the global anisotropy ratio and Péclet number, since these quantities appear in the elliptic regularity estimate. We close the chapter with a numerical illustration including two- and three-dimensional tests.

Chapter 4 addresses two additional topics on purely diffusive models: a posteriori error analysis and locally varying diffusion. The goal of a posteriori error analysis consists in estimating through a computable quantity the error between a known numerical approximation and the unknown exact solution. This information can be used, e.g., to drive local mesh refinement by identifying those elements where the error is larger. This is a crucial point to fully exploit the potential of high-order methods when the exact solution exhibits singularities, as is often the case for the complex geometries encountered in applications. Following the seminal ideas of Mikhlin [237] and Ladevèze [224] based on the Prager–Synge equality [252], we derive an upper bound for the error—defined as the difference between the potential reconstruction and the exact solution—in terms of three estimators measuring, respectively, the lack of conformity of the method, the residual of the equation in strong form, and the stabilisation. The upper bound is guaranteed, meaning that the estimators are fully computable from known quantities. The estimators are also locally efficient, that is, they provide local lower bounds of the error, making them suitable to drive mesh refinement. Their practical performance in this context is numerically demonstrated by solving three-dimensional singular problems with an adaptive HHO algorithm. Interestingly, the support of general polyhedral elements can be exploited in this case to perform local mesh coarsening instead of refinement, which does not require to generate a novel mesh at each iteration. Mesh coarsening as a means to reduce the computational cost while preserving geometric accuracy was first proposed in [38] in the context of Discontinuous Galerkin methods, and later pursued in [39, 40, 67] with particular focus on computational fluid mechanics.

The second section of Chap. 4 extends the results of the first section of Chap. 3 to models where the diffusion coefficient varies smoothly inside the mesh elements. A key difference with respect to the piecewise constant diffusion case is that we introduce here a richer reconstruction of the gradient using the full space of vector-valued polynomials of total degree k, instead of the gradients of polynomials of total degree $(k + 1)$. This strategy, different from the one adopted in [148], is a precursor to the developments of Chap. 6, where more complex nonlinear diffusion models are considered. The main result of this section is an energy-norm error estimate which shows that, when the diffusion coefficient varies inside mesh elements, the contribution to the error of every mesh element is proportional to a full power (as opposed to the half power found in Chap. 3) of the local anisotropy ratio. We also derive an L^2-norm estimate, showing that, also in this case, the HHO method enjoys a superconvergence property of the element-based unknowns towards the L^2-orthogonal projection of degree k of the exact solution.

In **Chap. 5** we study links between HHO methods and various other classical or modern methods. We start by presenting a variant of the HHO method of Chap. 2, in which the element unknowns are polynomials of degree $\ell = k - 1$ or $k + 1$ instead of $\ell = k$. We show that the principles behind the design of HHO methods can easily be adapted to this choice of unknowns and lead to numerical schemes that have the same $O(h^{k+1})$ convergence rates in energy norm as the standard HHO method corresponding to $\ell = k$. A particular treatment has to be made in the case $(k, \ell) = (0, -1)$, where the absent element unknowns have to be reconstructed by averaging face unknowns. In this case, the corresponding potential reconstruction is linked, through the interpolator of smooth functions, to a slightly different elliptic projector in which the closure equation involves a certain average over the element faces rather than the average over the element itself. The approximation properties of this modified elliptic projector are analysed, and turn out to be similar to the ones of the classical elliptic projector. An $O(h^{k+2})$ superconvergence rate in L^2-norm is established for the (k, ℓ)-variant of the HHO method, except in the case $(k, \ell) = (1, 0)$. Numerical tests show indeed that, if $(k, \ell) = (1, 0)$, the rate of convergence in L^2-norm stagnates to $O(h^{k+1}) = O(h^2)$, the same rate as in the energy norm.

Next, we consider two low-order methods and present their links with HHO. On matching simplicial meshes, it is shown that the matrix of the variant of the HHO method corresponding to $(k, \ell) = (0, -1)$ is identical to that of the non-conforming $\mathbb{P}^1$ finite element method; only their source terms differ. On generic polytopal meshes, we prove that the standard HHO scheme for $k = 0$ is a particular case of the Hybrid Mimetic Mixed method [176]. Hence, HHO schemes can be seen as high-order extensions on generic polytopal meshes of these two low-order methods.

We then analyse the links between the HHO method and the Mixed High-Order method of [149]. The latter discretises the mixed formulation of the Poisson problem, a saddle-point problem on the pair of unknowns potential–flux. The potential unknowns are broken polynomials of degree k on the mesh; the flux unknowns are gradients of polynomials of degree k inside the elements, and polynomials of degree k on each face—representing normal components of the fluxes. As for the HHO method, the design of the Mixed High-Order scheme relies on local reconstructions in the elements from the flux unknowns: a divergence in the space of polynomials of degree k and a flux in the space of gradients of polynomials of degree $(k+1)$. When composed with the natural interpolator, the former coincides with the L^2-orthogonal projector of the divergence on the local polynomial space of degree k, while the latter is polynomially consistent up to the degree $(k + 1)$. As usual in mixed methods, the global space accounts for the continuity of the normal fluxes across the interface. We present a hybridisation of the Mixed High-Order method in which this continuity condition is removed from the space and accounted for by introducing Lagrange multipliers, polynomials of degree k on the faces that can be interpreted as additional potential unknowns; this transforms the space of potential unknowns into the standard space of unknowns for the HHO method of degree k. After designing a potential-to-flux operator, we conclude this analysis by proving an algebraic equivalence between the Mixed High-Order scheme, its hybridised version, and the HHO scheme (when the stabilisation term is chosen in a form that involves the potential-to-flux operator).

The next link between the HHO method and other polytopal methods revolves around Virtual Element Methods: we give a presentation of the HHO scheme for the Poisson problem in the form of a Virtual Elements scheme. We start by identifying a space of broken H^1 functions on the mesh that is in one-to-one correspondence with the space of HHO unknowns. These functions do not have explicit representations, but are fully determined by the standard HHO degrees of freedom (L^2-projections on polynomials of degree k on the mesh elements, and L^2-projections on polynomials of degree k on the mesh faces). Through this correspondence, the potential reconstruction becomes the elliptic projector, and we can thus interpret the Galerkin contribution of the HHO scheme in terms of the elliptic projections of virtual functions. The stabilisation term is then seen as a penalisation of the defect between virtual functions and polynomials of degree $(k+1)$. The same process carried out on the (k, ℓ)-variant of the HHO method, with $\ell = k-1$, gives the Nonconforming Virtual Element Method of [27]. For the sake of completeness, we also address Conforming Virtual Element Methods in two space dimension, discussing the difference with respect to HHO methods, and providing a non-standard analysis inspired by that of HHO schemes; in particular, we study the approximation properties of the projector, relevant to Conforming Virtual Elements, in $W^{s,p}$-spaces with p possibly different from 2.

The final section of this chapter shows that HHO methods can be embedded into the Gradient Discretisation Method, a generic framework for the design and analysis of numerical schemes for diffusion problems [178]. It consists in designing and analysing schemes, for various elliptic and parabolic models, using three abstract discrete elements—a finite-dimensional space and two reconstruction operators (function and gradient). The accuracy of schemes written in the form of a Gradient Discretisation Method is assessed using three quantities only dependent on the choice of space and reconstruction operators: a discrete Poincaré inequality, a measure of approximability properties, and a measure of defect of conformity (how well a discrete integration by parts formula is satisfied). We show that a proper choice of the space and reconstruction operators lead to the HHO method, and we evaluate the three aforementioned quantities, showing that they behave as expected for a high-order scheme.

Applications to Advanced Models

In **Chap. 6** we consider the extension of HHO methods to fully nonlinear elliptic equations involving Leray–Lions operators [230], which contain the p-Laplacian as a special case. These operators appear in various physical models including, e.g., glacier motion [201], incompressible turbulent flows in porous media [164], and flow around airfoils [200]; they can also be regarded as a simplified version of the viscous term in power-law fluids. From a mathematical standpoint, Leray–Lions elliptic problems involve two important novelties compared to the models covered in Chaps. 2–5, namely the presence of nonlinearities and the fact that their weak formulation is naturally posed in a non-Hilbertian setting. An important consequence of the first point is that the convergence analysis cannot solely rely on error estimates. Such estimates are attainable for particular Leray–Lions operators (notably, the p-Laplacian), but impossible to prove for other models for which the solution may not be unique [178, Remark 2.42]. In order to circumvent this difficulty, we resort to compactness arguments inspired by the Finite Volumes literature; see, e.g., [188] and references therein. Compactness arguments have been traditionally employed also for the convergence analysis of conforming Finite Element Methods for nonlinear problems (an example is provided by [231]) and only recently applied to the nonconforming setting; see, e.g., [144] and references therein. Here, we further extend these techniques to the fully discrete high-order setting in which HHO methods are formulated. Specifically, after introducing a discrete version of the $W^{1,p}$-norm which enables us to emulate a Sobolev structure on the space of discrete unknowns introduced in Chap. 2, we show that the latter is continuously or compactly embedded into $L^q(\Omega)$ for suitable values of the exponent q. This embedding is established by comparing the order $k \geq 0$ HHO space and operators on a polytopal mesh with its order 0 version on a simplicial submesh and by invoking the discrete functional analysis results developed in an abstract low-order setting in [178]. Having established this new framework for the analysis, we follow the usual principles to formulate an HHO scheme for Leray–Lions elliptic operators. For a given integer $k \geq 0$, the local contribution revolves around the reconstruction, introduced in Chap. 4, of the gradient in the full space of vector-valued polynomials of total degree k. As originally observed in [163, Section 4.1], this choice is required here to preserve a scaling of the consistency error analogous to the one observed for linear problems. The second local ingredient is a non-Hilbertian version of the stabilisation term, which ensures stability with respect to the discrete $W^{1,p}$-norm. The convergence analysis is then carried out using two approaches: on the one hand, adapting the ideas of Appendix A, we establish convergence rates for regular solutions of the p-Laplace equation by deriving error estimates that generalise the ones derived in Chap. 2 for the linear case; on the other hand, we exploit the novel discrete functional analysis tools developed in the first part of this chapter to demonstrate convergence for more general Leray–Lions operators. The error estimates are illustrated by a numerical validation for various values of the index p.

Chapter 7 deals with the linear elasticity model describing the small deformations of a linear isotropic (Hookean) body under load. We consider the model in its primal formulation, where the unknown is a vector-valued function representing the displacement field. The main difference with respect to the scalar diffusion models considered in Chaps. 2 and 3 is that, in this case, the symmetric part of the gradient (representing the strain tensor) replaces the gradient in the constitutive law linking the stress and strain fields. A relevant consequence from the mathematical point of view is that the well-posedness of the continuous problem hinges on Korn's instead of Poincaré's inequality. To derive an HHO discretisation, we start by introducing a novel (strain) projector on local polynomial spaces obtained by minimising the difference with respect to the projected function measured by the L^2-norm of the symmetric part of the gradient. Optimal approximation properties for this projector follow from a discrete counterpart of Korn's second inequality valid inside mesh elements, with a constant independent of the element shape and with an explicit dependence on the meshsize. Following similar ideas as in Chap. 2 we show that, given an integer $k \geq 0$ and a smooth enough vector-valued function over a mesh element T, the strain projection of degree $(k + 1)$ of the function can be computed from its polynomial L^2-projections over T and its faces. On the one hand, this prompts us to introduce an HHO space where the discrete unknowns are vector-valued polynomials of total degree k over each mesh element and face; equipped with a discrete norm which mimics the L^2-norm of the strain tensor, this space satisfies a discrete version of Korn's first inequality. On the other hand, it suggests to define a local reconstruction of the displacement field which, composed with the natural interpolator, yields the strain projector of degree $(k + 1)$. This displacement reconstruction is used to formulate a stabilisation term whereas, for the consistent term, we use the symmetric part of a full gradient reconstruction defined in the spirit of Chap. 4. The latter choice enables the treatment of the full physical range for the Lamé parameters. The analysis for the resulting scheme is carried out using the abstract framework of Appendix A showing, for $k \geq 1$, convergence in h^{k+1} and h^{k+2} for the energy- and L^2-norms of the error, respectively. Crucially, the provided error estimates are robust in the quasi-incompressible limit corresponding to bodies which deform at constant volume. This is a crucial advantage with respect to, e.g., lowest-order H^1-conforming Finite Elements, which are known to provide unsatisfactory results in this case [31]. We next discuss a possible modification of the method which enables the use of the lowest-order version corresponding to $k = 0$. The key idea consists in adding a novel term penalising the jumps of the displacement reconstruction across interfaces. This new term allows us to recover stability, from which optimal error estimates can be inferred. A panel of numerical examples closes this chapter.

Chapters 8 and 9 deal with PDE problems arising in incompressible fluid mechanics. We start in **Chap. 8** with creeping flows of Newtonian, uniform density fluids modelled by the Stokes equations, which express the fundamental principles of momentum and mass conservation. In this setting, the conservation of mass takes the form of a zero-divergence constraint on the velocity, with the pressure acting as the corresponding Lagrange multiplier. Thus, the weak formulation has a saddle point structure, whose well-posedness hinges on the inf–sup stability of the pressure-velocity coupling. Deriving numerical approximations of saddle-point problems is not straightforward in the classical conforming Finite Elements framework since, unlike coercivity, inf–sup stability is not inherited by the discrete problem [57]. The HHO discretisation of the Stokes problem hinges on the space of vector-valued discrete unknowns of degree $k \geq 0$ introduced in Chap. 7 for the velocity and on the space of broken polynomials of the same degree for the pressure. Based on the local velocity unknowns, two reconstructions are introduced inside each element: (1) a velocity reconstruction of degree $(k+1)$ obtained mimicking the procedure introduced in Chap. 2 and (2) a divergence reconstruction of degree k which, applied to the interpolate of a local velocity field, yields the L^2-projection of the divergence of that field. These reconstructions are used to formulate the discrete counterparts of the viscous and pressure-velocity coupling terms, which are then assembled element by element. The choice of the discrete space for the pressure and the definition of the divergence reconstruction enable the use of the classical Fortin technique [194] to prove discrete stability and well-posedness. As for most of the other models considered in this book, it is possible to reformulate the HHO method for the Stokes problem in terms of conservative numerical fluxes, thus building a bridge with Finite Volume Methods. The analysis yields convergence in h^{k+1} for the energy norm of the error on the velocity and the L^2-norm of the error on the pressure. An improved estimate in h^{k+2} for the L^2-norm of the error on the velocity is derived under a suitable elliptic regularity assumption. To close this chapter, we address the topic of pressure-robustness. Specifically, we show that, by hacking the discretisation of the right-hand side, we can reproduce at the discrete level an important property of the continuous problem, namely that modifying the irrotational part of body forces only affects the pressure, leaving the velocity field unaltered. Mimicking this property at the discrete level yields error estimates for the velocity where the multiplicative constant in the right-hand side is independent of the pressure, whence the term pressure-robustness.

In **Chap. 9** we tackle the HHO discretisation of the full Navier–Stokes equations. The difference with respect to the Stokes problem considered in Chap. 8 is the presence of a nonlinear convective term, which is at the root of physically relevant phenomena such as turbulence. A key remark is that, when wall boundary conditions are enforced, this term does not contribute to the kinetic energy balance obtained taking the velocity as a test function in the momentum equation. This property, along with boundedness and consistency, turns out to be one of the key design assumptions on the discrete convective trilinear form associated with this nonlinear term. With these assumptions, we show well-posedness of the discrete problem under the usual small data condition, as well as a convergence estimate similar to the one obtained for the Stokes problem. For the sake of completeness, we also discuss the possibility of including a convective (upwind-like) stabilisation term, although numerical experiments indicate that this is typically not necessary in practice. We next discuss two discrete trilinear forms that match the design assumptions: the first is inspired by a skew-symmetric reformulation of the continuous trilinear form and hinges on a reconstruction of the gradient in the space of polynomials of degree $2k$; the second is inspired by Temam's modification [268] of the continuous trilinear form, originally considered in the context of Finite Element Methods. The advantage of this second trilinear form over the one inspired by the skew-symmetric approach is that it enables a flux formulation of the HHO scheme for the Navier–Stokes equations. We next examine the convergence of the method for general, possibly large, data. Using the compactness techniques introduced in Chap. 6, we show strong convergence (up to the extraction of a subsequence) of the velocity in $L^p(\Omega)^d$ for $p \in [1, \infty)$ if $d = 2$ and $p \in [1, 6)$ if $d = 3$, of the strain rate in $L^2(\Omega)^{d\times d}$, and of the pressure in $L^2(\Omega)$. These results classically extend to the whole sequence of discrete solutions when the continuous solution is unique. The numerical performance of the method is showcased on a panel of classical test cases, including the two- and three-dimensional lid-driven cavity problems.

Audience

This book is primarily intended for graduate students and researchers in applied mathematics and numerical analysis, who will find here valuable analysis tools of general scope. It can also prove a precious instrument for graduate students and researchers in engineering sciences and computational physics interested in the mathematical aspects underpinning HHO and polytopal methods. The book addresses both basic and advanced models encountered in these fields and pays particular attention to practically relevant issues such as robustness with respect to the model parameters. The reader is assumed to be familiar with the standard theory of conforming Finite Elements, including weak formulations of model problems and error analysis, and to have some acquaintance with the basic PDEs in continuum mechanics. Special care has been devoted to making the exposition as self-contained as possible. The material in the book has already been utilised by the authors to give lectures and courses at Université de Montpellier (France), Monash University (Australia), Institut Henri Poincaré (France), Università di Bergamo (Italy), Université Côte d'Azur (France), and several other prestigious institutions.

The general level of the book is best suited for specialised graduate-level courses. An introductory course can be designed based on Chaps. 1 and 2 together with Appendices A and B. A more advanced course would complement these with one of Chaps. 3 or 4. For an expert course, start with the introductory material and add either Chaps. 4 and 6 (to tackle strongly nonlinear problems), or Chaps. 7 and 8 (to sample linear solid and fluid mechanics), or Chaps. 8 and 9 (to focus on models of incompressible flows).

HHO Libraries

As of today, HHO methods have been implemented in several open-source codes including, on the academic side, the SpaFEDte (https://github.com/SpaFEDTe/spafedte.github.com), the HArD::Core (https://github.com/jdroniou/HArDCore) and the POLYPHO (http://www.comphys.com) libraries and, on the industrial side, the Code_Aster (https://www.code-aster.org) and Code_Saturne (https://www.code-saturne.org) simulators by EDF. Most of the schemes presented in the following chapters can be found in these libraries, which may help beginners as well as advanced users to get a practical grasp on HHO methods.

Compiled languages are typically best suited to match the computational requirements of polytopal methods. The above libraries are therefore written in C or C++. To facilitate practical initiation to the basics of HHO, an Octave/MATLAB implementation of the HHO scheme for the 2D Poisson problem can be found here https://github.com/jdroniou/HHO-Lapl-OM/. However, the intrinsic limitations of interpreted languages mean that only the simplest cases (relatively low polynomial degrees and small meshes) can be run with this code. A more serious usage of HHO schemes requires one of the other aforementioned libraries.

Acknowledgements

We are grateful to several colleagues for their constructive remarks that helped improve the manuscript. Our gratitude goes, in particular, to Lorenzo Botti (Università di Bergamo) and Neela Nataraj (Indian Institute of Technology Bombay) for carefully reading large portions of the manuscript and providing precious suggestions to make certain points more accessible. We also thank our students and collaborators for sharing their opinions on early versions of some chapters and for participating in the development of HHO libraries: Daniel Anderson (Carnegie Mellon University), Michele Botti (Politecnico di Milano), Daniel Castanon Quiroz (Université de Montpellier), Hanz Martin Cheng (Monash University), Lachlan Grose (Monash University), André Harnist (Université de Montpellier), Tom Lemaitre (Monash University).

Special thanks go to our colleagues and friends from the polytopal community around the world, with whom we have shared intense research moments during the IHP quarter "Numerical Methods for PDEs", the POEMS conference series, as well as several thematic sessions and minisymposia organised within the most important conferences on Numerical Analysis.

Finally, our deepest gratitude goes to our spouses and children (Margherita, Leone, and Ascanio for Daniele, Caroline and Anaïs for Jérôme) for their understanding and support during our long hours of work—often from home—on this book.

Montpellier, France — Daniele Antonio Di Pietro
Melbourne, Australia — Jérôme Droniou
November 2019

The original version of the book was revised: Author name has been updated. The erratum to the book is available at https://doi.org/10.1007/978-3-030-37203-3

Contents

Part II Applications to Advanced Models

Part I
Foundations

Chapter 1
Setting

In this chapter we introduce the setting for the development and analysis of Hybrid High-Order (HHO) methods. These methods are built upon general meshes possibly including polytopal elements and non-matching interfaces. In Sect. 1.1 we give a precise definition of polytopal mesh, and introduce the notion of regular sequence of h-refined polytopal meshes. In Sect. 1.2 we recall some basic notions on standard Lebesgue and Sobolev spaces, on the space $\boldsymbol{H}(\mathrm{div};\Omega)$, and on polynomial spaces. We next introduce the first building block of HHO methods, namely local polynomial spaces, and prove some fundamental results for the analysis including, in particular, the comparison of Lebesgue and Sobolev (semi)norms defined on such spaces, as well as local trace inequalities valid on regular mesh sequences. Section 1.3 is devoted to the second key ingredient in HHO methods: projectors on local polynomial spaces. After introducing the corresponding notion, we study their approximation properties in an abstract framework, then apply the abstract results to two particularly relevant instances, the L^2-orthogonal and elliptic projectors. Finally, Sect. 1.4 contains technical results required for non-star-shaped elements. All the results established in this chapter on generic polytopal elements and meshes apply to "usual" polytopes, such as triangles/tetrahedra or rectangles/hexahedra commonly encountered in Finite Element Methods.

1.1 Mesh

The starting point to write the HHO discretisation of a PDE problem is a suitable decomposition (mesh) of the domain in which the problem is set. The meshes supported by HHO methods are more general than those encountered in standard Finite Element Methods, and possibly include general polytopal elements and non-matching interfaces. The goal of this section is to introduce precise notions of mesh and h-refined mesh sequence suitable for the analysis.

D. A. Di Pietro, J. Droniou, *The Hybrid High-Order Method for Polytopal Meshes*, MS&A 19, https://doi.org/10.1007/978-3-030-37203-3_1

1.1.1 Polytopal Mesh

We start by defining the notion of simplex and polytopal set.

Definition 1.1 (Simplex and Polytopal Set) Let an integer $d \geq 2$ be fixed. Given a set of *vertices* $\mathcal{P} := \{\boldsymbol{P}_0, \ldots, \boldsymbol{P}_d\} \subset \mathbb{R}^d$ such that the family of vectors $\{\boldsymbol{P}_1 - \boldsymbol{P}_0, \ldots, \boldsymbol{P}_d - \boldsymbol{P}_0\}$ is linearly independent, the interior of the convex hull of $\mathcal{P}$ is a *simplex* of $\mathbb{R}^d$. For each integer $i \in \{0, \ldots, d\}$, the convex hull of $\mathcal{P} \setminus \{\boldsymbol{P}_i\}$ is a *simplicial face*.

A polytopal set (or polytope) is a connected set that is the interior of a finite union of closures of simplices.

According to the previous definition, if $d = 2$, a simplex is an open triangle and a polytope is an open polygonal set; if $d = 3$, a simplex is an open tetrahedron and a polytope is an open polyhedral set.

To identify the meshsize and express the shape regularity properties of a set, not necessarily polytopal, we introduce the following notions:

Definition 1.2 (Diameter and Inradius) Given an open bounded connected set $X \subset \mathbb{R}^d$, we define its *diameter* h_X as

$$h_X := \sup\{\mathrm{dist}(\boldsymbol{x}, \boldsymbol{y}) \ : \ \boldsymbol{x}, \boldsymbol{y} \in X\}. \tag{1.1}$$

The *inradius* r_X of X is the radius of the largest ball included in X.

Throughout the rest of this book, we make the following assumption, without necessarily recalling it at each occurrence, on the domain Ω over which the models are set. Note that this assumption implies in particular that Ω does not have any cracks, i.e., it lies on one side of its boundary $\partial\Omega$.

Assumption 1.3 (Domain Ω) *A space dimension $d \geq 2$ being fixed, Ω is a polytopal set of $\mathbb{R}^d$.*

The following definition of polytopal mesh, closely inspired by [178, Definition 7.2], enables the treatment of meshes as general as the ones depicted in Fig. 1.1:

Definition 1.4 (Polytopal Mesh) A *polytopal mesh* of Ω is a couple $\mathcal{M}_h = (\mathcal{T}_h, \mathcal{F}_h)$ where:

(i) The set of *mesh elements* (or *mesh cells*) $\mathcal{T}_h$ is a finite collection of nonempty disjoint polytopes T with boundary ∂T and diameter h_T such that the *meshsize* h satisfies

$$h = \max_{T \in \mathcal{T}_h} h_T$$

(continued)

Definition 1.4 (continued)
and it holds

$$\overline{\Omega} = \bigcup_{T \in \mathcal{T}_h} \overline{T}.$$

(ii) The set of *mesh faces* $\mathcal{F}_h$ is a finite collection of disjoint subsets of $\overline{\Omega}$ such that, for any $F \in \mathcal{F}_h$, F is a non-empty open connected subset of a hyperplane of $\mathbb{R}^d$ and the $(d-1)$-dimensional Hausdorff measure of its relative boundary $\overline{F} \backslash F$ is zero. We denote by h_F the diameter of F. Further assume that:

(a) For each $F \in \mathcal{F}_h$, either there exist distinct mesh elements $T_1, T_2 \in \mathcal{T}_h$ such that $F \subset \partial T_1 \cap \partial T_2$ and F is called an *interface*, or there exists one mesh element $T \in \mathcal{T}_h$ such that $F \subset \partial T \cap \partial\Omega$ and F is called a *boundary face*;

(b) The set of mesh faces is a partition of the mesh skeleton, i.e.,

$$\bigcup_{T \in \mathcal{T}_h} \partial T = \bigcup_{F \in \mathcal{F}_h} \overline{F}.$$

Interfaces are collected in the set $\mathcal{F}_h^{\mathrm{i}}$ and boundary faces in $\mathcal{F}_h^{\mathrm{b}}$, so that $\mathcal{F}_h = \mathcal{F}_h^{\mathrm{i}} \cup \mathcal{F}_h^{\mathrm{b}}$. For any mesh element $T \in \mathcal{T}_h$,

$$\mathcal{F}_T := \{F \in \mathcal{F}_h \ : \ F \subset \partial T\}$$

denotes the set of faces contained in ∂T. Symmetrically, for any mesh face $F \in \mathcal{F}_h$,

$$\mathcal{T}_F := \{T \in \mathcal{T}_h \ : \ F \subset \partial T\} \tag{1.2}$$

is the set containing the one or two mesh elements sharing F. Finally, for all $T \in \mathcal{T}_h$ and all $F \in \mathcal{F}_T$, $\boldsymbol{n}_{TF}$ denotes the unit normal vector to F pointing out of T.

Notice that, in the above definition, the terminology "mesh face" is preferred over "face" or "edge" when $d = 2$. The reason is twofold: on the one hand, this makes the discussion dimension-independent whenever possible; on the other hand, it emphasises the fact that the mesh faces do not necessarily coincide with the faces of the polytopal elements in $\mathcal{T}_h$. The latter fact provides the increased flexibility required, e.g., to handle nonconforming junctions such as the one depicted

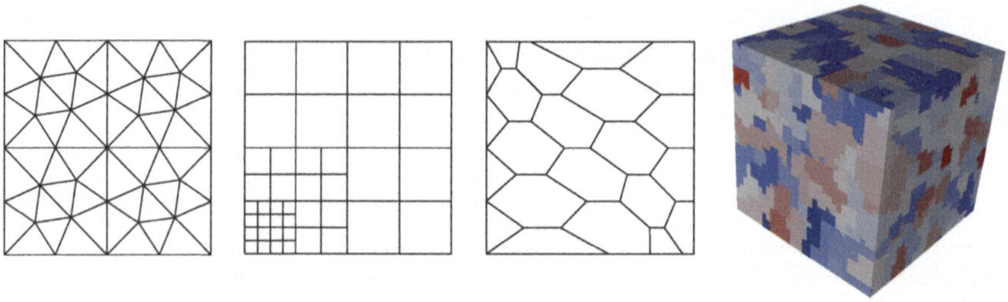

(a) (b) (c) (d)

Fig. 1.1 Examples of polytopal meshes in two and three space dimensions. The triangular and nonconforming meshes are taken from the FVCA5 benchmark [207], the polygonal mesh from [152, Section 4.2.3], and the agglomerated polyhedral mesh from [154]. (**a**) Matching triangular. (**b**) Nonconforming. (**c**) Polygonal. (**d**) Agglomerated

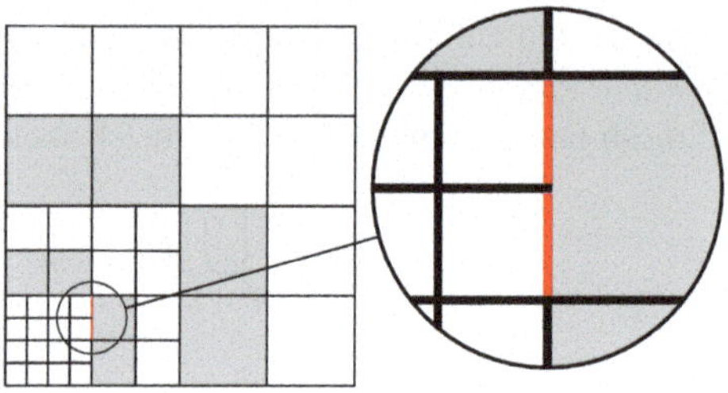

Fig. 1.2 Treatment of a nonconforming junction (red) as multiple coplanar faces. Gray elements are pentagons, white elements are squares

in Fig. 1.2: this case can be simply dealt with by treating each face containing hanging nodes as multiple coplanar mesh faces.

Remark 1.5 (Other Notions of Polytopal Meshes) In the context of Discontinuous Galerkin methods, the notion of mesh face for polytopal meshes proposed in [38, 144] and [145, Chapter 1] is different from the one introduced in Definition 1.4: in these references, interfaces and boundary faces are simply defined as, respectively, the intersection of the closures of two distinct mesh elements and the intersection of the closure of one mesh element with the domain boundary. As a consequence, mesh faces are possibly non-planar and even non-connected. This is possible because Discontinuous Galerkin methods feature only element-based unknowns, and links among elements are established through boundary terms involving their averages and jumps across faces. In HHO methods, on the other hand, transmission conditions are enforced via face-based discrete unknowns, which requires simpler face geometries.

Remark 1.6 (Curved Faces) Following [66], it is possible to construct optimally convergent HHO methods on meshes featuring curved faces that result from high-order geometric mappings. This requires to adapt the polynomial degree on mesh faces by accounting for the so-called effective mapping order; see also [65] and the precursor work [22] on this subject. We also refer the reader to [85] concerning the extension of the Mimetic Finite Difference method to meshes with curved faces, and to [53] on similar developments for Virtual Element Methods.

1.1.2 Regular Mesh Sequence

When studying the convergence of HHO methods with respect to the meshsize h, one needs to make assumptions on how the mesh is refined. The ones provided here are inspired by [145, Chapter 1], and refer to the case of *isotropic meshes* with *non-degenerate faces*. Isotropic means that we do not consider the case where elements become more and more stretched when refining. Non-degenerate faces means, on the other hand, that the diameter of each mesh face is uniformly comparable to that of the element(s) the face belongs to. To formulate the regularity assumptions, we need the notion of matching simplicial mesh, which corresponds to the standard one in the context of Finite Element Methods.

Definition 1.7 (Matching Simplicial Mesh) $\mathcal{M}_h = (\mathcal{T}_h, \mathcal{F}_h)$ is a *simplicial mesh* of Ω if, for all $T \in \mathcal{T}_h$, T is a simplex of $\mathbb{R}^d$. $\mathcal{M}_h$ is a *matching simplicial mesh* of Ω if it is a simplicial mesh and the following additional conditions hold: (i) For any $T, T' \in \mathcal{T}_h$ with $T' \neq T$, the set $\partial T \cap \partial T'$ is the convex hull of a (possibly empty) subset of the vertices of T; (ii) The set $\mathcal{F}_h$ is composed of the simplicial faces of the elements in $\mathcal{T}_h$.

The following definition introduces the notion, illustrated in Fig. 1.3, of matching simplicial submesh of a polytopal mesh.

Definition 1.8 (Matching Simplicial Submesh) Let $\mathcal{M}_h = (\mathcal{T}_h, \mathcal{F}_h)$ be a polytopal mesh of Ω. We say that $\mathfrak{M}_h = (\mathfrak{T}_h, \mathfrak{F}_h)$ is a *matching simplicial submesh* of $\mathcal{M}_h$ if: (i) $\mathfrak{M}_h$ is a matching simplicial mesh of Ω; (ii) for any simplex $\tau \in \mathfrak{T}_h$, there is a unique mesh element $T \in \mathcal{T}_h$ such that $\tau \subset T$; (iii) for any simplicial face $\sigma \in \mathfrak{F}_h$ and any mesh face $F \in \mathcal{F}_h$, either $\sigma \cap F = \emptyset$ or $\sigma \subset F$.

The regularity requirements for sequences of refined polytopal meshes are expressed in terms of a corresponding sequence of matching simplicial submeshes. We emphasise the fact that the simplicial submesh is merely a theoretical tool, and needs not be constructed in practice.

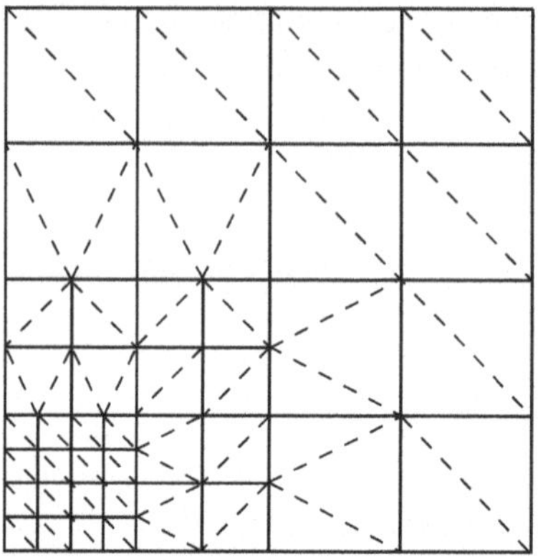

Fig. 1.3 Example of a matching simplicial submesh (dashed lines) of the non-conforming mesh in Fig. 1.1b

Definition 1.9 (Regular Mesh Sequence) Denote by $\mathcal{H} \subset (0, +\infty)$ a countable set of meshsizes having 0 as its unique accumulation point. A family of meshes $(\mathcal{M}_h)_{h\in\mathcal{H}} = (\mathcal{T}_h, \mathcal{F}_h)_{h\in\mathcal{H}}$ is said to be *regular* if there exists a real number $\varrho \in (0, 1)$, independent of h and called the *mesh regularity parameter*, such that, for all $h \in \mathcal{H}$, there exists a matching simplicial submesh $\mathfrak{M}_h = (\mathfrak{T}_h, \mathfrak{F}_h)$ of $\mathcal{M}_h$ that satisfies the following conditions:

(i) *Shape regularity.* For any simplex $\tau \in \mathfrak{T}_h$, denoting by h_τ its diameter and by r_τ its inradius, it holds

$$\varrho h_\tau \leq r_\tau; \tag{1.3}$$

(ii) *Contact regularity.* For any mesh element $T \in \mathcal{T}_h$ and any simplex $\tau \in \mathfrak{T}_T$, where $\mathfrak{T}_T := \{\tau \in \mathfrak{T}_h \ : \ \tau \subset T\}$ is the set of simplices contained in T, it holds

$$\varrho h_T \leq h_\tau. \tag{1.4}$$

Remark 1.10 (Matching Simplicial Mesh Sequences) If, for all $h \in \mathcal{H}$, $\mathcal{M}_h$ is matching simplicial, we can simply take $\mathfrak{M}_h = \mathcal{M}_h$. In this case, the contact regularity condition (1.4) is trivially verified for any $\varrho \in (0, 1)$, and the shape regularity requirement (1.3) coincides with the classical one for Finite Element Methods; see, e.g., [113, Eq. (3.1.43)] or [183, Definition 1.107].

Remark 1.11 (Degenerate Faces) A framework allowing for face degeneration has been proposed in [91] in the context of interior-penalty Discontinuous Galerkin methods, allowing one to use a discrete trace inequality sharper than (1.55) below; see also [18, 94]. In principle, one expects that this framework could be used herein with an appropriate adaptation of the penalty strategy. Notice, however, that the number of globally coupled unknowns in HHO methods is proportional to the number of mesh faces, so one should always make sure that the number of faces of each element stays bounded while refining. For this reason, we do not develop further this point here, and refer to the above references for details.

1.1.3 Geometric Bounds on Regular Mesh Sequences

We collect in the following lemma some useful geometric bounds that hold on regular mesh sequences. Here and throughout the book, $|X|_n$ denotes the n-dimensional Hausdorff measure of a set X (length if $n = 1$, area if $n = 2$, volume if $n = 3$).

Lemma 1.12 (Geometric Bounds on Regular Mesh Sequences) *Let $(\mathcal{M}_h)_{h\in\mathcal{H}}$ denote a regular mesh sequence in the sense of Definition 1.9. Then, the following results hold:*

(i) Bound on the number of faces. *There is an integer $N_\partial \geq d+1$, depending only on ϱ and d, such that*

$$\sup_{h\in\mathcal{H}} \max_{T\in\mathcal{T}_h} \mathrm{card}(\mathcal{F}_T) \leq N_\partial. \tag{1.5}$$

(ii) Comparison of element and face diameters. *For all $h \in \mathcal{H}$, all $T \in \mathcal{T}_h$, and all $F \in \mathcal{F}_T$, it holds that*

$$2\varrho^2 h_T \leq h_F \leq h_T. \tag{1.6}$$

(iii) Comparison of diameters and measures of elements and faces. *For all $h \in \mathcal{H}$, all $T \in \mathcal{T}_h$, and all $F \in \mathcal{F}_h$, it holds that*

$$\left(|\mathcal{B}_d|_d \varrho^{2d}\right) h_T^d \leq |T|_d \leq |\mathcal{B}_d|_d h_T^d \tag{1.7}$$

and

$$\left(|\mathcal{B}_{d-1}|_{d-1} \varrho^{2(d-1)}\right) h_F^{d-1} \leq |F|_{d-1} \leq |\mathcal{B}_{d-1}|_{d-1} h_F^{d-1}, \tag{1.8}$$

where, for $n = d$ or $n = d-1$, $\mathcal{B}_n$ is the unit ball in $\mathbb{R}^n$.

Proof (i) *Bound on the number of faces.* We start by proving that there is an integer $N \geq 0$, depending only on ϱ and d, such that

$$\sup_{h\in\mathcal{H}} \max_{T\in\mathcal{T}_h} \mathrm{card}(\mathfrak{T}_T) \leq N, \tag{1.9}$$

which means that every mesh element is decomposed into a number of submesh simplices that is bounded uniformly in h. Since each $T \in \mathcal{T}_h$ is contained in a ball of radius h_T and each $\tau \in \mathfrak{T}_h$ contains a ball of radius r_τ, we have that

$$
\begin{aligned}
|\mathcal{B}_d|_d h_T^d \geq |T|_d = \sum_{\tau \in \mathfrak{T}_T} |\tau|_d \geq \sum_{\tau \in \mathfrak{T}_T} |\mathcal{B}_d|_d r_\tau^d \\
&\geq \sum_{\tau \in \mathfrak{T}_T} |\mathcal{B}_d|_d \varrho^d h_\tau^d && \text{Eq. (1.3)} \\
&\geq \sum_{\tau \in \mathfrak{T}_T} |\mathcal{B}_d|_d \varrho^{2d} h_T^d && \text{Eq. (1.4)} \\
&= \operatorname{card}(\mathfrak{T}_T) |\mathcal{B}_d|_d \varrho^{2d} h_T^d,
\end{aligned}
\tag{1.10}
$$

and (1.9) follows with N the smallest natural number greater than ϱ^{-2d}. For all $h \in \mathcal{H}$ and all $T \in \mathcal{T}_h$, let now

$$
\mathfrak{F}_T := \{\sigma \in \mathfrak{F}_h \ : \ \sigma \subset \partial T\}
$$

denote the set of simplicial subfaces that lie on the boundary of T. Then, we have that

$$
\operatorname{card}(\mathcal{F}_T) \leq \operatorname{card}(\mathfrak{F}_T) \leq (d+1) \operatorname{card}(\mathfrak{T}_T) \leq (d+1)N, \tag{1.11}
$$

where we have used in the second bound the fact that a d-simplex has exactly $(d+1)$ faces, and (1.9) to conclude. The bound (1.5) follows with

$$
N_\partial = (d+1)N.
$$

(ii) *Comparison of element and face diameters.* Let a meshsize $h \in \mathcal{H}$, a mesh element $T \in \mathcal{T}_h$, and a mesh face $F \in \mathcal{F}_T$ be fixed. Since $F \subset \overline{T}$, it holds that $h_F \leq h_T$, which is the second inequality in (1.6). To prove the first inequality, let $\sigma \in \mathfrak{F}_h$ be such that $\sigma \subset F$, denote by h_σ its diameter, and let $\tau \in \mathfrak{T}_T$ be a simplex such that σ is contained in the boundary of τ. Let $\mathcal{B}_d(\boldsymbol{x}, r_\tau)$ be a ball of radius r_τ (the inradius of τ) centred at $\boldsymbol{x}$ and included in τ, and let β_σ be the intersection of this ball with the hyperplane parallel to σ and going through $\boldsymbol{x}$ (see Fig. 1.4). Then β_σ has diameter $2r_\tau$. Let H be the homothecy with centre the vertex $\boldsymbol{P}_\sigma$ of τ opposed to σ and that sends $\boldsymbol{x}$ on σ. H has ratio $\lambda > 1$, and sends β_σ onto $H(\beta_\sigma) \subset \sigma$ (this is due to the fact that τ is a simplex). Hence, the diameter $2\lambda r_\tau$ of $H(\beta_\sigma)$ is less than the diameter of σ, which shows that $h_\sigma \geq 2r_\tau$.

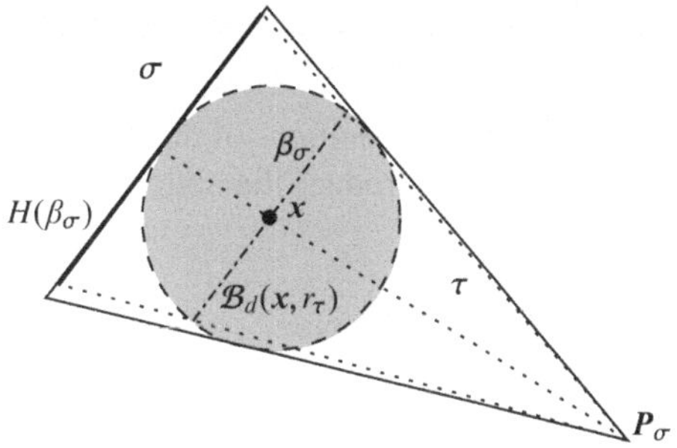

Fig. 1.4 Justification of $h_\sigma \geq 2r_\tau$ for the proof of (1.6)

It therefore holds that

$$h_F \geq h_\sigma \geq 2r_\tau \geq 2\varrho h_\tau \geq 2\varrho^2 h_T, \tag{1.12}$$

where we have used the shape (1.3) and contact regularity (1.4) conditions in the third and fourth inequalities, respectively.

(iii) *Comparison of diameters and measures of elements and faces.* The estimates (1.7) follow from (1.10), noticing that $\mathrm{card}(\mathfrak{T}_T) \geq 1$. The upper bound on $|F|_{d-1}$ in (1.8) is a simple consequence of the definition of h_F, which ensures that F is contained in a ball of radius h_F in the hyperplane (of dimension $d-1$) that it spans. For the lower bound, we recall that, with the notations used in the proof of Point (ii), F contains $\sigma \supset H(\beta_\sigma)$, where $H(\beta_\sigma)$ is a ball, in the hyperplane spanned by F, of radius $\lambda r_\tau \geq r_\tau$. Hence,

$$|F|_{d-1} \geq |\mathcal{B}_{d-1}|_{d-1} r_\tau^{d-1} \geq |\mathcal{B}_{d-1}|_{d-1} (\varrho^2 h_T)^{d-1},$$

where the conclusion follows from (1.12). The lower bound on $|F|_{d-1}$ stated in (1.8) then follows from (1.6). □

Remark 1.13 (Modification in Dimension $d = 1$) In dimension $d = 1$, Ω is an open interval and a mesh is a subdivision of Ω into intervals. The concept of regular mesh sequence in this situation is the following: There exists $\varrho > 0$ such that, for all $h \in \mathcal{H}$ and all intervals $T, T' \in \mathcal{T}_h$ sharing a common endpoint, $\varrho h_T \leq h_{T'}$. A "face" F is then just a node and therefore has zero diameter. The relation (1.6) no longer holds, but is also not necessary to the analysis. Also, whenever a length scale associated to a node $F \in \mathcal{F}_h$ is needed in this case, one can use, instead of $h_F = 0$, the average of the lengths of the intervals sharing F.

1.2 Function Spaces

The functional setting for the design and analysis of HHO methods is given by local and broken versions of the usual Lebesgue, Sobolev, $\boldsymbol{H}(\mathrm{div}; \Omega)$, and polynomial spaces on polytopal meshes. The corresponding notions are introduced in this section.

1.2.1 *Lebesgue and Sobolev Spaces*

We give here the definitions of the usual Lebesgue and Sobolev spaces, recall some basic facts, and introduce the notion of broken Sobolev space on a polytopal mesh.

1.2.1.1 Lebesgue Spaces

Let X denote an open bounded subset of $\mathbb{R}^n$, $n \geq 1$. In practice, the dimension n will usually be equal to d (e.g. when X is Ω or a mesh element), or $d-1$ (e.g. when X is a mesh face). We consider functions $v : X \to \mathbb{R}$ that are Lebesgue measurable and we denote by $\int_X v(\boldsymbol{x})\,\mathrm{d}\boldsymbol{x}$ the Lebesgue integral of v over X, when it exists (that is, v is non-negative or integrable). Whenever no ambiguity can arise, we omit both the dependence on $\boldsymbol{x}$ and on the measure in integrals, and simply write $\int_X v$. Let $p \in [1, \infty]$ be a real number. We set

$$\|v\|_{L^p(X)} := \begin{cases} \left(\int_X |v|^p\right)^{\frac{1}{p}} & \text{if } p \in [1, \infty), \\ \inf\{M \in \mathbb{R} \,:\, |v(\boldsymbol{x})| \leq M \text{ for a.e. } \boldsymbol{x} \in X\} & \text{if } p = \infty. \end{cases} \tag{1.13}$$

We define the *Lebesgue space*

$$L^p(X) := \left\{v \text{ Lebesgue measurable } : \|v\|_{L^p(X)} < \infty\right\}.$$

Equipped with the norm $\|\cdot\|_{L^p(X)}$, $L^p(X)$ is a Banach space (see, e.g., [186, p. 249] or [81, Proposition 9.1]). Moreover, if $p < +\infty$, the space $C_c^\infty(X)$ spanned by infinitely differentiable functions with compact support in X is dense in $L^p(X)$.

Remark 1.14 (The Case $p = 2$) For $p = 2$, $L^2(X)$ is a real Hilbert space when equipped with the scalar product

$$(v, w)_X := \int_X vw$$

and the associated norm $\|\cdot\|_X$. In what follows, we adopt the convention that the index X is omitted from both the inner product and the norm when $X = \Omega$. The same notation will be used for the spaces $L^2(X)^d$ and $L^2(X)^{d\times d}$ of vector- and tensor-valued, square-integrable functions. The use of a special notation for the case $p = 2$ is justified by the fact that the construction underlying HHO methods is inherently L^2-based, even when these methods are used for the approximation of problems posed in a non-Hilbertian setting such as the ones considered in Chap. 6.

A useful tool in Lebesgue spaces is the *Hölder inequality*: For all couples of conjugate Hölder exponents $(p, q) \in [1, \infty]^2$ such that $\frac{1}{p} + \frac{1}{q} = 1$, all $v \in L^p(X)$, and all $w \in L^q(X)$, there holds $vw \in L^1(X)$ and

$$\int_X |vw| \le \|v\|_{L^p(X)} \|w\|_{L^q(X)}.$$

The particular case $p = q = 2$ corresponds to the *Cauchy–Schwarz inequality*. The following generalisation of the Hölder inequality will also be useful to us: For all $(p, q, r) \in [1, \infty]^3$ such that $\frac{1}{p} + \frac{1}{q} + \frac{1}{r} = 1$, all $v \in L^p(X)$, $w \in L^q(X)$ and $z \in L^r(X)$, there holds $vwz \in L^1(X)$ and

$$\int_X |vwz| \le \|v\|_{L^p(X)} \|w\|_{L^q(X)} \|z\|_{L^r(X)}.$$

Most of the time, we will use this inequality with $(p, q, r) = (2, 2, \infty)$ and, for the analysis of Navier–Stokes equations in Chap. 9, with $(p, q, r) = (2, 4, 4)$. Straightforward generalisations of the Hölder inequality to products of four functions will also be occasionally required in Chap. 9 to estimate boundary contributions related to the convective term in the momentum equation.

1.2.1.2 Sobolev Spaces

Let X be as in the previous section and consider the Cartesian basis of $\mathbb{R}^n$ with coordinates $(x_1, \ldots, x_n)$. For $i \in \{1, \ldots, n\}$, we denote by ∂_i the distributional partial derivative with respect to x_i. For an n-tuple $\boldsymbol{\alpha} = (\alpha_1, \ldots, \alpha_n) \in \mathbb{N}^n$, $\partial^{\boldsymbol{\alpha}} v$ denotes the distributional derivative $\partial_1^{\alpha_1} \cdots \partial_n^{\alpha_n} v$ of v, with the convention that $\partial^{(0,\ldots,0)} v := v$. For all $p \in [1, \infty]$, we define the p-norm on $\mathbb{R}^n$ such that, for all $\boldsymbol{x} = (x_1, \ldots, x_n) \in \mathbb{R}^n$,

$$\|\boldsymbol{x}\|_p := \begin{cases} \left(\sum_{i=1}^n |x_i|^p \right)^{\frac{1}{p}} & \text{if } 1 \le p < \infty, \\ \max_{1 \le i \le n} |x_i| & \text{if } p = \infty. \end{cases} \tag{1.14}$$

Remark 1.15 (Notation for the Euclidean Norm) For the Euclidean norm obtained taking $p = 2$ in (1.14) we also use the less obtrusive notation $|\cdot|$ in the following chapters.

For any real number $p \in [1, \infty]$ and any integer $s \ge 0$, we define the *Sobolev space*

$$W^{s,p}(X) := \left\{ v \in L^p(X) \ : \ \partial^{\boldsymbol{\alpha}} v \in L^p(X) \quad \forall \boldsymbol{\alpha} \in A_n^s \right\},$$

with

$$A_n^s := \left\{\boldsymbol{\alpha} \in \mathbb{N}^n \ : \ \|\boldsymbol{\alpha}\|_1 \leq s\right\}. \tag{1.15}$$

We notice that

$$W^{0,p}(X) = L^p(X).$$

The Sobolev norm $\|\cdot\|_{W^{s,p}(X)}$ and seminorm $|\cdot|_{W^{s,p}(X)}$ are defined such that

$$\begin{aligned} \|v\|_{W^{s,p}(X)} &:= \sum_{\boldsymbol{\alpha}\in A_n^s} \|\partial^{\boldsymbol{\alpha}} v\|_{L^p(X)}, \\ |v|_{W^{s,p}(X)} &:= \sum_{\boldsymbol{\alpha}\in\mathbb{N}^n, \|\boldsymbol{\alpha}\|_1 = s} \|\partial^{\boldsymbol{\alpha}} v\|_{L^p(X)}. \end{aligned} \tag{1.16}$$

This choice enables a seamless treatment of the case $p = \infty$. Equipped with the norm $\|\cdot\|_{W^{s,p}(X)}$, $W^{s,p}(X)$ is a Banach space. The usual gradient operator $\nabla : W^{1,p}(X) \to L^p(X)^n$ is such that, for all $v \in W^{1,p}(X)$,

$$\nabla v := \begin{pmatrix} \partial_1 v \\ \vdots \\ \partial_n v \end{pmatrix}.$$

We classically denote by $W_0^{s,p}(X)$ the closure of $C_c^\infty(X)$ in $W^{s,p}(X)$.

Remark 1.16 (Hilbert Spaces) For $p = 2$, we introduce the special notations

$$H^s(X) := W^{s,2}(X) \quad \text{and} \quad H_0^s(X) := W_0^{s,2}(X).$$

These notations are reminiscent of the fact that $H^s(X)$ and $H_0^s(X)$ are Hilbert spaces when equipped with the scalar product

$$(v, w)_{H^s(X)} := \sum_{\boldsymbol{\alpha}\in A_n^s} (\partial^{\boldsymbol{\alpha}} v, \partial^{\boldsymbol{\alpha}} w)_X.$$

The corresponding norm is equivalent to (but not coincident with if $s \neq 0$) the one obtained setting $p = 2$ in (1.16).

In the context of diffusive PDE problems set on a d-dimensional domain Ω as in Assumption 1.3, the flux (see the introduction to Chap. 2 for this nomenclature) is a vector-valued function that belongs to the space

$$\boldsymbol{H}(\text{div}; \Omega) := \left\{\boldsymbol{\tau} = (\tau_1, \ldots, \tau_d) \in L^2(\Omega)^d \ : \ \sum_{i=1}^d \partial_i \tau_i \in L^2(\Omega)\right\}. \tag{1.17}$$

We classically denote the divergence $\sum_{i=1}^{d} \partial_i \tau_i$ of $\boldsymbol{\tau}$ by $\nabla\cdot\boldsymbol{\tau}$. For a given $p \in [1, \infty]$, we generalise $\boldsymbol{H}(\mathrm{div};\Omega)$ to the L^p setting

$$\boldsymbol{W}^p(\mathrm{div};\Omega) := \left\{ \boldsymbol{\tau} \in L^p(\Omega)^d \ :\ \nabla\cdot\boldsymbol{\tau} \in L^p(\Omega) \right\}. \tag{1.18}$$

Of course, $\boldsymbol{H}(\mathrm{div};\Omega) = \boldsymbol{W}^2(\mathrm{div};\Omega)$.

1.2.2 Broken Sobolev Spaces

Let $\mathcal{M}_h$ denote a polytopal mesh of Ω in the sense of Definition 1.4. With s and p as in the previous section, we define the broken Sobolev space

$$W^{s,p}(\mathcal{T}_h) := \left\{ v \in L^p(\Omega) \ :\ v_{|T} \in W^{s,p}(T) \quad \forall T \in \mathcal{T}_h \right\}.$$

The broken Sobolev norm and seminorm are defined similarly to (1.16): For $p < \infty$,

$$\begin{aligned} \|v\|_{W^{s,p}(\mathcal{T}_h)} &:= \sum_{\boldsymbol{\alpha}\in \boldsymbol{A}_n^s} \left(\sum_{T\in\mathcal{T}_h} \|\partial^{\boldsymbol{\alpha}} v\|_{L^p(T)}^p \right)^{\frac{1}{p}}, \\ |v|_{W^{s,p}(\mathcal{T}_h)} &:= \sum_{\boldsymbol{\alpha}\in\mathbb{N}^n,\, \|\boldsymbol{\alpha}\|_1 = s} \left(\sum_{T\in\mathcal{T}_h} \|\partial^{\boldsymbol{\alpha}} v\|_{L^p(T)}^p \right)^{\frac{1}{p}}. \end{aligned} \tag{1.19}$$

For $p = \infty$, we set

$$\begin{aligned} \|v\|_{W^{s,\infty}(\mathcal{T}_h)} &:= \sum_{\boldsymbol{\alpha}\in \boldsymbol{A}_n^s} \max_{T\in\mathcal{T}_h} \|\partial^{\boldsymbol{\alpha}} v\|_{L^\infty(T)}, \\ |v|_{W^{s,\infty}(\mathcal{T}_h)} &:= \sum_{\boldsymbol{\alpha}\in\mathbb{N}^n,\, \|\boldsymbol{\alpha}\|_1 = s} \max_{T\in\mathcal{T}_h} \|\partial^{\boldsymbol{\alpha}} v\|_{L^\infty(T)}. \end{aligned} \tag{1.20}$$

The case $s = 1$ will play an important role in the rest of this book, and deserves further discussion. Functions in $W^{1,p}(\mathcal{T}_h)$ do not admit a global weak gradient in general. We can, however, define the broken gradient operator $\nabla_h : W^{1,p}(\mathcal{T}_h) \to L^p(\Omega)^d$ such that, for all $v \in W^{1,p}(\mathcal{T}_h)$,

$$(\nabla_h v)_{|T} := \nabla(v_{|T}) \qquad \forall T \in \mathcal{T}_h. \tag{1.21}$$

Similarly, setting

$$\boldsymbol{W}^p(\mathrm{div};\mathcal{T}_h) := \left\{ \boldsymbol{\tau} \in L^p(\Omega)^d \ :\ \boldsymbol{\tau}_{|T} \in \boldsymbol{W}^p(\mathrm{div};T) \quad \forall T \in \mathcal{T}_h \right\}$$

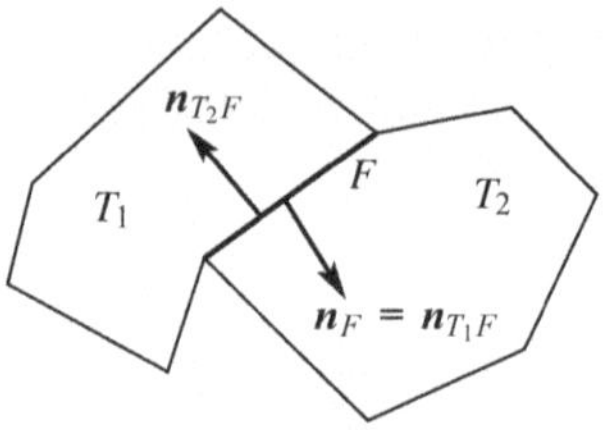

Fig. 1.5 Arbitrary numbering of the elements around a face F, and oriented normal $\boldsymbol{n}_F$

(where $\boldsymbol{W}^p(\text{div}; T)$ is defined by (1.18) with $\Omega = T$), the broken divergence $\boldsymbol{\nabla}_h\cdot$: $\boldsymbol{W}^p(\text{div}; \mathcal{T}_h) \to L^p(\Omega)$ is given by, for all $\boldsymbol{\tau} \in \boldsymbol{W}^p(\text{div}; \mathcal{T}_h)$,

$$(\boldsymbol{\nabla}_h\cdot\boldsymbol{\tau})_{|T} := \boldsymbol{\nabla}\cdot(\boldsymbol{\tau}_{|T}) \qquad \forall T \in \mathcal{T}_h.$$

For any $F \in \mathcal{F}_h^{\text{i}}$, denote by T_1 and T_2 the distinct elements of $\mathcal{T}_h$ such that $F \subset \partial T_1 \cap \partial T_2$. In what follows, we fix an arbitrary numbering of T_1 and T_2 and introduce the jump operator such that, for any function v smooth enough to admit a (possibly two-valued) trace on F,

$$[v]_F := \left(v_{|T_1}\right)_{|F} - \left(v_{|T_2}\right)_{|F}. \tag{1.22}$$

Throughout the rest of the book, in order to alleviate the notation, we will omit the restriction to F when using the definition (1.22) and simply write $[v]_F = v_{|T_1} - v_{|T_2}$. We also let (see Fig. 1.5)

$$\boldsymbol{n}_F := \boldsymbol{n}_{T_1F} = -\boldsymbol{n}_{T_2F}.$$

On boundary faces, we take $\boldsymbol{n}_F$ to be the unit normal vector to $\partial\Omega$ pointing out of Ω. An important characterisation of functions in $\boldsymbol{W}^p(\text{div}; \Omega)$ is contained in the following lemma.

Lemma 1.17 (Characterisation of $\boldsymbol{W}^p(\text{div}; \Omega)$) *Let a real number $p \in [1, \infty]$ be fixed, and let $\boldsymbol{\tau} \in W^{1,p}(\mathcal{T}_h)^d$. Then, $\boldsymbol{\tau} \in \boldsymbol{W}^p(\text{div}; \Omega)$ if and only if*

$$[\boldsymbol{\tau}]_F\cdot\boldsymbol{n}_F = 0 \qquad \forall F \in \mathcal{F}_h^{\text{i}}. \tag{1.23}$$

Remark 1.18 (Additional Regularity) The regularity $\boldsymbol{\tau} \in W^{1,p}(\mathcal{T}_h)^d$ can be weakened into $\boldsymbol{\tau} \in \boldsymbol{W}^p(\text{div}; \mathcal{T}_h)$, but the values of $\boldsymbol{\tau}$ on each side of $F \in \mathcal{F}_h^{\text{i}}$ have then to be understood in a weak sense (these “traces” might not be functions on the interfaces, see [145, Section 1.2.6] and references therein). This subtlety will however not be useful in this book.

Proof Let $\varphi \in C_c^\infty(\Omega)$. Integrating by parts element by element, and accounting for the fact that φ is smooth inside Ω and vanishes on $\partial\Omega$, we obtain

$$\begin{aligned}
\int_\Omega \boldsymbol{\tau}\cdot\nabla\varphi &= \sum_{T\in\mathcal{T}_h}\int_T \boldsymbol{\tau}_{|T}\cdot\nabla\varphi \\
&= -\sum_{T\in\mathcal{T}_h}\int_T (\nabla\cdot\boldsymbol{\tau}_{|T})\varphi + \sum_{T\in\mathcal{T}_h}\sum_{F\in\mathcal{F}_T}\int_F (\boldsymbol{\tau}_{|T}\cdot\boldsymbol{n}_{TF})\varphi \\
&= -\sum_{T\in\mathcal{T}_h}\int_T (\nabla\cdot\boldsymbol{\tau}_{|T})\varphi + \sum_{F\in\mathcal{F}_h}\int_F \left(\sum_{T\in\mathcal{T}_F}\boldsymbol{\tau}_{|T}\cdot\boldsymbol{n}_{TF}\right)\varphi \\
&= -\int_\Omega (\nabla_h\cdot\boldsymbol{\tau})\varphi + \sum_{F\in\mathcal{F}_h^{\mathrm{i}}}\int_F [\boldsymbol{\tau}]_F\cdot\boldsymbol{n}_F\varphi.
\end{aligned} \tag{1.24}$$

In the third line, we have exchanged the sums over faces and elements according to

$$\sum_{T\in\mathcal{T}_h}\sum_{F\in\mathcal{F}_T}\bullet = \sum_{F\in\mathcal{F}_h}\sum_{T\in\mathcal{T}_F}\bullet \tag{1.25}$$

and used the fact that φ is continuous across interfaces, while the conclusion holds using the definition of the jumps across internal faces and the fact that φ vanishes on boundary faces.

Assume that (1.23) holds. Then, (1.24) shows that

$$\int_\Omega \boldsymbol{\tau}\cdot\nabla\varphi = -\int_\Omega (\nabla_h\cdot\boldsymbol{\tau})\varphi, \tag{1.26}$$

which precisely states that $\nabla\cdot\boldsymbol{\tau} = \nabla_h\cdot\boldsymbol{\tau} \in L^p(\Omega)$, and thus that $\boldsymbol{\tau} \in \boldsymbol{W}^p(\mathrm{div};\Omega)$.

Conversely, if $\boldsymbol{\tau} \in \boldsymbol{W}^p(\mathrm{div};\Omega)$, then $\nabla\cdot\boldsymbol{\tau} = \nabla_h\cdot\boldsymbol{\tau}$ and thus (1.26) holds. Combined with (1.24), this gives

$$\sum_{F\in\mathcal{F}_h^{\mathrm{i}}}\int_F [\boldsymbol{\tau}]_F\cdot\boldsymbol{n}_F\varphi = 0.$$

Fix $F \in \mathcal{F}_h^{\mathrm{i}}$. Take $g \in C_c^\infty(F)$ and extend it into a function $\varphi \in C_c^\infty(\Omega)$ whose support does not intersect any other face. With this choice, the relation above shows that $\int_F [\boldsymbol{\tau}]_F\cdot\boldsymbol{n}_F g = 0$. Since this holds for any $g \in C_c^\infty(F)$, this shows that $[\boldsymbol{\tau}]_F\cdot\boldsymbol{n}_F = 0$ on F, and thus that (1.23) holds. □

The following corollary will often be invoked in the exposition.

Corollary 1.19 (Cancellation of Boundary Terms) *Take $p \in [1, \infty]$, and let $p' \in [1, \infty]$ be such that $\frac{1}{p} + \frac{1}{p'} = 1$. Let $\boldsymbol{\tau} \in \boldsymbol{W}^p(\mathrm{div}; \Omega) \cap W^{1,p}(\mathcal{T}_h)^d$ and $(\varphi_F)_{F \in \mathcal{F}_h}$ denote a family of functions such that $\varphi_F \in L^{p'}(F)$ for all $F \in \mathcal{F}_h$. Then, it holds*

$$\sum_{T \in \mathcal{T}_h} \sum_{F \in \mathcal{F}_T} \int_F (\boldsymbol{\tau}_{|T} \cdot \boldsymbol{n}_{TF})\, \varphi_F = \sum_{F \in \mathcal{F}_h^{\mathrm{b}}} \int_F (\boldsymbol{\tau} \cdot \boldsymbol{n}_F)\, \varphi_F. \tag{1.27}$$

In particular, if $\varphi_F = 0$ or $\boldsymbol{\tau} \cdot \boldsymbol{n}_F = 0$ for all $F \in \mathcal{F}_h^{\mathrm{b}}$,

$$\sum_{T \in \mathcal{T}_h} \sum_{F \in \mathcal{F}_T} \int_F (\boldsymbol{\tau}_{|T} \cdot \boldsymbol{n}_{TF})\, \varphi_F = 0. \tag{1.28}$$

Remark 1.20 (Regularity Assumptions) As in Remark 1.18, we notice that the results of Corollary 1.19 extend to certain situations where $\boldsymbol{\tau}$ does not belong to $W^{1,p}(\mathcal{T}_h)^d$. Essentially, if $\boldsymbol{\tau} \in \boldsymbol{W}^p(\mathrm{div}; \Omega)$ and the family $(\varphi_F)_{F \in \mathcal{F}_h}$ are such that $(\boldsymbol{\tau}_{|T} \cdot \boldsymbol{n}_{TF})\varphi_F$ makes sense, for all $F \in \mathcal{F}_h$, as an integrable function over F, then (1.27) and (1.28) hold.

Proof Exchanging the order of the summations over elements and faces according to (1.25), we can write

$$\begin{aligned}
\sum_{T \in \mathcal{T}_h} \sum_{F \in \mathcal{F}_T} \int_F (\boldsymbol{\tau}_{|T} \cdot \boldsymbol{n}_{TF})\, \varphi_F &= \sum_{F \in \mathcal{F}_h} \int_F \left(\sum_{T \in \mathcal{T}_F} \boldsymbol{\tau}_{|T} \cdot \boldsymbol{n}_{TF} \right) \varphi_F \\
&= \sum_{F \in \mathcal{F}_h^{\mathrm{i}}} \int_F \cancel{([\boldsymbol{\tau}]_F \cdot \boldsymbol{n}_F)}\, \varphi_F + \sum_{F \in \mathcal{F}_h^{\mathrm{b}}} \int_F (\boldsymbol{\tau} \cdot \boldsymbol{n}_F)\, \varphi_F,
\end{aligned}$$

where, in the second line, we have split the summation over $\mathcal{F}_h = \mathcal{F}_h^{\mathrm{i}} \cup \mathcal{F}_h^{\mathrm{b}}$, used the definition (1.22) of the jump operator for the first term, and invoked Lemma 1.17 to cancel the normal jumps of $\boldsymbol{\tau}$ across interfaces. □

A characterisation of $W^{1,p}(\Omega)$ can be deduced from the previous lemma.

Lemma 1.21 (Characterisation of $W^{1,p}(\Omega)$) *Let a real number $p \in [1, \infty]$ be fixed. Then, a function $v \in W^{1,p}(\mathcal{T}_h)$ is in $W^{1,p}(\Omega)$ if and only if*

$$[v]_F = 0 \qquad \forall F \in \mathcal{F}_h^{\mathrm{i}}. \tag{1.29}$$

Proof Let $(e^i)_{i=1,\dots,d}$ be the canonical basis of $\mathbb{R}^d$. For $i \in \{1,\dots,d\}$, consider $\tau^i = ve^i \in W^{1,p}(\mathcal{T}_h)^d$.

If $v \in W^{1,p}(\Omega)$, then $\tau^i \in W^{1,p}(\Omega)^d \subset \boldsymbol{W}^p(\mathrm{div};\Omega)$ and thus, by Lemma 1.17, for any $F \in \mathcal{F}_h^{\mathrm{i}}$, we have $0 = [\tau^i]_F\cdot n_F = [v]_F e^i\cdot n_F$. Since, for any $F \in \mathcal{F}_h^{\mathrm{i}}$, there is at least one i such that $e^i\cdot n_F \neq 0$, we deduce that $[v]_F = 0$ on F.

Conversely, if $[v]_F = 0$ on any $F \in \mathcal{F}_h^{\mathrm{i}}$, then $[\tau^i]_F = 0$ on all interfaces and Lemma 1.17 shows that $\tau^i \in \boldsymbol{W}^p(\mathrm{div};\Omega)$. Since $\nabla\cdot\tau^i = \partial_i v$, this partial derivative of v belongs to $L^p(\Omega)$. Being true for all $i \in \{0,\dots,d\}$, this shows that $v \in W^{1,p}(\Omega)$. □

1.2.3 Polynomial Spaces

The discrete unknowns in HHO methods are local polynomials over mesh elements and faces. The goal of this section is to make this notion precise and to recall some fundamental inequalities that hold on local polynomial spaces.

1.2.3.1 The Polynomial Space $\mathbb{P}_n^l$

Let $n \geq 1$ and $l \geq 0$ be two integers and, recalling (1.15), set

$$N_n^l := \mathrm{card}(\boldsymbol{A}_n^l) = \binom{l+n}{n}. \tag{1.30}$$

We define the space of n-variate polynomials of total degree l as

$$\mathbb{P}_n^l := \Big\{p : \mathbb{R}^n \to \mathbb{R} \,:\, \exists(\gamma_{\boldsymbol{\alpha}})_{\boldsymbol{\alpha}\in\boldsymbol{A}_n^l} \in \mathbb{R}^{N_n^l} \text{ such that } p(\boldsymbol{x}) = \sum_{\boldsymbol{\alpha}\in\boldsymbol{A}_n^l} \gamma_{\boldsymbol{\alpha}}\boldsymbol{x}^{\boldsymbol{\alpha}} \text{ for all } \boldsymbol{x}\in\mathbb{R}^n\Big\},$$

where, for a given multi-index $\boldsymbol{\alpha} \in \boldsymbol{A}_n^l$, we have set

$$\boldsymbol{x}^{\boldsymbol{\alpha}} := x_1^{\alpha_1}\cdots x_n^{\alpha_n}.$$

We recall that "polynomial of degree l" is actually an abuse of terminology, committed throughout the book, for "polynomial of degree l or less". The dimension of the vector space $\mathbb{P}_n^l$ is

$$\dim(\mathbb{P}_n^l) = N_n^l. \tag{1.31}$$

1.2.3.2 Local and Broken Polynomial Spaces

Definition 1.22 (Local Polynomial Space) Let $X \subset \mathbb{R}^n$, $n \geq 1$, be an open bounded connected set, and let an integer $l \geq 0$ be fixed. The *local (real-valued) polynomial space* $\mathbb{P}^l(X)$ is defined as the space spanned by the restrictions to X of functions in the polynomial space $\mathbb{P}^l_n$.

More generally, if V is a finite-dimensional vector space, the *V-valued local polynomial space* $\mathbb{P}^l(X; V)$ is the space of functions $f : X \to V$ such that the components of f on a basis of V belong to $\mathbb{P}^l(X)$; we note that this definition does not depend on the chosen basis (if the components in one basis are polynomial, then the components in any basis are polynomial).

HHO methods hinge on local polynomial spaces defined over mesh elements and faces.

Proposition 1.23 (Dimension of Local Polynomial Spaces on Mesh Elements and Faces) *Let $\mathcal{M}_h = (\mathcal{T}_h, \mathcal{F}_h)$ denote a polytopal mesh in the sense of Definition 1.4. Then it holds, for all $T \in \mathcal{T}_h$,*

$$\dim(\mathbb{P}^l(T)) = N^l_d \tag{1.32}$$

and, for all $F \in \mathcal{F}_h$,

$$\dim(\mathbb{P}^l(F)) = N^l_{d-1}. \tag{1.33}$$

Proof Relation (1.32) is an immediate consequence of (1.31), and of the fact that two polynomial functions that coincide on the set T, which is non-empty and open in $\mathbb{R}^d$, coincide everywhere, so that $\mathbb{P}^l_d \ni f \mapsto f_{|T} \in \mathbb{P}^l(T)$ is an isomorphism.

Relation (1.33), on the other hand, hinges on the assumption that faces are planar and non-degenerate: For any face F, there is a unique affine hyperplane H_F in $\mathbb{R}^d$ containing F, and in which F is open. Take an affine bijective mapping $\boldsymbol{T}_F : \mathbb{R}^{d-1} \to H_F$. The space $\mathbb{P}^l(H_F)$ can be described as $\mathbb{P}^l(H_F) = \mathbb{P}^l_{d-1} \circ \boldsymbol{T}_F^{-1}$, so that $\mathbb{P}^l(H_F)$ is isomorphic to $\mathbb{P}^l_{d-1}$ and has dimension N^l_{d-1}. Finally, with the same argument as for $\mathbb{P}^l(T)$, since F is non-empty and open in H_F, we see that $\mathbb{P}^l(F)$ is isomorphic to $\mathbb{P}^l(H_F)$ (hence also to $\mathbb{P}^l_{d-1}$). □

Definition 1.24 (Broken Polynomial Space) Let $\mathcal{M}_h = (\mathcal{T}_h, \mathcal{F}_h)$ denote a polytopal mesh of Ω in the sense of Definition 1.4, and let an integer $l \geq 0$ be given. We define the *broken polynomial space*

$$\mathbb{P}^l(\mathcal{T}_h) := \left\{ v_h \in L^1(\Omega) \ : \ v_{h|T} \in \mathbb{P}^l(T) \quad \forall T \in \mathcal{T}_h \right\}.$$

Crucially, the functions in $\mathbb{P}^l(\mathcal{T}_h)$ are possibly discontinuous at the interfaces in $\mathcal{F}^{\mathrm{i}}_h$.

1.2.4 Convention for Inequalities up to a Positive Constant

Throughout this book, many geometric or functional estimates are written in terms of inequalities that hold up to a multiplicative quantity depending on some parameters and independent of others. We therefore write

"*$A \lesssim B$ (resp. $A \gtrsim B$) with hidden constant depending only on X, Y, etc.*"

to mean that there exists C depending only on X, Y, etc. such that $A \le CB$ (resp. $A \ge CB$). Similarly,

"*$A \lesssim B$ (resp. $A \gtrsim B$) with hidden constant independent of X, Y, etc.*"

means that there exists C independent of X, Y, etc. such that $A \le CB$ (resp. $A \ge CB$). The notation

$$A \simeq B$$

is used as a shorthand for $A \lesssim B$ and $B \lesssim A$, with the prescribed dependency of the hidden constants. Unless otherwise specified, when these notations are used inside the proof of a certain estimate, it is assumed that the dependency of the hidden constants is the same as for the estimate itself.

1.2.5 Lebesgue and Sobolev Embeddings in Local Polynomial Spaces

The following result enables the comparison of Lebesgue norms on local polynomial spaces.

Lemma 1.25 (Direct and Inverse Lebesgue Embeddings in Local Polynomial Spaces) *Let $n \ge 1$ be a natural number and X be an open bounded connected subset of $\mathbb{R}^n$, with inradius r_X and diameter h_X. Let $\varrho > 0$ be a real number such that*

$$\varrho h_X \le r_X. \tag{1.34}$$

Let an integer $l \ge 0$ and two real numbers $q, m \in [1, \infty]$ be fixed. Then, for all $w \in \mathbb{P}^l(X)$, it holds that

$$\|w\|_{L^q(X)} \simeq |X|_n^{\frac{1}{q}-\frac{1}{m}} \|w\|_{L^m(X)}, \tag{1.35}$$

where we recall that $|X|_n$ denotes the Lebesgue measure of X in $\mathbb{R}^n$, and the hidden constants depend only on n, l, ϱ, q and m. The norm equivalence (1.35) *also holds, with the same dependency of the hidden constants, if X is an open bounded connected subset of $\mathbb{R}^n$ such that*

$$\overline{X} = \bigcup_{Y \in \mathfrak{T}_X} \overline{Y} \quad \text{and} \quad \operatorname{card}(\mathfrak{T}_X) \leq \varrho^{-1}, \tag{1.36a}$$

with $\mathfrak{T}_X$ denoting a family of open connected subsets of X that satisfy

$$\varrho h_Y \leq r_Y \text{ and } \varrho h_X \leq h_Y \qquad \forall Y \in \mathfrak{T}_X. \tag{1.36b}$$

Remark 1.26 (Inverse Embeddings) For $w \in \mathbb{P}^l(X)$ and $q \leq m$, the inequality

$$\|w\|_{L^q(X)} \lesssim |X|_n^{\frac{1}{q}-\frac{1}{m}} \|w\|_{L^m(X)}, \tag{1.37}$$

with hidden constant having the same dependency as in (1.35) (actually, in this case, the hidden constant is 1), is a classical direct Lebesgue embedding due to the Hölder inequality. If $m < q$, on the other hand, it holds solely because we consider polynomials. In this case, as $h_X \to 0$, we have $|X|_n \to 0$, and thus the scaling factor $|X|_n^{\frac{1}{q}-\frac{1}{m}}$ explodes since its exponent is negative.

Remark 1.27 (Conditions (1.36) *for Mesh Elements and Faces)* The geometrical conditions of Lemma 1.25 are met by X element or face of a polytopal mesh from a regular sequence (cf. Definition 1.9). When $X = T \in \mathcal{T}_h$, it suffices to take $\mathfrak{T}_X = \mathfrak{T}_T$, so that (1.36a) holds owing to (1.9) and replacing ϱ with $\min(\varrho, N^{-1})$, while the conditions (1.36b) coincide with (1.3) and (1.4), respectively. When $X = F \in \mathcal{F}_h$, on the other hand, we can take $\mathfrak{T}_X = \mathfrak{F}_F$, with $\mathfrak{F}_F$ denoting the set of simplicial subfaces $\sigma \subset F$. Property (1.36a) holds by (1.11). To check condition (1.36b), let $\sigma \in \mathfrak{F}_F$ be given, and let $T \in \mathcal{T}_F$ and $\tau \in \mathfrak{T}_T$ be such that $\sigma \subset \partial\tau$. Reasoning as in Point (ii) of the proof of Lemma 1.12, it is inferred that $h_\sigma \leq h_\tau \leq \varrho^{-1} r_\tau \leq \varrho^{-1} r_\sigma$. Moreover, we can write $h_F \leq h_T \leq \varrho^{-1} h_\tau \leq \frac{1}{2}\varrho^{-2} h_\sigma$, where we have used (1.4) and (1.12) in the second and third inequalities, respectively. Hence, the regularity conditions (1.36b) follow replacing ϱ with $\min(2\varrho^2, \varrho)$.

Proof (Lemma 1.25) (i) *Proof of* (1.35) *with X satisfying* (1.34). Since the indices m and q play symmetrical roles in (1.35), we only have to extend (1.37) to arbitrary q and m. For $\boldsymbol{x} \in \mathbb{R}^n$ and $r \geq 0$, let $\mathcal{B}_n(\boldsymbol{x}, r)$ denote the ball of centre $\boldsymbol{x}$ and radius r, and set, for the sake of brevity, $\mathcal{B}_n := \mathcal{B}_n(\mathbf{0}, 1)$. By (1.34) and the definitions of h_X and r_X, there is $\boldsymbol{x}_X \in X$ such that

$$\mathcal{B}_n(\boldsymbol{x}_X, \varrho h_X) \subset X \subset \mathcal{B}_n(\boldsymbol{x}_X, h_X). \tag{1.38}$$

Hence, $|\mathcal{B}_n|_n \varrho^n h_X^n \le |X|_n \le |\mathcal{B}_n|_n h_X^n$ and thus

$$|X|_n \simeq h_X^n. \tag{1.39}$$

Let

$$\widehat{X} := \frac{1}{h_X}(X - \boldsymbol{x}_X).$$

Using the linear change of variable $X \ni \boldsymbol{x} \mapsto \widehat{\boldsymbol{x}} = \frac{1}{h_X}(\boldsymbol{x} - \boldsymbol{x}_X) \in \widehat{X}$ and the relation (1.39), we see that, for any $s \in [1, \infty]$,

$$\|w\|_{L^s(X)} = h_X^{\frac{n}{s}} \|\widehat{w}\|_{L^s(\widehat{X})} \simeq |X|_n^{\frac{1}{s}} \|\widehat{w}\|_{L^s(\widehat{X})}, \tag{1.40}$$

where we have set $\widehat{w}(\widehat{\boldsymbol{x}}) := w(\boldsymbol{x}_X + h_X \widehat{\boldsymbol{x}})$ and, here, the hidden constants in $\simeq$ additionally depend on s. Assume now that, for all $\widehat{v} \in \mathbb{P}^l(\widehat{X})$,

$$\|\widehat{v}\|_{L^q(\widehat{X})} \lesssim \|\widehat{v}\|_{L^m(\widehat{X})}. \tag{1.41}$$

Then, combining this inequality with (1.40) for $s = q$ and $s = m$, since $\widehat{w} \in \mathbb{P}^l(\widehat{X})$, we have

$$\|w\|_{L^q(X)} \lesssim |X|_n^{\frac{1}{q}} \|\widehat{w}\|_{L^q(\widehat{X})} \lesssim |X|_n^{\frac{1}{q}} \|\widehat{w}\|_{L^m(\widehat{X})} \simeq |X|_n^{\frac{1}{q} - \frac{1}{m}} \|w\|_{L^m(X)},$$

and the lemma is proved.

It remains to establish (1.41). To this end we notice that, by (1.38), we have

$$\mathcal{B}_n(\boldsymbol{0}, \varrho) \subset \widehat{X} \subset \mathcal{B}_n. \tag{1.42}$$

Since $\|\cdot\|_{L^q(\mathcal{B}_n)}$ and $\|\cdot\|_{L^m(\mathcal{B}_n(\boldsymbol{0},\varrho))}$ are both norms on $\mathbb{P}^l(\mathbb{R}^n)$ (any polynomial that vanishes on a ball vanishes everywhere), they are equivalent on this finite dimensional space. Hence, for all $\widehat{v} \in \mathbb{P}^l(\widehat{X})$, considering $\widehat{v}$ as an element of $\mathbb{P}^l(\mathbb{R}^n)$,

$$\|\widehat{v}\|_{L^q(\mathcal{B}_n)} \lesssim \|\widehat{v}\|_{L^m(\mathcal{B}_n(\boldsymbol{0},\varrho))}, \tag{1.43}$$

with hidden constant depending only on n, q, l, m and ϱ. To prove (1.41), it then suffices to write, using (1.42),

$$\|\widehat{v}\|_{L^q(\widehat{X})} \le \|\widehat{v}\|_{L^q(\mathcal{B}_n)} \lesssim \|\widehat{v}\|_{L^m(\mathcal{B}_n(\boldsymbol{0},\varrho))} \le \|\widehat{v}\|_{L^m(\widehat{X})}.$$

(ii) *Proof of* (1.35) *with X satisfying* (1.36). In view of Remark 1.26, it suffices to prove (1.37) for $q > m$. By virtue of the first inequality in (1.36b) and Point (i) in this proof it holds, for any $Y \in \mathfrak{T}_X$,

$$\|w\|_{L^q(Y)} \lesssim |Y|_n^{\frac{1}{q}-\frac{1}{m}} \|w\|_{L^m(Y)}. \tag{1.44}$$

Using successively the definition (1.1) of h_X, the second and then first inequalities in (1.36b), and finally the definition of r_Y, we can estimate

$$|X|_n \leq |\mathcal{B}_n|_n h_X^n \leq |\mathcal{B}_n|_n h_Y^n \varrho^{-n} \leq |\mathcal{B}_n|_n r_Y^n \varrho^{-2n} \leq |Y|_n \varrho^{-2n},$$

so that, in particular, $|Y|_n^{\frac{1}{q}-\frac{1}{m}} \lesssim |X|_n^{\frac{1}{q}-\frac{1}{m}}$ since $\frac{1}{q} - \frac{1}{m} < 0$ having assumed $q > m$. Using $\|w\|_{L^m(Y)} \leq \|w\|_{L^m(X)}$ (a consequence of (1.36a)), we infer from (1.44) that

$$\|w\|_{L^q(Y)} \lesssim |X|_n^{\frac{1}{q}-\frac{1}{m}} \|w\|_{L^m(X)}. \tag{1.45}$$

We now distinguish two cases: $q < \infty$ and $q = \infty$. If $q < \infty$, taking the power q of the above inequality and summing over $Y \in \mathfrak{T}_X$, we infer

$$\|w\|_{L^q(X)}^q \leq \sum_{Y \in \mathfrak{T}_X} \|w\|_{L^q(Y)}^q \lesssim \operatorname{card}(\mathfrak{T}_X) |X|_n^{1-\frac{q}{m}} \|w\|_{L^m(X)}^q.$$

Using the bound on $\operatorname{card}(\mathfrak{T}_X)$ stated in (1.36a), estimate (1.37) follows. If $q = \infty$, on the other hand, we observe that $\|w\|_{L^\infty(X)} = \max_{Y \in \mathfrak{T}_X} \|w\|_{L^\infty(Y)}$ (a consequence of (1.36a)) and take the maximum over $Y \in \mathfrak{T}_X$ of (1.45) to obtain (1.37). □

In practice, one is also interested in the comparison of Sobolev seminorms of local polynomial functions. A key intermediate result in this direction is provided by the following lemma.

Lemma 1.28 (Discrete Inverse Inequality in Local Polynomial Spaces) *Let X be an open bounded connected subset of* $\mathbb{R}^n$ *that satisfies* (1.34) *or* (1.36). *Let an integer* $l \geq 0$ *and a real number* $p \in [1, \infty]$ *be fixed. Then, the following inverse inequality holds: For all* $v \in \mathbb{P}^l(X)$,

$$\|\nabla v\|_{L^p(X)^n} \lesssim h_X^{-1} \|v\|_{L^p(X)}, \tag{1.46}$$

with hidden constant depending only on n, ϱ*, l and p.*

Proof (i) *Proof of* (1.46) *with X satisfying* (1.34). We use the same notations and change of variable

$$X \ni \boldsymbol{x} \mapsto \widehat{\boldsymbol{x}} = \frac{1}{h_X}(\boldsymbol{x} - \boldsymbol{x}_X) \in \widehat{X}$$

as in Point (i) of the proof of Lemma 1.25. Since $\widehat{v}(\widehat{\boldsymbol{x}}) = v(\boldsymbol{x}_X + h_X\widehat{\boldsymbol{x}})$, we have

$$\widehat{\nabla v}(\widehat{\boldsymbol{x}}) := \nabla v(\boldsymbol{x}_X + h_X\widehat{\boldsymbol{x}}) = h_X^{-1}\widehat{\nabla}\widehat{v}(\widehat{\boldsymbol{x}}),$$

where ∇ is the gradient with respect to $\boldsymbol{x} \in X$ and $\widehat{\nabla}$ is the gradient with respect to $\widehat{\boldsymbol{x}} \in \widehat{X}$. Hence, applying (1.40) to $s = p$ and $w =$ components of ∇v, and using $\widehat{X} \subset \mathcal{B}_n$, we obtain

$$\begin{aligned}\|\nabla v\|_{L^p(X)^n} \simeq |X|_n^{\frac{1}{p}} \|\widehat{\nabla v}\|_{L^p(\widehat{X})^n} &= |X|_n^{\frac{1}{p}} h_X^{-1} \|\widehat{\nabla}\widehat{v}\|_{L^p(\widehat{X})^n} \\ &\le |X|_n^{\frac{1}{p}} h_X^{-1} \|\widehat{\nabla}\widehat{v}\|_{L^p(\mathcal{B}_n)^n}. \end{aligned} \tag{1.47}$$

Note that the polynomial $\widehat{\nabla}\widehat{v}$, originally defined on $\widehat{X}$, has been naturally extended into a polynomial on $\mathcal{B}_n$. Let us endow the spaces $\mathbb{P}^l(\mathcal{B}_n)$ and $\mathbb{P}^{l-1}(\mathcal{B}_n)^n$ with their respective L^p-norms. Since $\widehat{\nabla}$ is a linear mapping between these finite-dimensional spaces, it is continuous with a norm bounded above by a constant depending only on these spaces, that is, depending only on n, l and p. Hence, (1.47) can be continued writing

$$\|\nabla v\|_{L^p(X)^n} \lesssim |X|_n^{\frac{1}{p}} h_X^{-1} \|\widehat{v}\|_{L^p(\mathcal{B}_n)} \lesssim |X|_n^{\frac{1}{p}} h_X^{-1} \|\widehat{v}\|_{L^p(\mathcal{B}_n(\mathbf{0},\varrho))} \le |X|_n^{\frac{1}{p}} h_X^{-1} \|\widehat{v}\|_{L^p(\widehat{X})},$$

where the second inequality follows from (1.43) with $q = m = p$, and the conclusion from $\mathcal{B}_n(\mathbf{0}, \varrho) \subset \widehat{X}$. The proof of (1.46) with X satisfying (1.34) is completed invoking (1.40) with $s = p$ and $w = v$.

(ii) *Proof of* (1.46) *with X satisfying* (1.36). By virtue of the first inequality in (1.36b) together with Point (i) in this proof, it holds, for all $Y \in \mathfrak{T}_X$,

$$\|\nabla v\|_{L^p(Y)^n} \lesssim h_Y^{-1} \|v\|_{L^p(Y)} \lesssim h_X^{-1} \|v\|_{L^p(X)}, \tag{1.48}$$

where we have used, in the second bound, the second inequality in (1.36b) together with $Y \subset X$. We then conclude as in Point (ii) of the proof of Lemma 1.25. If $p < \infty$, taking the power p of (1.48), summing over $Y \in \mathfrak{T}_X$, using $\|\nabla v\|_{L^p(X)^n}^p \le \sum_{Y \in \mathfrak{T}_X} \|\nabla v\|_{L^p(Y)^n}^p$ (a consequence of (1.36a)), and taking the pth root of the resulting inequality proves (1.46) since $\mathrm{card}(\mathfrak{T}_X) \lesssim 1$ by (1.36a). If $p = \infty$, on the other hand, we take the maximum of (1.48) over $Y \in \mathfrak{T}_X$ and conclude observing that $\|\nabla v\|_{L^\infty(X)^n} = \max_{Y \in \mathfrak{T}_X} \|\nabla v\|_{L^\infty(Y)^n}$. □

The following corollary, whose proof results from a combination of Lemmas 1.25 and 1.28 and is left as an exercise to the reader, states inverse Sobolev embeddings valid in local polynomial spaces.

Corollary 1.29 (Inverse Sobolev Embeddings in Local Polynomial Spaces) *Let X be an open bounded connected subset of $\mathbb{R}^n$ that satisfies* (1.34) *or* (1.36). *Let three integers $l \geq 0$, $r \geq 0$, and $m \geq 0$ be given such that*

$$r \leq m, \tag{1.49}$$

as well as two real numbers $p, q \in [1, \infty]$. Then, for any $w \in \mathbb{P}^l(X)$,

$$|w|_{W^{m,p}(X)} \lesssim h_X^{r-m} |X|_n^{\frac{1}{p}-\frac{1}{q}} |w|_{W^{r,q}(X)} \tag{1.50}$$

with hidden multiplicative constant depending only on n, ϱ, l, r, m, p, and q.

Remark 1.30 (Condition (1.49)*)* Corollary 1.29 obviously cannot hold if $m < r$ and $m \leq l$. To check this point, consider for w a polynomial of degree exactly m (that is, w has a non-zero coefficient on at least one monomial of total degree m): the left-hand side of (1.50) does not vanish, while the right-hand side does if $r > m$.

1.2.6 Local Trace Inequalities on Regular Mesh Sequences

Trace inequalities enable the control of face norms through element norms. They play an important role in the analysis of HHO methods, which rely on a mixture of element-based and face-based terms. For a given mesh element $T \in \mathcal{T}_h$, the following lemma shows that the control of the L^p-norm of a (smooth enough) function on a face $F \in \mathcal{F}_T$ requires the control of the L^p-norm of both the function and its first derivatives inside T, and makes explicit the dependency of the constants with respect to h_T.

Lemma 1.31 (Continuous Local Trace Inequality) *Let $(\mathcal{M}_h)_{h\in\mathcal{H}}$ denote a regular mesh sequence in the sense of Definition 1.9, and let a real number $p \in [1, \infty]$ be fixed. Then, for all $h \in \mathcal{H}$, all $T \in \mathcal{T}_h$, all $F \in \mathcal{F}_T$, and all $v \in W^{1,p}(T)$,*

$$\|v\|_{L^p(F)} \lesssim h_T^{-\frac{1}{p}} \left(\|v\|_{L^p(T)} + h_T \|\nabla v\|_{L^p(T)^d}\right) \tag{1.51}$$

with hidden constant depending only on d, ϱ, and p.

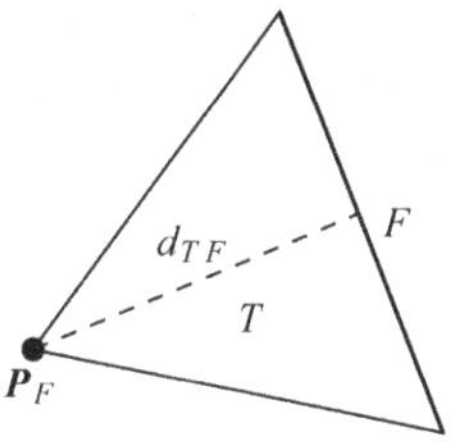

Fig. 1.6 Notation for the proof of Lemma 1.31

Proof We first consider the case $p < \infty$. Let $T \in \mathcal{T}_h$ and $F \in \mathcal{F}_T$. Assume first that T is simplicial. Since $C^1(\overline{T})$ is dense in $W^{1,p}(T)$ (see, e.g., [6, Theorem 3.22]), it suffices to prove (1.51) for $v \in C^1(\overline{T})$, the general case being then obtained approximating v by such smooth functions. For $v \in C^1(\overline{T})$, we notice that $|v|^p$ is at least Lipschitz-continuous (and thus in $W^{1,\infty}(T)$), with $\nabla|v|^p = p\,\text{sign}(v)|v|^{p-1}\nabla v$ (the function $|v|^p$ actually belongs to $C^1(\overline{T})$ if $p > 1$).

Consider the function $\boldsymbol{\varphi}_F : T \to \mathbb{R}^d$ such that, for all $\boldsymbol{x} \in T$,

$$\boldsymbol{\varphi}_F(\boldsymbol{x}) = \frac{|F|_{d-1}}{d|T|_d}(\boldsymbol{x} - \boldsymbol{P}_F),$$

where $\boldsymbol{P}_F$ denotes the vertex in the simplex T opposite to F; cf. Fig. 1.6. The function $\boldsymbol{\varphi}_F$ coincides with the lowest-order Raviart–Thomas–Nédélec shape function associated to the face F; it has normal component identically equal to one on F and identically equal to zero on the remaining faces in $\mathcal{F}_T$ (see, e.g., [57, Section 2.3.1], [56, Example 4], or [183, Section 1.2.7] for further details). Using this fact, we have that

$$\begin{aligned}\|v\|_{L^p(F)}^p &= \int_F |v|^p = \sum_{F' \in \mathcal{F}_T} \int_{F'} |v|^p (\boldsymbol{\varphi}_F \cdot \boldsymbol{n}_{TF'}) \\ &= \int_T \nabla\cdot(|v|^p \boldsymbol{\varphi}_F) = \int_T |v|^p (\nabla\cdot\boldsymbol{\varphi}_F) + \int_T p\,\text{sign}(v)|v|^{p-1}\boldsymbol{\varphi}_F\cdot\nabla v,\end{aligned}$$

where the third equality follows from the divergence theorem, valid since $|v|^p \in W^{1,\infty}(T)$. Since

$$\nabla\cdot\boldsymbol{\varphi}_F = \frac{|F|_{d-1}}{|T|_d} \quad \text{and} \quad \|\boldsymbol{\varphi}_F\|_{L^\infty(T)^d} \le \frac{|F|_{d-1}h_T}{d|T|_d},$$

we infer, using for the second term the generalised Hölder inequality with exponents (p', ∞, p), p' being such that $\frac{1}{p} + \frac{1}{p'} = 1$, that

$$\|v\|_{L^p(F)}^p \le \frac{|F|_{d-1}}{|T|_d}\|v\|_{L^p(T)}^p + \frac{|F|_{d-1}h_T}{d|T|_d}p\|v\|_{L^p(T)}^{p-1}\|\nabla v\|_{L^p(T)^d}. \tag{1.52}$$

Since T is simplicial, denoting by d_{TF} the orthogonal distance between $\boldsymbol{P}_F$ and F and by r_T the inradius of T, we have that

$$\frac{|T|_d}{|F|_{d-1}} = \frac{d_{TF}}{d} \geq \frac{r_T}{d} \geq \frac{\varrho h_T}{d}. \tag{1.53}$$

Combined with (1.52), this yields the following trace inequality on simplices:

$$\|v\|_{L^p(F)}^p \leq \varrho^{-1} h_T^{-1} \left(d\|v\|_{L^p(T)}^p + p\|v\|_{L^p(T)}^{p-1} h_T \|\nabla v\|_{L^p(T)^d} \right).$$

Using, if $p > 1$, the Young inequality $ab \leq \frac{a^p}{p} + \frac{b^{p'}}{p'}$ for the second term gives

$$\begin{aligned} \|v\|_{L^p(F)}^p &\leq \varrho^{-1} h_T^{-1} \left((d+p-1)\|v\|_{L^p(T)}^p + h_T^p \|\nabla v\|_{L^p(T)^d}^p \right) \\ &\leq (d+p-1)\varrho^{-1} h_T^{-1} \left(\|v\|_{L^p(T)} + h_T \|\nabla v\|_{L^p(T)^d} \right)^p. \end{aligned} \tag{1.54}$$

After taking the power $1/p$, (1.51) follows with hidden constant $C = (d + p - 1)^{\frac{1}{p}} \varrho^{-\frac{1}{p}}$.

Let us now turn to the general case where T belongs to a polytopal mesh. For each $\sigma \in \mathfrak{F}_F$, with $\mathfrak{F}_F$ collecting the simplicial subfaces of F, let $\tau_\sigma \in \mathfrak{T}_h$ denote the simplex contained in T of which σ is a face. Observing that we can apply the continuous trace inequality for simplices (1.54) to σ and τ_σ (this is possible owing to (1.3)), we obtain

$$\begin{aligned} \|v\|_{L^p(F)}^p &= \sum_{\sigma \in \mathfrak{F}_F} \|v\|_{L^p(\sigma)}^p \\ &\leq \varrho^{-1} \sum_{\sigma \in \mathfrak{F}_F} h_{\tau_\sigma}^{-1} \left((d+p-1)\|v\|_{L^p(\tau_\sigma)}^p + h_{\tau_\sigma}^p \|\nabla v\|_{L^p(\tau_\sigma)^d}^p \right) && \text{Eq. (1.54)} \\ &\leq \varrho^{-2} h_T^{-1} \sum_{\sigma \in \mathfrak{F}_F} \left((d+p-1)\|v\|_{L^p(\tau_\sigma)}^p + h_T^p \|\nabla v\|_{L^p(\tau_\sigma)^d}^p \right) && \text{Eq. (1.4)} \\ &\leq \varrho^{-2} h_T^{-1} \left((d+p-1)\|v\|_{L^p(T)}^p + h_T^p \|\nabla v\|_{L^p(T)^d}^p \right) && \bigcup_{\sigma \in \mathfrak{F}_F} \tau_\sigma \subset T, \end{aligned}$$

and (1.51) for $p < \infty$ follows with hidden constant $C = (d + p - 1)^{\frac{1}{p}} \varrho^{-\frac{2}{p}}$.

In the case $p = \infty$, a function $v \in W^{1,\infty}(T)$ is actually continuous over $\overline{T}$ and $\|v\|_{L^\infty(T)} = \max_{\boldsymbol{x} \in \overline{T}} |v(\boldsymbol{x})|$. We therefore simply have $\|v\|_{L^\infty(F)} \leq \|v\|_{L^\infty(T)}$, which implies (1.51) with hidden constant 1. □

The result of Lemma 1.31 can be simplified for local polynomial functions in view of the inverse inequality of Lemma 1.28, as made precise in the following lemma.

Lemma 1.32 (Discrete Local Trace Inequality) *Let $(\mathcal{M}_h)_{h\in\mathcal{H}}$ denote a regular mesh sequence, and let a real number $p \in [1,\infty]$ and an integer $l \geq 0$ be fixed. Then, for all $h \in \mathcal{H}$, all $T \in \mathcal{T}_h$, all $F \in \mathcal{F}_T$, and all $v \in \mathbb{P}^l(T)$,*

$$\|v\|_{L^p(F)} \lesssim h_T^{-\frac{1}{p}} \|v\|_{L^p(T)} \tag{1.55}$$

with hidden constant depending only on d, ϱ, p, and l.

Proof It suffices to combine the continuous local trace inequality (1.51) with the inverse inequality (1.46) applied to $X = T$, which is made possible by Remark 1.27. □

Remark 1.33 (Inequalities for Piecewise Polynomial Functions) The direct and inverse Lebesgue embeddings (Lemma 1.25), discrete inverse inequality (Lemma 1.28), inverse Sobolev embeddings (Corollary 1.29) and discrete local trace inequality (Lemma 1.32) also hold for functions that are piecewise polynomial on a subdivision $(S_i)_{i\in I}$ of the considered set X or T, provided that (a) each S_i has a diameter comparable (with constant ϱ) to the diameter of the set X or T, and (b) each S_i satisfies the geometric condition (1.34) or (1.36). This can be seen applying the same kind of argument as in Step (ii) of the proof of Lemma 1.25, using the fact that (a) and (b) above imply a bound on card(I) that only depends on ϱ.

1.3 Projectors on Local Polynomial Spaces

In this section we study projectors on local polynomial spaces, which play a key role in the design and analysis of HHO methods.

1.3.1 Definition and Examples

Definition 1.34 (Projector on a Local Polynomial Space) Let an integer $l \geq 0$ and an open bounded connected set $X \subset \mathbb{R}^n$, $n \geq 1$, be given. Let W be a vector space such that $\mathbb{P}^l(X) \subset W$. A linear mapping $\Pi_X^l : W \to \mathbb{P}^l(X)$ is a *projector on the local polynomial space* $\mathbb{P}^l(X)$ if it is onto and idempotent, i.e., $\Pi_X^l \circ \Pi_X^l = \Pi_X^l$.

The following proposition provides a simple condition to check that a projector meets the requirements of Definition 1.34, namely the invariance of polynomials under projection.

Proposition 1.35 (Characterisation of Projectors on Local Polynomial Spaces) *Let an integer $l \geq 0$ and an open bounded connected set $X \subset \mathbb{R}^n$, $n \geq 1$, be given. Let W be a vector space such that $\mathbb{P}^l(X) \subset W$. A linear mapping $\Pi^l_X : W \to \mathbb{P}^l(X)$ is a projector on the local polynomial space $\mathbb{P}^l(X)$ in the sense of Definition 1.34 if and only if, for any $v \in \mathbb{P}^l(X)$,*

$$\Pi^l_X v = v. \tag{1.56}$$

Proof Let us first assume that Π^l_X is onto and idempotent, and let us prove (1.56). Take $v \in \mathbb{P}^l(X)$. Since Π^l_X is onto, there exists $w \in W$ such that $v = \Pi^l_X w$. Taking the projection of this equality and using the idempotence property, we obtain

$$\Pi^l_X v = \Pi^l_X(\Pi^l_X w) = \Pi^l_X w = v,$$

which is (1.56).

Assume now (1.56). Then, since $\mathbb{P}^l(X) \subset W$, we have that

$$\mathbb{P}^l(X) = \Pi^l_X \mathbb{P}^l(X) \subset \Pi^l_X W \subset \mathbb{P}^l(X),$$

which shows that $\Pi^l_X W = \mathbb{P}^l(X)$, i.e., Π^l_X is onto. Moreover, using again the polynomial invariance (1.56) we have, for any $w \in W$, that $\Pi^l_X(\Pi^l_X w) = \Pi^l_X w$, which proves that Π^l_X is idempotent. □

We next discuss two key examples of projectors on local polynomial spaces: the L^2-orthogonal and elliptic projectors.

Definition 1.36 (The L^2-Orthogonal Projector) The L^2-orthogonal projector (in short, L^2-projector) $\pi^{0,l}_X : L^1(X) \to \mathbb{P}^l(X)$ is defined as follows: For all $v \in L^1(X)$, the polynomial $\pi^{0,l}_X v \in \mathbb{P}^l(X)$ satisfies

$$(\pi^{0,l}_X v - v, w)_X = 0 \qquad \forall w \in \mathbb{P}^l(X). \tag{1.57}$$

Existence and uniqueness of $\pi^{0,l}_X v$ immediately follow from the Riesz representation theorem in $\mathbb{P}^l(X)$ for the standard $L^2(X)$-inner product. Moreover, we have the following characterisation:

$$\pi^{0,l}_X v = \operatorname*{argmin}_{w \in \mathbb{P}^l(X)} \|w - v\|^2_X,$$

which is proved observing that (1.57) is the Euler equation for the above minimisation problem; see, e.g., [12, Chapter 10]. This means that $\pi^{0,l}_X v$ is the element of $\mathbb{P}^l(X)$ that minimises the distance from v in the $L^2(X)$-norm. To check that $\pi^{0,l}_X$ satisfies (1.56) (and hence, by Proposition 1.35, that it meets the conditions of Definition 1.34) it suffices to observe that, if $v \in \mathbb{P}^l(X)$, then (1.57) with $w = \pi^{0,l}_X v - v \in \mathbb{P}^l(X)$ implies $\pi^{0,l}_X v - v = 0$. It can also be checked that $\pi^{0,l}_X$ is a linear operator. The details are left to the reader.

Remark 1.37 (Case $l = 0$) The lowest-order L^2-projector satisfies

$$\pi_T^{0,0} v = \frac{1}{|T|_d} \int_T v = \frac{1}{|T|_d} (v, 1)_T \qquad \forall v \in L^1(T). \tag{1.58}$$

In the construction of HHO methods carried out in the following chapters, for a given polytopal mesh $\mathcal{M}_h = (\mathcal{T}_h, \mathcal{F}_h)$, we will need the L^2-projectors on $\mathbb{P}^l(T)$, $T \in \mathcal{T}_h$, and $\mathbb{P}^l(F)$, $F \in \mathcal{F}_h$. We will also need the vector and tensor versions of the L^2-projector, obtained by applying $\pi_X^{0,l}$ component-wise and denoted with the bold symbol $\boldsymbol{\pi}_X^{0,l}$. Finally, in some circumstances the following global (patched) version of the L^2-orthogonal projector will be useful.

Definition 1.38 (Global L^2-Orthogonal Projector on Broken Polynomial Spaces) Given a polynomial degree $l \geq 0$ and a polytopal mesh $\mathcal{M}_h = (\mathcal{T}_h, \mathcal{F}_h)$, we define the global L^2-orthogonal projector $\pi_h^{0,l} : L^1(\Omega) \to \mathbb{P}^l(\mathcal{T}_h)$ as follows: For all $v \in L^1(\Omega)$ and all $T \in \mathcal{T}_h$,

$$(\pi_h^{0,l} v)_{|T} = \pi_T^{0,l} v_{|T}. \tag{1.59}$$

Definition 1.39 (The Elliptic Projector) The elliptic projector $\pi_X^{1,l} : W^{1,1}(X) \to \mathbb{P}^l(X)$ is defined as follows: For all $v \in W^{1,1}(X)$, the polynomial $\pi_X^{1,l} v \in \mathbb{P}^l(X)$ satisfies

$$(\nabla(\pi_X^{1,l} v - v), \nabla w)_X = 0 \qquad \forall w \in \mathbb{P}^l(X) \tag{1.60a}$$

and

$$(\pi_X^{1,l} v - v, 1)_X = 0. \tag{1.60b}$$

By the Riesz representation theorem in $\nabla \mathbb{P}^l(X)$ for the $L^2(X)^n$-inner product, (1.60a) defines a unique element $\nabla \pi_X^{1,l} v \in \nabla \mathbb{P}^l(X)$, and thus a polynomial $\pi_X^{1,l} v$ up to an additive constant. This constant is fixed by (1.60b).

Notice that (1.60) is equivalent to requiring that

$$(\nabla(\pi_X^{1,l} v - v), \nabla w)_X + (\pi_X^{1,l} v - v, \pi_X^{0,0} w)_X = 0 \qquad \forall w \in \mathbb{P}^l(X). \tag{1.61}$$

This can be seen by adding (1.60a) to (1.60b) multiplied by $\pi_X^{0,0} w$ to get (1.61); conversely, apply (1.61) with $w - \pi_X^{0,0} w$ (resp. $w = 1$) to recover (1.60a) (resp. (1.60b)).

Observing that (1.60a) is trivially verified when $l = 0$, it follows from (1.60b) that $\pi_X^{1,0} = \pi_X^{0,0}$. The following characterisation holds:

$$\pi_X^{1,l} v = \underset{w \in \mathbb{P}^l(X),\ (w - v, 1)_X = 0}{\operatorname{argmin}} \|\nabla(w - v)\|_X^2,$$

which is proved observing that (1.60) is the Euler equation for the above minimisation problem; see, e.g., [12, Chapter 10]. Let us check that $\pi_X^{1,l}$ satisfies the polynomial invariance condition (1.56) (and hence, by Proposition 1.35, that it meets the requirements of Definition 1.34). Let $v \in \mathbb{P}^l(X)$ and observe that, by (1.60a) with $w = \pi_X^{1,l} v - v \in \mathbb{P}^l(X)$, $\nabla(\pi_X^{1,l} v - v) = \mathbf{0}$. As a result, $\pi_X^{1,l} v$ and v only differ by a constant, which must be zero in view of (1.60b). We leave it to the reader to check that $\pi_X^{1,l}$ is a linear operator.

When constructing the HHO approximation of the Poisson problem on a polytopal mesh $\mathcal{M}_h = (\mathcal{T}_h, \mathcal{F}_h)$ in Chap. 2, the elliptic projectors $\pi_T^{1,l}$, $T \in \mathcal{T}_h$, will play a key role. Other examples of projectors that will be encountered and studied in this book include: the oblique elliptic projector of Sect. 3.1.2, relevant in the discretisation of anisotropic diffusion; the modified elliptic projector of Sect. 5.1.2, with a closure equation involving an average on the boundary of the element (instead of the average (1.60b) on the element itself); and the strain projector of Sect. 7.2.2, used in the context of linear elasticity.

1.3.2 Approximation Properties of Bounded Projectors on Local Polynomial Spaces

We study in this section the approximation properties of projectors on local polynomial spaces. We start with an abstract result, which states that projectors that are bounded in a suitable (small) set of Sobolev seminorms have optimal approximation properties in all Sobolev seminorms. Optimal means here that the error committed approximating a smooth function v by its projection has the same scaling in the diameter h_X as the error with respect to the best approximation of v in the selected Sobolev seminorm.

Such approximation properties are established under geometrical assumptions on the sets. The simplest one is the following.

Definition 1.40 (Star-Shaped Set) A non-empty open set $X \subset \mathbb{R}^n$ of boundary ∂X is *star-shaped* with respect to a point $\boldsymbol{x} \in X$ if, for any $\boldsymbol{y} \in \partial X$, the segment $\{\alpha\boldsymbol{x} + (1-\alpha)\boldsymbol{y} : \alpha \in (0,1)\}$ is contained in X.

This is the notion used, e.g., in the main approximation results of [77, 179]. In the present context, however, we have to expand this notion in order to cover the case where the set X coincides with an element or face of a polytopal mesh from a regular sequence. We therefore introduce the following generalisation of the notion of star-shaped set:

Definition 1.41 (Set Connected by Star-Shaped Sets) An open bounded set X of $\mathbb{R}^n$ is *connected by star-shaped sets* with parameter $\theta > 0$ if X is connected and if there exists a family of open subsets $(X_i)_{i=1,\ldots,N}$ of X such that

$$\overline{X} = \bigcup_{i=1}^{N} \overline{X_i}, \quad N \le \theta^{-1}, \tag{1.62a}$$

$$\begin{aligned} &\forall i \in \{1, \ldots, N\}, X_i \text{ is star-shaped with respect to all points} \\ &\text{in a ball of radius } \theta h_{X_i}, \end{aligned} \tag{1.62b}$$

and

$$\begin{aligned} &\forall i \in \{2, \ldots, N\},\ \exists j \in \{1, \ldots, i-1\} \text{ such that} \\ &X_i \cap X_j \text{ contains a ball of radius } \theta h_X. \end{aligned} \tag{1.62c}$$

The main interest of this notion for us is that it covers the elements and faces in a polytopal mesh of a regular mesh sequence.

Lemma 1.42 (Mesh Elements and Faces are Connected by Shar-Shaped Sets) *Let $(\mathcal{M}_h)_{h\in\mathcal{H}} = (\mathcal{T}_h, \mathcal{F}_h)_{h\in\mathcal{H}}$ be a regular mesh sequence in the sense of Definition 1.9. Then, for all $h \in \mathcal{H}$, any $T \in \mathcal{T}_h$ and any $F \in \mathcal{F}_h$ is connected by star-shaped sets in the sense of Definition 1.41, with parameter θ depending only on d and ϱ in Definition 1.9.*

Proof See Sect. 1.4.2. □

Lemma 1.43 ($W^{s,p}$-Approximation for W-Bounded Projectors) *Assume that $X \subset \mathbb{R}^n$ is connected by star-shaped sets with parameter $\theta > 0$ in the sense of Definition 1.41. Let a real number $p \in [1, \infty]$ and four integers $l \ge 0$, $s \in \{0, \ldots, l+1\}$, and $q, m \in \{0, \ldots, s\}$ be fixed. Denote by $\Pi_X^{q,l} : W^{q,p}(X) \to \mathbb{P}^l(X)$ a projector on the local polynomial space $\mathbb{P}^l(X)$ in the sense of Definition 1.34. Assume that it holds, with hidden constant depending only on n, θ, l, s, q, m, and p: For all $v \in W^{q,p}(X)$,*

$$\textit{If } m < q, \quad |\Pi_X^{q,l} v|_{W^{m,p}(X)} \lesssim \sum_{r=m}^{q} h_X^{r-m} |v|_{W^{r,p}(X)}, \tag{1.63a}$$

$$\textit{If } m \ge q, \quad |\Pi_X^{q,l} v|_{W^{q,p}(X)} \lesssim |v|_{W^{q,p}(X)}. \tag{1.63b}$$

Then, for all $v \in W^{s,p}(X)$, with hidden constant having the same dependencies as above,

$$|v - \Pi_X^{q,l} v|_{W^{m,p}(X)} \lesssim h_X^{s-m} |v|_{W^{s,p}(X)}. \tag{1.64}$$

Proof We first notice that (1.62) in Definition 1.41 implies (1.36) with $\varrho = \theta$. Hence, the inverse Sobolev embeddings of Corollary 1.29 apply to X, with hidden constant having the same dependency as in (1.63).

We consider the following representation of v:

$$v = Q^s v + R^s v, \tag{1.65}$$

where $Q^s v \in \mathbb{P}^{s-1}(X) \subset \mathbb{P}^l(X)$ is an averaged Taylor polynomial, while the remainder $R^s v$ satisfies, for all $r \in \{0, \ldots, s\}$,

$$|R^s v|_{W^{r,p}(X)} \lesssim h_X^{s-r} |v|_{W^{s,p}(X)}. \tag{1.66}$$

A proof of this result for X star-shaped with respect to all points in a ball of radius θh_X is given in [77, Lemma 4.3.8]. Its extension to X connected by star-shaped sets is detailed in Sect. 1.4.1 below. Since $\Pi_X^{q,l}$ is a projector, it holds by Proposition 1.35 that $\Pi_X^{q,l}(Q^s v) = Q^s v$ so that, taking the projection of (1.65), it is inferred

$$\Pi_X^{q,l} v = Q^s v + \Pi_X^{q,l}(R^s v).$$

Subtracting this equation from (1.65), we arrive at $v - \Pi_X^{q,l} v = R^s v - \Pi_X^{q,l}(R^s v)$. Hence, passing to the seminorm and using a triangle inequality, we obtain

$$|v - \Pi_X^{q,l} v|_{W^{m,p}(X)} \leq |R^s v|_{W^{m,p}(X)} + |\Pi_X^{q,l}(R^s v)|_{W^{m,p}(X)}. \tag{1.67}$$

For the first term in the right-hand side, the estimate (1.66) with $r = m$ readily yields

$$|R^s v|_{W^{m,p}(X)} \lesssim h_X^{s-m} |v|_{W^{s,p}(X)}. \tag{1.68}$$

Let us estimate the second term in (1.67). If $m < q$, using the boundedness assumption (1.63a) followed by the estimate (1.66), it is inferred

$$\begin{aligned} |\Pi_X^{q,l}(R^s v)|_{W^{m,p}(X)} &\lesssim \sum_{r=m}^{q} h_X^{r-m} |R^s v|_{W^{r,p}(X)} \\ &\lesssim \sum_{r=m}^{q} h_X^{r-m} h_X^{s-r} |v|_{W^{s,p}(X)} \lesssim h_X^{s-m} |v|_{W^{s,p}(X)}. \end{aligned}$$

If, on the other hand, $m \geq q$, using the inverse Sobolev embeddings (1.50) followed by assumption (1.63b) and the estimate (1.66) with $r = q$, it is inferred that

$$\begin{aligned}|\Pi_X^{q,l}(R^s v)|_{W^{m,p}(X)} &\lesssim h_X^{q-m} |\Pi_X^{q,l}(R^s v)|_{W^{q,p}(X)} \\ &\lesssim h_X^{q-m} |R^s v|_{W^{q,p}(X)} \lesssim h_X^{s-m} |v|_{W^{s,p}(X)}.\end{aligned}$$

In conclusion we have, in either case $m < q$ or $m \geq q$,

$$|\Pi_X^{q,l}(R^s v)|_{W^{m,p}(X)} \lesssim h_X^{s-m} |v|_{W^{s,p}(X)}. \tag{1.69}$$

Using (1.68) and (1.69) to estimate the first and second terms in the right-hand side of (1.67), respectively, the conclusion follows. □

We close this section with a technical lemma which will play an important role in the study, carried out in the following section, of the approximation properties of the L^2-orthogonal and elliptic projectors.

Lemma 1.44 (L^p-Boundedness of L^2-Orthogonal Projectors on Local Polynomial Subspaces) *Let X denote an open bounded connected set of $\mathbb{R}^n$, $n \geq 1$, let two integers $l \geq 0$ and $m \geq 1$ be fixed, and let $\mathcal{P}$ be a subspace of $\mathbb{P}^l(X)^m$. We consider the L^2-orthogonal projector $\Pi_{\mathcal{P}} : L^1(X)^m \to \mathcal{P}$ such that, for all $\Phi \in L^1(X)^m$,*

$$(\Pi_{\mathcal{P}}\Phi - \Phi, \Psi)_X = 0 \qquad \forall \Psi \in \mathcal{P}. \tag{1.70}$$

Let a real number $p \in [1, \infty]$ be given and, if $p \neq 2$, assume that X satisfies (1.34) *or* (1.36)*. Then, for all $\Phi \in L^p(X)^m$,*

$$\|\Pi_{\mathcal{P}}\Phi\|_{L^p(X)^m} \lesssim \|\Phi\|_{L^p(X)^m} \tag{1.71}$$

with hidden constant equal to 1 *if $p = 2$ and depending only on n, l, m, ϱ and p otherwise.*

Proof (i) *The case $p = 2$.* Using (1.70) with $\Psi = \Pi_{\mathcal{P}}\Phi$ and the Cauchy–Schwarz inequality, it is inferred that

$$\|\Pi_{\mathcal{P}}\Phi\|_X^2 = (\Phi, \Pi_{\mathcal{P}}\Phi)_X \leq \|\Phi\|_X \|\Pi_{\mathcal{P}}\Phi\|_X,$$

and thus, simplifying by $\|\Pi_{\mathcal{P}}\Phi\|_X$,

$$\|\Pi_{\mathcal{P}}\Phi\|_X \leq \|\Phi\|_X. \tag{1.72}$$

(ii) *The case* $p > 2$. Using the inverse Lebesgue embeddings on local polynomial spaces of Lemma 1.25 followed by (1.72) and the Hölder inequality (with functions Φ, 1 and exponents $\frac{p}{2}$, $\frac{p}{p-2}$), it is inferred that

$$\|\Pi_{\mathcal{P}}\Phi\|_{L^p(X)^m} \lesssim |X|_n^{\frac{1}{p}-\frac{1}{2}} \|\Pi_{\mathcal{P}}\Phi\|_X \le |X|_n^{\frac{1}{p}-\frac{1}{2}} \|\Phi\|_X \lesssim \|\Phi\|_{L^p(X)^m},$$

which proves (1.71) for $p > 2$.

(iii) *The case* $p < 2$. We first observe that, using the definition (1.70) of $\Pi_{\mathcal{P}}$ twice, for all $\Phi, \Psi \in L^1(X)^m$ we have that

$$\int_X (\Pi_{\mathcal{P}}\Phi)\cdot\Psi = \int_X (\Pi_{\mathcal{P}}\Phi)\cdot(\Pi_{\mathcal{P}}\Psi) = \int_X \Phi\cdot(\Pi_{\mathcal{P}}\Psi).$$

Hence, with p' such that $\frac{1}{p} + \frac{1}{p'} = 1$, it holds

$$\begin{aligned}
\|\Pi_{\mathcal{P}}\Phi\|_{L^p(X)^m} &= \sup_{\Psi \in L^{p'}(X)^m, \|\Psi\|_{L^{p'}(X)^m}=1} \int_X (\Pi_{\mathcal{P}}\Phi)\cdot\Psi \\
&= \sup_{\Psi \in L^{p'}(X)^m, \|\Psi\|_{L^{p'}(X)^m}=1} \int_X \Phi\cdot(\Pi_{\mathcal{P}}\Psi) \\
&\le \sup_{\Psi \in L^{p'}(X)^m, \|\Psi\|_{L^{p'}(X)^m}=1} \|\Phi\|_{L^p(X)^m} \|\Pi_{\mathcal{P}}\Psi\|_{L^{p'}(X)^m}, \qquad (1.73)
\end{aligned}$$

where we have used the Hölder inequality to conclude. Use (1.71) for $p' > 2$ to write $\|\Pi_{\mathcal{P}}\Psi\|_{L^{p'}(X)^m} \lesssim \|\Psi\|_{L^{p'}(X)^m} = 1$ and plug this bound into (1.73) to conclude the proof of (1.71). □

1.3.3 Approximation Properties of the Local L^2-Orthogonal and Elliptic Projectors

Theorem 1.45 (Approximation Properties of the L^2-Orthogonal Projector on Elements and Faces) *Let $(\mathcal{M}_h)_{h\in\mathcal{H}} = (\mathcal{T}_h, \mathcal{F}_h)_{h\in\mathcal{H}}$ be a regular mesh sequence in the sense of Definition 1.9. Let a polynomial degree $l \ge 0$, an integer $s \in \{0, \ldots, l+1\}$, and a real number $p \in [1, \infty]$ be given. Then, for any X element or face of $\mathcal{M}_h$, all $v \in W^{s,p}(X)$, and all $m \in \{0, \ldots, s\}$,*

$$|v - \pi_X^{0,l} v|_{W^{m,p}(X)} \lesssim h_X^{s-m} |v|_{W^{s,p}(X)}. \qquad (1.74)$$

(continued)

Theorem 1.45 (continued)
Moreover, if $s \geq 1$, for all $T \in \mathcal{T}_h$, all $v \in W^{s,p}(T)$, all $F \in \mathcal{F}_T$, and all $m \in \{0, \ldots, s-1\}$, it holds that

$$h_T^{\frac{1}{p}} |v - \pi_T^{0,l} v|_{W^{m,p}(F)} \lesssim h_T^{s-m} |v|_{W^{s,p}(T)}. \tag{1.75}$$

In (1.74) *and* (1.75)*, the hidden constants depend only on d, ϱ, l, s, and p.*

Proof Let X denote an element or face of $\mathcal{M}_h$. Using Lemma 1.44 with $\mathcal{P} = \mathbb{P}^l(X)$ (which is possible in view of Remark 1.27), we have the following boundedness property for $\pi_X^{0,l}$: For all $v \in L^p(X)$,

$$\|\pi_X^{0,l} v\|_{L^p(X)} \lesssim \|v\|_{L^p(X)}.$$

The estimate (1.74) is then an immediate consequence of Lemma 1.43 with $q = 0$ and $\Pi_X^{0,l} = \pi_X^{0,l}$ (notice that Lemma 1.43 applies to X as a consequence of Lemma 1.42).

Let us now turn to the trace approximation property (1.75). Take $\boldsymbol{\alpha} \in \mathbb{N}^{d-1}$ such that $\|\boldsymbol{\alpha}\|_1 = m$. Apply the continuous trace inequality (1.51) with v replaced by $\partial^{\boldsymbol{\alpha}}(v - \pi_T^{0,l} v)$ (the derivative being taken with respect to Cartesian coordinates along the hyperplane spanned by F) to get

$$\begin{aligned} h_T^{\frac{1}{p}} \|\partial^{\boldsymbol{\alpha}}(v - \pi_T^{0,l} v)\|_{L^p(F)} &\lesssim \|\partial^{\boldsymbol{\alpha}}(v - \pi_T^{0,l} v)\|_{L^p(T)} + h_T \|\nabla \partial^{\boldsymbol{\alpha}}(v - \pi_T^{0,l} v)\|_{L^p(T)} \\ &\lesssim |v - \pi_T^{0,l} v|_{W^{m,p}(T)} + h_T |v - \pi_T^{0,l} v|_{W^{m+1,p}(T)}. \end{aligned}$$

Invoke then (1.74) for $X = T$ twice, first with m and then with $(m+1)$ instead of m (note that, by assumption, $m + 1 \leq s$), to deduce

$$\begin{aligned} h_T^{\frac{1}{p}} \|\partial^{\boldsymbol{\alpha}}(v - \pi_T^{0,l} v)\|_{L^p(F)} &\lesssim h_T^{s-m} |v|_{W^{s,p}(T)} + h_T h_T^{s-m-1} |v|_{W^{s,p}(T)} \\ &\lesssim h_T^{s-m} |v|_{W^{s,p}(T)}. \end{aligned}$$

Estimate (1.75) follows summing over $\boldsymbol{\alpha} \in \mathbb{N}^{d-1}$ such that $\|\boldsymbol{\alpha}\|_1 = m$. □

Some remarks are in order to highlight relevant consequences of Theorem 1.45.

Remark 1.46 (Local Poincaré–Wirtinger Inequality) From (1.58) and (1.74) with $p = 2$, $l = 0$, $s = 1$, and $m = 0$, we infer a local Poincaré–Wirtinger inequality, which will often be invoked in the following chapters: For any $T \in \mathcal{T}_h$ and any $v \in H^1(T)$ such that $\int_T v = 0$,

$$\|v\|_T \lesssim h_T \|\nabla v\|_T \tag{1.76}$$

with hidden constant depending only on d and ϱ.

Remark 1.47 ($W^{s,p}$-Boundedness of L^2-Orthogonal Projectors on Elements and Faces) For any X element or face of a mesh $\mathcal{M}_h$ and any $v \in W^{s,p}(X)$, it holds with hidden constant depending only on d, ϱ, s, and p:

$$|\pi_X^{0,l} v|_{W^{s,p}(X)} \lesssim |v|_{W^{s,p}(X)}, \tag{1.77}$$

which expresses the fact that the L^2-orthogonal projector on $\mathbb{P}^l(X)$ is bounded in any Sobolev seminorm. To prove (1.77), it suffices to use the triangle inequality to write

$$|\pi_X^{0,l} v|_{W^{s,p}(X)} \le |\pi_X^{0,l} v - v|_{W^{s,p}(X)} + |v|_{W^{s,p}(X)}$$

and conclude using (1.74) with $m = s$ for the first term. We notice, in passing, that the hidden constant in (1.77) is equal to 1 if $s = 0$ and $p = 2$ (see Point (i) in the proof of Lemma 1.44).

Theorem 1.48 (Approximation Properties of the Elliptic Projector on Elements) *Let $(\mathcal{M}_h)_{h\in\mathcal{H}} = (\mathcal{T}_h, \mathcal{F}_h)_{h\in\mathcal{H}}$ be a regular mesh sequence in the sense of Definition 1.9. Let a polynomial degree $l \ge 0$, an integer $s \in \{1, \ldots, l+1\}$, and a real number $p \in [1, \infty]$ be given. Then, for all $T \in \mathcal{T}_h$, all $v \in W^{s,p}(T)$, and all $m \in \{0, \ldots, s\}$,*

$$|v - \pi_T^{1,l} v|_{W^{m,p}(T)} \lesssim h_T^{s-m} |v|_{W^{s,p}(T)}. \tag{1.78}$$

Moreover, if $m \le s-1$, for all $F \in \mathcal{F}_T$,

$$h_T^{\frac{1}{p}} |v - \pi_T^{1,l} v|_{W^{m,p}(F)} \lesssim h_T^{s-m} |v|_{W^{s,p}(T)}. \tag{1.79}$$

The hidden constants in (1.78) *and* (1.79) *depend only on d, ϱ, l, s, and p.*

Proof The proof of (1.78) is obtained applying Lemma 1.43 (this is possible in view of Lemma 1.42) with $q = 1$ and $\Pi_T^{1,l} = \pi_T^{1,l}$, provided that we can establish (1.63). Combining this result with the continuous trace inequality (1.51) yields (1.79), in a similar way as in the proof of Theorem 1.45. To prove that the condition (1.63) holds, we distinguish two cases corresponding, respectively, to $m \geq 1$ and $m = 0$.

(i) *The case* $m \geq 1$. We need to show that (1.63b) holds, i.e., for all $v \in W^{1,p}(T)$,

$$|\pi_T^{1,l} v|_{W^{1,p}(T)} \lesssim |v|_{W^{1,p}(T)}. \tag{1.80}$$

By definition (1.60a) of $\pi_T^{1,l}$, it holds, for all $v \in W^{1,1}(T)$,

$$\nabla \pi_T^{1,l} v = \Pi_{\nabla \mathbb{P}^l(T)} \nabla v, \tag{1.81}$$

where $\Pi_{\nabla \mathbb{P}^l(T)}$ denotes the L^2-orthogonal projector on $\nabla \mathbb{P}^l(T) \subset \mathbb{P}^{l-1}(T)^d$. Then, (1.80) is proved observing that, by definition (1.16) of the $|\cdot|_{W^{1,p}(T)}$-seminorm, and invoking (1.81) and the L^p-boundedness of $\Pi_{\nabla \mathbb{P}^l(T)}$ resulting from (1.71) with $\mathcal{P} = \nabla \mathbb{P}^l(T)$, we have

$$\begin{aligned} |\pi_T^{1,l} v|_{W^{1,p}(T)} \lesssim \|\nabla \pi_T^{1,l} v\|_{L^p(T)^d} = \|\Pi_{\nabla \mathbb{P}^l(T)} \nabla v\|_{L^p(T)^d} &\lesssim \|\nabla v\|_{L^p(T)^d} \\ &\lesssim |v|_{W^{1,p}(T)}. \end{aligned}$$

(ii) *The case* $m = 0$. We need to prove that (1.63a) holds, i.e., for all $v \in W^{1,p}(T)$,

$$\|\pi_T^{1,l} v\|_{L^p(T)} \lesssim \|v\|_{L^p(T)} + h_T |v|_{W^{1,p}(T)}. \tag{1.82}$$

Let $v \in W^{1,p}(T)$ and set $\overline{v} := \pi_T^{0,0} v$. By (1.58) and the closure condition (1.60b) in the definition of the elliptic projector, we also have that $\overline{v} = \pi_T^{0,0}(\pi_T^{1,l} v)$. The approximation property (1.74) of the L^2-projector (applied with $m = 0$ and $s = 1$ to $\pi_T^{1,l} v$ instead of v) therefore gives

$$\|\pi_T^{1,l} v - \overline{v}\|_{L^p(T)} \lesssim h_T |\pi_T^{1,l} v|_{W^{1,p}(T)}.$$

We infer

$$\begin{aligned} \|\pi_T^{1,l} v\|_{L^p(T)} &\leq \|\pi_T^{1,l} v - \overline{v}\|_{L^p(T)} + \|\overline{v}\|_{L^p(T)} \\ &\lesssim h_T |\pi_T^{1,l} v|_{W^{1,p}(T)} + \|\pi_T^{0,0} v\|_{L^p(T)} \\ &\lesssim h_T |v|_{W^{1,p}(T)} + \|v\|_{L^p(T)}, \end{aligned}$$

where we have inserted $\pm\overline{v}$ inside the norm and used the triangle inequality in the first line, while the terms in the second line have been estimated using (1.80) for the first one and (1.77) with $(l, s) = (0, 0)$ and $X = T$ for the second one. This establishes (1.82) and concludes the proof. □

Remark 1.49 (Estimates in Fractional Sobolev Spaces) Lemma 1.43 and Theorems 1.45 and 1.48 have been stated for simplicity in integral Sobolev spaces, that is, considering only the case where s and m are integers. However, using standard interpolation techniques (see, e.g., [233, Theorem 5.1]), it is easily deduced from the integer case that the estimates in these theorems also hold for non-integer s and m within the admissible bounds (that is, $s \in [0, l+1]$ or $s \in [1, l+1]$, and $m \in [0, s]$ or $m \in [0, s-1]$).

1.4 Technical Results on Sets that are Connected by Star-Shaped Sets

1.4.1 Approximation by Local Polynomials

As announced in the proof of Lemma 1.43, we prove in this section that the decomposition (1.65)–(1.66) of functions in $W^{s,p}(X)$ holds when X is connected by star-shaped sets, in the sense of Definition 1.41.

The ideas developed here are inspired by [179, Section 7], in which it is shown that a polynomial approximation property holds on a connected finite union of open sets provided that it holds on each set. However, this setting does not enable a proper tracking of the dependency of the constants involved in the estimates. We will show that the notion of set connected by star-shaped sets enables such a tracking.

Theorem 1.50 (Local Polynomial Approximations of $W^{s,p}$-Functions) *Let $X \subset \mathbb{R}^n$ be connected by star-shaped sets with parameter θ, in the sense of Definition 1.41, and take an integer $s \geq 0$ and a real number $p \in [1, \infty]$. Let $v \in W^{s,p}(X)$. Then, there exists $Q^s v \in \mathbb{P}^{s-1}(X)$ such that, for all $r \in \{0, \ldots, s\}$,*

$$|v - Q^s v|_{W^{r,p}(X)} \lesssim h_X^{s-r} |v|_{W^{s,p}(X)}, \tag{1.83}$$

with hidden constant depending only on n, s, p, r and θ.

Proof Let $(X_i)_{i=1,\ldots,N}$ be the sets given by Definition 1.41. By [77, Chapter 4] and (1.62b), for each $i \in \{1, \ldots, N\}$ there exists $Q^s_{X_i} v \in \mathbb{P}^{s-1}(X_i)$ such that, for all $r \in \{0, \ldots, s\}$,

$$|v - Q^s_{X_i} v|_{W^{r,p}(X_i)} \lesssim h_{X_i}^{s-r} |v|_{W^{s,p}(X_i)} \lesssim h_X^{s-r} |v|_{W^{s,p}(X)}, \tag{1.84}$$

the second inequality following from $X_i \subset X$. Each polynomial function $Q^s_{X_i} v$ can obviously be uniquely extended into a polynomial in $\mathbb{P}^{s-1}(X)$. We will prove that $Q^s v := Q^s_{X_1} v$ satisfies (1.83).

Let us start with a preliminary estimate. Using the same arguments as in Point (i) of the proof of Lemma 1.25, that is to say a translation and scaling argument, and the equivalence of norms on $\mathbb{P}^{s-1}(\mathbb{R}^n)$, we see that, for any ball $\mathcal{B}_n(\boldsymbol{x}, \theta h_X)$ contained in X, there holds

$$\|w\|_{L^p(X)} \lesssim \|w\|_{L^p(\mathcal{B}_n(\boldsymbol{x},\theta h_X))} \qquad \forall w \in \mathbb{P}^{s-1}(X),$$

with hidden constant depending only on n, s, p and θ. Take $w \in \mathbb{P}^{s-1}(X)$ and apply this estimate to its derivatives $\partial^{\boldsymbol{\alpha}} w \in \mathbb{P}^{s-r-1}(X) \subset \mathbb{P}^{s-1}(X)$, with $\boldsymbol{\alpha} \in \mathbb{N}^n$ such that $\|\boldsymbol{\alpha}\|_1 = r$, to get

$$|w|_{W^{r,p}(X)} \lesssim |w|_{W^{r,p}(\mathcal{B}_n(\boldsymbol{x},\theta h_X))} \qquad \forall w \in \mathbb{P}^{s-1}(X). \tag{1.85}$$

We now turn to the proof that $Q^s_{X_1} v$ satisfies (1.83). By (1.62c), $X_1 \cap X_2$ contains a ball $\mathcal{B}_n(\boldsymbol{x}, \theta h_X)$. Applying (1.85) to this ball and $w = Q^s_{X_1} v - Q^s_{X_2} v$ yields

$$\begin{aligned}
|Q^s_{X_1} v - Q^s_{X_2} v|_{W^{r,p}(X)} &\lesssim |Q^s_{X_1} v - Q^s_{X_2} v|_{W^{r,p}(\mathcal{B}_n(\boldsymbol{x},\theta h_X))} \\
&\lesssim |Q^s_{X_1} v - v|_{W^{r,p}(\mathcal{B}_n(\boldsymbol{x},\theta h_X))} + |v - Q^s_{X_2} v|_{W^{r,p}(\mathcal{B}_n(\boldsymbol{x},\theta h_X))} \\
&\lesssim |Q^s_{X_1} v - v|_{W^{r,p}(X_1)} + |v - Q^s_{X_2} v|_{W^{r,p}(X_2)} \\
&\lesssim h_X^{s-r} |v|_{W^{s,p}(X)},
\end{aligned}$$

where we have inserted $\pm v$ into the seminorm and used the triangle inequality to pass to the second line, used the fact that $\mathcal{B}_n(\boldsymbol{x}, \theta h_X) \subset X_1 \cap X_2$ in the third line, and concluded invoking (1.84) with $i = 1, 2$.

Following similar arguments we obtain, for all $i \in \{1, \ldots, N\}$,

$$|Q^s_{X_1} v - Q^s_{X_i} v|_{W^{r,p}(X)} \lesssim h_X^{s-r} |v|_{W^{s,p}(X)}. \tag{1.86}$$

This estimate is established by strong induction on i. In the inductive step, which assumes that (1.86) holds with i replaced by any $j \in \{1, \ldots, i-1\}$, we use (1.62c) to estimate $|Q^s_{X_i} v - Q^s_{X_j} v|_{W^{s,p}(X)}$ for some $j < i$ (as for $|Q^s_{X_1} v - Q^s_{X_2} v|_{W^{r,p}(X)}$ above), invoke the induction hypothesis to estimate $|Q^s_{X_1} v - Q^s_{X_j} v|_{W^{r,p}(X)}$, and conclude by triangle inequality.

As a consequence, for all $i \in \{1, \ldots, N\}$,

$$\begin{aligned}
|Q^s_{X_1} v - v|_{W^{r,p}(X_i)} &\lesssim |Q^s_{X_1} v - Q^s_{X_i} v|_{W^{r,p}(X_i)} + |Q^s_{X_i} v - v|_{W^{r,p}(X_i)} \\
&\lesssim h_X^{s-r} |v|_{W^{s,p}(X)},
\end{aligned} \tag{1.87}$$

where we have inserted $\pm Q^s_{X_i} v$ into the seminorm and used the triangle inequality in the first line, and invoked (1.86) and (1.84) to conclude.

The proof for $p < \infty$ is completed using (1.62a) and invoking (1.87) to write

$$|Q^s_{X_1}v - v|_{W^{r,p}(X)} \leq \left(\sum_{i=1}^{N} |Q^s_{X_1}v - v|^p_{W^{r,p}(X_i)}\right)^{1/p} \lesssim h_X^{s-r}|v|_{W^{s,p}(X)}.$$

For $p = \infty$, we conclude the proof by taking the maximum of (1.87) over $i \in \{1, \ldots, N\}$. □

1.4.2 The Case of Mesh Elements and Faces

In this section, we prove Lemma 1.42, that is, we show that elements and faces of a regular mesh sequence are connected by star-shaped sets.

Proof (Lemma 1.42) We only present the proof for a mesh element, the case of a face being similar. Let $h \in \mathcal{H}$ and $\mathfrak{M}_h = (\mathfrak{T}_h, \mathfrak{F}_h)$ be a matching simplicial submesh of $\mathcal{M}_h$ given by Definition 1.9. In this proof, the hidden constants in $\lesssim$ depend only on d and ϱ.

(i) *Preliminary result on simplices.* Let $\tau \in \mathfrak{T}_h$ and $\sigma \in \mathfrak{F}_\tau$ be a face of τ. Let

$$\widehat{\tau} := \left\{ \boldsymbol{x} = (x_1, \ldots, x_d) \in \mathbb{R}^d \;:\; x_i > 0 \quad \forall i = 1, \ldots, d\,,\; \sum_{i=1}^{d} x_i < 1 \right\}$$

be the reference simplex and

$$\widehat{\sigma}_0 := \left\{ \boldsymbol{x} = (x_1, \ldots, x_d) \in \mathbb{R}^d \;:\; x_i > 0 \quad \forall i = 1, \ldots, d\,,\; \sum_{i=1}^{d} x_i = 1 \right\}$$

be the face of $\widehat{\tau}$ opposite to $\mathbf{0}$ (see Fig. 1.7 for an illustration of these sets and of the arguments to follow).

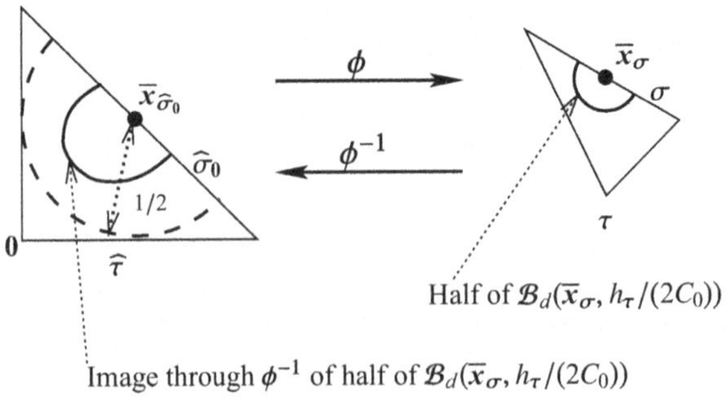

Fig. 1.7 Illustration of Point (i) in the proof of Lemma 1.42

Since τ contains a ball of radius $\gtrsim h_\tau$, by [178, Lemma 8.8] there exists an affine mapping $\boldsymbol{\phi}(\cdot) = \boldsymbol{x}_0 + M\cdot$ of $\mathbb{R}^d$, with M an invertible matrix, such that $\boldsymbol{\phi}(\widehat{\tau}) = \tau$ and $\|M^{-1}\|_2 \le C_0 h_\tau^{-1}$ with C_0 depending only on d and ϱ (here, $\|\cdot\|_2$ is the norm induced on the space of matrices by the Euclidean norm $|\cdot|$ of $\mathbb{R}^d$). Possibly upon a permutation of the vertices of τ, we can assume that $\boldsymbol{\phi}$ sends $\widehat{\sigma}_{\mathbf{0}}$ onto σ.

Since the centre of mass $\overline{\boldsymbol{x}}_\sigma$ of σ is the barycentre of its vertices, it is sent by $\boldsymbol{\phi}^{-1}$ to the centre of mass $\overline{\boldsymbol{x}}_{\widehat{\sigma}_0}$ of $\widehat{\sigma}_{\mathbf{0}}$. Let $\boldsymbol{y} \in \mathbb{R}^d$ be on the same side of σ as τ and such that

$$|\boldsymbol{y} - \overline{\boldsymbol{x}}_\sigma| < \frac{1}{2C_0} h_\tau.$$

Then, applying $\boldsymbol{\phi}^{-1}$,

$$\begin{aligned}|\boldsymbol{\phi}^{-1}(\boldsymbol{y}) - \overline{\boldsymbol{x}}_{\widehat{\sigma}_0}| = |\boldsymbol{\phi}^{-1}(\boldsymbol{y}) - \boldsymbol{\phi}^{-1}(\overline{\boldsymbol{x}}_\sigma)| &= |M^{-1}(\boldsymbol{y} - \overline{\boldsymbol{x}}_\sigma)| \\ &< \|M^{-1}\|_2 \frac{1}{2C_0} h_\tau \le \frac{1}{2}.\end{aligned}$$

Hence, $\boldsymbol{\phi}^{-1}(\boldsymbol{y}) \in \widehat{\tau}$ since $\widehat{\tau}$ is the reference simplex, and therefore contains the half ball of radius $\frac{1}{2}$ centred at $\overline{\boldsymbol{x}}_{\widehat{\sigma}_0}$ and on the same side of $\widehat{\sigma}_{\mathbf{0}}$ as $\widehat{\tau}$. This proves that $\boldsymbol{y} = \boldsymbol{\phi}(\boldsymbol{\phi}^{-1}(\boldsymbol{y})) \in \boldsymbol{\phi}(\widehat{\tau}) = \tau$. In other words, we have proved what is illustrated on the right of Fig. 1.7:

$$\begin{aligned}&\tau \text{ contains a half ball, centred at } \overline{\boldsymbol{x}}_\sigma \text{ and of radius } \gtrsim h_\tau, \\ &\text{that lies on the same side of } \sigma \text{ as } \tau.\end{aligned} \tag{1.88}$$

(ii) *Construction of the family* $(X_i)_{i=1,\ldots,N}$. Let $T \in \mathcal{T}_h$ and $\tau, \tau' \in \mathfrak{T}_h$ be two simplices in T that share a common face $\sigma \in \mathfrak{F}_h$. Owing to (1.88) and (1.4), each of τ and τ' contains a half ball centred at $\overline{\boldsymbol{x}}_\sigma$ and of radius $\gtrsim h_T$. Let r_σ be the smallest of the radii of these two half balls, so that $r_\sigma \gtrsim h_T$, and set $\mathfrak{B}_\sigma := \mathcal{B}_d(\overline{\boldsymbol{x}}_\sigma, r_\sigma)$.

A family $(X_i)_{i=1,\ldots,N}$ satisfying (1.62) for $X = T$ is constructed the following way. We first list the simplices $\tau_1, \ldots, \tau_r$ of $\mathfrak{T}_T = \{\tau \in \mathfrak{T}_h \ :\ \tau \subset T\}$ in such a way that, for any $i \in \{2, \ldots, r\}$, there exists $j < i$ such that τ_i and τ_j share a face. Then, denoting by $\mathfrak{F}_\tau^{\mathrm{i}}$ the set of internal faces of $\tau \in \mathfrak{T}_T$ (that is, the faces shared by another simplex in $\mathfrak{T}_T$), the union in (1.62a) is written as

$$\overline{T} = \overline{\tau_1} \cup \bigcup_{\sigma \in \mathfrak{F}_{\tau_1}^{\mathrm{i}}} \overline{\mathfrak{B}_\sigma} \cup \overline{\tau_2} \cup \bigcup_{\sigma \in \mathfrak{F}_{\tau_2}^{\mathrm{i}}} \overline{\mathfrak{B}_\sigma} \cup \cdots \cup \overline{\tau_{r-1}} \cup \bigcup_{\sigma \in \mathfrak{F}_{\tau_{r-1}}^{\mathrm{i}}} \overline{\mathfrak{B}_\sigma} \cup \overline{\tau_r}. \tag{1.89}$$

The order in which the τ_j and $\mathfrak{F}_{\tau_j}^{\mathrm{i}}$ are listed is important, but the order in which we list each $\overline{\mathfrak{B}_\sigma}$ in $\bigcup_{\sigma \in \mathfrak{F}_{\tau_j}^{\mathrm{i}}} \overline{\mathfrak{B}_\sigma}$ is not. This union satisfies (1.62a) since the number of its elements is bounded above by $(d+2)\,\mathrm{card}(\mathfrak{T}_T)$ (each simplex $\tau \in \mathfrak{T}_T$ has at most $(d + 1)$ internal faces), and $\mathrm{card}(\mathfrak{T}_T) \lesssim 1$ by (1.9). It obviously satisfies (1.62b) by the regularity assumption (1.3) on the simplices, and the fact that all the balls in the family have radius $\gtrsim h_T$ and are star-shaped with respect to all their points.

Finally, (1.62c) comes from the order chosen on the simplices. Indeed, considering first the case of a simplex τ_i, $i \geq 2$, in the list (1.89), τ_i has a face σ in common with some τ_j for $j < i$. The ball $\mathfrak{B}_\sigma$ corresponding to this face σ therefore appears in the list before τ_i, in the union over $\mathfrak{F}^{\mathrm{i}}_{\tau_j}$, and by construction $\tau_i \cap \mathfrak{B}_\sigma$ contains half of $\mathfrak{B}_\sigma$. This ball has radius $\gtrsim h_T$, so its half contains a ball of radius $\gtrsim h_T/2 \gtrsim h_T$. If we consider now the case of a ball $\mathfrak{B}_\sigma$ in the union over some $\mathfrak{F}^{\mathrm{i}}_{\tau_j}$ in the list (1.89), then the simplex τ_j appears before $\mathfrak{B}_\sigma$ in the list and $\mathfrak{B}_\sigma \cap \tau_j$ contains half of $\mathfrak{B}_\sigma$ which, as above, contains a ball of radius $\gtrsim h_T$. This completes the proof of the lemma. □

Remark 1.51 (Optimality of the Choice of $(X_i)_{i=1,\ldots,N}$) The sets constructed in this proof are far from being optimal in terms of their numbers, or the sizes of the balls contained in their pairwise intersections. For a given explicit polytopal set T, a simple inspection usually identifies a small family made of two or three sets. However, establishing a generic proof of the existence of these $(X_i)_{i=1,\ldots,N}$ requires to only rely on the definition of an element in a regular family of meshes, which is what we did above.

Chapter 2
Basic Principles of Hybrid High-Order Methods: The Poisson Problem

In this chapter we introduce the main ideas underlying HHO methods, using to this purpose the Poisson problem: Find $u : \Omega \to \mathbb{R}$ such that

$$-\Delta u = f \qquad \text{in } \Omega, \tag{2.1a}$$

$$u = 0 \qquad \text{on } \partial\Omega, \tag{2.1b}$$

where Ω is an open bounded polytopal subset of $\mathbb{R}^n$, $n \geq 2$, with boundary $\partial\Omega$ and $f : \Omega \to \mathbb{R}$ is a given volumetric source term, assumed to be in $L^2(\Omega)$. Recalling the notation introduced in Remark 1.14 for L^2-products, the starting point to devise an HHO discretisation is the following classical weak formulation of problem (2.1): Find $u \in H_0^1(\Omega)$ such that

$$a(u, v) = (f, v) \qquad \forall v \in H_0^1(\Omega), \tag{2.2}$$

where the bilinear form $a : H^1(\Omega) \times H^1(\Omega) \to \mathbb{R}$ is such that

$$a(u, v) := (\nabla u, \nabla v). \tag{2.3}$$

In what follows, the quantities u and $-\nabla u$ will be referred to, respectively, as the *potential* and the *flux*.

Throughout the rest of this chapter, $(\mathcal{M}_h)_{h\in\mathcal{H}} = (\mathcal{T}_h, \mathcal{F}_h)_{h\in\mathcal{H}}$ denotes a regular mesh sequence in the sense of Definition 1.9. This fact is explicitly recalled only in the statements of central results for the sake of easy consultation. HHO methods are based on discrete unknowns that are broken polynomials over mesh elements and faces, and rely on two key ingredients: (1) problem-dependent local reconstructions obtained by solving small, trivially parallel problems on each mesh element $T \in \mathcal{T}_h$, and (2) stabilisation terms penalising high-order differences. These ingredients are combined to formulate local contributions, which are then assembled as in Finite

D. A. Di Pietro, J. Droniou, *The Hybrid High-Order Method for Polytopal Meshes*, MS&A 19, https://doi.org/10.1007/978-3-030-37203-3_2

Element Methods. The reconstruction is usually conceived so that its composition with the interpolator coincides with a projector on a local polynomial space.

In Sect. 2.1 we decline these general ideas for the Poisson problem, starting with a key remark: For any mesh element $T \in \mathcal{T}_h$, the elliptic projection of degree $(k+1)$ of a function can be computed from its L^2-orthogonal projections of degree k on T and on each of its faces. This remark motivates:

1. the choice of a local space of discrete unknowns $\underline{U}_T^k$ spanned by polynomials of degree $k \geq 0$ on T and on each of its faces, the latter possibly discontinuous at face boundaries;
2. the choice of a local interpolator from smooth functions to $\underline{U}_T^k$ constructed from the L^2-orthogonal projections of degree k on T and on each of its faces;
3. the introduction of a local potential reconstruction operator $\mathrm{p}_T^{k+1} : \underline{U}_T^k \to \mathbb{P}^{k+1}(T)$ whose composition with the interpolator on $\underline{U}_T^k$ coincides with the elliptic projector of degree $(k+1)$.

From these ingredients, we devise the local approximation a_T of the continuous bilinear form a defined by (2.3). The local bilinear form a_T is composed of two terms: a standard Galerkin contribution and a stabilisation term. The latter is conceived so that stability and boundedness hold with respect to a suitable $H^1(T)$-like seminorm, and that polynomial consistency up to degree $(k+1)$ is verified.

In Sect. 2.2 we introduce the global space of discrete unknowns with single-valued interface values (meaning that the interface unknowns match from one element to the adjacent one), as well as a global bilinear form a_h obtained by element by element assembly of the local contributions a_T, $T \in \mathcal{T}_h$. Based on these ingredients, we formulate the discrete problem and discuss its well-posedness. We close this section by showing that the HHO method is locally conservative on each element, and identify a computable expression for the normal trace of the numerical flux. This interpretation highlights the different roles of the equations associated to element and face unknowns: the former correspond to local balances inside each element, whereas the latter enforce the continuity of fluxes. These balance and continuity equations lead to an interpretation of HHO schemes as high-order Finite Volume schemes.

In Sect. 2.3 we carry out an exhaustive error analysis of the method based on the abstract framework of Appendix A. Specifically, we show that the approximation error measured in the energy norm converges as h^{k+1}. A similar convergence rate is then proved for the global potential reconstruction in $\mathbb{P}^{k+1}(\mathcal{T}_h)$, obtained by glueing together the local reconstructions, and for its jumps. Finally, under the usual elliptic regularity assumption, we show that improved convergence in h^{k+2} holds for the L^2-norm of the error. The latter result hinges on a key feature of HHO methods, namely the superconvergence of element-based unknowns.

In Sect. 2.4 we briefly discuss other boundary conditions. Finally, in Sect. 2.5, we illustrate the theoretical results with numerical examples in two and three space dimensions.

2.1 Local Construction

Let a polynomial degree $k \geq 0$ and a mesh element $T \in \mathcal{T}_h$ be given. We introduce the local ingredients underlying the HHO construction: the discrete unknowns, the interpolator, the potential reconstruction operator, and the local approximation of the continuous bilinear form defined by (2.3).

2.1.1 Computing the Local Elliptic Projection from L^2-Projections

Consider a function $v \in W^{1,1}(T)$. We note the following integration by parts formula, valid for all $w \in C^\infty(\overline{T})$:

$$(\nabla v, \nabla w)_T = -(v, \Delta w)_T + \sum_{F \in \mathcal{F}_T} (v, \nabla w \cdot \boldsymbol{n}_{TF})_F. \tag{2.4}$$

Let us specialise (2.4) to $w \in \mathbb{P}^{k+1}(T)$. Observing that $\Delta w \in \mathbb{P}^{k-1}(T) \subset \mathbb{P}^k(T)$ and using the definition (1.57) of $\pi_T^{0,k}$, we can write $(\pi_T^{0,k} v, \Delta w)_T$ instead of $(v, \Delta w)_T$ in the right-hand side. Moreover, for all $F \in \mathcal{F}_T$, we have that $(\nabla w)_{|F} \in \mathbb{P}^k(F)^d$ by Definition 1.22 of local polynomial spaces and that $\boldsymbol{n}_{TF} \in \mathbb{P}^0(F)^d$ by the planarity of faces (see Definition 1.4), so that $(\nabla w)_{|F} \cdot \boldsymbol{n}_{TF} \in \mathbb{P}^k(F)$. Hence, invoking the definition (1.57) of $\pi_F^{0,k}$, we can further replace $(v, \nabla w \cdot \boldsymbol{n}_{TF})_F$ by $(\pi_F^{0,k} v, \nabla w \cdot \boldsymbol{n}_{TF})_F$ in the right-hand side. Finally, using the definition (1.60a) of the elliptic projector $\pi_T^{1,k+1}$, we can write $(\nabla \pi_T^{1,k+1} v, \nabla w)_T$ instead of $(\nabla v, \nabla w)_T$ in the left-hand side. In conclusion, we have that

$$(\nabla \pi_T^{1,k+1} v, \nabla w)_T = -(\pi_T^{0,k} v, \Delta w)_T + \sum_{F \in \mathcal{F}_T} (\pi_F^{0,k} v, \nabla w \cdot \boldsymbol{n}_{TF})_F. \tag{2.5a}$$

Notice that, here and in what follows, it is understood that the L^2-projectors over faces act on the traces of the considered functions. On the other hand, using again the definition (1.57) of $\pi_T^{0,k}$, we have that

$$0 = (\pi_T^{1,k+1} v - v, 1)_T = (\pi_T^{1,k+1} v - \pi_T^{0,k} v, 1)_T. \tag{2.5b}$$

The relations (2.5) show that computing the elliptic projection $\pi_T^{1,k+1} v$ does not require the full knowledge of the function v. All that is required is

1. $\pi_T^{0,k} v$, the L^2-projection of v on the local polynomial space $\mathbb{P}^k(T)$;
2. for all $F \in \mathcal{F}_T$, $\pi_F^{0,k} v$, the L^2-projection of the trace of v on $\mathbb{P}^k(F)$.

Remark 2.1 (Choice of the Polynomial Degree for the Element-Based L^2-Projector) In (2.5a), since $\Delta w \in \mathbb{P}^{k-1}(T)$, we could have replaced $\pi_T^{0,k}$ with $\pi_T^{0,\ell}$ for any $\ell \geq \max(0, k-1)$. The same holds in (2.5b). This choice leads to variations of the method, discussed in Sect. 5.1.

2.1.2 Local Space of Discrete Unknowns

The discussion in the previous section motivates the introduction of the following space of discrete unknowns (see Fig. 2.1):

$$\underline{U}_T^k := \left\{ \underline{v}_T = (v_T, (v_F)_{F\in\mathcal{F}_T}) \ :\ v_T \in \mathbb{P}^k(T) \text{ and } v_F \in \mathbb{P}^k(F) \quad \forall F \in \mathcal{F}_T \right\}. \tag{2.6}$$

On $\underline{U}_T^k$, we define the H^1-like seminorm $\|{\cdot}\|_{1,T}$ such that, for all $\underline{v}_T \in \underline{U}_T^k$,

$$\begin{aligned} \|\underline{v}_T\|_{1,T} &:= \left(\|\nabla v_T\|_T^2 + |\underline{v}_T|_{1,\partial T}^2 \right)^{\frac{1}{2}}, \\ |\underline{v}_T|_{1,\partial T} &:= \left(\sum_{F\in\mathcal{F}_T} h_F^{-1} \|v_F - v_T\|_F^2 \right)^{\frac{1}{2}}, \end{aligned} \tag{2.7}$$

where h_F denotes the diameter of F. The negative power of h_F in the second term ensures that both contributions have the same scaling. The discrete unknowns corresponding to a smooth function $v \in W^{1,1}(T)$ are obtained via the local interpolator $\underline{I}_T^k : W^{1,1}(T) \to \underline{U}_T^k$ such that

$$\underline{I}_T^k v := (\pi_T^{0,k} v, (\pi_F^{0,k} v)_{F\in\mathcal{F}_T}). \tag{2.8}$$

The next proposition states a boundedness property of this interpolator that will be instrumental to the analysis of the HHO method.

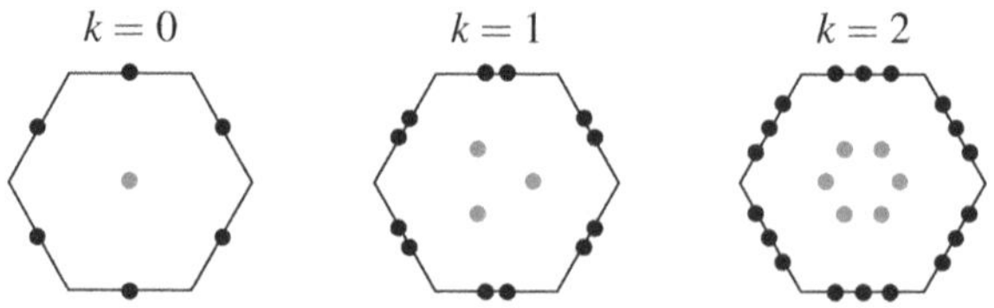

Fig. 2.1 Discrete unknowns in $\underline{U}_T^k$ for $k \in \{0, 1, 2\}$. The dots represent the number of unknowns attached to an element or face, in dimension $d = 2$. When writing the HHO scheme (2.48), the discrete unknowns attached to elements (in grey) can be eliminated by static condensation; see Sect. B.3.2 for further details

Proposition 2.2 (Boundedness of the Local Interpolator) *For all $T \in \mathcal{T}_h$ and all $v \in H^1(T)$,*

$$\|\underline{I}_T^k v\|_{1,T} \lesssim |v|_{H^1(T)}, \tag{2.9}$$

where the hidden constant depends only on d, ϱ and k.

Proof Expanding the local seminorm according to its definition (2.7) and recalling the definition (2.8) of $\underline{I}_T^k$, we get

$$\begin{aligned}\|\underline{I}_T^k v\|_{1,T}^2 &= \|\nabla \pi_T^{0,k} v\|_T^2 + \sum_{F\in\mathcal{F}_T} h_F^{-1}\|\pi_F^{0,k} v - \pi_T^{0,k} v\|_F^2 \\ &\lesssim \|\nabla v\|_T^2 + \sum_{F\in\mathcal{F}_T} h_F^{-1}\|\pi_F^{0,k} v - \pi_T^{0,k} v\|_F^2,\end{aligned} \tag{2.10}$$

where the inequality follows from the boundedness properties of $\pi_T^{0,k}$ resulting from (1.77) with $X = T$, $l = k$, $p = 2$, and $s = 1$. To bound the second term in (2.10), we observe that, using the linearity and idempotency of $\pi_F^{0,k}$ followed by its L^2-boundedness expressed by (1.77) with $X = F$, $l = k$, $p = 2$, and $s = 0$, we have

$$\|\pi_F^{0,k} v - \pi_T^{0,k} v\|_F = \|\pi_F^{0,k}(v - \pi_T^{0,k} v)\|_F \le \|v - \pi_T^{0,k} v\|_F \lesssim h_T^{\frac{1}{2}}|v|_{H^1(T)},$$

where the conclusion follows from the trace approximation property (1.75) of $\pi_T^{0,k}$ with $l = k$, $p = 2$, $s = 1$, and $m = 0$. Using the above estimate together with $\frac{h_T}{h_F} \le \frac{1}{2\varrho^2}$ (see (1.6)) and the uniform bound (1.5) on the number of faces of T, we obtain

$$\sum_{F\in\mathcal{F}_T} h_F^{-1}\|\pi_F^{0,k} v - \pi_T^{0,k} v\|_F^2 \lesssim \sum_{F\in\mathcal{F}_T} \frac{h_T}{h_F}|v|_{H^1(T)}^2 \lesssim |v|_{H^1(T)}^2.$$

Plugged into (2.10), this concludes the proof of (2.9) after observing that, by definition (1.16) of the Sobolev seminorm with $X = T$, $s = 1$, and $p = 2$, $\|\nabla v\|_T \lesssim |v|_{H^1(T)}$. □

2.1.3 Potential Reconstruction Operator

Inspired by (2.5), we introduce the potential reconstruction operator $\mathrm{p}_T^{k+1} : \underline{U}_T^k \to \mathbb{P}^{k+1}(T)$ such that, for all $\underline{v}_T \in \underline{U}_T^k$,

$$(\nabla \mathrm{p}_T^{k+1}\underline{v}_T, \nabla w)_T = -(v_T, \Delta w)_T + \sum_{F\in\mathcal{F}_T}(v_F, \nabla w\cdot\boldsymbol{n}_{TF})_F \quad \forall w \in \mathbb{P}^{k+1}(T) \tag{2.11a}$$

and

$$(\mathrm{p}_T^{k+1}\underline{v}_T - v_T, 1)_T = 0. \tag{2.11b}$$

For future use, we also note the following equivalent statement of (2.11a), obtained integrating by parts the first term in the right-hand side: for all $w \in \mathbb{P}^{k+1}(T)$,

$$(\nabla \mathrm{p}_T^{k+1}\underline{v}_T, \nabla w)_T = (\nabla v_T, \nabla w)_T + \sum_{F\in\mathcal{F}_T} (v_F - v_T, \nabla w{\cdot}\boldsymbol{n}_{TF})_F. \tag{2.12}$$

Remark 2.3 (Equivalent Definition of p_T^{k+1}*)* Letting $\lambda_T \neq 0$, it is useful to notice that Eqs. (2.11) can be equivalently reformulated as: For all $w \in \mathbb{P}^{k+1}(T)$,

$$\begin{aligned}(\nabla \mathrm{p}_T^{k+1}\underline{v}_T, \nabla w)_T &+ \lambda_T(\mathrm{p}_T^{k+1}\underline{v}_T, \pi_T^{0,0} w)_T \\ &= (\nabla v_T, \nabla w)_T + \sum_{F\in\mathcal{F}_T} (v_F - v_T, \nabla w{\cdot}\boldsymbol{n}_{TF})_F + \lambda_T(v_T, \pi_T^{0,0} w)_T.\end{aligned} \tag{2.13}$$

This can be seen summing (2.12) (equivalent to (2.11a)) and (2.11b) multiplied by $\lambda_T \pi_T^{0,0} w$ to get (2.13); conversely, applying (2.13) with $w - \pi_T^{0,0} w$ (resp. $w = 1/\lambda_T$), we recover (2.12) (resp. (2.11b)). In the practical implementation, the parameter λ_T can be tuned so as to improve the conditioning of the local problem matrix; see Sect. B.2.1.

The local reconstruction $\mathrm{p}_T^{k+1}\underline{v}_T$ is a polynomial function on T one degree higher than the element-based discrete unknown v_T. Comparing (2.5) and (2.11) shows that, for all $v \in W^{1,1}(T)$,

$$\mathrm{p}_T^{k+1}\underline{I}_T^k v = \pi_T^{1,k+1} v, \tag{2.14}$$

i.e., the composition of the reconstruction operator with the interpolator gives the elliptic projector of degree $(k+1)$. This commutation property is illustrated in Fig. 2.2.

An immediate consequence of (2.14) together with Theorem 1.48 is that $\mathrm{p}_T^{k+1}\underline{I}_T^k$ has optimal approximation properties in $\mathbb{P}^{k+1}(T)$.

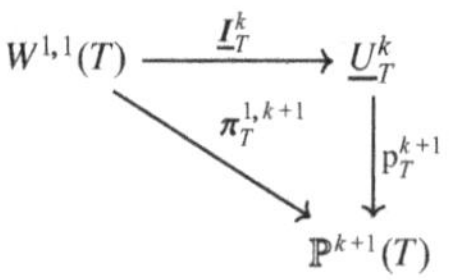

Fig. 2.2 Illustration of the commutation property (2.14) of p_T^{k+1}

2.1.4 Local Contribution

Inside T, we approximate the continuous bilinear form a defined by (2.3) by the discrete bilinear form $\mathrm{a}_T : \underline{U}_T^k \times \underline{U}_T^k \to \mathbb{R}$ such that, for all $\underline{u}_T, \underline{v}_T \in \underline{U}_T^k$,

$$\mathrm{a}_T(\underline{u}_T, \underline{v}_T) := (\nabla \mathrm{p}_T^{k+1} \underline{u}_T, \nabla \mathrm{p}_T^{k+1} \underline{v}_T)_T + \mathrm{s}_T(\underline{u}_T, \underline{v}_T), \tag{2.15}$$

where the first term in the right-hand side is responsible for consistency, while $\mathrm{s}_T : \underline{U}_T^k \times \underline{U}_T^k \to \mathbb{R}$ is a local stabilisation bilinear form, whose role is to ensure the coercivity of the discrete problem defined by (2.48) below. The following design conditions have been originally proposed in [56]:

Assumption 2.4 (Local Stabilisation Bilinear Form s_T) *The local stabilisation bilinear form $\mathrm{s}_T : \underline{U}_T^k \times \underline{U}_T^k \to \mathbb{R}$ satisfies the following properties:*

(S1) Symmetry and positivity. *s_T is symmetric and positive semidefinite;*
(S2) Stability and boundedness. *There is a real number $\eta > 0$ independent of h and T such that, for all $\underline{v}_T \in \underline{U}_T^k$,*

$$\eta^{-1} \|\underline{v}_T\|_{1,T}^2 \le \mathrm{a}_T(\underline{v}_T, \underline{v}_T) \le \eta \|\underline{v}_T\|_{1,T}^2; \tag{2.16}$$

(S3) Polynomial consistency. *For all $w \in \mathbb{P}^{k+1}(T)$ and all $\underline{v}_T \in \underline{U}_T^k$, it holds*

$$\mathrm{s}_T(\underline{I}_T^k w, \underline{v}_T) = 0. \tag{2.17}$$

Some comments are in order.

Remark 2.5 (Local Stabilisation Bilinear Form s_T) Condition (S1) is a natural requirement reflecting the fact that, at the discrete level, we wish to preserve both the symmetry and the positive semidefinite nature of a. Condition (S2) implies, in particular, that a_T vanishes if one of its arguments is the interpolate of a constant function, and that the converse is also true:

$$\mathrm{a}_T(\underline{w}_T, \underline{v}_T) = 0 \quad \forall \underline{v}_T \in \underline{U}_T^k \iff \text{there exists } c \in \mathbb{R} \text{ such that } \underline{w}_T = \underline{I}_T^k c. \tag{2.18}$$

While it can be checked that the above condition is indeed sufficient to ensure the uniqueness of the solution to the discrete problem (2.48) below, in (S2) we also stipulate that the equivalence between $\|\cdot\|_{1,T}$ and the seminorm induced by a_T is uniform with respect to the meshsize h. This fact plays a key role in the proof of the uniform a priori bound on the exact solution in Lemma 2.19 and, in conjunction with (S3), in the derivation of an optimal estimate for the consistency error in Lemma 2.18. For further insight into this point, we refer the reader to the proof of the Third Strang Lemma A.7 and to Remark A.8.

The requirements in Assumption 2.4 suggest that s_T can be obtained penalising in a least square sense quantities that vanish for interpolates of polynomial functions in $\mathbb{P}^{k+1}(T)$. Paradigmatic examples of such quantities are obtained through the operators $\delta_T^k : \underline{U}_T^k \to \mathbb{P}^k(T)$ and, for all $F \in \mathcal{F}_T$, $\delta_{TF}^k : \underline{U}_T^k \to \mathbb{P}^k(F)$ such that, for all $\underline{v}_T \in \underline{U}_T^k$,

$$\delta_T^k \underline{v}_T := \pi_T^{0,k}(\mathrm{p}_T^{k+1}\underline{v}_T - v_T), \qquad \delta_{TF}^k \underline{v}_T := \pi_F^{0,k}(\mathrm{p}_T^{k+1}\underline{v}_T - v_F) \quad \forall F \in \mathcal{F}_T. \tag{2.19}$$

Recalling the definition (2.8) of the local interpolator, it is a simple matter to check that

$$(\delta_T^k \underline{v}_T, (\delta_{TF}^k \underline{v}_T)_{F\in\mathcal{F}_T}) = \underline{I}_T^k \mathrm{p}_T^{k+1}\underline{v}_T - \underline{v}_T. \tag{2.20}$$

Proposition 2.6 (Polynomial Consistency of the Difference Operators) *It holds, for all $T \in \mathcal{T}_h$ and all $w \in \mathbb{P}^{k+1}(T)$,*

$$\delta_T^k \underline{I}_T^k w = 0 \quad \textit{and} \quad \delta_{TF}^k \underline{I}_T^k w = 0 \quad \forall F \in \mathcal{F}_T. \tag{2.21}$$

Proof Let us check that δ_T^k vanishes when its argument is of the form $\underline{I}_T^k w$ with $w \in \mathbb{P}^{k+1}(T)$. By definition of δ_T^k and $\underline{I}_T^k w = (\pi_T^{0,k} w, (\pi_F^{0,k} w)_{F\in\mathcal{F}_T})$, we have that

$$\delta_T^k \underline{I}_T^k w = \pi_T^{0,k}(\mathrm{p}_T^{k+1}\underline{I}_T^k w - \pi_T^{0,k} w).$$

Using the relation (2.14) to replace $\mathrm{p}_T^{k+1}\underline{I}_T^k$ by $\pi_T^{1,k+1}$ and the fact that $\pi_T^{0,k} w \in \mathbb{P}^k(T)$ together with the linearity and polynomial invariance (1.56) for $\pi_T^{0,k}$ to remove the latter projector from the second term in parentheses, we get

$$\delta_T^k \underline{I}_T^k w = \pi_T^{0,k}(\pi_T^{1,k+1} w - w).$$

Using again the polynomial invariance (1.56), this time for $\pi_T^{1,k+1}$, we conclude that

$$\delta_T^k \underline{I}_T^k w = \pi_T^{0,k}(w - w) = 0.$$

Similar arguments can be used to prove the second identity in (2.21). The details are left as an exercise to the reader. □

Relevant examples of stabilisation bilinear forms obtained by penalising, in a least square sense, the differences defined in (2.19) are discussed in what follows.

Example 2.7 (Original HHO Stabilisation) The original HHO stabilisation of [158] is obtained setting

$$s_T(\underline{u}_T, \underline{v}_T) := \sum_{F\in\mathcal{F}_T} h_F^{-1}((\delta_{TF}^k - \delta_T^k)\underline{u}_T, (\delta_{TF}^k - \delta_T^k)\underline{v}_T)_F. \tag{2.22}$$

In this case, only quantities on faces are penalised, and the factor h_F^{-1} ensures dimensional homogeneity with the consistency term in (2.15). The proof that this stabilisation bilinear form satisfies Assumption 2.4 is provided in Proposition 2.13 below. Another important example of a stabilisation bilinear form used in the HHO literature can be found in [9, Eq. (3.24)]. This expression results from the hybridisation of the Mixed High-Order method of [149]; see Sect. 5.4 for further details.

Example 2.8 (A Stabilisation Inspired by Virtual Elements) An expression for the stabilisation term inspired by the Virtual Elements literature [45] is obtained setting

$$\mathrm{s}_T(\underline{u}_T, \underline{v}_T) := h_T^{-2}(\delta_T^k \underline{u}_T, \delta_T^k \underline{v}_T)_T + \sum_{F\in\mathcal{F}_T} h_F^{-1}(\delta_{TF}^k \underline{u}_T, \delta_{TF}^k \underline{v}_T)_F. \tag{2.23}$$

Unlike in (2.22), both volumetric and boundary contributions are present. The negative powers of the element and face diameters in each term are again selected so as to ensure dimensional homogeneity with the consistency term.

Remark 2.9 (Original Stabilisation in Hybridisable Discontinuous Galerkin Methods) The following stabilisation bilinear form is used in the original Hybridisable Discontinuous Galerkin method of [100, 127]:

$$\mathrm{s}_T(\underline{u}_T, \underline{v}_T) = \sum_{F\in\mathcal{F}_T} \alpha h_F^{-1}(u_F - u_T, v_F - v_T)_F,$$

where $\alpha > 0$ denotes a user-dependent penalty parameter. While this choice obviously satisfies the properties (S1)–(S2), it fails to satisfy (S3) (it is only consistent for polynomials of degree up to k). As a result, up to one order of convergence is lost with respect to the estimates of Theorems 2.28 and 2.32 below. For a discussion including fixes that restore optimal orders of convergence, see Sect. 5.1.6 and also [129].

Remark 2.10 (Modification in Dimension $d = 1$) In the case of spatial dimension $d = 1$, each "face" F is a point, $\mathbb{P}^k(F)$ is identified with $\mathbb{R}$, the integral over F boils down to taking the value of the function at F, and the scaling factors h_F^{-1} in the semi-norm $|\cdot|_{1,\partial T}$ and the stabilisation terms s_T above have to be replaced by h_T^{-1}.

Following up on the previous remarks, the next lemma shows that consistent stabilisation bilinear forms are inevitably constructed from the difference operators (2.19).

Lemma 2.11 (Dependency of s_T) *Let $T \in \mathcal{T}_h$ and let $\mathrm{s}_T : \underline{U}_T^k \times \underline{U}_T^k \to \mathbb{R}$ be a symmetric bilinear form. Then, s_T satisfies the polynomial consistency (S3) in Assumption 2.4 if and only if it depends on its arguments only via the difference operators* (2.19).

Proof If s_T only depends on its arguments through the difference operators (2.19), then (S3) follows from the polynomial consistency (2.21) of these difference operators.

Conversely, assume that s_T satisfies (S3) and take $\underline{u}_T, \underline{v}_T \in \underline{U}_T^k$. Using the bilinearity and symmetry of s_T, and applying (S3) first with $(w, \underline{v}_T) = (\mathrm{p}_T^{k+1}\underline{u}_T, \underline{v}_T)$, then with $(w, \underline{v}_T) = (\mathrm{p}_T^{k+1}\underline{v}_T, \underline{u}_T - \underline{I}_T^k \mathrm{p}_T^{k+1}\underline{u}_T)$, we get

$$\mathrm{s}_T(\underline{u}_T, \underline{v}_T) = \mathrm{s}_T(\underline{u}_T - \underline{I}_T^k \mathrm{p}_T^{k+1}\underline{u}_T, \underline{v}_T - \underline{I}_T^k \mathrm{p}_T^{k+1}\underline{v}_T).$$

The conclusion follows from (2.20) which shows that both $(\underline{u}_T - \underline{I}_T^k \mathrm{p}_T^{k+1}\underline{u}_T)$ and $(\underline{v}_T - \underline{I}_T^k \mathrm{p}_T^{k+1}\underline{v}_T)$ depend only on the difference operators (2.19) applied, respectively, to $\underline{u}_T$ and $\underline{v}_T$. □

Remark 2.12 (On the Choice of the Difference Operators) An inspection of the proof above shows that we could have used, instead of p_T^{k+1}, any polynomial reconstruction $R_T^{k+1} : \underline{U}_T^k \to \mathbb{P}^{k+1}(T)$ consistent for polynomials of degree $(k+1)$. It would have given a dependency of s_T in terms of the differences $\delta_{R,T}^k$ and $\delta_{R,TF}^k$, $F \in \mathcal{F}_T$, defined such that, for all $\underline{v}_T \in \underline{U}_T^k$,

$$(\delta_{R,T}^k \underline{v}_T, (\delta_{R,TF}^k \underline{v}_T)_{F \in \mathcal{F}_T}) := \underline{I}_T^k R_T^{k+1} \underline{v}_T - \underline{v}_T. \tag{2.24}$$

However, designing a stabilisation term s_T from the above dependencies for an arbitrary choice of R_T^{k+1} makes it difficult to ensure that (S2) holds. As shown in Proposition 2.13, the difference operators (2.19) seem to be the natural choice to build a stabilisation bilinear form that satisfies the stability and boundedness property (S2).

We now prove that the original HHO stabilisation form satisfies Assumption 2.4.

Proposition 2.13 (Original HHO Stabilisation) *The original HHO stabilisation bilinear form* s_T *defined by* (2.22) *satisfies Assumption 2.4.*

Proof The bilinear form s_T is clearly symmetric and positive semidefinite, so that property (S1) holds. Property (S3), on the other hand, is a consequence of Lemma 2.11. It only remains to prove property (S2). Throughout the rest of the proof, we let $\underline{v}_T$ be a generic element of $\underline{U}_T^k$. For the sake of brevity, we also set

$$\check{v}_T := \mathrm{p}_T^{k+1}\underline{v}_T$$

and the notation $\lesssim$ is understood with hidden constant independent of h, T, and $\underline{v}_T$. We first estimate the volumetric components in $\|\underline{v}_T\|_{1,T}^2$ and $\mathrm{a}_T(\underline{v}_T, \underline{v}_T)$, and then establish (2.16).

(i) *Volumetric components*. We prove here that

$$\|\nabla v_T\|_T \lesssim \|\nabla \check{v}_T\|_T + |\underline{v}_T|_{1,\partial T} \tag{2.25}$$

and

$$\|\nabla \check{v}_T\|_T \lesssim \|\underline{v}_T\|_{1,T}. \tag{2.26}$$

Letting $w = v_T$ in (2.12) and using Cauchy–Schwarz and generalised Hölder inequalities with exponents $(2, 2, \infty)$ along with $\|\boldsymbol{n}_{TF}\|_{L^\infty(F)^d} = 1$, we have that

$$\begin{aligned}
\|\nabla v_T\|_T^2 &= (\nabla \check{v}_T, \nabla v_T)_T - \sum_{F\in\mathcal{F}_T} (v_F - v_T, \nabla v_T \cdot \boldsymbol{n}_{TF})_F. \\
&\le \|\nabla \check{v}_T\|_T \|\nabla v_T\|_T + \sum_{F\in\mathcal{F}_T} \|v_F - v_T\|_F \|\nabla v_T\|_F \|\boldsymbol{n}_{TF}\|_{L^\infty(F)^d} \\
&\le \|\nabla \check{v}_T\|_T \|\nabla v_T\|_T \\
&\quad + \left(\sum_{F\in\mathcal{F}_T} h_F^{-1} \|v_F - v_T\|_F^2\right)^{\frac{1}{2}} \left(\sum_{F\in\mathcal{F}_T} h_F \|\nabla v_T\|_F^2\right)^{\frac{1}{2}},
\end{aligned} \tag{2.27}$$

where, in the third line, we have multiplied the boundary term by $h_F^{-\frac{1}{2}} h_F^{\frac{1}{2}} = 1$ and used a discrete Cauchy–Schwarz inequality on the sum over the faces to write

$$\begin{aligned}
\sum_{F\in\mathcal{F}_T} \|v_F - v_T\|_F \|\nabla v_T\|_F &= \sum_{F\in\mathcal{F}_T} h_F^{-\frac{1}{2}} \|v_F - v_T\|_F \, h_F^{\frac{1}{2}} \|\nabla v_T\|_F \\
&\le \left(\sum_{F\in\mathcal{F}_T} h_F^{-1} \|v_F - v_T\|_F^2\right)^{\frac{1}{2}} \left(\sum_{F\in\mathcal{F}_T} h_F \|\nabla v_T\|_F^2\right)^{\frac{1}{2}}.
\end{aligned}$$

We continue from (2.27) using the discrete trace inequality (1.55) with $p = 2$ on the components of ∇v_T, the bound $h_F \le h_T$, and $\mathrm{card}(\mathcal{F}_T) \lesssim 1$ (see (1.5)) to write

$$\|\nabla v_T\|_T^2 \lesssim \|\nabla \check{v}_T\|_T \|\nabla v_T\|_T + |\underline{v}_T|_{1,\partial T} \|\nabla v_T\|_T.$$

Simplifying by $\|\nabla v_T\|_T$ leads to (2.25).

We now estimate $\|\nabla \check{v}_T\|_T$. Make $w = \check{v}_T$ in (2.12) and, following similar arguments as above, use a Cauchy–Schwarz inequality for the volumetric terms and generalised Hölder inequalities with exponents $(2, 2, \infty)$ along with $\|\boldsymbol{n}_{TF}\|_{L^\infty(F)^d} = 1$ for the boundary terms to infer

$$\|\nabla \check{v}_T\|_T^2 \le \|\nabla v_T\|_T \|\nabla \check{v}_T\|_T + |\underline{v}_T|_{1,\partial T} \left(\sum_{F\in\mathcal{F}_T} h_F \|\nabla \check{v}_T\|_F^2\right)^{\frac{1}{2}} \lesssim \|\underline{v}_T\|_{1,T} \|\nabla \check{v}_T\|_T,$$

where we have invoked the discrete trace inequality (1.55) with $p = 2$ on the components of $\nabla\check{v}_T$ together with the bound $h_F \le h_T$ and (1.5) to conclude. Simplifying by $\|\nabla\check{v}_T\|_T$, we arrive at (2.26).

(ii) *Proof of (S2)*. Let

$$\underline{z}_T := \underline{I}_T^k\check{v}_T - \underline{v}_T = (\delta_T^k\underline{v}_T, (\delta_{TF}^k\underline{v}_T)_{F\in\mathcal{F}_T}), \tag{2.28}$$

where the second equality follows from (2.20). Using the definition (2.15) of a_T together with the choice (2.22) of s_T and the expression (2.28) of $\underline{z}_T$ in terms of difference operators, we have

$$\mathrm{a}_T(\underline{v}_T, \underline{v}_T) = \|\nabla\check{v}_T\|_T^2 + |\underline{z}_T|_{1,\partial T}^2. \tag{2.29}$$

Use (2.28) to write $\underline{v}_T = \underline{I}_T^k\check{v}_T - \underline{z}_T$, invoke the boundedness (2.9) of $\underline{I}_T^k$ with $v = \check{v}_T$ to get

$$|\underline{I}_T^k\check{v}_T|_{1,\partial T} \lesssim |\check{v}_T|_{H^1(T)} \lesssim \|\nabla\check{v}_T\|_T, \tag{2.30}$$

and use finally (2.29) to obtain, by triangle inequality,

$$\begin{aligned}|\underline{v}_T|_{1,\partial T}^2 = |\underline{I}_T^k\check{v}_T - \underline{z}_T|_{1,\partial T}^2 &\le 2|\underline{I}_T^k\check{v}_T|_{1,\partial T}^2 + 2|\underline{z}_T|_{1,\partial T}^2\\ &\lesssim \|\nabla\check{v}_T\|_T^2 + |\underline{z}_T|_{1,\partial T}^2 = \mathrm{a}_T(\underline{v}_T, \underline{v}_T).\end{aligned}$$

Combining this estimate and (2.25) yields

$$\|\underline{v}_T\|_{1,T}^2 = \|\nabla v_T\|_T^2 + |\underline{v}_T|_{1,\partial T}^2 \lesssim \|\nabla\check{v}_T\|_T^2 + |\underline{v}_T|_{1,\partial T}^2 \lesssim \mathrm{a}_T(\underline{v}_T, \underline{v}_T),$$

which proves the first estimate in (2.16). To establish the second, we start from (2.29) and substitute $\underline{z}_T = \underline{I}_T^k\check{v}_T - \underline{v}_T$ to write

$$\begin{aligned}\mathrm{a}_T(\underline{v}_T, \underline{v}_T) &\le \|\nabla\check{v}_T\|_T^2 + 2|\underline{I}_T^k\check{v}_T|_{1,\partial T}^2 + 2|\underline{v}_T|_{1,\partial T}^2\\ &\lesssim \|\nabla\check{v}_T\|_T^2 + |\underline{v}_T|_{1,\partial T}^2\\ &\lesssim \|\underline{v}_T\|_{1,T}^2,\end{aligned}$$

where the first line follows from a triangle inequality, the second line is a consequence of (2.30), and the conclusion is obtained invoking (2.26). The proof of (S2) is complete. □

To close this section, we study the consistency properties of s_T when its arguments are interpolates of smooth functions.

Proposition 2.14 (Consistency of s_T) *Let $T \in \mathcal{T}_h$ and let s_T denote a stabilisation bilinear form satisfying Assumption 2.4. Let $r \in \{0, \ldots, k\}$. Then, for all $v \in$*

$H^{r+2}(T)$,

$$s_T(\underline{I}_T^k v, \underline{I}_T^k v)^{\frac{1}{2}} \lesssim h_T^{r+1} |v|_{H^{r+2}(T)}, \tag{2.31}$$

where the hidden constant is independent of h, T and v.

Proof Using (S3) with $w = \pi_T^{0,k+1} v \in \mathbb{P}^{k+1}(T)$ and (S2), we infer that

$$\begin{aligned}
s_T(\underline{I}_T^k v, \underline{I}_T^k v)^{\frac{1}{2}} &= s_T(\underline{I}_T^k (v - \pi_T^{0,k+1} v), \underline{I}_T^k (v - \pi_T^{0,k+1} v))^{\frac{1}{2}} \\
&\le \eta^{\frac{1}{2}} \|\underline{I}_T^k (v - \pi_T^{0,k+1} v)\|_{1,T} \\
&\lesssim |v - \pi_T^{0,k+1} v|_{H^1(T)} \\
&\lesssim h_T^{r+1} |v|_{H^{r+2}(T)},
\end{aligned}$$

where the third line follows from the boundedness (2.9) of $\underline{I}_T^k$ with $(v - \pi_T^{0,k+1} v)$ instead of v, and the conclusion is obtained applying the approximation property (1.74) of the orthogonal projector with $X = T$, $l = k+1$, $p = 2$, $s = r+2$, and $m = 1$. □

2.2 Discrete Problem

In this section we formulate the discrete problem based on the local contributions introduced in the previous section.

2.2.1 Global Space of Discrete Unknowns

We define the following global space with single-valued interface unknowns:

$$\underline{U}_h^k := \Big\{ \underline{v}_h = ((v_T)_{T \in \mathcal{T}_h}, (v_F)_{F \in \mathcal{F}_h}) \; : \\
v_T \in \mathbb{P}^k(T) \quad \forall T \in \mathcal{T}_h \text{ and } v_F \in \mathbb{P}^k(F) \quad \forall F \in \mathcal{F}_h \Big\}. \tag{2.32}$$

The restriction of a generic element $\underline{v}_h \in \underline{U}_h^k$ to $T \in \mathcal{T}_h$ is denoted by $\underline{v}_T = (v_T, (v_F)_{F \in \mathcal{F}_T}) \in \underline{U}_T^k$. We also define the broken polynomial function $v_h \in \mathbb{P}^k(\mathcal{T}_h)$ such that

$$(v_h)_{|T} := v_T \qquad \forall T \in \mathcal{T}_h. \tag{2.33}$$

The discrete unknowns corresponding to a smooth function $v \in W^{1,1}(\Omega)$ are obtained via the global interpolator $\underline{I}_h^k : W^{1,1}(\Omega) \to \underline{U}_h^k$ such that

$$\underline{I}_h^k v := ((\pi_T^{0,k} v)_{T\in\mathcal{T}_h}, (\pi_F^{0,k} v)_{F\in\mathcal{F}_h}). \tag{2.34}$$

We define on $\underline{U}_h^k$ the global seminorm $\|\cdot\|_{1,h}$ such that, for all $\underline{v}_h \in \underline{U}_h^k$,

$$\|\underline{v}_h\|_{1,h} := \left(\sum_{T\in\mathcal{T}_h} \|\underline{v}_T\|_{1,T}^2 \right)^{\frac{1}{2}}, \tag{2.35}$$

with local seminorm $\|\cdot\|_{1,T}$ defined by (2.7). To account for the homogeneous Dirichlet boundary condition (2.1b) in a strong manner, we introduce the subspace

$$\underline{U}_{h,0}^k := \left\{ \underline{v}_h \in \underline{U}_h^k \ : \ v_F = 0 \quad \forall F \in \mathcal{F}_h^{\rm b} \right\}, \tag{2.36}$$

where we recall that $\mathcal{F}_h^{\rm b}$ gathers all the faces that lie on $\partial\Omega$ (see Definition 1.4). It is a simple matter to check that $\underline{I}_h^k$ maps functions in $H_0^1(\Omega)$ on vectors of discrete unknowns in $\underline{U}_{h,0}^k$.

2.2.2 *A Discrete Poincaré Inequality*

In the following lemma, we establish a discrete version of the Poincaré inequality that will be used to prove that $\|\cdot\|_{1,h}$ is a norm on $\underline{U}_{h,0}^k$ (see Corollary 2.16), as well as to establish the uniform a priori bound (2.49) on the discrete solution.

Lemma 2.15 (Discrete Poincaré Inequality) *There exists $C_{\rm P} > 0$ depending only on Ω, d, and ϱ such that, for all $\underline{v}_h \in \underline{U}_{h,0}^k$,*

$$\|v_h\| \le C_{\rm P} \|\underline{v}_h\|_{1,h}. \tag{2.37}$$

Proof Let $\underline{v}_h \in \underline{U}_{h,0}^k$. Since the divergence operator $\nabla\cdot : H^1(\Omega)^d \to L^2(\Omega)$ is onto (see Lemma 8.3 in Chap. 8 for a proof), there exists $\boldsymbol{\tau}_{v_h} \in H^1(\Omega)^d$ such that

$$\nabla\cdot\boldsymbol{\tau}_{v_h} = v_h \text{ and } \|\boldsymbol{\tau}_{v_h}\|_{H^1(\Omega)^d} \lesssim \|v_h\|. \tag{2.38}$$

Here and in the rest of the proof, the hidden constants in $\lesssim$ depend only on Ω, d and ϱ. Using the first relation in (2.38) and integrating by parts element by element, we obtain

$$\|v_h\|^2 = (v_h, \nabla\cdot\boldsymbol{\tau}_{v_h}) = -\sum_{T\in\mathcal{T}_h} \left((\nabla v_T, \boldsymbol{\tau}_{v_h})_T + \sum_{F\in\mathcal{F}_T} (v_F - v_T, \boldsymbol{\tau}_{v_h}\cdot\boldsymbol{n}_{TF})_F \right),$$

where we have used Corollary 1.19 with $p = 2$, $\boldsymbol{\tau} = \boldsymbol{\tau}_{v_h} \in H^1(\Omega)^d \subset \boldsymbol{H}(\text{div}; \Omega) \cap H^1(\mathcal{T}_h)^d$, and $(\varphi_F)_{F \in \mathcal{F}_h} = (v_F)_{F \in \mathcal{F}_h}$ to insert v_F into the boundary term. Using Cauchy–Schwarz inequalities for the first term and generalised Hölder inequalities with exponents $(2, 2, \infty)$ along with $\|\boldsymbol{n}_{TF}\|_{L^\infty(F)^d} = 1$ for the second, we can go on writing

$$
\begin{aligned}
\|v_h\|^2 &\le \sum_{T \in \mathcal{T}_h} \left(\|\nabla v_T\|_T \|\boldsymbol{\tau}_{v_h}\|_T + \sum_{F \in \mathcal{F}_T} h_F^{-\frac{1}{2}} \|v_F - v_T\|_F \, h_F^{\frac{1}{2}} \|\boldsymbol{\tau}_{v_h}\|_F \right) \\
&\le \|\underline{v}_h\|_{1,h} \left[\sum_{T \in \mathcal{T}_h} \left(\|\boldsymbol{\tau}_{v_h}\|_T^2 + h_T \|\boldsymbol{\tau}_{v_h}\|_{\partial T}^2 \right) \right]^{\frac{1}{2}} \\
&\lesssim \|\underline{v}_h\|_{1,h} \|\boldsymbol{\tau}_{v_h}\|_{H^1(\Omega)^d},
\end{aligned}
$$

where we have recalled the bound $h_F \le h_T$ (for $F \in \mathcal{F}_T$) together with the definition (2.35) of the global H^1-like seminorm after using discrete Cauchy–Schwarz inequalities on the sums to pass to the second line, and used the trace inequality (1.51) with $p = 2$ along with the fact that, for all $T \in \mathcal{T}_h$, $h_T \le h_\Omega$ (with h_Ω denoting the diameter of Ω) to conclude. Using the second condition in (2.38) to further bound the second factor in the right-hand side, we arrive at

$$
\|v_h\|^2 \lesssim \|\underline{v}_h\|_{1,h} \|v_h\|,
$$

which yields the conclusion after simplifying by $\|v_h\|$. □

Corollary 2.16 (Norm $\|\cdot\|_{1,h}$) *The map $\|\cdot\|_{1,h}$ defines a norm on $\underline{U}_{h,0}^k$.*

Proof The seminorm property is evident. It therefore suffices to prove that, for all $\underline{v}_h \in \underline{U}_{h,0}^k$, $\|\underline{v}_h\|_{1,h} = 0$ implies $\underline{v}_h = \underline{0}$. Let $\underline{v}_h \in \underline{U}_{h,0}^k$ be such that $\|\underline{v}_h\|_{1,h} = 0$. By the Poincaré inequality (2.37), we have $\|v_h\| = 0$, hence $v_T = 0$ for all $T \in \mathcal{T}_h$. From the definition (2.7) of the norm $\|\cdot\|_{1,T}$, we also have that $\|v_F - v_T\|_F = 0$, hence $v_F = v_T = 0$ on F, for all $T \in \mathcal{T}_h$ and all $F \in \mathcal{F}_T$. Since any mesh face belongs to a set of faces $\mathcal{F}_T$ for at least one mesh element $T \in \mathcal{T}_h$, this concludes the proof. □

Remark 2.17 (Discrete Poincaré Inequalities on Broken Spaces) The discrete Poincaré inequality (2.37) on HHO spaces can also be proved starting from the corresponding result on broken polynomial spaces. This strategy is adopted in [142, Proposition 5.4], based on the results of [144, Theorem 6.1] and [145, Theorem 5.3]. In the latter references, Sobolev embeddings on broken polynomial spaces are proved using arguments inspired by the recent Finite Volumes literature; see, in particular, [190, Section 5], [178, Appendix B], and also Sects. 6.5 and 6.6 in Chap. 6. In the nonconforming Finite Elements literature, Poincaré inequalities on broken spaces are proved, e.g., in [20, 76], where stronger assumptions on the mesh are needed.

2.2.3 Global Bilinear Form

We define the global bilinear forms $\mathrm{a}_h : \underline{U}_h^k \times \underline{U}_h^k \to \mathbb{R}$ and $\mathrm{s}_h : \underline{U}_h^k \times \underline{U}_h^k \to \mathbb{R}$ by element by element assembly: For all $\underline{u}_h, \underline{v}_h \in \underline{U}_h^k$,

$$\mathrm{a}_h(\underline{u}_h, \underline{v}_h) := \sum_{T\in\mathcal{T}_h} \mathrm{a}_T(\underline{u}_T, \underline{v}_T), \qquad \mathrm{s}_h(\underline{u}_h, \underline{v}_h) := \sum_{T\in\mathcal{T}_h} \mathrm{s}_T(\underline{u}_T, \underline{v}_T). \tag{2.39}$$

For future use, we also define the stabilisation seminorm $|\cdot|_{\mathrm{s},h}$ such that, for all $\underline{v}_h \in \underline{U}_h^k$,

$$|\underline{v}_h|_{\mathrm{s},h} := \mathrm{s}_h(\underline{v}_h, \underline{v}_h)^{\frac{1}{2}}. \tag{2.40}$$

Lemma 2.18 (Properties of a_h) *The bilinear form a_h enjoys the following properties:*

(i) Stability and boundedness. *For all $\underline{v}_h \in \underline{U}_{h,0}^k$, it holds with η as in* (2.16) *that*

$$\eta^{-1}\|\underline{v}_h\|_{1,h}^2 \le \|\underline{v}_h\|_{\mathrm{a},h}^2 \le \eta\|\underline{v}_h\|_{1,h}^2 \text{ with } \|\underline{v}_h\|_{\mathrm{a},h} := \mathrm{a}_h(\underline{v}_h, \underline{v}_h)^{\frac{1}{2}}. \tag{2.41}$$

(ii) Consistency. *It holds for all $r \in \{0, \ldots, k\}$ and all $w \in H_0^1(\Omega) \cap H^{r+2}(\mathcal{T}_h)$ such that $\Delta w \in L^2(\Omega)$,*

$$\sup_{\underline{v}_h\in\underline{U}_{h,0}^k, \|\underline{v}_h\|_{\mathrm{a},h}=1} |\mathcal{E}_h(w; \underline{v}_h)| \lesssim h^{r+1}|w|_{H^{r+2}(\mathcal{T}_h)}, \tag{2.42}$$

where the hidden constant is independent of w and h, and the linear form $\mathcal{E}_h(w; \cdot) : \underline{U}_{h,0}^k \to \mathbb{R}$ representing the consistency error is such that, for all $\underline{v}_h \in \underline{U}_{h,0}^k$,

$$\mathcal{E}_h(w; \underline{v}_h) := -(\Delta w, v_h) - \mathrm{a}_h(\underline{I}_h^k w, \underline{v}_h). \tag{2.43}$$

Proof (i) *Stability and boundedness.* Summing inequalities (2.16) over $T \in \mathcal{T}_h$, (2.41) follows.

(ii) *Consistency.* Let $\underline{v}_h \in \underline{U}_{h,0}^k$ be such that $\|\underline{v}_h\|_{\mathrm{a},h} = 1$. Throughout the proof, the hidden constant in $A \lesssim B$ is independent of both w and h. For the sake of brevity, we also let, for all $T \in \mathcal{T}_h$,

$$\check{w}_T := \mathrm{p}_T^{k+1}\underline{I}_T^k w = \pi_T^{1,k+1} w,$$

where the equality is a consequence of the commutation property (2.14). Integrating by parts element by element and using Corollary 1.19 with $p = 2$, $\boldsymbol{\tau} = \nabla w$ and

$(\varphi_F)_{F\in\mathcal{F}_h} = (v_F)_{F\in\mathcal{F}_h}$ to insert v_F into the boundary term after noticing that, by the assumed regularity, $\boldsymbol{\nabla} w \in \boldsymbol{H}(\mathrm{div};\Omega)\cap H^1(\mathcal{T}_h)^d$, we infer

$$
\begin{aligned}
-(\Delta w, v_h) &= \sum_{T\in\mathcal{T}_h}\left((\boldsymbol{\nabla} w, \boldsymbol{\nabla} v_T)_T - \sum_{F\in\mathcal{F}_T}(\boldsymbol{\nabla} w\cdot\boldsymbol{n}_{TF}, v_T)_F\right) \\
&= \sum_{T\in\mathcal{T}_h}\left((\boldsymbol{\nabla} w, \boldsymbol{\nabla} v_T)_T + \sum_{F\in\mathcal{F}_T}(\boldsymbol{\nabla} w\cdot\boldsymbol{n}_{TF}, v_F - v_T)_F\right).
\end{aligned}
\tag{2.44}
$$

On the other hand, using the definitions (2.39) of a_h and s_h, and (2.15) of a_T, and expanding $\mathrm{p}_T^{k+1}\underline{v}_T$ according to (2.12) with $w = \check{w}_T$, it is inferred that

$$
\begin{aligned}
&\mathrm{a}_h(\underline{I}_h^k w, \underline{v}_h) \\
&\quad= \sum_{T\in\mathcal{T}_h}(\boldsymbol{\nabla}\check{w}_T, \boldsymbol{\nabla}\mathrm{p}_T^{k+1}\underline{v}_T)_T + \mathrm{s}_h(\underline{I}_h^k w, \underline{v}_h) \\
&\quad= \sum_{T\in\mathcal{T}_h}\left((\boldsymbol{\nabla}\check{w}_T, \boldsymbol{\nabla} v_T)_T + \sum_{F\in\mathcal{F}_T}(\boldsymbol{\nabla}\check{w}_T\cdot\boldsymbol{n}_{TF}, v_F - v_T)_F\right) + \mathrm{s}_h(\underline{I}_h^k w, \underline{v}_h).
\end{aligned}
\tag{2.45}
$$

Subtracting (2.45) from (2.44), taking absolute values, and using the definition (1.60a) of $\pi_T^{1,k+1}$ to cancel the first term in parentheses, we get

$$
\begin{aligned}
&|\mathcal{E}_h(w;\underline{v}_h)| \\
&= \left|\sum_{T\in\mathcal{T}_h}\left(\cancel{(\boldsymbol{\nabla}(w-\check{w}_T), \boldsymbol{\nabla} v_T)_T} + \sum_{F\in\mathcal{F}_T}(\boldsymbol{\nabla}(w-\check{w}_T)\cdot\boldsymbol{n}_{TF}, v_F - v_T)_F\right)\right. \\
&\qquad \left. - \mathrm{s}_h(\underline{I}_h^k w, \underline{v}_h)\right| \\
&\le \sum_{T\in\mathcal{T}_h}\sum_{F\in\mathcal{F}_T} h_F^{\frac12}\|\boldsymbol{\nabla}(w-\check{w}_T)\|_F\, h_F^{-\frac12}\|v_F - v_T\|_F + |\mathrm{s}_h(\underline{I}_h^k w, \underline{v}_h)|,
\end{aligned}
\tag{2.46}
$$

where we have used a generalised Hölder inequality with exponents $(2,\infty,2)$ together with $\|\boldsymbol{n}_{TF}\|_{L^\infty(F)^d} = 1$ to conclude. The Cauchy–Schwarz inequality on the positive semidefinite bilinear form s_h (see (S1)) gives, on the other hand,

$$
|\mathrm{s}_h(\underline{I}_h^k w, \underline{v}_h)| \le \mathrm{s}_h(\underline{I}_h^k w, \underline{I}_h^k w)^{\frac12}\, \mathrm{s}_h(\underline{v}_h, \underline{v}_h)^{\frac12}.
$$

Hence, since $h_F \leq h_T$ whenever $F \in \mathcal{F}_T$, applying Cauchy–Schwarz inequalities on the sums and recalling the definition (2.7) of $|\cdot|_{1,\partial T}$, we get

$$|\mathcal{E}_h(w; \underline{v}_h)| \leq \left(\sum_{T\in\mathcal{T}_h} h_T \|\nabla(w - \check{w}_T)\|_{\partial T}^2\right)^{\frac{1}{2}} \left(\sum_{T\in\mathcal{T}_h} |\underline{v}_T|_{1,\partial T}^2\right)^{\frac{1}{2}} + \mathrm{s}_h(\underline{I}_h^k w, \underline{I}_h^k w)^{\frac{1}{2}}\, \mathrm{s}_h(\underline{v}_h, \underline{v}_h)^{\frac{1}{2}}.$$

Using the trace approximation properties (1.79) of the elliptic projector with $l = k+1$, $p = 2$, $s = r+2$, and $m = 1$ to estimate $\|\nabla(w - \check{w}_T)\|_{\partial T} = \|\nabla(w - \pi_T^{1,k+1} w)\|_{\partial T}$ and (2.31) to estimate $\mathrm{s}_h(\underline{I}_h^k w, \underline{I}_h^k w) = \sum_{T\in\mathcal{T}_h} \mathrm{s}_T(\underline{I}_T^k w, \underline{I}_T^k w)$, we infer

$$|\mathcal{E}_h(w; \underline{v}_h)| \lesssim h^{r+1} |w|_{H^{r+2}(\mathcal{T}_h)} \left[\left(\sum_{T\in\mathcal{T}_h} |\underline{v}_T|_{1,\partial T}^2\right)^{\frac{1}{2}} + |\underline{v}_h|_{\mathrm{s},h}\right]. \tag{2.47}$$

Recalling the definition (2.39) of a_h, the coercivity property (2.41), and that $\|\underline{v}_h\|_{\mathrm{a},h} = 1$, the second factor in the right-hand side of (2.47) is bounded by a constant independent of h, and (2.42) follows. □

2.2.4 Discrete Problem and Well-Posedness

The HHO scheme for the approximation of problem (2.2) reads: Find $\underline{u}_h \in \underline{U}_{h,0}^k$ such that

$$\mathrm{a}_h(\underline{u}_h, \underline{v}_h) = (f, v_h) \qquad \forall \underline{v}_h \in \underline{U}_{h,0}^k. \tag{2.48}$$

Lemma 2.19 (Well-Posedness of Problem (2.48)**)** *Problem* (2.48) *is well-posed, and we have the following a priori bound for the unique discrete solution* $\underline{u}_h \in \underline{U}_{h,0}^k$:

$$\|\underline{u}_h\|_{\mathrm{a},h} \leq \eta^{\frac{1}{2}} C_{\mathrm{P}} \|f\|, \tag{2.49}$$

where C_{P} *denotes the constant of the discrete Poincaré inequality* (2.37) *and* η *is as in* (2.16).

The proof hinges on the Lax–Milgram Lemma [226], which we recall hereafter.

Lemma 2.20 (Lax–Milgram) *Let* U *be a real Hilbert space, let* $\mathrm{a} : \mathrm{U} \times \mathrm{U} \to \mathbb{R}$ *denote a bounded bilinear form, and let* $\mathrm{f} \in \mathrm{U}^\star$, *with* $\mathrm{U}^\star$ *denoting the dual space of* U. *Further assume that the bilinear form* a *is* U-*coercive, i.e., there exists a real* *number $C > 0$ such that, for all $v \in \mathtt{U}$,*

$$C\|v\|_{\mathtt{U}}^2 \le \mathtt{a}(v, v),$$

where $\|{\cdot}\|_{\mathtt{U}}$ denotes the norm induced by the inner product in $\mathtt{U}$. Then, the problem: Find $u \in \mathtt{U}$ such that

$$\mathtt{a}(u, v) = \langle \mathtt{f}, v\rangle_{\mathtt{U}^\star,\mathtt{U}} \qquad \forall v \in \mathtt{U},$$

is well-posed, i.e., it admits a unique solution for which the following a priori bound holds:

$$\|u\|_{\mathtt{U}} \le C^{-1}\|\mathtt{f}\|_{\mathtt{U}^\star}.$$

Proof (Lemma 2.19) We check the assumptions of the Lax–Milgram Lemma with $\mathtt{U} = \underline{U}_{h,0}^k$, $\mathtt{a} = \mathrm{a}_h$, and $\langle \mathtt{f}, \underline{v}_h\rangle_{\mathtt{U}^\star,\mathtt{U}} = (f, v_h)$. Clearly, $\underline{U}_{h,0}^k$ equipped with the inner product norm $\|{\cdot}\|_{\mathrm{a},h}$ is a Hilbert space. The bilinear form a_h is also clearly $\mathtt{U}$-coercive with coercivity constant equal to 1. To conclude the proof, it suffices to observe that, owing to the discrete Poincaré inequality (2.37) and to the global norm equivalence (2.41), it holds that

$$|(f, v_h)| \le \|f\|\ \|v_h\| \le C_{\mathrm{P}}\|f\|\ \|\underline{v}_h\|_{1,h} \le \eta^{\frac{1}{2}} C_{\mathrm{P}}\|f\|\ \|\underline{v}_h\|_{\mathrm{a},h},$$

which implies, in particular, that the dual norm of the linear form $\mathtt{f} : \underline{v}_h \mapsto (f, v_h)$ is bounded above by $\eta^{\frac{1}{2}} C_{\mathrm{P}}\|f\|$. □

2.2.5 Flux Formulation

In this section we reformulate the HHO scheme (2.48) in terms of numerical fluxes, and show that the latter satisfy local balances and are continuous across interfaces. These features are relevant from both the engineering and mathematical points of view, and can be exploited, for example, to design schemes for coupled systems with advection terms [14], or to derive a posteriori error estimators by equilibration techniques (see, e.g, [184]).

We start by showing that local balances with continuous fluxes hold inside each element for the continuous solution. We next identify conditions under which similar relations hold for an abstract HHO scheme. Finally, we show that the HHO scheme (2.48) for the Poisson problem meets these conditions.

2.2.5.1 Local Balances and Continuity of the Flux for the Continuous Problem

Let $u \in H_0^1(\Omega)$ solve (2.2) and further assume, for the sake of simplicity, that $u \in H^2(\mathcal{T}_h)$. Let a mesh element $T \in \mathcal{T}_h$ be fixed. Using the fact that Eq. (2.1a) holds almost everywhere in Ω, multiplying it by a function $v_T \in \mathbb{P}^k(T)$ and using the assumed regularity to integrate by parts, we infer the following local balance:

$$(\nabla u, \nabla v_T)_T - \sum_{F\in\mathcal{F}_T} (\nabla u{\cdot}\boldsymbol{n}_{TF}, v_T)_F = (f, v_T)_T. \tag{2.50a}$$

The first term in the left-hand side of this relation accounts for the redistribution inside the element T, the second for the exchanges through its boundary ∂T, while the term in the right-hand side represents the generation (or depletion) through the volumetric source term. Taking $v_T \equiv 1$, we have the classical local balance

$$-\sum_{F\in\mathcal{F}_T} \int_F \nabla u{\cdot}\boldsymbol{n}_{TF} = \int_T f,$$

which is an underlying principle of Finite Volume Methods [171]. Crucially, since $\nabla u \in \boldsymbol{H}(\mathrm{div};\Omega) \cap H^1(\mathcal{T}_h)^d$, by virtue of Lemma 1.17 the normal traces of the flux are continuous, that is, for all $F \in \mathcal{F}_h^{\mathrm{i}}$ such that $F \subset \partial T_1 \cap \partial T_2$ with distinct mesh elements $T_1, T_2 \in \mathcal{T}_h$, it holds

$$(\nabla u)_{|T_1}{\cdot}\boldsymbol{n}_{T_1F} + (\nabla u)_{|T_2}{\cdot}\boldsymbol{n}_{T_2F} = 0. \tag{2.50b}$$

This relation shows that the flux exiting T_1 through F enters T_2 and vice-versa. Notice that, in the spirit of Remark 1.18, the relations (2.50) can be formulated with weaker regularity on the exact solution, but we do not further develop this point here as it is not relevant to our discussion.

2.2.5.2 Flux Formulation for an Abstract HHO Scheme

The following lemma identifies conditions under which an abstract HHO scheme admits a flux formulation which mimics the relations (2.50). It will serve as a starting point to derive the flux formulation corresponding to the scheme (2.48) for the Poisson problem.

Lemma 2.21 (Flux Formulation for an Abstract HHO Scheme) *Let* $\mathrm{a}_h : \underline{U}_{h,0}^k \times \underline{U}_{h,0}^k \to \mathbb{R}$ *denote a function such that, for all* $\underline{u}_h, \underline{v}_h \in \underline{U}_{h,0}^k$,

$$\mathrm{a}_h(\underline{u}_h, \underline{v}_h) = \sum_{T\in\mathcal{T}_h} \left(\mathrm{a}_{\mathrm{V},T}(\underline{u}_T, v_T) - \sum_{F\in\mathcal{F}_T} (\Phi_{TF}(\underline{u}_T), v_F - v_T)_F \right), \tag{2.51}$$

where

(i) *For all* $T \in \mathcal{T}_h$, *the* volumetric contribution $\mathrm{a}_{\mathrm{V},T} : \underline{U}_T^k \times \mathbb{P}^k(T) \to \mathbb{R}$ *is linear in its second argument;*
(ii) *For all* $T \in \mathcal{T}_h$ *and all* $F \in \mathcal{F}_T$, $\Phi_{TF} : \underline{U}_T^k \to \mathbb{P}^k(F)$ *represents the* numerical normal trace of the flux.

Let $f \in L^2(\Omega)$. *Then,* $\underline{u}_h \in \underline{U}_{h,0}^k$ *is such that*

$$\mathrm{a}_h(\underline{u}_h, \underline{v}_h) = (f, v_h) \qquad \forall \underline{v}_h \in \underline{U}_{h,0}^k \tag{2.52}$$

if and only if the following properties hold:

(i) Local balance. *For all* $T \in \mathcal{T}_h$ *and all* $v_T \in \mathbb{P}^k(T)$, *it holds*

$$\mathrm{a}_{\mathrm{V},T}(\underline{u}_T, v_T) + \sum_{F \in \mathcal{F}_T} (\Phi_{TF}(\underline{u}_T), v_T)_F = (f, v_T). \tag{2.53a}$$

(ii) Continuity of the numerical normal traces of the flux. *For any interface* $F \in \mathcal{F}_h^{\mathrm{i}}$ *such that* $F \subset \partial T_1 \cap \partial T_2$ *with distinct mesh elements* $T_1, T_2 \in \mathcal{T}_h$, *it holds*

$$\Phi_{T_1 F}(\underline{u}_{T_1}) + \Phi_{T_2 F}(\underline{u}_{T_2}) = 0. \tag{2.53b}$$

Problem (2.53) *is called the* flux formulation *of problem* (2.52).

Remark 2.22 (Forcing Term) Crucial to obtain the flux continuity relation (2.53b) is the fact that face-based unknowns do not appear in the discretisation of the right-hand side.

Remark 2.23 (Balance of Fluxes) If we moreover assume that $\mathrm{a}_{\mathrm{V},T}(\underline{w}_T, 1) = 0$ for all $T \in \mathcal{T}_h$ and all $\underline{w}_T \in \underline{U}_T^k$, then making $v_T = 1$ in (2.53a) shows that the following low-order balance of numerical normal traces of the fluxes, essential component of Finite Volume Methods, holds:

$$\sum_{F \in \mathcal{F}_T} \int_F \Phi_{TF}(\underline{u}_T) = \int_T f \qquad \forall T \in \mathcal{T}_h.$$

Proof (Lemma 2.21) Since both contributions in the right-hand side of (2.51) are linear in their second arguments, so is the case for a_h. Hence, it is sufficient to test (2.52) for $\underline{v}_h$ in a basis of $\underline{U}_{h,0}^k$. Such a basis can be obtained by selecting, for each $T \in \mathcal{T}_h$, vectors such that v_T spans $\mathbb{P}^k(T)$ whilst $v_{T'} = 0$ for $T' \in \mathcal{T}_h \backslash \{T\}$ and $v_F = 0$ for all $F \in \mathcal{F}_h$, and then, for each interface $F \in \mathcal{F}_h^{\mathrm{i}}$, vectors such that v_F spans $\mathbb{P}^k(F)$ whilst $v_{F'} = 0$ for all $F' \in \mathcal{F}_h \backslash \{F\}$ and $v_T = 0$ for all $T \in \mathcal{T}_h$. The first type of basis function simplifies (2.52) into (2.53a). Using the second type of basis function in (2.52) gives (2.53b) since both $\Phi_{T_1 F}(\underline{u}_{T_1})$ and $\Phi_{T_2 F}(\underline{u}_{T_2})$ belong to $\mathbb{P}^k(F)$. □

2.2.5.3 Flux Formulation for the HHO Approximation of the Poisson Problem

Lemma 2.21 indicates that, to recast the HHO scheme (2.48) for the Poisson problem in flux formulation, it suffices to show that the bilinear form a_h defined by (2.39) admits the reformulation (2.51). Plugging the definition (2.15) of a_T into (2.39) and using, for all $T \in \mathcal{T}_h$, the property (2.12) of $\mathrm{p}_T^{k+1}\underline{v}_T$ with $w = \mathrm{p}_T^{k+1}\underline{u}_T$, we can write

$$\begin{aligned}&\mathrm{a}_h(\underline{u}_h, \underline{v}_h)\\&= \sum_{T\in\mathcal{T}_h}\left((\nabla\mathrm{p}_T^{k+1}\underline{u}_T, \nabla v_T)_T + \sum_{F\in\mathcal{F}_T}(\nabla\mathrm{p}_T^{k+1}\underline{u}_T\cdot\boldsymbol{n}_{TF}, v_F - v_T)_F + \mathrm{s}_T(\underline{u}_T, \underline{v}_T)\right).\end{aligned} \tag{2.54}$$

Clearly, the first term inside the summation over $T \in \mathcal{T}_h$ can be incorporated into $\mathrm{a}_{\mathrm{V},T}$ in (2.51), while the second reveals that the consistent contribution to $\Phi_{TF}(\underline{u}_T)$ is $-\nabla\mathrm{p}_T^{k+1}\underline{u}_T\cdot\boldsymbol{n}_{TF}$.

We next prove that the stabilisation term can also be incorporated into $\Phi_{TF}(\underline{u}_T)$ in two steps: first, we show that the stabilisation can be interpreted as acting on boundary differences; second, based on this reformulation, we define the boundary residual operator which constitutes the contribution of the stabilisation bilinear form to $\Phi_{TF}(\underline{u}_T)$. Let a mesh element $T \in \mathcal{T}_h$ be fixed. We define the space

$$\underline{D}_{\partial T}^k := \left\{\underline{\alpha}_{\partial T} = (\alpha_{TF})_{F\in\mathcal{F}_T} \ : \ \alpha_{TF} \in \mathbb{P}^k(F) \quad \forall F \in \mathcal{F}_T\right\} \tag{2.55}$$

and the boundary difference operator $\underline{\Delta}_{\partial T}^k : \underline{U}_T^k \to \underline{D}_{\partial T}^k$ such that, for all $\underline{v}_T \in \underline{U}_T^k$,

$$\underline{\Delta}_{\partial T}^k\underline{v}_T := (v_F - v_T)_{F\in\mathcal{F}_T}. \tag{2.56}$$

A useful remark is that, for all $\underline{v}_T = (v_T, (v_F)_{F\in\mathcal{F}_T}) \in \underline{U}_T^k$, it holds

$$\underline{v}_T - \underline{I}_T^k v_T = (v_T - \pi_T^{0,k} v_T, (v_F - \pi_F^{0,k} v_T)_{F\in\mathcal{F}_T}) = (0, \underline{\Delta}_{\partial T}^k\underline{v}_T), \tag{2.57}$$

where the conclusion follows using the polynomial invariance property (1.56) of $\pi_T^{0,k}$ and $\pi_F^{0,k}$ to infer, for all $T \in \mathcal{T}_h$, $\pi_T^{0,k} v_T = v_T$ (since $v_T \in \mathbb{P}^k(T)$) and, for all $F \in \mathcal{F}_T$, $\pi_F^{0,k} v_T = (v_T)_{|F}$ (since $(v_T)_{|F} \in \mathbb{P}^k(F)$).

Proposition 2.24 (Reformulation of the Stabilisation Bilinear Form) *Let $T \in \mathcal{T}_h$, and let $\mathrm{s}_T : \underline{U}_T^k \times \underline{U}_T^k \to \mathbb{R}$ be a symmetric bilinear form that satisfies (S3) in Assumption 2.4. Then, it holds, for all $\underline{u}_T, \underline{v}_T \in \underline{U}_T^k$, that*

$$\mathrm{s}_T(\underline{u}_T, \underline{v}_T) = \mathrm{s}_T\big(\underline{u}_T, (0, \underline{\Delta}_{\partial T}^k\underline{v}_T)\big) = \mathrm{s}_T\big((0, \underline{\Delta}_{\partial T}^k\underline{u}_T), (0, \underline{\Delta}_{\partial T}^k\underline{v}_T)\big). \tag{2.58}$$

Proof Let $\underline{v}_T \in \underline{U}_T^k$. The polynomial consistency (2.21) of the difference operators applied to $w = v_T \in \mathbb{P}^k(T)$, together with (2.57), shows that $\delta_T^k \underline{v}_T = \delta_T^k(\underline{v}_T - \underline{I}_T^k v_T) = \delta_T^k(0, \underline{\Delta}_{\partial T}^k \underline{v}_T)$ and, for all $F \in \mathcal{F}_T$, $\delta_{TF}^k \underline{v}_T = \delta_{TF}^k(\underline{v}_T - \underline{I}_T^k v_T) = \delta_{TF}^k(0, \underline{\Delta}_{\partial T}^k \underline{v}_T)$. Since s_T only depends on its arguments through the difference operators (see Lemma 2.11), these two relations establish (2.58). □

Let now the boundary residual operator $\underline{R}_{\partial T}^k : \underline{U}_T^k \to \underline{D}_{\partial T}^k$ be such that, for all $\underline{v}_T \in \underline{U}_T^k$, the vector of polynomials

$$\underline{R}_{\partial T}^k \underline{v}_T := (R_{TF}^k \underline{v}_T)_{F \in \mathcal{F}_T}$$

satisfies, for all $\underline{\alpha}_{\partial T} = (\alpha_{TF})_{F \in \mathcal{F}_T} \in \underline{D}_{\partial T}^k$,

$$-\sum_{F \in \mathcal{F}_T} (R_{TF}^k \underline{v}_T, \alpha_{TF})_F = s_T((0, \underline{\Delta}_{\partial T}^k \underline{v}_T), (0, \underline{\alpha}_{\partial T})). \tag{2.59}$$

By the Riesz representation theorem in $\underline{D}_{\partial T}^k$ endowed with the $L^2(\partial T)$-inner product, problem (2.59) is well-posed; computing $R_{TF}^k \underline{v}_T$ only requires to invert the boundary mass matrix, which has a block-diagonal structure with each block corresponding to a face in $\mathcal{F}_T$.

Lemma 2.25 (Flux Formulation) *Let $\mathcal{M}_h$ denote a polytopal mesh in the sense of Definition 1.4. Let $\underline{u}_h \in \underline{U}_{h,0}^k$ and, for all $T \in \mathcal{T}_h$, let s_T satisfy Assumption 2.4. Define, for all $F \in \mathcal{F}_T$, the numerical normal trace of the flux*

$$\Phi_{TF}(\underline{u}_T) := -\nabla p_T^{k+1} \underline{u}_T \cdot \boldsymbol{n}_{TF} + R_{TF}^k \underline{u}_T$$

with R_{TF}^k given by (2.59).

Then, $\underline{u}_h$ is the unique solution of problem (2.48) if and only if the following two properties hold:

(i) Local balance. *For all $T \in \mathcal{T}_h$ and all $v_T \in \mathbb{P}^k(T)$, it holds*

$$(\nabla p_T^{k+1} \underline{u}_T, \nabla v_T)_T + \sum_{F \in \mathcal{F}_T} (\Phi_{TF}(\underline{u}_T), v_T)_F = (f, v_T)_T. \tag{2.60a}$$

(ii) Continuity of the numerical normal traces of the flux. *For any interface $F \in \mathcal{F}_h^{\rm i}$ such that $F \subset \partial T_1 \cap \partial T_2$ with distinct mesh elements $T_1, T_2 \in \mathcal{T}_h$, it holds*

$$\Phi_{T_1 F}(\underline{u}_{T_1}) + \Phi_{T_2 F}(\underline{u}_{T_2}) = 0. \tag{2.60b}$$

Proof Let $\underline{v}_h \in \underline{U}_{h,0}^k$. Using the reformulation (2.58) of s_T together with the definition (2.59) of $\underline{R}_{\partial T}^k$, we can write

$$\mathrm{s}_T(\underline{u}_T, \underline{v}_T) = -\sum_{F \in \mathcal{F}_T} (R_{TF}^k \underline{u}_T, v_F - v_T)_F \qquad \forall T \in \mathcal{T}_h. \tag{2.61}$$

Recalling (2.54), we infer that the bilinear form a_h defined by (2.39) admits the reformulation (2.51) with, for all $T \in \mathcal{T}_h$, $\mathrm{a}_{\nabla,T}(\underline{u}_T, v_T) = (\nabla \mathrm{p}_T^{k+1} \underline{u}_T, \nabla v_T)_T$ for all $(\underline{u}_T, v_T) \in \underline{U}_T^k \times \mathbb{P}^k(T)$ and, for all $\underline{u}_T \in \underline{U}_T^k$ and all $F \in \mathcal{F}_T$, $\Phi_{TF}(\underline{u}_T) = -\nabla \mathrm{p}_T^{k+1} \underline{u}_T \cdot \boldsymbol{n}_{TF} + R_{TF}^k \underline{u}_T$. The conclusion is an immediate consequence of Lemma 2.21. □

Remark 2.26 (Interpretation of the Discrete Problem) Lemma 2.25 provides further insight into the structure of the discrete problem (2.48), which consists of the local balances (2.60a) (corresponding to the algebraic subproblem (B.13a)) and the global transmission condition (2.60b) enforcing the continuity of numerical fluxes (corresponding to the algebraic subproblem (B.13b)).

2.3 Error Analysis

Having proved that the discrete problem (2.48) is well-posed, it remains to determine the convergence of the discrete solution towards the exact solution, which is precisely the goal of this section. The main results here are established using the generic analysis framework presented in Sect. A.1 of Appendix A, and the reader should familiarise themselves with this framework before continuing further.

2.3.1 Energy Error Estimate

We start by deriving a convergence result in a discrete energy norm, using the interpolate of the solution to the continuous problem.

Theorem 2.27 (Discrete Energy Error Estimate) *Let $(\mathcal{M}_h)_{h \in \mathcal{H}}$ denote a regular mesh sequence in the sense of Definition 1.9. Let a polynomial degree $k \geq 0$ be fixed. Let $u \in H_0^1(\Omega)$ denote the unique solution to (2.2), for which we assume the additional regularity $u \in H^{r+2}(\mathcal{T}_h)$ for some $r \in \{0, \ldots, k\}$. For all $h \in \mathcal{H}$, let $\underline{u}_h \in \underline{U}_{h,0}^k$ denote the unique solution to (2.48) with stabilisation bilinear forms s_T, $T \in \mathcal{T}_h$, in (2.15) satisfying Assumption 2.4.*

(continued)

Theorem 2.27 (continued)
Then,

$$\|\underline{u}_h - \underline{I}_h^k u\|_{a,h} \lesssim h^{r+1}|u|_{H^{r+2}(\mathcal{T}_h)}, \tag{2.62}$$

where $\|\cdot\|_{a,h}$ *is defined in* (2.41) *and the hidden constant is independent of* h *and* u.

Proof We invoke the Third Strang Lemma A.7 with $\mathrm{U} = H_0^1(\Omega)$, $\mathrm{a}(u, v) = (\nabla u, \nabla v)$, $\mathrm{l}(v) = (f, v)$, $\mathrm{U}_h = \underline{U}_{h,0}^k$ endowed with the norm $\|\cdot\|_{a,h}$, $\mathrm{a}_h = \mathrm{a}_h$, $\mathrm{l}_h(\underline{v}_h) = (f, v_h)$, and $\mathrm{I}_h u = \underline{I}_h^k u$. We notice that a_h is obviously coercive for $\|\cdot\|_{a,h}$ with constant 1 and, since $-\Delta u = f$, the consistency error (A.5) is exactly (2.43) with $w = u$. Hence, (2.62) follows plugging (2.42) into (A.6). □

From this convergence result in a discrete norm, we now deduce an estimate for the error measured as the difference between the exact solution and the global reconstruction obtained from the discrete solution through the operator $\mathrm{p}_h^{k+1} : \underline{U}_h^k \to \mathbb{P}^{k+1}(\mathcal{T}_h)$ such that, for all $\underline{v}_h \in \underline{U}_h^k$,

$$(\mathrm{p}_h^{k+1}\underline{v}_h)_{|T} := \mathrm{p}_T^{k+1}\underline{v}_T \qquad \forall T \in \mathcal{T}_h. \tag{2.63}$$

Theorem 2.28 (Energy Error Estimate for the Reconstructed Approximate Solution) *Under the assumptions and notations of Theorem 2.27, it holds that*

$$\|\nabla_h(\mathrm{p}_h^{k+1}\underline{u}_h - u)\| + |\underline{u}_h|_{s,h} \lesssim h^{r+1}|u|_{H^{r+2}(\mathcal{T}_h)}, \tag{2.64}$$

where the hidden constant is independent of h *and* u, *and the* $|\cdot|_{s,h}$ *seminorm is defined by* (2.40).

Proof For the sake of brevity, let $\underline{\hat{u}}_h := \underline{I}_h^k u$. Inserting $\pm\nabla_h \mathrm{p}_h^{k+1}\underline{\hat{u}}_h$ into the first term, $\pm\underline{\hat{u}}_h$ into the second, and using the triangle inequality, it is readily inferred that

$$\begin{aligned}
&\|\nabla_h(\mathrm{p}_h^{k+1}\underline{u}_h - u)\| + |\underline{u}_h|_{s,h} \\
&\le \underbrace{\|\nabla_h \mathrm{p}_h^{k+1}(\underline{u}_h - \underline{\hat{u}}_h)\| + |\underline{u}_h - \underline{\hat{u}}_h|_{s,h}}_{\mathfrak{T}_1} + \underbrace{\|\nabla_h(\mathrm{p}_h^{k+1}\underline{\hat{u}}_h - u)\| + |\underline{\hat{u}}_h|_{s,h}}_{\mathfrak{T}_2}.
\end{aligned} \tag{2.65}$$

By definition (2.15) of a_T, (2.39) of a_h, and (2.41) of $\|\cdot\|_{\mathrm{a},h}$, we deduce from (2.62) that

$$\mathfrak{T}_1 \lesssim \|\underline{u}_h - \hat{\underline{u}}_h\|_{\mathrm{a},h} \lesssim h^{r+1}|u|_{H^{r+2}(\mathcal{T}_h)}.$$

Clearly, $(\mathrm{p}_h^{k+1}\hat{\underline{u}}_h)_{|T} = \pi_T^{1,k+1}u$ for all $T \in \mathcal{T}_h$ by virtue of (2.14). Then, the approximation properties (1.78) of the elliptic projector with $l = k+1$, $p = 2$, $s = r+2$, and $m = 1$ together with the consistency (2.31) of s_T yield

$$\mathfrak{T}_2 \lesssim h^{r+1}|u|_{H^{r+2}(\mathcal{T}_h)}.$$

Plugging the above bounds into (2.65) concludes the proof of (2.64). □

Remark 2.29 (Estimates in Fractional Sobolev Spaces) Remark 1.49 and the proofs above easily show that (2.62) and (2.64) also hold for fractional $r \in [0, k]$.

Remark 2.30 (p- and hp-Error Analysis) In this manuscript, we do not address the p- or hp-versions of the HHO method, where convergence is attained by increasing the polynomial degree rather than reducing the meshsize (p-version) or by combining these two strategies (hp-version). The key points are, in this case: (1) an accurate tracking of the dependence on the polynomial degree of the constants appearing in discrete inverse and trace inequalities and (2) hp-approximation results for local polynomial spaces. These issues are treated in [91, Section 4] based on classical results for simplicial meshes. Similar results, but with a proof based on a direct extension of the classical hp-approximation results of [29] to regular mesh sequences in arbitrary space dimension, can be found in [10, Lemma 2.3]; see also [51] for the two-dimensional case. On this subject, the interested reader can also consult the recent monograph [94], which focuses on hp-Discontinuous Galerkin methods on meshes with a (possibly) unbounded number of faces.

The hp-analysis for HHO methods applied to a pure diffusion problem can be found in [10], where the option of letting the polynomial degree vary locally is also contemplated. Specialised to the present setting, Theorem 3.3 therein asserts that, assuming the regularity $u \in H^{k+2}(\mathcal{T}_h)$ for the solution to (2.2) and denoting by $\underline{u}_h \in \underline{U}_{h,0}^k$ the solution to the HHO scheme (2.48), it holds

$$\|\underline{u}_h - \underline{I}_h^k u\|_{\mathrm{a},h} \lesssim \frac{h^{k+1}}{(k+1)^k}|u|_{H^{k+2}(\mathcal{T}_h)}$$

with hidden constant independent of h, k, and u.

2.3.2 Convergence of the Jumps

Functions in $H^1(\mathcal{T}_h)$ are in $H^1_0(\Omega)$ if their jumps vanish a.e. on the interfaces and if their trace is zero a.e. on $\partial\Omega$. Thus, a measure of the nonconformity "up to degree k" in $H^1_0(\Omega)$ is provided by the jump seminorm $|\cdot|_{\mathrm{J},h}$ such that, for all $v \in H^1(\mathcal{T}_h)$,

$$|v|_{\mathrm{J},h}^2 := \sum_{F\in\mathcal{F}_h} h_F^{-1}\|\pi_F^{0,k}[v]_F\|_F^2 \tag{2.66}$$

with jump operator $[\cdot]_F$ defined by (1.22) if $F \in \mathcal{F}_h^{\mathrm{i}}$, and by $[v]_F := v_F$ if $F \in \mathcal{F}_h^{\mathrm{b}}$. A natural question is whether the jump seminorm of $\mathrm{p}_h^{k+1}\underline{u}_h$ converges to zero. The answer is provided by the following lemma.

Lemma 2.31 (Convergence of the Jumps) *Under the assumptions and notations of Theorem 2.27, and further supposing, for the sake of simplicity, that, for all $T \in \mathcal{T}_h$, the local stabilisation bilinear form s_T is given by (2.23), it holds, with hidden constant independent of h and u, that*

$$|\mathrm{p}_h^{k+1}\underline{u}_h|_{\mathrm{J},h} \lesssim h^{r+1}|u|_{H^{r+2}(\mathcal{T}_h)}. \tag{2.67}$$

Proof For $F \in \mathcal{F}_h^{\mathrm{i}}$ with bordering elements T_1 and T_2, write

$$[\mathrm{p}_h^{k+1}\underline{u}_h]_F = p_{T_1}^{k+1}\underline{u}_{T_1} - p_{T_2}^{k+1}\underline{u}_{T_2} = (p_{T_1}^{k+1}\underline{u}_{T_1} - u_F) + (u_F - p_{T_2}^{k+1}\underline{u}_{T_2}).$$

For $F \in \mathcal{F}_h^{\mathrm{b}}$ with bordering element T, write $[\mathrm{p}_h^{k+1}\underline{u}_h]_F = \mathrm{p}_T^{k+1}\underline{u}_T - u_F$, owing to the fact that $u_F = 0$. Using the triangle inequality and gathering the sum by elements as per (1.25), it is then inferred that

$$\begin{aligned}
\sum_{F\in\mathcal{F}_h} h_F^{-1}\|\pi_F^{0,k}[\mathrm{p}_h^{k+1}\underline{u}_h]_F\|_F^2 &\le 2\sum_{F\in\mathcal{F}_h}\sum_{T\in\mathcal{T}_F} h_F^{-1}\|\pi_F^{0,k}(\mathrm{p}_T^{k+1}\underline{u}_T - u_F)\|_F^2\\
&\le 2\sum_{T\in\mathcal{T}_h}\sum_{F\in\mathcal{F}_T} h_F^{-1}\|\pi_F^{0,k}(\mathrm{p}_T^{k+1}\underline{u}_T - u_F)\|_F^2\\
&= 2\sum_{T\in\mathcal{T}_h}\sum_{F\in\mathcal{F}_T} h_F^{-1}\|\delta_{TF}^k\underline{u}_T\|_F^2\\
&\le 2|\underline{u}_h|_{\mathrm{s},h}^2.
\end{aligned}$$

Using (2.64) to bound the right-hand side yields (2.67). □

2.3.3 L^2-Error Estimate

We next study the convergence of the error in the L^2-norm. Optimal error estimates in this context require further regularity for the continuous problem. More precisely, we assume that, for all $g \in L^2(\Omega)$, the unique solution of the dual problem: Find $z_g \in H_0^1(\Omega)$ such that

$$a(v, z_g) = (g, v) \qquad \forall v \in H_0^1(\Omega) \tag{2.68}$$

satisfies the a priori estimate

$$\|z_g\|_{H^2(\Omega)} \leq C\|g\|, \tag{2.69}$$

with real number C depending only on Ω. This property, called *elliptic regularity*, holds when the domain Ω is convex; see, e.g., [205]. Notice that, for the Poisson equation, the dual problem coincides with the primal problem (2.2) owing to the symmetry of a.

Theorem 2.32 (L^2-Error Estimate) *Let $(\mathcal{M}_h)_{h\in\mathcal{H}}$ denote a regular mesh sequence in the sense of Definition 1.9. Let a polynomial degree $k \geq 0$ be fixed. Let $u \in H_0^1(\Omega)$ denote the unique solution of (2.2), for which we assume the additional regularity $u \in H^{r+2}(\mathcal{T}_h)$ for some $r \in \{0, \ldots, k\}$. For all $h \in \mathcal{H}$, let $\underline{u}_h \in \underline{U}_{h,0}^k$ denote the unique solution to (2.48) with stabilisation bilinear forms s_T, $T \in \mathcal{T}_h$, in (2.15) satisfying Assumption 2.4. Further assuming elliptic regularity and that $f \in H^1(\mathcal{T}_h)$ if $k = 0$, it holds*

$$\|\mathrm{p}_h^{k+1}\underline{u}_h - u\| \lesssim \begin{cases} h^2\|f\|_{H^1(\mathcal{T}_h)} & \text{if } k = 0, \\ h^{r+2}|u|_{H^{r+2}(\mathcal{T}_h)} & \text{if } k \geq 1, \end{cases} \tag{2.70}$$

with hidden constant independent of both h and u.

The proof of Theorem 2.32 hinges on the following lemma, which shows that the element-based unknowns of the HHO solution are very close to the L^2-orthogonal projection of u on the broken polynomial space $\mathbb{P}^k(\mathcal{T}_h)$. Since this corresponds, when $r = k$, to an error estimate of higher degree than the approximation properties of the discrete space, we speak of *superconvergence*.

Lemma 2.33 (Superconvergence of Element Unknowns) *Under the assumptions and notations of Theorem 2.32, it holds that*

$$\|u_h - \pi_h^{0,k}u\| \lesssim \begin{cases} h^2\|f\|_{H^1(\mathcal{T}_h)} & \text{if } k = 0, \\ h^{r+2}|u|_{H^{r+2}(\mathcal{T}_h)} & \text{if } k \geq 1, \end{cases} \tag{2.71}$$

where the hidden constant is independent of both h and u, and the global L^2-orthogonal projection $\pi_h^{0,k}u$ is defined according to (1.59), i.e., $(\pi_h^{0,k}u)_{|T} = \pi_T^{0,k}u_{|T}$ for all $T \in \mathcal{T}_h$.

Proof (Theorem 2.32) Let, for the sake of brevity, $\underline{\hat{u}}_h := \underline{I}_h^k u$ and $\check{u}_h := \mathrm{p}_h^{k+1}\underline{\hat{u}}_h$ so that, by the commutation property (2.14), $(\check{u}_h)_{|T} = \pi_T^{1,k+1}u$ for all $T \in \mathcal{T}_h$. Inserting $0 = \mathrm{p}_h^{k+1}\underline{\hat{u}}_h - \check{u}_h$ inside the norm and using the triangle inequality, we have that

$$\|\mathrm{p}_h^{k+1}\underline{u}_h - u\| \le \|u - \check{u}_h\| + \|\mathrm{p}_h^{k+1}(\underline{\hat{u}}_h - \underline{u}_h)\| =: \mathfrak{T}_1 + \mathfrak{T}_2. \tag{2.72}$$

Using inside each element $T \in \mathcal{T}_h$ the approximation properties (1.78) of the elliptic projector with $l = k+1$, $p = 2$, $s = r+2$, and $m = 0$ readily gives for the first term:

$$\mathfrak{T}_1 \lesssim h^{r+2}|u|_{H^{r+2}(\mathcal{T}_h)}. \tag{2.73}$$

For the second term, on the other hand, we observe that

$$\begin{aligned}
\mathfrak{T}_2^2 &= \sum_{T\in\mathcal{T}_h} \|\mathrm{p}_T^{k+1}(\underline{\hat{u}}_T - \underline{u}_T)\|_T^2 \\
&\lesssim \sum_{T\in\mathcal{T}_h} \left(h_T^2\|\nabla\mathrm{p}_T^{k+1}(\underline{\hat{u}}_T - \underline{u}_T)\|_T^2 + \|\pi_T^{0,0}(\hat{u}_T - u_T)\|_T^2\right) \\
&\le h^2\|\underline{\hat{u}}_h - \underline{u}_h\|_{\mathrm{a},h}^2 + \|\hat{u}_h - u_h\|^2,
\end{aligned} \tag{2.74}$$

where the second line follows writing

$$\mathrm{p}_T^{k+1}(\underline{\hat{u}}_T - \underline{u}_T) = \left[\mathrm{p}_T^{k+1}(\underline{\hat{u}}_T - \underline{u}_T) - \pi_T^{0,0}(\hat{u}_T - u_T)\right] + \pi_T^{0,0}(\hat{u}_T - u_T)$$

and using the triangle inequality followed by the local Poincaré–Wirtinger inequality (1.76) on $\mathrm{p}_T^{k+1}(\underline{\hat{u}}_T - \underline{u}_T) - \pi_T^{0,0}(\hat{u}_T - u_T)$, while the conclusion in (2.74) is obtained invoking the definition (2.41) of the $\|{\cdot}\|_{\mathrm{a},h}$-norm together with the L^2-boundedness of $\pi_T^{0,0}$. Using (2.62) and (2.71) to bound respectively the first and second terms in (2.74), and plugging the resulting inequality together with (2.73) into (2.72), the estimate (2.70) follows. □

To complete the proof of the L^2-error estimate, it only remains to prove Lemma 2.33, which we do next.

Proof (Lemma 2.33) The result follows from the Aubin–Nitsche Lemma A.10 in Appendix A, with the same setting as in the proof of Theorem 2.27, that is: $\mathtt{U} = H_0^1(\Omega)$, $\mathtt{a}(u,v) = (\nabla u, \nabla v)$, $\mathtt{l}(v) = (f, v)$, $\mathtt{U}_h = \underline{U}_{h,0}^k$, $\|{\cdot}\|_{\mathtt{U}_h} = \|{\cdot}\|_{\mathrm{a},h}$, $\mathtt{a}_h = \mathrm{a}_h$, $\mathtt{l}_h(\underline{v}_h) = (f, v_h)$ and $\mathtt{I}_h u = \underline{I}_h^k u$. Additionally, we take $\mathtt{L} = L^2(\Omega)$ and $\mathtt{r}_h$:

$\underline{U}_{h,0}^k \to L^2(\Omega)$ defined by $\mathrm{r}_h \underline{v}_h = v_h$. In what follows, the hidden constants in the inequalities $A \lesssim B$ do not depend on h, f, u, or g in the dual problem (2.68).

With this setting, (A.10) is identical to (2.68) and, by choice of r_h, since the bilinear form a is symmetric, the dual consistency error $\mathcal{E}_h^{\mathrm{d}}(z_g;\cdot)$ is equal to the primal consistency error $\mathcal{E}_h(z_g;\cdot)$ defined by (2.43). By definition of r_h and $\underline{I}_h^k$, denoting by $\|\cdot\|_{\mathrm{a},h,\star}$ the dual norm of $\|\cdot\|_{\mathrm{a},h}$, the Aubin–Nitsche Lemma A.10 therefore shows that

$$\begin{aligned}\|u_h - \pi_h^{0,k} u\| \le{}& \underbrace{\|\underline{u}_h - \underline{I}_h^k u\|_{\mathrm{a},h} \sup_{g\in L^2(\Omega),\ \|g\|\le 1} \|\mathcal{E}_h(z_g;\cdot)\|_{\mathrm{a},h,\star}}_{\mathfrak{T}_1} \\ &+ \underbrace{\sup_{g\in L^2(\Omega),\ \|g\|\le 1} |\mathcal{E}_h(u;\underline{I}_h^k z_g)|}_{\mathfrak{T}_2}.\end{aligned} \tag{2.75}$$

(i) *Estimate of* $\mathfrak{T}_1$. Since $z_g \in H_0^1(\Omega)\cap H^2(\Omega)$, using the definition of the dual norm $\|\cdot\|_{\mathrm{a},h,\star}$ followed by the estimates (2.42) with $r = 0$ and (2.69) yields

$$\|\mathcal{E}_h(z_g;\cdot)\|_{\mathrm{a},h,\star} = \sup_{\underline{v}_h\in \underline{U}_{h,0}^k,\ \|\underline{v}_h\|_{\mathrm{a},h}=1} |\mathcal{E}_h(z_g;\underline{v}_h)| \lesssim h|z_g|_{H^2(\Omega)} \lesssim h\|g\|.$$

Combining this result with (2.62), the first term in the right-hand side of (2.75) is estimated as

$$\mathfrak{T}_1 \lesssim h^{r+2}|u|_{H^{r+2}(\mathcal{T}_h)}. \tag{2.76}$$

(ii) *Estimate of* $\mathfrak{T}_2$. To estimate the second term in the right-hand side of (2.75), we treat the cases $k \ge 1$ and $k = 0$ separately.

(ii.A) *The case* $k \ge 1$. Applying (2.47) to $w = u$ and $\underline{v}_h = \underline{I}_h^k z_g$ yields

$$|\mathcal{E}_h(u;\underline{I}_h^k z_g)| \lesssim h^{r+1}|u|_{H^{r+2}(\mathcal{T}_h)}\left[\left(\sum_{T\in\mathcal{T}_h} |\underline{I}_T^k z_g|_{1,\partial T}^2\right)^{\frac{1}{2}} + |\underline{I}_h^k z_g|_{\mathrm{s},h}\right]. \tag{2.77}$$

Using, for all $T \in \mathcal{T}_h$, the consistency property (2.31) of the local stabilisation bilinear form with $r = 0$, we see that

$$|\underline{I}_h^k z_g|_{\mathrm{s},h} \lesssim h|z_g|_{H^2(\Omega)}.$$

On the other hand, recalling the definition (2.7) of $|\cdot|_{1,\partial T}$, we can write for any $T \in \mathcal{T}_h$

$$
\begin{aligned}
|\underline{I}_T^k z_g|_{1,\partial T}^2 &= \sum_{F\in\mathcal{F}_T} h_F^{-1}\|\pi_F^{0,k} z_g - \pi_T^{0,k} z_g\|_F^2 \\
&= \sum_{F\in\mathcal{F}_T} h_F^{-1}\|\pi_F^{0,k}(z_g - \pi_T^{0,k} z_g)\|_F^2 \\
&\le \sum_{F\in\mathcal{F}_T} h_F^{-1}\|z_g - \pi_T^{0,k} z_g\|_F^2 \lesssim h_T^2 |z_g|_{H^2(T)}^2,
\end{aligned}
\tag{2.78}
$$

where we have used the definition of $\underline{I}_T^k z_g$ in the first equality, followed by the linearity and polynomial invariance (1.56) of $\pi_F^{0,k}$ in the second equality, its L^2-boundedness in the third line, and concluded by the trace approximation property (1.75) with $l = k$, $p = 2$, $m = 0$ and $s = 2$ (we have $s \le l + 1$ since, here, $k \ge 1$), along with the uniform equivalence of face and element diameters (1.6). Plugging the above bounds into (2.77) and recalling the elliptic regularity estimate (2.69), we infer that $|\mathcal{E}_h(u; \underline{I}_h^k z_g)| \lesssim h^{r+2}|u|_{H^{r+2}(\mathcal{T}_h)}\|g\|$, hence

$$
\mathfrak{T}_2 \lesssim h^{r+2}|u|_{H^{r+2}(\mathcal{T}_h)}.
$$

Plugging this estimate together with (2.76) into (2.75) concludes the proof of (2.71) in the case $k \ge 1$.

(ii.B) *The case* $k = 0$. Substituting $f = -\Delta u$ in the expression (2.43) of the consistency error $\mathcal{E}_h(u; \underline{I}_h^0 z_g)$, using the definitions (2.39) and (2.15) of the bilinear forms a_h and a_T to expand the quantity $\mathrm{a}_h(\underline{I}_h^k u, \underline{I}_h^k z_g)$, and invoking the property $\mathrm{p}_T^1 \underline{I}_T^0 = \pi_T^{1,1}$ (see (2.14)), we have that

$$
\begin{aligned}
\mathcal{E}_h(u; \underline{I}_h^0 z_g) = &\sum_{T\in\mathcal{T}_h} (f, \pi_T^{0,0} z_g)_T - \sum_{T\in\mathcal{T}_h} (\nabla \pi_T^{1,1} u, \nabla \pi_T^{1,1} z_g)_T \\
&- \mathrm{s}_h(\underline{I}_h^0 u, \underline{I}_h^0 z_g).
\end{aligned}
\tag{2.79}
$$

The orthogonality property of $\pi_T^{0,0}$ and the fact that $(f, z_g) = (\nabla u, \nabla z_g)$ (see (2.2)) justify the following algebra:

$$
\begin{aligned}
\sum_{T\in\mathcal{T}_h} (f, \pi_T^{0,0} z_g)_T &= \sum_{T\in\mathcal{T}_h} (\pi_T^{0,0} f, z_g)_T \\
&= \sum_{T\in\mathcal{T}_h} (\pi_T^{0,0} f - f, z_g)_T + (f, z_g) \\
&= \sum_{T\in\mathcal{T}_h} (\pi_T^{0,0} f - f, z_g - \pi_T^{0,0} z_g)_T + (\nabla u, \nabla z_g).
\end{aligned}
\tag{2.80}
$$

The Cauchy–Schwarz inequality and the approximation property (1.74) applied to f and z_g with $p = 2$, $l = k = 0$, and $s = 1$ yield

$$\begin{aligned}\Big|\sum_{T\in\mathcal{T}_h}(\pi_T^{0,0}f - f, z_g - \pi_T^{0,0}z_g)_T\Big| &\le \sum_{T\in\mathcal{T}_h}\|\pi_T^{0,0}f - f\|_T\|z_g - \pi_T^{0,0}z_g\|_T\\ &\lesssim \sum_{T\in\mathcal{T}_h} h|f|_{H^1(T)}\, h|z_g|_{H^1(T)}\\ &\lesssim h^2|f|_{H^1(\mathcal{T}_h)}\|g\|,\end{aligned}$$

where the conclusion follows by the Cauchy–Schwarz inequality on the sum and the standard stability estimate $\|z_g\|_{H^1(\Omega)} \lesssim \|g\|$. Using the Cauchy–Schwarz inequality on the positive semidefinite form s_h, the consistency estimate (2.31) with $k = r = 0$ gives

$$|\mathrm{s}_h(\underline{I}_h^0 u, \underline{I}_h^0 z_g)| \le |\underline{I}_h^0 u|_{\mathrm{s},h}\, |\underline{I}_h^0 z_g|_{\mathrm{s},h} \lesssim h|u|_{H^2(\Omega)}\, h|z_g|_{H^2(\Omega)}. \tag{2.81}$$

Hence, coming back to (2.79) and invoking the elliptic regularity estimate (2.69) for both z_g and u, we find

$$\begin{aligned}|\mathcal{E}_h(u; \underline{I}_h^0 z_g)| \lesssim{}& \left|\sum_{T\in\mathcal{T}_h}(\nabla u, \nabla z_g)_T - (\nabla\pi_T^{1,1}u, \nabla\pi_T^{1,1}z_g)_T\right|\\ &+ h^2\|f\|_{H^1(\mathcal{T}_h)}\|g\|.\end{aligned}$$

For all $T \in \mathcal{T}_h$, introducing $\pm(\nabla\pi_T^{1,1}u, \nabla z_g)_T$ and using the definition (1.60a) of $\pi_T^{1,1}$ with $(v, w) = (z_g, \pi_T^{1,1}u)$ and $(v, w) = (u, \pi_T^{1,1}z_g)$, we write

$$\begin{aligned}&|(\nabla u, \nabla z_g)_T - (\nabla\pi_T^{1,1}u, \nabla\pi_T^{1,1}z_g)_T|\\ &\quad= |(\nabla(u - \pi_T^{1,1}u), \nabla z_g)_T + (\nabla\pi_T^{1,1}u, \nabla(z_g - \pi_T^{1,1}z_g))_T|\\ &\quad= |(\nabla(u - \pi_T^{1,1}u), \nabla(z_g - \pi_T^{1,1}z_g))_T|\\ &\quad\lesssim h_T|u|_{H^2(T)}\, h_T|z_g|_{H^2(T)},\end{aligned} \tag{2.82}$$

where we have used the Cauchy–Schwarz inequality and the approximation property (1.78) of $\pi_T^{1,1}$ with $l = 1$, $p = 2$, $s = 2$, and $m = 1$ to conclude. Hence, using again the elliptic regularity estimate (2.69) for u and z_g,

$$|\mathcal{E}_h(u; \underline{I}_h^0 z_g)| \lesssim h^2\|f\|_{H^1(\mathcal{T}_h)}\|g\|. \tag{2.83}$$

The proof of (2.71) for $k = 0$ is completed by plugging this estimate and (2.76) into (2.75). □

2.4 Other Boundary Conditions

We hint in this section at the treatment of more general boundary conditions. For the sake of simplicity, we consider mixed boundary conditions under the assumption that they do not degenerate into the pure Neumann case (the adaptation to the pure Neumann case is addressed in detail in Chap. 6 for more general, possibly nonlinear diffusion problems). Let $\Gamma_{\rm D}$ denote a relatively open subset of $\partial\Omega$ with non-zero $(d-1)$-dimensional Hausdorff measure, and set $\Gamma_{\rm N} := \partial\Omega \setminus \Gamma_{\rm D}$. Let $g_{\rm D} := (u_{\rm D})_{|\Gamma_{\rm D}}$ with $u_{\rm D} \in H^1(\Omega)$, $g_{\rm N} \in L^2(\Gamma_{\rm N})$, and consider the problem: Find $u : \Omega \to \mathbb{R}$ such that

$$
\begin{aligned}
-\Delta u &= f && \text{in } \Omega,\\
u &= g_{\rm D} && \text{on } \Gamma_{\rm D},\\
\nabla u{\cdot}\boldsymbol{n}_\Omega &= g_{\rm N} && \text{on } \Gamma_{\rm N},
\end{aligned}
\tag{2.84}
$$

where $\boldsymbol{n}_\Omega$ denotes the outer unit normal to Ω on $\partial\Omega$. Denote by $H^1_{\rm D}(\Omega)$ the space of functions in $H^1(\Omega)$ which vanish (in the sense of traces) on $\Gamma_{\rm D}$. Classically, a weak solution to Problem (2.84) can be obtained as $u = u_0 + u_{\rm D}$ where $u_0 \in H^1_{\rm D}(\Omega)$ is such that

$$
(\nabla u_0, \nabla v) = (f, v) - (\nabla u_{\rm D}, \nabla v) + (g_{\rm N}, v)_{\Gamma_{\rm N}} \qquad \forall v \in H^1_{\rm D}(\Omega). \tag{2.85}
$$

In order to write the HHO discretisation of problem (2.85), we consider a polytopal mesh $\mathcal{M}_h$ in the sense of Definition 1.4 for which we make the following assumption.

Assumption 2.34 (Boundary Datum-Compliant Mesh) *We assume that the interior of every boundary face $F \in \mathcal{F}_h^{\rm b}$ is contained either in $\Gamma_{\rm D}$ (the set of all such F is denoted by $\mathcal{F}_h^{\rm D}$) or in $\Gamma_{\rm N}$ (the set of all such F is denoted by $\mathcal{F}_h^{\rm N}$).*

We next introduce the space

$$
\underline{U}_{h,\rm D}^k := \left\{ \underline{v}_h \in \underline{U}_h^k \;:\; u_F = 0 \quad \forall F \in \mathcal{F}_h^{\rm D} \right\},
$$

and we let $\underline{u}_{h,\mathrm{D}} := ((u_{T,\mathrm{D}})_{T\in\mathcal{T}_h}, (u_{F,\mathrm{D}})_{F\in\mathcal{F}_h}) \in \underline{U}_h^k$ be such that

$$u_{T,\mathrm{D}} = 0 \quad \forall T \in \mathcal{T}_h, \quad u_{F,\mathrm{D}} = \pi_F^{0,k} g_{\mathrm{D}} \quad \forall F \in \mathcal{F}_h^{\mathrm{D}}, \quad u_{F,\mathrm{D}} = 0 \quad \forall F \in \mathcal{F}_h^{\not{\mathrm{D}}}, \tag{2.86}$$

where we have introduced the set of non-Dirichlet faces

$$\mathcal{F}_h^{\not{\mathrm{D}}} := \mathcal{F}_h \setminus \mathcal{F}_h^{\mathrm{D}} = \mathcal{F}_h^{\mathrm{i}} \cup \mathcal{F}_h^{\mathrm{N}}. \tag{2.87}$$

Then, the HHO solution $\underline{u}_h \in \underline{U}_h^k$ is obtained as $\underline{u}_h = \underline{u}_{h,0} + \underline{u}_{h,\mathrm{D}}$ with $\underline{u}_{h,0} \in \underline{U}_{h,\mathrm{D}}^k$ such that

$$\mathrm{a}_h(\underline{u}_{h,0}, \underline{v}_h) = (f, v_h) - \mathrm{a}_h(\underline{u}_{h,\mathrm{D}}, \underline{v}_h) + \sum_{F\in\mathcal{F}_h^{\mathrm{N}}} (g_{\mathrm{N}}, v_F)_F \qquad \forall \underline{v}_h \in \underline{U}_{h,\mathrm{D}}^k. \tag{2.88}$$

2.5 Numerical Examples

We illustrate the numerical performance of the HHO method on a set of model problems.

2.5.1 Two-Dimensional Test Case

The first test case, taken from [158], aims at illustrating the demonstrated orders of convergence in two space dimensions. We solve the Dirichlet problem in the unit square $\Omega = (0, 1)^2$ with

$$u(x_1, x_2) = \sin(\pi x_1)\sin(\pi x_2), \tag{2.89}$$

and corresponding right-hand side $f = 2\pi^2 \sin(\pi x_1)\sin(\pi x_2)$, on families of triangular and polygonal meshes, an instance of each being described in Fig. 1.1a and c, respectively. Figure 2.3 displays convergence results for both mesh families and polynomial degrees up to 4. By virtue of (2.62) and (2.71) (both with $r = k$), we can measure the energy- and L^2-errors through the quantities $\|\underline{I}_h^k u - \underline{u}_h\|_{\mathrm{a},h}$ and $\|\pi_h^{0,k} u - u_h\|$, respectively. In all cases, the numerical results show asymptotic convergence rates that match those predicted by the theory.

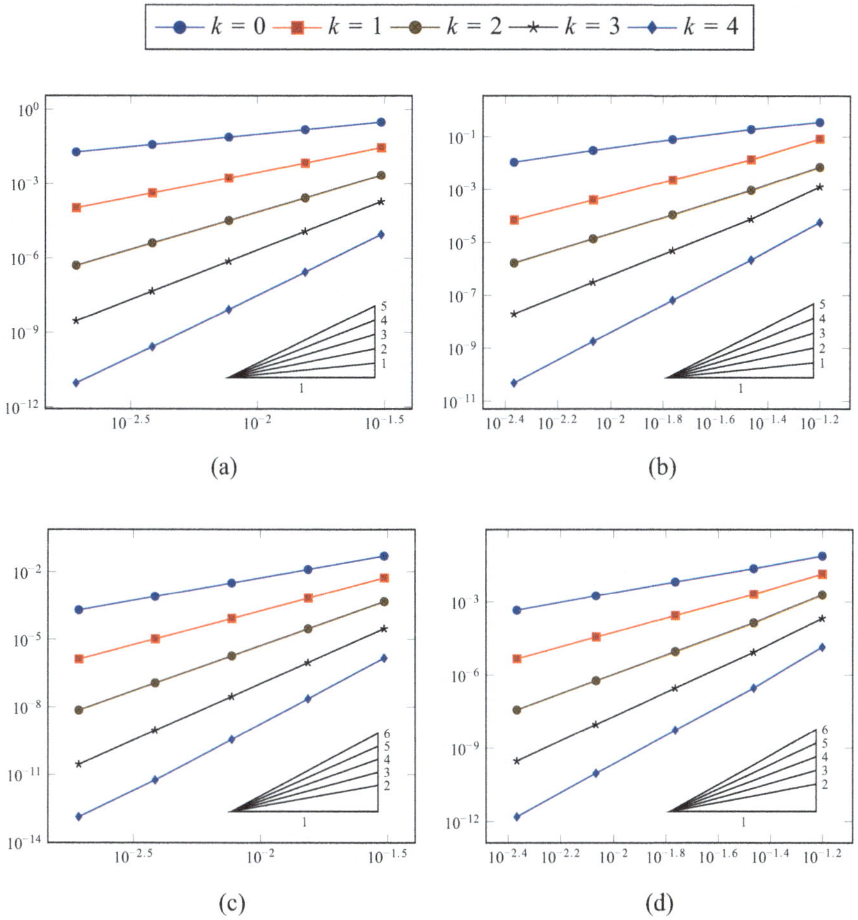

Fig. 2.3 Error vs. h for the test case of Sect. 2.5.1. The reference slopes refer to the expected order of convergence for each polynomial degree $k \in \{0, \ldots, 4\}$. **(a)** $\|\underline{I}_h^k u - \underline{u}_h\|_{a,h}$ vs. h, triangular mesh. **(b)** $\|\underline{I}_h^k u - \underline{u}_h\|_{a,h}$ vs. h, polygonal mesh. **(c)** $\|\pi_h^{0,k} u - u_h\|$ vs. h, triangular mesh. **(d)** $\|\pi_h^{0,k} u - u_h\|$ vs. h, polygonal mesh

2.5.2 Three-Dimensional Test Case

The second test case, taken from [154], demonstrates the orders of convergence in three space dimensions. We solve the Dirichlet problem in the unit cube $\Omega = (0, 1)^3$ with

$$u(x_1, x_2, x_3) = \sin(\pi x_1) \sin(\pi x_2) \sin(\pi x_3),$$

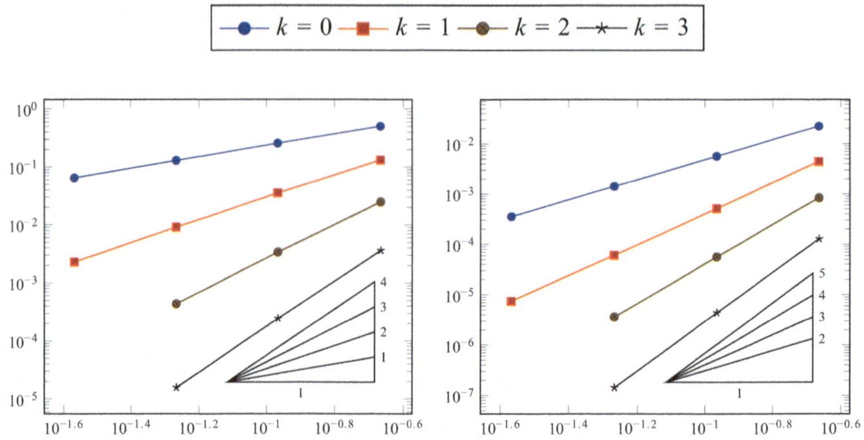

(a) (b)

Fig. 2.4 Error vs. h for the test case of Sect. 2.5.2. The reference slopes refer to the expected order of convergence for each polynomial degree $k \in \{0, \ldots, 3\}$. (**a**) $\|\nabla_h(\mathrm{p}_h^{k+1}\underline{u}_h - u)\|$ vs. h. (**b**) $\|\mathrm{p}_h^{k+1}\underline{u}_h - u\|$ vs. h

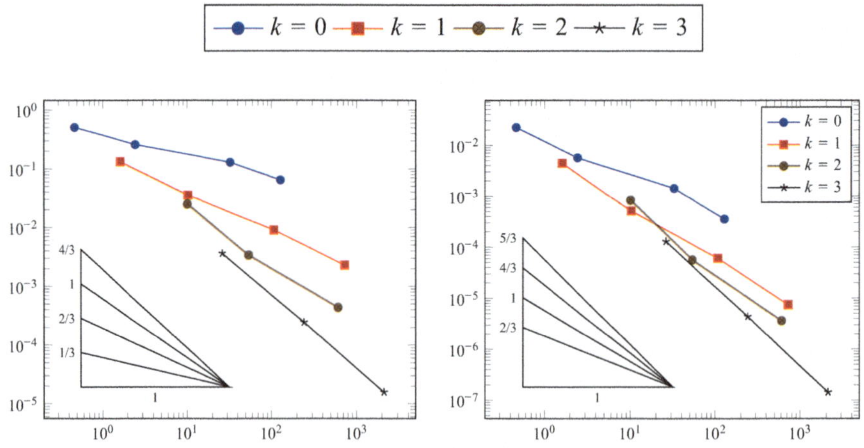

(a) (b)

Fig. 2.5 Error vs. total computational time (in seconds) for the test case of Sect. 2.5.2. The reference slopes refer to the optimal scaling for each polynomial degree $k \in \{0, \ldots, 3\}$. (**a**) $\|\nabla_h(\mathrm{p}_h^{k+1}\underline{u}_h - u)\|$ vs. t_{tot}. (**b**) $\|\mathrm{p}_h^{k+1}\underline{u}_h - u\|$ vs. t_{tot}

and corresponding right-hand side $f(x_1, x_2, x_3) = 3\pi^2 \sin(\pi x_1)\sin(\pi x_2)\sin(\pi x_3)$, on a matching simplicial mesh family and for polynomial degrees k up to 3. The numerical results displayed in Fig. 2.4 show asymptotic convergence rates that match those predicted by (2.64) and (2.70), both with $r = k$. In Fig. 2.5 we display the error versus the total computational time t_{tot} (including the pre-processing, solution, and post-processing), in seconds. It can be seen that the energy- and L^2-errors scale as $t_{\text{tot}}^{\frac{(k+1)}{d}}$ and $t_{\text{tot}}^{\frac{(k+2)}{d}}$ (with $d = 3$), respectively.

Chapter 3
Variable Diffusion and Diffusion–Advection–Reaction

In this chapter we extend the HHO method to the scalar diffusion–advection–reaction problem: Find $u : \Omega \to \mathbb{R}$ such that

$$\begin{aligned} \nabla\cdot(-\boldsymbol{K}\nabla u + \boldsymbol{\beta} u) + \mu u &= f &&\text{in } \Omega, \\ u &= 0 &&\text{on } \partial\Omega, \end{aligned} \tag{3.1}$$

where $\boldsymbol{K} : \Omega \to \mathbb{R}^{d\times d}_{\rm sym}$ (with $\mathbb{R}^{d\times d}_{\rm sym}$ denoting the space of symmetric $d \times d$ matrices) is the spatially varying and possibly anisotropic diffusion coefficient, $\boldsymbol{\beta} : \Omega \to \mathbb{R}^d$ is the velocity, and $\mu : \Omega \to \mathbb{R}$ is the reaction coefficient.

We first consider in Sect. 3.1 the pure diffusion case, that is, we take $\boldsymbol{\beta} = \boldsymbol{0}$ and $\mu = 0$. A key point is to design a method robust with respect to the variations of $\boldsymbol{K}$. We start by introducing and studying the local oblique elliptic projector, which modifies the elliptic projector of Definition 1.39 by including a dependence on the diffusion coefficient. For this projector, we prove approximation properties in both weighted and standard Sobolev seminorms. We next introduce the key ingredient of the local construction, namely a diffusion-dependent potential reconstruction inspired by the oblique elliptic projector, and formulate the local contribution, the global bilinear form, and the discrete problem. Finally, we prove energy error estimates that are fully robust with respect to the heterogeneity of the diffusion coefficient, and have only a moderate dependence on its local anisotropy ratio.

In Sect. 3.2, we then consider the full diffusion–advection–reaction model. The main novel ingredient introduced in this section is the robust HHO discretisation of first-order terms. Problem (3.1) is characterised by the presence of spatially varying coefficients, which can give rise to different regimes in different regions of the domain. In practice, one is typically interested in numerical methods that handle in a robust way locally dominant advection, corresponding to large values of a local Péclet number (a measure of the relative magnitude of the advective and diffusive processes in the model; see (3.82) for a precise definition in the present context). As pointed out in [157], this requires that the discrete counterpart of the bilinear

D. A. Di Pietro, J. Droniou, *The Hybrid High-Order Method for Polytopal Meshes*, MS&A 19, https://doi.org/10.1007/978-3-030-37203-3_3

form corresponding to the terms $\nabla\cdot(\boldsymbol{\beta} u) + \mu u$ satisfies a stability condition that guarantees well-posedness even in the absence of diffusion. This stability property is achieved here combining a reconstruction of the advective derivative, obtained in the HHO spirit, with an upwind stabilisation that penalises the differences between face- and element-based discrete unknowns. The material in Sect. 3.2 is inspired by Di Pietro et al. [159], with some noticeable differences. In particular, an important addition in the present setting are improved error estimates in the L^2-norm.

3.1 Variable Diffusion

In this section, we consider pure diffusion problems with spatially varying coefficients. Denote by $\mathbb{R}^{d\times d}_{\rm sym}$ the space of symmetric $d \times d$ real matrices, and let $\boldsymbol{K} : \Omega \to \mathbb{R}^{d\times d}_{\rm sym}$ be a diffusion coefficient which we assume uniformly elliptic, i.e., such that it holds, for every $\boldsymbol{\xi} \in \mathbb{R}^d$ and a.e. $\boldsymbol{x} \in \Omega$,

$$\underline{K}|\boldsymbol{\xi}|^2 \le \boldsymbol{K}(\boldsymbol{x})\boldsymbol{\xi}\cdot\boldsymbol{\xi} \le \overline{K}|\boldsymbol{\xi}|^2 \tag{3.2}$$

for two given real numbers $0 < \underline{K} \le \overline{K}$. We make the following additional assumption concerning the spatial dependence of $\boldsymbol{K}$.

Assumption 3.1 (Piecewise Constant Diffusion Coefficient) *The diffusion coefficient $\boldsymbol{K}$ is piecewise constant on a finite collection $P_\Omega := \{\Omega_i\}_{i\in I}$ of disjoint polytopes such that $\overline{\Omega} = \bigcup_{i\in I} \overline{\Omega_i}$, i.e.,*

$$\boldsymbol{K}_{|\Omega_i} \in \mathbb{P}^0(\Omega_i; \mathbb{R}^{d\times d}_{\rm sym}) \qquad \forall i \in I.$$

The case of diffusion coefficients that are not piecewise constant is covered in Sect. 4.2.

For a given volumetric source term $f \in L^2(\Omega)$, we consider the problem that consists in seeking $u : \Omega \to \mathbb{R}$ such that

$$-\nabla\cdot(\boldsymbol{K}\nabla u) = f \qquad \text{in } \Omega, \tag{3.3a}$$

$$u = 0 \qquad \text{on } \partial\Omega. \tag{3.3b}$$

Recalling the notation introduced in Remark 1.14 for L^2-products, the weak formulation of problem (3.3) reads: Find $u \in H^1_0(\Omega)$ such that

$$a_{\boldsymbol{K}}(u, v) = (f, v) \qquad \forall v \in H^1_0(\Omega), \tag{3.4}$$

with bilinear form $a_{\boldsymbol{K}} : H^1(\Omega) \times H^1(\Omega) \to \mathbb{R}$ defined by

$$a_{\boldsymbol{K}}(u, v) := (\boldsymbol{K}\nabla u, \nabla v).$$

The key idea to extend the HHO method discussed in Chap. 2 to the variable diffusion problem (3.4) consists in modifying the local potential reconstruction so as to incorporate a dependence on the diffusion tensor. This diffusion-dependent potential reconstruction is designed so that its composition with the local interpolator coincides with a modified elliptic projector that we call the *oblique elliptic projector*. A local contribution including a high-order stabilisation term is then designed based on this reconstruction, following similar principles as in Sect. 2.1.4.

The material is organised as described in what follows. In Sect. 3.1.1 we introduce the notion of mesh sequence compliant with the diffusion coefficient, meaning that jumps of $\boldsymbol{K}$ do not occur inside mesh elements. We note here the advantage of using polyhedral meshes over more common (e.g. triangular) meshes, as this usually enables the construction of compliant meshes using fewer elements. In Sect. 3.1.2 we define the oblique elliptic projector and study its approximation properties. An important point when dealing with variable diffusion problems is to ensure robustness with respect to both the *heterogeneity* (i.e., the spatial variations) and the *anisotropy* (i.e., the directional dependence) of the diffusion coefficient. In the derivation of robust error estimates, the approximation properties of the oblique elliptic projector in both diffusion-weighted and standard Sobolev norms play a key role. Their study is the purpose of Theorem 3.3 and Corollary 3.6. In Sect. 3.1.3, we describe the local construction underlying the HHO method and introduce novel abstract assumptions on the stabilisation term. Finally, in Sect. 3.1.4 we formulate the discrete problem, study the stability, boundedness, and approximation properties of the discrete bilinear form, and derive an error estimate in the energy norm. The robustness of this error estimate is discussed in Remark 3.20.

3.1.1 Compliant Mesh Sequence

In what follows, we consider a regular mesh sequence $(\mathcal{M}_h)_{h\in\mathcal{H}}$ in the sense of Definition 1.9, without necessarily recalling this fact at each occurrence. The role of the following assumption (and, in particular, of its consequence (3.5)) in the design and analysis of the HHO method is discussed in Remarks 3.5 and 3.8.

Assumption 3.2 (Compliant Mesh Sequence) *For all $h \in \mathcal{H}$, we assume that $\mathcal{M}_h$ is* compliant *with the partition P_Ω introduced in Assumption 3.1, in the sense that, for all $T \in \mathcal{T}_h$, there exists a unique index $i \in I$ such that $T \subset \Omega_i$.*

Assumption 3.2 is typically satisfied, e.g., in the modelling of petroleum reservoirs, where the mesh reflects the structure of the subsoil resulting from the petrophysical analysis. On a compliant mesh, jumps of the diffusion coefficient can occur at interfaces but not inside elements. Accounting for Assumption 3.1 this implies, in particular, that

$$\boldsymbol{K} \in \mathbb{P}^0(\mathcal{T}_h; \mathbb{R}^{d\times d}_{\text{sym}}). \tag{3.5}$$

For any mesh element $T \in \mathcal{T}_h$, we let $\boldsymbol{K}_T := \boldsymbol{K}_{|T}$ denote the constant value of the diffusion coefficient inside T, and introduce the *local anisotropy ratio*

$$\alpha_T := \frac{\overline{K}_T}{\underline{K}_T}, \tag{3.6}$$

where $\overline{K}_T$ and $\underline{K}_T$ denote, respectively, the largest and smallest eigenvalues of $\boldsymbol{K}_T$. For any mesh element $T \in \mathcal{T}_h$ and any face $F \in \mathcal{F}_T$, we also define the following positive real number:

$$K_{TF} := \boldsymbol{K}_T \boldsymbol{n}_{TF} \cdot \boldsymbol{n}_{TF}.$$

Finally, for any $T \in \mathcal{T}_h$, we denote by $\boldsymbol{K}_T^{\frac{1}{2}} \in \mathbb{R}^{d \times d}_{\rm sym}$ the unique symmetric positive definite matrix such that

$$\boldsymbol{K}_T = \boldsymbol{K}_T^{\frac{1}{2}} \boldsymbol{K}_T^{\frac{1}{2}},$$

and let $\boldsymbol{K}^{\frac{1}{2}}$ be the piecewise constant matrix-valued field whose restriction to any mesh element $T \in \mathcal{T}_h$ coincides with $\boldsymbol{K}_T^{\frac{1}{2}}$.

3.1.2 The Oblique Elliptic Projector

The starting point to devise an HHO discretisation of problem (3.4) is a modified version of the elliptic projector, introduced in Definition 1.39, that accounts for the presence of the diffusion coefficient. Its definition and the study of its approximation properties make the topic of this section.

Let a mesh element $T \in \mathcal{T}_h$ and an integer $l \geq 0$ be fixed. We define the *oblique elliptic projector* $\pi^{1,l}_{\boldsymbol{K},T} : W^{1,1}(T) \to \mathbb{P}^l(T)$ such that, for all $v \in W^{1,1}(T)$,

$$(\boldsymbol{K}_T \nabla(\pi^{1,l}_{\boldsymbol{K},T} v - v), \nabla w)_T = 0 \qquad \forall w \in \mathbb{P}^l(T). \tag{3.7a}$$

By the Riesz representation theorem in $\nabla\mathbb{P}^l(T)$ for the $\boldsymbol{K}_T$-weighted L^2-inner product, this relation defines a unique element $\nabla\pi^{1,l}_{\boldsymbol{K},T} v$, and thus a polynomial $\pi^{1,l}_{\boldsymbol{K},T} v$ up to an additive constant. This constant is fixed by imposing

$$(\pi^{1,l}_{\boldsymbol{K},T} v - v, 1)_T = 0. \tag{3.7b}$$

Using similar arguments as to pass from (1.60) to (1.61), the conditions (3.7) can be alternatively formulated as follows:

$$(\boldsymbol{K}_T\nabla(\pi_{\boldsymbol{K},T}^{1,l}v-v),\nabla w)_T+(\pi_{\boldsymbol{K},T}^{1,l}v-v,\pi_T^{0,0}w)_T=0 \qquad \forall w\in\mathbb{P}^l(T). \tag{3.8}$$

We have the following characterisation:

$$\pi_{\boldsymbol{K},T}^{1,l}v=\underset{w\in\mathbb{P}^l(T),\,(w-v,1)_T=0}{\operatorname{argmin}}\ \|\boldsymbol{K}_T^{\frac{1}{2}}\nabla(w-v)\|_T^2, \tag{3.9}$$

as can be easily checked by observing that (3.7a) is the Euler equation for the minimisation problem (3.9). Comparing (3.7) with (1.60), we see that $\pi_{\boldsymbol{K},T}^{1,l}$ coincides with the standard elliptic projector $\pi_T^{1,l}$ when $\boldsymbol{K}_T=\boldsymbol{I}_d$, where $\boldsymbol{I}_d$ denotes the identity matrix of $\mathbb{R}^{d\times d}$. To check that $\pi_{\boldsymbol{K},T}^{1,l}$ satisfies the polynomial invariance requirement (1.56) (and hence, by Proposition 1.35, that it meets the conditions of Definition 1.34) it suffices to observe that, if $v\in\mathbb{P}^l(T)$, then making $w=\pi_{\boldsymbol{K},T}^{1,l}v-v$ in (3.7a) implies $\nabla(\pi_{\boldsymbol{K},T}^{1,l}v-v)=0$ and thus that $(\pi_{\boldsymbol{K},T}^{1,l}v-v)$ is constant on T. Using (3.7b), we deduce that $\pi_{\boldsymbol{K},T}^{1,l}v-v=0$.

We next study the approximation properties of the oblique elliptic projector in diffusion-weighted seminorms. Such seminorms are required in the analysis to obtain error estimates with a sharp dependence on the local anisotropy ratio α_T; see Remark 3.20 below. We focus on the Hilbertian case since, as will be made clear in Chap. 6, a different construction is required to treat nonlinear diffusion problems in a non-Hilbertian setting.

Theorem 3.3 (Approximation Properties of the Oblique Elliptic Projector in Diffusion-Weighted Seminorms) *Let $(\mathcal{M}_h)_{h\in\mathcal{H}}=(\mathcal{T}_h,\mathcal{F}_h)_{h\in\mathcal{H}}$ be a regular mesh sequence in the sense of Definition 1.9. Let Assumptions 3.1 and 3.2 hold true. For a given polynomial degree $l\ge 0$, let an integer $s\in\{1,\ldots,l+1\}$ be given. Then, for all $T\in\mathcal{T}_h$, all $v\in H^s(T)$, and all $m\in\{0,\ldots,s-1\}$, it holds*

$$|\boldsymbol{K}_T^{\frac{1}{2}}\nabla(v-\pi_{\boldsymbol{K},T}^{1,l}v)|_{H^m(T)^d}\lesssim \overline{K}_T^{\frac{1}{2}}h_T^{s-m-1}|v|_{H^s(T)} \tag{3.10}$$

with hidden constant independent of h, T, v, m and $\boldsymbol{K}$, but possibly depending on d, ϱ, l and s. If, additionally, $m\le s-2$ (which enforces $s\ge 2$), then, for all $F\in\mathcal{F}_T$,

$$h_T^{\frac{1}{2}}|\boldsymbol{K}_T^{\frac{1}{2}}\nabla(v-\pi_{\boldsymbol{K},T}^{1,l}v)|_{H^m(F)^d}\lesssim \overline{K}_T^{\frac{1}{2}}h_T^{s-m-1}|v|_{H^s(T)}, \tag{3.11}$$

where the hidden constant has the same dependencies as in (3.10).

Remark 3.4 (Approximation Estimates in Diffusion-Weighted Seminorms) The crucial point in estimates (3.10) and (3.11) is that the right-hand side does not depend on the local anisotropy ratio α_T.

Proof We adapt the arguments of the proofs of Lemma 1.43 and Theorem 1.48. Consider the following representation of v:

$$v = Q^s v + R^s v, \tag{3.12}$$

where $Q^s v \in \mathbb{P}^{s-1}(T) \subset \mathbb{P}^l(T)$ is the averaged Taylor polynomial, while the remainder $R^s v$ satisfies, for all $r \in \{0, \ldots, s\}$ (cf. [77, Lemma 4.3.8] for star-shaped elements, and Theorem 1.50 together with Lemma 1.42 for general elements),

$$|R^s v|_{H^r(T)} \lesssim h_T^{s-r} |v|_{H^s(T)}. \tag{3.13}$$

By definition of the oblique elliptic projector, it holds, for any $\phi \in H^1(T)$,

$$\|\boldsymbol{K}_T^{\frac{1}{2}} \nabla \pi_{\boldsymbol{K},T}^{1,l} \phi\|_T \leq \|\boldsymbol{K}_T^{\frac{1}{2}} \nabla \phi\|_T, \tag{3.14}$$

as can be inferred selecting $w = \pi_{\boldsymbol{K},T}^{1,l} \phi$ as a test function in (3.7a) and using the Cauchy–Schwarz inequality. Taking the projection of (3.12), and using the polynomial invariance (1.56) for $\pi_{\boldsymbol{K},T}^{1,l}$ along with $l \geq s - 1$ to write $\pi_{\boldsymbol{K},T}^{1,l} Q^s v = Q^s v$, it is inferred that

$$\pi_{\boldsymbol{K},T}^{1,l} v = Q^s v + \pi_{\boldsymbol{K},T}^{1,l}(R^s v).$$

Subtracting this equation from (3.12), we obtain $v - \pi_{\boldsymbol{K},T}^{1,l} v = R^s v - \pi_{\boldsymbol{K},T}^{1,l}(R^s v)$. Applying the operator $\boldsymbol{K}_T^{1/2} \nabla$ to this expression, taking the seminorm, and using the triangle inequality, we arrive at

$$|\boldsymbol{K}_T^{\frac{1}{2}} \nabla (v - \pi_{\boldsymbol{K},T}^{1,l} v)|_{H^m(T)^d} \leq \underbrace{|\boldsymbol{K}_T^{\frac{1}{2}} \nabla R^s v|_{H^m(T)^d}}_{\mathfrak{T}_1} + \underbrace{|\boldsymbol{K}_T^{\frac{1}{2}} \nabla \pi_{\boldsymbol{K},T}^{1,l}(R^s v)|_{H^m(T)^d}}_{\mathfrak{T}_2}. \tag{3.15}$$

For the first term, it is readily inferred that $\mathfrak{T}_1 \lesssim \overline{K}_T^{\frac{1}{2}} |R^s v|_{H^{m+1}(T)}$ which, combined with (3.13) for $r = m + 1$, gives

$$\mathfrak{T}_1 \lesssim \overline{K}_T^{\frac{1}{2}} h^{s-m-1} |v|_{H^s(T)}. \tag{3.16}$$

For the second term, on the other hand, we can proceed as follows:

$$
\begin{aligned}
\mathfrak{T}_2 &\lesssim h_T^{-m} \|\boldsymbol{K}_T^{\frac{1}{2}} \nabla(\pi_{\boldsymbol{K},T}^{1,l} R^s v)\|_T && \text{Eq. (1.50) with } (p,q,r) = (2,2,0) \\
&\lesssim h_T^{-m} \|\boldsymbol{K}_T^{\frac{1}{2}} \nabla R^s v\|_T && \text{Eq. (3.14) with } \phi = R^s v \\
&\lesssim \overline{K}_T^{\frac{1}{2}} h_T^{-m} |R^s v|_{H^1(T)} \\
&\lesssim \overline{K}_T^{\frac{1}{2}} h_T^{s-m-1} |v|_{H^s(T)}. && \text{Eq. (3.13) with } r = 1
\end{aligned}
\tag{3.17}
$$

Plugging the bounds (3.16) and (3.17) into (3.15), (3.10) follows. To prove (3.11), it suffices to combine (3.10) with the continuous trace inequality (1.51) as in Theorem 1.45. □

Remark 3.5 (Role of Assumptions 3.1 and 3.2 in the Proof of Theorem 3.3) As already pointed out in Sect. 3.1.1, Assumption 3.1 and 3.2 combined imply that the diffusion coefficient is constant inside each mesh element. This fact is used in the first line of the estimate (3.17) of $\mathfrak{T}_2$ to apply the inverse Sobolev embeddings (1.50) to the polynomial function $\boldsymbol{K}_T^{\frac{1}{2}} \nabla(\pi_{\boldsymbol{K},T}^{1,l} R^s v) \in \mathbb{P}^{l-1}(T)^d$, as well as in the applications of the continuous trace inequality (1.51) to obtain (3.11).

In the analysis, we will also need the following result concerning the approximation properties of the oblique elliptic projector in standard Sobolev seminorms. Unlike the estimates of Theorem 3.3, the multiplicative constant in the right-hand side depends, in this case, on the square root of the local anisotropy ratio.

Corollary 3.6 (Approximation Properties of the Oblique Elliptic Projector in Standard Sobolev Seminorms) *Let $(\mathcal{M}_h)_{h\in\mathcal{H}} = (\mathcal{T}_h, \mathcal{F}_h)_{h\in\mathcal{H}}$ be a regular mesh sequence in the sense of Definition 1.9. Let Assumptions 3.1 and 3.2 hold true. For a given polynomial degree $l \geq 0$, let an integer $s \in \{1, \ldots, l+1\}$ be given. Then, for all $T \in \mathcal{T}_h$, all $v \in H^s(T)$, and all $m \in \{0, \ldots, s\}$,*

$$
|v - \pi_{\boldsymbol{K},T}^{1,l} v|_{H^m(T)} \lesssim \alpha_T^{\frac{1}{2}} h_T^{s-m} |v|_{H^s(T)},
\tag{3.18}
$$

with α_T defined by (3.6) *and hidden constant independent of h, T, v, m and $\boldsymbol{K}$, but possibly depending on d, ϱ, l and s. If, additionally, $s \geq 1$ and $m \in \{0, \ldots, s-1\}$, then, for all $F \in \mathcal{F}_T$,*

$$
h_T^{\frac{1}{2}} |v - \pi_{\boldsymbol{K},T}^{1,l} v|_{H^m(F)} \lesssim \alpha_T^{\frac{1}{2}} h_T^{s-m} |v|_{H^s(T)},
\tag{3.19}
$$

with hidden constant having the same dependency as in (3.18).

Proof We first remark that (3.19) is an immediate consequence of (3.18) combined with the continuous trace inequality (1.51), as in Theorem 1.45. To prove (3.18), we distinguish two cases.

(i) *The case* $m = 0$. Recalling that $H^0(T) = L^2(T)$, and using (3.7b) together with the local Poincaré–Wirtinger inequality (1.76), we infer that

$$\|v - \pi_{\boldsymbol{K},T}^{1,l} v\|_T \lesssim h_T \|\nabla(v - \pi_{\boldsymbol{K},T}^{1,l} v)\|_T \le \frac{h_T}{\underline{K}_T^{\frac{1}{2}}} \|\boldsymbol{K}_T^{\frac{1}{2}} \nabla(v - \pi_{\boldsymbol{K},T}^{1,l} v)\|_T$$
$$\lesssim \alpha_T^{\frac{1}{2}} h_T^s |v|_{H^s(T)},$$

where we have used (3.10) to conclude.

(ii) *The case* $m \ge 1$. We have that

$$|v - \pi_{\boldsymbol{K},T}^{1,l} v|_{H^m(T)} \lesssim \frac{1}{\underline{K}_T^{\frac{1}{2}}} |\boldsymbol{K}_T^{\frac{1}{2}} \nabla(v - \pi_{\boldsymbol{K},T}^{1,l} v)|_{H^{m-1}(T)^d} \lesssim \alpha_T^{\frac{1}{2}} h_T^{s-m} |v|_{H^s(T)},$$

where we have used the definition (1.16) of the Sobolev seminorms (with $s = m$ and $p = 2$) in the first inequality and concluded invoking (3.10) with m replaced by $(m - 1)$. □

3.1.3 *Local Construction*

In this section we describe the local construction underlying the HHO method for problem (3.4).

3.1.3.1 Diffusion-Dependent Local Potential Reconstruction

For the variable diffusion problem (3.3), the relevant integration by parts formula is the following: For all $v \in W^{1,1}(T)$ and all $w \in C^\infty(\overline{T})$,

$$(\boldsymbol{K}_T \nabla v, \nabla w)_T = -(v, \nabla\cdot(\boldsymbol{K}_T \nabla w))_T + \sum_{F \in \mathcal{F}_T} (v, \boldsymbol{K}_T \nabla w \cdot \boldsymbol{n}_{TF})_F. \tag{3.20}$$

Let a polynomial degree $k \ge 0$ be fixed. Specialising (3.20) to $w \in \mathbb{P}^{k+1}(T)$, and using the fact that $\boldsymbol{K}_T$ is constant inside T owing to (3.5) to infer $\nabla\cdot(\boldsymbol{K}_T \nabla w) \in \mathbb{P}^{k-1}(T) \subset \mathbb{P}^k(T)$ and $(\boldsymbol{K}_T \nabla w)_{|F} \cdot \boldsymbol{n}_{TF} \in \mathbb{P}^k(F)$ for all $F \in \mathcal{F}_T$, we obtain the following relation, which will inspire the definition of the local potential reconstruction:

$$(\boldsymbol{K}_T\nabla\pi_{\boldsymbol{K},T}^{1,k+1}v,\nabla w)_T=-(\pi_T^{0,k}v,\nabla\cdot(\boldsymbol{K}_T\nabla w))_T+\sum_{F\in\mathcal{F}_T}(\pi_F^{0,k}v,\boldsymbol{K}_T\nabla w\cdot\boldsymbol{n}_{TF})_F,\tag{3.21}$$

where we have used the definitions (3.7a) to insert the elliptic projector $\pi_{\boldsymbol{K},T}^{1,k+1}$ into the left-hand side and (1.57) to insert the L^2-orthogonal projectors $\pi_T^{0,k}$ and $\pi_F^{0,k}$ into the right-hand side. Notice that, as for the Poisson problem, we could have replaced $\pi_T^{0,k}$ by $\pi_T^{0,k-1}$, but this would have required a separate treatment for the case $k=0$; see Sect. 5.1 for variants of HHO with enriched or depleted element unknowns.

Let $\underline{U}_T^k$ denote the local space of discrete unknowns defined by (2.6), which we recall hereafter for the sake of convenience:

$$\underline{U}_T^k:=\left\{\underline{v}_T=(v_T,(v_F)_{F\in\mathcal{F}_T})\ :\ v_T\in\mathbb{P}^k(T)\text{ and }v_F\in\mathbb{P}^k(F)\quad\forall F\in\mathcal{F}_T\right\}.$$

By principles similar to those illustrated in Sect. 2.1.3, inspired by (3.21), we define the diffusion-dependent local potential reconstruction operator $\mathrm{p}_{\boldsymbol{K},T}^{k+1}:\underline{U}_T^k\to\mathbb{P}^{k+1}(T)$ such that, for all $\underline{v}_T\in\underline{U}_T^k$ and all $w\in\mathbb{P}^{k+1}(T)$,

$$(\boldsymbol{K}_T\nabla\mathrm{p}_{\boldsymbol{K},T}^{k+1}\underline{v}_T,\nabla w)_T=-(v_T,\nabla\cdot(\boldsymbol{K}_T\nabla w))_T+\sum_{F\in\mathcal{F}_T}(v_F,\boldsymbol{K}_T\nabla w\cdot\boldsymbol{n}_{TF})_F\tag{3.22a}$$

and

$$(\mathrm{p}_{\boldsymbol{K},T}^{k+1}\underline{v}_T-v_T,1)_T=0.\tag{3.22b}$$

For future use, we note the following equivalent reformulation of (3.22a), obtained integrating by parts the first term in the right-hand side:

$$(\boldsymbol{K}_T\nabla\mathrm{p}_{\boldsymbol{K},T}^{k+1}\underline{v}_T,\nabla w)_T=(\nabla v_T,\boldsymbol{K}_T\nabla w)_T+\sum_{F\in\mathcal{F}_T}(v_F-v_T,\boldsymbol{K}_T\nabla w\cdot\boldsymbol{n}_{TF})_F.\tag{3.23}$$

Remark 3.7 (Equivalent Definition of $\mathrm{p}_{\boldsymbol{K},T}^{k+1}$*)* As in Remark 2.3, fixing $\lambda_T\neq 0$ we note that the conditions (3.22) can be equivalently reformulated as: For all $w\in\mathbb{P}^{k+1}(T)$,

$$\begin{aligned}(\boldsymbol{K}_T\nabla\mathrm{p}_{\boldsymbol{K},T}^{k+1}\underline{v}_T,\nabla w)_T&+\lambda_T(\mathrm{p}_T^{k+1}\underline{v}_T,\pi_T^{0,0}w)_T\\&=(\nabla v_T,\boldsymbol{K}_T\nabla w)_T+\sum_{F\in\mathcal{F}_T}(v_F-v_T,\boldsymbol{K}_T\nabla w\cdot\boldsymbol{n}_{TF})_F+\lambda_T(v_T,\pi_T^{0,0}w)_T.\end{aligned}$$

Comparing (3.22) with (2.11), we see that $\mathrm{p}_{\boldsymbol{K},T}^{k+1} = \mathrm{p}_T^{k+1}$ when $\boldsymbol{K}_T = \boldsymbol{I}_d$. Additionally, recalling the definition (2.8) of the local interpolator $\underline{I}_T^k$, the following crucial relation follows accounting for (3.21) : For all $v \in W^{1,1}(T)$,

$$(\mathrm{p}_{\boldsymbol{K},T}^{k+1} \circ \underline{I}_T^k)v = \pi_{\boldsymbol{K},T}^{1,k+1} v, \tag{3.24}$$

which means that the composition of the diffusion-dependent reconstruction operator with the interpolator gives the oblique elliptic projector on $\mathbb{P}^{k+1}(T)$.

Remark 3.8 (Non-piecewise Constant Diffusion Coefficients) The definition (3.7) of the oblique elliptic projector extends in a straightforward manner to non-constant (but still uniformly elliptic) $\boldsymbol{K}_T$. In this case, however, we can no longer introduce the L^2-projectors $\pi_T^{0,k}$ and $\pi_F^{0,k}$ into the right-hand side of (3.21). The reason is that the functions $\nabla{\cdot}(\boldsymbol{K}_T\nabla w)$ and $(\boldsymbol{K}_T\nabla w)_{|F}{\cdot}\boldsymbol{n}_{TF}$ are no longer in $\mathbb{P}^{k-1}(T)$ and $\mathbb{P}^k(F)$, respectively (in fact, they are possibly not even polynomials). A consequence of this fact is that (3.24) will no longer hold in general. For this reason, in the case of non-piecewise constant diffusion, we consider a different approach not based on the oblique elliptic projector; see Sect. 4.2 on this topic.

3.1.3.2 Local Contribution

We define on $\underline{U}_T^k$ the diffusion-dependent local seminorm such that, for all $\underline{v}_T \in \underline{U}_T^k$,

$$\|\underline{v}_T\|_{1,\boldsymbol{K},T} := \left(\|\boldsymbol{K}_T^{\frac{1}{2}}\nabla v_T\|_T^2 + \sum_{F\in\mathcal{F}_T} \frac{K_{TF}}{h_F}\|v_F - v_T\|_F^2 \right)^{\frac{1}{2}}, \tag{3.25}$$

and we let $\mathrm{a}_{\boldsymbol{K},T} : \underline{U}_T^k \times \underline{U}_T^k \to \mathbb{R}$ be the bilinear form such that, for all $\underline{u}_T, \underline{v}_T \in \underline{U}_T^k$,

$$\mathrm{a}_{\boldsymbol{K},T}(\underline{u}_T, \underline{v}_T) := (\boldsymbol{K}_T\nabla \mathrm{p}_{\boldsymbol{K},T}^{k+1}\underline{u}_T, \nabla \mathrm{p}_{\boldsymbol{K},T}^{k+1}\underline{v}_T)_T + \mathrm{s}_{\boldsymbol{K},T}(\underline{u}_T, \underline{v}_T). \tag{3.26}$$

Here, $\mathrm{s}_{\boldsymbol{K},T} : \underline{U}_T^k \times \underline{U}_T^k \to \mathbb{R}$ is a stabilisation bilinear form that satisfies the design conditions summarised in the following assumption.

Assumption 3.9 (Local Stabilisation Bilinear Form $\mathrm{s}_{\boldsymbol{K},T}$) *The local stabilisation bilinear form $\mathrm{s}_{\boldsymbol{K},T} : \underline{U}_T^k \times \underline{U}_T^k \to \mathbb{R}$ satisfies the following properties:*

(SK1) Symmetry and positivity. *$\mathrm{s}_{\boldsymbol{K},T}$ is symmetric and positive semidefinite;*
(SK2) Stability and boundedness. *There is a real number $\eta > 0$ independent of h, T, and $\boldsymbol{K}$ such that, for all $\underline{v}_T \in \underline{U}_T^k$,*

$$(\alpha_T\eta)^{-1}\|\underline{v}_T\|_{1,\boldsymbol{K},T}^2 \le \mathrm{a}_{\boldsymbol{K},T}(\underline{v}_T, \underline{v}_T) \le \alpha_T\eta\|\underline{v}_T\|_{1,\boldsymbol{K},T}^2; \tag{3.27}$$

(SK3) Polynomial consistency. *For all $w \in \mathbb{P}^{k+1}(T)$ and all $\underline{v}_T \in \underline{U}_T^k$, it holds*

$$\mathrm{s}_{\boldsymbol{K},T}(\underline{I}_T^k w, \underline{v}_T) = 0. \tag{3.28}$$

The following lemma collects a few significant properties enjoyed by stabilisation bilinear forms that satisfy Assumption 3.9.

Lemma 3.10 (Properties of $\mathrm{s}_{\boldsymbol{K},T}$) *Let $T \in \mathcal{T}_h$ and let $\mathrm{s}_{\boldsymbol{K},T}$ denote a stabilisation bilinear form satisfying Assumption 3.9. Then, the following properties hold:*

(i) Dependence through difference operators. *$\mathrm{s}_{\boldsymbol{K},T}$ depends on its arguments only through the diffusion-dependent difference operators $\delta^k_{\boldsymbol{K},T} : \underline{U}_T^k \to \mathbb{P}^k(T)$ and $\delta^k_{\boldsymbol{K},TF} : \underline{U}_T^k \to \mathbb{P}^k(F)$, $F \in \mathcal{F}_T$, such that, for all $\underline{v}_T \in \underline{U}_T^k$,*

$$\begin{aligned} \delta^k_{\boldsymbol{K},T}\underline{v}_T &:= \pi_T^{0,k}(\mathrm{p}^{k+1}_{\boldsymbol{K},T}\underline{v}_T - v_T), \\ \delta^k_{\boldsymbol{K},TF}\underline{v}_T &:= \pi_F^{0,k}(\mathrm{p}^{k+1}_{\boldsymbol{K},T}\underline{v}_T - v_F) \quad \forall F \in \mathcal{F}_T. \end{aligned} \tag{3.29}$$

(ii) Boundary difference reformulation. *For all $\underline{u}_T, \underline{v}_T \in \underline{U}_T^k$, it holds that*

$$\mathrm{s}_{\boldsymbol{K},T}(\underline{u}_T, \underline{v}_T) = \mathrm{s}_{\boldsymbol{K},T}((0, \underline{\Delta}^k_{\partial T}\underline{u}_T), (0, \underline{\Delta}^k_{\partial T}\underline{v}_T)), \tag{3.30}$$

where $\underline{\Delta}^k_{\partial T}$ is the boundary difference operator defined by (2.56).

(iii) Consistency for smooth functions. *For all $T \in \mathcal{T}_h$, all $r \in \{0, \ldots, k\}$, and all $v \in H^{r+2}(T)$, it holds that*

$$\mathrm{s}_{\boldsymbol{K},T}(\underline{I}_T^k v, \underline{I}_T^k v)^{\frac{1}{2}} \lesssim \overline{K}_T^{\frac{1}{2}} \alpha_T^{\frac{1}{2}} h_T^{r+1} |v|_{H^{r+2}(T)} \tag{3.31}$$

with hidden constant independent of h, T, v, r and $\boldsymbol{K}$.

Remark 3.11 (Diffusion-Dependent Difference Operators and Interpolator) The diffusion-dependent difference operators (3.29) satisfy the following relation:

$$(\delta^k_{\boldsymbol{K},T}\underline{v}_T, (\delta^k_{\boldsymbol{K},TF}\underline{v}_T)_{F \in \mathcal{F}_T}) = \underline{I}_T^k \mathrm{p}^{k+1}_{\boldsymbol{K},T}\underline{v}_T - \underline{v}_T. \tag{3.32}$$

Proof (i) *Dependence through difference operators.* This property follows from (SK3) reasoning as in Lemma 2.11 with $\mathrm{p}^{k+1}_{\boldsymbol{K},T}$ instead of p^{k+1}_T. See also Remark 2.12.

(ii) *Boundary difference reformulation.* The commutation property (3.24) and the polynomial invariance of $\pi^{1,k+1}_{\boldsymbol{K},T}$ easily show, as in the proof of Proposition 2.6, that the diffusion-dependent difference operators (3.29) satisfy the polynomial consistency (2.21), that is to say: For all $w \in \mathbb{P}^{k+1}(T)$,

$$\delta^k_{\boldsymbol{K},T}\underline{I}_T^k w = 0, \qquad \delta^k_{\boldsymbol{K},TF}\underline{I}_T^k w = 0 \quad \forall F \in \mathcal{F}_T. \tag{3.33}$$

Using this and Item (i), the proof of (3.30) is done as that of Proposition 2.24, with the diffusion-dependent difference operators instead of $(\delta_T^k, (\delta_{TF}^k)_{F\in\mathcal{F}_T})$.

(iii) *Consistency for smooth functions.* We set, for the sake of brevity, $\check{v}_T := \pi_T^{0,k+1}v$, and we start by observing that

$$
\begin{aligned}
\mathrm{s}_{\boldsymbol{K},T}(\underline{I}_T^k v, \underline{I}_T^k v)^{\frac{1}{2}} &= \mathrm{s}_{\boldsymbol{K},T}(\underline{I}_T^k(v-\check{v}_T), \underline{I}_T^k(v-\check{v}_T))^{\frac{1}{2}} && \text{Eq. (3.28)}\\
&= \mathrm{s}_{\boldsymbol{K},T}((0,\underline{\Delta}_{\partial T}^k\underline{I}_T^k(v-\check{v}_T)),(0,\underline{\Delta}_{\partial T}^k\underline{I}_T^k(v-\check{v}_T)))^{\frac{1}{2}}. && \text{Eq. (3.30)}
\end{aligned}
$$

The property (SK2) then yields

$$
\mathrm{s}_{\boldsymbol{K},T}(\underline{I}_T^k v, \underline{I}_T^k v)^{\frac{1}{2}} \le \alpha_T^{\frac{1}{2}}\eta^{\frac{1}{2}}\|(0,\underline{\Delta}_{\partial T}^k\underline{I}_T^k(v-\check{v}_T))\|_{1,\boldsymbol{K},T}. \tag{3.34}
$$

Recalling the definitions (3.25) of $\|{\cdot}\|_{1,\boldsymbol{K},T}$ and (2.56) of $\underline{\Delta}_{\partial T}^k$, we have that

$$
\begin{aligned}
\|(0,\underline{\Delta}_{\partial T}^k\underline{I}_T^k(v-\check{v}_T))\|_{1,\boldsymbol{K},T}^2 &= \sum_{F\in\mathcal{F}_T}\frac{K_{TF}}{h_F}\|\pi_F^{0,k}(v-\check{v}_T)-\pi_T^{0,k}(v-\check{v}_T)\|_F^2\\
&\le \overline{K}_T\sum_{F\in\mathcal{F}_T}h_F^{-1}\|\pi_F^{0,k}(v-\check{v}_T)-\pi_T^{0,k}(v-\check{v}_T)\|_F^2\\
&= \overline{K}_T|\underline{I}_T^k(v-\check{v}_T)|_{1,\partial T}^2,
\end{aligned} \tag{3.35}
$$

where we have used $K_{TF} \le \overline{K}_T$ in the second line, and the definitions (2.7) and (2.8) of the local seminorm $|{\cdot}|_{1,\partial T}$ and of the local interpolator $\underline{I}_T^k$ to conclude. We then invoke the boundedness (2.9) of $\underline{I}_T^k$ together with the approximation property (1.74) of $\check{v}_T = \pi_T^{0,k+1}v$ with $s = r+2$, $m = 1$ and $p = 2$ to deduce

$$
\|(0,\underline{\Delta}_{\partial T}^k\underline{I}_T^k(v-\check{v}_T))\|_{1,\boldsymbol{K},T} \lesssim \overline{K}_T^{\frac{1}{2}}|v-\check{v}_T|_{H^1(T)} \lesssim \overline{K}_T^{\frac{1}{2}}h_T^{r+1}|v|_{H^{r+2}(T)},
$$

which, combined with (3.34), yields (3.31). □

Remark 3.12 (On the Choice of the Difference Operators) As observed in Remark 2.12 for the Poisson problem, we could have used, in the first point of Lemma 3.10, difference operators defined as in (2.24) starting from any polynomial reconstruction $R_T^{k+1} : \underline{U}_T^k \to \mathbb{P}^{k+1}(T)$ consistent for polynomials of degree $(k+1)$. The specific choice (3.29) is, however, the natural one to prove (SK2); see Proposition 3.13.

An important example of a stabilisation bilinear form that matches the requirements in Assumption 3.9 is given in the following proposition. This stabilisation is a generalisation of the original HHO stabilisation s_T defined by (2.22),

Proposition 3.13 (Diffusion-Dependent HHO Stabilisation) *The bilinear form defined, for all $\underline{u}_T, \underline{v}_T \in \underline{U}_T^k$, by*

$$\mathrm{s}_{\boldsymbol{K},T}(\underline{u}_T, \underline{v}_T) := \sum_{F\in\mathcal{F}_T} \frac{K_{TF}}{h_F}((\delta^k_{\boldsymbol{K},TF} - \delta^k_{\boldsymbol{K},T})\underline{u}_T, (\delta^k_{\boldsymbol{K},TF} - \delta^k_{\boldsymbol{K},T})\underline{v}_T)_F \qquad (3.36)$$

satisfies properties (SK1)–(SK3).

Proof Property (SK1) can be checked by simple inspection. Property (SK3) is a consequence of the fact that the difference operators defined by (3.29) satisfy the polynomial consistency (3.33). It only remains to prove (SK2). In the following, the multiplicative constants in $\lesssim$ do not depend on h, T, v or $\boldsymbol{K}$, and we set, for the sake of brevity,

$$\check{v}_T := \mathrm{p}^{k+1}_{\boldsymbol{K},T}\underline{v}_T.$$

(i) *Coercivity.* Let $\underline{w}_T := \underline{I}^k_T \check{v}_T = (\delta^k_{\boldsymbol{K},T}\underline{v}_T, (\delta^k_{\boldsymbol{K},TF}\underline{v}_T)_{F\in\mathcal{F}_T}) + \underline{v}_T$ (see (3.32)). The triangle inequality gives

$$\|v_F - v_T\|_F \lesssim \|(\delta^k_{\boldsymbol{K},TF} - \delta^k_{\boldsymbol{K},T})\underline{v}_T\|_F + \|w_F - w_T\|_F.$$

Raising to the square, multiplying by K_{TF}/h_F, summing over $F \in \mathcal{F}_T$, and using $K_{TF} \le \overline{K}_T$ and the definition of $\underline{w}_T$ leads to

$$\begin{aligned}\sum_{F\in\mathcal{F}_T} \frac{K_{TF}}{h_F}\|v_F - v_T\|_F^2 &\lesssim \mathrm{s}_{\boldsymbol{K},T}(\underline{v}_T, \underline{v}_T) + \overline{K}_T|\underline{w}_T|^2_{1,\partial T} \\ &\lesssim \mathrm{s}_{\boldsymbol{K},T}(\underline{v}_T, \underline{v}_T) + \overline{K}_T|\underline{I}^k_T\check{v}_T|^2_{1,\partial T} \\ &\lesssim \mathrm{s}_{\boldsymbol{K},T}(\underline{v}_T, \underline{v}_T) + \overline{K}_T\|\nabla\check{v}_T\|^2_T,\end{aligned} \qquad (3.37)$$

where we have used the boundedness (2.9) of $\underline{I}^k_T$ with $v = \check{v}_T$ to conclude. We then notice that $\underline{K}_T\|\nabla\check{v}_T\|^2_T \le \|\boldsymbol{K}_T^{\frac{1}{2}}\nabla\check{v}_T\|^2_T$ to infer

$$\begin{aligned}\sum_{F\in\mathcal{F}_T} \frac{K_{TF}}{h_F}\|v_F - v_T\|_F^2 &\lesssim \mathrm{s}_{\boldsymbol{K},T}(\underline{v}_T, \underline{v}_T) + \alpha_T\|\boldsymbol{K}_T^{\frac{1}{2}}\nabla\check{v}_T\|^2_T \\ &\le \alpha_T\left(\mathrm{s}_{\boldsymbol{K},T}(\underline{v}_T, \underline{v}_T) + \|\boldsymbol{K}_T^{\frac{1}{2}}\nabla\check{v}_T\|^2_T\right) \\ &= \alpha_T\mathrm{a}_{\boldsymbol{K},T}(\underline{v}_T, \underline{v}_T),\end{aligned} \qquad (3.38)$$

where we have used the fact that $\alpha_T \ge 1$ to pass to the second line. We now estimate the volumetric term in $\|\underline{v}_T\|_{1,\boldsymbol{K},T}$. To this purpose, write (3.23) with $w = v_T$ and use Cauchy–Schwarz inequalities to see that

$$\begin{aligned}
&\|\boldsymbol{K}_T^{\frac{1}{2}}\nabla v_T\|_T^2 \\
&\quad = (\boldsymbol{K}_T\nabla \check{v}_T, \nabla v_T)_T - \sum_{F\in\mathcal{F}_T}(v_F - v_T, \boldsymbol{K}_T\nabla v_T{\cdot}\boldsymbol{n}_{TF})_F \\
&\quad \le \|\boldsymbol{K}_T^{\frac{1}{2}}\nabla \check{v}_T\|_T\|\boldsymbol{K}_T^{\frac{1}{2}}\nabla v_T\|_T + \sum_{F\in\mathcal{F}_T}\||\boldsymbol{K}_T^{\frac{1}{2}}\boldsymbol{n}_{TF}|(v_F - v_T)\|_F\|\boldsymbol{K}_T^{\frac{1}{2}}\nabla v_T\|_F,
\end{aligned} \tag{3.39}$$

where we have used the symmetry of $\boldsymbol{K}_T^{\frac{1}{2}}$ to write $(\boldsymbol{K}_T\nabla \check{v}_T, \nabla v_T)_T = (\boldsymbol{K}_T^{\frac{1}{2}}\nabla \check{v}_T, \boldsymbol{K}_T^{\frac{1}{2}}\nabla v_T)_T$ and

$$\begin{aligned}
(v_F - v_T, \boldsymbol{K}_T\nabla v_T{\cdot}\boldsymbol{n}_{TF})_F &= (v_F - v_T, \boldsymbol{K}_T^{\frac{1}{2}}\nabla v_T{\cdot}\boldsymbol{K}_T^{\frac{1}{2}}\boldsymbol{n}_{TF})_F \\
&= ((v_F - v_T)\boldsymbol{K}_T^{\frac{1}{2}}\boldsymbol{n}_{TF}, \boldsymbol{K}_T^{\frac{1}{2}}\nabla v_T)_F.
\end{aligned} \tag{3.40}$$

Invoking the discrete trace inequality (1.55) on $\boldsymbol{K}_T^{\frac{1}{2}}\nabla v_T \in \mathbb{P}^{k-1}(T)^d$ and the bounds $h_F \le h_T$, whenever $F \in \mathcal{F}_T$, and $\mathrm{card}(\mathcal{F}_T) \lesssim 1$ (see (1.5)), we deduce

$$\begin{aligned}
\|\boldsymbol{K}_T^{\frac{1}{2}}\nabla v_T\|_T^2 \lesssim{}& \|\boldsymbol{K}_T^{\frac{1}{2}}\nabla \check{v}_T\|_T\|\boldsymbol{K}_T^{\frac{1}{2}}\nabla v_T\|_T \\
&+ \left(\sum_{F\in\mathcal{F}_T} h_F^{-1}\||\boldsymbol{K}_T^{\frac{1}{2}}\boldsymbol{n}_{TF}|(v_F - v_T)\|_F^2\right)^{\frac{1}{2}}\|\boldsymbol{K}_T^{\frac{1}{2}}\nabla v_T\|_T.
\end{aligned}$$

Extracting the constant scalar $|\boldsymbol{K}_T^{\frac{1}{2}}\boldsymbol{n}_{TF}|$ from the norm, noticing that

$$|\boldsymbol{K}_T^{\frac{1}{2}}\boldsymbol{n}_{TF}|^2 = \boldsymbol{K}_T^{\frac{1}{2}}\boldsymbol{n}_{TF}{\cdot}\boldsymbol{K}_T^{\frac{1}{2}}\boldsymbol{n}_{TF} = \boldsymbol{K}_T\boldsymbol{n}_{TF}{\cdot}\boldsymbol{n}_{TF} = K_{TF} \tag{3.41}$$

and simplifying by $\|\boldsymbol{K}_T^{\frac{1}{2}}\nabla v_T\|_T$ yields

$$\|\boldsymbol{K}_T^{\frac{1}{2}}\nabla v_T\|_T \lesssim \|\boldsymbol{K}_T^{\frac{1}{2}}\nabla \check{v}_T\|_T + \left(\sum_{F\in\mathcal{F}_T}\frac{K_{TF}}{h_F}\|v_F - v_T\|_F^2\right)^{\frac{1}{2}}.$$

Raising to the square and estimating the second term using (3.38), we infer that $\|\boldsymbol{K}_T^{\frac{1}{2}}\nabla v_T\|_T^2 \lesssim \alpha_T \mathrm{a}_{\boldsymbol{K},T}(\underline{v}_T, \underline{v}_T)$. Adding together this estimate and (3.38) yields the first inequality in (3.27).

(ii) *Boundedness.* We now prove the second inequality in (3.27). Use $w = \check{v}_T$ in (3.23), Cauchy–Schwarz inequalities, the discrete trace inequality (1.55) and the symmetry of $\boldsymbol{K}_T^{\frac{1}{2}}$ as above to write

$$\|\boldsymbol{K}_T^{\frac{1}{2}}\nabla\check{v}_T\|_T^2 \lesssim \|\boldsymbol{K}_T^{\frac{1}{2}}\nabla v_T\|_T\|\boldsymbol{K}_T^{\frac{1}{2}}\nabla\check{v}_T\|_T + \left(\sum_{F\in\mathcal{F}_T}\frac{K_{TF}}{h_F}\|v_F - v_T\|_F^2\right)^{\frac{1}{2}}\|\boldsymbol{K}_T^{\frac{1}{2}}\nabla\check{v}_T\|_T.$$

Simplifying by $\|\boldsymbol{K}_T^{\frac{1}{2}}\nabla\check{v}_T\|_T$ and raising to the square yields

$$\|\boldsymbol{K}_T^{\frac{1}{2}}\nabla\check{v}_T\|_T^2 \lesssim \|\underline{v}_T\|_{1,\boldsymbol{K},T}^2. \tag{3.42}$$

To estimate the stabilisation term in $\mathrm{a}_{\boldsymbol{K},T}$, set $\underline{w}_T = \underline{I}_T^k\check{v}_T$ and recall (3.32) to write

$$\|(\delta_{\boldsymbol{K},TF}^k - \delta_{\boldsymbol{K},T}^k)\underline{v}_T\|_F \lesssim \|w_F - w_T\|_F + \|v_F - v_T\|_F.$$

Raise to the square, multiply by K_{TF}/h_F, use $K_{TF} \le \overline{K}_T$, sum over $F \in \mathcal{F}_T$ and use the boundedness (2.9) of $\underline{I}_T^k$ with $v = \check{v}_T$ to get

$$\begin{aligned}
\mathrm{s}_{\boldsymbol{K},T}(\underline{v}_T, \underline{v}_T) &\le \overline{K}_T|\underline{w}_T|_{1,\partial T}^2 + \sum_{F\in\mathcal{F}_T}\frac{K_{TF}}{h_F}\|v_F - v_T\|_F^2 \\
&\lesssim \overline{K}_T\|\nabla\check{v}_T\|_T^2 + \sum_{F\in\mathcal{F}_T}\frac{K_{TF}}{h_F}\|v_F - v_T\|_F^2.
\end{aligned}$$

Since $\|\boldsymbol{K}_T^{\frac{1}{2}}\nabla\check{v}_T\|_T^2 \ge \underline{K}_T\|\nabla\check{v}_T\|_T^2$, this shows that

$$\mathrm{s}_{\boldsymbol{K},T}(\underline{v}_T, \underline{v}_T) \lesssim \alpha_T\|\boldsymbol{K}_T^{\frac{1}{2}}\nabla\check{v}_T\|_T^2 + \sum_{F\in\mathcal{F}_T}\frac{K_{TF}}{h_F}\|v_F - v_T\|_F^2 \lesssim \alpha_T\|\underline{v}_T\|_{1,\boldsymbol{K},T}^2,$$

the conclusion following from (3.42) and $\alpha_T \ge 1$. Adding together this estimate and (3.42) yields the second inequality in (3.27). □

Remark 3.14 (Comparison with Poisson) The proof of Proposition 3.13 highlights two important differences with respect to the Poisson problem treated in Chap. 2; see Proposition 2.13.

First, a slightly different treatment of the boundary terms is required to have a sharp dependence on the local anisotropy ratio, which involves the rewriting (3.40) in order to use Cauchy–Schwarz instead of generalised Hölder inequalities to estimate the boundary terms in (3.39) (compare with (2.27) in Proposition 2.13).

Second, the use of the diffusion-dependent difference operators (3.29) in $s_{\boldsymbol{K},T}$ over the diffusion-independent ones defined by (2.19) is justified by the fact that they result in the correct volumetric term involving the gradient of $\mathrm{p}_{\boldsymbol{K},T}^{k+1}\underline{v}_T$ instead of $\mathrm{p}_T^{k+1}\underline{v}_T$ in (3.37).

3.1.4 *Discrete Problem and Convergence*

In this section, we formulate the global problem and prove energy norm error estimates.

3.1.4.1 Global Bilinear Form

Recall the definition (2.32) of the global HHO space $\underline{U}_h^k$ with single-valued interface unknowns:

$$\underline{U}_h^k := \Big\{\underline{v}_h = ((v_T)_{T\in\mathcal{T}_h}, (v_F)_{F\in\mathcal{F}_h}) : \\ v_T \in \mathbb{P}^k(T) \quad \forall T \in \mathcal{T}_h \text{ and } v_F \in \mathbb{P}^k(F) \quad \forall F \in \mathcal{F}_h\Big\},$$

and its subspace accounting for homogeneous Dirichlet boundary conditions

$$\underline{U}_{h,0}^k := \Big\{\underline{v}_h \in \underline{U}_h^k \ : \ v_F = 0 \quad \forall F \in \mathcal{F}_h^{\mathrm{b}}\Big\}. \tag{3.43}$$

Let $\mathrm{a}_{\boldsymbol{K},h} : \underline{U}_h^k \times \underline{U}_h^k \to \mathbb{R}$ denote the global bilinear form obtained by element by element assembly setting, for $\underline{u}_h, \underline{v}_h \in \underline{U}_h^k$,

$$\mathrm{a}_{\boldsymbol{K},h}(\underline{u}_h, \underline{v}_h) := \sum_{T\in\mathcal{T}_h} \mathrm{a}_{\boldsymbol{K},T}(\underline{u}_T, \underline{v}_T). \tag{3.44}$$

Stability and boundedness are expressed with respect to the $\boldsymbol{K}$-weighted seminorm on $\underline{U}_h^k$ such that, for all $\underline{v}_h \in \underline{U}_h^k$,

$$\|\underline{v}_h\|_{1,\boldsymbol{K},h} := \left(\sum_{T\in\mathcal{T}_h} \|\underline{v}_T\|_{1,\boldsymbol{K},T}^2 \right)^{\frac{1}{2}} \tag{3.45}$$

with $\|\cdot\|_{1,\boldsymbol{K},T}$ defined by (3.25). The fact that $\|\cdot\|_{1,\boldsymbol{K},h}$ defines a norm on $\underline{U}_{h,0}^k$ follows from Corollary 2.16 after observing that $\|\underline{v}_h\|_{1,h} \le \underline{K}^{-\frac{1}{2}} \|\underline{v}_h\|_{1,\boldsymbol{K},h}$ for all $\underline{v}_h \in \underline{U}_{h,0}^k$.

Lemma 3.15 (Properties of $\mathrm{a}_{\boldsymbol{K},h}$) *The bilinear form $\mathrm{a}_{\boldsymbol{K},h}$ enjoys the following properties:*

(i) Stability and boundedness. *For all $\underline{v}_h \in \underline{U}_{h,0}^k$ it holds*

$$\begin{aligned} &(\alpha\eta)^{-1} \|\underline{v}_h\|_{1,\boldsymbol{K},h}^2 \le \|\underline{v}_h\|_{\mathrm{a},\boldsymbol{K},h}^2 \le \alpha\eta \|\underline{v}_h\|_{1,\boldsymbol{K},h}^2 \\ &\textit{with } \|\underline{v}_h\|_{\mathrm{a},\boldsymbol{K},h} := \mathrm{a}_{\boldsymbol{K},h}(\underline{v}_h, \underline{v}_h)^{\frac{1}{2}}, \end{aligned} \tag{3.46}$$

where η is as in (3.27) *and*

$$\alpha := \max_{T\in\mathcal{T}_h} \alpha_T \tag{3.47}$$

denotes the global anisotropy ratio.

(ii) Consistency. *It holds for all $r \in \{0, \ldots, k\}$ and all $w \in H_0^1(\Omega) \cap H^{r+2}(\mathcal{T}_h)$ such that $\nabla\cdot(\boldsymbol{K}\nabla w) \in L^2(\Omega)$,*

$$\sup_{\underline{v}_h \in \underline{U}_{h,0}^k, \|\underline{v}_h\|_{\mathrm{a},\boldsymbol{K},h}=1} |\mathcal{E}_{\boldsymbol{K},h}(w; \underline{v}_h)| \lesssim \left(\sum_{T\in\mathcal{T}_h} \overline{K}_T \alpha_T h_T^{2(r+1)} |w|_{H^{r+2}(T)}^2 \right)^{\frac{1}{2}}, \tag{3.48}$$

where the hidden constant is independent of w, h and $\boldsymbol{K}$, and the linear form $\mathcal{E}_{\boldsymbol{K},h}(w; \cdot) : \underline{U}_{h,0}^k \to \mathbb{R}$ representing the consistency error is such that, for all $\underline{v}_h \in \underline{U}_{h,0}^k$,

$$\mathcal{E}_{\boldsymbol{K},h}(w; \underline{v}_h) := -(\nabla\cdot(\boldsymbol{K}\nabla w), v_h) - \mathrm{a}_{\boldsymbol{K},h}(\underline{I}_h^k w, \underline{v}_h). \tag{3.49}$$

Proof The proof uses similar arguments as that of Lemma 2.18.

(i) *Stability and boundedness.* Summing the inequalities (3.27) over $T \in \mathcal{T}_h$, and observing that $\alpha_T \le \alpha$ for all $T \in \mathcal{T}_h$, (3.46) follows.

(ii) *Consistency.* Let $\underline{v}_h \in \underline{U}_{h,0}^k$ be such that $\|\underline{v}_h\|_{a,\boldsymbol{K},h} = 1$. For the sake of brevity, we set $\check{w}_T := \mathrm{p}_{\boldsymbol{K},T}^{k+1} \underline{I}_T^k w = \pi_{\boldsymbol{K},T}^{1,k+1} w$ (cf. (3.24)) for all $T \in \mathcal{T}_h$. Integrating by parts element by element and using Corollary 1.19 with $\boldsymbol{\tau} = \boldsymbol{K}\nabla w$ and $(\varphi_F)_{F\in\mathcal{F}_h} = (v_F)_{F\in\mathcal{F}_h}$ to insert v_F into the boundary term after noticing that, by the assumed regularity on w and $\boldsymbol{K}$, $\boldsymbol{K}\nabla w \in \boldsymbol{H}(\mathrm{div};\Omega) \cap H^1(\mathcal{T}_h)^d$, we infer that

$$-(\nabla\cdot(\boldsymbol{K}\nabla w), v_h) = \sum_{T\in\mathcal{T}_h} \left((\boldsymbol{K}_T \nabla w, \nabla v_T)_T + \sum_{F\in\mathcal{F}_T} (\boldsymbol{K}_T \nabla w_{|T}\cdot\boldsymbol{n}_{TF}, v_F - v_T)_F \right). \tag{3.50}$$

On the other hand, plugging the definition (3.26) of $\mathrm{a}_{\boldsymbol{K},T}$ into (3.44) and expanding, for all $T \in \mathcal{T}_h$, $\mathrm{p}_{\boldsymbol{K},T}^{k+1}\underline{v}_T$ according to (3.23) with $w = \check{w}_T$, it is inferred that

$$\begin{aligned}\mathrm{a}_{\boldsymbol{K},h}(\underline{I}_h^k w, \underline{v}_h) &= \sum_{T\in\mathcal{T}_h} \left((\boldsymbol{K}_T \nabla \check{w}_T, \nabla v_T)_T + \sum_{F\in\mathcal{F}_T} (\boldsymbol{K}_T \nabla \check{w}_T\cdot\boldsymbol{n}_{TF}, v_F - v_T)_F \right) \\ &\quad + \sum_{T\in\mathcal{T}_h} \mathrm{s}_{\boldsymbol{K},T}(\underline{I}_T^k w, \underline{v}_T).\end{aligned} \tag{3.51}$$

Subtracting (3.51) from (3.50), taking absolute values, and using the definition (3.7a) of $\check{w}_T = \pi_{\boldsymbol{K},T}^{1,k+1} w$ to cancel the first terms inside the summation leads to

$$\begin{aligned}&|\mathcal{E}_{\boldsymbol{K},h}(w; \underline{v}_h)| \\ &= \left| \sum_{T\in\mathcal{T}_h} \sum_{F\in\mathcal{F}_T} (\boldsymbol{K}_T \nabla(w_{|T} - \check{w}_T)\cdot\boldsymbol{n}_{TF}, v_F - v_T)_F - \sum_{T\in\mathcal{T}_h} \mathrm{s}_{\boldsymbol{K},T}(\underline{I}_T^k w, \underline{v}_T) \right| \\ &\lesssim \sum_{T\in\mathcal{T}_h} \left(\sum_{F\in\mathcal{F}_T} h_F \|\boldsymbol{K}_T^{\frac{1}{2}} \nabla(w_{|T} - \check{w}_T)\|_F^2 \right)^{\frac{1}{2}} \left(\sum_{F\in\mathcal{F}_T} \frac{K_{TF}}{h_F} \|v_F - v_T\|_F^2 \right)^{\frac{1}{2}} \\ &\quad + \sum_{T\in\mathcal{T}_h} \mathrm{s}_{\boldsymbol{K},T}(\underline{I}_T^k w, \underline{I}_T^k w)^{\frac{1}{2}} \mathrm{s}_{\boldsymbol{K},T}(\underline{v}_T, \underline{v}_T)^{\frac{1}{2}},\end{aligned}$$

where the inequality follows using the same algebraic manipulations as in (3.40) and (3.41), and Cauchy–Schwarz inequalities (including on the positive semidefinite forms $\mathrm{s}_{\boldsymbol{K},T}$). Using the trace approximation properties (3.11) of the oblique elliptic projector with $l = k+1$, $s = r+2$, and $m = 0$ together with the consistency property (3.31) of $\mathrm{s}_{\boldsymbol{K},T}$, we continue with

$$|\mathcal{E}_{\boldsymbol{K},h}(w; \underline{v}_h)| \lesssim \sum_{T \in \mathcal{T}_h} \overline{K}_T^{\frac{1}{2}} h_T^{r+1} |w|_{H^{r+2}(T)} \left(\sum_{F \in \mathcal{F}_T} \frac{K_{TF}}{h_F} \|v_F - v_T\|_F^2 \right)^{\frac{1}{2}} + \sum_{T \in \mathcal{T}_h} \overline{K}_T^{\frac{1}{2}} \alpha_T^{\frac{1}{2}} h_T^{r+1} |w|_{H^{r+2}(T)} \mathrm{s}_{\boldsymbol{K},T}(\underline{v}_T, \underline{v}_T)^{\frac{1}{2}}. \tag{3.52}$$

We then use again Cauchy–Schwarz inequalities to infer

$$|\mathcal{E}_{\boldsymbol{K},h}(w; \underline{v}_h)| \lesssim \left(\sum_{T \in \mathcal{T}_h} \alpha_T \overline{K}_T h_T^{2(r+1)} |w|_{H^{r+2}(T)}^2 \right)^{\frac{1}{2}} \times \left\{ \left(\sum_{T \in \mathcal{T}_h} \alpha_T^{-1} \sum_{F \in \mathcal{F}_T} \frac{K_{TF}}{h_F} \|v_F - v_T\|_F^2 \right)^{\frac{1}{2}} + \left(\sum_{T \in \mathcal{T}_h} \mathrm{s}_{\boldsymbol{K},T}(\underline{v}_T, \underline{v}_T) \right)^{\frac{1}{2}} \right\}.$$

The proof is completed by noticing that the local seminorm equivalence (3.27) together with the fact that $\|\underline{v}_h\|_{\mathrm{a},\boldsymbol{K},h} = 1$ imply

$$\sum_{T \in \mathcal{T}_h} \alpha_T^{-1} \sum_{F \in \mathcal{F}_T} \frac{K_{TF}}{h_F} \|v_F - v_T\|_F^2 \lesssim \sum_{T \in \mathcal{T}_h} \mathrm{a}_{\boldsymbol{K},T}(\underline{v}_T, \underline{v}_T) = 1$$

and

$$\sum_{T \in \mathcal{T}_h} \mathrm{s}_{\boldsymbol{K},T}(\underline{v}_T, \underline{v}_T) \le \sum_{T \in \mathcal{T}_h} \mathrm{a}_{\boldsymbol{K},T}(\underline{v}_T, \underline{v}_T) = 1.$$

□

3.1.4.2 Discrete Problem

The HHO discretisation of problem (3.4) reads: Find $\underline{u}_h \in \underline{U}_{h,0}^k$ such that

$$\mathrm{a}_{\boldsymbol{K},h}(\underline{u}_h, \underline{v}_h) = (f, v_h) \qquad \forall \underline{v}_h \in \underline{U}_{h,0}^k, \tag{3.53}$$

where we remind the reader that, according to (2.33), the broken polynomial function $v_h \in \mathbb{P}^k(\mathcal{T}_h)$ is obtained from $\underline{v}_h$ setting $(v_h)_{|T} := v_T$ for all $T \in \mathcal{T}_h$. The proof of the following lemma is a straightforward variation of that of Lemma 2.19 and is left as an exercise to the reader.

Lemma 3.16 (Well-Posedness of Problem (3.53)**)** *Problem* (3.53) *is well-posed, and we have the following a priori bound for the unique discrete solution* $\underline{u}_h \in \underline{U}_{h,0}^k$*:*

$$\|\underline{u}_h\|_{1,\boldsymbol{K},h} \leq \frac{\alpha\eta C_{\mathrm{P}}}{\underline{K}^{\frac{1}{2}}}\|f\|,$$

where C_{P} *denotes the constant of the discrete Poincaré inequality* (2.37) *and* $\underline{K}$ *is as in* (3.2).

3.1.4.3 Flux Formulation

The following lemma contains a reformulation of the discrete problem (3.53) in terms of numerical fluxes. Its proof is a straightforward adaptation of that of Lemma 2.25, and is left as an exercise to the reader. The main difference with respect to the Poisson problem considered in Chap. 2 is that, for the variable diffusion problem (3.3), the flux whose normal component is continuous across interfaces is $-\boldsymbol{K}\nabla u$.

We recall that the boundary difference space $\underline{D}_{\partial T}^k$ and operator $\underline{\Delta}_{\partial T}^k$ are defined by (see (2.55) and (2.56)):

$$\underline{D}_{\partial T}^k := \left\{\underline{\alpha}_{\partial T} = (\alpha_{TF})_{F\in\mathcal{F}_T} \;:\; \alpha_{TF} \in \mathbb{P}^k(F) \quad \forall F \in \mathcal{F}_T\right\}$$

and

$$\underline{\Delta}_{\partial T}^k \underline{v}_T = (\Delta_{TF}^k \underline{v}_T)_{F\in\mathcal{F}_T} := (v_F - v_T)_{F\in\mathcal{F}_T}.$$

Lemma 3.17 (Flux Formulation) *Let* $\mathcal{M}_h$ *denote a polytopal mesh in the sense of Definition 1.4, and let Assumptions 3.1 and 3.2 hold true. For all* $T \in \mathcal{T}_h$*, let* $\mathrm{s}_{\boldsymbol{K},T}$ *be a bilinear form satisfying Assumption 3.9, and define the boundary residual operator* $\underline{R}_{\boldsymbol{K},\partial T}^k : \underline{U}_T^k \to \underline{D}_{\partial T}^k$ *such that, for all* $\underline{v}_T \in \underline{U}_T^k$*,*

$$\underline{R}_{\boldsymbol{K},\partial T}^k \underline{v}_T := (R_{\boldsymbol{K},TF}^k \underline{v}_T)_{F\in\mathcal{F}_T}$$

and, for all $\underline{\alpha}_{\partial T} = (\alpha_{TF})_{F\in\mathcal{F}_T} \in \underline{D}_{\partial T}^k$*,*

$$-\sum_{F\in\mathcal{F}_T} (R_{\boldsymbol{K},TF}^k \underline{v}_T, \alpha_{TF})_F = \mathrm{s}_{\boldsymbol{K},T}((0, \underline{\Delta}_{\partial T}^k \underline{v}_T), (0, \underline{\alpha}_{\partial T})).$$

(continued)

Lemma 3.17 (continued)
Let $\underline{u}_h \in \underline{U}_{h,0}^k$ *and, for all* $T \in \mathcal{T}_h$ *and all* $F \in \mathcal{F}_T$*, define the numerical normal trace of the flux*

$$\Phi_{\boldsymbol{K},TF}(\underline{u}_T) := -\boldsymbol{K}_T \nabla \mathrm{p}_{\boldsymbol{K},T}^{k+1} \underline{u}_T \cdot \boldsymbol{n}_{TF} + R_{\boldsymbol{K},TF}^k \underline{u}_T. \tag{3.54}$$

Then, $\underline{u}_h$ *is the unique solution of problem* (3.53) *if and only if the following two properties hold:*

(i) Local balance. *For all* $T \in \mathcal{T}_h$ *and all* $v_T \in \mathbb{P}^k(T)$*, it holds*

$$(\boldsymbol{K}_T \nabla \mathrm{p}_{\boldsymbol{K},T}^{k+1} \underline{u}_T, \nabla v_T)_T + \sum_{F \in \mathcal{F}_T} (\Phi_{\boldsymbol{K},TF}(\underline{u}_T), v_T)_F = (f, v_T)_T.$$

(ii) Continuity of the numerical normal traces of the fluxes. *For any interface* $F \in \mathcal{F}_h^{\mathrm{i}}$ *such that* $F \subset \partial T_1 \cap \partial T_2$ *with distinct mesh elements* $T_1, T_2 \in \mathcal{T}_h$*, it holds*

$$\Phi_{\boldsymbol{K},T_1F}(\underline{u}_{T_1}) + \Phi_{\boldsymbol{K},T_2F}(\underline{u}_{T_2}) = 0.$$

3.1.4.4 Energy Error Estimate

In this section we study the convergence of the discrete solution of the HHO problem (3.53) towards the solution of problem (3.4). As in Sect. 2.3.1, we first state a convergence result for the energy norm of the error measured as the difference between the solution to the HHO scheme and the interpolate of the exact solution. This discrete energy error estimate is the starting point to prove an estimate for the error measured as the difference between the exact solution and the global reconstruction obtained through the operator $\mathrm{p}_{\boldsymbol{K},h}^{k+1} : \underline{U}_h^k \to \mathbb{P}^{k+1}(\mathcal{T}_h)$ such that, for all $\underline{v}_h \in \underline{U}_h^k$,

$$(\mathrm{p}_{\boldsymbol{K},h}^{k+1} \underline{v}_h)_{|T} := \mathrm{p}_{\boldsymbol{K},T}^{k+1} \underline{v}_T \qquad \forall T \in \mathcal{T}_h. \tag{3.55}$$

Theorem 3.18 (Discrete Energy Error Estimate) *Let $(\mathcal{M}_h)_{h\in\mathcal{H}}$ denote a regular mesh sequence in the sense of Definition 1.9. Let Assumptions 3.1 and 3.2 hold true. Let a polynomial degree $k \geq 0$ be fixed. Denote by $u \in H^1_0(\Omega)$ the unique solution to (3.4), for which we assume the additional regularity $u \in H^{r+2}(\mathcal{T}_h)$ for some $r \in \{0,\ldots,k\}$. For all $h \in \mathcal{H}$, let $\underline{u}_h \in \underline{U}^k_{h,0}$ denote the unique solution to (3.53) with stabilisation bilinear forms $\mathrm{s}_{\boldsymbol{K},T}$, $T \in \mathcal{T}_h$, in (3.26) satisfying Assumptions 3.9. Then, it holds that*

$$\|\underline{u}_h - \underline{I}^k_h u\|_{\mathrm{a},\boldsymbol{K},h} \lesssim \left(\sum_{T\in\mathcal{T}_h} \overline{K}_T \alpha_T h_T^{2(r+1)} |u|^2_{H^{r+2}(T)} \right)^{\frac{1}{2}}, \tag{3.56}$$

where the norm $\|{\cdot}\|_{\mathrm{a},\boldsymbol{K},h}$ is defined in (3.46) and the hidden constant is independent of h, u and $\boldsymbol{K}$.

Proof Identical to the proof of Theorem 2.27, using the Third Strang Lemma A.7 and the consistency estimate (3.48) (noting that $\mathrm{a}_{\boldsymbol{K},h}$ is obviously coercive with constant $\gamma = 1$ for the norm $\|{\cdot}\|_{\mathrm{a},\boldsymbol{K},h}$ on $\underline{U}^k_{h,0}$). □

Theorem 3.19 (Energy Error Estimate for the Reconstructed Approximate Solution) *Under the assumptions of Theorem 3.18, it holds that*

$$\|\boldsymbol{K}^{\frac{1}{2}}\nabla_h(\mathrm{p}^{k+1}_{\boldsymbol{K},h}\underline{u}_h - u)\| + |\underline{u}_h|_{\mathrm{s},\boldsymbol{K},h} \lesssim \left(\sum_{T\in\mathcal{T}_h} \overline{K}_T \alpha_T h_T^{2(r+1)} |u|^2_{H^{r+2}(T)} \right)^{\frac{1}{2}}, \tag{3.57}$$

where the hidden constant is independent of h, u and $\boldsymbol{K}$ and, for all $\underline{v}_h \in \underline{U}^k_{h,0}$, we have set

$$|\underline{v}_h|_{\mathrm{s},\boldsymbol{K},h} := \left(\sum_{T\in\mathcal{T}_h} \mathrm{s}_{\boldsymbol{K},T}(\underline{v}_T, \underline{v}_T) \right)^{\frac{1}{2}}.$$

Remark 3.20 (Robustness of the Estimates) The estimates (3.56) and (3.57) are (i) *fully robust* with respect to the heterogeneity of the diffusion coefficient in that they do not depend on the jumps of $\boldsymbol{K}$ across mesh elements; (ii) *partially robust* with respect to the anisotropy of the diffusion coefficient, meaning that the multiplicative constants in the right-hand sides do not depend on the global anisotropy ratio α defined by (3.47), but only on the square roots of the local anisotropy ratios α_T, $T \in \mathcal{T}_h$.

Proof (Theorem 3.19) Let, for the sake of brevity, $\underline{\hat{u}}_h := \underline{I}_h^k u$ and $\check{u}_h := \mathrm{p}_{\boldsymbol{K},h}^{k+1}\underline{\hat{u}}_h$. Clearly, $(\check{u}_h)_{|T} = \pi_{\boldsymbol{K},T}^{1,k+1} u$ for all $T \in \mathcal{T}_h$ by virtue of (3.24). Using the triangle and Cauchy–Schwarz inequalities, it is readily inferred that

$$\begin{aligned}\|\boldsymbol{K}^{\frac{1}{2}}\nabla_h(\mathrm{p}_{\boldsymbol{K},h}^{k+1}\underline{u}_h - u)\| + |\underline{u}_h|_{\mathrm{s},\boldsymbol{K},h} \\ \le \|\underline{u}_h - \underline{\hat{u}}_h\|_{\mathrm{a},\boldsymbol{K},h} + \left(\|\boldsymbol{K}^{\frac{1}{2}}\nabla_h(\check{u}_h - u)\|^2 + |\underline{\hat{u}}_h|_{\mathrm{s},\boldsymbol{K},h}^2\right)^{\frac{1}{2}}. \end{aligned} \tag{3.58}$$

Using the approximation properties (3.10) of $\check{u}_T$ with $l = k+1$, $s = r+2$, and $m = 0$, the consistency estimate (3.31) of $\mathrm{s}_{\boldsymbol{K},T}$, and the fact that $\alpha_T \ge 1$ for all $T \in \mathcal{T}_h$, we have

$$\left(\|\boldsymbol{K}^{\frac{1}{2}}\nabla_h(\check{u}_h - u)\|^2 + |\underline{\hat{u}}_h|_{\mathrm{s},\boldsymbol{K},h}^2\right)^{\frac{1}{2}} \lesssim \left(\sum_{T\in\mathcal{T}_h} \overline{K}_T\alpha_T h_T^{2(r+1)}|u|_{H^{r+2}(T)}^2\right)^{\frac{1}{2}}. \tag{3.59}$$

Using (3.56) and (3.59) to bound the right-hand side of (3.58), (3.57) follows. □

Remark 3.21 (Error Estimate in L^2-Norm) For the Poisson problem, improved estimates for the L^2-norm of the error have been established in Theorem 2.32. These estimates require the elliptic regularity of the adjoint problem which, in that case, is the same as the original problem. In case of a varying diffusion coefficient, the adjoint problem is also the original problem (3.3). The elliptic regularity for this model is only known if Ω is convex and $\boldsymbol{K}$ is Lipschitz continuous. Combined with Assumption 3.1, this enforces $\boldsymbol{K}$ constant over the entire domain, which means that (3.3) essentially reduces to the Poisson problem (up to a linear transformation of the coordinates). The L^2-error estimate is then a straightforward consequence of Theorem 2.32. In Sect. 4.2, we tackle the issue of a diffusion equation with non-piecewise constant diffusion coefficient $\boldsymbol{K}$, and we establish improved L^2-error estimates for the HHO approximation of this equation (these estimates are not direct consequences of Theorem 2.32 since, when allowed to vary inside the mesh elements, $\boldsymbol{K}$ can be Lipschitz-continuous without being constant).

3.2 Diffusion–Advection–Reaction

We consider in this section the full diffusion–advection–reaction model

$$\begin{aligned} \nabla\cdot(-\boldsymbol{K}\nabla u + \boldsymbol{\beta} u) + \mu u &= f && \text{in } \Omega, \\ u &= 0 && \text{on } \partial\Omega, \end{aligned} \tag{3.60}$$

where the volumetric source term f belongs to $L^2(\Omega)$, the diffusion tensor $\boldsymbol{K}$ satisfies the same assumptions as in Sect. 3.1 (that is, uniform coercivity and Assumption 3.1), and the velocity field $\boldsymbol{\beta} : \Omega \to \mathbb{R}^d$ and reaction coefficient $\mu : \Omega \to \mathbb{R}$ satisfy the regularity requirements formulated hereafter.

Assumption 3.22 (Velocity and Reaction Coefficient) *The advection field is such that $\boldsymbol{\beta} \in \mathrm{Lip}(\Omega)^d$ (or, in other words, $\boldsymbol{\beta} \in W^{1,\infty}(\Omega)^d$), the reaction coefficient satisfies $\mu \in L^\infty(\Omega)$, and there is a real number $\mu_0 > 0$ such that $\frac{1}{2}\nabla\cdot\boldsymbol{\beta} + \mu \geq \mu_0$ a.e. in Ω.*

Having assumed $\boldsymbol{K}$ uniformly elliptic, the following weak formulation is well-posed: Find $u \in H^1_0(\Omega)$ such that

$$a_{\boldsymbol{K},\boldsymbol{\beta},\mu}(u,v) = (f,v) \qquad \forall v \in H^1_0(\Omega), \tag{3.61}$$

where the bilinear form $a_{\boldsymbol{K},\boldsymbol{\beta},\mu} : H^1(\Omega) \times H^1(\Omega) \to \mathbb{R}$ is such that

$$a_{\boldsymbol{K},\boldsymbol{\beta},\mu}(u,v) := a_{\boldsymbol{K}}(u,v) + a_{\boldsymbol{\beta},\mu}(u,v),$$

and the diffusive and advective–reactive contributions are respectively defined by

$$\begin{aligned} a_{\boldsymbol{K}}(u,v) &:= (\boldsymbol{K}\nabla u, \nabla v), \\ a_{\boldsymbol{\beta},\mu}(u,v) &:= \tfrac{1}{2}(\boldsymbol{\beta}\cdot\nabla u, v) - \tfrac{1}{2}(u, \boldsymbol{\beta}\cdot\nabla v) + \left(\left[\tfrac{1}{2}\nabla\cdot\boldsymbol{\beta} + \mu\right] u, v\right). \end{aligned} \tag{3.62}$$

Remark 3.23 (Continuous Advection–Reaction Bilinear Form) The usage of the bilinear form $a_{\boldsymbol{\beta},\mu}(u,v)$ in the weak formulation (3.61) is justified by the following algebra, based on a splitting of the advection term, the expansion of $\nabla\cdot(\boldsymbol{\beta}u)$ and an integration by parts:

$$\begin{aligned} (\nabla\cdot(\boldsymbol{\beta}u), v) &= \tfrac{1}{2}(\nabla\cdot(\boldsymbol{\beta}u), v) + \tfrac{1}{2}(\nabla\cdot(\boldsymbol{\beta}u), v) \\ &= \tfrac{1}{2}(\boldsymbol{\beta}\cdot\nabla u, v) + \tfrac{1}{2}((\nabla\cdot\boldsymbol{\beta})u, v) - \tfrac{1}{2}(\boldsymbol{\beta}u, \nabla v). \end{aligned}$$

The discretisation of $a_{\boldsymbol{K}}$ was discussed in Sect. 3.1. We therefore focus, in the next section, on the discretisation of the advective–reactive bilinear form $a_{\boldsymbol{\beta},\mu}$.

3.2.1 Discretisation of Advective Terms with Upwind Stabilisation

We introduce the ingredients for the discretisation of first-order terms: a local advective derivative reconstruction and an upwind stabilisation term penalizing the differences between face- and element-based discrete unknowns. In the following, we consider meshes that are compliant with the discontinuities of $\boldsymbol{K}$, that is, that satisfy Assumption 3.2.

3.2.1.1 Reconstructed Advective Derivative

Let a mesh element $T \in \mathcal{T}_h$ be fixed. Our objective is to reconstruct, from a local vector of unknowns $\underline{v}_T \in \underline{U}_T^k$, an advective derivative $\mathrm{G}_{\boldsymbol{\beta},T}^k \underline{v}_T$ that approximates $\boldsymbol{\beta}\cdot\nabla v$, when $\underline{v}_T = \underline{I}_T^k v$ for a function $v \in W^{1,1}(T)$. For such a function, and w smooth, integrating by parts shows that

$$(\boldsymbol{\beta}\cdot\nabla v, w)_T = -(v, \nabla\cdot(\boldsymbol{\beta} w))_T + \sum_{F\in\mathcal{F}_T} ((\boldsymbol{\beta}\cdot\boldsymbol{n}_{TF})v, w)_F.$$

Approximating v by $\pi_T^{0,k} v$ inside T and by $\pi_F^{0,k} v$ on $F \in \mathcal{F}_T$, we see that

$$(\boldsymbol{\beta}\cdot\nabla v, w)_T \approx -(\pi_T^{0,k} v, \nabla\cdot(\boldsymbol{\beta} w))_T + \sum_{F\in\mathcal{F}_T} ((\boldsymbol{\beta}\cdot\boldsymbol{n}_{TF})\pi_F^{0,k} v, w)_F. \tag{3.63}$$

Specialising this formula to $w \in \mathbb{P}^k(T)$ leads to the following definition of the local discrete advective derivative reconstruction $\mathrm{G}_{\boldsymbol{\beta},T}^k : \underline{U}_T^k \to \mathbb{P}^k(T)$: For all $\underline{v}_T \in \underline{U}_T^k$,

$$(\mathrm{G}_{\boldsymbol{\beta},T}^k \underline{v}_T, w)_T = -(v_T, \nabla\cdot(\boldsymbol{\beta} w))_T + \sum_{F\in\mathcal{F}_T} ((\boldsymbol{\beta}\cdot\boldsymbol{n}_{TF})v_F, w)_F \quad \forall w \in \mathbb{P}^k(T). \tag{3.64}$$

The existence of a unique $\mathrm{G}_{\boldsymbol{\beta},T}^k \underline{v}_T$ satisfying the equation above follows from the Riesz representation theorem in $\mathbb{P}^k(T)$ endowed with the $L^2(T)$-inner product. Note that, using an integration by parts on the first term in the right-hand side of (3.64), we also have, for all $w \in \mathbb{P}^k(T)$,

$$(\mathrm{G}_{\boldsymbol{\beta},T}^k \underline{v}_T, w)_T = (\boldsymbol{\beta}\cdot\nabla v_T, w)_T + \sum_{F\in\mathcal{F}_T} ((\boldsymbol{\beta}\cdot\boldsymbol{n}_{TF})(v_F - v_T), w)_F. \tag{3.65}$$

In Sect. 2.1.1, we passed from (2.4) to (2.5a) by specifying the test function w to be a polynomial function, and replacing v with its projections on local polynomial spaces using their orthogonality properties. This was possible because, if w is a polynomial function inside T, Δw and, for all $F \in \mathcal{F}_T$, $(\nabla w)_{|F}\cdot\boldsymbol{n}_{TF}$ are also polynomial. As a consequence, the potential reconstruction satisfies the commutation property (2.14). Here, the terms $\nabla\cdot(\boldsymbol{\beta} w)$ and $(\boldsymbol{\beta}\cdot\boldsymbol{n}_{TF})w$ are not necessarily polynomial functions, even if w is polynomial. Hence, introducing the projections of v on local polynomial spaces over T and $F \in \mathcal{F}_T$ leads to the *approximate* Eq. (3.63): unless $\boldsymbol{\beta}$ is constant over T, this is not an exact relation, and no exact commutation property can be stated in general for $(\mathrm{G}_{\boldsymbol{\beta},T}^k \circ \underline{I}_T^k)$. That being said, we can nonetheless establish approximation properties for this operator. To state them, we introduce the reference velocity on T, defined by

$$\hat{\beta}_T := \|\boldsymbol{\beta}\|_{L^\infty(T)^d}. \tag{3.66}$$

Lemma 3.24 (Approximation Properties for $(\mathrm{G}^k_{\boldsymbol{\beta},T} \circ \underline{I}^k_T)$) *If $r \in \{0, \dots, k\}$ and $v \in H^{r+1}(T)$, then*

$$\|\mathrm{G}^k_{\boldsymbol{\beta},T}\underline{I}^k_T v - \pi^{0,k}_T(\boldsymbol{\beta}\cdot\nabla v)\|_T \lesssim \hat{\beta}_T h^r_T |v|_{H^{r+1}(T)}. \tag{3.67}$$

As a consequence, if $\boldsymbol{\beta} \in W^{r,\infty}(T)^d$, then

$$\|\mathrm{G}^k_{\boldsymbol{\beta},T}\underline{I}^k_T v - \boldsymbol{\beta}\cdot\nabla v\|_T \lesssim h^r_T \left(\hat{\beta}_T |v|_{H^{r+1}(T)} + |\boldsymbol{\beta}\cdot\nabla v|_{H^r(T)}\right). \tag{3.68}$$

In the above estimates, the hidden constants are independent of h, T, v, r and $\boldsymbol{\beta}$ (but may depend on d, ϱ and k).

Proof For any $w \in \mathbb{P}^k(T)$, subtracting $(\boldsymbol{\beta}\cdot\nabla v, w)_T$ from (3.65) with $\underline{v}_T = \underline{I}^k_T v = (\pi^{0,k}_T v, (\pi^{0,k}_F v)_{F\in\mathcal{F}_T})$, we have

$$\begin{aligned}(\mathrm{G}^k_{\boldsymbol{\beta},T}\underline{I}^k_T v - \boldsymbol{\beta}\cdot\nabla v, w)_T &= (\boldsymbol{\beta}\cdot\nabla(\pi^{0,k}_T v - v), w)_T \\ &\quad + \sum_{F\in\mathcal{F}_T} ((\boldsymbol{\beta}\cdot\boldsymbol{n}_{TF})(\pi^{0,k}_F v - \pi^{0,k}_T v), w)_F.\end{aligned}$$

We can insert the orthogonal projector $\pi^{0,k}_T$ into the left-hand side using its definition (1.57) with $X = T$ and $l = k$ to obtain

$$\begin{aligned}&(\mathrm{G}^k_{\boldsymbol{\beta},T}\underline{I}^k_T v - \pi^{0,k}_T(\boldsymbol{\beta}\cdot\nabla v), w)_T \\ &\quad \le \hat{\beta}_T \|\nabla(\pi^{0,k}_T v - v)\|_T \|w\|_T + \sum_{F\in\mathcal{F}_T} \hat{\beta}_T \|\pi^{0,k}_F v - \pi^{0,k}_T v\|_F \|w\|_F \\ &\quad \lesssim \hat{\beta}_T \|\nabla(\pi^{0,k}_T v - v)\|_T \|w\|_T + \hat{\beta}_T \sum_{F\in\mathcal{F}_T} \|v - \pi^{0,k}_T v\|_F h_T^{-\frac{1}{2}} \|w\|_T \\ &\quad \lesssim \hat{\beta}_T h^r_T |v|_{H^{r+1}(T)} \|w\|_T,\end{aligned} \tag{3.69}$$

where the first line follows from a generalised Hölder inequality with exponents $(\infty, 2, 2)$ along with the definition (3.66) of $\hat{\beta}_T$, the second line from the linearity, idempotency, and $L^2(F)$-boundedness of $\pi^{0,k}_F$ (which yield $\|\pi^{0,k}_F v - \pi^{0,k}_T v\|_F = \|\pi^{0,k}_F(v - \pi^{0,k}_T v)\|_F \le \|v - \pi^{0,k}_T v\|_F$) and the discrete trace inequality (1.55), while the conclusion is obtained by invoking the approximation properties (1.74) and (1.75) of $\pi^{0,k}_T$ with $p = 2$, $s = r + 1$, and $m = 1$ and $m = 0$ for the volumetric and boundary terms, respectively. The proof of (3.67) is completed by taking $w = \mathrm{G}^k_{\boldsymbol{\beta},T}\underline{v}_T - \pi^{0,k}_T(\boldsymbol{\beta}\cdot\nabla v) \in \mathbb{P}^k(T)$ in (3.69) and by simplifying by $\|w\|_T$.

Estimate (3.68) follows inserting $\pm\pi_T^{0,k}(\boldsymbol{\beta}\cdot\nabla v)$ inside the norm in the left-hand side, using a triangle inequality to write

$$\|\mathrm{G}_{\boldsymbol{\beta},T}^k \underline{I}_T^k v - \boldsymbol{\beta}\cdot\nabla v\| \leq \|\mathrm{G}_{\boldsymbol{\beta},T}^k \underline{I}_T^k v - \pi_T^{0,k}(\boldsymbol{\beta}\cdot\nabla v)\| + \|\boldsymbol{\beta}\cdot\nabla v - \pi_T^{0,k}(\boldsymbol{\beta}\cdot\nabla v)\|,$$

and using (3.67) to estimate the first term and the approximation properties (1.74) of the L^2-orthogonal projector with $l = k$, $p = 2$, $s = r$ and $m = 0$ to estimate the second (notice that $\boldsymbol{\beta}\cdot\nabla v \in H^r(T)$ whenever $\boldsymbol{\beta} \in W^{r,\infty}(T)^d$ and $v \in H^{r+1}(T)$). □

The following global discrete integration by parts formula will also be useful.

Lemma 3.25 (Integration by Parts for the Reconstructed Advective Derivative) *For all $\underline{u}_h, \underline{v}_h \in \underline{U}_{h,0}^k$, it holds*

$$\begin{aligned}\sum_{T\in\mathcal{T}_h} (\mathrm{G}_{\boldsymbol{\beta},T}^k \underline{u}_T, v_T)_T = &-\sum_{T\in\mathcal{T}_h} (u_T, \mathrm{G}_{\boldsymbol{\beta},T}^k \underline{v}_T)_T - \sum_{T\in\mathcal{T}_h} ((\nabla\cdot\boldsymbol{\beta})u_T, v_T)_T \\ &- \sum_{T\in\mathcal{T}_h}\sum_{F\in\mathcal{F}_T} ((\boldsymbol{\beta}\cdot\boldsymbol{n}_{TF})(u_F - u_T), (v_F - v_T))_F.\end{aligned} \tag{3.70}$$

Proof Let us first work on an element $T \in \mathcal{T}_h$. Use the definition (3.64) of $\mathrm{G}_{\boldsymbol{\beta},T}^k \underline{u}_T$ with $w = v_T$, develop $\nabla\cdot(\boldsymbol{\beta} v_T)$, and invoke (3.65) with $w = u_T$ to see that

$$\begin{aligned}(\mathrm{G}_{\boldsymbol{\beta},T}^k \underline{u}_T, v_T)_T &= -(u_T, \nabla\cdot(\boldsymbol{\beta} v_T))_T + \sum_{F\in\mathcal{F}_T} ((\boldsymbol{\beta}\cdot\boldsymbol{n}_{TF})u_F, v_T)_F \\ &= -(u_T, (\nabla\cdot\boldsymbol{\beta})v_T)_T - (u_T, \boldsymbol{\beta}\cdot\nabla v_T)_T + \sum_{F\in\mathcal{F}_T} ((\boldsymbol{\beta}\cdot\boldsymbol{n}_{TF})u_F, v_T)_F \\ &= -(u_T, (\nabla\cdot\boldsymbol{\beta})v_T)_T - (\mathrm{G}_{\boldsymbol{\beta},T}^k \underline{v}_T, u_T)_T \\ &\quad + \sum_{F\in\mathcal{F}_T} ((\boldsymbol{\beta}\cdot\boldsymbol{n}_{TF})(v_F - v_T), u_T)_F + \sum_{F\in\mathcal{F}_T} ((\boldsymbol{\beta}\cdot\boldsymbol{n}_{TF})u_F, v_T)_F.\end{aligned}$$

Subtracting and adding u_F in the first boundary sum in the right-hand side, and noticing that

$$\sum_{F\in\mathcal{F}_T} ((\boldsymbol{\beta}\cdot\boldsymbol{n}_{TF})(v_F - v_T), u_F)_F + \sum_{F\in\mathcal{F}_T} ((\boldsymbol{\beta}\cdot\boldsymbol{n}_{TF})u_F, v_T)_F = \sum_{F\in\mathcal{F}_T} ((\boldsymbol{\beta}\cdot\boldsymbol{n}_{TF})v_F, u_F)_F,$$

we infer, after summation over $T \in \mathcal{T}_h$, that

$$\begin{aligned}
\sum_{T\in\mathcal{T}_h} (\mathrm{G}^k_{\boldsymbol{\beta},T}\underline{u}_T, v_T)_T
&= -\sum_{T\in\mathcal{T}_h} (u_T, (\nabla\cdot\boldsymbol{\beta})v_T)_T - \sum_{T\in\mathcal{T}_h} (\mathrm{G}^k_{\boldsymbol{\beta},T}\underline{v}_T, u_T)_T \\
&\quad + \sum_{T\in\mathcal{T}_h}\sum_{F\in\mathcal{F}_T} ((\boldsymbol{\beta}\cdot\boldsymbol{n}_{TF})(v_F - v_T), (u_T - u_F))_F + \sum_{T\in\mathcal{T}_h}\sum_{F\in\mathcal{F}_T} ((\boldsymbol{\beta}\cdot\boldsymbol{n}_{TF})u_F, v_F)_F.
\end{aligned}$$

Recalling Assumption 3.22, Corollary 1.19 with $p = \infty$, $\boldsymbol{\tau} = \boldsymbol{\beta}$ and $(\varphi_F)_{F\in\mathcal{F}_h} = (u_F v_F)_{F\in\mathcal{F}_h}$ gives

$$\sum_{T\in\mathcal{T}_h}\sum_{F\in\mathcal{F}_T} ((\boldsymbol{\beta}\cdot\boldsymbol{n}_{TF})u_F, v_F)_F = 0 \tag{3.71}$$

and the proof is complete. □

3.2.1.2 Local Advective–Reactive Bilinear Form

Let a mesh element $T \in \mathcal{T}_h$ be fixed. Inspired by the definition (3.62) of the continuous advective–reactive bilinear form $a_{\boldsymbol{\beta},\mu}$, we define its local discrete counterpart $\mathrm{a}_{\boldsymbol{\beta},\mu,T} : \underline{U}^k_T \times \underline{U}^k_T \to \mathbb{R}$ based on the reconstructed advective derivative $\mathrm{G}^k_{\boldsymbol{\beta},T}$ as follows: for all $\underline{u}_T, \underline{v}_T \in \underline{U}^k_T$,

$$\begin{aligned}
\mathrm{a}_{\boldsymbol{\beta},\mu,T}(\underline{u}_T, \underline{v}_T) &:= \frac{1}{2}(\mathrm{G}^k_{\boldsymbol{\beta},T}\underline{u}_T, v_T)_T - \frac{1}{2}(u_T, \mathrm{G}^k_{\boldsymbol{\beta},T}\underline{v}_T)_T \\
&\quad + \left(\left[\tfrac{1}{2}\nabla\cdot\boldsymbol{\beta} + \mu\right] u_T, v_T\right)_T + \frac{1}{2}\mathrm{s}_{\boldsymbol{\beta},T}(\underline{u}_T, \underline{v}_T),
\end{aligned} \tag{3.72}$$

where the bilinear form $\mathrm{s}_{\boldsymbol{\beta},T} : \underline{U}^k_T \times \underline{U}^k_T \to \mathbb{R}$ is such that

$$\mathrm{s}_{\boldsymbol{\beta},T}(\underline{u}_T, \underline{v}_T) := \sum_{F\in\mathcal{F}_T} (|\boldsymbol{\beta}\cdot\boldsymbol{n}_{TF}|(u_F - u_T), v_F - v_T)_F. \tag{3.73}$$

This form can be interpreted as an upwind stabilisation term; see Remark 3.31.

Remark 3.26 (Element-Face Upwind Stabilisation) Upwinding is realised in (3.73) by penalizing the difference between face- and element-based discrete unknowns. This is a relevant difference with respect to classical (element-based) Finite Volumes and Discontinuous Galerkin methods, where jumps of element-based discrete unknowns are considered instead (see, e.g., [145, Chapter 2] and [84] for an interpretation of upwind stabilisation as a jump penalisation). With the choice (3.73) for the stabilisation term, the stencil remains the same as for a pure diffusion problem, and static condensation of element-based discrete unknowns is possible, in the spirit of Sect. B.3.2. In the context of the low-order Hybrid Mimetic Mixed methods, face-element upwind terms have been considered in [44], and shown on numerical examples to be more accurate, in the advection-dominated regime, than element-based upwinding.

To express the stability properties of $\mathrm{a}_{\boldsymbol{\beta},\mu,T}$, we define the local seminorm such that, for all $\underline{v}_T \in \underline{U}_T^k$,

$$\|\underline{v}_T\|_{\boldsymbol{\beta},\mu,T}^2 := \frac{1}{2} \sum_{F\in\mathcal{F}_T} \||\boldsymbol{\beta}\cdot\boldsymbol{n}_{TF}|^{\frac{1}{2}}(v_F - v_T)\|_F^2 + \mu_0\|v_T\|_T^2. \tag{3.74}$$

Notice that the map $\|\cdot\|_{\boldsymbol{\beta},\mu,T}$ is actually a norm on $\underline{U}_T^k$ provided that, for each $F \in \mathcal{F}_T$, $\boldsymbol{\beta}\cdot\boldsymbol{n}_{TF}$ is non-zero on a set of positive $(d-1)$-measure in F. Letting $\underline{u}_T = \underline{v}_T$ in (3.72) and recalling that $\frac{1}{2}\nabla\cdot\boldsymbol{\beta} + \mu \geq \mu_0 > 0$ (see Assumption 3.22) yields the following coercivity property:

$$\|\underline{v}_T\|_{\boldsymbol{\beta},\mu,T}^2 \leq \mathrm{a}_{\boldsymbol{\beta},\mu,T}(\underline{v}_T, \underline{v}_T) \qquad \forall \underline{v}_T \in \underline{U}_T^k. \tag{3.75}$$

3.2.1.3 Global Advective–Reactive Bilinear Form

The global advective–reactive bilinear form $\mathrm{a}_{\boldsymbol{\beta},\mu,h}(\underline{u}_h, \underline{v}_h) : \underline{U}_h^k \times \underline{U}_h^k \to \mathbb{R}$ is obtained by assembling the elementary contributions in the usual way: For all $\underline{u}_h, \underline{v}_h \in \underline{U}_h^k$,

$$\mathrm{a}_{\boldsymbol{\beta},\mu,h}(\underline{u}_h, \underline{v}_h) := \sum_{T\in\mathcal{T}_h} \mathrm{a}_{\boldsymbol{\beta},\mu,T}(\underline{u}_T, \underline{v}_T). \tag{3.76}$$

Expanding the definition (3.72) of $\mathrm{a}_{\boldsymbol{\beta},\mu,T}$ in the previous expression leads to the formula

$$\begin{aligned}\mathrm{a}_{\boldsymbol{\beta},\mu,h}(\underline{u}_h, \underline{v}_h) &= \frac{1}{2} \sum_{T\in\mathcal{T}_h} (\mathrm{G}_{\boldsymbol{\beta},T}^k \underline{u}_T, v_T)_T - \frac{1}{2} \sum_{T\in\mathcal{T}_h} (u_T, \mathrm{G}_{\boldsymbol{\beta},T}^k \underline{v}_T)_T \\ &\quad + \sum_{T\in\mathcal{T}_h} \left(\left[\tfrac{1}{2}\nabla\cdot\boldsymbol{\beta} + \mu\right] u_T, v_T\right)_T + \frac{1}{2} \sum_{T\in\mathcal{T}_h} \mathrm{s}_{\boldsymbol{\beta},T}(\underline{u}_T, \underline{v}_T).\end{aligned} \tag{3.77}$$

Using the discrete integration by parts formula (3.70) to substitute the first (resp. second) term in this relation, we obtain the following two equivalent reformulations of $\mathrm{a}_{\boldsymbol{\beta},\mu,h}$ on $\underline{U}_{h,0}^k \times \underline{U}_{h,0}^k$: For any $\underline{u}_h, \underline{v}_h \in \underline{U}_{h,0}^k$

$$
\begin{aligned}
\mathrm{a}_{\boldsymbol{\beta},\mu,h}(\underline{u}_h, \underline{v}_h) = &- \sum_{T\in\mathcal{T}_h} (u_T, \mathrm{G}_{\boldsymbol{\beta},T}^k \underline{v}_T)_T + \sum_{T\in\mathcal{T}_h} \mathrm{s}_{\boldsymbol{\beta},T}^-(\underline{u}_T, \underline{v}_T) \\
&+ \sum_{T\in\mathcal{T}_h} (\mu u_T, v_T)_T,
\end{aligned}
\tag{3.78}
$$

and

$$
\begin{aligned}
\mathrm{a}_{\boldsymbol{\beta},\mu,h}(\underline{u}_h, \underline{v}_h) = &\sum_{T\in\mathcal{T}_h} (\mathrm{G}_{\boldsymbol{\beta},T}^k \underline{u}_T, v_T)_T + \sum_{T\in\mathcal{T}_h} \mathrm{s}_{\boldsymbol{\beta},T}^+(\underline{u}_T, \underline{v}_T) \\
&+ \sum_{T\in\mathcal{T}_h} ([\nabla\cdot\boldsymbol{\beta} + \mu]\, u_T, v_T)_T,
\end{aligned}
\tag{3.79}
$$

where, denoting by $x^{\pm} := \frac{1}{2}(|x| \pm x) = \max(\pm x, 0)$ the positive and negative parts of a real number x, we have introduced the local bilinear forms $\mathrm{s}_{\boldsymbol{\beta},T}^{\pm} : \underline{U}_T^k \times \underline{U}_T^k \to \mathbb{R}$ such that

$$
\mathrm{s}_{\boldsymbol{\beta},T}^{\pm}(\underline{u}_T, \underline{v}_T) := \sum_{F\in\mathcal{F}_T} ((\boldsymbol{\beta}\cdot\boldsymbol{n}_{TF})^{\pm}(u_F - u_T), v_F - v_T)_F. \tag{3.80}
$$

The formulation (3.77) decomposes the global advective–reactive bilinear form into a skew-symmetric part (first line) and a symmetric semidefinite positive part (second line), and is adapted to studying its stability properties (as already seen in (3.75) for the local bilinear form). The formulation (3.78), on the other hand, clearly separates the advective component (first line) and the reactive component (second line), and is appropriate for the consistency analysis. Before proceeding, a few remarks are in order.

Remark 3.27 (Comparison with [159]) The formulation (3.78) corresponds to [159, Eq. (16)], when the upwind stabilisation discussed in Sect. 4.2 therein is used, and the boundary conditions are enforced strongly. This formulation also has a more familiar look for the reader accustomed to upwind stabilisation terms; see also Remark 3.31 on this subject.

Remark 3.28 (Other Approaches for the Discretisation of Convective Terms) Approaches different from the one proposed here can be found in the literature on hybrid and polytopal element methods. In [126], the authors devise and numerically investigate a Hybridisable Discontinuous Galerkin method for the diffusion-dominated regime based on a mixed formulation where an approximation for the total advective-diffusive flux is sought. A convergence analysis for a variable degree Hybridisable Discontinuous Galerkin method on semimatching nonconforming simplicial meshes is carried out in [108], where the impact of mesh nonconformity on the supercloseness of the potential is also investigated. The formulation differs from [126] in that the flux variable approximates the diffusive component only. In the context of Virtual Element Methods, diffusion-advection problems are considered in [50], where the discretisation of the advective terms hinges on a projection of the gradient of virtual functions on full polynomial spaces. Mixed Virtual Element Methods, on the other hand, are considered in [49]. In both cases, the analysis is mainly tailored to the diffusion-dominated regime.

Define the global advective–reactive seminorm such that, for all $\underline{v}_h \in \underline{U}_h^k$,

$$\|\underline{v}_h\|_{\boldsymbol{\beta},\mu,h} := \left(\sum_{T\in\mathcal{T}_h} \|\underline{v}_T\|_{\boldsymbol{\beta},\mu,T}^2 \right)^{\frac{1}{2}}.$$

In a similar way as Lemmas 2.18 and 3.15 for the diffusion bilinear forms, we now prove stability and consistency properties for $\mathrm{a}_{\boldsymbol{\beta},\mu,h}$. The consistency estimates are established in a norm that gathers this advective–reactive norm and the $\boldsymbol{K}$-weighted diffusive norm defined in (3.46):

$$\|\underline{v}_h\|_{\flat,h} := \left(\|\underline{v}_h\|_{\mathrm{a},\boldsymbol{K},h}^2 + \|\underline{v}_h\|_{\boldsymbol{\beta},\mu,h}^2 \right)^{\frac{1}{2}}. \tag{3.81}$$

Note that $\|\cdot\|_{\flat,h}$ is indeed a norm on $\underline{U}_{h,0}^k$, not just a semi-norm, since $\|\cdot\|_{\mathrm{a},\boldsymbol{K},h}$ is a norm on this space. In order to state consistency estimates that are robust across the various possible regimes (diffusion-dominated, advection-dominated, or in between), we introduce, for each mesh element $T \in \mathcal{T}_h$, the local Péclet number such that

$$\mathrm{Pe}_T := \max_{F\in\mathcal{F}_T} \frac{h_F \|\boldsymbol{\beta}\cdot\boldsymbol{n}_{TF}\|_{L^\infty(F)}}{K_{TF}}. \tag{3.82}$$

For the mesh elements where diffusion dominates we have $\mathrm{Pe}_T \le h_T$, for those where advection dominates we have $\mathrm{Pe}_T \ge 1$, while intermediate regimes correspond to $\mathrm{Pe}_T \in (h_T, 1)$.

In the statement of the following lemma, we also make use of the quantity

$$\hat{\tau}_T := \frac{1}{\max(\|\mu\|_{L^\infty(T)}, \mathrm{L}_{\boldsymbol{\beta},T})}, \tag{3.83}$$

where $\mathrm{L}_{\boldsymbol{\beta},T}$ is the Lipschitz constant of $\boldsymbol{\beta}_{|T}$. If the steady model (3.60) is regarded as a time-stepping for a transient diffusive–advective–reactive equation, then $\hat{\tau}_T$ has the dimension of a time and can be interpreted as a reference time.

Lemma 3.29 (Properties of $\mathrm{a}_{\beta,\mu,h}$) *The bilinear form $\mathrm{a}_{\boldsymbol{\beta},\mu,h}$ enjoys the following properties:*

(i) Stability. *For all $\underline{v}_h \in \underline{U}_h^k$ it holds*

$$\|\underline{v}_h\|_{\boldsymbol{\beta},\mu,h}^2 \le \mathrm{a}_{\boldsymbol{\beta},\mu,h}(\underline{v}_h, \underline{v}_h). \tag{3.84}$$

(ii) Consistency. *It holds for all $r \in \{0, \ldots, k\}$ and all $w \in H_0^1(\Omega) \cap H^{r+2}(\mathcal{T}_h)$,*

$$\sup_{\underline{v}_h \in \underline{U}_{h,0}^k, \|\underline{v}_h\|_{\flat,h}=1} |\mathcal{E}_{\boldsymbol{\beta},\mu,h}(w; \underline{v}_h)|$$
$$\lesssim \left\{ \sum_{T\in\mathcal{T}_h} \hat{\tau}_T^{-2} \mu_0^{-1} h_T^{2(r+1)} |w|_{H^{r+1}(T)}^2 + \hat{\beta}_T \left[\min(1, \mathrm{Pe}_T)^{\frac{1}{2}} h_T^{r+\frac{1}{2}} \right]^2 |w|_{H^{r+1}(T)}^2 \right\}^{\frac{1}{2}}, \tag{3.85}$$

where the hidden constant is independent of h, w, r, $\boldsymbol{\beta}$ and μ, and the linear form $\mathcal{E}_{\boldsymbol{\beta},\mu,h}(w; \cdot) : \underline{U}_{h,0}^k \to \mathbb{R}$ representing the consistency error is such that, for all $\underline{v}_h \in \underline{U}_{h,0}^k$,

$$\mathcal{E}_{\boldsymbol{\beta},\mu,h}(w; \underline{v}_h) := (\nabla\cdot(\boldsymbol{\beta} w) + \mu w, v_h) - \mathrm{a}_{\boldsymbol{\beta},\mu,h}(\underline{I}_h^k w, \underline{v}_h).$$

Proof

(i) *Stability.* The stability property (3.84) is obtained summing up (3.75) over $T \in \mathcal{T}_h$.

(ii) *Consistency.* To prove the consistency estimate, we split the consistency error in its reactive and advective components, using the representation (3.78) of $\mathrm{a}_{\boldsymbol{\beta},\mu,h}(\underline{I}_h^k w, \underline{v}_h)$:

$$\mathcal{E}_{\boldsymbol{\beta},\mu,h}(w; \underline{v}_h) = \mathcal{E}_{\mu,h}(w; \underline{v}_h) + \mathcal{E}_{\boldsymbol{\beta},h}(w; \underline{v}_h)$$

with, setting $\underline{\hat{w}}_h := \underline{I}_h^k w$,

$$\mathcal{E}_{\mu,h}(w; \underline{v}_h) := (\mu w, v_h) - \sum_{T\in\mathcal{T}_h} (\mu \hat{w}_T, v_T)_T,$$
$$\mathcal{E}_{\boldsymbol{\beta},h}(w; \underline{v}_h) := (\nabla\cdot(\boldsymbol{\beta} w), v_h) + \sum_{T\in\mathcal{T}_h} (\hat{w}_T, \mathrm{G}_{\boldsymbol{\beta},T}^k \underline{v}_T)_T - \sum_{T\in\mathcal{T}_h} \mathrm{s}_{\boldsymbol{\beta},T}^-(\underline{\hat{w}}_T, \underline{v}_T). \tag{3.86}$$

Let us first deal with the reactive consistency error. For any $\underline{v}_h \in \underline{U}_{h,0}^k$, by definition (2.33) of v_h, we have $(\mu w, v_h) = \sum_{T\in\mathcal{T}_h}(\mu w, v_T)_T$ and thus

$$|\mathcal{E}_{\mu,h}(w; \underline{v}_h)| = \left|\sum_{T\in\mathcal{T}_h}(\mu(w - \pi_T^{0,k} w), v_T)_T\right|$$
$$\lesssim \sum_{T\in\mathcal{T}_h} \|\mu\|_{L^\infty(T)} h_T^{r+1} |w|_{H^{r+1}(T)} \mu_0^{-\frac{1}{2}} \|\underline{v}_T\|_{\boldsymbol{\beta},\mu,T},$$

where the estimate follows from a generalised Hölder inequality with exponents $(\infty, 2, 2)$, the approximation properties (1.74) of the L^2-orthogonal projector (with $l = k$, $p = 2$, $m = 0$, and $s = r + 1$), and the definition (3.74) of $\|\cdot\|_{\boldsymbol{\beta},\mu,T}$. Since $\|\mu\|_{L^\infty(T)} \le \hat{\tau}_T^{-1}$ (cf. (3.83)), a Cauchy–Schwarz inequality on the sum over $T \in \mathcal{T}_h$ yields

$$|\mathcal{E}_{\mu,h}(w; \underline{v}_h)| \lesssim \left(\sum_{T\in\mathcal{T}_h} \hat{\tau}_T^{-2} \mu_0^{-1} h_T^{2(r+1)} |w|^2_{H^{r+1}(T)}\right)^{\frac{1}{2}} \|\underline{v}_h\|_{\boldsymbol{\beta},\mu,h}. \tag{3.87}$$

Let us now turn to $\mathcal{E}_{\boldsymbol{\beta},h}(w; \underline{v}_h)$. The definition (3.64) of $\mathrm{G}_{\boldsymbol{\beta},T}^k \underline{v}_T$ (with $w = \hat{w}_T$) and element-wise integrations by parts yield

$$\begin{aligned}
&(\nabla\cdot(\boldsymbol{\beta} w), v_h) + \sum_{T\in\mathcal{T}_h}(\hat{w}_T, \mathrm{G}_{\boldsymbol{\beta},T}^k \underline{v}_T)_T \\
&\quad= \sum_{T\in\mathcal{T}_h}\left((\nabla\cdot(\boldsymbol{\beta} w), v_T)_T - (\nabla\cdot(\boldsymbol{\beta}\hat{w}_T), v_T)_T + \sum_{F\in\mathcal{F}_T}(\hat{w}_T, (\boldsymbol{\beta}\cdot\boldsymbol{n}_{TF})v_F)_F\right) \\
&\quad= \sum_{T\in\mathcal{T}_h}(\hat{w}_T - w, \boldsymbol{\beta}\cdot\nabla v_T)_T \\
&\qquad+ \sum_{T\in\mathcal{T}_h}\sum_{F\in\mathcal{F}_T}\Big[((\boldsymbol{\beta}\cdot\boldsymbol{n}_{TF})(w - \hat{w}_T), v_T)_F + (\hat{w}_T, (\boldsymbol{\beta}\cdot\boldsymbol{n}_{TF})v_F)_F\Big].
\end{aligned} \tag{3.88}$$

Corollary 1.19 with $\boldsymbol{\tau} = \boldsymbol{\beta}$ and $(\varphi_F)_{F\in\mathcal{F}_h} = (v_F w)_{F\in\mathcal{F}_h}$ gives

$$\sum_{T\in\mathcal{T}_h}\sum_{F\in\mathcal{F}_T}(w, (\boldsymbol{\beta}\cdot\boldsymbol{n}_{TF})v_F)_F = 0.$$

Subtracting this quantity from (3.88) and combining it with the last addend in this equation leads to

$$(\nabla\cdot(\boldsymbol{\beta}w), v_h) + \sum_{T\in\mathcal{T}_h}(\hat{w}_T, \mathrm{G}^k_{\boldsymbol{\beta},T}\underline{v}_T)_T$$
$$= \sum_{T\in\mathcal{T}_h}(\hat{w}_T - w, \boldsymbol{\beta}\cdot\nabla v_T)_T + \sum_{T\in\mathcal{T}_h}\sum_{F\in\mathcal{F}_T}((\boldsymbol{\beta}\cdot\boldsymbol{n}_{TF})(w-\hat{w}_T), v_T - v_F)_F.$$

Hence, recalling the definition (3.80) of $\mathrm{s}^-_{\boldsymbol{\beta},T}$,

$$\begin{aligned}\mathcal{E}_{\boldsymbol{\beta},h}(w;\underline{v}_h) &= \left.\sum_{T\in\mathcal{T}_h}(\hat{w}_T - w, \boldsymbol{\beta}\cdot\nabla v_T)_T\right\}\mathfrak{T}_1(\underline{v}_h)\\ &\left.\begin{aligned}&+\sum_{T\in\mathcal{T}_h}\sum_{F\in\mathcal{F}_T}((\boldsymbol{\beta}\cdot\boldsymbol{n}_{TF})(w-\hat{w}_T), v_T - v_F)_F\\ &-\sum_{T\in\mathcal{T}_h}\sum_{F\in\mathcal{F}_T}((\boldsymbol{\beta}\cdot\boldsymbol{n}_{TF})^-(\hat{w}_F-\hat{w}_T), v_F - v_T)_F.\end{aligned}\right\}\mathfrak{T}_2(\underline{v}_h).\end{aligned}\tag{3.89}$$

To estimate $\mathfrak{T}_1(\underline{v}_h)$, observe that $(\pi_T^{0,0}\boldsymbol{\beta})\cdot\nabla v_T \in \mathbb{P}^{k-1}(T) \subset \mathbb{P}^k(T)$ and recall the orthogonality property (1.57) of $\pi_T^{0,k}$ to write, since $\hat{w}_T = \pi_T^{0,k}w$,

$$\mathfrak{T}_1(\underline{v}_h) = \sum_{T\in\mathcal{T}_h}(\pi_T^{0,k}w - w, (\boldsymbol{\beta} - \pi_T^{0,0}\boldsymbol{\beta})\cdot\nabla v_T)_T.$$

Hence, using a generalised Hölder inequality with exponents $(2, \infty, 2)$, we get

$$\begin{aligned}|\mathfrak{T}_1(\underline{v}_h)| &\lesssim \sum_{T\in\mathcal{T}_h}\|w - \pi_T^{0,k}w\|_T\|\boldsymbol{\beta} - \pi_T^{0,0}\boldsymbol{\beta}\|_{L^\infty(T)^d}\|\nabla v_T\|_T\\ &\lesssim \sum_{T\in\mathcal{T}_h} h_T^{r+1}|w|_{H^{r+1}(T)}\hat{\tau}_T^{-1}h_T\|\nabla v_T\|_T,\end{aligned}\tag{3.90}$$

where the second inequality is obtained using the fact that $\boldsymbol{\beta}$ is Lipschitz continuous (see Assumption 3.22) together with the definition (3.83) of $\hat{\tau}_T$ to infer $\|\boldsymbol{\beta} - \pi_T^{0,0}\boldsymbol{\beta}\|_{L^\infty(T)^d} \lesssim \mathrm{L}_{\boldsymbol{\beta},T}h_T \lesssim \hat{\tau}_T^{-1}h_T$, along with the approximation properties (1.74) of the L^2-orthogonal projector with $l = k$, $p = 2$, $s = r+1$, and $m = 0$. The inverse inequality (1.46) yields $h_T \|\nabla v_T\|_T \lesssim \|v_T\|_T$ so, by a Cauchy–Schwarz inequality on the sum over $T \in \mathcal{T}_h$ and the definition (3.74) of $\|\cdot\|_{\boldsymbol{\beta},\mu,T}$,

$$
\begin{aligned}
|\mathfrak{T}_1(\underline{v}_h)| &\lesssim \left(\sum_{T\in\mathcal{T}_h} \hat{\tau}_T^{-2}\mu_0^{-1}h_T^{2(r+1)}|w|^2_{H^{r+1}(T)}\right)^{\frac{1}{2}} \left(\sum_{T\in\mathcal{T}_h} \mu_0\|v_T\|_T^2\right)^{\frac{1}{2}} \\
&\lesssim \left(\sum_{T\in\mathcal{T}_h} \hat{\tau}_T^{-2}\mu_0^{-1}h_T^{2(r+1)}|w|^2_{H^{r+1}(T)}\right)^{\frac{1}{2}} \|\underline{v}_h\|_{\boldsymbol{\beta},\mu,h}. \qquad (3.91)
\end{aligned}
$$

Let us now turn to $\mathfrak{T}_2(\underline{v}_h)$. We first observe that, for all $T \in \mathcal{T}_h$ and all $F \in \mathcal{F}_T$, the following holds:

$$
\|\hat{w}_F - \hat{w}_T\|_F = \|\pi_F^{0,k}(w - \hat{w}_T)\|_F \le \|w - \hat{w}_T\|_F \lesssim h_T^{r+\frac{1}{2}}|w|_{H^{r+1}(T)}, \qquad (3.92)
$$

where we have used the fact that $\pi_F^{0,k}$ is linear and idempotent, and invoked the boundedness property (1.72) (with $X = F$ and $\mathcal{P} = \mathbb{P}^k(F)$) of $\pi_F^{0,k}$ and the trace approximation property (1.75) of $\pi_T^{0,k}$ with $l = k$, $p = 2$, $s = r + 1$ and $m = 0$. This estimate shows that

$$
\begin{aligned}
|\mathfrak{T}_2(\underline{v}_h)| &\lesssim \sum_{T\in\mathcal{T}_h}\sum_{F\in\mathcal{F}_T} h_T^{r+\frac{1}{2}}|w|_{H^{r+1}(T)}\||\boldsymbol{\beta}\cdot\boldsymbol{n}_{TF}|(v_F - v_T)\|_F \qquad (3.93) \\
&\lesssim \sum_{T\in\mathcal{T}_h,\mathrm{Pe}_T\le 1}(\cdots) + \sum_{T\in\mathcal{T}_h,\mathrm{Pe}_T> 1}(\cdots) =: \mathfrak{T}_{2,\mathrm{d}}(\underline{v}_h) + \mathfrak{T}_{2,\mathrm{a}}(\underline{v}_h). \qquad (3.94)
\end{aligned}
$$

The quantity $\||\boldsymbol{\beta}\cdot\boldsymbol{n}_{TF}|(v_F - v_T)\|_F$ in these terms is estimated using two different norms: the diffusive norm for $\mathfrak{T}_{2,\mathrm{d}}(\underline{v}_h)$, the advective–reactive (semi)norm for $\mathfrak{T}_{2,\mathrm{a}}(\underline{v}_h)$.

Let us start with $\mathfrak{T}_{2,\mathrm{d}}(\underline{v}_h)$. The definitions (3.82) and (3.66) of Pe_T and $\hat{\beta}_T$ show that, if $\mathrm{Pe}_T \le 1$ (so that $\mathrm{Pe}_T = \min(1, \mathrm{Pe}_T)$), a.e. on $F \in \mathcal{F}_T$ it holds

$$
|\boldsymbol{\beta}\cdot\boldsymbol{n}_{TF}| = |\boldsymbol{\beta}\cdot\boldsymbol{n}_{TF}|^{\frac{1}{2}}|\boldsymbol{\beta}\cdot\boldsymbol{n}_{TF}|^{\frac{1}{2}} \le \hat{\beta}_T^{\frac{1}{2}}\left(\mathrm{Pe}_T\frac{K_{TF}}{h_F}\right)^{\frac{1}{2}} = (\hat{\beta}_T\min(1,\mathrm{Pe}_T))^{\frac{1}{2}}\left(\frac{K_{TF}}{h_F}\right)^{\frac{1}{2}}.
$$

Hence,

$$
\begin{aligned}
&\mathfrak{T}_{2,\mathrm{d}}(\underline{v}_h) \\
&\lesssim \sum_{T\in\mathcal{T}_h}\sum_{F\in\mathcal{F}_T}(\hat{\beta}_T\min(1,\mathrm{Pe}_T))^{\frac{1}{2}}h_T^{r+\frac{1}{2}}|w|_{H^{r+1}(T)}\left(\frac{K_{TF}}{h_F}\right)^{\frac{1}{2}}\|v_F-v_T\|_F \\
&\lesssim\left(\sum_{T\in\mathcal{T}_h}\hat{\beta}_T\min(1,\mathrm{Pe}_T)h_T^{2(r+\frac{1}{2})}|w|^2_{H^{r+1}(T)}\right)^{\frac{1}{2}}\left(\sum_{T\in\mathcal{T}_h}\sum_{F\in\mathcal{F}_T}\frac{K_{TF}}{h_F}\|v_F-v_T\|_F^2\right)^{\frac{1}{2}} \\
&\le\left(\sum_{T\in\mathcal{T}_h}\hat{\beta}_T\min(1,\mathrm{Pe}_T)h_T^{2(r+\frac{1}{2})}|w|^2_{H^{r+1}(T)}\right)^{\frac{1}{2}}\|\underline{v}_h\|_{1,\boldsymbol{K},h},
\end{aligned}
\tag{3.95}
$$

where the second bound follows from a Cauchy–Schwarz inequality on the sums over $T\in\mathcal{T}_h$ and $F\in\mathcal{F}_T$ along with $\mathrm{card}(\mathcal{F}_T)\lesssim 1$ (see (1.5)), and the conclusion is a consequence of the definitions (3.45) of $\|{\cdot}\|_{1,\boldsymbol{K},h}$ and (3.25) of $\|{\cdot}\|_{1,\boldsymbol{K},T}$.

To estimate $\mathfrak{T}_{2,\mathrm{a}}(\underline{v}_h)$ we simply observe that, whenever $\mathrm{Pe}_T>1$ (so that $1=\min(1,\mathrm{Pe}_T)$), a.e. on $F\in\mathcal{F}_T$ it holds

$$
|\boldsymbol{\beta}{\cdot}\boldsymbol{n}_{TF}|\le\hat{\beta}_T^{\frac{1}{2}}|\boldsymbol{\beta}{\cdot}\boldsymbol{n}_{TF}|^{\frac{1}{2}}=(\hat{\beta}_T\min(1,\mathrm{Pe}_T))^{\frac{1}{2}}|\boldsymbol{\beta}{\cdot}\boldsymbol{n}_{TF}|^{\frac{1}{2}}
$$

and thus, by the Cauchy–Schwarz inequality and the definition (3.74) of $\|{\cdot}\|_{\boldsymbol{\beta},\mu,T}$,

$$
\begin{aligned}
&\mathfrak{T}_{2,\mathrm{a}}(\underline{v}_h) \\
&\lesssim\sum_{T\in\mathcal{T}_h}\sum_{F\in\mathcal{F}_T}(\hat{\beta}_T\min(1,\mathrm{Pe}_T))^{\frac{1}{2}}h_T^{r+\frac{1}{2}}|w|_{H^{r+1}(T)}\||\boldsymbol{\beta}{\cdot}\boldsymbol{n}_{TF}|^{\frac{1}{2}}(v_F-v_T)\|_F \\
&\lesssim\left(\sum_{T\in\mathcal{T}_h}\hat{\beta}_T\min(1,\mathrm{Pe}_T)h_T^{2(r+\frac{1}{2})}|w|^2_{H^{r+1}(T)}\right)^{\frac{1}{2}}\|\underline{v}_h\|_{\boldsymbol{\beta},\mu,h}.
\end{aligned}
\tag{3.96}
$$

Gathering the above estimates on $\mathfrak{T}_{2,\mathrm{d}}(\underline{v}_h)$ and $\mathfrak{T}_{2,\mathrm{a}}(\underline{v}_h)$ into (3.94), and plugging the resulting estimate on $\mathfrak{T}_2(\underline{v}_h)$ together with (3.91) into (3.89), we obtain, recalling the definition (3.81) of $\|{\cdot}\|_{\flat,h}$ and using $a^{\frac{1}{2}}+b^{\frac{1}{2}}\le 2(a+b)^{\frac{1}{2}}$,

$$
\begin{aligned}
&|\mathcal{E}_{\boldsymbol{\beta},h}(w;\underline{v}_h)| \\
&\lesssim\left(\sum_{T\in\mathcal{T}_h}\hat{\tau}_T^{-2}\mu_0^{-1}h_T^{2(r+1)}|w|^2_{H^{r+1}(T)}+\hat{\beta}_T\min(1,\mathrm{Pe}_T)h_T^{2(r+\frac{1}{2})}|w|^2_{H^{r+1}(T)}\right)^{\frac{1}{2}}\|\underline{v}_h\|_{\flat,h}.
\end{aligned}
$$

The proof of (3.85) is completed by summing up this estimate and (3.87), and by taking the supremum over $\underline{v}_h \in \underline{U}_{h,0}^k$ such that $\|\underline{v}_h\|_{\flat,h} = 1$. □

3.2.2 *Discrete Problem and Initial Convergence Result*

We formulate in this section the discrete problem, discuss local conservation, and prove an initial convergence result.

3.2.2.1 Discrete Problem

We define the global bilinear form $\mathrm{a}_{\boldsymbol{K},\boldsymbol{\beta},\mu,h} : \underline{U}_h^k \times \underline{U}_h^k \to \mathbb{R}$ combining the diffusive and advective–reactive contributions:

$$\mathrm{a}_{\boldsymbol{K},\boldsymbol{\beta},\mu,h}(\underline{u}_h, \underline{v}_h) := \mathrm{a}_{\boldsymbol{K},h}(\underline{u}_h, \underline{v}_h) + \mathrm{a}_{\boldsymbol{\beta},\mu,h}(\underline{u}_h, \underline{v}_h), \tag{3.97}$$

where $\mathrm{a}_{\boldsymbol{K},h}$ is defined by (3.44) with local contributions given by (3.26), while $\mathrm{a}_{\boldsymbol{\beta},\mu,h}$ is defined by (3.76) with local contributions given by (3.72). The HHO approximation of problem (3.61) then reads: Find $\underline{u}_h \in \underline{U}_{h,0}^k$ such that, for all $\underline{v}_h \in \underline{U}_{h,0}^k$,

$$\mathrm{a}_{\boldsymbol{K},\boldsymbol{\beta},\mu,h}(\underline{u}_h, \underline{v}_h) = (f, v_h), \tag{3.98}$$

where we remind the reader that the broken polynomial function $v_h \in \mathbb{P}^k(\mathcal{T}_h)$ is obtained from $\underline{v}_h$ setting $(v_h)_{|T} := v_T$ for all $T \in \mathcal{T}_h$.

The well-posedness of this problem, together with estimates on the solution in $\|\cdot\|_{\flat,h}$ norm, follows easily from the stability property (3.84). We leave the details as an exercise to the reader.

3.2.2.2 Flux Formulation

The following lemma shows that the solution to the HHO scheme (3.98) satisfies local balances inside each elements, with numerical fluxes that have continuous normal trace across interfaces.

Lemma 3.30 (Flux Formulation) *Let the assumptions and notations of Lemma 3.17 hold, and further suppose that Assumption 3.22 is verified. Let* $\underline{u}_h \in \underline{U}^k_{h,0}$. *For all* $T \in \mathcal{T}_h$ *and all* $F \in \mathcal{F}_T$, *let the normal trace of the diffusive flux* $\Phi_{K,TF}(\underline{u}_T)$ *be defined by (3.54), and additionally define the normal trace of the advective flux as follows:*

$$\Phi_{\beta,TF}(\underline{u}_T) := \pi_F^{0,k}\left((\beta\cdot n_{TF})^+ u_T - (\beta\cdot n_{TF})^- u_F\right). \tag{3.99}$$

Then, $\underline{u}_h$ *is the unique solution of problem (3.98) if and only if the following two properties hold:*

(i) Local balance. *For all* $T \in \mathcal{T}_h$ *and all* $v_T \in \mathbb{P}^k(T)$,

$$\begin{aligned}(K_T \nabla \mathrm{p}^{k+1}_{K,T}\underline{u}_T, \nabla v_T)_T - (u_T, \beta\cdot\nabla v_T)_T + (\mu u_T, v_T)_T \\ + \sum_{F\in\mathcal{F}_T} (\Phi_{K,TF}(\underline{u}_T) + \Phi_{\beta,TF}(\underline{u}_T), v_T)_F = (f, v_T)_T.\end{aligned} \tag{3.100}$$

(ii) Continuity of the numerical normal traces of the fluxes. *For any interface* $F \in \mathcal{F}_h^{\mathrm{i}}$ *such that* $F \subset \partial T_1 \cap \partial T_2$ *with distinct mesh elements* $T_1, T_2 \in \mathcal{T}_h$, *it holds*

$$\left(\Phi_{K,T_1F}(\underline{u}_{T_1}) + \Phi_{\beta,T_1F}(\underline{u}_{T_1})\right) + \left(\Phi_{K,T_2F}(\underline{u}_{T_2}) + \Phi_{\beta,T_2F}(\underline{u}_{T_2})\right) = 0. \tag{3.101}$$

Remark 3.31 (Upwind Stabilisation) The expression (3.99) of the advective flux reveals that the stabilisation term introduces upwinding in the scheme. As a matter of fact, (3.99) is equivalent to

$$\Phi_{\beta,TF}(\underline{u}_T) = \pi_F^{0,k}\left((\beta\cdot n_{TF})u_{TF}^{\uparrow}\right) \text{ with } u_{TF}^{\uparrow} := \begin{cases} u_T & \text{if } \beta\cdot n_{TF} \geq 0, \\ u_F & \text{otherwise.}\end{cases}$$

Here, $u_{TF}^{\uparrow}$ represents the upwind value of the advected quantity u: if the flow exits T (that is, recalling that n_{TF} points out of T, $\beta\cdot n_{TF} \geq 0$), $u_{TF}^{\uparrow}$ is equal to the trace of u_T on F; if, on the other hand, the flow enters T (that is, $\beta\cdot n_{TF} < 0$), $u_{TF}^{\uparrow}$ is equal to the face value u_F.

Proof We use Lemma 2.21 by showing that the bilinear form $\mathrm{a}_{\boldsymbol{K},\boldsymbol{\beta},\mu,h}$ defined by (3.97) admits the reformulation (2.51).

Working as in the proof of Lemma 2.25, we can write for the diffusive bilinear form defined by (3.44):

$$\begin{aligned}\mathrm{a}_{\boldsymbol{K},h}(\underline{u}_h,\underline{v}_h) = &\sum_{T\in\mathcal{T}_h}(\boldsymbol{K}_T\nabla \mathrm{p}^{k+1}_{\boldsymbol{K},T}\underline{u}_T,\nabla v_T)_T\\&-\sum_{T\in\mathcal{T}_h}\sum_{F\in\mathcal{F}_T}(\Phi_{\boldsymbol{K},TF}(\underline{u}_T),v_F-v_T)_F.\end{aligned}\tag{3.102}$$

Let us now reformulate the advection–reaction bilinear form. The starting point is (3.79), which we recall here for the sake of readability:

$$\begin{aligned}\mathrm{a}_{\boldsymbol{\beta},\mu,h}(\underline{u}_h,\underline{v}_h) &= \sum_{T\in\mathcal{T}_h}(\mathrm{G}^k_{\boldsymbol{\beta},T}\underline{u}_T,v_T)_T+\sum_{T\in\mathcal{T}_h}\mathrm{s}^+_{\boldsymbol{\beta},T}(\underline{u}_T,\underline{v}_T)+\sum_{T\in\mathcal{T}_h}([\nabla{\cdot}\boldsymbol{\beta}+\mu]u_T,v_T)_T\\&=: \mathfrak{T}_1+\mathfrak{T}_2+\mathfrak{T}_3.\end{aligned}$$

For the sum of the first and third term, expanding, for all $T\in\mathcal{T}_h$, $\mathrm{G}^k_{\boldsymbol{\beta},T}$ according to its definition (3.64) with $\underline{v}_T=\underline{u}_T$ and $w=v_T$, and applying the chain rule to write $\nabla{\cdot}(\boldsymbol{\beta}v_T)=(\nabla{\cdot}\boldsymbol{\beta})v_T+\boldsymbol{\beta}{\cdot}\nabla v_T$, we obtain

$$\mathfrak{T}_1+\mathfrak{T}_3=\sum_{T\in\mathcal{T}_h}[-(u_T,\boldsymbol{\beta}{\cdot}\nabla v_T)_T+(\mu u_T,v_T)_T]-\sum_{T\in\mathcal{T}_h}\sum_{F\in\mathcal{F}_T}((\boldsymbol{\beta}{\cdot}\boldsymbol{n}_{TF})u_F,v_F-v_T)_F.$$

To insert v_F into the boundary integral, we have used (3.71). On the other hand, expanding $(\boldsymbol{\beta}{\cdot}\boldsymbol{n}_{TF})^+$ in $\mathrm{s}^+_{\boldsymbol{\beta},T}$ (see (3.80)), we can write for the second term:

$$\mathfrak{T}_2=\sum_{T\in\mathcal{T}_h}\sum_{F\in\mathcal{F}_T}\left(\frac{\boldsymbol{\beta}{\cdot}\boldsymbol{n}_{TF}+|\boldsymbol{\beta}{\cdot}\boldsymbol{n}_{TF}|}{2}(u_F-u_T),v_F-v_T\right)_F.$$

In conclusion, recalling the definition (3.99) of the normal trace of the advective flux and, since $(v_F-v_T)_{|F}\in\mathbb{P}^k(F)$, using (1.57) to insert $\pi^{0,k}_F$ in front of the first argument of boundary integrals, we get

$$\begin{aligned}\mathrm{a}_{\boldsymbol{\beta},\mu,h}(\underline{u}_h,\underline{v}_h) = &\sum_{T\in\mathcal{T}_h}[-(u_T,\boldsymbol{\beta}{\cdot}\nabla v_T)_T+(\mu u_T,v_T)_T]\\&-\sum_{T\in\mathcal{T}_h}\sum_{F\in\mathcal{F}_T}(\Phi_{\boldsymbol{\beta},TF}(\underline{u}_T),v_F-v_T)_F.\end{aligned}\tag{3.103}$$

Combining (3.102) and (3.103), we see that the reformulation (2.51) holds for $\mathrm{a}_{\boldsymbol{K},\boldsymbol{\beta},\mu,h}$ with, for all $T \in \mathcal{T}_h$,

$$\mathrm{a}_{\mathrm{V},T}(\underline{u}_T, v_T) = (\boldsymbol{K}_T \nabla \mathrm{p}_{\boldsymbol{K},T}^{k+1} \underline{u}_T, \nabla v_T)_T - (u_T, \boldsymbol{\beta} \cdot \nabla v_T)_T + (\mu u_T, v_T)_T$$
$$\forall (\underline{u}_T, v_T) \in \underline{U}_T^k \times \mathbb{P}^k(T)$$

and, for all $\underline{u}_T \in \underline{U}_T^k$ and all $F \in \mathcal{F}_T$, $\Phi_{TF}(\underline{u}_T) = \Phi_{\boldsymbol{K},TF}(\underline{u}_T) + \Phi_{\boldsymbol{\beta},TF}(\underline{u}_T)$. □

3.2.2.3 Initial Convergence Result

We next investigate the convergence of the method in the $\|\cdot\|_{\flat,h}$-norm.

Theorem 3.32 (Discrete Energy Error Estimate) *Let $(\mathcal{M}_h)_{h \in \mathcal{H}}$ denote a regular mesh sequence in the sense of Definition 1.9. Let Assumptions 3.1, 3.2 and 3.22 hold true. Let a polynomial degree $k \geq 0$ be fixed. Denote by $u \in H_0^1(\Omega)$ the unique solution to* (3.61)*, for which we assume the additional regularity $u \in H^{r+2}(\mathcal{T}_h)$ for some $r \in \{0, \ldots, k\}$. For all $h \in \mathcal{H}$, let $\underline{u}_h \in \underline{U}_{h,0}^k$ denote the unique solution to* (3.98) *with local stabilisation bilinear forms $\mathrm{s}_{\boldsymbol{K},T}$, $T \in \mathcal{T}_h$, in* (3.26) *satisfying Assumptions 3.9. Then, it holds that*

$$\|\underline{u}_h - \underline{I}_h^k u\|_{\flat,h} \lesssim \left\{ \sum_{T \in \mathcal{T}_h} \left(\overline{K}_T \alpha_T |u|_{H^{r+2}(T)}^2 + \hat{\tau}_T^{-2} \mu_0^{-1} |u|_{H^{r+1}(T)}^2 \right) h_T^{2(r+1)} \right.$$
$$\left. + \sum_{T \in \mathcal{T}_h} \beta_T \left[\min(1, \mathrm{Pe}_T)^{\frac{1}{2}} h_T^{r+\frac{1}{2}} \right]^2 |u|_{H^{r+1}(T)}^2 \right\}^{\frac{1}{2}}$$
$$=: E_{\boldsymbol{K},\boldsymbol{\beta},\mu,h}(u), \tag{3.104}$$

where the hidden constant is independent of h, u, $\boldsymbol{K}$, $\boldsymbol{\beta}$ and μ, and $\|\cdot\|_{\flat,h}$ is defined by (3.81).

Remark 3.33 (Robustness of the Estimate (3.104)*)* As in the pure diffusion case (see Remark 3.20), the estimate (3.104) is fully robust with respect to the heterogeneity of $\boldsymbol{K}$ and partially robust with respect to its anisotropy. Additionally, it is fully robust with respect to the local Péclet number: (1) diffusion-dominated elements (for which $\mathrm{Pe}_T \leq h_T$) contribute with a term in $\mathcal{O}(h_T^{r+1})$ (as for a pure diffusion problem); (2) convection-dominated elements (for which $\mathrm{Pe}_T \geq 1$) contribute with a term in $O(h_T^{r+\frac{1}{2}})$ (as for pure advection problem, see [159] where it is recovered as a special case); (3) elements in an intermediate regime (that is, $\mathrm{Pe}_T \in (h_T, 1)$) contribute with the intermediate rate $O(\mathrm{Pe}_T^{\frac{1}{2}} h_T^{r+\frac{1}{2}})$, which transitions continuously from the advection-dominated rate of convergence to the diffusion-dominated rate of convergence. We note that, in most analyses of numerical methods for advection–diffusion equations, the intermediate rates are usually not made explicit, and only the extreme regimes are fully studied.

Proof We invoke the Third Strang Lemma A.7 with $\mathtt{U} = H_0^1(\Omega)$, $\mathtt{a} = a_{\boldsymbol{K},\boldsymbol{\beta},\mu}$, $\mathtt{l}(v) = (f, v)$, $\mathtt{U}_h = \underline{U}_{h,0}^k$ endowed with the norm $\|\cdot\|_{\flat,h}$, $\mathtt{a}_h = \mathrm{a}_{\boldsymbol{K},\boldsymbol{\beta},\mu,h}$, $\mathtt{l}_h(\underline{v}_h) = (f, v_h)$, and $\mathtt{I}_h u = \underline{I}_h^k u$. Owing to (3.84), the global bilinear form $\mathrm{a}_{\boldsymbol{K},\boldsymbol{\beta},\mu,h}$ is coercive with respect to the norm $\|\cdot\|_{\flat,h}$, with coercivity constant equal to 1. An inspection shows that the consistency error $\mathcal{E}_h(u;\cdot)$ is the sum of $\mathcal{E}_{\boldsymbol{K},h}(u;\cdot)$ and $\mathcal{E}_{\boldsymbol{\beta},\mu,h}(u;\cdot)$. Since $\|\cdot\|_{\mathrm{a},\boldsymbol{K},h} \le \|\cdot\|_{\flat,h}$ by definition (see (3.81)), the estimates (3.48) and (3.85) give a bound on the dual norm of $\mathcal{E}_h(u;\cdot)$ with respect to $\|\cdot\|_{\flat,h}$. Plugged into (A.6), this bound establishes (3.104). □

As a consequence of the fully discrete estimate (3.104), we can state an error estimate between the exact solution and the reconstructed approximate solution obtained through the operator $\mathrm{p}_{\boldsymbol{K},h}^{k+1}$ defined by (3.55). The following result is the pendant of Theorem 3.19 for diffusion–advection–reaction equations.

Theorem 3.34 (Energy Error Estimate for the Reconstructed Approximate Solution) *Under the assumptions and notations in Theorem 3.32, it holds that*

$$
\begin{aligned}
&\left(\|\boldsymbol{K}^{\frac{1}{2}}\nabla_h(\mathrm{p}_{\boldsymbol{K},h}^{k+1}\underline{u}_h - u)\| + |\underline{u}_h|_{\mathrm{s},\boldsymbol{K},h} + \mu_0^{\frac{1}{2}}\|u_h - u\|\right) \\
&\qquad\lesssim E_{\boldsymbol{K},\boldsymbol{\beta},\mu,h}(u) + \left(\sum_{T\in\mathcal{T}_h} \mu_0 h_T^{2(r+1)} |u|_{H^{r+1}(T)}^2\right)^{\frac{1}{2}},
\end{aligned}
\tag{3.105}
$$

where the hidden constant is independent of h, u, $\boldsymbol{K}$, $\boldsymbol{\beta}$ and μ, and $E_{\boldsymbol{K},\boldsymbol{\beta},\mu,h}(u)$ is defined by (3.104).

Proof Since $\|\cdot\|_{a,\boldsymbol{K},h} \le \|\cdot\|_{b,h}$ (see (3.81)), (3.104) gives an estimate on the error $\|\underline{u}_h - \underline{I}_h^k u\|_{a,\boldsymbol{K},h}$; reasoning exactly as in the proof of Theorem 3.19 then yields the estimate on $\|\boldsymbol{K}^{\frac{1}{2}}\nabla_h(\mathrm{p}_{\boldsymbol{K},h}^{k+1}\underline{u}_h - u)\| + |\underline{u}_h|_{s,\boldsymbol{K},h}$. On the other hand, recalling the definition (3.74) of $\|\cdot\|_{\boldsymbol{\beta},\mu,T}$ and using again (3.104), we infer that

$$\mu_0^{\frac{1}{2}}\|u_h - \pi_h^{0,k}u\| \lesssim E_{\boldsymbol{K},\boldsymbol{\beta},\mu,h}(u).$$

The estimate on $\mu_0^{\frac{1}{2}}\|u_h - u\|$ follows from this bound, the triangle inequality, and the approximation properties (1.74) of the L^2-orthogonal projector with $p = 2$, $l = k$, $s = r + 1$ and $m = 0$, which imply

$$\mu_0\|\pi_h^{0,k}u - u\|^2 = \sum_{T\in\mathcal{T}_h}\mu_0\|\pi_T^{0,k}u - u\|_T^2 \lesssim \sum_{T\in\mathcal{T}_h}\mu_0 h_T^{2(r+1)}|u|_{H^{r+1}(T)}^2.$$

□

3.2.3 Robust Convergence Including the Advective Derivative

Although the estimates (3.104) and (3.105) enjoy some robustness properties (see Remark 3.33), a non-desirable phenomenon occurs in the limit $\boldsymbol{K} \to 0$: all approximation properties on derivatives of u are lost, and only an approximation property on u itself remains (through the term $\mu_0^{\frac{1}{2}}\|u_h - u\|$). However, even for very small diffusion tensor $\boldsymbol{K}$, the model (3.60) still contains some stable information on a derivative of u, namely, the advective derivative $\nabla\cdot(\boldsymbol{\beta}u)$ (or, equivalently, $\boldsymbol{\beta}\cdot\nabla u$). This information is not made readily available in the estimate (3.105).

The purpose of this section is to present a more robust estimate, in a discrete norm that involves some stable information on the reconstructed advective derivative. To this end, we will need the following additional assumption:

Assumption 3.35 (Small Damköhler Number) *It holds*

$$h_T\mu_0 \le \hat{\beta}_T \qquad \forall T \in \mathcal{T}_h.$$

Since $\mu_0 > 0$ by Assumption 3.22, this means in particular that there is no element in which the velocity is identically zero.

Remark 3.36 (Assumption 3.35) Given a mesh element $T \in \mathcal{T}_h$, the quantity $\mathrm{Da}_T := \frac{h_T\mu_0}{\hat{\beta}_T}$ can be interpreted as a local Damköhler number, measuring the relative importance of reactive and advective phenomena. Under Assumption 3.35, we have $\mathrm{Da}_T \le 1$ for all $T \in \mathcal{T}_h$, which means that we are not concerned with dominant reaction.

The improved discrete estimate is established in the following norm, which is stronger than $\|\cdot\|_{\flat,h}$ and accounts for the discrete advective derivative:

$$\|\underline{v}_h\|_{\sharp,h} := \left(\|\underline{v}_h\|_{\flat,h}^2 + \alpha^{-1} \sum_{T\in\mathcal{T}_h} h_T \alpha_T^{-\frac{1}{2}} \hat{\beta}_T^{-1} \|\mathrm{G}_{\boldsymbol{\beta},T}^k \underline{v}_T\|_T^2 \right)^{\frac{1}{2}}, \tag{3.106}$$

where α_T is the local anisotropy ratio defined by (3.6), while α is the global counterpart defined by (3.47). The global bilinear form of the HHO scheme (3.98) satisfies an inf–sup condition with respect to this improved norm.

Lemma 3.37 (Inf-sup Stability of $\mathrm{a}_{\boldsymbol{K},\boldsymbol{\beta},\mu,h}$**)** *Under Assumption 3.35, for all* $\underline{w}_h \in \underline{U}_{h,0}^k$ *it holds that*

$$\chi \|\underline{w}_h\|_{\sharp,h} \lesssim \sup_{\underline{v}_h\in\underline{U}_{h,0}^k\setminus\{\underline{0}_h\}} \frac{\mathrm{a}_{\boldsymbol{K},\boldsymbol{\beta},\mu,h}(\underline{w}_h,\underline{v}_h)}{\|\underline{v}_h\|_{\sharp,h}} \quad \textit{with } \chi := \min_{T\in\mathcal{T}_h}(1, \hat{\tau}_T\mu_0), \tag{3.107}$$

where the hidden constant is independent of h, $\underline{w}_h$, $\boldsymbol{K}$, $\boldsymbol{\beta}$, μ *and* μ_0*, but possibly depends on* d, ϱ*, and* k.

Proof Denote by $\mathcal{S}_{\sharp,h}$ the supremum in the right-hand side of (3.107). We first notice that

$$\chi \|\underline{w}_h\|_{\flat,h} \le \|\underline{w}_h\|_{\flat,h} \le \mathcal{S}_{\sharp,h}, \tag{3.108}$$

which follows from the coercivity of $\mathrm{a}_{\boldsymbol{K},\boldsymbol{\beta},\mu,h}$ with respect to $\|\cdot\|_{\flat,h}$ (see the proof of Theorem 3.32). The main difficulty of the proof is therefore to bound the term $\alpha^{-1}\sum_{T\in\mathcal{T}_h} h_T\alpha^{-\frac{1}{2}}\hat{\beta}_T^{-1}\|\mathrm{G}_{\boldsymbol{\beta},T}^k\underline{w}_T\|_T^2$. In order to do so, remark the following: if, for all $T\in\mathcal{T}_h$, we take $v_T = h_T\alpha_T^{-\frac{1}{2}}\hat{\beta}_T^{-1}\mathrm{G}_{\boldsymbol{\beta},T}^k\underline{w}_T$, then this term naturally appears in the expression (3.79) of $\mathrm{a}_{\boldsymbol{\beta},\mu,h}(\underline{w}_h,\underline{v}_h)$. This idea, which consists in using the scaled advective derivative as a test function, can be found, e.g., in [215] in the context of Discontinuous Galerkin methods, and was extended to HHO methods in [159]. We therefore define, for a given $\underline{w}_h\in\underline{U}_{h,0}^k$, the element $\underline{v}_h\in\underline{U}_{h,0}^k$ such that

$$v_T = h_T\alpha_T^{-\frac{1}{2}}\hat{\beta}_T^{-1}\mathrm{G}_{\boldsymbol{\beta},T}^k\underline{w}_T \quad \forall T\in\mathcal{T}_h, \qquad v_F\equiv 0 \quad \forall F\in\mathcal{F}_h. \tag{3.109}$$

Step 1. We prove that

$$\|\underline{v}_h\|_{\sharp,h} \lesssim \alpha\|\underline{w}_h\|_{\sharp,h}. \tag{3.110}$$

(i) *Diffusive contribution.* We first establish an estimate on $\|\mathrm{G}^k_{\boldsymbol{\beta},T}\underline{w}_T\|_T$ for a generic mesh element $T \in \mathcal{T}_h$. Using the characterisation (3.65) of the reconstructed advective derivative we infer that, for all $\phi \in \mathbb{P}^k(T)$ with $\|\phi\|_T \le 1$,

$$
\begin{aligned}
(\mathrm{G}^k_{\boldsymbol{\beta},T}\underline{w}_T, \phi)_T &= (\boldsymbol{\beta}\cdot\nabla w_T, \phi)_T + \sum_{F\in\mathcal{F}_T} ((\boldsymbol{\beta}\cdot\boldsymbol{n}_{TF})(w_F - w_T), \phi)_F \\
&\le \hat{\beta}_T \left(\|\nabla w_T\|_T \|\phi\|_T + \sum_{F\in\mathcal{F}_T} h_F^{-\frac{1}{2}} \|w_F - w_T\|_F \, h_F^{\frac{1}{2}} \|\phi\|_F \right) \\
&\lesssim \hat{\beta}_T \left(\|\nabla w_T\|_T^2 + \sum_{F\in\mathcal{F}_T} h_F^{-1} \|w_F - w_T\|_F^2 \right)^{\frac{1}{2}} \\
&\lesssim \hat{\beta}_T \underline{K}_T^{-\frac{1}{2}} \|\underline{w}_T\|_{1,\boldsymbol{K},T}.
\end{aligned}
$$

where we have used generalised Hölder inequalities with exponents $(\infty, 2, 2)$ together with the definition (3.66) of $\hat{\beta}_T$ to pass to the second line, the discrete trace inequality (1.55) (with $v = \phi$ and $p = 2$) together with $\|\phi\|_T \le 1$, $h_F \le h_T$, and a discrete Cauchy–Schwarz inequality to pass to the third line, and we have concluded recalling the definitions of $\underline{K}_T$ and of $\|\cdot\|_{1,\boldsymbol{K},T}$ (see (3.25)). Since $\mathrm{G}^k_{\boldsymbol{\beta},T}\underline{w}_T \in \mathbb{P}^k(T)$, taking the supremum over all $\phi \in \mathbb{P}^k(T)$ such that $\|\phi\|_T \le 1$ gives an estimate on $\|\mathrm{G}^k_{\boldsymbol{\beta},T}\underline{w}_T\|_T$, from which we deduce, multiplying by $\overline{K}_T^{\frac{1}{2}}$ and using the definition (3.6) of α_T,

$$
\overline{K}_T^{\frac{1}{2}} \|\mathrm{G}^k_{\boldsymbol{\beta},T}\underline{w}_T\|_T \lesssim \hat{\beta}_T \alpha_T^{\frac{1}{2}} \|\underline{w}_T\|_{1,\boldsymbol{K},T}, \tag{3.111}
$$

where the hidden multiplicative constant is additionally independent of T.

Let us now turn to estimating the diffusive contributions in $\|\underline{v}_h\|_{\sharp,h}$. By the definitions (3.25) of $\|\cdot\|_{1,\boldsymbol{K},T}$ and (3.109) of $\underline{v}_T$,

$$
\begin{aligned}
\|\underline{v}_T\|_{1,\boldsymbol{K},T}^2 &= h_T^2 \alpha_T^{-1} \hat{\beta}_T^{-2} \|\boldsymbol{K}_T^{\frac{1}{2}} \nabla \mathrm{G}^k_{\boldsymbol{\beta},T}\underline{w}_T\|_T^2 + h_T^2 \alpha_T^{-1} \hat{\beta}_T^{-2} \sum_{F\in\mathcal{F}_T} \frac{K_{TF}}{h_F} \|\mathrm{G}^k_{\boldsymbol{\beta},T}\underline{w}_T\|_F^2 \\
&\lesssim \alpha_T^{-1} \hat{\beta}_T^{-2} \overline{K}_T \|\mathrm{G}^k_{\boldsymbol{\beta},T}\underline{w}_T\|_T^2 \\
&\lesssim \|\underline{w}_T\|_{1,\boldsymbol{K},T}^2,
\end{aligned} \tag{3.112}
$$

where the second line follows from the discrete inverse inequality (1.46) and the discrete trace inequality (1.55), both applied with $p = 2$ to $v = \mathrm{G}^k_{\boldsymbol{\beta},T}\underline{w}_T$, from (1.6) to write $h_T/h_F \le \varrho^{-2}/2 \lesssim 1$, and from the uniform bound (1.5) on the number of faces of T, while the conclusion is a consequence of (3.111). Summing over $T \in \mathcal{T}_h$, using the norm equivalence (3.46) and the estimate (3.112), and recalling that, by (3.81) and (3.106), $\|\cdot\|_{\mathrm{a},\boldsymbol{K},h} \le \|\cdot\|_{\flat,h} \le \|\cdot\|_{\sharp,h}$, we find

$$\|\underline{v}_h\|_{\mathrm{a},\boldsymbol{K},h}^2 \lesssim \alpha\|\underline{v}_h\|_{1,\boldsymbol{K},h}^2 \lesssim \alpha\|\underline{w}_h\|_{1,\boldsymbol{K},h}^2 \le \alpha^2\|\underline{w}_h\|_{\mathrm{a},\boldsymbol{K},h}^2 \le \alpha^2\|\underline{w}_h\|_{\sharp,h}^2. \tag{3.113}$$

(ii) *Advective-reactive contribution.* Let a mesh element $T \in \mathcal{T}_h$ be fixed. By definition (3.74) of $\|\cdot\|_{\boldsymbol{\beta},\mu,T}$ and (3.66) of $\hat{\beta}_T$, the discrete trace inequality (1.55) with $p = 2$ and the uniform bound (1.5) on the number of faces of T yield

$$\begin{aligned}
\|\underline{v}_T\|_{\boldsymbol{\beta},\mu,T}^2 &\lesssim \hat{\beta}_T h_T^{-1}\|v_T\|_T^2 + \mu_0\|v_T\|_T^2 \\
&= h_T\alpha_T^{-1}\hat{\beta}_T^{-1}\|\mathrm{G}_{\boldsymbol{\beta},T}^k\underline{w}_T\|_T^2 + \mu_0 h_T^2\alpha_T^{-1}\hat{\beta}_T^{-2}\|\mathrm{G}_{\boldsymbol{\beta},T}^k\underline{w}_T\|_T^2 \\
&\le 2h_T\alpha_T^{-1}\hat{\beta}_T^{-1}\|\mathrm{G}_{\boldsymbol{\beta},T}^k\underline{w}_T\|_T^2,
\end{aligned}$$

the conclusion being a consequence of Assumption 3.35. Since $\alpha_T \ge 1$, we have $\alpha_T^{-1} \le \alpha_T^{-\frac{1}{2}}$ and, after summing over $T \in \mathcal{T}_h$, the previous estimate therefore leads to

$$\|\underline{v}_h\|_{\boldsymbol{\beta},\mu,h}^2 \lesssim \sum_{T\in\mathcal{T}_h} h_T\alpha_T^{-\frac{1}{2}}\hat{\beta}_T^{-1}\|\mathrm{G}_{\boldsymbol{\beta},T}^k\underline{w}_T\|_T^2 \le \alpha\|\underline{w}_h\|_{\sharp,h}^2. \tag{3.114}$$

The definition (3.65) of $\mathrm{G}_{\boldsymbol{\beta},T}^k$ shows that

$$\begin{aligned}
\|\mathrm{G}_{\boldsymbol{\beta},T}^k\underline{v}_T\|_T &= \sup_{\phi\in\mathbb{P}^k(T),\ \|\phi\|_T\le 1} (\mathrm{G}_{\boldsymbol{\beta},T}^k\underline{v}_T, \phi)_T \\
&= \sup_{\phi\in\mathbb{P}^k(T),\ \|\phi\|_T\le 1} \left((\boldsymbol{\beta}\cdot\nabla v_T, \phi)_T - \sum_{F\in\mathcal{F}_T}((\boldsymbol{\beta}\cdot\boldsymbol{n}_{TF})v_T, \phi)_F\right) \\
&\lesssim h_T^{-1}\hat{\beta}_T\|v_T\|_T = \alpha_T^{-\frac{1}{2}}\|\mathrm{G}_{\boldsymbol{\beta},T}^k\underline{w}_T\|_T \le \|\mathrm{G}_{\boldsymbol{\beta},T}^k\underline{w}_T\|_T,
\end{aligned}$$

where, to pass to the third line, we have used generalised Hölder inequalities with exponents $(\infty, 2, 2)$, the discrete inverse and trace inequalities (1.46) and (1.55), both with $p = 2$, the bound $\|\phi\|_T \le 1$, the definition (3.109) of v_T, and $\alpha_T \ge 1$. Squaring this estimate, multiplying by $\alpha^{-1}h_T\alpha_T^{-\frac{1}{2}}\hat{\beta}_T^{-1}$, and summing over $T \in \mathcal{T}_h$, we obtain

$$\alpha^{-1}\sum_{T\in\mathcal{T}_h} h_T\alpha_T^{-\frac{1}{2}}\hat{\beta}_T^{-1}\|\mathrm{G}_{\boldsymbol{\beta},T}^k\underline{v}_T\|_T^2 \lesssim \alpha^{-1}\sum_{T\in\mathcal{T}_h} h_T\alpha_T^{-\frac{1}{2}}\hat{\beta}_T^{-1}\|\mathrm{G}_{\boldsymbol{\beta},T}^k\underline{w}_T\|_T^2 \le \|\underline{w}_h\|_{\sharp,h}^2.$$

Combine this bound with (3.113) and (3.114). Since $1 \leq \alpha \leq \alpha^2$, this completes the proof of (3.110).

Step 2. We establish (3.107). Using the test function $\underline{v}_h$, defined by (3.109), in (3.79) with $\underline{u}_h = \underline{w}_h$, and recalling the definition (3.97) of $\mathrm{a}_{\boldsymbol{K},\boldsymbol{\beta},\mu,h}$, it is inferred that

$$\begin{aligned}
&\sum_{T\in\mathcal{T}_h} h_T \alpha_T^{-\frac{1}{2}} \hat{\beta}_T^{-1} \|\mathrm{G}_{\boldsymbol{\beta},T}^k \underline{w}_T\|_T^2 \\
&\quad = \mathrm{a}_{\boldsymbol{\beta},\mu,h}(\underline{w}_h, \underline{v}_h) - \sum_{T\in\mathcal{T}_h} ([\nabla\cdot\boldsymbol{\beta} + \mu] w_T, v_T)_T - \sum_{T\in\mathcal{T}_h} \mathrm{s}_{\boldsymbol{\beta},T}^+(\underline{w}_T, \underline{v}_T) \\
&\quad = \mathrm{a}_{\boldsymbol{K},\boldsymbol{\beta},\mu,h}(\underline{w}_h, \underline{v}_h) - \mathrm{a}_{\boldsymbol{K},h}(\underline{w}_h, \underline{v}_h) - \sum_{T\in\mathcal{T}_h} ([\nabla\cdot\boldsymbol{\beta} + \mu] w_T, v_T)_T \\
&\qquad - \sum_{T\in\mathcal{T}_h} \mathrm{s}_{\boldsymbol{\beta},T}^+(\underline{w}_T, \underline{v}_T).
\end{aligned} \tag{3.115}$$

Denote by $\mathfrak{T}_1, \ldots, \mathfrak{T}_4$ the addends in the right-hand side of (3.115). By definition of $\mathcal{S}_{\sharp,h}$ and (3.110), we have

$$\mathfrak{T}_1 \leq \mathcal{S}_{\sharp,h} \|\underline{v}_h\|_{\sharp,h} \lesssim \alpha \mathcal{S}_{\sharp,h} \|\underline{w}_h\|_{\sharp,h}. \tag{3.116}$$

The definition (3.46) of $\|\cdot\|_{\mathrm{a},\boldsymbol{K},h}$ together with the Cauchy–Schwarz inequality followed by the definition (3.81) of $\|\cdot\|_{\flat,h}$ and (3.110) yield

$$\mathfrak{T}_2 \leq \|\underline{w}_h\|_{\mathrm{a},\boldsymbol{K},h} \|\underline{v}_h\|_{\mathrm{a},\boldsymbol{K},h} \lesssim \|\underline{w}_h\|_{\flat,h} \alpha \|\underline{w}_h\|_{\sharp,h}. \tag{3.117}$$

Since $|(\nabla\cdot\boldsymbol{\beta} + \mu)_{|T}| \leq d\mathrm{L}_{\boldsymbol{\beta},T} + \|\mu\|_{L^\infty(T)} \leq (d+1)\hat{\tau}_T^{-1} \leq (d+1)\mu_0 \chi^{-1}$ (recall the definitions (3.83) of $\hat{\tau}_T$ and (3.107) of χ), by definition (3.74) of $\|\cdot\|_{\boldsymbol{\beta},\mu,T}$ and the Cauchy–Schwarz inequality, the estimate on $\mathfrak{T}_3$ is

$$\mathfrak{T}_3 \lesssim \chi^{-1} \|\underline{w}_h\|_{\boldsymbol{\beta},\mu,h} \|\underline{v}_h\|_{\boldsymbol{\beta},\mu,h} \lesssim \chi^{-1} \|\underline{w}_h\|_{\flat,h} \alpha \|\underline{w}_h\|_{\sharp,h}. \tag{3.118}$$

For $\mathfrak{T}_4$, writing

$$\begin{aligned}
|((\boldsymbol{\beta}\cdot\boldsymbol{n}_{TF})^+(w_F - w_T), v_F - v_T)_F| &\leq (|\boldsymbol{\beta}\cdot\boldsymbol{n}_{TF}|\,|w_F - w_T|, |v_F - v_T|)_F \\
&= (|\boldsymbol{\beta}\cdot\boldsymbol{n}_{TF}|^{\frac{1}{2}}|w_F - w_T|, |\boldsymbol{\beta}\cdot\boldsymbol{n}_{TF}|^{\frac{1}{2}}|v_F - v_T|)_F
\end{aligned}$$

and using the Cauchy–Schwarz inequality, the definition (3.74) of $\|\cdot\|_{\beta,\mu,T}$, and (3.110), we have

$$\begin{aligned}\mathfrak{T}_4 &\le \left(\sum_{T\in\mathcal{T}_h}\sum_{F\in\mathcal{F}_T} \||\boldsymbol{\beta}\cdot\boldsymbol{n}_{TF}|^{\frac{1}{2}}(w_F - w_T)\|_F^2\right)^{\frac{1}{2}}\left(\sum_{T\in\mathcal{T}_h}\sum_{F\in\mathcal{F}_T} \||\boldsymbol{\beta}\cdot\boldsymbol{n}_{TF}|^{\frac{1}{2}}(v_F - v_T)\|_F^2\right)^{\frac{1}{2}}\\ &\lesssim \|\underline{w}_h\|_{\boldsymbol{\beta},\mu,h}\|\underline{v}_h\|_{\boldsymbol{\beta},\mu,h} \lesssim \|\underline{w}_h\|_{\flat,h}\alpha\|\underline{w}_h\|_{\sharp,h}.\end{aligned} \tag{3.119}$$

Hence, plugging (3.116)–(3.119) into (3.115), we obtain

$$\begin{aligned}\chi\alpha^{-1}\sum_{T\in\mathcal{T}_h} h_T\alpha_T^{-\frac{1}{2}}\hat{\beta}_T^{-1}\|\mathrm{G}^k_{\boldsymbol{\beta},T}\underline{w}_T\|_T^2 &\lesssim \mathcal{S}_{\sharp,h}\chi\,\|\underline{w}_h\|_{\sharp,h} + (1+\chi)\|\underline{w}_h\|_{\flat,h}\|\underline{w}_h\|_{\sharp,h}\\ &\lesssim \mathcal{S}_{\sharp,h}\|\underline{w}_h\|_{\sharp,h},\end{aligned} \tag{3.120}$$

where the conclusion follows recalling (3.108) and observing that, by definition, $\chi \le 1$. Adding

$$\chi\,\|\underline{w}_h\|_{\flat,h}^2 \le \chi\,\|\underline{w}_h\|_{\flat,h}\|\underline{w}_h\|_{\sharp,h}$$

to (3.120) and using again (3.108), we infer $\chi\,\|\underline{w}_h\|_{\sharp,h}^2 \lesssim \mathcal{S}_{\sharp,h}\|\underline{w}_h\|_{\sharp,h}$ and the proof of (3.107) is completed simplifying by $\|\underline{w}_h\|_{\sharp,h}$. □

Since $\|\cdot\|_{\sharp,h}$ is stronger than $\|\cdot\|_{\flat,h}$ (and thus than $\|\cdot\|_{\mathrm{a},\boldsymbol{K},h}$), the consistency estimates (3.48) and (3.85) still hold when calculated using $\|\cdot\|_{\sharp,h}$. Hence, owing to Lemma 3.37, the Third Strang Lemma A.7 directly yields the following discrete error estimate.

Theorem 3.38 (Energy Error Estimate) *Under the assumptions and notations in Theorem 3.32 together with Assumption 3.35, we have, with χ defined in* (3.107),

$$\|\underline{u}_h - \underline{I}_h^k u\|_{\sharp,h} \lesssim \chi^{-1}E_{\boldsymbol{K},\boldsymbol{\beta},\mu,h}(u), \tag{3.121}$$

where the hidden constant is independent of h, u, $\boldsymbol{K}$, $\boldsymbol{\beta}$, μ, and μ_0, and $E_{\boldsymbol{K},\boldsymbol{\beta},\mu,h}(u)$ is defined by (3.104).

Remark 3.39 (Extension to Locally Vanishing Diffusion) In [159], an analysis similar to the previous one but using slightly different norms is extended to the singular limit corresponding to locally vanishing diffusion. Considering this singular limit entails some additional difficulties, including the fact that the solution may exhibit jumps across the interface between diffusive and non-diffusive regions;

see, e.g., the discussion in [195] for the one-dimensional case and [157] for the multi-dimensional case. In this situation, estimates similar to the ones in Theorem 3.38 are obtained, with the convention that, in $E_{\boldsymbol{K},\boldsymbol{\beta},\mu,h}(u)$, $\mathrm{Pe}_T = \infty$ for any element $T \in \mathcal{T}_h$ such that $K_{TF} = 0$ for some $F \in \mathcal{F}_T$. In the context of Discontinuous Galerkin methods, a convergence analysis for locally vanishing diffusion with advection is carried out in [157]; see also [26, 213].

As a consequence of the estimate in Theorem 3.38, we can establish the following approximation property of $\mathrm{G}^k_{\boldsymbol{\beta},T}\underline{u}_T$ which, contrary to (3.105), is fully robust with respect to the Péclet number (it persists in the limit $\boldsymbol{K} \to 0$, provided that the anisotropy ratios remain bounded above).

Theorem 3.40 (Error Estimate for the Advective Derivative) *Under the assumptions of Theorem 3.38 we have, with χ defined in* (3.107),

$$\left(\alpha^{-1}\sum_{T\in\mathcal{T}_h} h_T\alpha_T^{-\frac{1}{2}}\hat{\beta}_T^{-1}\|\mathrm{G}^k_{\boldsymbol{\beta},T}\underline{u}_T - \pi_T^{0,k}(\boldsymbol{\beta}\cdot\nabla u)\|_T^2\right)^{\frac{1}{2}}$$

$$\lesssim \chi^{-1}E_{\boldsymbol{K},\boldsymbol{\beta},\mu,h}(u) + \left(\alpha^{-1}\sum_{T\in\mathcal{T}_h}\alpha_T^{-\frac{1}{2}}\hat{\beta}_T h_T^{2r+1}|u|^2_{H^{r+1}(T)}\right)^{\frac{1}{2}}, \tag{3.122}$$

where the hidden constant is independent of h, u, $\boldsymbol{K}$, $\boldsymbol{\beta}$, μ and μ_0. If, additionally, we suppose that $\boldsymbol{\beta} \in W^{r,\infty}(\mathcal{T}_h)^d$, then

$$\left(\alpha^{-1}\sum_{T\in\mathcal{T}_h} h_T\alpha_T^{-\frac{1}{2}}\hat{\beta}_T^{-1}\|\mathrm{G}^k_{\boldsymbol{\beta},T}\underline{u}_T - \boldsymbol{\beta}\cdot\nabla u\|_T^2\right)^{\frac{1}{2}} \lesssim \chi^{-1}E_{\boldsymbol{K},\boldsymbol{\beta},\mu,h}(u)$$

$$+\left(\alpha^{-1}\sum_{T\in\mathcal{T}_h}\alpha_T^{-\frac{1}{2}}\hat{\beta}_T^{-1}h_T^{2r+1}\left(\hat{\beta}_T|u|_{H^{r+1}(T)} + |\boldsymbol{\beta}\cdot\nabla u|_{H^r(T)}\right)^2\right)^{\frac{1}{2}}. \tag{3.123}$$

Remark 3.41 (Robustness and Dominating Term) For the sake of simplicity, let us consider the case of a quasi-uniform mesh (that is, $h \lesssim h_T$ for all $T \in \mathcal{T}_h$), and of a uniformly bounded anisotropy ratio α. The estimate (3.105) provides an $O(h^{r+\frac{1}{2}})$ (or even $O(h^{r+1})$ in diffusion-dominated regime) approximation of the complete gradient ∇u of the solution. This estimate is, however, not robust in the limit $\boldsymbol{K} \to 0$ since the term involving ∇u in left-hand side then vanishes.

On the contrary, the estimate (3.123) yields an approximation of order $O(h^r)$ of the advective derivative $\boldsymbol{\beta}\cdot\nabla u$, which is a reduced order and only deals with part of the gradient of the solution, but this estimate is fully robust in the limit $\boldsymbol{K} \to 0$.

Proof Squaring the approximation property (3.67) with $v = u$, multiplying by $\alpha^{-1} h_T \alpha_T^{-\frac{1}{2}} \hat{\beta}_T^{-1}$ and summing over $T \in \mathcal{T}_h$ shows that

$$\left(\alpha^{-1} \sum_{T\in\mathcal{T}_h} h_T \alpha_T^{-\frac{1}{2}} \hat{\beta}_T^{-1} \|\mathrm{G}^k_{\boldsymbol{\beta},T} \underline{I}^k_T u - \pi_T^{0,k}(\boldsymbol{\beta}\cdot\nabla u)\|_T^2\right)^{\frac{1}{2}}$$
$$\lesssim \left(\alpha^{-1} \sum_{T\in\mathcal{T}_h} \alpha_T^{-\frac{1}{2}} \hat{\beta}_T h_T^{2r+1} |u|^2_{H^{r+1}(T)}\right)^{\frac{1}{2}}. \tag{3.124}$$

On the other hand, the definition (3.106) of $\|\cdot\|_{\sharp,h}$ and the estimate (3.121) yield

$$\left(\alpha^{-1} \sum_{T\in\mathcal{T}_h} h_T \alpha_T^{-\frac{1}{2}} \hat{\beta}_T^{-1} \|\mathrm{G}^k_{\boldsymbol{\beta},T} \underline{u}_T - \mathrm{G}^k_{\boldsymbol{\beta},T} \underline{I}^k_T u\|_T^2\right)^{\frac{1}{2}} \lesssim \chi^{-1} E_{\boldsymbol{K},\boldsymbol{\beta},\mu,h}(u). \tag{3.125}$$

The bound (3.122) follows from (3.124), (3.125) and the triangle inequality. The estimate (3.123) is obtained similarly, replacing (3.124) with

$$\left(\alpha^{-1} \sum_{T\in\mathcal{T}_h} h_T \alpha_T^{-\frac{1}{2}} \hat{\beta}_T^{-1} \|\mathrm{G}^k_{\boldsymbol{\beta},T} \underline{I}^k_T u - \boldsymbol{\beta}\cdot\nabla u\|_T^2\right)^{\frac{1}{2}}$$
$$\lesssim \left(\alpha^{-1} \sum_{T\in\mathcal{T}_h} \alpha_T^{-\frac{1}{2}} \hat{\beta}_T^{-1} h_T^{2r+1} \left(\hat{\beta}_T |u|_{H^{r+1}(T)} + |\boldsymbol{\beta}\cdot\nabla u|_{H^r(T)}\right)^2\right)^{\frac{1}{2}},$$

which follows from (3.68) with $v = u$. □

3.2.4 L^2-Error Estimate

We conclude our series of estimates by considering the L^2-norm of the error. An estimate for this quantity is already available in (3.104) via the following terms composing $\|\underline{u}_h - \underline{I}^k_h u\|_{\flat,h}$: $\mu_0^{\frac{1}{2}} \|u_h - \pi_h^{0,k} u\|$ and, through a discrete Poincaré inequality, $\|\underline{u}_h - \underline{I}^k_h u\|_{\mathrm{a},\boldsymbol{K},h}$. Our goal here is to see whether these estimates can be improved. As for the Poisson problem, this requires to assume full elliptic regularity of the dual problem. Here, given $g \in L^2(\Omega)$, the dual problem of (3.60) is

$$\begin{aligned} \nabla\cdot(-\boldsymbol{K}\nabla z_g) - \boldsymbol{\beta}\cdot\nabla z_g + \mu z_g &= g && \text{in } \Omega, \\ z_g &= 0 && \text{on } \partial\Omega. \end{aligned} \tag{3.126}$$

As usual, the solution is understood in the weak sense, that is, we consider the problem obtained by switching the test and trial functions in the bilinear form in the left-hand side of (3.61): Find $z_g \in H_0^1(\Omega)$ such that

$$a_{\boldsymbol{K},\boldsymbol{\beta},\mu}(v, z_g) = (g, v) \qquad \forall v \in H_0^1(\Omega). \tag{3.127}$$

Full elliptic regularity on this problem entails that

$$\exists C_{\text{ell}} > 0 \text{ such that } \|z_g\|_{H^2(\Omega)} \leq C_{\text{ell}}\|g\| \quad \forall g \in L^2(\Omega). \tag{3.128}$$

As mentioned in Remark 3.21, full elliptic regularity with a varying diffusion tensor requires Ω convex together with the Lipschitz-continuity of that tensor, which means, in view of Assumption 3.1, that it should be constant over Ω. Under this assumption and Assumption 3.22, z_g is the solution of $\nabla\cdot(-\boldsymbol{K}\nabla z_g) = g + \boldsymbol{\beta}\cdot\nabla z_g - \mu z_g \in L^2(\Omega)$ with homogeneous Dirichlet boundary conditions, and full elliptic regularity easily follows from the same regularity for the pure elliptic model. Note, however, that C_{ell} additionally depends in this case on $\|\boldsymbol{\beta}\|_{L^\infty(\Omega)^d}$ and $\|\mu\|_{L^\infty(\Omega)}$.

Using $v = z_g$ in (3.127), noticing that

$$(-\boldsymbol{\beta}\cdot\nabla z_g, z_g) + (\mu z_g, z_g) = \int_\Omega \left(\mu + \frac{1}{2}\nabla\cdot\boldsymbol{\beta}\right) z_g^2 \geq 0,$$

and invoking the following Poincaré inequality from [250]:

$$\|z_g\| \leq \frac{h_\Omega}{\pi}|z|_{H^1(\Omega)}, \tag{3.129}$$

we infer the following H^1-stability result:

$$|z_g|_{H^1(\Omega)} \leq \frac{h_\Omega}{\pi \underline{K}}\|g\|. \tag{3.130}$$

Combining this estimate with the Poincaré inequality (3.129), we have the L^2-estimate

$$\|z_g\| \leq \frac{h_\Omega^2}{\pi^2 \underline{K}}\|g\|. \tag{3.131}$$

Theorem 3.42 (L^2-Error Estimate) *Under the assumptions and notations in Theorem 3.32, assume furthermore* (3.128), *that* $\mu \in W^{1,\infty}(\mathcal{T}_h)$, *and that*

$$k \geq 1 \text{ or } \big(k = 0 \text{ and } \nabla\cdot(\boldsymbol{K}\nabla u) \in H^1(\mathcal{T}_h)\big).$$

Then, it holds, with hidden constant independent of h, u, $\boldsymbol{K}$, $\boldsymbol{\beta}$ *and* μ,

$$\|u_h - \pi_h^{0,k} u\| \lesssim E_h^{(1)} E_{\boldsymbol{K},\boldsymbol{\beta},\mu,h}(u) + E_h^{(2)}, \tag{3.132}$$

where $E_{\boldsymbol{K},\boldsymbol{\beta},\mu,h}(u)$ *is defined in* (3.104),

$$E_h^{(1)} := C_{\rm ell} \max_{T\in\mathcal{T}_h}\left(\overline{K}_T^{\frac12}\alpha_T^{\frac12} h_T\right) + \frac{h_\Omega}{\pi \underline{K}} \max_{T\in\mathcal{T}_h}\left(\hat{\beta}_T^{\frac12}\Big[\min(1,\mathrm{Pe}_T)h_T\Big]^{\frac12}\right)$$
$$+ \frac{h_\Omega \mu_0^{-\frac12}}{\pi \underline{K}} \max_{T\in\mathcal{T}_h}\left(\max\big[\|\mu + \nabla\cdot\boldsymbol{\beta}\|_{L^\infty(T)}, \mathrm{L}_{\boldsymbol{\beta},T}\big]h_T\right)$$

and:

(i) If $k \geq 1$,

$$E_h^{(2)} := C_{\rm ell}\left(\sum_{T\in\mathcal{T}_h} \overline{K}_T^2 \alpha_T^2 h_T^{2(r+2)} |u|^2_{H^{r+2}(T)}\right)^{\frac12}$$
$$+ C_{\rm ell}\left(\sum_{T\in\mathcal{T}_h} \hat{\beta}_T^2 h_T^{2(r+2)} |u|^2_{H^{r+1}(T)}\right)^{\frac12}$$
$$+ \frac{h_\Omega^2}{\pi^2 \underline{K}}\left(\sum_{T\in\mathcal{T}_h} \|\nabla \mu\|^2_{L^\infty(T)^d} h_T^{2(r+2)} |u|^2_{H^{r+1}(T)}\right)^{\frac12}$$
$$+ \frac{h_\Omega}{\pi \underline{K}}\left(\sum_{T\in\mathcal{T}_h} \hat{\tau}_T^{-2} h_T^{2(r+2)} |u|^2_{H^{r+1}(T)}\right)^{\frac12}.$$

(continued)

Theorem 3.42 (continued)
(ii) If $k = 0$,

$$E_h^{(2)} := \frac{h_\Omega}{\pi \underline{K}} \left(\sum_{T\in\mathcal{T}_h} h_T^4 |\nabla\cdot(\boldsymbol{K}\nabla u)|^2_{H^1(T)} \right)^{\frac{1}{2}}$$

$$+ C_{\rm ell} \left(\sum_{T\in\mathcal{T}_h} \overline{K}_T^2 \alpha_T^2 h_T^4 |u|^2_{H^2(T)} \right)^{\frac{1}{2}}$$

$$+ \frac{h_\Omega}{\pi \underline{K}} \left(\sum_{T\in\mathcal{T}_h} \hat{\beta}_T^2 h_T^2 |u|^2_{H^1(T)} \right)^{\frac{1}{2}}$$

$$+ \frac{h_\Omega^2}{\pi^2 \underline{K}} \left(\sum_{T\in\mathcal{T}_h} \|\nabla \mu\|^2_{L^\infty(T)^d} h_T^4 |u|^2_{H^1(T)} \right)^{\frac{1}{2}}.$$

Remark 3.43 (Rates of Convergence) Considering only global rates of convergence, we have $E_{\boldsymbol{K},\boldsymbol{\beta},\mu,h}(u) = \mathcal{O}(h^{r+1})$ if the diffusion dominates globally (that is, $\mathrm{Pe}_T = \mathcal{O}(h_T)$ in every $T \in \mathcal{T}_h$), and $E_{\boldsymbol{K},\boldsymbol{\beta},\mu,h}(u) = \mathcal{O}(h^{r+\frac{1}{2}})$ if advection is dominant (in which case $\min(1, \mathrm{Pe}_T) = 1$ for all $T \in \mathcal{T}_h$). These are the rates of convergence in energy norm provided by Theorem 3.32 (see also Remark 3.33).

Regarding the quantities introduced in Theorem 3.42:

- *if* $k \geq 1$ *or* $\boldsymbol{\beta} = \boldsymbol{0}$: in any regime we have $E_h^{(2)} = \mathcal{O}(h^{r+2})$. For dominating diffusion, $E_h^{(1)} = \mathcal{O}(h)$ and (3.132) then provides an $\mathcal{O}(h^{r+2})$ rate of convergence in L^2-norm. If, on the contrary, advection dominates then $E_h^{(1)} = \mathcal{O}(h^{\frac{1}{2}})$ and the rate of convergence given by (3.132) is $\mathcal{O}(h^{r+1})$.
- *if* $k = 0$ *and* $\boldsymbol{\beta} \neq \boldsymbol{0}$: $E_h^{(2)} = \mathcal{O}(h)$ and $E_h^{(1)} = \mathcal{O}(h^{\frac{1}{2}})$ (at worst). The rate of convergence given by (3.132) is thus $\mathcal{O}(h) = \mathcal{O}(h^{r+1})$ (since $r = 0$ whenever $k = 0$).

This demonstrates that Theorem 3.42 indeed yields an improved order of convergence in L^2-norm, compared to the energy estimate in Theorem 3.32. For $k \geq 1$, the improvement is one full power of h if diffusion dominates, and one-half power of h in advection-dominated regimes. For $k = 0$, the improvement is one full power of h in absence of advection, and one half-power of h if advection is dominant.

Proof (Theorem 3.42) We apply the Aubin–Nitsche Lemma A.10 with $\mathrm{U}_h = \underline{U}_{h,0}^k$ endowed with the norm $\|\cdot\|_{\flat,h}$ defined by (3.81), $L = L^2(\Omega)$ and $\mathrm{r}_h : \underline{U}_{h,0}^k \to L^2(\Omega)$ defined by $\mathrm{r}_h \underline{v}_h = v_h$ for all $\underline{v}_h \in \underline{U}_{h,0}^k$. Owing to (3.104), the estimate (3.132) holds if we can, for all $g \in L^2(\Omega)$ with $\|g\| \le 1$, bound the dual consistency error $\|\mathcal{E}_h^{\mathrm{d}}(z_g; \cdot)\|_{\flat,h,\star}$ (with $\|\cdot\|_{\flat,h,\star}$ denoting the norm dual to $\|\cdot\|_{\flat,h}$) by $E_h^{(1)}$ and the primal-dual consistency error $\mathcal{E}_h(u; \mathrm{I}_h z_g)$ by $E_h^{(2)}$.

(i) *Dual consistency error.* The definition (3.77) of $\mathrm{a}_{\boldsymbol{\beta},\mu,h}$ together with the property $\mathrm{G}_{-\boldsymbol{\beta},T}^k = -\mathrm{G}_{\boldsymbol{\beta},T}^k$ (see (3.64)) shows that $\mathrm{a}_{\boldsymbol{K},\boldsymbol{\beta},\mu,h}(\underline{v}_h, \underline{I}_h^k z) = \mathrm{a}_{\boldsymbol{K},\widetilde{\boldsymbol{\beta}},\widetilde{\mu},h}(\underline{I}_h^k z, \underline{v}_h)$ where $\widetilde{\boldsymbol{\beta}} = -\boldsymbol{\beta}$ and $\widetilde{\mu} = \mu + \boldsymbol{\nabla}{\cdot}\boldsymbol{\beta}$. Hence, the dual consistency error $\mathcal{E}_h^{\mathrm{d}}(z_g; \cdot)$ as in Definition A.9 is nothing else but the consistency error for the primal problem (3.60) with $(\boldsymbol{\beta}, \mu)$ replaced by $(\widetilde{\boldsymbol{\beta}}, \widetilde{\mu})$. The relation $\frac{1}{2}\boldsymbol{\nabla}{\cdot}\widetilde{\boldsymbol{\beta}} + \widetilde{\mu} = \frac{1}{2}\boldsymbol{\nabla}{\cdot}\boldsymbol{\beta} + \mu \ge \mu_0$ shows that $(\widetilde{\boldsymbol{\beta}}, \widetilde{\mu})$ satisfies Assumption 3.22. Estimates (3.48) and (3.85) thus give a bound on $\|\mathcal{E}_h^{\mathrm{d}}(z_g; \cdot)\|_{\flat,h,\star}$. When replacing $(\boldsymbol{\beta}, \mu)$ by $(\widetilde{\boldsymbol{\beta}}, \widetilde{\mu})$, the reference velocity and Péclet numbers are still $\hat{\beta}_T$ and Pe_T, while the inverse of the reference time becomes $\max\big[\|\mu + \boldsymbol{\nabla}{\cdot}\boldsymbol{\beta}\|_{L^\infty(T)}, \mathrm{L}_{\boldsymbol{\beta},T}\big]$. Hence, since $z_g \in H^2(\Omega)$, (3.48) and (3.85) with $r = 0$ yield

$$
\begin{aligned}
&\|\mathcal{E}_h^{\mathrm{d}}(z_g; \cdot)\|_{\flat,h,\star}\\
&\lesssim \Bigg\{ \sum_{T\in\mathcal{T}_h} \Big(\overline{K}_T \alpha_T |z_g|_{H^2(T)}^2 + \max\big[\|\mu + \boldsymbol{\nabla}{\cdot}\boldsymbol{\beta}\|_{L^\infty(T)}, \mathrm{L}_{\boldsymbol{\beta},T}\big]^2 \mu_0^{-1} |z_g|_{H^1(T)}^2 \Big) h_T^2\\
&\qquad + \sum_{T\in\mathcal{T}_h} \hat{\beta}_T |z_g|_{H^1(T)}^2 \Big[\min(1, \mathrm{Pe}_T)^{\frac{1}{2}} h_T^{\frac{1}{2}} \Big]^2 \Bigg\}^{\frac{1}{2}}\\
&\lesssim \Bigg\{ |z_g|_{H^2(\Omega)}^2 \max_{T\in\mathcal{T}_h} \Big(\overline{K}_T \alpha_T h_T^2 \Big)\\
&\qquad + \mu_0^{-1} |z_g|_{H^1(\Omega)}^2 \max_{T\in\mathcal{T}_h} \big(\max\big[\|\mu + \boldsymbol{\nabla}{\cdot}\boldsymbol{\beta}\|_{L^\infty(T)}, \mathrm{L}_{\boldsymbol{\beta},T}\big] h_T \big)^2\\
&\qquad + |z_g|_{H^1(\Omega)}^2 \max_{T\in\mathcal{T}_h} \Big(\hat{\beta}_T \big[\min(1, \mathrm{Pe}_T) h_T \big] \Big) \Bigg\}^{\frac{1}{2}}.
\end{aligned}
$$

Invoking (3.128) and (3.130) and recalling that $\|g\| \le 1$ leads to

$$
\|\mathcal{E}_h^{\mathrm{d}}(z_g; \cdot)\|_{\flat,h,\star} \lesssim E_h^{(1)},
$$

which is the required estimate on the dual consistency error.

(ii.A) *Primal-dual consistency error, case $k \geq 1$.* Recalling the definitions (3.49) and (3.86) of $\mathcal{E}_{\boldsymbol{K},h}(u;\cdot)$, $\mathcal{E}_{\mu,h}(u;\cdot)$ and $\mathcal{E}_{\boldsymbol{\beta},h}(u;\cdot)$, we decompose the primal-dual consistency error into

$$\mathcal{E}_h(u; \underline{I}_h^k z_g) = \mathcal{E}_{\boldsymbol{K},h}(u; \underline{I}_h^k z_g) + \mathcal{E}_{\mu,h}(u; \underline{I}_h^k z_g) + \mathcal{E}_{\boldsymbol{\beta},h}(u; \underline{I}_h^k z_g). \tag{3.133}$$

Let $\hat{\underline{z}}_h = \underline{I}_h^k z_g$. To estimate the diffusive contribution, we use as a starting point (3.52) with $w = u$ and $\underline{v}_h = \hat{\underline{z}}_h$. The consistency property (3.31) with $r = 0$ shows that

$$\mathrm{s}_{\boldsymbol{K},T}(\hat{\underline{z}}_T, \hat{\underline{z}}_T)^{\frac{1}{2}} \lesssim \overline{K}_T^{\frac{1}{2}} \alpha_T^{\frac{1}{2}} h_T |z_g|_{H^2(T)}.$$

Invoking (2.78) (which requires $k \geq 1$), we also see that

$$\sum_{F \in \mathcal{F}_T} \frac{K_{TF}}{h_F} \|\hat{z}_F - \hat{z}_T\|_F^2 \leq \overline{K}_T \sum_{F \in \mathcal{F}_T} h_F^{-1} \|\hat{z}_F - \hat{z}_T\|_F^2 \lesssim |z_g|_{H^2(T)}^2 h_T^2 \overline{K}_T.$$

Plug these estimates into (3.52) with $w = u$ and $\underline{v}_h = \hat{\underline{z}}_h$. Using $1 \leq \alpha_T$, a Cauchy–Schwarz inequality, (3.128), and $\|g\| \leq 1$, we infer

$$\begin{aligned}
|\mathcal{E}_{\boldsymbol{K},h}(u; \hat{\underline{z}}_h)| &\lesssim \sum_{T \in \mathcal{T}_h} \overline{K}_T h_T^{r+2} |u|_{H^{r+2}(T)} |z_g|_{H^2(T)} \\
&\quad + \sum_{T \in \mathcal{T}_h} \overline{K}_T \alpha_T h_T^{r+2} |u|_{H^{r+2}(T)} |z_g|_{H^2(T)} \\
&\lesssim C_{\mathrm{ell}} \left(\sum_{T \in \mathcal{T}_h} \overline{K}_T^2 \alpha_T^2 h_T^{2(r+2)} |u|_{H^{r+2}(T)}^2 \right)^{\frac{1}{2}}.
\end{aligned} \tag{3.134}$$

We now estimate the reactive contribution to the primal-dual consistency error. Since $(\pi_T^{0,0}\mu)\hat{z}_T \in \mathbb{P}^k(T)$, the orthogonality property of the projector $\pi_T^{0,k}$ yields $(u - \pi_T^{0,k} u, (\pi_T^{0,0}\mu)\hat{z}_T)_T = 0$ and thus

$$\begin{aligned}
\mathcal{E}_{\mu,h}(u; \hat{\underline{z}}_h) &= \sum_{T \in \mathcal{T}_h} (\mu u, \hat{z}_T)_T - (\mu \pi_T^{0,k} u, \hat{z}_T)_T \\
&= \sum_{T \in \mathcal{T}_h} (u - \pi_T^{0,k} u, \mu \hat{z}_T - (\pi_T^{0,0}\mu)\hat{z}_T)_T.
\end{aligned}$$

Hence, using generalised Hölder inequalities with exponents $(2, \infty, 2)$, we obtain

$$
\begin{aligned}
|\mathcal{E}_{\mu,h}(u; \hat{\underline{z}}_h)| &\le \sum_{T\in\mathcal{T}_h} \|u - \pi_T^{0,k} u\|_T \|\mu - \pi_T^{0,0}\mu\|_{L^\infty(T)} \|\pi_T^{0,k} z_g\|_T \\
&\lesssim \left(\sum_{T\in\mathcal{T}_h} \|\nabla\mu\|^2_{L^\infty(T)^d} h_T^{2(r+2)} |u|^2_{H^{r+1}(T)} \right)^{\frac{1}{2}} \frac{h_\Omega^2}{\pi^2 \underline{K}},
\end{aligned}
\tag{3.135}
$$

where we have used the approximation property (1.74) of the L^2-orthogonal projectors on T with $(l, p, s, m, v) = (k, 2, r+1, 0, u)$ and $(l, p, s, m, v) = (0, \infty, 1, 0, \mu)$, followed by the L^2-stability $\|\pi_T^{0,k} z_g\|_T \le \|z_g\|_T$, a discrete Cauchy–Schwarz inequality on the sum over $T \in \mathcal{T}_h$, the a priori estimate (3.131), and $\|g\| \le 1$. Note that we did not use $k \ge 1$ to establish the estimate (3.135), which is therefore also valid for $k = 0$.

Finally, we estimate the advective contribution $\mathcal{E}_{\boldsymbol{\beta},h}(u; \hat{\underline{z}}_h)$ by estimating each term $\mathfrak{T}_1(\hat{\underline{z}}_h)$ and $\mathfrak{T}_2(\hat{\underline{z}}_h)$ in (3.89) with $w = u$ and $\underline{v}_h = \hat{\underline{z}}_h$. The boundedness property (1.77) of $\pi_T^{0,k}$ with $p = 2$ and $s = 1$ shows that $\|\nabla \hat{z}_T\|_T = \|\nabla \pi_T^{0,k} z_g\|_T \lesssim \|\nabla z_g\|_T$. Estimate (3.90) therefore gives

$$
\begin{aligned}
|\mathfrak{T}_1(\hat{\underline{z}}_h)| &\le \sum_{T\in\mathcal{T}_h} \hat{\tau}_T^{-1} h_T^{r+1} |u|_{H^{r+1}(T)} h_T \|\nabla z_g\|_T \\
&\le \left(\sum_{T\in\mathcal{T}_h} \hat{\tau}_T^{-2} h_T^{2(r+2)} |u|^2_{H^{r+1}(T)} \right)^{\frac{1}{2}} |z_g|_{H^1(\Omega)}.
\end{aligned}
$$

To estimate $|\mathfrak{T}_2(\hat{z}_h)|$ we use (3.93) and

$$
\|\hat{z}_F - \hat{z}_T\|_F = \|\pi_F^{0,k}(z - \pi_T^{0,k} z)\|_F \le \|z - \pi_T^{0,k} z\|_F \lesssim h_T^{\frac{3}{2}} |z_g|_{H^2(T)},
\tag{3.136}
$$

which follows from (1.75) with $m = 0$, $p = 2$, $l = k$, $s = 2$ (this s is valid since $k \ge 1$ here), to write

$$
\begin{aligned}
|\mathfrak{T}_2(\hat{z}_h)| &\lesssim \sum_{T\in\mathcal{T}_h} \hat{\beta}_T h_T^{r+2} |u|_{H^{r+1}(T)} |z_g|_{H^2(T)} \\
&\lesssim \left(\sum_{T\in\mathcal{T}_h} \hat{\beta}_T^2 h_T^{2(r+2)} |u|^2_{H^{r+1}(T)} \right)^{\frac{1}{2}} |z_g|_{H^2(\Omega)}.
\end{aligned}
$$

These estimates on $\mathfrak{T}_1(\hat{z}_h)$ and $\mathfrak{T}_2(\hat{z}_h)$ lead to

$$|\mathcal{E}_{\boldsymbol{\beta},h}(u;\underline{\hat{z}}_h)| \lesssim$$

$$\left(\sum_{T\in\mathcal{T}_h} \hat{\tau}_T^{-2} h_T^{2(r+2)} |u|^2_{H^{r+1}(T)}\right)^{\frac{1}{2}} \frac{h_\Omega}{\pi \underline{K}} + \left(\sum_{T\in\mathcal{T}_h} \hat{\beta}_T^2 h_T^{2(r+2)} |u|^2_{H^{r+1}(T)}\right)^{\frac{1}{2}} C_{\mathrm{ell}},$$

where we have used the elliptic regularity (3.128) and the a priori H^1-estimate (3.130) together with $\|g\| \le 1$. Plugging the above estimate together with (3.134) and (3.135) into (3.133), we infer

$$|\mathcal{E}_h(u;\underline{I}_h^k z_g)| \lesssim E_h^{(2)}.$$

(ii.B) *Primal-dual consistency error, case* $k = 0$. As made explicit in the above arguments, the primal-dual consistency error coming from the reaction term can still be estimated by (3.135) if $k = 0$. We therefore only have to analyse the consistency error coming from the advective and diffusive contributions.

Let us first consider $\mathcal{E}_{\boldsymbol{\beta},h}(u;\underline{\hat{z}}_h) = \mathfrak{T}_1(\underline{\hat{z}}_h) + \mathfrak{T}_2(\underline{\hat{z}}_h)$, where $\mathfrak{T}_1(\underline{\hat{z}}_h)$ and $\mathfrak{T}_2(\underline{\hat{z}}_h)$ are defined in (3.89) with $w = u$. Since $k = 0$, $\hat{z}_T = \pi_T^{0,0} z \in \mathbb{P}^0(T)$ and thus $\mathfrak{T}_1(\underline{\hat{z}}_h) = 0$. For $\mathfrak{T}_2(\underline{\hat{z}}_h)$ we notice that, owing to (1.75) with $(p, l, m, s) = (2, 0, 0, 1)$, the estimate (3.136) is transformed into

$$\|\hat{z}_F - \hat{z}_T\|_F \le \|z - \pi_T^{0,0} z\|_F \lesssim h_T^{\frac{1}{2}} |z_g|_{H^1(T)}.$$

Hence, recalling that $r = 0$ here (since $k = 0$), the estimate (3.93), Cauchy–Schwarz inequalities and the stability estimate (3.130) lead to

$$\begin{aligned} |\mathcal{E}_{\boldsymbol{\beta},h}(u;\underline{\hat{z}}_h)| = |\mathfrak{T}_2(\underline{\hat{z}}_h)| &\lesssim \left(\sum_{T\in\mathcal{T}_h} \hat{\beta}_T^2 h_T^2 |u|^2_{H^1(T)}\right)^{\frac{1}{2}} |z_g|_{H^1(\Omega)} \\ &\lesssim \left(\sum_{T\in\mathcal{T}_h} \hat{\beta}_T^2 h_T^2 |u|^2_{H^1(T)}\right)^{\frac{1}{2}} \frac{h_\Omega}{\pi \underline{K}} \|g\|. \end{aligned} \tag{3.137}$$

For $\mathcal{E}_{\boldsymbol{K},h}(u;\underline{I}_h^k z_g)$, we use the same ideas as in (ii.B) in the proof of Lemma 2.33. The definitions (3.49) of $\mathcal{E}_{\boldsymbol{K},h}(u;\underline{I}_h^0 z_g)$ and (3.26) of $\mathrm{a}_{\boldsymbol{K},T}$, together with the property $\mathrm{p}^1_{\boldsymbol{K},T}\underline{I}_T^0 = \pi^{1,1}_{\boldsymbol{K},T}$ (see (3.24)), give

$$\begin{aligned} \mathcal{E}_{\boldsymbol{K},h}(u;\underline{I}_h^0 z_g) = \sum_{T\in\mathcal{T}_h} (-\nabla\cdot(\boldsymbol{K}\nabla u), \pi_T^{0,0} z_g)_T - \sum_{T\in\mathcal{T}_h} (\boldsymbol{K}_T \nabla \pi^{1,1}_{\boldsymbol{K},T} u, \nabla \pi^{1,1}_{\boldsymbol{K},T} z_g)_T \\ - \sum_{T\in\mathcal{T}_h} \mathrm{s}_{\boldsymbol{K},T}(\underline{I}_T^0 u, \underline{I}_T^0 z_g). \end{aligned}$$

Let $f_{\boldsymbol{K}} = -\nabla\cdot(\boldsymbol{K}\nabla u) \in H^1(\mathcal{T}_h)$. Then $(f_{\boldsymbol{K}}, z_g) = (\boldsymbol{K}\nabla u, \nabla z_g)$ and thus

$$\begin{aligned}\sum_{T\in\mathcal{T}_h}(-\nabla\cdot(\boldsymbol{K}\nabla u), \pi_T^{0,0} z_g)_T &= \sum_{T\in\mathcal{T}_h}(\pi_T^{0,0} f_{\boldsymbol{K}}, z_g)_T \\ &= \sum_{T\in\mathcal{T}_h}(\pi_T^{0,0} f_{\boldsymbol{K}} - f_{\boldsymbol{K}}, z_g)_T + (\boldsymbol{K}\nabla u, \nabla z_g) \\ &= \sum_{T\in\mathcal{T}_h}(\pi_T^{0,0} f_{\boldsymbol{K}} - f_{\boldsymbol{K}}, z_g - \pi_T^{0,0} z_g)_T + (\boldsymbol{K}\nabla u, \nabla z_g).\end{aligned}$$

Hence,

$$\begin{aligned}|\mathcal{E}_{\boldsymbol{K},h}(u; \underline{I}_h^k z_g)| \le &\sum_{T\in\mathcal{T}_h} \|\pi_T^{0,0} f_{\boldsymbol{K}} - f_{\boldsymbol{K}}\|_T \|z_g - \pi_T^{0,0} z_g\|_T \\ &+ \underbrace{\left|\sum_{T\in\mathcal{T}_h} (\boldsymbol{K}_T\nabla u, \nabla z_g)_T - (\boldsymbol{K}_T\nabla\pi_{\boldsymbol{K},T}^{1,1} u, \nabla\pi_{\boldsymbol{K},T}^{1,1} z_g)_T\right|}_{\mathfrak{T}_{\boldsymbol{K}}} \\ &+ \sum_{T\in\mathcal{T}_h} \mathrm{s}_{\boldsymbol{K},T}(\underline{I}_T^0 u, \underline{I}_T^0 u)^{\frac{1}{2}} \mathrm{s}_{\boldsymbol{K},T}(\underline{I}_T^0 z_g, \underline{I}_T^0 z_g)^{\frac{1}{2}} \\ \lesssim &\left(\sum_{T\in\mathcal{T}_h} h_T^4 |f_{\boldsymbol{K}}|_{H^1(T)}^2\right)^{\frac{1}{2}} |z_g|_{H^1(\Omega)} + \mathfrak{T}_{\boldsymbol{K}} \\ &+ \left(\sum_{T\in\mathcal{T}_h} \overline{K}_T^2 \alpha_T^2 h_T^4 |u|_{H^2(T)}^2\right)^{\frac{1}{2}} |z_g|_{H^2(\Omega)},\end{aligned} \tag{3.138}$$

where the conclusion follows from the approximation properties (1.74) of $\pi_T^{0,0}$, the consistency property (3.31) of $\mathrm{s}_{\boldsymbol{K},T}$, and Cauchy–Schwarz inequalities. To estimate $\mathfrak{T}_{\boldsymbol{K}}$, we write

$$\begin{aligned}&(\boldsymbol{K}_T\nabla u, \nabla z_g)_T - (\boldsymbol{K}_T\nabla\pi_{\boldsymbol{K},T}^{1,1} u, \nabla\pi_{\boldsymbol{K},T}^{1,1} z_g)_T \\ &\quad = (\boldsymbol{K}_T(\nabla u - \nabla\pi_{\boldsymbol{K},T}^{1,1} u), \nabla z_g)_T + \underbrace{(\nabla\pi_{\boldsymbol{K},T}^{1,1} u, \boldsymbol{K}_T(\nabla z_g - \nabla\pi_{\boldsymbol{K},T}^{1,1} z_g))_T}_{=0 \text{ by (3.7a) with } v = z_g \text{ and } w = \pi_{\boldsymbol{K},T}^{1,1} u} \\ &\quad = (\boldsymbol{K}_T(\nabla u - \nabla\pi_{\boldsymbol{K},T}^{1,1} u), (\nabla z_g - \nabla\pi_{\boldsymbol{K},T}^{1,1} z_g))_T + \underbrace{(\boldsymbol{K}_T(\nabla u - \nabla\pi_{\boldsymbol{K},T}^{1,1} u), \nabla\pi_{\boldsymbol{K},T}^{1,1} z_g)_T}_{=0 \text{ by (3.7a) with } v = u \text{ and } w = \pi_{\boldsymbol{K},T}^{1,1} z_g}.\end{aligned}$$

Hence, Cauchy–Schwarz inequalities and the approximation property (3.10) of $\pi^{1,1}_{\boldsymbol{K},T}$ yield

$$\mathfrak{T}_{\boldsymbol{K}} \leq \sum_{T\in\mathcal{T}_h} \|\boldsymbol{K}_T^{\frac{1}{2}}(\nabla u - \nabla\pi^{1,1}_{\boldsymbol{K},T}u)\|_T \|\boldsymbol{K}_T^{\frac{1}{2}}(\nabla z_g - \nabla\pi^{1,1}_{\boldsymbol{K},T}z_g)\|_T$$

$$\lesssim \left(\sum_{T\in\mathcal{T}_h} \overline{K}_T^2 h_T^4 |u|^2_{H^2(T)}\right)^{\frac{1}{2}} |z_g|_{H^2(\Omega)}.$$

Plugged into (3.138) and using (3.130) and (3.128) together with $\|g\| \leq 1$, this shows, since $\alpha_T \geq 1$, that

$$|\mathcal{E}_{\boldsymbol{K},h}(u; \underline{I}_h^k z_g)|$$

$$\lesssim \frac{h_\Omega}{\pi\underline{K}}\left(\sum_{T\in\mathcal{T}_h} h_T^4|\nabla\cdot(\boldsymbol{K}\nabla u)|^2_{H^1(T)}\right)^{\frac{1}{2}} + C_{\rm ell}\left(\sum_{T\in\mathcal{T}_h} \overline{K}_T^2\alpha_T^2 h_T^4 |u|^2_{H^2(T)}\right)^{\frac{1}{2}}.$$

This estimate, combined with (3.135) and (3.137), shows that the primal-dual consistency error is bounded above by $E_h^{(2)}$, as defined in the theorem in the case $k = 0$. This concludes the proof. □

3.3 Numerical Examples

We provide in this section numerical examples to illustrate the performance of the HHO method for the diffusion–advection–reaction model (3.1).

3.3.1 Two-Dimensional Test Case

In the first test case, we solve in the unit square $\Omega = (0, 1)^2$ the Dirichlet problem (3.60) corresponding to the solution

$$u(\boldsymbol{x}) = \sin(\pi x_1)\sin(\pi x_2)$$

with $\boldsymbol{\beta}(\boldsymbol{x}) = (\frac{1}{2} - x_2, x_1 - \frac{1}{2})$, $\mu = 1$, and a uniform diffusion coefficient $\boldsymbol{K} = \nu\boldsymbol{I}_d$ with real number ν taking values in {1e-03, 1, 1e+03}, corresponding to the advection-dominated, intermediate and diffusion-dominated regimes. The domain is discretised by means of a refined sequence of unstructured triangular meshes. The first four refinements are depicted in Fig. 3.1a. We monitor the energy- and L^2-norms of the error, whose value is normalised with respect to the corresponding

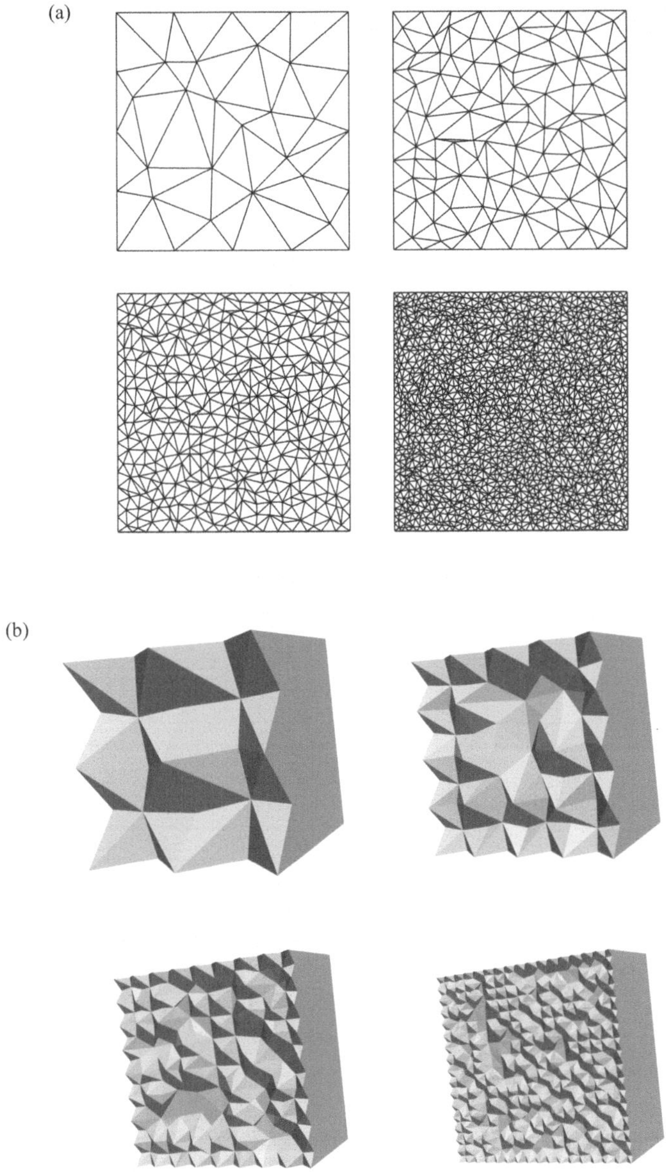

Fig. 3.1 Meshes for the numerical tests of Sect. 3.3. (**a**) First four refinements of the mesh sequence used in the test of Sect. 3.3.1. (**b**) Cuts of the first four refinements of the mesh sequence used in the test of Sect. 3.3.2

Table 3.1 Convergence results for the two-dimensional test case of Sect. 3.3.1, $\nu = 1$

$N_{\text{dof},h}$	$\Vert \underline{u}_h - \underline{\hat{u}}_h \Vert_{\flat,h}$	EOC	$\Vert u_h - \pi_h^{0,k} u \Vert$	EOC
$k = 0$				
83	3.51e−01	–	1.29e−01	–
319	1.78e−01	0.98	3.08e−02	2.06
1247	8.73e−02	1.03	6.76e−03	2.19
4765	4.48e−02	0.96	1.33e−03	2.35
19,280	2.23e−02	1.01	1.98e−04	2.74
75,181	1.13e−02	0.98	1.18e−04	0.75
$k = 1$				
166	5.75e−02	–	1.56e−02	–
638	1.48e−02	1.96	2.13e−03	2.87
2494	3.76e−03	1.97	2.57e−04	3.05
9530	9.76e−04	1.95	3.47e−05	2.89
38,560	2.43e−04	2.01	4.32e−06	3.01
150,362	6.20e−05	1.97	5.61e−07	2.94
$k = 2$				
249	6.62e−03	–	1.72e−03	–
957	8.65e−04	2.94	1.05e−04	4.03
3741	1.09e−04	2.99	6.78e−06	3.95
14,295	1.39e−05	2.97	4.33e−07	3.97
57,840	1.74e−06	3.00	2.72e−08	4.00
225,543	2.28e−07	2.94	1.80e−09	3.92
$k = 3$				
332	6.41e−04	–	1.51e−04	–
1276	3.72e−05	4.11	4.28e−06	5.14
4988	2.37e−06	3.97	1.34e−07	4.99
19,060	1.62e−07	3.87	4.68e−09	4.84
77,120	9.77e−09	4.05	1.42e−10	5.04
300,724	6.38e−10	3.94	1.66e−11	3.10

norm of $\underline{I}_h^k u$. Each error measure is accompanied by the Estimated Order of Convergence (EOC) which, denoting by e_i an error on the ith mesh refinement, is computed as

$$\text{EOC} = \frac{\log e_i - \log e_{i+1}}{\log h_i - \log h_{i+1}}.$$

The convergence results collected in Tables 3.1, 3.2, and 3.3 show that the energy norm converges as h^{k+1} when $\nu = 1$ or $\nu =$ 1e+03 while, as predicted, a loss of about a half order of convergence is observed when $\nu =$ 1e-03. The L^2-norm of the error, on the other hand, converges as h^{k+2} when $\nu = 1$ or $\nu =$ 1e+03 while, when $\nu =$ 1e-03, the observed convergence is in h for $k = 0$ and between h^{k+1} and $h^{k+\frac{3}{2}}$ for $k \geq 1$. The apparent loss of convergence in Table 3.1 on the last mesh for

Table 3.2 Convergence results for the two-dimensional test case of Sect. 3.3.1, $\nu = $ 1e-03

$N_{\mathrm{dof},h}$	$\Vert \underline{u}_h - \underline{\hat{u}}_h \Vert_{\flat,h}$	EOC	$\Vert u_h - \pi_h^{0,k} u \Vert$	EOC
$k = 0$				
83	2.95e−01	–	1.61e−01	–
319	2.11e−01	0.48	8.45e−02	0.93
1247	1.48e−01	0.51	4.65e−02	0.86
4765	1.04e−01	0.51	2.44e−02	0.93
19,280	6.84e−02	0.61	1.16e−02	1.07
75,181	4.22e−02	0.70	5.48e−03	1.09
$k = 1$				
166	6.33e−02	–	2.56e−02	–
638	2.36e−02	1.42	6.86e−03	1.90
2494	8.34e−03	1.50	1.62e−03	2.08
9530	2.87e−03	1.54	3.52e−04	2.20
38,560	9.34e−04	1.62	7.07e−05	2.32
150,362	2.99e−04	1.64	1.35e−05	2.39
$k = 2$				
249	7.24e−03	–	2.54e−03	–
957	1.24e−03	2.55	3.15e−04	3.01
3741	2.18e−04	2.51	3.49e−05	3.17
14,295	4.00e−05	2.45	4.18e−06	3.06
57,840	6.30e−06	2.67	3.90e−07	3.42
225,543	1.00e−06	2.65	3.68e−08	3.41
$k = 3$				
332	5.59e−04	–	1.81e−04	–
1276	5.19e−05	3.43	1.17e−05	3.95
4988	5.14e−06	3.34	7.62e−07	3.94
19,060	4.32e−07	3.57	3.96e−08	4.27
77,120	3.41e−08	3.66	1.91e−09	4.37
300,724	2.83e−09	3.59	9.95e−11	4.27

$k = 0$ in L^2-norm could correspond to an adjustment of the error, after the superconvergence that occurred for the previous meshes; the overall rate remains close to h^2. On the contrary, the reduced rate on the last mesh for $k = 3$ is due to rounding errors that start to become perceptible at these magnitudes.

These results corroborate overall the findings of Theorems 3.32 and 3.42 (see also Remark 3.43), with a notable exception for $k = 0$ when $\nu = 1$ or $\nu = $ 1e+03; in these cases, Tables 3.1 and 3.3 seem to suggest that the rate of convergence in L^2-norm is actually not impacted by the presence of advection when said advection is not dominant.

Table 3.3 Convergence results for the two-dimensional test case of Sect. 3.3.1, ν = 1e+03

$N_{\text{dof},h}$	$\Vert \underline{u}_h - \underline{\hat{u}}_h \Vert_{\flat,h}$	EOC	$\Vert u_h - \pi_h^{0,k} u \Vert$	EOC
$k = 0$				
83	3.66e−01	–	1.45e−01	–
319	1.84e−01	0.99	3.56e−02	2.03
1247	8.97e−02	1.04	8.54e−03	2.06
4765	4.59e−02	0.97	2.18e−03	1.97
19,280	2.29e−02	1.01	5.38e−04	2.02
75,181	1.16e−02	0.98	1.38e−04	1.97
$k = 1$				
166	5.99e−02	–	1.62e−02	–
638	1.52e−02	1.97	2.16e−03	2.91
2494	3.87e−03	1.98	2.59e−04	3.06
9530	1.00e−03	1.95	3.48e−05	2.90
38,560	2.49e−04	2.01	4.33e−06	3.01
150,362	6.35e−05	1.97	5.61e−07	2.95
$k = 2$				
249	6.89e−03	–	1.78e−03	–
957	8.96e−04	2.94	1.07e−04	4.06
3741	1.12e−04	3.00	6.82e−06	3.97
14,295	1.43e−05	2.97	4.35e−07	3.97
57,840	1.79e−06	3.00	2.72e−08	4.00
225,543	2.33e−07	2.94	1.80e−09	3.92
$k = 3$				
332	6.67e−04	–	1.55e−04	–
1276	3.81e−05	4.13	4.31e−06	5.17
4988	2.44e−06	3.97	1.36e−07	4.99
19,060	1.66e−07	3.88	4.68e−09	4.86
77,120	1.00e−08	4.05	1.42e−10	5.05
300,724	6.54e−10	3.94	2.40e−11	2.56

3.3.2 *Three-Dimensional Test Case*

In the second test case, we solve in the unit cube $\Omega = (0,1)^3$ the Dirichlet problem (3.60) corresponding to the solution

$$u(\boldsymbol{x}) = \sin(\pi x_1)\sin(\pi x_2)\sin(\pi x_3)$$

with $\boldsymbol{\beta}(\boldsymbol{x}) = (x_2 - x_3, x_3 - x_1, x_1 - x_2)$, $\mu = 1$, and a uniform diffusion coefficient $\boldsymbol{K} = \nu \boldsymbol{I}_d$ with real number ν taking values in {1e-03, 1, 1e+03}. The domain is discretised by means of a refined sequence of unstructured simplicial meshes; see Fig. 3.1b. Similar considerations as for the two-dimensional test case hold for the corresponding results collected in Tables 3.4, 3.5, and 3.6.

Table 3.4 Convergence results for the three-dimensional test case of Sect. 3.3.2, $\nu = 1$

$N_{\mathrm{dof},h}$	$\|\underline{u}_h - \underline{\hat{u}}_h\|_{\flat,h}$	EOC	$\|u_h - \pi_h^{0,k} u\|$	EOC
$k = 0$				
432	3.89e−01	–	1.40e−01	–
3264	1.93e−01	1.01	2.97e−02	2.23
25,344	9.70e−02	0.99	6.24e−03	2.25
199,680	4.87e−02	0.99	9.67e−04	2.69
1,585,152	2.44e−02	1.00	2.28e−04	2.08
$k = 1$				
1296	1.11e−01	–	3.12e−02	–
9792	2.95e−02	1.91	4.07e−03	2.94
76,032	7.53e−03	1.97	5.17e−04	2.97
599,040	1.89e−03	1.99	6.54e−05	2.98
4,755,456	4.75e−04	2.00	8.22e−06	2.99
$k = 2$				
2592	2.13e−02	–	5.31e−03	–
19,584	2.78e−03	2.94	3.61e−04	3.88
152,064	3.55e−04	2.97	2.31e−05	3.96
1,198,080	4.50e−05	2.98	1.47e−06	3.97
9,510,912	5.65e−06	2.99	9.28e−08	3.99

Table 3.5 Convergence results for the three-dimensional test case of Sect. 3.3.2, ν = 1e-03

$N_{\mathrm{dof},h}$	$\|\underline{u}_h - \underline{\hat{u}}_h\|_{\flat,h}$	EOC	$\|u_h - \pi_h^{0,k} u\|$	EOC
$k = 0$				
432	4.15e−01	–	2.55e−01	–
3264	3.47e−01	0.26	1.77e−01	0.53
25,344	2.48e−01	0.49	9.41e−02	0.91
199,680	1.69e−01	0.55	4.70e−02	1.00
1,585,152	1.10e−01	0.62	2.32e−02	1.02
$k = 1$				
1296	1.57e−01	–	6.81e−02	–
9792	5.94e−02	1.40	1.70e−02	2.00
76,032	2.11e−02	1.49	4.05e−03	2.07
599,040	7.23e−03	1.54	8.74e−04	2.21
4,755,456	2.39e−03	1.60	1.72e−04	2.34
$k = 2$				
2592	2.75e−02	–	9.45e−03	–
19,584	5.47e−03	2.33	1.38e−03	2.77
152,064	9.56e−04	2.52	1.56e−04	3.14
1,198,080	1.63e−04	2.55	1.64e−05	3.26
9,510,912	2.68e−05	2.60	1.58e−06	3.37

Table 3.6 Convergence results for the three-dimensional test case of Sect. 3.3.2, $\nu = 1e{+}03$

$N_{\mathrm{dof},h}$	$\|\underline{u}_h - \underline{\hat{u}}_h\|_{\flat,h}$	EOC	$\|u_h - \pi_h^{0,k} u\|$	EOC
$k = 0$				
432	3.89e−01	–	1.40e−01	–
3264	1.93e−01	1.01	2.97e−02	2.23
25,344	9.70e−02	0.99	6.24e−03	2.25
199,680	4.87e−02	0.99	9.67e−04	2.69
1,585,152	2.44e−02	1.00	2.28e−04	2.08
$k = 1$				
1296	1.11e−01	–	3.12e−02	–
9792	2.95e−02	1.91	4.07e−03	2.94
76,032	7.53e−03	1.97	5.17e−04	2.97
599,040	1.89e−03	1.99	6.54e−05	2.98
4,755,456	4.75e−04	2.00	8.22e−06	2.99
$k = 2$				
2592	2.22e−02	–	5.53e−03	–
19,584	2.85e−03	2.96	3.68e−04	3.91
152,064	3.62e−04	2.98	2.33e−05	3.98
1,198,080	4.58e−05	2.98	1.48e−06	3.98
9,510,912	5.75e−06	2.99	9.29e−08	3.99

Chapter 4
Complements on Pure Diffusion

This chapter covers two unrelated topics on HHO methods for linear diffusion problems: an a posteriori error analysis for the Poisson problem and the extension of the HHO method to the case of a diffusion tensor that varies inside each element. These topics build up on Chaps. 1 and 2, and can be used in a short introductory course to present more advanced notions on HHO.

In Sect. 4.1 we derive a posteriori error estimates for the HHO discretisation (2.48) of the Poisson problem. We prove a reliable (and, in the case of simplicial meshes, also guaranteed and fully computable) upper bound of the error in terms of residual-based estimators. We next show that the estimate is both locally and globally efficient, meaning that the estimators are in turn controlled by the discretisation error. We finally investigate the numerical performance of an adaptive algorithm where local mesh refinement is driven by the estimators derived in the previous sections.

Section 4.2 considers the case of a diffusion model where, contrary to the situation covered in Sect. 3.1, the diffusion coefficient is allowed to vary inside each cell. Designing an HHO scheme for such a model requires a different approach from the one taken in Sect. 3.1, using a new gradient reconstruction operator. After introducing the HHO scheme based on this reconstruction, we prove optimal error estimates, in both the energy- and L^2-norms. Numerical validation is provided solving a test case with a locally variable and highly anisotropic diffusion tensor.

4.1 A Posteriori Error Analysis

A priori error estimates such as (2.62) are useful to assess the rate of convergence of the method, but the bound they provide is not computable as it involves the (unknown) exact solution u. It is often useful to establish computable estimates of the error between the approximate and exact solution, which is precisely the goal of

D. A. Di Pietro, J. Droniou, *The Hybrid High-Order Method for Polytopal Meshes*,
MS&A 19, https://doi.org/10.1007/978-3-030-37203-3_4

a posteriori error analysis. A particularly important application of a posteriori error estimates is mesh adaptation, as briefly discussed hereafter. For smooth enough exact solutions, increasing the polynomial degree yields a corresponding increase in the convergence rate; see, e.g., the a priori error estimates proved in Sect. 2.3 and the numerical tests in Sect. 2.5. However, when the regularity of the exact solution is insufficient, the order of convergence is limited by the latter instead of the polynomial degree (see for example the tests in Sect. 4.1.3 below). To improve this situation (and, possibly, restore optimal orders of convergence in terms of error vs. the number of degrees of freedom; see, e.g., [97]), one can resort to local mesh adaptation. This is typically carried out by using local a posteriori error estimators to mark the elements where the error is larger, and by refining the computational mesh based on this information.

In this section, we present energy-norm a posteriori error estimates for the HHO discretisation (2.48) of the Poisson problem (2.2) recalled hereafter: Find $u \in H_0^1(\Omega)$ such that

$$a(u, v) := (\nabla u, \nabla v) = (f, v) \qquad \forall v \in H_0^1(\Omega), \tag{4.1}$$

where $f \in L^2(\Omega)$ denotes a given source term. We follow the residual-based approach of [154]; see also [155, Section 3.4]. Our goal here is to show how this classical approach (see [218]) can be applied to HHO schemes.

The rest of this section is organised as follows: in Sect. 4.1.1 we prove a reliable (and, in the case of simplicial meshes, guaranteed and fully computable) upper bound on the discretisation error; in Sect. 4.1.2 we prove local and global efficiency by showing that the error estimator is (locally or globally) bounded by the discretisation error; finally, in Sect. 4.1.3 we numerically demonstrate the performance of an automatic mesh adaptation algorithm driven by our a posteriori error estimators.

4.1.1 Energy Error Upper Bound

Denote by u the unique solution of the Poisson problem (4.1), and by $\underline{u}_h \in \underline{U}_{h,0}^k$ its HHO approximation obtained solving (2.48). Recalling the definition (2.63) of the global reconstruction operator p_h^{k+1}, the goal of this section is to prove an upper bound of the discretisation error of the form

$$\|\nabla_h(\mathrm{p}_h^{k+1}\underline{u}_h - u)\| \lesssim \varepsilon, \tag{4.2}$$

where the hidden constant is independent of the meshsize and of the problem data, while the quantity ε is computable in terms of the discrete solution and of the problem data only. An a posteriori error estimator that satisfies property (4.2) is said to be *reliable*. At least for simplicial meshes, we will aim at a stronger property than reliability, namely we want to obtain an estimate of the form

$$\|\nabla_h(\mathrm{p}_h^{k+1}\underline{u}_h - u)\| \le \varepsilon,$$

where the difference with respect to (4.2) is that no undetermined constant appears in the right-hand side, so that the bound is *guaranteed* and *fully computable*. To this purpose, we recall the following local Poincaré–Wirtinger (see Remark 1.46) and Friedrichs inequalities (consequence of (1.75)), valid for all $T \in \mathcal{T}_h$ and all $\varphi \in H^1(T)$:

$$\|\varphi - \pi_T^{0,0}\varphi\|_T \le C_{\mathrm{P},T} h_T \|\nabla\varphi\|_T, \tag{4.3}$$

$$\|\varphi - \pi_T^{0,0}\varphi\|_{\partial T} \le C_{\mathrm{F},T}^{\frac{1}{2}} h_T^{\frac{1}{2}} \|\nabla\varphi\|_T. \tag{4.4}$$

It was proved in [42, 250] that the real number $C_{\mathrm{P},T}$ in (4.3) can be taken equal to π^{-1} if T is convex. The real number $C_{\mathrm{F},T}$ in (4.4), on the other hand, can be defined by $C_{\mathrm{F},T} := C_{\mathrm{P},T}(h_T|\partial T|_{d-1}/|T|_d)(2/d + C_{\mathrm{P},T})$ if T is a simplex (see [145, Section 5.6.2.2]).

We extend the continuous bilinear form a to $H^1(\mathcal{T}_h) \times H^1(\mathcal{T}_h)$ by replacing the standard gradient operator ∇ with its broken counterpart ∇_h defined by (1.21). For any integer $l \ge 1$ and any broken polynomial function $v_h \in \mathbb{P}^l(\mathcal{T}_h)$, we then define the residual $\mathcal{R}(v_h) \in H^{-1}(\Omega)$ such that, for all $\varphi \in H_0^1(\Omega)$,

$$\langle \mathcal{R}(v_h), \varphi\rangle_{-1,1} := a(u - v_h, \varphi) = (f, \varphi) - a(v_h, \varphi), \tag{4.5}$$

where $\langle\cdot,\cdot\rangle_{-1,1}$ denotes the duality pairing between $H^{-1}(\Omega)$ and $H_0^1(\Omega)$. Our starting point to derive a guaranteed and fully computable upper bound on the discretisation error is contained in the following lemma, inspired by [218, Lemma 4.4]; see also [145, Lemma 5.44] and [154, Lemma 7].

Lemma 4.1 (Abstract Estimate) *Let $u \in H_0^1(\Omega)$ solve* (4.1). *Then, for any integer $l \ge 1$ and any broken polynomial function $v_h \in \mathbb{P}^l(\mathcal{T}_h)$, it holds that*

$$\|\nabla_h(u - v_h)\|^2 \le \inf_{\varphi \in H_0^1(\Omega)} \|\nabla_h(\varphi - v_h)\|^2 + \left(\sup_{\varphi \in H_0^1(\Omega), \|\nabla\varphi\|=1} \langle \mathcal{R}(v_h), \varphi\rangle_{-1,1}\right)^2. \tag{4.6}$$

Remark 4.2 (Nonconformity and Residual Terms) In (4.6), the difference between the exact solution u of the Poisson problem and a generic broken polynomial function v_h is estimated by two terms: the first one measures the nonconformity of v_h (i.e., the difference between v_h and the closest element of $H_0^1(\Omega)$); the second measures how far v_h is from being a solution to the Poisson problem in terms of the dual norm of the residual (4.5).

Proof Let $\psi \in H_0^1(\Omega)$ be such that

$$a(\psi, \varphi) = a(v_h, \varphi) \qquad \forall \varphi \in H_0^1(\Omega). \tag{4.7}$$

We observe that ψ is well-defined since the bilinear form a and the linear form $H_0^1(\Omega) \ni \varphi \mapsto a(v_h, \varphi) \in \mathbb{R}$ satisfy the assumptions of the Lax–Milgram Lemma 2.20. We have the following characterisation of ψ:

$$\|\nabla_h(v_h - \psi)\|^2 = \inf_{\varphi \in H_0^1(\Omega)} \|\nabla_h(v_h - \varphi)\|^2, \tag{4.8}$$

which can be inferred observing that (4.7) is the Euler equation for the minimisation problem (4.8); see, e.g., [12, Chapter 10]. Additionally, by definition (4.1) of the bilinear form $a(\cdot,\cdot)$, it holds that

$$\begin{aligned} \|\nabla(u-\psi)\| = \frac{a(u-\psi, u-\psi)}{\|\nabla(u-\psi)\|} &\le \sup_{\varphi \in H_0^1(\Omega), \|\nabla\varphi\|=1} a(u-\psi, \varphi) \\ &= \sup_{\varphi \in H_0^1(\Omega), \|\nabla\varphi\|=1} \langle \mathcal{R}(v_h), \varphi\rangle_{-1,1}, \end{aligned} \tag{4.9}$$

where the conclusion follows using the linearity of a in its first argument together with the definitions (4.7) of ψ and (4.5) of $\mathcal{R}(v_h)$. Finally, since $(v_h - \psi)$ is by definition a-orthogonal to the functions in $H_0^1(\Omega)$, using the Pythagorean theorem we have that

$$\|\nabla_h(u - v_h)\|^2 = \|\nabla_h(v_h - \psi)\|^2 + \|\nabla(u - \psi)\|^2.$$

To conclude, it suffices to use (4.8) and (4.9) to bound the terms in the right-hand side. □

We are now ready to prove the upper bound on the discretisation error.

Theorem 4.3 (A Posteriori Error Upper Bound) *Let $\mathcal{M}_h$ denote a polytopal mesh in the sense of Definition 1.4, and let an integer $k \ge 0$ be fixed. Let $u \in H_0^1(\Omega)$ and $\underline{u}_h \in \underline{U}_{h,0}^k$ denote the unique solutions to problems* (2.2) *and* (2.48)*, respectively, with local stabilisation bilinear forms s_T, $T \in \mathcal{T}_h$, satisfying Assumption 2.4. Then, it holds that*

$$\|\nabla_h(\mathrm{p}_h^{k+1}\underline{u}_h - u)\| \le \varepsilon := \left[\sum_{T\in\mathcal{T}_h} \left(\varepsilon_{\mathrm{nc},T}^2 + (\varepsilon_{\mathrm{res},T} + \varepsilon_{\mathrm{sta},T})^2\right)\right]^{\frac{1}{2}}, \tag{4.10}$$

(continued)

Theorem 4.3 (continued)
with local nonconformity, residual, *and* stabilisation *estimators such that, for all $T \in \mathcal{T}_h$,*

$$\varepsilon_{\mathrm{nc},T} := \|\nabla(\mathrm{p}_T^{k+1}\underline{u}_T - u_h^*)\|_T, \tag{4.11a}$$

$$\varepsilon_{\mathrm{res},T} := C_{\mathrm{P},T} h_T \|(f + \Delta \mathrm{p}_T^{k+1}\underline{u}_T) - \pi_T^{0,0}(f + \Delta \mathrm{p}_T^{k+1}\underline{u}_T)\|_T, \tag{4.11b}$$

$$\varepsilon_{\mathrm{sta},T} := C_{\mathrm{F},T}^{\frac{1}{2}} h_T^{\frac{1}{2}} \left(\sum_{F \in \mathcal{F}_T} \|R_{TF}^k \underline{u}_T\|_F^2 \right)^{\frac{1}{2}}, \tag{4.11c}$$

where u_h^* is an arbitrary function in $H_0^1(\Omega)$ and, for all $F \in \mathcal{F}_T$, the boundary residual operator R_{TF}^k is defined by (2.59).

Remark 4.4 (Nonconformity Estimator) To compute the nonconformity estimator $\varepsilon_{\mathrm{nc},T}$, we can obtain an $H_0^1(\Omega)$-conforming function u_h^* from the HHO solution $\underline{u}_h$ by applying the node-averaging operator, described hereafter, to the global potential reconstruction $\mathrm{p}_h^{k+1}\underline{u}_h$ (see (2.63)). The node-averaging operator has been considered in the context of a posteriori error estimates for nonconforming Finite Element Methods in, among others, [5, 216].

Let an integer $l \geq 1$ be fixed. When $\mathcal{M}_h = (\mathcal{T}_h, \mathcal{F}_h)$ is a matching simplicial mesh in the sense of Definition 1.7, the node-averaging operator $\mathcal{I}_{\mathrm{av},h}^l : \mathbb{P}^l(\mathcal{T}_h) \to \mathbb{P}^l(\mathcal{T}_h) \cap H_0^1(\Omega)$ is defined by setting, for each Lagrange interpolation node $\boldsymbol{N}$ (see, e.g., [183, Section 1.2.3] or [77, Section 3.2]),

$$(\mathcal{I}_{\mathrm{av},h}^l v_h)(\boldsymbol{N}) := \begin{cases} \dfrac{1}{\mathrm{card}(\mathcal{T}_{\boldsymbol{N}})} \displaystyle\sum_{T \in \mathcal{T}_{\boldsymbol{N}}} (v_h)_{|T}(\boldsymbol{N}) & \text{if } \boldsymbol{N} \in \Omega, \\ 0 & \text{if } \boldsymbol{N} \in \partial\Omega, \end{cases}$$

where the set $\mathcal{T}_{\boldsymbol{N}} \subset \mathcal{T}_h$ collects the simplices to which $\boldsymbol{N}$ belongs. We then set

$$u_h^* := \mathcal{I}_{\mathrm{av},h}^{k+1} \mathrm{p}_h^{k+1} \underline{u}_h. \tag{4.12}$$

The generalisation to polytopal meshes can be realised applying the node averaging operator to $\mathrm{p}_h^{k+1}\underline{u}_h$ on a matching simplicial submesh of $\mathcal{T}_h$ (whose existence is guaranteed for regular mesh sequences, see Definition 1.9).

For future use, we note the following result proved in [216] for matching simplicial meshes: For all $v_h \in \mathbb{P}^l(\mathcal{T}_h)$ and all $T \in \mathcal{T}_h$, denoting by $\mathcal{F}_{\boldsymbol{N},T}$ the set of mesh faces that have at least one vertex in common with T (see (4.22) below),

$$\|v_h - \mathcal{I}_{\mathrm{av},h}^l v_h\|_T^2 \lesssim \sum_{F \in \mathcal{F}_{\boldsymbol{N},T}} h_F \|[v_h]_F\|_F^2, \tag{4.13}$$

with hidden constant independent of h and T, but possibly depending on d, ϱ, and l, and jump operator defined by (1.22) and extended to boundary faces $F \in \mathcal{F}_h^{\mathrm{b}}$ setting $[v_h]_F := v_h$. Following [145, Section 5.5.2], (4.13) still holds on regular polyhedral meshes when the node-averaging interpolator is defined on the matching simplicial submesh of Definition 1.9; see Sect. 7.3.2.

Proof (Theorem 4.3) Set, for the sake of brevity,

$$\mathcal{R} := \mathcal{R}(\mathrm{p}_h^{k+1}\underline{u}_h).$$

It follows from Lemma 4.1 that it holds for all $u_h^* \in H_0^1(\Omega)$,

$$\|\nabla_h(\mathrm{p}_h^{k+1}\underline{u}_h - u)\|^2 \leq \|\nabla_h(\mathrm{p}_h^{k+1}\underline{u}_h - u_h^*)\|^2 + \left(\sup_{\varphi \in H_0^1(\Omega), \|\nabla\varphi\|=1} \langle \mathcal{R}, \varphi \rangle_{-1,1}\right)^2, \tag{4.14}$$

where we have used the fact that u_h^* is arbitrary in $H_0^1(\Omega)$ to estimate the infimum in the right-hand side of (4.6). We denote by $\mathfrak{T}_1$ and $\mathfrak{T}_2$ the addends in the right-hand side of (4.14) and proceed to bound them.

(i) *Bound of* $\mathfrak{T}_1$. Recalling the definition (4.11a) of the nonconformity estimator, it is readily inferred that

$$\mathfrak{T}_1 = \sum_{T \in \mathcal{T}_h} \varepsilon_{\mathrm{nc},T}^2. \tag{4.15}$$

(ii) *Bound of* $\mathfrak{T}_2$. We estimate the argument of the supremum in $\mathfrak{T}_2$ for a generic function $\varphi \in H_0^1(\Omega)$ such that $\|\nabla\varphi\| = 1$. Using an element by element integration by parts for the second term in the right-hand side of (4.5) with $v_h = \mathrm{p}_h^{k+1}\underline{u}_h$, we obtain

$$\langle \mathcal{R}, \varphi \rangle_{-1,1} = \sum_{T \in \mathcal{T}_h} \left((f + \Delta \mathrm{p}_T^{k+1}\underline{u}_T, \varphi)_T - \sum_{F \in \mathcal{F}_T} (\nabla \mathrm{p}_T^{k+1}\underline{u}_T \cdot \boldsymbol{n}_{TF}, \varphi)_F \right). \tag{4.16}$$

Let now $\underline{\varphi}_h \in \underline{U}_{h,0}^k$ be such that $\varphi_T = \pi_T^{0,0}\varphi$ for all $T \in \mathcal{T}_h$ and $\varphi_F = \pi_F^{0,k}\varphi_{|F}$ for all $F \in \mathcal{F}_h$ (note the usage of two different polynomial degrees in the elements and on the faces). We have that

$$\begin{aligned}
&\sum_{T\in\mathcal{T}_h} (\pi_T^{0,0}(f + \Delta \mathrm{p}_T^{k+1}\underline{u}_T), \varphi)_T \\
&\quad= \sum_{T\in\mathcal{T}_h} (f + \Delta \mathrm{p}_T^{k+1}\underline{u}_T, \pi_T^{0,0}\varphi)_T \\
&\quad= \sum_{T\in\mathcal{T}_h} (f + \Delta \mathrm{p}_T^{k+1}\underline{u}_T, \varphi_T)_T \\
&\quad= \sum_{T\in\mathcal{T}_h} \left(\mathrm{a}_T(\underline{u}_T, \underline{\varphi}_T) + \sum_{F\in\mathcal{F}_T} (\nabla \mathrm{p}_T^{k+1}\underline{u}_T \cdot \boldsymbol{n}_{TF}, \varphi_T)_F\right),
\end{aligned}$$

where we have used the definition (1.57) of $\pi_T^{0,0}$ in the first equality and the discrete problem (2.48) with $\underline{v}_h = \underline{\varphi}_h$ together with an element by element integration by parts and the fact that $\nabla\varphi_T = 0$ for all $T \in \mathcal{T}_h$ to conclude. Expanding a_T according to its definition (2.15) and using the definition (2.11a) of p_T^{k+1} with $\underline{v}_T = \underline{\varphi}_T$ and $w = \mathrm{p}_T^{k+1}\underline{u}_T$ for the consistency term, we obtain

$$\begin{aligned}
&\sum_{T\in\mathcal{T}_h} (\pi_T^{0,0}(f + \Delta \mathrm{p}_T^{k+1}\underline{u}_T), \varphi)_T \\
&\qquad= \sum_{T\in\mathcal{T}_h} \left(\mathrm{s}_T(\underline{u}_T, \underline{\varphi}_T) + \sum_{F\in\mathcal{F}_T} (\nabla \mathrm{p}_T^{k+1}\underline{u}_T \cdot \boldsymbol{n}_{TF}, \varphi)_F\right),
\end{aligned} \tag{4.17}$$

where we have used the fact that $(\nabla \mathrm{p}_T^{k+1}\underline{u}_T)_{|F}\cdot\boldsymbol{n}_{TF} \in \mathbb{P}^k(F)$ together with the definition (1.57) of $\pi_F^{0,k}$ to write φ instead of $\varphi_F = \pi_F^{0,k}\varphi_{|F}$ in the boundary term. Sum (4.17) and (4.16), and rearrange the terms, to arrive at

$$\begin{aligned}
&\langle \mathcal{R}, \varphi\rangle_{-1,1} \\
&\quad= \sum_{T\in\mathcal{T}_h} \Big((f + \Delta \mathrm{p}_T^{k+1}\underline{u}_T - \pi_T^{0,0}(f + \Delta \mathrm{p}_T^{k+1}\underline{u}_T), \varphi - \varphi_T)_T + \mathrm{s}_T(\underline{u}_T, \underline{\varphi}_T)\Big) \\
&\quad=: \sum_{T\in\mathcal{T}_h} \big(\mathfrak{T}_{2,1}(T) + \mathfrak{T}_{2,2}(T)\big),
\end{aligned} \tag{4.18}$$

where the definition (1.57) of $\pi_T^{0,0}$ was used to insert $\varphi_T = \pi_T^{0,0}\varphi$ into the first term. Let us estimate the addends inside the summation. Using the Cauchy–Schwarz and local Poincaré (4.3) inequalities, and recalling the definition (4.11b) of the residual estimator, we readily infer that, for all $T \in \mathcal{T}_h$,

$$|\mathfrak{T}_{2,1}(T)| \le \varepsilon_{\mathrm{res},T} \|\nabla\varphi\|_T. \tag{4.19}$$

On the other hand, recalling the reformulation (2.61) of the local stabilisation bilinear form s_T we have, for all $T \in \mathcal{T}_h$,

$$|\mathfrak{T}_{2,2}(T)| = \left| \sum_{F\in\mathcal{F}_T} (R^k_{TF}\underline{u}_T, \varphi - \varphi_T)_F \right| \le \varepsilon_{\mathrm{sta},T} \|\nabla\varphi\|_T, \tag{4.20}$$

where we have used the fact that $R^k_{TF}\underline{u}_T \in \mathbb{P}^k(F)$ together with the definition (1.57) of $\pi^{0,k}_F$ to write φ instead of $\varphi_F = \pi^{0,k}_F \varphi_{|F}$ inside the boundary term, and the Cauchy–Schwarz and local Friedrichs (4.4) inequalities followed by definition (4.11c) of the stability estimator to conclude. Using (4.19) and (4.20) to estimate the right-hand side of (4.18) followed by a Cauchy–Schwarz inequality on the sum over $T \in \mathcal{T}_h$ and $\|\nabla\varphi\| = 1$, and plugging the resulting bound inside the supremum in $\mathfrak{T}_2$, we arrive at

$$\mathfrak{T}_2 \le \sum_{T\in\mathcal{T}_h} (\varepsilon_{\mathrm{res},T} + \varepsilon_{\mathrm{sta},T})^2. \tag{4.21}$$

(iii) *Conclusion.* Plug (4.15) and (4.21) into (4.14). □

4.1.2 Energy Error Lower Bounds

In practice, we want to make sure that the error estimators are able to correctly localise the error (for use, e.g., in adaptive mesh refinement) and that they do not overestimate it. The goal of this section is precisely to show that the error estimators defined in Theorem 4.3 are *efficient*, i.e., that they are controlled by the error. We start by proving *local efficiency*, meaning that the error estimators on a given mesh element $T \in \mathcal{T}_h$ are bounded by the approximation error on a patch of elements surrounding T. This shows that the a posteriori estimate is suitable to drive local mesh refinement. We next show *global efficiency*, expressed by an inequality of the form

$$\varepsilon \lesssim \|\nabla_h(\mathrm{p}^{k+1}_h\underline{u}_h - u)\| + |\underline{u}_h|_{\mathrm{s},h}$$

with seminorm $|\cdot|_{\mathrm{s},h}$ defined by (2.40) and hidden constant independent of both the meshsize and the problem data. This inequality guarantees that the estimated error does not depart from the actual discretisation error.

Let a mesh element $T \in \mathcal{T}_h$ be fixed, and define the following sets of elements and faces sharing at least one node with T:

$$\mathcal{T}_{N,T} := \left\{T' \in \mathcal{T}_h \,:\, \overline{T}' \cap \overline{T} \neq \emptyset\right\}, \qquad \mathcal{F}_{N,T} := \left\{F \in \mathcal{F}_h \,:\, \overline{F} \cap \partial T \neq \emptyset\right\}. \tag{4.22}$$

The following proposition contains useful geometric bounds for these sets.

Proposition 4.5 (Geometric Bounds for $\mathcal{T}_{N,T}$ and $\mathcal{F}_{N,T}$) *Let $(\mathcal{M}_h)_{h\in\mathcal{H}}$ denote a regular sequence of polytopal meshes in the sense of Definition 1.9. Then, the following bounds hold with hidden constants depending only on d and ϱ:*

(i) Number of elements and faces sharing a node with T. *For all $h \in \mathcal{H}$ and all $T \in \mathcal{T}_h$,*

$$\max\left(\mathrm{card}(\mathcal{T}_{N,T}), \mathrm{card}(\mathcal{F}_{N,T})\right) \lesssim 1. \tag{4.23}$$

(ii) Diameter of the faces sharing a node with T. *For all $h \in \mathcal{H}$ and all $T \in \mathcal{T}_h$,*

$$h_F \lesssim h_T \qquad \forall F \in \mathcal{F}_{N,T}. \tag{4.24}$$

Proof (i) *Number of elements and faces sharing a node with T.* Let us first assume that, for all $h \in \mathcal{H}$, $\mathcal{M}_h = (\mathcal{T}_h, \mathcal{F}_h)$ is a matching simplicial mesh in the sense of Definition 1.7. Then, the shape regularity condition (1.3) implies that, for all $h \in \mathcal{H}$ and all $T \in \mathcal{T}_h$, the smallest solid angle of T is bounded from below by a real number depending only on the mesh regularity parameter ϱ. As a consequence, for each vertex $\boldsymbol{a}$ of T, the cardinality of the set of mesh elements to which $\boldsymbol{a}$ belongs is bounded from above uniformly in h, i.e., $\mathrm{card}\{T' \in \mathcal{T}_h \,:\, \boldsymbol{a} \in \overline{T'}\} \lesssim 1$. Since each mesh element in $\mathcal{T}_{N,T}$ shares at least one vertex with T, and T has a finite number (equal to $d+1$) of vertices, this means that $\mathrm{card}(\mathcal{T}_{N,T}) \lesssim 1$. On the other hand, since $\mathrm{card}(\mathcal{F}_{N,T}) \leq (d+1)\,\mathrm{card}(\mathcal{T}_{N,T})$, this also implies $\mathrm{card}(\mathcal{F}_{N,T}) \lesssim 1$.

Let us now turn to the case when $\mathcal{M}_h$ belongs to a regular polytopal mesh sequence and, for every $h \in \mathcal{H}$, denote by $\mathfrak{M}_h$ the corresponding matching simplicial submesh in the sense of Definition 1.8. Let $h \in \mathcal{H}$ and $T \in \mathcal{T}_h$ be fixed. Then, for all $\tau \in \mathfrak{T}_T$ (with $\mathfrak{T}_T$ denoting the set of simplices contained in T), owing to the uniform bounds for matching simplicial meshes there holds

$$\mathrm{card}(\mathfrak{T}_{N,\tau}) \lesssim 1 \quad \text{with} \quad \mathfrak{T}_{N,\tau} := \left\{\tau' \in \mathfrak{T}_h \,:\, \overline{\tau'} \cap \overline{\tau} \neq \emptyset\right\}. \tag{4.25}$$

The uniform bound on card($\mathcal{T}_{N,T}$) then follows observing that

$$\text{card}(\mathcal{T}_{N,T}) \le \sum_{\tau\in\mathfrak{T}_T} \text{card}(\mathfrak{T}_{N,\tau}) \le \text{card}(\mathfrak{T}_T) \max_{\tau\in\mathfrak{T}_T} \text{card}(\mathfrak{T}_{N,\tau}) \lesssim 1,$$

where, to conclude, we have bounded the first factor using (1.9) and the second one using (4.25). A similar reasoning, whose details are left to the reader, yields the uniform bound on card($\mathcal{F}_{N,T}$).

(ii) *Diameter of the faces sharing a node with T.* Let $h \in \mathcal{H}$, $T \in \mathcal{T}_h$, and $F \in \mathcal{F}_{N,T}$ be fixed. Then, either $F \in \mathcal{F}_T$ and (4.24) is trivial, or there exist two finite sequences $(F_i)_{0\le i\le n} \subset (\mathcal{F}_{N,T})^{n+1}$ and $(T_i)_{1\le i\le n} \subset (\mathcal{T}_{N,T})^{n+1}$ with no repeated elements such that $T_0 = T$, $F_0 \in \mathcal{F}_T$, $F_n = F$ and, for all $0 \le i \le n-1$, $F_i \in \mathcal{F}_{T_i} \cap \mathcal{F}_{T_{i+1}}$. Recalling (1.6), we have, for all $0 \le i \le n-1$,

$$h_{F_{i+1}} \le h_{T_{i+1}} \le \frac{1}{2\varrho^2} h_{F_i}.$$

Iterating this inequality, we infer that

$$h_F \le \left(\frac{1}{2\varrho^2}\right)^n h_{F_0} \le \left(\frac{1}{2\varrho^2}\right)^n h_T,$$

where the last inequality follows from $F_0 \in \mathcal{F}_T$. Since $n \le \text{card}(\mathcal{T}_{N,T}) \lesssim 1$ owing to the bounds proved in Point (i), (4.24) follows. □

We also note the following technical result.

Proposition 4.6 (Estimate of Boundary Oscillations) *For all $T \in \mathcal{T}_h$, all $\varphi \in H^1(T)$ and all $F \in \mathcal{F}_T$, it holds that*

$$h_F^{-\frac{1}{2}} \|\varphi - \pi_F^{0,k}\varphi\|_F \lesssim \|\nabla\varphi\|_T, \tag{4.26}$$

with hidden constant independent of h, T, φ and F, but possibly depending on d, ϱ, and k.

Proof We write

$$\|\varphi - \pi_F^{0,k}\varphi\|_F \le \|\varphi - \pi_T^{0,k}\varphi\|_F + \|\pi_T^{0,k}\varphi - \pi_F^{0,k}\varphi\|_F \lesssim h_F^{\frac{1}{2}}\|\nabla\varphi\|_T, \tag{4.27}$$

where we have inserted $\pm\pi_T^{0,k}\varphi$ and used the triangle inequality to obtain the first bound, and we have concluded invoking the trace approximation property (1.75) of $\pi_T^{0,k}$ (with $p = 2$, $m = 0$ and $s = 1$), the property $h_T \lesssim h_F$ (see (1.6)), and the boundedness (2.9) of the interpolator $\underline{I}_T^k$, after noticing that $\|\pi_T^{0,k}\varphi - \pi_F^{0,k}\varphi\|_F \le h_F^{\frac{1}{2}}\|\underline{I}_T^k\varphi\|_{1,T}$. □

The following theorem, whose proof is inspired by classical bubble function techniques (see, e.g., [274]), states the optimality of the a posteriori indicators.

Theorem 4.7 (A Posteriori Local Error Lower Bound) *We let the assumptions of Theorem 4.3 hold and further assume, for the sake of simplicity, that*

(i) For all $T \in \mathcal{T}_h$, the stabilisation bilinear form s_T is given by (2.23);
(ii) The H^1_0-conforming reconstruction u_h^* is obtained as described in Remark 4.4, using the node-averaging operator on the matching simplicial submesh $\mathfrak{M}_h = (\mathfrak{T}_h, \mathfrak{F}_h)$ of $\mathcal{M}_h = (\mathcal{T}_h, \mathcal{F}_h)$ of Definition 1.9;
(iii) We have, for the forcing term, $f \in \mathbb{P}^{k+1}(\mathcal{T}_h)$.

Then, it holds, for all $T \in \mathcal{T}_h$,

$$\varepsilon_{\mathrm{nc},T} \lesssim \left(\|\nabla_h(\mathrm{p}_h^{k+1}\underline{u}_h - u)\|_{\mathcal{N},T} + |\underline{u}_h|_{\mathrm{s},\mathcal{N},T}\right), \tag{4.28a}$$

$$\varepsilon_{\mathrm{res},T} \lesssim \|\nabla(\mathrm{p}_T^{k+1}\underline{u}_T - u)\|_T, \tag{4.28b}$$

$$\varepsilon_{\mathrm{sta},T} \lesssim |\underline{u}_T|_{\mathrm{s},T}, \tag{4.28c}$$

with hidden constants possibly depending on d, ϱ, and on k, but independent of h, T, and u. For all $T \in \mathcal{T}_h$, $\|\cdot\|_{\mathcal{N},T}$ denotes the L^2-norm on the union of the elements in $\mathcal{T}_{\mathcal{N},T}$ and we have set, with stabilisation seminorm $|\cdot|_{\mathrm{s},T'}$, for $T' \in \mathcal{T}_{\mathcal{N},T}$, such that, for all $\underline{v}_{T'} \in \underline{U}_{T'}^k$, $|\underline{v}_{T'}|_{\mathrm{s},T'}^2 := \mathrm{s}_{T'}(\underline{v}_{T'}, \underline{v}_{T'})$,

$$|\underline{u}_h|_{\mathrm{s},\mathcal{N},T} := \left(\sum_{T' \in \mathcal{T}_{\mathcal{N},T}} |\underline{u}_{T'}|_{\mathrm{s},T'}^2\right)^{\frac{1}{2}}.$$

Proof Let a mesh element $T \in \mathcal{T}_h$ be fixed.

(i) *Bound* (4.28a) *on the nonconformity estimator.* Using the inverse Sobolev embedding (1.50) with $X = T$, $p = 2$, $m = 1$, and $r = 0$ together with the estimates (4.13) and (4.24), we infer from (4.11a) that

$$\varepsilon_{\mathrm{nc},T}^2 \lesssim h_T^{-2}\|\mathrm{p}_T^{k+1}\underline{u}_T - u_h^*\|_T^2 \lesssim \sum_{F \in \mathcal{F}_{\mathcal{N},T}} h_F^{-1}\|[\mathrm{p}_h^{k+1}\underline{u}_h]_F\|_F^2. \tag{4.29}$$

Using the fact that $[u]_F = 0$ for all $F \in \mathcal{F}_h$ (use Lemma 1.21 for $F \in \mathcal{F}_h^{\mathrm{i}}$ and recall that $[u]_F = u_{|F} = 0$ for all $F \in \mathcal{F}_h^{\mathrm{b}}$ since $u \in H^1_0(\Omega)$) to write $[\mathrm{p}_h^{k+1}\underline{u}_h]_F =$

$[\mathrm{p}_h^{k+1}\underline{u}_h - u]_F$, inserting $\pi_F^{0,k}[\mathrm{p}_h^{k+1}\underline{u}_h]_F - \pi_F^{0,k}[\mathrm{p}_h^{k+1}\underline{u}_h - u]_F = 0$ inside the norm, and using the triangle inequality, we have for all $F \in \mathcal{F}_{N,T}$,

$$\begin{aligned}
\|[\mathrm{p}_h^{k+1}\underline{u}_h]_F\|_F &\le \|[\mathrm{p}_h^{k+1}\underline{u}_h - u]_F - \pi_F^{0,k}[\mathrm{p}_h^{k+1}\underline{u}_h - u]_F\|_F \\
&\quad + \|\pi_F^{0,k}[\mathrm{p}_h^{k+1}\underline{u}_h]_F\|_F \\
&\le \sum_{T'\in\mathcal{T}_F} \|(\mathrm{p}_{T'}^{k+1}\underline{u}_{T'} - u) - \pi_F^{0,k}(\mathrm{p}_{T'}^{k+1}\underline{u}_{T'} - u)\|_F \\
&\quad + \|\pi_F^{0,k}[\mathrm{p}_h^{k+1}\underline{u}_h]_F\|_F,
\end{aligned}$$

where we have expanded the jump according to its definition (1.22) and used a triangle inequality to pass to the second line. Plugging the above bound into (4.29), and using multiple times (4.26) with $\varphi = (\mathrm{p}_{T'}^{k+1}\underline{u}_{T'} - u)$ for $T' \in \mathcal{T}_{N,T}$, we arrive at

$$\varepsilon_{\mathrm{nc},T}^2 \lesssim \|\nabla_h(\mathrm{p}_h^{k+1}\underline{u}_h - u)\|_{N,T}^2 + \sum_{F\in\mathcal{F}_{N,T}} h_F^{-1}\|\pi_F^{0,k}[\mathrm{p}_h^{k+1}\underline{u}_h]_F\|_F^2.$$

To conclude, we proceed as in the proof of Lemma 2.31 to prove that the last term is bounded by $|\underline{u}_h|_{\mathrm{s},N,T}^2$ up to a constant independent of h.

(ii) *Bound* (4.28b) *on the residual estimator.* For the sake of brevity, we let $r_T := f_{|T} + \Delta\mathrm{p}_T^{k+1}\underline{u}_T$ and recall that $\mathfrak{T}_T := \{\tau \in \mathfrak{T}_h : \tau \subset T\}$ denotes the set of simplices contained in T. For all $\tau \in \mathfrak{T}_T$, let $b_\tau \in H_0^1(\tau)$ be the element bubble function equal to the product of barycentric coordinates of τ and rescaled so as to take the value 1 at the centre of mass of τ. Letting $\psi_\tau := b_\tau r_T$ for all $\tau \in \mathfrak{T}_T$, the following properties hold [274]:

$$\psi_\tau = 0 \text{ on } \partial\tau, \tag{4.30a}$$

$$\|r_T\|_\tau^2 \lesssim (r_T, \psi_\tau)_\tau, \tag{4.30b}$$

$$\|\psi_\tau\|_\tau \le \|r_T\|_\tau. \tag{4.30c}$$

We have that

$$\begin{aligned}
\|r_T\|_T^2 &= \sum_{\tau\in\mathfrak{T}_T} \|r_T\|_\tau^2 \lesssim \sum_{\tau\in\mathfrak{T}_T} (r_T, \psi_\tau)_\tau \\
&= \sum_{\tau\in\mathfrak{T}_T} (\nabla(u - \mathrm{p}_T^{k+1}\underline{u}_T), \nabla\psi_\tau)_\tau \\
&\le \|\nabla(u - \mathrm{p}_T^{k+1}\underline{u}_T)\|_T \left(\sum_{\tau\in\mathfrak{T}_T} h_\tau^{-2}\|\psi_\tau\|_\tau^2\right)^{\frac{1}{2}} \\
&\lesssim h_T^{-1}\|\nabla(u - \mathrm{p}_T^{k+1}\underline{u}_T)\|_T\|r_T\|_T,
\end{aligned} \tag{4.31}$$

where we have used property (4.30b) in the first line, the fact that $f = -\Delta u$ a.e. in Ω together with an integration by parts and property (4.30a) to pass to the second line, the Cauchy–Schwarz inequality together with the inverse Sobolev embedding (1.50) with $X = \tau$, $p = 2$, $m = 1$, and $r = 0$ to pass to the third line, and (4.30c) together with the fact that $h_\tau^{-1} \le (\varrho h_T)^{-1}$ for all $\tau \in \mathfrak{T}_T$ (see Definition 1.9) to conclude. Recalling the definition (4.11b) of the residual estimator, observing that $\|r_T - \pi_T^{0,0} r_T\|_T \le 2\|r_T\|_T$ as a result of the triangle inequality followed by the $L^2(T)$-boundedness of $\pi_T^{0,0}$ expressed by (1.72) with $X = T$, $\mathcal{P} = \mathbb{P}^0(T)$ and $m = 1$, and using (4.31), the bound (4.28b) follows.

(iii) *Bound* (4.28c) *on the stabilisation estimator.* Using the definition (2.59) of the boundary residual operator $\underline{R}_{\partial T}^k$ with $\underline{v}_T = \underline{u}_T$ and $\underline{\alpha}_{\partial T} = -h_T \underline{R}_{\partial T}^k \underline{u}_T = (-h_T R_{TF}^k \underline{u}_T)_{F\in\mathcal{F}_T}$ together with the property (2.58) with $\underline{v}_T = (0, (-h_T R_{TF}^k \underline{u}_T)_{F\in\mathcal{F}_T})$, the stabilisation estimator (4.11c) can be bounded as follows:

$$\varepsilon_{\mathrm{sta},T}^2 = C_{\mathrm{F},T} s_T(\underline{u}_T, (0, -h_T \underline{R}_{\partial T}^k \underline{u}_T)) \lesssim |\underline{u}_T|_{\mathrm{s},T} |(0, -h_T \underline{R}_{\partial T}^k \underline{u}_T)|_{\mathrm{s},T}. \tag{4.32}$$

On the other hand, from the seminorm equivalence (2.16), the fact that $h_F^{-1} \lesssim h_T^{-1}$ (cf. (1.6)), and the definition (4.11c) of $\varepsilon_{\mathrm{sta},T}$, it is inferred that

$$|(0, -h_T \underline{R}_{\partial T}^k \underline{u}_T)|_{\mathrm{s},T} \le \eta^{\frac{1}{2}} \left(\sum_{F\in\mathcal{F}_T} h_F^{-1} \|h_T R_{TF}^k \underline{u}_T\|_F^2 \right)^{\frac{1}{2}} \lesssim \varepsilon_{\mathrm{sta},T}.$$

Using this estimate to bound the right-hand side of (4.32) and simplifying, (4.28c) follows. □

An immediate consequence of the local lower bounds is that the following global lower bound holds.

Corollary 4.8 (Global Lower Bound) *Under the assumptions of Theorem 4.7, it holds that*

$$\left[\sum_{T\in\mathcal{T}_h} \left(\varepsilon_{\mathrm{nc},T}^2 + (\varepsilon_{\mathrm{res},T} + \varepsilon_{\mathrm{sta},T})^2 \right) \right]^{\frac{1}{2}} \lesssim \left(\|\nabla_h (\mathrm{p}_h^{k+1} \underline{u}_h - u)\| + |\underline{u}_h|_{\mathrm{s},h} \right),$$

with hidden constant independent of h and f, but possibly depending on d, ϱ and k.

4.1.3 Numerical Examples: A Posteriori-Driven Mesh Adaptivity

In this section we illustrate the performance of the adaptive procedure described in Algorithm 1 and based on the error estimators of Theorem 4.3. This adaptive algorithm follows the usual "solve–estimate–mark–refine" process.

Algorithm 1 Pseudocode of the automatic mesh adaptation procedure

Set a tolerance tol > 0 and a maximum number of iterations $N_{\max}$
Generate an initial coarse mesh $\mathcal{T}_h^{(0)}$, set $n \leftarrow 0$, and let $\mathcal{T}_h^{(n)} \leftarrow \mathcal{T}_h^{(0)}$
repeat
  Solve the HHO problem (2.48) on $\mathcal{T}_h^{(n)}$
  for $T \in \mathcal{T}_h^{(n)}$ **do**
    Compute and store the local estimator $\varepsilon_T := \left[\varepsilon_{\mathrm{nc},T}^2 + (\varepsilon_{\mathrm{res},T} + \varepsilon_{\mathrm{sta},T})^2\right]^{\frac{1}{2}}$
  end for
  for $T \in \mathcal{T}_h^{(n)}$ **do**
    if T is among the 5% elements with the largest local estimator **then**
      Set a target diameter of $h_T/2$
    else
      Set a target diameter of h_T
    end if
  end for
  Set $n \leftarrow n + 1$ and generate a novel mesh $\mathcal{T}_h^{(n)}$ by using the target element diameters
until $\varepsilon <$ tol **or** $n > N_{\max}$

We consider in what follows a numerical test case taken from [154] and based on the exact solution of [193] on the etched three-dimensional domain $\Omega = (-1, 1)^3 \setminus [0, 1]^3$:

$$u(\boldsymbol{x}) = \sqrt[4]{x_1^2 + x_2^2 + x_3^2},$$

with right-hand side

$$f = -\frac{3}{4}\left(x_1^2 + x_2^2 + x_3^2\right)^{-\frac{3}{4}}.$$

In this case, the gradient of the solution has a singularity at the origin which prevents the method from attaining optimal convergence rates even for $k = 0$ (since the regularity requirements detailed in Theorems 2.28 and 2.32 are not matched). The stabilisation bilinear form used in the computations results from the hybridisation of the Mixed High-Order method of [149], see (5.94) in Sect. 5.4.

4.1.3.1 Adaptively Refined Matching Tetrahedral Meshes

We first consider matching tetrahedral meshes obtained using the open source software Netgen [258]. At each refinement iteration, a new mesh is generated by specifying the target local meshsize at the barycentres of the elements.

In Fig. 4.1 we plot the numerical error versus the number of degrees of freedom (DOFs) N_{dof} (cf. (B.13c)) on uniformly and adaptively refined mesh sequences for polynomial degrees up to 2. Notice that, when considering adaptive mesh refinement, we evaluate convergence in terms of error versus the number of DOFs since the global meshsize h may not vary from one refinement level to another. The convergence curves for uniformly refined mesh sequences show that the order of convergence is clearly limited by the solution regularity. When using adaptively refined mesh sequences, on the other hand, we recover the optimal orders of convergence $N_{\mathrm{dof}}^{\frac{(k+1)}{d}}$ and $N_{\mathrm{dof}}^{\frac{(k+2)}{d}}$ (with $d = 3$) for the energy- and L^2-norms of the error, respectively. This shows that the adaptive Algorithm 1 is capable of restoring optimal orders of convergence.

In Fig. 4.2 we display the energy error vs. the total computational wall time in seconds, including the pre-processing (mesh generation and creation of the connectivity), the assembly of the sparse matrix (including static condensation, see Sect. B.3.2), the solution of the linear system, and the post-processing (reconstruction of the element unknowns and error computation). These computations were run on a laptop equipped with an Intel Core i7-3720QM processor clocked at 2.60 GHz and 16 Gb of RAM. The global linear systems are solved with the algebraic multigrid solver AGMG [248], and the iterations are stopped once the

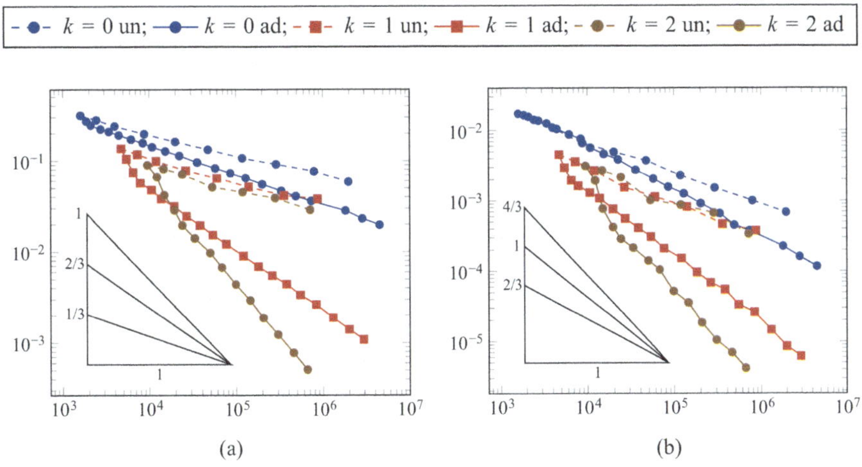

Fig. 4.1 Error vs. N_{dof} for the test case of Sect. 4.1.3.1. The triangles represent reference slopes corresponding to the optimal convergence rates. "un" = uniformly refined meshes, "ad" = adaptively refined meshes. (**a**) Energy-error vs. N_{dof}. (**b**) L^2-error vs. N_{dof}

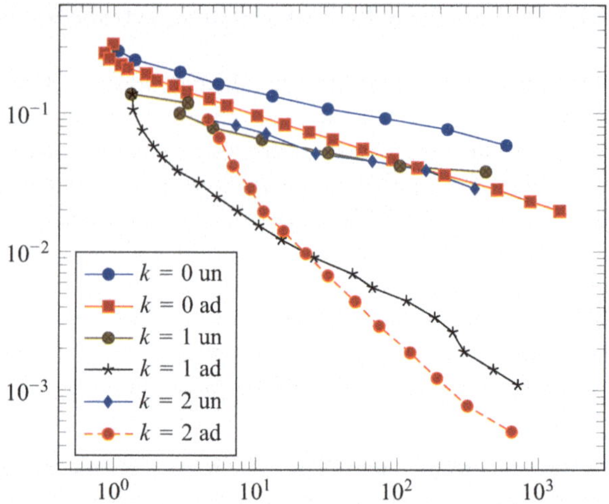

Fig. 4.2 Energy error vs. computational wall time

relative residual reaches 10^{-8}. While the displayed times obviously depend on both the implementation and the machine used to run the tests, the plot gives a clear indication that the combined use of high polynomial orders and a posteriori-driven mesh adaptation leads to a better use of the computational resources.

4.1.3.2 Adaptive Mesh Coarsening

We next consider adaptively coarsened meshes in the spirit of [17, 38]. The idea here consists in starting from a fine mesh (chosen, e.g., to accurately represent the geometric features of the domain or to capture the finest scales in the solution) and in solving the problem on a coarsened mesh obtained by selectively merging the fine elements into polyhedral conglomerates; see Fig. 4.3. If the coarsening procedure is well-designed, one can achieve a precision comparable to that of a computation on the fine mesh, but for a smaller number of DOFs.

In our case, we start from a tetrahedral mesh consisting of 51,534 nodes and 2.72×10^5 tetrahedra, and we create an initial agglomerated mesh with a modified version of MGridGen[1] [239] by setting a typical agglomeration target of 1024 tetrahedra per coarse element. An adaptive mesh sequence is then generated by locally reducing the meshsize using the same procedure as for the standard meshes considered in the previous section. Figure 4.3a shows on the left the initial polyhedral mesh and on the right the one obtained at the final refinement step. Figure 4.3b displays the fields u_h^* defined as in Remark 4.4 (obtained by an averaging

[1]The authors are grateful to Lorenzo Botti and Alessandro Colombo (Università di Bergamo) for providing this modified version of MGridGen.

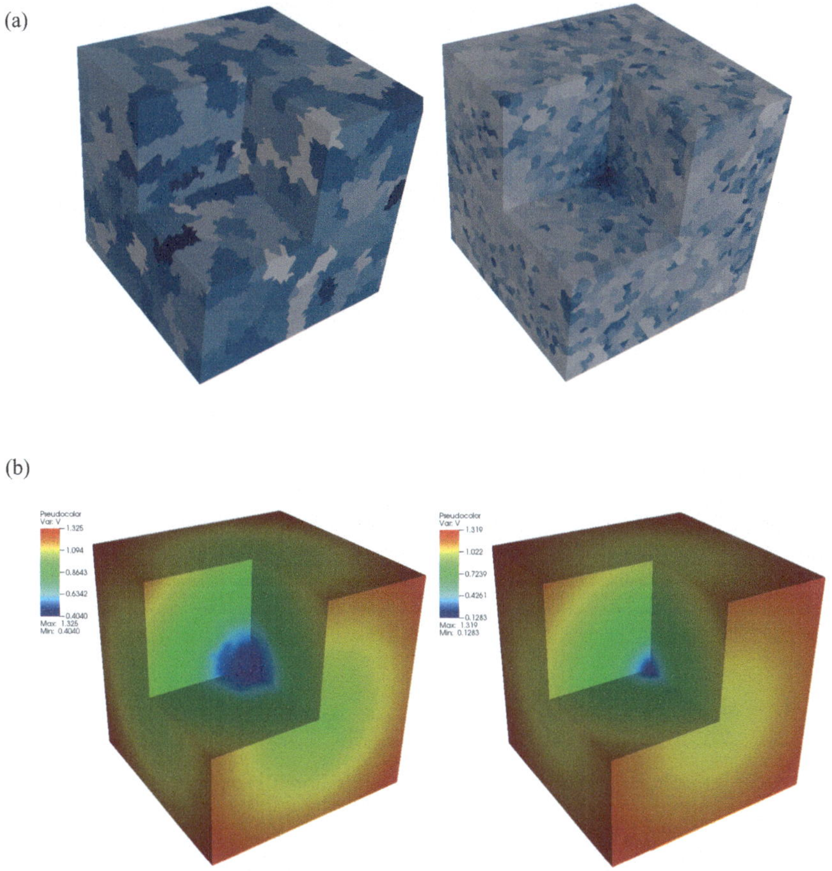

Fig. 4.3 Numerical results with adaptive mesh coarsening for $k = 0$ taken from [154]. (**a**) Left. Initial coarse mesh with agglomerated elements containing about 1024 tetrahedral elements each. Right. Final adaptive mesh. (**b**) Left. Interpolated potential u_h^* (cf. Remark 4.4) on the first mesh of (**a**). Right. Potential u_h^* on the second mesh of (**a**)

interpolation on the matching simplicial submesh) corresponding to the two meshes of Fig. 4.3a.

In order to assess the performance of the adaptive coarsening procedure, we compare the results with a sequence of meshes obtained by uniform coarsening of the same initial mesh (i.e., the size of the agglomerated elements is not adapted in accordance with the distribution of the error). Figure 4.4 shows the energy- and L^2-errors as functions of the number of DOFs on both the adaptively and uniformly agglomerated mesh sequences. In Fig. 4.4a, one can see that a similar precision in the energy-norm as the one achievable on the fine mesh is obtained for less than half the number of DOFs. The gain is even larger when considering the L^2-norm. For both norms, the error stagnates when the accuracy made possible by the initial mesh

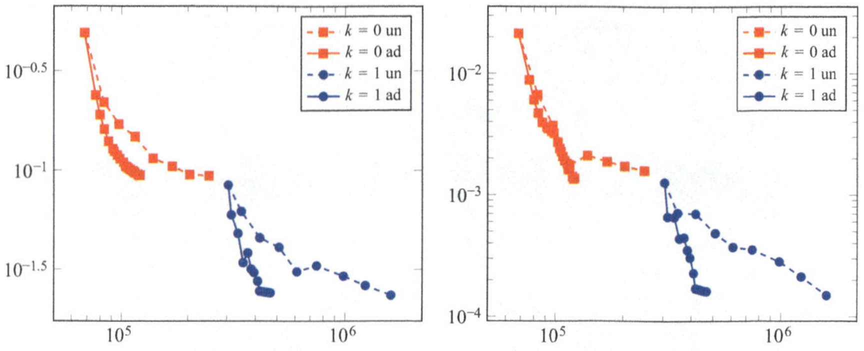

(a) (b)

Fig. 4.4 Error vs. $N_{\rm dof}$ for the test case of Sect. 4.1.3.2 on adaptively coarsened meshes ("ad"), and comparison with uniformly coarsened meshes ("un"). (**a**) Energy-error vs. $N_{\rm dof}$. (**b**) L^2-error vs. $N_{\rm dof}$

is approached. To further reduce the error, one would need to adaptively refine the initial mesh.

4.2 Locally Variable Diffusion

We consider here the same model as (3.3), which we recall for the sake of legibility:

$$-\nabla\cdot(\boldsymbol{K}\nabla u) = f \qquad \text{in } \Omega, \tag{4.33a}$$

$$u = 0 \qquad \text{on } \partial\Omega. \tag{4.33b}$$

The source term f belongs to $L^2(\Omega)$, and the diffusion tensor $\boldsymbol{K}:\Omega \to \mathbb{R}^{d\times d}_{\rm sym}$ is measurable, uniformly bounded and elliptic, that is, there are two strictly positive real numbers $\underline{K}$ and $\overline{K}$ such that

$$\underline{K} \le \boldsymbol{K}(\boldsymbol{x})\boldsymbol{\xi}\cdot\boldsymbol{\xi} \le \overline{K}. \tag{4.34}$$

Contrary to Sect. 3.1, however, we do not assume that $\boldsymbol{K}$ is piecewise constant on a partition of Ω. We recall that the weak formulation of problem (4.33) is: Find $u \in H_0^1(\Omega)$ such that

$$a_{\boldsymbol{K}}(u, v) = (f, v) \qquad \forall v \in H_0^1(\Omega), \tag{4.35}$$

with bilinear form $a_{\boldsymbol{K}} : H^1(\Omega) \times H^1(\Omega) \to \mathbb{R}$ defined by

$$a_{\boldsymbol{K}}(u, v) := (\boldsymbol{K}\nabla u, \nabla v).$$

4.2.1 Discrete Gradient

As seen in previous chapters, HHO schemes are built from local bilinear forms made of two components: a consistent contribution, and a stabilisation term. Considering the case of the Poisson equation, for example, the consistent component $(\nabla \mathrm{p}_T^{k+1}\underline{v}_T, \nabla \mathrm{p}_T^{k+1}\underline{v}_T)_T$ in (2.15) is constructed using the gradient of the local reconstruction p_T^{k+1}. The definition (2.11a) of $\nabla \mathrm{p}_T^{k+1}\underline{v}_T$ through gradients of functions in $\mathbb{P}^{k+1}(T)$ is essential, in the analysis of the consistency error, to cancel out volumetric terms by recasting $\mathrm{a}_h(\underline{I}_h^k w, \underline{v}_h)$ under the form (2.45). This recasting is made possible precisely because, in the consistent component of $\mathrm{a}_T(\underline{I}_T^k w, \underline{v}_T)$, $\nabla \mathrm{p}_T^{k+1}\underline{v}_T$ appears in a scalar product with $\nabla \mathrm{p}_T^{k+1}\underline{I}_T^k w$, which lies in $\nabla\mathbb{P}^{k+1}(T)$.

In Sect. 3.1 we considered diffusion equations with a diffusion coefficient that could be anisotropic in each cell. Similar considerations as above led us to define an oblique reconstruction operator $\mathrm{p}_{\boldsymbol{K},T}^{k+1}$ through (3.22), to ensure a complete elimination of the volumetric terms between (3.50) and (3.51) in the consistency analysis. The definition (3.22) is inspired by the formula (3.21), to ensure the commutativity property (3.24) and thus the optimal approximation properties of the HHO scheme. The formula (3.21) is however only exact if $\boldsymbol{K}$ is constant in the cell T—otherwise, the L^2-orthogonal projectors on $\mathbb{P}^k(T)$ and $\mathbb{P}^k(F)$ cannot be introduced in the right-hand side of (3.20).

Hence, if $\boldsymbol{K}$ is allowed to vary inside each cell, the approach in Sect. 3.1 has to be revised. The main ingredient for the HHO discretisation in this case is a reconstructed gradient in the space $\mathbb{P}^k(T)^d$, instead of its subspace $\nabla\mathbb{P}^{k+1}(T)$.

Let a mesh element $T \in \mathcal{T}_h$ be fixed. As usual, we start with an inspiring remark, in a similar way as at the beginning of Sects. 2.1.1 and 3.1.3.1. This time, however, instead of showing that the elliptic or oblique elliptic projector of a function v can be computed using its L^2-projections on $\mathbb{P}^k(T)$ and $\mathbb{P}^k(F)$, for $F \in \mathcal{F}_T$, we show that the $L^2(T)^d$-projection of ∇v can be computed using these projections of v. If $v \in W^{1,1}(T)$ and $\boldsymbol{\tau} \in C^\infty(\overline{T})^d$, an integration by parts gives

$$(\nabla v, \boldsymbol{\tau})_T = -(v, \nabla\cdot\boldsymbol{\tau})_T + \sum_{F\in\mathcal{F}_T} (v, \boldsymbol{\tau}\cdot\boldsymbol{n}_{TF})_F.$$

Specialising this relation to $\boldsymbol{\tau} \in \mathbb{P}^k(T)^d$, we can introduce the orthogonal projectors using their definitions (1.57) and the fact that $\nabla\cdot\boldsymbol{\tau} \in \mathbb{P}^{k-1}(T) \subset \mathbb{P}^k(T)$ and $\boldsymbol{\tau}_{|F}\cdot\boldsymbol{n}_{TF} \in \mathbb{P}^k(F)$ for all $F \in \mathcal{F}_T$:

$$(\boldsymbol{\pi}_T^{0,k}\nabla v, \boldsymbol{\tau})_T = -(\pi_T^{0,k} v, \nabla\cdot\boldsymbol{\tau})_T + \sum_{F\in\mathcal{F}_T} (\pi_F^{0,k} v, \boldsymbol{\tau}\cdot\boldsymbol{n}_{TF})_F. \tag{4.36}$$

As announced, this gives a formula for the L^2-orthogonal projection of ∇v on $\mathbb{P}^k(T)^d$ using the orthogonal projections of v on $\mathbb{P}^k(T)$ and of $v_{|F}$ on $\mathbb{P}^k(F)$ for all $F \in \mathcal{F}_T$. Following similar principles as in Sect. 2.1.1, and recalling that the unknowns in $\underline{U}_T^k$ precisely play the role of such projections, this leads to defining the local gradient reconstruction $\mathbf{G}_T^k : \underline{U}_T^k \to \mathbb{P}^k(T)^d$ such that, for all $\underline{v}_T \in \underline{U}_T^k$,

$$(\mathbf{G}_T^k \underline{v}_T, \boldsymbol{\tau})_T = -(v_T, \nabla\cdot\boldsymbol{\tau})_T + \sum_{F\in\mathcal{F}_T} (v_F, \boldsymbol{\tau}\cdot\boldsymbol{n}_{TF})_F \qquad \forall \boldsymbol{\tau} \in \mathbb{P}^k(T)^d. \tag{4.37}$$

For future use, we notice that an integration by parts in the first term in the right-hand side gives the following alternative definition of $\mathbf{G}_T^k \underline{v}_T$:

$$(\mathbf{G}_T^k \underline{v}_T, \boldsymbol{\tau})_T = (\nabla v_T, \boldsymbol{\tau})_T + \sum_{F\in\mathcal{F}_T} (v_F - v_T, \boldsymbol{\tau}\cdot\boldsymbol{n}_{TF})_F \qquad \forall \boldsymbol{\tau} \in \mathbb{P}^k(T)^d. \tag{4.38}$$

Remark 4.9 (Relation Between $\mathbf{G}_T^k$ and p_T^{k+1}) Taking $\boldsymbol{\tau} = \nabla w$ with $w \in \mathbb{P}^{k+1}(T)$ in (4.37) and comparing with (2.11a), it is readily inferred that

$$(\mathbf{G}_T^k \underline{v}_T - \nabla \mathrm{p}_T^{k+1} \underline{v}_T, \nabla w)_T = 0 \qquad \forall w \in \mathbb{P}^{k+1}(T). \tag{4.39}$$

In other words, $\nabla \mathrm{p}_T^{k+1} \underline{v}_T$ is the L^2-orthogonal projection of $\mathbf{G}_T^k \underline{v}_T$ on $\nabla \mathbb{P}^{k+1}(T) \subset \mathbb{P}^k(T)^d$. In the case $k = 0$, since $\nabla \mathbb{P}^1(T) = \mathbb{P}^0(T)^d$, (4.39) implies that $\mathbf{G}_T^0 \underline{v}_T = \nabla \mathrm{p}_T^1 \underline{v}_T$.

Equations (4.36) and (4.37) show that $\mathbf{G}_T^k$ and $\underline{I}_T^k$ enjoy the following commutation property (which differs from the one obtained taking the gradient of (2.14)):

$$\mathbf{G}_T^k \underline{I}_T^k v = \boldsymbol{\pi}_T^{0,k}(\nabla v) \qquad \forall v \in W^{1,1}(T). \tag{4.40}$$

Figure 4.5 illustrates this commutation property.

The analysis of the HHO scheme for (4.33) will require us to consider the $L^2(T)^d$-inner product of $\mathbf{G}_T^k \underline{v}_T$ against functions $\boldsymbol{\tau}$ that are not necessarily in $\mathbb{P}^k(T)^d$ (typically because $\boldsymbol{\tau} = \boldsymbol{K}_{|T}\boldsymbol{\gamma}$ with $\boldsymbol{\gamma} \in \mathbb{P}^k(T)^d$ and $\boldsymbol{K}_{|T}$ is not assumed constant). The formula in the following lemma will be essential to manage these terms.

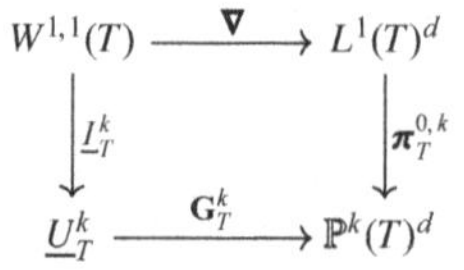

Fig. 4.5 Illustration of the commutation property (4.40) of $\mathbf{G}_T^k$

Lemma 4.10 *For all $\underline{v}_T \in \underline{U}_T^k$ and all $\boldsymbol{\tau} \in L^1(T)^d$ it holds*

$$(\mathbf{G}_T^k \underline{v}_T, \boldsymbol{\tau})_T = (\nabla v_T, \boldsymbol{\tau})_T + \sum_{F \in \mathcal{T}_h} (v_F - v_T, (\boldsymbol{\pi}_T^{0,k} \boldsymbol{\tau}) \cdot \boldsymbol{n}_{TF})_F. \tag{4.41}$$

Proof Apply (4.38) to $\boldsymbol{\pi}_T^{0,k} \boldsymbol{\tau} \in \mathbb{P}^k(T)^d$ and notice that, since $\mathbf{G}_T^k \underline{v}_T$ and ∇v_T both belong to $\mathbb{P}^k(T)^d$, $(\mathbf{G}_T^k \underline{v}_T, \boldsymbol{\pi}_T^{0,k} \boldsymbol{\tau})_T = (\mathbf{G}_T^k \underline{v}_T, \boldsymbol{\tau})_T$ and $(\nabla v_T, \boldsymbol{\pi}_T^{0,k} \boldsymbol{\tau})_T = (\nabla v_T, \boldsymbol{\tau})_T$. □

4.2.2 Local and Global Bilinear Forms

We make the following assumption.

Assumption 4.11 (Piecewise Continuity of $\boldsymbol{K}$) *For each $T \in \mathcal{T}_h$, $\boldsymbol{K}_{|T}$ can be extended into a continuous function over $\overline{T}$, also denoted by $\boldsymbol{K}_{|T}$.*

For $T \in \mathcal{T}_h$, we denote by $\overline{K}_T$ and $\underline{K}_T$, respectively, the maximal and minimal eigenvalues of $\boldsymbol{K}_{|T}$, and we define the local anisotropy-heterogeneity ratio as

$$\alpha_T = \frac{\overline{K}_T}{\underline{K}_T}. \tag{4.42}$$

We also set, for all $T \in \mathcal{T}_h$ and all $F \in \mathcal{F}_T$,

$$K_{TF} := \|\boldsymbol{K}_{|T} \boldsymbol{n}_{TF} \cdot \boldsymbol{n}_{TF}\|_{L^\infty(F)} = \|\boldsymbol{K}_{|T}^{\frac{1}{2}} \boldsymbol{n}_{TF}\|_{L^\infty(F)^d}^2, \tag{4.43}$$

where the second equality follows from $\boldsymbol{K}_{|T} \boldsymbol{n}_{TF} \cdot \boldsymbol{n}_{TF} = \boldsymbol{K}_{|T}^{\frac{1}{2}} \boldsymbol{n}_{TF} \cdot \boldsymbol{K}_{|T}^{\frac{1}{2}} \boldsymbol{n}_{TF} = |\boldsymbol{K}_{|T}^{\frac{1}{2}} \boldsymbol{n}_{TF}|^2$.

The discretisation space $\underline{U}_{h,0}^k$ defined in (2.36) is endowed with the following norm, similar to the one used in Sect. 3.1: For all $\underline{v}_h \in \underline{U}_{h,0}^k$,

$$\begin{aligned} \|\underline{v}_h\|_{1,\boldsymbol{K},h} &:= \left(\sum_{T \in \mathcal{T}_h} \|\underline{v}_T\|_{1,\boldsymbol{K},T}^2 \right)^{\frac{1}{2}}, \\ \|\underline{v}_T\|_{1,\boldsymbol{K},T} &= \left(\|\boldsymbol{K}_T^{\frac{1}{2}} \nabla v_T\|_T^2 + \sum_{F \in \mathcal{F}_T} \frac{K_{TF}}{h_F} \|v_F - v_T\|_F^2 \right)^{\frac{1}{2}} \quad \forall T \in \mathcal{T}_h. \end{aligned} \tag{4.44}$$

The local bilinear form $\mathrm{a}_{\boldsymbol{K},T} : \underline{U}_T^k \times \underline{U}_T^k \to \mathbb{R}$ is obtained as the sum of a consistent term, based on the discrete gradient $\mathbf{G}_T^k$, and of a stabilisation term, similar to the one used in Sect. 3.1: For all $\underline{u}_T, \underline{v}_T \in \underline{U}_T^k$,

$$\mathrm{a}_{\boldsymbol{K},T}(\underline{u}_T, \underline{v}_T) := (\boldsymbol{K}_{|T}\mathbf{G}_T^k\underline{u}_T, \mathbf{G}_T^k\underline{v}_T)_T + \mathrm{s}_{\boldsymbol{K},T}(\underline{u}_T, \underline{v}_T), \tag{4.45}$$

with stabilisation bilinear form defined as in (2.22), but with a scaling accounting for the local diffusion strength and orientation:

$$\mathrm{s}_{\boldsymbol{K},T}(\underline{u}_T, \underline{v}_T) := \sum_{F\in\mathcal{F}_T} \frac{K_{TF}}{h_F}((\delta_{TF}^k - \delta_T^k)\underline{u}_T, (\delta_{TF}^k - \delta_T^k)\underline{v}_T)_F. \tag{4.46}$$

We recall that the difference operators δ_T^k and δ_{TF}^k are given by (2.19). Other choices of stabilisation terms could be made, following similar design conditions as in Assumption 3.9.

As usual, the global bilinear form $\mathrm{a}_{\boldsymbol{K},h} : \underline{U}_{h,0}^k \times \underline{U}_{h,0}^k \to \mathbb{R}$ is obtained by assembling the local forms: For all $\underline{u}_h, \underline{v}_h \in \underline{U}_{h,0}^k$,

$$\mathrm{a}_{\boldsymbol{K},h}(\underline{u}_h, \underline{v}_h) := \sum_{T\in\mathcal{T}_h} \mathrm{a}_{\boldsymbol{K},T}(\underline{u}_T, \underline{v}_T). \tag{4.47}$$

Associated to this bilinear form, we define the norm

$$|||\underline{v}_h|||_{\mathrm{a},\boldsymbol{K},h} := \mathrm{a}_{\boldsymbol{K},h}(\underline{v}_h, \underline{v}_h)^{\frac{1}{2}} \qquad \forall \underline{v}_h \in \underline{U}_{h,0}^k. \tag{4.48}$$

Note that, even though this norm is still defined from $\mathrm{a}_{\boldsymbol{K},h}$, it is different from the norm $\|\cdot\|_{\mathrm{a},\boldsymbol{K},h}$ used in Sect. 3.1 even when $\boldsymbol{K}$ is piecewise constant, a difference we highlight by using a triple-bar notation. There, the bilinear form $\mathrm{a}_{\boldsymbol{K},h}$ was constructed using $\nabla \mathrm{p}_{\boldsymbol{K},T}^{k+1}$ in the volumetric terms; here, $\mathrm{a}_{\boldsymbol{K},h}$ is built from $\mathbf{G}_T^k$. The following lemma is the equivalent, for the bilinear form $\mathrm{a}_{\boldsymbol{K},h}$ defined above, of Lemma 3.15.

Lemma 4.12 (Properties of $\mathrm{a}_{\boldsymbol{K},h}$) *Under Assumption 4.11, the bilinear form $\mathrm{a}_{\boldsymbol{K},h}$ enjoys the following properties:*

(i) Estimate of boundary terms in $\|\cdot\|_{1,\boldsymbol{K},T}$. *It holds: For all $T \in \mathcal{T}_h$ and all $\underline{v}_T \in \underline{U}_T^k$,*

$$\sum_{F\in\mathcal{F}_T} \frac{K_{TF}}{h_F}\|v_F - v_T\|_F^2 \lesssim \alpha_T \mathrm{a}_{\boldsymbol{K},T}(\underline{v}_T, \underline{v}_T), \tag{4.49}$$

with hidden constant independent of h, T, $\underline{v}_T$ and $\boldsymbol{K}$.

(ii) Consistency. *It holds, for all* $r \in \{0, \ldots, k\}$ *and all* $w \in H_0^1(\Omega) \cap H^{r+2}(\mathcal{T}_h)$ *such that* $\nabla\cdot(\boldsymbol{K}\nabla w) \in L^2(\Omega)$,

$$\sup_{\underline{v}_h \in \underline{U}_{h,0}^k, \|\underline{v}_h\|_{\mathrm{a},\boldsymbol{K},h}=1} |\mathcal{E}_{\boldsymbol{K},h}(w;\underline{v}_h)| \lesssim \left(\sum_{T\in\mathcal{T}_h} \overline{K}_T h_T^{2(r+1)} |w|_{H^{r+2}(T)}^2\right)^{\frac{1}{2}}$$
$$+ \left(\sum_{T\in\mathcal{T}_h} \alpha_T \sum_{F\in\mathcal{F}_T} h_F \|(\boldsymbol{K}^{\frac{1}{2}}\nabla w)_{|T} - \boldsymbol{K}_{|T}^{-\frac{1}{2}} \boldsymbol{\pi}_T^{0,k}(\boldsymbol{K}\nabla w)\|_F^2\right)^{\frac{1}{2}}, \tag{4.50}$$

where the hidden constant is independent of w, h *and* $\boldsymbol{K}$, *and the consistency error* $\mathcal{E}_{\boldsymbol{K},h}(w;\cdot) : \underline{U}_{h,0}^k \to \mathbb{R}$ *is such that, for all* $\underline{v}_h \in \underline{U}_{h,0}^k$,

$$\mathcal{E}_{\boldsymbol{K},h}(w;\underline{v}_h) := -(\nabla\cdot(\boldsymbol{K}\nabla w), v_h) - \mathrm{a}_{\boldsymbol{K},h}(\underline{I}_h^k w, \underline{v}_h). \tag{4.51}$$

Remark 4.13 (Estimate on $\|\nabla v_T\|_T^2$ *by* $\mathrm{a}_{\boldsymbol{K},T}(\underline{v}_T,\underline{v}_T)$*)* In Propositions 2.13 and 3.13, a full coercivity of the HHO bilinear form is established with respect to the corresponding norms on $\underline{U}_{h,0}^k$. In particular, not only the boundary terms are estimated as in (4.49), but an estimate of the volumetric term is obtained. Such an estimate is also possible here, but scales with α_T^2 instead of α_T itself (compare with (4.49)): For all $T \in \mathcal{T}_h$ and all $\underline{v}_T \in \underline{U}_T^k$,

$$\|\boldsymbol{K}_{|T}^{\frac{1}{2}}\nabla v_T\|_T^2 \lesssim \alpha_T^2 \mathrm{a}_{\boldsymbol{K},T}(\underline{v}_T,\underline{v}_T). \tag{4.52}$$

Proof (i) *Estimate of boundary terms in* $\|\cdot\|_{1,\boldsymbol{K},T}$. Setting $\underline{\hat{v}}_T := \underline{I}_T^k \mathrm{p}_T^{k+1}\underline{v}_T$, we have $\underline{\hat{v}}_T - \underline{v}_T = (\delta_T^k \underline{v}_T, (\delta_{TF}^k \underline{v}_T)_{F\in\mathcal{F}_T})$ (see (2.20)), and thus

$$\|v_F - v_T\|_F \lesssim \|\delta_{TF}^k \underline{v}_T - \delta_T^k \underline{v}_T\|_F + \|\hat{v}_F - \hat{v}_T\|_F.$$

Square this inequality, multiply by K_{TF}/h_F, sum over $F \in \mathcal{F}_T$, and use $K_{TF} \le \overline{K}_T$ to get

$$\sum_{F\in\mathcal{F}_T} \frac{K_{TF}}{h_F}\|v_F - v_T\|_F^2 \lesssim \mathrm{s}_{\boldsymbol{K},T}(\underline{v}_T,\underline{v}_T) + \overline{K}_T|\underline{\hat{v}}_T|_{1,\partial T}^2$$
$$\lesssim \mathrm{s}_{\boldsymbol{K},T}(\underline{v}_T,\underline{v}_T) + \overline{K}_T\|\nabla \mathrm{p}_T^{k+1}\underline{v}_T\|_T^2, \tag{4.53}$$

where the conclusion follows by the boundedness (2.9) of $\underline{I}_T^k$ with $v = \mathrm{p}_T^{k+1}\underline{v}_T$. As seen in Remark 4.9, $\nabla \mathrm{p}_T^{k+1}\underline{v}_T$ is an $L^2(T)^d$-orthogonal projection of $\mathbf{G}_T^k\underline{v}_T$. Hence, $\|\nabla \mathrm{p}_T^{k+1}\underline{v}_T\|_T \le \|\mathbf{G}_T^k\underline{v}_T\|_T$ and

$$\sum_{F\in\mathcal{F}_T} \frac{K_{TF}}{h_F}\|v_F - v_T\|_F^2 \lesssim \mathrm{s}_{\boldsymbol{K},T}(\underline{v}_T,\underline{v}_T) + \overline{K}_T\|\mathbf{G}_T^k\underline{v}_T\|_T^2$$
$$\lesssim \mathrm{s}_{\boldsymbol{K},T}(\underline{v}_T,\underline{v}_T) + \alpha_T\|\boldsymbol{K}^{\frac12}\mathbf{G}_T^k\underline{v}_T\|_T^2,$$

where we have used $\|\boldsymbol{K}^{\frac12}\mathbf{G}_T^k\underline{v}_T\|_T^2 \ge \underline{K}_T\|\mathbf{G}_T^k\underline{v}_T\|_T^2$ and the definition (4.42) of α_T in the second line. Recalling the definition (4.45) of $\mathrm{a}_{\boldsymbol{K},T}$ and that $\alpha_T \ge 1$, this completes the proof of (4.49).

(ii) *Consistency.* Let $\underline{v}_h \in \underline{U}_{h,0}^k$ be such that $|||\underline{v}_h|||_{\mathrm{a},\boldsymbol{K},h} = 1$. Using element-wise integrations by parts (justified because $\boldsymbol{K}\nabla w \in \boldsymbol{H}(\mathrm{div};\Omega)$, $\nabla w \in H^1(\mathcal{T}_h)^d$ and $\boldsymbol{K}_{|T}$ is continuous) and (4.41) with $\boldsymbol{\tau} = (\boldsymbol{K}\nabla w)_{|T}$, we write

$$\begin{aligned}
-(\nabla\cdot(\boldsymbol{K}\nabla w), v_h) &= \sum_{T\in\mathcal{T}_h}\left((\boldsymbol{K}\nabla w, \nabla v_T)_T + \sum_{F\in\mathcal{F}_T}((\boldsymbol{K}\nabla w)_{|T}\cdot\boldsymbol{n}_{TF}, v_F - v_T)_F\right)\\
&= \sum_{T\in\mathcal{T}_h}(\boldsymbol{K}\nabla w, \mathbf{G}_T^k\underline{v}_T)_T\\
&\quad + \sum_{T\in\mathcal{T}_h}\sum_{F\in\mathcal{F}_T}([(\boldsymbol{K}\nabla w)_{|T} - \boldsymbol{\pi}_T^{0,k}(\boldsymbol{K}\nabla w)]\cdot\boldsymbol{n}_{TF}, v_F - v_T)_F,
\end{aligned} \tag{4.54}$$

where the introduction of v_F in the first line is justified by Corollary 1.19 with $\boldsymbol{\tau} = \boldsymbol{K}\nabla w \in \boldsymbol{H}(\mathrm{div};\Omega)$ and $(\varphi_F)_{F\in\mathcal{F}_h} = (v_F)_{F\in\mathcal{F}_h}$ (see also Remark 1.20 and notice that the regularities of $\boldsymbol{K}_{|T}$ and ∇w recalled above ensure that $(\boldsymbol{K}\nabla w)_{|T}$ has an L^2-trace on the faces of T), while the second line is obtained invoking Lemma 4.10 with $\boldsymbol{\tau} = \boldsymbol{K}\nabla w$. We then write, by definition of $\mathrm{a}_{\boldsymbol{K},T}$ and using $\mathbf{G}_T^k\underline{I}_T^k w = \boldsymbol{\pi}_T^{0,k}(\nabla w)$ (see (4.40)),

$$\mathrm{a}_{\boldsymbol{K},h}(\underline{I}_h^k w, \underline{v}_h) = \sum_{T\in\mathcal{T}_h}(\boldsymbol{K}\boldsymbol{\pi}_T^{0,k}(\nabla w), \mathbf{G}_T^k\underline{v}_T)_T + \sum_{T\in\mathcal{T}_h}\mathrm{s}_{\boldsymbol{K},T}(\underline{I}_T^k w, \underline{v}_T). \tag{4.55}$$

Subtracting (4.55) from (4.54) we infer

$$
\begin{aligned}
\mathcal{E}_{\boldsymbol{K},h}(w;\underline{v}_h) = &\underbrace{\sum_{T\in\mathcal{T}_h}(\boldsymbol{K}\nabla w - \boldsymbol{K}\boldsymbol{\pi}_T^{0,k}(\nabla w), \mathbf{G}_T^k\underline{v}_T)_T}_{\mathfrak{T}_1(\underline{v}_h)} \\
&+ \underbrace{\sum_{T\in\mathcal{T}_h}\sum_{F\in\mathcal{F}_T}\left(\left[(\boldsymbol{K}\nabla w)_{|T} - \boldsymbol{\pi}_T^{0,k}(\boldsymbol{K}\nabla w)\right]\cdot\boldsymbol{n}_{TF}, v_F - v_T\right)_F}_{\mathfrak{T}_2(\underline{v}_h)} \\
&- \underbrace{\sum_{T\in\mathcal{T}_h} \mathrm{s}_{\boldsymbol{K},T}(\underline{I}_T^k w, \underline{v}_T).}_{\mathfrak{T}_3(\underline{v}_h)}
\end{aligned}
\tag{4.56}
$$

We estimate $\mathfrak{T}_1(\underline{v}_h)$ starting with Cauchy–Schwarz inequalities, recalling the definition of $|||\cdot|||_{\mathrm{a},\boldsymbol{K},h}$ together with the fact that $|||\underline{v}_h|||_{\mathrm{a},\boldsymbol{K},h} = 1$ to estimate the term involving $\mathbf{G}_T^k\underline{v}_T$, and applying the approximation property (1.74) of $\boldsymbol{\pi}_T^{0,k}$ with $l = k$, $p = 2$, $s = r+1$, $m = 0$ and $v =$ the components of ∇w:

$$
\begin{aligned}
|\mathfrak{T}_1(\underline{v}_h)| &\le \left(\sum_{T\in\mathcal{T}_h}\|\boldsymbol{K}^{\frac{1}{2}}(\nabla w - \boldsymbol{\pi}_T^{0,k}(\nabla w))\|_T^2\right)^{\frac{1}{2}}\left(\sum_{T\in\mathcal{T}_h}\|\boldsymbol{K}^{\frac{1}{2}}\mathbf{G}_T^k\underline{v}_T\|_T^2\right)^{\frac{1}{2}} \\
&\le \left(\sum_{T\in\mathcal{T}_h}\overline{K}_T\|\nabla w - \boldsymbol{\pi}_T^{0,k}(\nabla w)\|_T^2\right)^{\frac{1}{2}} \\
&\lesssim \left(\sum_{T\in\mathcal{T}_h}\overline{K}_T h_T^{2(r+1)}|\nabla w|^2_{H^{r+1}(T)^d}\right)^{\frac{1}{2}}.
\end{aligned}
\tag{4.57}
$$

For $\mathfrak{T}_2(\underline{v}_h)$, notice first that

$$
\begin{aligned}
&\left|\left(\left[(\boldsymbol{K}\nabla w)_{|T} - \boldsymbol{\pi}_T^{0,k}(\boldsymbol{K}\nabla w)\right]\cdot\boldsymbol{n}_{TF}, v_F - v_T\right)_F\right| \\
&\qquad = \left|\left(\left[(\boldsymbol{K}^{\frac{1}{2}}\nabla w)_{|T} - \boldsymbol{K}_{|T}^{-\frac{1}{2}}\boldsymbol{\pi}_T^{0,k}(\boldsymbol{K}\nabla w)\right]\cdot\boldsymbol{K}_{|T}^{\frac{1}{2}}\boldsymbol{n}_{TF}, v_F - v_T\right)_F\right| \\
&\qquad \le \|(\boldsymbol{K}^{\frac{1}{2}}\nabla w)_{|T} - \boldsymbol{K}_{|T}^{-\frac{1}{2}}\boldsymbol{\pi}_T^{0,k}(\boldsymbol{K}\nabla w)\|_F K_{TF}^{\frac{1}{2}}\|v_F - v_T\|_F,
\end{aligned}
$$

where the first line is obtained by symmetry of $\boldsymbol{K}_{|T}^{\frac{1}{2}}$, and the second line follows from a generalised Hölder inequality with exponents $(2, \infty, 2)$ together with the definition (4.43) of K_{TF}. By Cauchy–Schwarz inequalities on the sums, we then obtain the bound

$$|\mathfrak{T}_2(\underline{v}_h)| \leq \left(\sum_{T\in\mathcal{T}_h} \alpha_T \sum_{F\in\mathcal{F}_T} h_F \|(\boldsymbol{K}^{\frac{1}{2}}\nabla w)_{|T} - \boldsymbol{K}_{|T}^{-\frac{1}{2}}\boldsymbol{\pi}_T^{0,k}(\boldsymbol{K}\nabla w)\|_F^2\right)^{\frac{1}{2}}$$

$$\times\left(\sum_{T\in\mathcal{T}_h} \alpha_T^{-1} \sum_{F\in\mathcal{F}_T} \frac{K_{TF}}{h_F}\|v_F - v_T\|_F^2\right)^{\frac{1}{2}} \tag{4.58}$$

$$\leq \left(\sum_{T\in\mathcal{T}_h} \alpha_T \sum_{F\in\mathcal{F}_T} h_F \|(\boldsymbol{K}^{\frac{1}{2}}\nabla w)_{|T} - \boldsymbol{K}_{|T}^{-\frac{1}{2}}\boldsymbol{\pi}_T^{0,k}(\boldsymbol{K}\nabla w)\|_F^2\right)^{\frac{1}{2}}, \tag{4.59}$$

the conclusion being a consequence of (4.49) and $|||\underline{v}_h|||_{\mathrm{a},\boldsymbol{K},h} = 1$.

Letting s_T be the stabilisation defined for the Poisson problem by (2.22), the bound $K_{TF} \leq \overline{K}_T$, the definition (4.46) of $\mathrm{s}_{\boldsymbol{K},T}$, and the consistency property (2.31) of s_T yield

$$\mathrm{s}_{\boldsymbol{K},T}(\underline{I}_T^k w, \underline{I}_T^k w) \leq \overline{K}_T \mathrm{s}_T(\underline{I}_T^k w, \underline{I}_T^k w) \lesssim \overline{K}_T h_T^{2(r+1)} |w|_{H^{r+2}(T)}^2. \tag{4.60}$$

This enables the following bound of $\mathfrak{T}_3(\underline{v}_h)$, based on a Cauchy–Schwarz inequality on the symmetric positive semidefinite form $\mathrm{s}_{\boldsymbol{K},T}$, and in which we also use $\sum_{T\in\mathcal{T}_h} \mathrm{s}_{\boldsymbol{K},T}(\underline{v}_T, \underline{v}_T) \leq |||\underline{v}_h|||_{\mathrm{a},\boldsymbol{K},h}^2 = 1$:

$$|\mathfrak{T}_3(\underline{v}_h)| \leq \left(\sum_{T\in\mathcal{T}_h} \mathrm{s}_{\boldsymbol{K},T}(\underline{I}_T^k w, \underline{I}_T^k w)\right)^{\frac{1}{2}} \left(\sum_{T\in\mathcal{T}_h} \mathrm{s}_{\boldsymbol{K},T}(\underline{v}_T, \underline{v}_T)\right)^{\frac{1}{2}} \tag{4.61}$$

$$\lesssim \left(\sum_{T\in\mathcal{T}_h} \overline{K}_T h_T^{2(r+1)} |w|_{H^{r+2}(T)}^2\right)^{\frac{1}{2}}. \tag{4.62}$$

Plugging (4.57), (4.59), and (4.62) into (4.56) concludes the proof of (4.50). □

Remark 4.14 (Approach to the Consistency Error Estimate) The reconstructed gradient $\mathbf{G}_T^k \underline{v}_T$ does not account for the anisotropic diffusion tensor. As a consequence, and contrary to what happens for the Poisson problem in the proof of Lemma 2.18 or for piecewise-constant diffusion in the proof of Lemma 3.15, one cannot expect here the volumetric term to cancel out when creating $\mathcal{E}_{\boldsymbol{K},h}(w; \underline{v}_h)$. As seen in

Remark 4.13, estimating $\|\boldsymbol{K}_{|T}^{1/2}\nabla v_T\|_T^2$ by $\mathrm{a}_{\boldsymbol{K},T}(\underline{v}_T, \underline{v}_T)$ introduces a local constant that scales like the square of the local anisotropy-heterogeneity ratio, and would lead to α_T being replaced by α_T^2 in (4.50), which is much worse than what has been obtained for piecewise constant diffusion in (3.48). To avoid this issue, we had to adopt a slightly different approach to estimate the dual norm of $\mathcal{E}_{\boldsymbol{K},h}(w; \underline{v}_h)$, by getting rid of all the terms ∇v_T, using (4.41) to replace them by $\mathbf{G}_T^k \underline{v}_T$ (see (4.54)).

For the same reason, we do not directly estimate the second addend, in the right-hand side of (4.50), in terms of approximability properties and Sobolev semi-norms. Such an estimate is obtained in Theorem 4.16 below (see (4.71)), but it is sometimes sub-optimal in terms of dependency with respect to the anisotropy-heterogeneity ratio. Preserving the last term in (4.50) provides the flexibility required to recover better estimates in certain circumstances, as demonstrated in (4.72).

4.2.3 Discrete Problem and Flux Formulation

The HHO scheme for (4.33) is obtained using the global bilinear form $\mathrm{a}_{\boldsymbol{K},h}$ in a classical way: Find $\underline{u}_h \in \underline{U}_{h,0}^k$ such that

$$\mathrm{a}_{\boldsymbol{K},h}(\underline{u}_h, \underline{v}_h) = (f, v_h) \qquad \forall \underline{v}_h \in \underline{U}_{h,0}^k. \tag{4.63}$$

As for the Poisson problem, the case of piecewise constant diffusion, and the case of diffusion–advection–reaction treated in Sects. 2.2.5, 3.1.4.3 and 3.2, respectively, this HHO scheme can be reformulated in terms of numerical fluxes that satisfy local balance and continuity properties. We recall that the boundary difference space $\underline{D}_{\partial T}^k$ and operator $\underline{\Delta}_{\partial T}^k$ are respectively defined by (2.55) and (2.56).

Lemma 4.15 (Flux Formulation) *Let $\mathcal{M}_h$ denote a polytopal mesh in the sense of Definition 1.4, and let Assumption 4.11 hold true. For all $T \in \mathcal{T}_h$, let $\mathrm{s}_{\boldsymbol{K},T}$ be defined by (4.46), and define the boundary residual operator $\underline{R}_{\boldsymbol{K},\partial T}^k : \underline{U}_T^k \to \underline{D}_{\partial T}^k$ such that, for all $\underline{v}_T \in \underline{U}_T^k$,*

$$\underline{R}_{\boldsymbol{K},\partial T}^k \underline{v}_T := (R_{\boldsymbol{K},TF}^k \underline{v}_T)_{F\in\mathcal{F}_T}$$

and, for all $\underline{\alpha}_{\partial T} = (\alpha_{TF})_{F\in\mathcal{F}_T} \in \underline{D}_{\partial T}^k$,

$$-\sum_{F\in\mathcal{F}_T} (R_{\boldsymbol{K},TF}^k \underline{v}_T, \alpha_{TF})_F = \mathrm{s}_{\boldsymbol{K},T}((0, \underline{\Delta}_{\partial T}^k \underline{v}_T), (0, \underline{\alpha}_{\partial T})). \tag{4.64}$$

(continued)

Lemma 4.15 (continued)
For $\underline{u}_h \in \underline{U}_{h,0}^k$, $T \in \mathcal{T}_h$ *and* $F \in \mathcal{F}_T$, *define the numerical normal trace of the flux*

$$\Phi_{\boldsymbol{K},TF}(\underline{u}_T) := -\boldsymbol{\pi}_T^{0,k}(\boldsymbol{K}_{|T}\mathbf{G}_T^k\underline{u}_T)\cdot\boldsymbol{n}_{TF} + R_{\boldsymbol{K},TF}^k\underline{u}_T. \tag{4.65}$$

Then $\underline{u}_h$ *is a solution of* (4.63) *if and only if the following two properties hold:*

(i) Local balance. *For all* $T \in \mathcal{T}_h$ *and all* $v_T \in \mathbb{P}^k(T)$,

$$(\boldsymbol{K}_T\mathbf{G}_T^k\underline{u}_T, \nabla v_T)_T + \sum_{F\in\mathcal{F}_T}(\Phi_{\boldsymbol{K},TF}(\underline{u}_T), v_T)_F = (f, v_T)_T. \tag{4.66}$$

(ii) Continuity of fluxes. *For any interface* $F \in \mathcal{F}_h^{\mathrm{i}}$ *such that* $F \subset \partial T_1 \cap \partial T_2$ *with distinct mesh elements* $T_1, T_2 \in \mathcal{T}_h$, *the numerical normal traces of the fluxes are continuous, i.e.,*

$$\Phi_{\boldsymbol{K},T_1F}(\underline{u}_{T_1}) + \Phi_{\boldsymbol{K},T_2F}(\underline{u}_{T_2}) = 0. \tag{4.67}$$

Proof The proof is similar to that of Lemma 2.25, but we provide it nonetheless as it requires a slight twist using (4.41).

Let $\underline{u}_h, \underline{v}_h \in \underline{U}_{h,0}^k$. Since $\mathrm{s}_{\boldsymbol{K},T}$ depends only on its arguments through the difference operators δ_{TF}^k and δ_T^k, Proposition 2.24 and the definition (4.64) show that

$$\mathrm{s}_{\boldsymbol{K},T}(\underline{u}_T, \underline{v}_T) = -\sum_{F\in\mathcal{F}_T}(R_{\boldsymbol{K},TF}^k\underline{u}_T, v_F - v_T)_F. \tag{4.68}$$

Applying (4.41) with $\boldsymbol{\tau} = \boldsymbol{K}_{|T}\mathbf{G}_T^k\underline{u}_T$ shows that

$$\begin{aligned}(\boldsymbol{K}_{|T}\mathbf{G}_T^k\underline{u}_T, \mathbf{G}_T^k\underline{v}_T)_T &= (\boldsymbol{K}_{|T}\mathbf{G}_T^k\underline{u}_T, \nabla v_T)_T \\ &\quad + \sum_{F\in\mathcal{F}_T}(\boldsymbol{\pi}_T^{0,k}(\boldsymbol{K}_{|T}\mathbf{G}_T^k\underline{u}_T)\cdot\boldsymbol{n}_{TF}, v_F - v_T)_F.\end{aligned} \tag{4.69}$$

Plugging (4.68) and (4.69) into the definition (4.45) of the local bilinear form $\mathrm{a}_{\boldsymbol{K},T}$, and the resulting relation into the definition (4.47) of the global bilinear form $\mathrm{a}_{\boldsymbol{K},h}$, we see that the latter admits the reformulation (2.51) with, for all $T \in \mathcal{T}_h$, $\mathrm{a}_{\mathrm{V},T}(\underline{u}_T, v_T) = (\boldsymbol{K}_{|T}\mathbf{G}_T^k\underline{u}_T, \nabla v_T)_T$ for all $(\underline{u}_T, v_T) \in \underline{U}_T^k \times \mathbb{P}^k(T)$ and, for all $\underline{u}_T \in \underline{U}_T^k$ and all $F \in \mathcal{F}_T$, $\Phi_{TF}(\underline{u}_T) = \Phi_{\boldsymbol{K},TF}(\underline{u}_T)$. The conclusion is an immediate consequence of Lemma 2.21. □

4.2.4 Energy Error Estimate

Using the Third Strang Lemma A.7, the following error estimates in energy norm for the HHO scheme (4.63) are easy consequences of the consistency estimate on $\mathrm{a}_{\boldsymbol{K},h}$ in Lemma 4.12.

Theorem 4.16 (Discrete Energy Error Estimate) *Let $(\mathcal{M}_h)_{h\in\mathcal{H}}$ denote a regular mesh sequence in the sense of Definition 1.9. Let Assumption 4.11 hold true and let a polynomial degree $k \geq 0$ be fixed. Denote by $u \in H_0^1(\Omega)$ the unique solution to (4.35), for which we assume the additional regularity $u \in H^{r+2}(\mathcal{T}_h)$ for some $r \in \{0, \ldots, k\}$. For all $h \in \mathcal{H}$, let $\underline{u}_h \in \underline{U}_{h,0}^k$ denote the unique solution to (4.63) with $\mathrm{a}_{\boldsymbol{K},h}$ defined by (4.45)–(4.47). Then, it holds that*

$$|||\underline{u}_h - \underline{I}_h^k u|||_{\mathrm{a},\boldsymbol{K},h} \lesssim \left(\sum_{T\in\mathcal{T}_h} \overline{K}_T h_T^{2(r+1)} |u|_{H^{r+2}(T)}^2\right)^{\frac{1}{2}} + \left(\sum_{T\in\mathcal{T}_h} \alpha_T \sum_{F\in\mathcal{F}_T} h_F \|(\boldsymbol{K}^{\frac{1}{2}}\nabla u)_{|T} - \boldsymbol{K}_{|T}^{-\frac{1}{2}} \boldsymbol{\pi}_T^{0,k}(\boldsymbol{K}\nabla u)\|_F^2\right)^{\frac{1}{2}}, \tag{4.70}$$

where the norm $|||\cdot|||_{\mathrm{a},\boldsymbol{K},h}$ is defined by (4.48) and the hidden constant is independent of h, u and $\boldsymbol{K}$. As a consequence,

(i) If $\boldsymbol{K}$ is constant on each element $T \in \mathcal{T}_h$ then

$$|||\underline{u}_h - \underline{I}_h^k u|||_{\mathrm{a},\boldsymbol{K},h} \lesssim \left(\sum_{T\in\mathcal{T}_h} \overline{K}_T \alpha_T h_T^{2(r+1)} |u|_{H^{r+2}(T)}^2\right)^{\frac{1}{2}}. \tag{4.71}$$

(ii) If $\boldsymbol{K}\nabla u \in H^{r+1}(\mathcal{T}_h)^d$ then

$$|||\underline{u}_h - \underline{I}_h^k u|||_{\mathrm{a},\boldsymbol{K},h} \lesssim \left(\sum_{T\in\mathcal{T}_h} \overline{K}_T h_T^{2(r+1)} |u|_{H^{r+2}(T)^d}^2\right)^{\frac{1}{2}} + \left(\sum_{T\in\mathcal{T}_h} \underline{K}_T^{-1} \alpha_T h_T^{2(r+1)} |\boldsymbol{K}\nabla u|_{H^{r+1}(T)^d}^2\right)^{\frac{1}{2}}. \tag{4.72}$$

A few remarks are in order.

Remark 4.17 (Combining (4.71) *and* (4.72)*)* Both estimates (4.71) and (4.72) are direct consequences of (4.70), based on particular treatments of the last term in this estimate. The proof shows that the estimates can easily be combined in the case where $\boldsymbol{K}$ is constant in some elements $T \in \mathcal{T}_{h,1}$, and $\boldsymbol{K}\nabla w \in H^{r+1}(T)$ for the elements $T \in \mathcal{T}_{h,2} := \mathcal{T}_h \backslash \mathcal{T}_{h,1}$. In this case, the upper bound on $|||u_h - \underline{I}_h^k u|||_{\mathrm{a},\boldsymbol{K},h}$ consists in the right-hand side of (4.71) with a sum limited to $T \in \mathcal{T}_{h,1}$ plus the right-hand side of (4.72) with sums limited to $T \in \mathcal{T}_{h,2}$.

Remark 4.18 (Rates and Dependency with Respect to the Anisotropy-Heterogeneity Ratio) Both estimates (4.71) and (4.72) give a global estimate $|||\underline{u}_h - \underline{I}_h^k u|||_{\mathrm{a},\boldsymbol{K},h} \lesssim h^{r+1}$ (with hidden constant depending on $\boldsymbol{K}$ and u). The difference lies in the dependency with respect to the anisotropy-heterogeneity ratio. Applied to $\boldsymbol{K}$ that is constant in each element, (4.72) yields

$$|||\underline{u}_h - \underline{I}_h^k u|||_{\mathrm{a},\boldsymbol{K},h} \lesssim \left(\sum_{T\in\mathcal{T}_h} \overline{K}_T \alpha_T^2 h_T^{2(r+1)} |u|^2_{H^{r+2}(T)^d} \right)^{\frac{1}{2}},$$

with hidden constant not depending on h, u or $\boldsymbol{K}$. This estimate is worse than (4.71), in which only the power one of α_T appears. This justifies keeping the last term in (4.70) in this form, as it can lead, in certain cases, to better estimates than a bound purely based on regularity assumptions, such as (4.72).

Remark 4.19 (Piecewise Constant Diffusion) For a piecewise constant diffusion coefficient, the estimate (4.71) is identical to the one obtained using the HHO method in Sect. 3.1 (see (3.56)). This shows that the method developed here enjoys similar error estimates as the one developed in that section, assuming at the onset that the diffusion was piecewise constant on the mesh. However, the method in Sect. 3.1 can be computationally slightly less expensive for $k \geq 1$, depending on the implementation, because the gradient $\nabla \mathrm{p}_{\boldsymbol{K},T}^{k+1} \underline{v}_T$ used in the consistent contribution to the local bilinear form only has to be constructed in the space $\nabla \mathbb{P}^{k+1}(T)$, whereas $\mathbf{G}_T^k \underline{v}_T$ defined by (4.37) is constructed in the larger space $\mathbb{P}^k(T)^d$.

Proof (Theorem 4.16) The estimate (4.70) follows combining the consistency error estimate (4.50) and the Third Strang Lemma A.7, as in the proofs of Theorems 2.27 and 3.18.

Let us now consider the case where $\boldsymbol{K}_{|T}$ is constant for each $T \in \mathcal{T}_h$. Then $\boldsymbol{\pi}_T^{0,k}(\boldsymbol{K}\nabla u) = \boldsymbol{K}_{|T}\boldsymbol{\pi}_T^{0,k}(\nabla u)$ and thus, for $F \in \mathcal{F}_T$,

$$\begin{aligned} h_F \|(\boldsymbol{K}^{\frac{1}{2}}\nabla u)_{|T} - \boldsymbol{K}_{|T}^{-\frac{1}{2}}\boldsymbol{\pi}_T^{0,k}(\boldsymbol{K}\nabla u)\|_F^2 &= h_F \|\boldsymbol{K}_{|T}^{\frac{1}{2}}[(\nabla u)_{|T} - \boldsymbol{\pi}_T^{0,k}(\nabla u)]\|_F^2 \\ &\leq \overline{K}_T h_F \|(\nabla u)_{|T} - \boldsymbol{\pi}_T^{0,k}(\nabla u)\|_F^2 \\ &\lesssim \overline{K}_T h_T^{2(r+1)} |\nabla u|^2_{H^{r+1}(T)^d}, \end{aligned}$$

the conclusion being a consequence of $h_F \leq h_T$ and of the trace approximation property (1.75) of $\pi_T^{0,k}$ applied to $l = k$, $s = r+1$, $m = 0$, $p = 2$ and $v =$ components of ∇u. Plugging this estimate into (4.70) yields (4.71) after recalling that, by definition, $\alpha_T \geq 1$ for all $T \in \mathcal{T}_h$.

We now assume that $\boldsymbol{K}$ can vary inside each element, but that $(\boldsymbol{K}\nabla u)_{|T} \in H^{r+1}(T)$ for all $T \in \mathcal{T}_h$. Write

$$(\boldsymbol{K}^{\frac{1}{2}}\nabla u)_{|T} - \boldsymbol{K}_{|T}^{-\frac{1}{2}}\boldsymbol{\pi}_T^{0,k}(\boldsymbol{K}\nabla u) = \boldsymbol{K}_{|T}^{-\frac{1}{2}}\left[(\boldsymbol{K}\nabla u)_{|T} - \boldsymbol{\pi}_T^{0,k}(\boldsymbol{K}\nabla u)\right]$$

and thus, by definition of $\underline{K}_T$ and by the same approximation property of $\pi_T^{0,k}$ as above but applied to $v =$ components of $\boldsymbol{K}\nabla u$, for all $F \in \mathcal{F}_T$,

$$\begin{aligned}h_F\|(\boldsymbol{K}^{\frac{1}{2}}\nabla u)_{|T} - \boldsymbol{K}_{|T}^{-\frac{1}{2}}\boldsymbol{\pi}_T^{0,k}(\boldsymbol{K}\nabla u)\|_F^2 &\leq \underline{K}_T^{-1}h_F\|(\boldsymbol{K}\nabla u)_{|T} - \boldsymbol{\pi}_T^{0,k}(\boldsymbol{K}\nabla u)\|_F^2\\ &\leq \underline{K}_T^{-1}h_T^{2(r+1)}|\boldsymbol{K}\nabla u|_{H^{r+1}(T)^d}^2. \qquad (4.73)\end{aligned}$$

Plugged into (4.70), this proves (4.72). □

In a similar way as in Theorem 3.19, this energy error estimate gives an estimate on a reconstructed approximate solution.

Corollary 4.20 (Energy Error Estimate for an Approximate Reconstructed Solution) *Under the assumptions of Point (ii) in Theorem 4.16, and recalling the definition* (2.63) *of* p_h^{k+1}, *it holds that*

$$\begin{aligned}&\underline{K}^{\frac{1}{2}}\|\nabla_h(\mathrm{p}_h^{k+1}\underline{u}_h - u)\| + |\underline{u}_h|_{\mathrm{s},\boldsymbol{K},h}\\ &\lesssim \left(\sum_{T\in\mathcal{T}_h}\overline{K}_T h_T^{2(r+1)}|u|_{H^{r+2}(T)^d}^2\right)^{\frac{1}{2}} + \left(\sum_{T\in\mathcal{T}_h}\underline{K}_T^{-1}\alpha_T h_T^{2(r+1)}|\boldsymbol{K}\nabla u|_{H^{r+1}(T)^d}^2\right)^{\frac{1}{2}},\end{aligned}$$

where the hidden constant is independent of h, u and $\boldsymbol{K}$ *and, for all* $\underline{v}_h \in \underline{U}_{h,0}^k$, *we have set*

$$|\underline{v}_h|_{\mathrm{s},\boldsymbol{K},h} := \left(\sum_{T\in\mathcal{T}_h}\mathrm{s}_{\boldsymbol{K},T}(\underline{v}_T,\underline{v}_T)\right)^{\frac{1}{2}}.$$

Proof We define the global operator $\mathbf{G}_h^k : \underline{U}_h^k \to \mathbb{P}^k(\mathcal{T}_h)^d$ such that, for all $\underline{v}_h \in \underline{U}_h^k$,

$$(\mathbf{G}_h^k\underline{v}_h)_{|T} := \mathbf{G}_T^k\underline{v}_T \qquad \forall T \in \mathcal{T}_h. \qquad (4.74)$$

Let $\underline{\hat{u}}_h = \underline{I}_h^k u$. Since, for all $T \in \mathcal{T}_h$, $\nabla \mathrm{p}_T^{k+1}(\underline{u}_T - \underline{\hat{u}}_T)$ is the $L^2(T)^d$-projection of $\mathbf{G}_T^k(\underline{u}_T - \underline{\hat{u}}_T)$ on $\nabla \mathbb{P}^{k+1}(T)$ (see Remark 4.9), we have

$$\underline{K}^{\frac{1}{2}} \|\nabla_h \mathrm{p}_h^{k+1}(\underline{u}_h - \underline{\hat{u}}_h)\| \le \underline{K}^{\frac{1}{2}} \|\mathbf{G}_h^k(\underline{u}_h - \underline{\hat{u}}_h)\| \le \|\boldsymbol{K}^{\frac{1}{2}} \mathbf{G}_h^k(\underline{u}_h - \underline{\hat{u}}_h)\|.$$

Hence,

$$\underline{K}^{\frac{1}{2}} \|\nabla_h \mathrm{p}_h^{k+1}(\underline{u}_h - \underline{\hat{u}}_h)\| + |\underline{u}_h - \underline{\hat{u}}_h|_{\mathrm{s},\boldsymbol{K},h} \le 2 |\!|\!| \underline{u}_h - \underline{\hat{u}}_h |\!|\!|_{\mathrm{a},\boldsymbol{K},h}.$$

The conclusion follows from this estimate as in the proof of Theorem 2.28, by using the triangle inequality, the property $\underline{K} \le \overline{K}_T$ for all $T \in \mathcal{T}_h$, the discrete energy estimate (4.72), and the consistency estimate (4.60) on $\mathrm{s}_{\boldsymbol{K},T}$. □

4.2.5 L^2-Error Estimate

As usual, the L^2-error estimates are obtained under an elliptic regularity assumption on the dual problem to (4.35) which, due to the symmetry of $\boldsymbol{K}$, is (4.35) itself. Specifically, we assume that there exists $C_{\mathrm{ell}} \ge 0$ such that, for any $g \in L^2(\Omega)$, letting z_g be the solution to (4.35) with $f = g$, we have $z_g \in H^2(\Omega)$, $\boldsymbol{K}\nabla z_g \in H^1(\Omega)^d$, and

$$\|z_g\|_{H^2(\Omega)} + \|\boldsymbol{K}\nabla z_g\|_{H^1(\Omega)^d} \le C_{\mathrm{ell}} \|g\|. \tag{4.75}$$

We recall that such an elliptic regularity assumption is known if Ω is convex and $\boldsymbol{K}$ is Lipschitz-continuous. Contrary to the case of a piecewise-constant diffusion coefficient (cf. Remark 3.21), under Assumption 4.11 we can have $\boldsymbol{K}$ Lipschitz-continuous and not globally constant. However, the dependency on C_{ell} is usually global in terms of $\boldsymbol{K}$, and no local estimates on the H^2-norm of z_g is known. For this reason, we make no attempt at tracking precise dependency of the error estimates in terms of the local behaviour of $\boldsymbol{K}$.

Theorem 4.21 (L^2-Error Estimate) *Let $(\mathcal{M}_h)_{h\in\mathcal{H}}$ denote a regular mesh sequence in the sense of Definition 1.9. Let Assumption 4.11 hold true, assume elliptic regularity, and let a polynomial degree $k \ge 0$ be fixed. Denote by $u \in H_0^1(\Omega)$ the unique solution to (4.35), for which we assume the additional regularity $u \in H^{r+2}(\mathcal{T}_h)$ and $\boldsymbol{K}\nabla u \in H^{r+1}(\mathcal{T}_h)^d$ for some $r \in \{0, \ldots, k\}$. If $k = 0$, we further assume that $f \in H^1(\mathcal{T}_h)$ and that $\boldsymbol{K} \in W^{1,\infty}(\mathcal{T}_h)^{d\times d}$. For all $h \in \mathcal{H}$, let $\underline{u}_h \in \underline{U}_{h,0}^k$ denote the unique solution to (4.63) with $\mathrm{a}_{\boldsymbol{K},h}$*

(continued)

Theorem 4.21 (continued)
defined by (4.45)–(4.47). *Then it holds, with a hidden constant independent of* h, u, *and* f, *but depending on* $\boldsymbol{K}$, ϱ, *and* k:

$$\|u_h - \pi_h^{0,k} u\| \lesssim \begin{cases} h^2 \|f\|_{H^1(\mathcal{T}_h)} \left(1 + |\boldsymbol{K}|_{W^{1,\infty}(\mathcal{T}_h)^{d\times d}}\right) & \text{if } k = 0, \\ h^{r+2} \left(|u|_{H^{r+2}(\mathcal{T}_h)} + |\boldsymbol{K}\nabla u|_{H^{r+1}(\mathcal{T}_h)^d}\right) & \text{if } k \geq 1. \end{cases} \tag{4.76}$$

Proof The theorem hinges on the abstract Aubin–Nitsche Lemma A.10, with a similar setting as in the proof of Lemma 2.33: $\mathtt{U} = H_0^1(\Omega)$, $\mathtt{a}(u, v) = (\boldsymbol{K}\nabla u, \nabla v)$, $\mathtt{l}(v) = (f, v)$, $\mathtt{U}_h = \underline{U}_{h,0}^k$, $\|\cdot\|_{\mathtt{U}_h} = |\!|\!|\cdot|\!|\!|_{\mathrm{a},\boldsymbol{K},h}$, $\mathtt{a}_h = \mathrm{a}_{\boldsymbol{K},h}$, $\mathtt{l}_h(\underline{v}_h) = (f, v_h)$, $\mathtt{I}_h u = \underline{I}_h^k u$, $\mathtt{L} = L^2(\Omega)$, and $\mathtt{r}_h : \underline{U}_{h,0}^k \to L^2(\Omega)$ defined by $\mathtt{r}_h \underline{v}_h = v_h$.

We first notice, as for the Poisson problem, that the dual consistency error $\mathcal{E}_h^{\mathrm{d}}(z_g; \cdot)$ is identical to the primal consistency error $\mathcal{E}_{\boldsymbol{K},h}(z_g; \cdot)$. Hence, (4.50) with $r = 0$ and the bound on the second term obtained for this r in Point (ii) of the proof of Theorem 4.16 show that $\|\mathcal{E}_h^{\mathrm{d}}(z_g; \cdot)\|_{\mathtt{U}_h^\star} \lesssim h \left(|z_g|_{H^2(\Omega)} + |\boldsymbol{K}\nabla z_g|_{H^1(\Omega)^d}\right) \lesssim h\|g\|$ (we have used (4.75) to conclude). By (4.72),

$$\|u_h - \mathtt{I}_h u\|_{\mathtt{U}_h} \lesssim h^{r+1} \left(|u|_{H^{r+2}(\mathcal{T}_h)} + |\boldsymbol{K}\nabla u|_{H^{r+1}(\mathcal{T}_h)^d}\right),$$

and, in the case $k = 0$ (which enforces $r = 0$), this right-hand side is bounded above by $h\|f\|$ by (4.75). This shows that the term $\|u_h - \mathtt{I}_h u\|_{\mathtt{U}_h} \sup_{g \in \mathtt{L}^*} \|\mathcal{E}_h^{\mathrm{d}}(z_g; \cdot)\|_{\mathtt{U}_h}$ in (A.11) is bounded above by the right-hand side in (4.76).

The proof is complete if we show a similar bound for the primal-dual consistency error $\mathcal{E}_{\boldsymbol{K},h}(u; \hat{\underline{z}}_h)$, with $\hat{\underline{z}}_h := \underline{I}_h^k z_g$ for $g \in L^2(\Omega)$ such that $\|g\| \leq 1$. We study the cases $k \geq 1$ and $k = 0$ separately.

(i) *Case* $k \geq 1$. As for the Poisson problem (proof of Lemma 2.33) and the diffusion–advection–reaction model (proof of Theorem 3.42), we re-visit the estimates done on the consistency error $\mathcal{E}_{\boldsymbol{K},h}(u; \underline{v}_h)$, by considering the special case $\underline{v}_h = \hat{\underline{z}}_h$. Here, we have to examine the terms $\mathfrak{T}_1(\hat{\underline{z}}_h)$, $\mathfrak{T}_2(\hat{\underline{z}}_h)$ and $\mathfrak{T}_3(\hat{\underline{z}}_h)$ in (4.56) with $w = u$.

Let us start with $\mathfrak{T}_1(\hat{\underline{z}}_h)$. Recalling that $\mathbf{G}_T^k\hat{\underline{z}}_T = \boldsymbol{\pi}_T^{0,k}(\nabla z_g)$ (see (4.40)) and inserting $\pm\nabla z_g$, we write

$$\begin{aligned}\mathfrak{T}_1(\hat{\underline{z}}_h) &= \sum_{T\in\mathcal{T}_h}(\boldsymbol{K}\nabla u - \boldsymbol{K}\boldsymbol{\pi}_T^{0,k}(\nabla u), \boldsymbol{\pi}_T^{0,k}(\nabla z_g) - \nabla z_g)_T \\ &\quad + \sum_{T\in\mathcal{T}_h}(\nabla u - \boldsymbol{\pi}_T^{0,k}(\nabla u), \boldsymbol{K}\nabla z_g)_T \\ &= \sum_{T\in\mathcal{T}_h}(\boldsymbol{K}[\nabla u - \boldsymbol{\pi}_T^{0,k}(\nabla u)], \boldsymbol{\pi}_T^{0,k}(\nabla z_g) - \nabla z_g)_T \\ &\quad + \sum_{T\in\mathcal{T}_h}(\nabla u - \boldsymbol{\pi}_T^{0,k}(\nabla u), \boldsymbol{K}\nabla z_g - \boldsymbol{\pi}_T^{0,0}(\boldsymbol{K}\nabla z_g))_T,\end{aligned}$$

the introduction of $\boldsymbol{\pi}_T^{0,0}(\boldsymbol{K}\nabla z_g) \in \mathbb{P}^0(T)^d$ being justified by the fact that $\nabla u - \boldsymbol{\pi}_T^{0,k}(\nabla u)$ is $L^2(T)^d$-orthogonal to $\mathbb{P}^k(T)^d \supset \mathbb{P}^0(T)^d$. We then use generalised Hölder and Cauchy–Schwarz inequalities, the approximation property (1.74) of the local L^2-projector with $m = 0$, $p = 2$ and $(l, s, v) = (k, r+1, (\nabla u)_i)$, $(l, s, v) = (k, 1, (\nabla z_g)_i)$ and $(l, s, v) = (0, 1, (\boldsymbol{K}\nabla z_g)_i)$ for $i \in \{1, \ldots, d\}$, and invoke (4.75) to obtain

$$\begin{aligned}|\mathfrak{T}_1(\hat{\underline{z}}_h)| &\lesssim \left(\sum_{T\in\mathcal{T}_h}\|\nabla u - \boldsymbol{\pi}_T^{0,k}(\nabla u)\|_T^2\right)^{\frac{1}{2}}\left(\sum_{T\in\mathcal{T}_h}\|\boldsymbol{\pi}_T^{0,k}(\nabla z_g) - \nabla z_g\|_T^2\right)^{\frac{1}{2}} \\ &\quad + \left(\sum_{T\in\mathcal{T}_h}\|\nabla u - \boldsymbol{\pi}_T^{0,k}(\nabla u)\|_T^2\right)^{\frac{1}{2}}\left(\sum_{T\in\mathcal{T}_h}\|\boldsymbol{K}\nabla z_g - \boldsymbol{\pi}_T^{0,0}(\boldsymbol{K}\nabla z_g)\|_T^2\right)^{\frac{1}{2}} \\ &\lesssim h^{r+1}|\nabla u|_{H^{r+1}(\mathcal{T}_h)^d} h|\nabla z_g|_{H^1(\Omega)^d} + h^{r+1}|\nabla u|_{H^{r+1}(\mathcal{T}_h)^d} h|\boldsymbol{K}\nabla z_g|_{H^1(\Omega)^d} \\ &\lesssim h^{r+2}|u|_{H^{r+2}(\mathcal{T}_h)}. \end{aligned} \tag{4.77}$$

To estimate $\mathfrak{T}_2(\hat{\underline{z}}_h)$, we start from (4.58) with $w = u$ and $\underline{v}_h = \hat{\underline{z}}_h$. Since $\hat{z}_F = \pi_F^{0,k} z_g$ and $\hat{z}_T = \pi_T^{0,k} z_g$, the estimates in (2.78) together with (4.75) show that

$$\left(\sum_{T\in\mathcal{T}_h}\alpha_T^{-1}\sum_{F\in\mathcal{F}_T}\frac{K_{TF}}{h_F}\|\hat{z}_F - \hat{z}_T\|_F^2\right)^{\frac{1}{2}} \lesssim h|z_g|_{H^2(\mathcal{T}_h)} \lesssim h.$$

On the other hand, by (4.73),

$$\left(\sum_{T\in\mathcal{T}_h}\alpha_T\sum_{F\in\mathcal{F}_T}h_F\|(\boldsymbol{K}^{\frac{1}{2}}\boldsymbol{\nabla}u)_{|T}-\boldsymbol{K}_{|T}^{-\frac{1}{2}}\boldsymbol{\pi}_T^{0,k}(\boldsymbol{K}\boldsymbol{\nabla}u)\|_F^2\right)^{\frac{1}{2}}\lesssim h^{r+1}|\boldsymbol{K}\boldsymbol{\nabla}u|_{H^{r+1}(\mathcal{T}_h)^d}.$$

Plugging these two estimates into (4.58) yields

$$|\mathfrak{T}_2(\hat{\underline{z}}_h)|\lesssim h^{r+2}|\boldsymbol{K}\boldsymbol{\nabla}u|_{H^{r+1}(\mathcal{T}_h)^d}. \tag{4.78}$$

For $\mathfrak{T}_3(\hat{\underline{z}}_h)$, (4.61) with $\underline{v}_h=\hat{\underline{z}}_h$ and (4.60) with $(w,r)=(u,r)$ and $(w,r)=(z_g,0)$ yield, after using (4.75),

$$|\mathfrak{T}_3(\hat{\underline{z}}_h)|\lesssim h^{r+1}|u|_{H^{r+2}(\mathcal{T}_h)}h|z_g|_{H^2(\Omega)}\lesssim h^{r+2}|u|_{H^{r+2}(\mathcal{T}_h)}. \tag{4.79}$$

Notice that this estimate on $\mathfrak{T}_3(\hat{\underline{z}}_h)$ is also valid for $k=0$.

Plugging (4.77)–(4.79) into (4.56) shows that $|\mathcal{E}_{\boldsymbol{K},h}(u;\hat{\underline{z}}_h)|$ is bounded above by the right-hand side of (4.76), which concludes the proof in this case $k\ge 1$.

(ii) *Case* $k=0$. We start from the definition of $\mathcal{E}_{\boldsymbol{K},h}(u;\hat{\underline{z}}_h)$ and work in a similar way as in the proof of Lemma 2.33. Using (4.40),

$$\mathcal{E}_{\boldsymbol{K},h}(u;\hat{\underline{z}}_h)=\sum_{T\in\mathcal{T}_h}(f,\pi_T^{0,0}z_g)_T-\sum_{T\in\mathcal{T}_h}(\boldsymbol{K}\boldsymbol{\pi}_T^{0,0}(\boldsymbol{\nabla}u),\boldsymbol{\pi}_T^{0,0}(\boldsymbol{\nabla}z_g))_T+\mathfrak{T}_3(\hat{\underline{z}}_h), \tag{4.80}$$

where $\mathfrak{T}_3(\hat{\underline{z}}_h)=-\sum_{T\in\mathcal{T}_h}s_{\boldsymbol{K},T}(\underline{I}_T^k u,\hat{\underline{z}}_T)$ is the same as in the case $k\ge 1$ and can be estimated by (4.79) with $r=0$. The first term in the right-hand side of (4.80) is manipulated as follows:

$$\begin{aligned}\sum_{T\in\mathcal{T}_h}(f,\pi_T^{0,0}z_g)_T&=\sum_{T\in\mathcal{T}_h}(\pi_T^{0,0}f,z_g)_T\\&=\sum_{T\in\mathcal{T}_h}(\pi_T^{0,0}f-f,z_g)_T+(f,z_g)\\&=\sum_{T\in\mathcal{T}_h}(\pi_T^{0,0}f-f,z_g-\pi_T^{0,0}z_g)_T+(\boldsymbol{K}\boldsymbol{\nabla}u,\boldsymbol{\nabla}z_g),\end{aligned}$$

where the first line follows from the orthogonality property of $\pi_T^{0,0}$, the second line is obtained by inserting $\pm f$, and we have used, in the third line, (4.35) together with the orthogonality property of $\pi_T^{0,0}$. Plugging this into (4.80), using Cauchy–Schwarz inequalities, the bound (4.79) for $\mathfrak{T}_3(\hat{\underline{z}}_h)$ and the approximation properties of $\pi_T^{0,0}$, we deduce, recalling the estimate (4.75) (for both z_g and u) and the choice $\|g\| \le 1$,

$$|\mathcal{E}_{\boldsymbol{K},h}(u; \hat{\underline{z}}_h)| \lesssim h|f|_{H^1(\mathcal{T}_h)} h|z_g|_{H^1(\mathcal{T}_h)} + |\mathfrak{T}_4| + h^2|u|_{H^2(\mathcal{T}_h)}$$
$$\lesssim h^2\|f\|_{H^1(\mathcal{T}_h)} + |\mathfrak{T}_4|, \tag{4.81}$$

where

$$\mathfrak{T}_4 := \sum_{T\in\mathcal{T}_h} (\boldsymbol{K}\nabla u, \nabla z_g)_T - (\boldsymbol{K}\boldsymbol{\pi}_T^{0,0}(\nabla u), \boldsymbol{\pi}_T^{0,0}(\nabla z_g))_T.$$

This term is then rearranged as

$$\mathfrak{T}_4 = \sum_{T\in\mathcal{T}_h} (\boldsymbol{K}(\nabla u - \boldsymbol{\pi}_T^{0,0}(\nabla u)), \nabla z_g)_T + (\boldsymbol{K}\boldsymbol{\pi}_T^{0,0}(\nabla u), \nabla z_g - \boldsymbol{\pi}_T^{0,0}(\nabla z_g))_T$$
$$= \sum_{T\in\mathcal{T}_h} (\nabla u - \boldsymbol{\pi}_T^{0,0}(\nabla u), \boldsymbol{K}\nabla z_g - \boldsymbol{\pi}_T^{0,0}(\boldsymbol{K}\nabla z_g))_T$$
$$+ \sum_{T\in\mathcal{T}_h} (\boldsymbol{K}\boldsymbol{\pi}_T^{0,0}(\nabla u) - \boldsymbol{\pi}_T^{0,0}(\boldsymbol{K}\boldsymbol{\pi}_T^{0,0}(\nabla u)), \nabla z_g - \boldsymbol{\pi}_T^{0,0}(\nabla z_g))_T,$$

where we have inserted $\pm(\boldsymbol{K}\boldsymbol{\pi}_T^{0,0}(\nabla u), \nabla z_g)_T$ in the addends in the first line, and used the symmetry of $\boldsymbol{K}$ together with the $L^2(T)^d$-orthogonality properties of $\boldsymbol{\pi}_T^{0,0}$ to conclude. Since $\boldsymbol{\pi}_T^{0,0}(\nabla u)$ is constant, we have $\boldsymbol{\pi}_T^{0,0}(\boldsymbol{K}\boldsymbol{\pi}_T^{0,0}(\nabla u)) = (\boldsymbol{\pi}_T^{0,0}\boldsymbol{K})(\boldsymbol{\pi}_T^{0,0}(\nabla u))$ (where $\boldsymbol{\pi}_T^{0,0}\boldsymbol{K}$ denotes the component-wise $L^2(T)$-projection of $\boldsymbol{K}$), and Cauchy–Schwarz and generalised Hölder inequalities lead to

$$|\mathfrak{T}_4| \lesssim h|\nabla u|_{H^1(\mathcal{T}_h)^d} h|\boldsymbol{K}\nabla z_g|_{H^1(\mathcal{T}_h)^d} + h|\boldsymbol{K}|_{W^{1,\infty}(\mathcal{T}_h)^{d\times d}}\|\nabla u\| h|\nabla z_g|_{H^1(\mathcal{T}_h)^d}.$$

Plugging this estimate into (4.81) and recalling (4.75) shows that $|\mathcal{E}_{\boldsymbol{K},h}(u; \hat{\underline{z}}_h)|$ is bounded above by the right-hand side of (4.76), which concludes the proof. □

4.2.6 Numerical Tests

This section presents numerical tests obtained with the scheme (4.63), on the domain $\Omega = (0,1)^2$ and for the exact solution and diffusion tensors given by:

$$u(x_1, x_2) = \sin(\pi x_1)\sin(\pi x_2), \quad \boldsymbol{K}(x_1, x_2) = \begin{bmatrix} \epsilon\bar{x}_1^2 + \bar{x}_2^2 & (\epsilon-1)\bar{x}_1\bar{x}_2 \\ (\epsilon-1)\bar{x}_1\bar{x}_2 & \bar{x}_1^2 + \epsilon\bar{x}_2^2 \end{bmatrix},$$

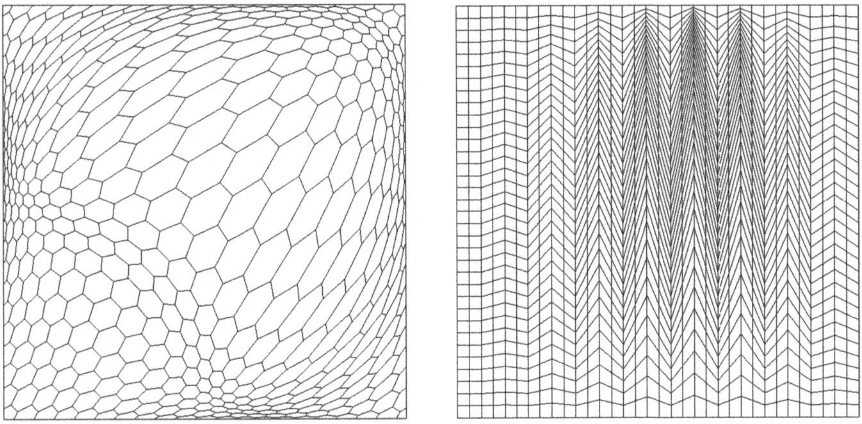

Fig. 4.6 Examples of meshes for the tests of Sect. 4.2.6: hexagonal mesh (left); Kershaw mesh (right)

where $\epsilon = 10^{-5}$ and $(\bar{x}_1, \bar{x}_2) := (x_1, x_2) + (0.1, 0.1)$. The tensor $\boldsymbol{K}(x_1, x_2)$ has eigendirections $(\bar{x}_1, \bar{x}_2)$ and $(\bar{x}_1, \bar{x}_2)^\perp$, and an anisotropy ratio equal to $\epsilon^{-1} = 10^5$. This tensor, which is taken from [172], is a modification of the one proposed in [227]. Two families of meshes are considered for the numerical tests: a family of (mostly) hexagonal meshes, and a family of highly distorted Kershaw meshes from [207]. A representative of each of these families is shown in Fig. 4.6.

The numerical results for $k \in \{0, \ldots, 3\}$ are presented in Fig. 4.7. They are in perfect agreement with the theoretical rates predicted by Theorems 4.16 and 4.21: the convergence in energy norm is in $\mathcal{O}(h^{k+1})$, while we observe an L^2-norm convergence in $\mathcal{O}(h^{k+2})$. Comparing Fig. 4.7a and c with the results for the HHO method (2.48) on the Laplace equation (see Fig. 2.3b and d), we notice that the HHO scheme (4.63) for locally variable diffusion tensor does not seem here to be very sensitive to the high anisotropy ratio of $\boldsymbol{K}$: the absolute errors in each norm are of similar magnitude for these two schemes.

The results on the Kershaw meshes (Fig. 4.7b and d) show some impact of the mesh distortion: an order of magnitude up to 3 is lost when compared to the results on hexagonal meshes. Despite this loss, the rates of convergence are preserved, showing a certain robustness of the HHO method for locally variable diffusion tensors, even with severe anisotropy and mesh distortion. We mention that the tests on the Kershaw meshes have been run using orthonormalised local basis functions; on this test case, the use of more naive monomial basis functions leads, starting from $k = 3$, to round-off errors that deteriorate the convergence (see Sect. B.1.1 in Appendix B for more on this topic).

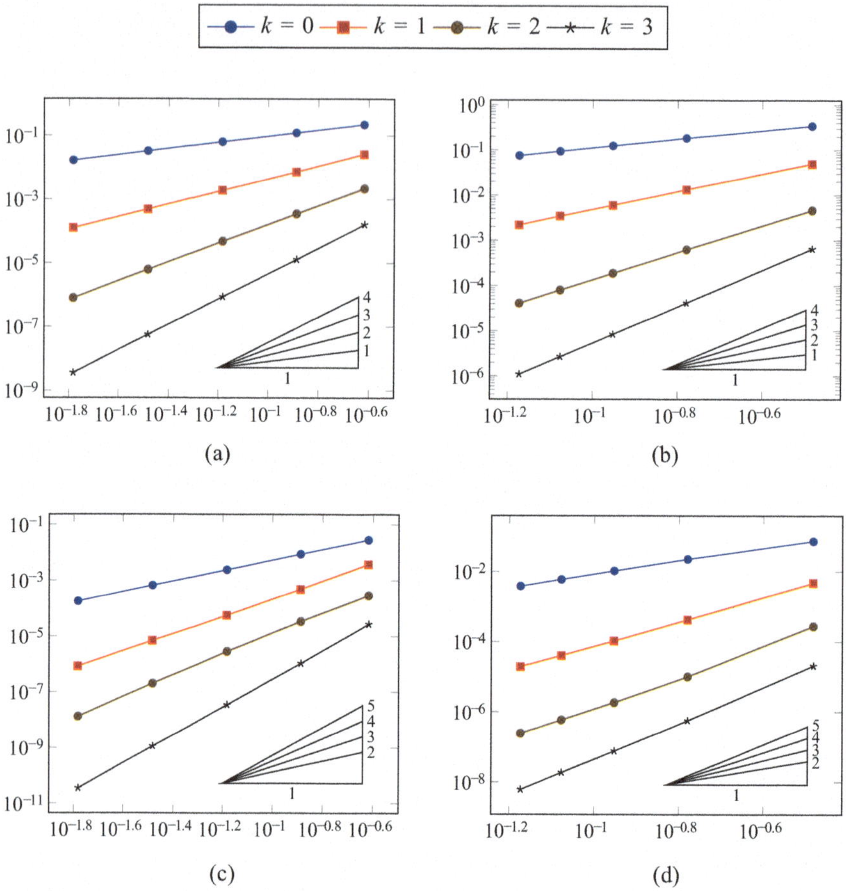

Fig. 4.7 Error vs. h for the test cases of Sect. 4.2.6. The reference slopes refer to the expected order of convergence for each polynomial degree $k \in \{0, \ldots, 3\}$. (**a**) $\|\underline{I}_h^k u - \underline{u}_h\|_{a,h}$ vs. h, hexagonal meshes. (**b**) $\|\underline{I}_h^k u - \underline{u}_h\|_{a,h}$ vs. h, Kershaw meshes. (**c**) $\|\pi_h^{0,k} u - u_h\|$ vs. h, hexagonal meshes. (**d**) $\|\pi_h^{0,k} u - u_h\|$ vs. h, Kershaw meshes

Chapter 5
Variations and Comparison with Other Methods

In this chapter we explore variations of the Hybrid High-Order method and establish links with other polytopal methods. Specifically, in Sect. 5.1 we consider the possibility of enriching or depleting element unknowns. Section 5.2 establishes a link with the nonconforming $\mathbb{P}^1$ Finite Element method on matching simplicial meshes, which can be regarded as a variation of the lowest-order depleted HHO method with a modified discretisation of the right-hand side. We next show, in Sect. 5.3, that the lowest-order version of the standard HHO method on generic polytopal meshes is intimately linked to Hybrid Mimetic Mixed methods. In Sect. 5.4 we discuss the Mixed High-Order method, which is developed using as a starting point the mixed version of the Poisson problem, and show that the HHO method corresponds to its hybridised version. Section 5.5 establishes a link between the HHO and the Nonconforming Virtual Element method. For the sake of completeness, we also discuss the Conforming Virtual Element method and prove key results for its analysis using HHO-inspired norms, which extend to the non-Hilbertian setting. Finally, we develop in Sect. 5.6 a Gradient Discretisation Method inspired by HHO. We focus on the Poisson problem (2.1), except in Sect. 5.6 where we consider the locally variable diffusion problem (4.33).

5.1 Enrichment and Depletion of Element Unknowns

The core of the HHO methods are formulas (2.5), which express the local elliptic projection of a function in terms of its L^2-orthogonal polynomial projections on an element and on its faces. These formulas drove, in particular, the choice of the local space of unknowns (2.6), the vectors of which are made of a polynomial of degree k in the element and polynomials of degree k on each face. However, following Remark 2.1, the polynomial degree of the element unknowns could be reduced from k to $k-1$ (at least for $k \geq 1$). In this section, we explore variants of the HHO method

D. A. Di Pietro, J. Droniou, *The Hybrid High-Order Method for Polytopal Meshes*, MS&A 19, https://doi.org/10.1007/978-3-030-37203-3_5

in which the polynomial degree of element unknowns possibly differs from that of the face unknowns.

Throughout the section, we denote by $k \geq 0$ and $\ell \in \{k-1, k, k+1\}$ two integers corresponding to the polynomial degrees of element- and face-based unknowns, respectively. The case $(k, \ell) = (0, -1)$ is allowed, and we recall that $\mathbb{P}^{-1}(T) = \{0\}$. The limit $\ell \geq k - 1$ comes from the desire to preserve optimal approximation properties (see discussion above) whereas, as we will see, the assumption $\ell \leq k+1$ is required to prove the stability of the method (see the proof of Proposition 5.10).

5.1.1 Local Space and Interpolator

For $\ell \geq 0$, we define the local space of unknowns as

$$\underline{U}_T^{k,\ell} := \left\{ \underline{v}_T = (v_T, (v_F)_{F\in\mathcal{F}_T}) \;:\; v_T \in \mathbb{P}^\ell(T) \text{ and } v_F \in \mathbb{P}^k(F) \quad \forall F \in \mathcal{F}_T \right\}. \tag{5.1a}$$

In the case $\ell = -1$, which can only occur if $k = 0$, this would lead to a space in which all element unknowns are 0, since $\mathbb{P}^{-1}(T) = \{0\}$. We therefore modify the definition for $(k, \ell) = (0, -1)$ and set

$$\begin{aligned} \underline{U}_T^{0,-1} := \Bigg\{ \underline{v}_T = (v_T, (v_F)_{F\in\mathcal{F}_T}) \;:\; & v_F \in \mathbb{P}^0(F) \quad \forall F \in \mathcal{F}_T\,, \\ & v_T = \frac{1}{|T|_d} \sum_{F\in\mathcal{F}_T} \omega_{TF} |F|_{d-1} v_F \Bigg\}, \end{aligned} \tag{5.1b}$$

where we have fixed weights $(\omega_{TF})_{F\in\mathcal{F}_T}$ such that

$$\omega_{TF} \geq 0 \qquad \forall F \in \mathcal{F}_T, \tag{5.2a}$$

$$\sum_{F\in\mathcal{F}_T} \omega_{TF} (q, 1)_F = (q, 1)_T \qquad \forall q \in \mathbb{P}^0(T). \tag{5.2b}$$

Remark 5.1 (Assumption on the Weights) The condition (5.2b) is equivalent to

$$\sum_{F\in\mathcal{F}_T} \omega_{TF} |F|_{d-1} = |T|_d. \tag{5.3}$$

By the geometric bounds (1.6)–(1.8), it holds $|T|_d \lesssim h_T |F|_{d-1}$ for all $F \in \mathcal{F}_T$, with hidden constant depending only on d and the mesh regularity parameter ϱ. Relation (5.3) together with the positivity of the weights therefore implies

$$|\omega_{TF}| \lesssim h_T \qquad \forall F \in \mathcal{F}_T. \tag{5.4}$$

The positivity condition (5.2a) is actually formally useful, in the following analysis, only because it implies (5.4), and the latter estimate could therefore be used in lieu of (5.2a). However, positive weights are often preferable for better stability of the scheme.

To obtain superconvergence in the L^2-norm, as highlighted in [234], the condition (5.2b) should be strengthen into

$$\sum_{F\in\mathcal{F}_T} \omega_{TF}(q,1)_F = (q,1)_T \qquad \forall q \in \mathbb{P}^1(T). \tag{5.5}$$

As shown in [234, Appendix A], if the cell T is star-shaped with respect to its centre of mass $\overline{\boldsymbol{x}}_T$, a set of weights that satisfy (5.5) is given by $\omega_{TF} = \mathrm{dist}(\overline{\boldsymbol{x}}_T, H_F)/d$, where H_F is the hyperplane spanned by F.

Remark 5.2 (Barycentric Elimination) In the wording of [178], the replacement in $\underline{U}_T^{0,-1}$ of the free unknown v_T by a convex combination of the other unknowns $(v_F)_{F\in\mathcal{F}_T}$ is called a *barycentric elimination.*

The space $\underline{U}_T^{k,\ell}$ is endowed with the norm $\|{\cdot}\|_{1,T}$ still formally defined by (2.7). The local interpolator associated with $\underline{U}_T^{k,\ell}$ is $\underline{I}_T^{k,\ell} : W^{1,1}(T) \to \underline{U}_T^{k,\ell}$ such that, for all $v \in W^{1,1}(T)$, if $\ell \geq 0$,

$$\underline{I}_T^{k,\ell} v := (\pi_T^{0,\ell} v, (\pi_F^{0,k} v)_{F\in\mathcal{F}_T}) \tag{5.6a}$$

and, if $\ell = -1$,

$$\underline{I}_T^{0,-1} v := (v_T, (\pi_F^{0,0} v)_{F\in\mathcal{F}_T}) \text{ with } v_T = \frac{1}{|T|_d} \sum_{F\in\mathcal{F}_T} \omega_{TF} |F|_{d-1} \pi_F^{0,0} v. \tag{5.6b}$$

As in the case $k = \ell$ covered in Chap. 2, the boundedness of the local interpolator will be instrumental to the analysis of the HHO scheme for $k \neq \ell$.

Proposition 5.3 (Boundedness of the Local Interpolator $\underline{I}_T^{k,\ell}$) *For all $v \in H^1(T)$,*

$$\|\underline{I}_T^{k,\ell} v\|_{1,T} \lesssim |v|_{H^1(T)}, \tag{5.7}$$

where the hidden constant depends only on d, ϱ, k, and ℓ.

Proof Let $\underline{I}_T^{k,\ell} v = (v_T, (v_F)_{F\in\mathcal{F}_T})$. Recalling the definition (2.7) of $\|{\cdot}\|_{1,T}$, we have

$$\|\underline{I}_T^{k,\ell} v\|_{1,T}^2 \lesssim \|\nabla v\|_T^2 + \sum_{F\in\mathcal{F}_T} h_F^{-1} \|v_F - v_T\|_F^2, \tag{5.8}$$

where, if $\ell \geq 0$, we have used the boundedness property (1.77) of $\pi_T^{0,\ell}$ with $s = 1$ and $p = 2$ to remove this projector from the volumetric term $\|\nabla v_T\|_T =$

$\|\nabla\pi_T^{0,\ell}v\|_T$ while, if $\ell = -1$, we have written $\|\nabla v_T\|_T = 0 \le \|\nabla v\|_T$. We now deal with the boundary term in (5.8). Notice first that, by the idempotency of $\pi_F^{0,k}$, its $L^2(F)$-boundedness expressed by (1.77) with $X = F$, $s = 0$, and $p = 2$, and the trace approximation property (1.75) of $\pi_T^{0,0}$ (with $m = 0$, $s = 1$, and $p = 2$),

$$\|\pi_F^{0,k}v - \pi_T^{0,0}v\|_F = \|\pi_F^{0,k}(v - \pi_T^{0,0}v)\|_F \lesssim \|v - \pi_T^{0,0}v\|_F \lesssim h_T^{\frac{1}{2}}\|\nabla v\|_T. \tag{5.9}$$

Hence, introducing $\pm\pi_T^{0,0}v$ and using the triangle inequality, we have

$$\begin{aligned}\|v_F - v_T\|_F &\le \|\pi_F^{0,k}v - \pi_T^{0,0}v\|_F + \|\pi_T^{0,0}v - v_T\|_F \\ &\lesssim h_T^{\frac{1}{2}}\|\nabla v\|_T + h_T^{-\frac{1}{2}}\|\pi_T^{0,0}v - v_T\|_T,\end{aligned} \tag{5.10}$$

where we have used, to pass to the second line, the estimate (5.9) and the discrete trace inequality (1.55) with $p = 2$ and $\pi_T^{0,0}v - v_T$ instead of v.

We now consider the second term in the right-hand side of (5.10). Given the different definitions of v_T if $\ell \ge 0$ or if $\ell = -1$, we have to treat each of these cases separately. If $\ell \ge 0$, then $v_T = \pi_T^{0,\ell}v$ and the idempotency, boundedness (1.77) (with $X = T$, $l = \ell$, $s = 0$, and $p = 2$) and approximation property (1.74) (with $l = 0$, $m = 0$, $s = 1$ and $p = 2$) of the orthogonal projector yield

$$\|\pi_T^{0,0}v - v_T\|_T = \|\pi_T^{0,\ell}(\pi_T^{0,0}v - v)\|_T \lesssim \|\pi_T^{0,0}v - v\|_T \lesssim h_T\|\nabla v\|_T. \tag{5.11}$$

Consider now $\ell = -1$. Recalling the definition of v_T in (5.6b) and the property (5.3) of the weights, we write

$$\pi_T^{0,0}v - v_T = \frac{1}{|T|_d}\sum_{F\in\mathcal{F}_T}\omega_{TF}|F|_{d-1}(\pi_T^{0,0}v - \pi_F^{0,0}v). \tag{5.12}$$

We have, by Cauchy–Schwarz inequality and (5.9),

$$\begin{aligned}\big||F|_{d-1}(\pi_T^{0,0}v - \pi_F^{0,0}v)\big| &= |(\pi_T^{0,0}v - \pi_F^{0,0}v, 1)_F| \\ &\le |F|_{d-1}^{\frac{1}{2}}\|\pi_T^{0,0}v - \pi_F^{0,0}v\|_F \\ &\lesssim |F|_{d-1}^{\frac{1}{2}}h_T^{\frac{1}{2}}\|\nabla v\|_T \lesssim |T|_d^{\frac{1}{2}}\|\nabla v\|_T,\end{aligned}$$

the conclusion following from the geometric bounds (1.6)–(1.8). Starting from (5.12) and using the triangle inequality and the above estimate to bound the right-hand side, we infer

$$\|\pi_T^{0,0}v - v_T\|_T = |T|_d^{\frac{1}{2}}|\pi_T^{0,0}v - v_T| \lesssim \sum_{F\in\mathcal{F}_T} \omega_{TF}\|\nabla v\|_T \lesssim h_T\|\nabla v\|_T, \quad (5.13)$$

where we have used (5.4) and $\mathrm{card}(\mathcal{F}_T) \lesssim 1$ (see (1.5)) to conclude.

Combining (5.10) and (5.11) (if $\ell \geq 0$) or (5.13) (if $\ell = -1$), we find that $\|v_F - v_T\|_F \lesssim h_T^{\frac{1}{2}}\|\nabla v\|_T$. Raise to the square, multiply by h_F^{-1}, use $h_T \lesssim h_F$ (see (1.6)), sum over $F \in \mathcal{F}_T$, and use $\mathrm{card}(\mathcal{F}_T) \lesssim 1$ again to deduce

$$\sum_{F\in\mathcal{F}_T} h_F^{-1}\|v_F - v_T\|_F^2 \lesssim \|\nabla v\|_T^2.$$

Plugged into (5.8), this concludes the proof of (5.7). □

5.1.2 *Modified Elliptic Projector*

Because of the particular choices in (5.1b) and (5.6b) for the element unknown when $\ell = -1$, the relevant projector for the analysis in the case $(k, \ell) = (0, -1)$ is not the elliptic projector given by Definition 1.39 but a modified version thereof, in which the closure equation (1.60b) is replaced by an equation that fixes a weighted integral on the boundary of the element.

Definition 5.4 (The Modified Elliptic Projector) Let $\mathcal{M}_h$ be a polytopal mesh, $T \in \mathcal{T}_h$, and $l \geq 0$ be a polynomial degree. Take weights $(\omega_{TF})_{F\in\mathcal{F}_T}$ satisfying (5.2). The modified elliptic projector $\tilde{\pi}_T^{1,l} : W^{1,1}(T) \to \mathbb{P}^l(T)$ is defined as follows: For all $v \in W^{1,1}(T)$, the polynomial $\tilde{\pi}_T^{1,l}v \in \mathbb{P}^l(T)$ satisfies

$$(\nabla(\tilde{\pi}_T^{1,l}v - v), \nabla w)_T = 0 \qquad \forall w \in \mathbb{P}^l(T) \quad (5.14\text{a})$$

and

$$\sum_{F\in\mathcal{F}_T} \omega_{TF}(\tilde{\pi}_T^{1,l}v - v, 1)_F = 0. \quad (5.14\text{b})$$

We note that $\tilde{\pi}_T^{1,l}$ is indeed a projector onto $\mathbb{P}^l(T)$: if v belongs to this space, then it obviously satisfies the constitutive equations (5.14a)–(5.14b) that define $\tilde{\pi}_T^{1,l}v$, and thus $\tilde{\pi}_T^{1,l}v = v$.

Remark 5.5 (Choice of the Weights and Closure Equation) If all the weights are identical, that is, according to (5.3), $\omega_{TF} = \frac{|T|_d}{|\partial T|_{d-1}}$, then the closure equation (5.14b) becomes

$$(\tilde{\pi}_T^{1,l} v - v, 1)_{\partial T} = 0.$$

In the case $l = 1$, if the weights satisfy the improved property (5.5), then using this property with $q = \tilde{\pi}_T^{1,1} v$ shows that the closure equation (5.14b) consists in fixing the average of $\tilde{\pi}_T^{1,1} v$ in the cell to

$$(\tilde{\pi}_T^{1,1} v, 1)_T = \sum_{F \in \mathcal{F}_T} \omega_{TF}(v, 1)_F. \tag{5.15}$$

Remark 5.6 (Case $l = 1$, Computing the Modified Elliptic Projector from L^2-Projections on the Faces) Let us consider the case $l = 1$. Taking $w \in \mathbb{P}^1(T)$ and integrating by parts, we have from (5.14a) and since $\Delta w = 0$,

$$(\nabla \tilde{\pi}_T^{1,1} v, \nabla w)_T = (\nabla v, \nabla w)_T = \sum_{F \in \mathcal{F}_T} (v, \nabla w \cdot \boldsymbol{n}_{TF})_F = \sum_{F \in \mathcal{F}_T} (\pi_F^{0,0} v, \nabla w \cdot \boldsymbol{n}_{TF})_F, \tag{5.16a}$$

where the introduction of the L^2-projector on $\mathbb{P}^0(F)$ is justified since $\nabla w \cdot \boldsymbol{n}_{TF}$ is constant on F. Likewise, (5.14b) can be recast as

$$\sum_{F \in \mathcal{F}_T} \omega_{TF}(\tilde{\pi}_T^{1,1} v, 1)_F = \sum_{F \in \mathcal{F}_T} \omega_{TF}(\pi_F^{0,0} v, 1)_F. \tag{5.16b}$$

Equations (5.16) show that $\tilde{\pi}_T^{1,1} v$ is computable from $(\pi_F^{0,0} v)_{F \in \mathcal{F}_T}$.

The approximation properties of this modified elliptic projector are similar to those of the standard elliptic projector (Theorem 1.48), as proved in the following theorem.

Theorem 5.7 (Approximation Properties of the Modified Elliptic Projector) *Let $(\mathcal{M}_h)_{h \in \mathcal{H}} = (\mathcal{T}_h, \mathcal{F}_h)_{h \in \mathcal{H}}$ be a regular mesh sequence in the sense of Definition 1.9. For a given polynomial degree $l \geq 0$, let an integer $s \in \{1, \ldots, l+1\}$ and a real number $p \in [1, \infty]$ be given. Then, for all $T \in \mathcal{T}_h$, all $v \in W^{s,p}(T)$, and all $m \in \{0, \ldots, s\}$,*

$$|v - \tilde{\pi}_T^{1,l} v|_{W^{m,p}(T)} \lesssim h_T^{s-m} |v|_{W^{s,p}(T)}. \tag{5.17}$$

Moreover, if $m \leq s - 1$, then, for all $F \in \mathcal{F}_T$,

(continued)

Theorem 5.7 (continued)

$$h_T^{\frac{1}{p}} |v - \widetilde{\pi}_T^{1,l} v|_{W^{m,p}(F)} \lesssim h_T^{s-m} |v|_{W^{s,p}(T)}. \tag{5.18}$$

The hidden constants above depend only on d, ϱ, l, s, p, and m.

Proof The proof relies on the approximation results of Lemma 1.43. As in the proof of Theorem 1.48, we have to consider two cases: $m \geq 1$ and $m = 0$. In the former case, the proof is identical to the proof made in Theorem 1.48 for $\pi_T^{1,l}$, since $\widetilde{\pi}_T^{1,l}$ also satisfies (1.60a) (see (5.14a)), which was shown to imply (1.80), that is,

$$\|\nabla \widetilde{\pi}_T^{1,l} v\|_{L^p(T)^d} \lesssim \|\nabla v\|_{L^p(T)^d}. \tag{5.19}$$

Let us now consider $m = 0$. According to Lemma 1.43, the approximation property (5.17) holds if we prove that, for all $v \in W^{1,1}(T)$,

$$\|\widetilde{\pi}_T^{1,l} v\|_{L^p(T)} \lesssim \|v\|_{L^p(T)} + h_T |v|_{W^{1,p}(T)}. \tag{5.20}$$

To this purpose, we first notice that, by the approximation properties (1.74) of the L^2-projector with $X = T$, $l = 0$, $m = 0$ and $s = 1$,

$$\|\widetilde{\pi}_T^{1,l} v - \pi_T^{0,0}(\widetilde{\pi}_T^{1,l} v)\|_{L^p(T)} \lesssim h_T \|\nabla \widetilde{\pi}_T^{1,l} v\|_{L^p(T)^d} \lesssim h_T \|\nabla v\|_{L^p(T)^d}, \tag{5.21}$$

where the conclusion follows invoking (5.19).

We now estimate $\pi_T^{0,0}(\widetilde{\pi}_T^{1,l} v)$, starting with

$$\begin{aligned}
&\left| \sum_{F \in \mathcal{F}_T} \omega_{TF} (v, 1)_F - \sum_{F \in \mathcal{F}_T} \omega_{TF} \big(\pi_T^{0,0}(\widetilde{\pi}_T^{1,l} v), 1\big)_F \right| \\
&\qquad = \left| \sum_{F \in \mathcal{F}_T} \omega_{TF} \big(\widetilde{\pi}_T^{1,l} v - \pi_T^{0,0}(\widetilde{\pi}_T^{1,l} v), 1\big)_F \right| \\
&\qquad \leq \sum_{F \in \mathcal{F}_T} \omega_{TF} |F|_{d-1}^{\frac{1}{p'}} \|\widetilde{\pi}_T^{1,l} v - \pi_T^{0,0}(\widetilde{\pi}_T^{1,l} v)\|_{L^p(F)} \\
&\qquad \lesssim \sum_{F \in \mathcal{F}_T} \omega_{TF} |F|_{d-1}^{\frac{1}{p'}} h_T^{\frac{1}{p'}} \|\nabla \widetilde{\pi}_T^{1,l} v\|_{L^p(T)^d} \\
&\qquad \lesssim h_T |T|_d^{\frac{1}{p'}} \|\nabla \widetilde{\pi}_T^{1,l} v\|_{L^p(T)^d},
\end{aligned} \tag{5.22}$$

where we have used (5.14b) to write the equality, a Hölder inequality in the second line, the trace approximation properties (1.75) of $\pi_T^{0,0}$ with $s = 1$ and $m = 0$ together with the property $1 - \frac{1}{p} = \frac{1}{p'}$ in the third line, and the estimates (5.4), $\mathrm{card}(\mathcal{F}_T) \lesssim 1$ and $|F|_{d-1}h_T \lesssim |T|_d$ (see Lemma 1.12) in the conclusion. We deduce that

$$
\begin{aligned}
|T|_d \left|\pi_T^{0,0}(\widetilde{\pi}_T^{1,l} v)\right| &= \left|\sum_{F\in\mathcal{F}_T} \omega_{TF}|F|_{d-1}\pi_T^{0,0}(\widetilde{\pi}_T^{1,l} v)\right| \\
&= \left|\sum_{F\in\mathcal{F}_T} \omega_{TF}\left(\pi_T^{0,0}(\widetilde{\pi}_T^{1,l} v), 1\right)_F\right| \\
&\lesssim h_T|T|_d^{\frac{1}{p'}} \|\nabla\widetilde{\pi}_T^{1,l} v\|_{L^p(T)^d} + \left|\sum_{F\in\mathcal{F}_T} \omega_{TF}(v,1)_F\right| \qquad (5.23)\\
&\lesssim |T|_d^{\frac{1}{p'}} \left(\|v\|_{L^p(T)} + h_T\|\nabla v\|_{L^p(T)^d}\right), \qquad (5.24)
\end{aligned}
$$

where we have used the property (5.3) of the weights in the first line, the fact that $\pi_T^{0,0}(\widetilde{\pi}_T^{1,l} v)$ is constant over F in the second line, and a triangle inequality (introducing $\pm\sum_{F\in\mathcal{F}_T}\omega_{TF}(v,1)_F$) together with (5.22) to pass to the third line. The fourth line is obtained invoking (5.19) to remove $\widetilde{\pi}_T^{1,l}$ together with a Hölder inequality, the continuous trace inequality (1.51), the property (5.4) of the weights, the relation $1 - \frac{1}{p} = \frac{1}{p'}$, and the estimate $h_T|F|_{d-1} \lesssim |T|_d$ for all $F \in \mathcal{F}_T$ (see (1.6)–(1.8)) to write

$$
\begin{aligned}
\left|\sum_{F\in\mathcal{F}_T} \omega_{TF}(v,1)_F\right| &\le \sum_{F\in\mathcal{F}_T} \omega_{TF}|F|_{d-1}^{\frac{1}{p'}}\|v\|_{L^p(F)} \\
&\lesssim \left(\sum_{F\in\mathcal{F}_T} h_T|F|_{d-1}^{\frac{1}{p'}} h_T^{-\frac{1}{p}}\right)\left(\|v\|_{L^p(T)} + h_T\|\nabla v\|_{L^p(T)^d}\right) \\
&\lesssim |T|_d^{\frac{1}{p'}}\left(\|v\|_{L^p(T)} + h_T\|\nabla v\|_{L^p(T)^d}\right).
\end{aligned}
$$

We infer from (5.24) and $\frac{1}{p'} + \frac{1}{p} = 1$ that it holds

$$
\|\pi_T^{0,0}(\widetilde{\pi}_T^{1,l} v)\|_{L^p(T)} = |T|_d^{\frac{1}{p}}\left|\pi_T^{0,0}(\widetilde{\pi}_T^{1,l} v)\right| \lesssim \|v\|_{L^p(T)} + h_T\|\nabla v\|_{L^p(T)^d}. \qquad (5.25)
$$

Combining (5.21), (5.25), and a triangle inequality shows that (5.20) holds, which concludes the proof of (5.17).

The estimate (5.18) follows from (5.17) and the continuous trace inequality (1.51), as in the proof of Theorem 1.45. □

5.1.3 Potential Reconstruction

The potential reconstruction is formally defined as in Sect. 2.1.3 if $\ell \geq 0$, and with a modified closure equation if $(k, \ell) = (0, -1)$. Specifically, we define $\tilde{\mathrm{p}}_T^{k+1} : \underline{U}_T^{k,\ell} \to \mathbb{P}^{k+1}(T)$ such that, for all $\underline{v}_T \in \underline{U}_T^{k,\ell}$ and $w \in \mathbb{P}^{k+1}(T)$,

$$(\nabla \tilde{\mathrm{p}}_T^{k+1} \underline{v}_T, \nabla w)_T = -(v_T, \Delta w)_T + \sum_{F \in \mathcal{F}_T} (v_F, \nabla w \cdot \boldsymbol{n}_{TF})_F \tag{5.26a}$$

$$= (\nabla v_T, \nabla w)_T + \sum_{F \in \mathcal{F}_T} (v_F - v_T, \nabla w \cdot \boldsymbol{n}_{TF})_F \tag{5.26b}$$

(the second equality coming from an integration by parts) and

$$\text{if } \ell \geq 0, \qquad (\tilde{\mathrm{p}}_T^{k+1} \underline{v}_T - v_T, 1)_T = 0, \tag{5.26c}$$

$$\text{if } (k, \ell) = (0, -1), \qquad \sum_{F \in \mathcal{F}_T} \omega_{TF} (\tilde{\mathrm{p}}_T^1 \underline{v}_T - v_F, 1)_F = 0. \tag{5.26d}$$

Remark 5.8 (Closure Equation if $\ell = -1$ and the Weights Satisfy (5.5)*)* Consider the case $(k, \ell) = (0, -1)$ and assume that the weights satisfy the property (5.5). Then, taking $\underline{v}_T \in \underline{U}_T^{0,-1}$, applying (5.5) to $q = \tilde{\mathrm{p}}_T^1 \underline{v}_T$ and recalling the choice of v_T in (5.1b), we see that (5.26d) is equivalent to

$$(\tilde{\mathrm{p}}_T^1 \underline{v}_T - v_T, 1)_T = 0.$$

In other words, we recover the closure equation (5.26c) based on averages in the cell.

If $\ell \geq 0$, by Remark 2.1, the relations (2.5) hold with $\pi_T^{0,k}$ replaced with $\pi_T^{0,\ell}$. Hence, recalling the definition (5.6a) of $\underline{I}_T^{k,\ell}$ and comparing (2.5) and (5.26a)–(5.26d), we see that

$$\text{if } \ell \geq 0, \quad \tilde{\mathrm{p}}_T^{k+1} \underline{I}_T^{k,\ell} v = \pi_T^{1,k+1} v \qquad \forall v \in W^{1,1}(T). \tag{5.27a}$$

On the other hand, if $(k, \ell) = (0, -1)$, taking $v \in W^{1,1}(T)$, applying (5.26a) (in which $\Delta w = 0$) and (5.26d) to $\underline{v}_T = \underline{I}_T^{0,-1} v$ (with $\underline{I}_T^{0,-1}$ defined by (5.6b)), and comparing with (5.16), we find that

$$\text{if } (k, \ell) = (0, -1), \quad \tilde{\mathrm{p}}_T^1 \underline{I}_T^{0,-1} v = \tilde{\pi}_T^{1,1} v \qquad \forall v \in W^{1,1}(T). \tag{5.27b}$$

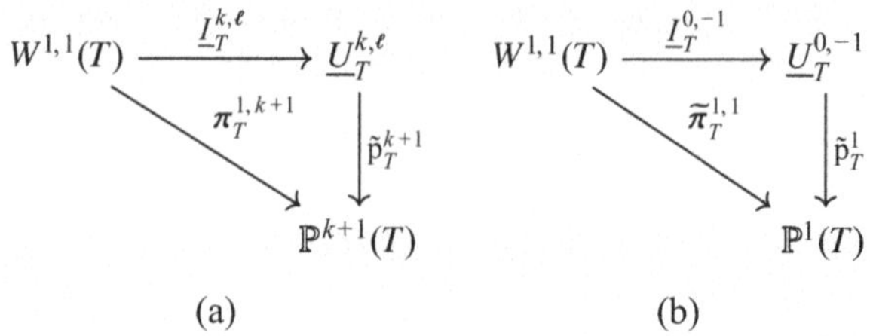

Fig. 5.1 Illustration of the commutation properties (5.27) of $\tilde{\mathrm{p}}_T^{k+1}$. (**a**) $\ell \geq 0$. (**b**) $(k, \ell) = (0, -1)$

These commutation properties are illustrated in Fig. 5.1.

The commutation properties (5.27) together with the approximation properties of the standard or modified elliptic projector (expressed by Theorems 1.48 and 5.7, respectively) ensure that $(\tilde{\mathrm{p}}_T^{k+1} \circ \underline{I}_T^{k,\ell})$ has optimal approximation properties.

5.1.4 *Local Bilinear Form*

The local bilinear form $\mathrm{a}_T : \underline{U}_T^{k,\ell} \times \underline{U}_T^{k,\ell} \to \mathbb{R}$ is defined in a similar way as in (2.15), using the modified potential reconstruction:

$$\mathrm{a}_T(\underline{u}_T, \underline{v}_T) := (\nabla \tilde{\mathrm{p}}_T^{k+1} \underline{u}_T, \nabla \tilde{\mathrm{p}}_T^{k+1} \underline{v}_T)_T + \mathrm{s}_T(\underline{u}_T, \underline{v}_T). \tag{5.28}$$

Here, the stabilisation term s_T is assumed to satisfy Assumption 2.4 in which the pair $(\underline{U}_T^k, \underline{I}_T^k)$ is replaced by $(\underline{U}_T^{k,\ell}, \underline{I}_T^{k,\ell})$, that is:

Assumption 5.9 (Local Stabilisation Bilinear Form s_T, $k \neq \ell$) *The local stabilisation bilinear form $\mathrm{s}_T : \underline{U}_T^{k,\ell} \times \underline{U}_T^{k,\ell} \to \mathbb{R}$ satisfies the following properties:*

(S1) Symmetry and positivity. *s_T is symmetric and positive semidefinite;*
(S2) Stability and boundedness. *There is a real number $\eta > 0$ independent of h and of T such that, for all $\underline{v}_T \in \underline{U}_T^{k,\ell}$,*

$$\eta^{-1} \|\underline{v}_T\|_{1,T}^2 \leq \mathrm{a}_T(\underline{v}_T, \underline{v}_T) \leq \eta \|\underline{v}_T\|_{1,T}^2; \tag{5.29}$$

(S3) Polynomial consistency. *For all $w \in \mathbb{P}^{k+1}(T)$ and all $\underline{v}_T \in \underline{U}_T^{k,\ell}$, it holds*

$$\mathrm{s}_T(\underline{I}_T^{k,\ell} w, \underline{v}_T) = 0. \tag{5.30}$$

Hereafter we describe a possible choice of stabilisation term, inspired by (2.22), that satisfies this assumption.

Proposition 5.10 (Example of Local Stabilisation Bilinear Form) *Let $\tilde{\delta}_{TF}^k : \underline{U}_T^{k,\ell} \to \mathbb{P}^k(F)$ and $\tilde{\delta}_T^\ell : \underline{U}_T^{k,\ell} \to \mathbb{P}^\ell(T)$ be the difference operators such that,* *for all* $\underline{v}_T \in \underline{U}_T^{k,\ell}$,

$$\tilde{\delta}_{TF}^k \underline{v}_T := \pi_F^{0,k}(\tilde{\mathrm{p}}_T^{k+1}\underline{v}_T - v_F) \qquad \tilde{\delta}_T^\ell := \pi_T^{0,\ell}(\tilde{\mathrm{p}}_T^{k+1}\underline{v}_T - v_T). \tag{5.31}$$

Then, the following bilinear form satisfies Assumption 5.9:

$$\mathrm{s}_T(\underline{u}_T, \underline{v}_T) := \sum_{F \in \mathcal{F}_T} h_F^{-1}((\tilde{\delta}_{TF}^k - \tilde{\delta}_T^\ell)\underline{u}_T, (\tilde{\delta}_{TF}^k - \tilde{\delta}_T^\ell)\underline{v}_T)_F. \tag{5.32}$$

Proof We follow the lines of the proof of Proposition 2.13. A simple inspection proves (S1). To establish (S3) we first notice that, by the commutation property (5.27) and idempotency of $\pi_T^{1,k+1}$ or $\tilde{\pi}_T^{1,k+1}$, if $w \in \mathbb{P}^{k+1}(T)$ then $\tilde{\mathrm{p}}_T^{k+1}\underline{I}_T^{k,\ell} w = w$. Hence, $\tilde{\delta}_{TF}^k \underline{I}_T^{k,\ell} w = \pi_F^{0,k}(w - \pi_F^{0,k} w) = 0$ for all $F \in \mathcal{F}_T$ and, if $\ell \geq 0$, $\tilde{\delta}_T^\ell \underline{I}_T^{k,\ell} w = \pi_T^{0,\ell}(w - \pi_T^{0,\ell} w) = 0$. If $\ell = -1$, then $\tilde{\delta}_T^\ell \equiv 0$. This establishes the polynomial consistency of the difference operators and, by construction of s_T, proves (S3).

It remains to check (S2). The estimates (2.25) and (2.26) on the volumetric terms in a_T and $\|\cdot\|_{1,T}$ are established exactly as in the proof of Proposition 2.13 (note that the assumption $\ell \leq k+1$ is needed in order to plug $w = v_T \in \mathbb{P}^\ell(T) \subset \mathbb{P}^{k+1}(T)$ into the definition (5.26b) of the potential reconstruction). To estimate the boundary terms, we set $\check{v}_T := \tilde{\mathrm{p}}_T^{k+1}\underline{v}_T$ and (compare with (2.28))

$$\underline{z}_T := \underline{I}_T^{k,\ell}\check{v}_T - \underline{v}_T.$$

If $\ell \geq 0$, the definition (5.6a) of $\underline{I}_T^{k,\ell}$ shows that

$$\underline{z}_T = (\tilde{\delta}_T^\ell \underline{v}_T, (\tilde{\delta}_{TF}^k \underline{v}_T)_{F \in \mathcal{F}_T}). \tag{5.33}$$

If $\ell = -1$, we write

$$\begin{aligned} |T|_d z_T &= \sum_{F \in \mathcal{F}_T} \omega_{TF}|F|_{d-1}\pi_F^{0,0}(\tilde{\mathrm{p}}_T^1 \underline{v}_T) - \sum_{F \in \mathcal{F}_T} \omega_{TF}|F|_{d-1} v_F \\ &= \sum_{F \in \mathcal{F}_T} \omega_{TF}(\tilde{\mathrm{p}}_T^1 \underline{v}_T, 1)_F - \sum_{F \in \mathcal{F}_T} \omega_{TF}(v_F, 1)_F \\ &= 0, \end{aligned}$$

where we have used in the first line the definitions of $\underline{I}_T^{0,-1}$ and v_T (see (5.6b) and (5.1b)), we passed to the second line by writing $|F|_{d-1}\pi_F^{0,0}(\tilde{\mathrm{p}}_T^1 \underline{v}_T) = (\pi_F^{0,0}(\tilde{\mathrm{p}}_T^1 \underline{v}_T), 1)_F = (\tilde{\mathrm{p}}_T^1 \underline{v}_T, 1)_F$, and we concluded invoking the closure condition (5.26d) on $\tilde{\mathrm{p}}_T^1 \underline{v}_T$. Noting that $\tilde{\delta}_T^{-1} \equiv 0$, this shows that (5.33) also holds if $\ell = -1$.

The conclusion of the proof of (S2) then follows exactly as in the proof of Proposition 2.13, using (5.33) and the boundedness property (5.7) of $\underline{I}_T^{k,\ell}$. □

The consistency properties, on interpolates of smooth functions, for a stabilisation bilinear form s_T satisfying Assumption 5.9 are stated in the following proposition.

Proposition 5.11 (Consistency of s_T, $k \neq \ell$) *Let $T \in \mathcal{T}_h$ and let s_T denote a stabilisation bilinear form satisfying Assumption 5.9. Let $r \in \{0, \ldots, k\}$. Then, for all $v \in H^{r+2}(T)$,*

$$s_T(\underline{I}_T^{k,\ell} v, \underline{I}_T^{k,\ell} v)^{\frac{1}{2}} \lesssim h_T^{r+1} |v|_{H^{r+2}(T)}, \tag{5.34}$$

where the hidden constant is independent of h, T, and v.

Proof Identical to the proof of Proposition 2.14, replacing $\underline{I}_T^k$ with $\underline{I}_T^{k,\ell}$ and using the boundedness (5.7) of the latter operator. □

5.1.5 Discrete Problem and Energy Error Estimate

The construction of the global space, norm, and interpolator is done by patching the corresponding local objects. We therefore define

$$\begin{aligned}\underline{U}_h^{k,\ell} := \Big\{&\underline{v}_h = ((v_T)_{T \in \mathcal{T}_h}, (v_F)_{F \in \mathcal{F}_h}) \,:\, v_F \in \mathbb{P}^k(F) \quad \forall F \in \mathcal{F}_h,\\ &v_T \in \mathbb{P}^\ell(T) \text{ if } \ell \geq 0,\, v_T = \frac{1}{|T|_d} \sum_{F \in \mathcal{F}_T} \omega_{TF} |F|_{d-1} v_F \text{ if } \ell = -1 \quad \forall T \in \mathcal{T}_h\Big\},\end{aligned} \tag{5.35}$$

that we endow with the seminorm $\|{\cdot}\|_{1,h}$ still formally defined by (2.35), and $\underline{I}_h^{k,\ell} : W^{1,1}(\Omega) \to \underline{U}_h^{k,\ell}$ such that, for $v \in W^{1,1}(\Omega)$,

$$\begin{aligned}&\underline{I}_h^{k,\ell} v := ((v_T)_{T \in \mathcal{T}_h}, (\pi_F^{0,k} v)_{F \in \mathcal{F}_h}) \text{ with, for all } T \in \mathcal{T}_h,\\ &v_T := \begin{cases} \pi_T^{0,\ell} v & \text{if } \ell \geq 0,\\ \frac{1}{|T|_d} \sum_{F \in \mathcal{F}_T} \omega_{TF} |F|_{d-1} \pi_F^{0,k} v & \text{if } \ell = -1.\end{cases}\end{aligned} \tag{5.36}$$

The subspace of $\underline{U}_h^{k,\ell}$ with strongly enforced homogeneous Dirichlet boundary conditions is

$$\underline{U}_{h,0}^{k,\ell} := \left\{\underline{v}_h \in \underline{U}_h^{k,\ell} \,:\, v_F = 0 \quad \forall F \in \mathcal{F}_h^{\mathrm{b}}\right\}, \tag{5.37}$$

and we notice that $\underline{I}_h^{k,\ell}$ maps functions in $W_0^{1,1}(\Omega)$ on vectors of discrete unknowns in $\underline{U}_{h,0}^{k,\ell}$.

Finally, for $\underline{v}_h \in \underline{U}_h^{k,\ell}$, we define $v_h \in L^2(\Omega)$ as in (2.33), that is:

$$(v_h)_{|T} = v_T \qquad \forall T \in \mathcal{T}_h. \tag{5.38}$$

Using these definitions, the HHO scheme for the Poisson problem (2.1) reads: Find $\underline{u}_h \in \underline{U}_{h,0}^{k,\ell}$ such that

$$\mathrm{a}_h(\underline{u}_h, \underline{v}_h) = (f, v_h) \qquad \forall \underline{v}_h \in \underline{U}_{h,0}^{k,\ell}, \tag{5.39a}$$

where the global bilinear form $\mathrm{a}_h : \underline{U}_h^{k,\ell} \times \underline{U}_h^{k,\ell} \to \mathbb{R}$ is classically obtained by assembling local contributions:

$$\mathrm{a}_h(\underline{u}_h, \underline{v}_h) := \sum_{T \in \mathcal{T}_h} \mathrm{a}_T(\underline{u}_T, \underline{v}_T) \text{ with, for all } T \in \mathcal{T}_h,\ \mathrm{a}_T \text{ given by (5.28).} \tag{5.39b}$$

To differentiate it from the standard HHO scheme based on the same polynomial degrees in the elements and on the faces, we will call this scheme the HHO(k, ℓ) scheme.

Owing to Assumption 5.9, the commutation property (5.27), the approximation properties (1.78)–(1.79) of $\pi_T^{1,k+1}$ or (5.17)–(5.18) of $\tilde{\pi}_T^{1,1}$, and the consistency property (5.34) of s_T, we can reproduce the proof of Lemma 2.18 to see that the bilinear form a_h defined by (5.39b) satisfies similar stability, boundedness, and consistency properties as the bilinear form a_h considered in Sect. 2.2.3, with obvious substitutions of space and interpolator. In particular, the following consistency property holds: For $r \in \{0, \ldots, k\}$ and $w \in H_0^1(\Omega) \cap H^{r+2}(\mathcal{T}_h)$ such that $\Delta w \in L^2(\Omega)$,

$$\sup_{\underline{v}_h \in \underline{U}_{h,0}^{k,\ell}, \|\underline{v}_h\|_{\mathrm{a},h}=1} |\mathcal{E}_h(w; \underline{v}_h)| \lesssim h^{r+1} |w|_{H^{r+2}(\mathcal{T}_h)}, \tag{5.40}$$

where the hidden constant does not depend on w or h, and we have set

$$\|\underline{v}_h\|_{\mathrm{a},h} := \mathrm{a}_h(\underline{v}_h, \underline{v}_h)^{\frac{1}{2}}. \tag{5.41}$$

As a consequence, using the Lax–Milgram Lemma 2.20 and following the same arguments as in the proof of Theorem 2.27, we obtain the following well-posedness and error estimate result for (5.39).

Theorem 5.12 (Well-posedness and Discrete Energy Error Estimate, $k \neq \ell$) *Let $(\mathcal{M}_h)_{h\in\mathcal{H}}$ denote a regular mesh sequence in the sense of Definition 1.9. Let polynomial degrees $k \geq 0$ and $\ell \in \{k-1, k, k+1\}$ be fixed and assume that, for all $h \in \mathcal{H}$, the stabilisation bilinear forms s_T, $T \in \mathcal{T}_h$, satisfy Assumption 5.9. Then, for all $h \in \mathcal{H}$, the discrete problem (5.39) has a unique solution $\underline{u}_h$.*

Moreover, if the solution $u \in H_0^1(\Omega)$ to the weak formulation (2.2) of the Poisson problem enjoys the additional regularity $u \in H^{r+2}(\mathcal{T}_h)$ for some $r \in \{0, \dots, k\}$, then

$$\|\underline{u}_h - \underline{I}_h^{k,\ell} u\|_{\mathrm{a},h} \lesssim h^{r+1} |u|_{H^{r+2}(\mathcal{T}_h)}, \tag{5.42}$$

where the hidden constant is independent of both h and u.

A few remarks are in order before proceeding.

Remark 5.13 (Choice of ℓ) Different considerations can drive the choice of the polynomial degree ℓ. For simple diffusion problems such as the one considered in this section, the choice $\ell = k - 1$ leads to a reduction in the number of element-based unknowns, but requires a separate treatment for the case $k = 0$; moreover, in this case, problem (5.39) no longer admits a flux reformulation (see Remark 5.14). Additionally, the case $(k, \ell) = (1, 0)$ fails to deliver the usual superconvergence in L^2-norm; see Remark 5.17 below. The choice $\ell = k$, on the other hand, permits a unified implementation including the case $k = 0$ and leads to L^2-superconvergence for any $k \geq 0$. The choice $\ell = k + 1$ is sometimes necessary when more complex problems are considered; see, e.g., [105, Remark 6] concerning the Cahn–Hilliard equation.

It is worth emphasising that the number of element-based unknowns has in fact a minor impact on the overall computational cost when static condensation is used (see Sect. B.3.2 in Appendix B). As a matter of fact, the global matrix after static condensation (see (B.13b)) only contains face-based unknowns, and therefore has the same size irrespective of the value of ℓ. Similarly, the size of the local matrix to invert in order to compute the local reconstruction operator (see (B.7)) does not depend on ℓ. The computation of the local reconstruction operator typically represents the larger cost in local computations.

Remark 5.14 (Flux Formulation) In a similar way as in Lemma 2.11, it can be seen, using (S3), that the bilinear stabilisation forms $(\mathrm{s}_T)_{T\in\mathcal{T}_h}$ depend on their arguments only through the difference operators (5.31). In the case $\ell \geq 0$, a reformulation of problem (5.39) in terms of numerical fluxes similar to the one in Lemma 2.25 can then be proved. The only difference is that the operator $\underline{\Delta}_{\partial T}^k$ defined by (2.56) has to be replaced with $\underline{\Delta}_{\partial T}^{k,\ell} : \underline{U}_T^{k,\ell} \to \underline{D}_{\partial T}^k$ such that, for all $\underline{v}_T \in \underline{U}_T^{k,\ell}$,

$$\underline{\Delta}_{\partial T}^{k,\ell}\underline{v}_T = (\Delta_{TF}^{k,\ell}\underline{v}_T)_{F\in\mathcal{F}_T} := \left(\pi_F^{0,k}\left(v_F - (v_T)_{|F}\right)\right)_{F\in\mathcal{F}_T}.$$

For $(k, \ell) = (0, -1)$, no flux reformulation can be obtained. The bilinear form a_h can still be recast as (2.51), but the term v_T inside cannot be chosen independently of the boundary values $(v_F)_{F\in\mathcal{F}_T}$ (see the definition (5.1b) of $\underline{U}_T^{0,-1}$), and the proof of Lemma 2.21 therefore cannot be reproduced.

Remark 5.15 (Energy Error Estimate for the Reconstruction) Using Theorems 5.12 and 1.48 (if $\ell \geq 0$) or Theorem 5.7 (if $\ell = -1$), and following the proof of Theorem 2.28, one can see that the HHO(k, ℓ) scheme (5.39) satisfies the estimate (2.64) with p_h^{k+1} replaced by $\tilde{\mathrm{p}}_h^{k+1}$ defined by: For all $\underline{v}_h \in \underline{U}_h^{k,\ell}$,

$$(\tilde{\mathrm{p}}_h^{k+1}\underline{v}_h)_{|T} = \tilde{\mathrm{p}}_T^{k+1}\underline{v}_T \qquad T \in \mathcal{T}_h. \tag{5.43}$$

5.1.6 Link with Hybridisable Discontinuous Galerkin Methods

It was shown in [129] that the choice $\ell = k + 1$ is linked to Hybridisable Discontinuous Galerkin methods. As pointed out in Remark 2.9, the original version of Hybridisable Discontinuous Galerkin methods may display reduced convergence orders in some circumstances. A possible improvement, proposed in [228, Remark 1.2.4] and analysed in [249], consists in using the local bilinear form $\mathrm{a}_T^{\mathrm{hdg}} : \underline{U}_T^{k,k+1} \times \underline{U}_T^{k,k+1} \to \mathbb{R}$ such that

$$\mathrm{a}_T^{\mathrm{hdg}}(\underline{u}_T, \underline{v}_T) := (\mathbf{G}_T^k\underline{u}_T, \mathbf{G}_T^k\underline{v}_T)_T + \mathrm{s}_T^{\mathrm{hdg}}(\underline{u}_T, \underline{v}_T),$$

where $\mathbf{G}_T^k : \underline{U}_T^{k,k+1} \to \mathbb{P}^k(T)^d$ is formally defined as in (4.37) but with $\underline{U}_T^k$ replaced by $\underline{U}_T^{k,k+1}$, while the stabilisation bilinear form $\mathrm{s}_T^{\mathrm{hdg}} : \underline{U}_T^{k,k+1} \times \underline{U}_T^{k,k+1} \to \mathbb{R}$ is such that, possibly up to a strictly positive coefficient,

$$\mathrm{s}_T^{\mathrm{hdg}}(\underline{u}_T, \underline{v}_T) := \sum_{F\in\mathcal{F}_T} h_F^{-1}(\pi_F^{0,k}(u_T - u_F), \pi_F^{0,k}(v_T - v_F))_F. \tag{5.44}$$

This stabilisation can be expressed in terms of the difference operators defined by (5.31) by observing that, for all $\underline{v}_T \in \underline{U}_T^{k,k+1}$ and all $F \in \mathcal{F}_T$, $\tilde{\delta}_T^{k+1}\underline{v}_T = \pi_T^{0,k+1}(\mathrm{p}_T^{k+1}\underline{v}_T - v_T) = \mathrm{p}_T^{k+1}\underline{v}_T - v_T$, so that

$$\pi_F^{0,k}(v_T - v_F) = \pi_F^{0,k}\left(\mathrm{p}_T^{k+1}\underline{v}_T - v_F - (\mathrm{p}_T^{k+1}\underline{v}_T - v_T)\right) = \pi_F^{0,k}(\tilde{\delta}_{TF}^k\underline{v}_T - \tilde{\delta}_T^{k+1}\underline{v}_T)$$

and

$$\mathrm{s}_T^{\mathrm{hdg}}(\underline{u}_T, \underline{v}_T) := \sum_{F\in\mathcal{F}_T} h_F^{-1}(\pi_F^{0,k}(\tilde{\delta}_{TF}^k\underline{u}_T - \tilde{\delta}_T^{k+1}\underline{u}_T), \pi_F^{0,k}(\tilde{\delta}_{TF}^k\underline{v}_T - \tilde{\delta}_T^{k+1}\underline{v}_T))_F. \tag{5.45}$$

Comparing with (5.32) written for $\ell = k + 1$, the difference is the presence of the projector $\pi_F^{0,k}$ in front of $(\tilde{\delta}_{TF}^k - \tilde{\delta}_T^{k+1})$.

Let us briefly show that this alternative stabilisation form satisfies Assumption 5.9. Given the polynomial consistency of the difference operators $\tilde{\delta}_T^{k+1}$ and $(\tilde{\delta}_{TF}^k)_{F\in\mathcal{F}_T}$, it is clear that $\mathrm{s}_T^{\mathrm{hdg}}$ satisfies (S1) and (S3). To show the stability property (S2), we first notice that, by the formula (5.45) and the L^2-boundedness property (1.77) of $\pi_F^{0,k}$, we have $\mathrm{s}_T^{\mathrm{hdg}}(\underline{v}_T, \underline{v}_T) \leq \mathrm{s}_T(\underline{v}_T, \underline{v}_T)$ with s_T defined by (5.32). Together with Proposition 5.10, this proves the upper bound in (5.29). To establish the lower bound, notice that

$$\begin{aligned}\sum_{F\in\mathcal{F}_T} h_F^{-1}\|\pi_F^{0,k}v_T - v_T\|_F^2 &= \sum_{F\in\mathcal{F}_T} h_F^{-1}\|\pi_F^{0,k}v_T - \pi_T^{0,k+1}v_T\|_F^2 \\ &= |\underline{I}_T^{k,k+1}v_T|_{1,\partial T}^2 \lesssim \|\nabla v_T\|_T^2,\end{aligned}$$

where the last inequality follows from the boundedness property (5.7) of the interpolator $\underline{I}_T^{k,k+1}$ applied to $v = v_T$. A triangle inequality and the definition (5.44) of $\mathrm{s}_T^{\mathrm{hdg}}$ then show that

$$\begin{aligned}|\underline{v}_T|_{1,\partial T}^2 &\lesssim \sum_{F\in\mathcal{F}_T} h_F^{-1}\|v_F - \pi_F^{0,k}v_T\|_F^2 + \sum_{F\in\mathcal{F}_T} h_F^{-1}\|\pi_F^{0,k}v_T - v_T\|_F^2 \\ &\lesssim \sum_{F\in\mathcal{F}_T} h_F^{-1}\|\pi_F^{0,k}(v_F - v_T)\|_F^2 + \|\nabla v_T\|_T^2 \\ &= \mathrm{s}_T^{\mathrm{hdg}}(\underline{v}_T, \underline{v}_T) + \|\nabla v_T\|_T^2,\end{aligned}$$

and the lower bound in (5.29) follows using similar arguments as in the proofs of Propositions 5.10 and 2.13.

5.1.7 L^2-Error Analysis

We present here improved error estimates in L^2-norm, under the standard elliptic regularity property (2.69) for the dual problem (which is actually nothing but the Poisson problem).

Theorem 5.16 (L^2-Error Estimate) *Let $(\mathcal{M}_h)_{h\in\mathcal{H}}$ denote a regular mesh sequence in the sense of Definition 1.9. Let polynomial degrees $k \geq 0$ and $\ell \in \{k-1, k, k+1\}$ be fixed. Let $u \in H_0^1(\Omega)$ denote the unique solution of (2.2), for which we assume the additional regularity $u \in H^{r+2}(\mathcal{T}_h)$ for some $r \in \{0, \ldots, k\}$. For all $h \in \mathcal{H}$, let $\underline{u}_h \in \underline{U}_{h,0}^{k,\ell}$ denote the unique solution to (5.39) with stabilisation bilinear forms s_T, $T \in \mathcal{T}_h$, in (5.28) satisfying Assumption 5.9. We further assume elliptic regularity and that*

- *if $(k,\ell) = (0,0)$, $f \in H^1(\mathcal{T}_h)$,*
- *if $(k,\ell) = (0,-1)$, $f \in H^1(\mathcal{T}_h)$ and, for all $T \in \mathcal{T}_h$, the weights $(\omega_{TF})_{F\in\mathcal{F}_T}$ satisfy the improved quadrature rule (5.5).*

Then, recalling the definition (5.43) of $\tilde{\mathrm{p}}_h^{k+1}$,

$$\|\tilde{\mathrm{p}}_h^{k+1}\underline{u}_h - u\| \lesssim \begin{cases} h^2\|f\|_{H^1(\mathcal{T}_h)} & \text{if } k = 0 \text{ and } \ell \leq 0, \\ h^{r+2}|u|_{H^{r+2}(\mathcal{T}_h)} & \text{if } \ell \geq 1, \end{cases} \tag{5.46}$$

with hidden constant independent of both h and u.

Remark 5.17 (The Case $(k,\ell) = (1,0)$) The case $(k,\ell) = (1,0)$ is not covered by (5.46). Actually, for this choice of polynomial degrees, numerical tests presented in Sect. 5.1.8 show that, on some mesh families, the rate of convergence in L^2-norm is not better than the $O(h^2)$ rate obtained in energy norm (see (5.42) with $r = k = 1$). A possible way to recover an improved L^2-convergence would be to change, in the scheme (5.39), the discretisation of the source term into $(f, \tilde{\mathrm{p}}_h^1\underline{v}_h)$ (where $\tilde{\mathrm{p}}_h^1$ has formally the same definition, but its domain is changed into $\underline{U}_h^{1,0}$). This, however, leads to a method that is more complex to implement, and for which the flux formulation mentioned in Remark 5.14 is no longer valid, see Remark 2.22 (on this problematic of loosing the flux formulation when a higher-order reconstruction is used in the source term, see also [174, Section A.3.2] in the context of Hybrid Mimetic Mixed methods—which contain the case $(k,\ell) = (0,0)$ of HHO methods, as shown in Sect. 5.3 below). If an L^2-superconvergence is required, a better choice for $k = 1$ is to simply take $\ell = 1$; as explained in Remark 5.13, the added computational cost corresponding to choosing $\ell = 1$ instead of $\ell = 0$ is minimal owing to the possibility of statically condensing the system to locally eliminate the element unknowns.

Proof As in the case $k = \ell$ covered in Sect. 2.3.3, the estimate (5.46) is a consequence of the following superconvergence result for element unknowns (see Lemma 2.33): Letting $\underline{\hat{u}}_h := \underline{I}_h^{k,\ell} u$ and defining $\hat{u}_h$ via (5.38),

$$\|u_h - \hat{u}_h\| \lesssim \begin{cases} h^2\|f\|_{H^1(\mathcal{T}_h)} & \text{if } k = 0 \text{ and } \ell \leq 0, \\ h^{r+2}|u|_{H^{r+2}(\mathcal{T}_h)} & \text{if } \ell \geq 1. \end{cases} \tag{5.47}$$

The proof of this estimate follows the lines of the proof of Lemma 2.33. We sketch here the ideas, focusing on the arguments that require a specific adaptation in the case $\ell \neq k$.

An estimate analogous to (2.75) for $\|u_h - \hat{u}_h\|$ is obtained applying the Aubin–Nitsche Lemma A.10 to $\mathrm{U} = H_0^1(\Omega)$, $\mathrm{a}(u, v) = (\nabla u, \nabla v)$, $\mathrm{l}(v) = (f, v)$, $\mathrm{U}_h = \underline{U}_{h,0}^{k,\ell}$, $\|\cdot\|_{\mathrm{U}_h} = \|\cdot\|_{\mathrm{a},h}$ defined by (5.41), $\mathrm{a}_h = \mathrm{a}_h$ defined by (5.39b), $\mathrm{l}_h(\underline{v}_h) = (f, v_h)$ and $\mathrm{I}_h u = \underline{I}_h^{k,\ell} u$, $\mathrm{L} = L^2(\Omega)$ and $\mathrm{r}_h : \underline{U}_{h,0}^{k,\ell} \to L^2(\Omega)$ defined by $\mathrm{r}_h \underline{v}_h = v_h$. Specifically, we have that

$$
\begin{aligned}
\|u_h - \hat{u}_h\| \le \underbrace{\|\underline{u}_h - \underline{I}_h^{k,\ell} u\|_{\mathrm{a},h} \sup_{g \in L^2(\Omega),\ \|g\| \le 1} \|\mathcal{E}_h(z_g; \cdot)\|_{\mathrm{a},h,\star}}_{\mathfrak{T}_1} \\
+ \underbrace{\sup_{g \in L^2(\Omega),\ \|g\| \le 1} |\mathcal{E}_h(u; \underline{I}_h^{k,\ell} z_g)|}_{\mathfrak{T}_2},
\end{aligned}
\tag{5.48}
$$

where the consistency error is still formally defined by (2.43). We now estimate the terms $\mathfrak{T}_1$ and $\mathfrak{T}_2$ in the right-hand side of (5.48).

(i) *Estimate of* $\mathfrak{T}_1$. This term is estimated exactly as in the proof of Lemma 2.33, using the symmetry of the problem and applying the consistency property (5.40) to $r = 0$ and $w = z_g$.

(ii) *Estimate of* $\mathfrak{T}_2$. We separate the cases $k \ge 1$ and $k = 0$.

(ii.A) *Case* $k \ge 1$ *and* $\ell \ge 1$. We still have (2.77), with $\underline{I}_h^{k,\ell}$ instead of $\underline{I}_h^k$, that is

$$
|\mathcal{E}_h(u; \underline{I}_h^{k,\ell} z_g)| \lesssim h^{r+1} |u|_{H^{r+2}(\mathcal{T}_h)} \left[\left(\sum_{T \in \mathcal{T}_h} |\underline{I}_h^{k,\ell} z_g|_{1,\partial T}^2 \right)^{\frac{1}{2}} + |\underline{I}_h^{k,\ell} z_g|_{\mathrm{s},h} \right].
$$

The term $|\underline{I}_h^{k,\ell} z_g|_{\mathrm{s},h}$ is estimated as in the proof of Lemma 2.33, using the consistency property (5.34) of the stabilisation bilinear form with $v = z_g$ and $r = 0$. To estimate the boundary terms $|\underline{I}_h^{k,\ell} z_g|_{1,\partial T}^2$, we insert $\pm \pi_T^{0,1} z_g$ and use the triangle inequality to write

$$
\begin{aligned}
|\underline{I}_T^{k,\ell} z_g|_{1,\partial T}^2 &= \sum_{F \in \mathcal{F}_T} h_F^{-1} \|\pi_F^{0,k} z_g - \pi_T^{0,\ell} z_g\|_F^2 \\
&\le 2 \sum_{F \in \mathcal{F}_T} h_F^{-1} \|\pi_F^{0,k} z_g - \pi_T^{0,1} z_g\|_F^2 \\
&\quad + 2 \sum_{F \in \mathcal{F}_T} h_F^{-1} \|\pi_T^{0,1} z_g - \pi_T^{0,\ell} z_g\|_F^2 \\
&= 2\mathfrak{T}_{\partial T,a} + 2\mathfrak{T}_{\partial T,b}.
\end{aligned}
\tag{5.49}
$$

The term $\mathfrak{T}_{\partial T,a}$ is estimated using the same arguments as for (2.78): polynomial invariance of $\pi_F^{0,k}$ (recall that $k \geq 1$ here), $L^2(F)$-boundedness of this projector, and trace approximation property (1.75) with $(l, p, m, s) = (1, 2, 0, 2)$:

$$\begin{aligned}\mathfrak{T}_{\partial T,a} &= \sum_{F\in\mathcal{F}_T} h_F^{-1} \|\pi_F^{0,k}(z_g - \pi_T^{0,1} z_g)\|_F^2 \\ &\leq \sum_{F\in\mathcal{F}_T} h_F^{-1} \|z_g - \pi_T^{0,1} z_g\|_F^2 \lesssim h_T^2 |z_g|_{H^2(T)}^2.\end{aligned}$$

To estimate $\mathfrak{T}_{\partial T,b}$, we write

$$\begin{aligned}\mathfrak{T}_{\partial T,b} &\lesssim \sum_{F\in\mathcal{F}_T} h_F^{-1} h_T^{-1} \|\pi_T^{0,1} z_g - \pi_T^{0,\ell} z_g\|_T^2 \\ &\lesssim h_T^{-2} \|\pi_T^{0,\ell}(\pi_T^{0,1} z_g - z_g)\|_T^2 \\ &\lesssim h_T^{-2} \|\pi_T^{0,1} z_g - z_g\|_T^2 \lesssim h_T^2 |z_g|_{H^2(T)}^2,\end{aligned}$$

where we have used the discrete trace inequality (1.55) with $p = 2$ in the first line, followed by the estimate $h_T \lesssim h_F$ (see (1.6)) and the bound (1.5) on card$(\mathcal{F}_T)$ together with the linearity and polynomial invariance of $\pi_T^{0,\ell}$ (recall that $\ell \geq 1$) to pass to the second line, and the $L^2(T)$-boundedness of this projector together with the approximation property (1.74) of the L^2-orthogonal projector with $X = T$ and $(l, p, m, s) = (1, 2, 0, 2)$ to conclude.

Plugging the above bounds into (5.49) shows that $|\underline{I}_T^{k,\ell} z_g|_{1,\partial T}^2$ satisfies the estimate (2.78) and thus, as in the proof of Lemma 2.33, that $\mathfrak{T}_2 \lesssim h^{r+2}|u|_{H^{r+2}(\mathcal{T}_h)}$.

(ii.B) *Case* $k = 0$. Letting $\underline{z}_h = \underline{I}_h^{0,\ell} z_g$, the relation (2.79) holds with $\underline{I}_h^0$ replaced by $\underline{I}_h^{0,\ell}$, $\pi_T^{0,0} z_g$ replaced by z_T and, if $\ell = -1$, $\pi_T^{1,1} z_g$ replaced with $\widetilde{\pi}_T^{1,1} z_g$ (due to (5.27b)), that is:

$$\mathcal{E}_h(u; \underline{I}_h^{0,\ell} z_g) = \sum_{T\in\mathcal{T}_h} (f, z_T)_T - \sum_{T\in\mathcal{T}_h} (\nabla\widehat{\pi}_T^{1,1} u, \nabla\widehat{\pi}_T^{1,1} z_g)_T - s_h(\underline{I}_h^{0,\ell} u, \underline{I}_h^{0,\ell} z_g), \tag{5.50}$$

where $\widehat{\pi}_T^{1,1} = \pi_T^{1,1}$ if $\ell \geq 0$ and $\widehat{\pi}_T^{1,1} = \widetilde{\pi}_T^{1,1}$ if $\ell = -1$. The case $k = \ell = 0$ is covered in Lemma 2.33, so it remains to consider the cases $\ell = 1$ and $\ell = -1$.

(ii.B.1) *Case* $k = 0$ *and* $\ell = 1$. We have

$$\sum_{T\in\mathcal{T}_h}(f, z_T)_T = \sum_{T\in\mathcal{T}_h}(f, \pi_T^{0,1} z_g)_T = \underbrace{\sum_{T\in\mathcal{T}_h}(f, \pi_T^{0,1} z_g - z_g)_T}_{\mathfrak{T}_{2,1}} + (f, z_g). \tag{5.51}$$

By Cauchy–Schwarz inequalities and the approximation property (1.74) of the local polynomial projector with $l = 1$, $p = 2$, $m = 0$ and $s = 2$, we have

$$|\mathfrak{T}_{2,1}| \le \sum_{T\in\mathcal{T}_h} h_T^2 \|f\|_T |z_g|_{H^2(T)} \lesssim h^2 \|f\| \|g\|, \tag{5.52}$$

the conclusion following from a Cauchy–Schwarz inequality on the sum together with the elliptic regularity assumption (2.69). Plugging (5.51) and (5.52) into (5.50) and using $(f, z_g) = (\nabla u, \nabla z_g)$ leads to

$$|\mathcal{E}_h(u; \underline{I}_h^{0,1} z_g)| \lesssim \sum_{T\in\mathcal{T}_h} \left|(\nabla u, \nabla z_g)_T - (\nabla \pi_T^{1,1} u, \nabla \pi_T^{1,1} z_g)_T\right|$$
$$+ |\mathrm{s}_h(\underline{I}_h^{0,1} u, \underline{I}_h^{0,1} z_g)| + h^2 \|f\| \|g\|.$$

The first two terms in this right-hand side can be manipulated as in Point (ii.B) of the proof of Lemma 2.33 (see (2.81) and (2.82)), leading to

$$|\mathcal{E}_h(u; \underline{I}_h^{0,1} z_g)| \lesssim h^2 |u|_{H^2(\mathcal{T}_h)} |z_g|_{H^2(\mathcal{T}_h)} + h^2 \|f\| \|g\| \lesssim h^2 |u|_{H^2(\mathcal{T}_h)} \|g\|,$$

which is the estimate required to conclude.
(ii.B.2) *Case* $k = 0$ *and* $\ell = -1$. We write

$$\sum_{T\in\mathcal{T}_h}(f, z_T)_T = \sum_{T\in\mathcal{T}_h}(f, \pi_T^{0,0} z_g)_T + \underbrace{\sum_{T\in\mathcal{T}_h}(f, z_T - \pi_T^{0,0} z_g)_T}_{\mathfrak{T}'_{2,1}}.$$

The first addend in the right-hand side is then manipulated as in (2.80), which leads to

$$|\mathcal{E}_h(u; \underline{I}_h^{0,-1} z_g)| \lesssim h^2 \|f\|_{H^1(\mathcal{T}_h)} \|g\| + |\mathfrak{T}'_{2,1}|. \tag{5.53}$$

To estimate $\mathfrak{T}'_{2,1}$, we use the assumption (5.5) on the weights. This assumption and the definition (5.6b) of z_T for $\underline{z}_h = \underline{I}_h^{0,-1} z_g$ show that, if z_g is polynomial of degree 1 in an element T, then $z_T = \pi_T^{0,0} z_g$ and the contribution of T in $\mathfrak{T}'_{2,1}$ is zero. Estimating $\mathfrak{T}'_{2,1}$ thus consists in approximating, in each element $T \in \mathcal{T}_h$, z_g by an element of $\mathbb{P}^1(T)$.

Fix $T \in \mathcal{T}_h$ and let $i_T : H^1(T) \to \mathbb{P}^0(T)$ be the interpolator defined by the first component of $\underline{I}_T^{0,-1}$, that is: For $w \in H^1(T)$,

$$i_T w := \frac{1}{|T|_d} \sum_{F \in \mathcal{F}_T} \omega_{TF} |F|_{d-1} \pi_F^{0,0} w.$$

We have

$$\begin{aligned}
\|i_T(z_g - \pi_T^{0,1} z_g)\|_T &= |T|_d^{\frac{1}{2}} |i_T(z_g - \pi_T^{0,1} z_g)| \\
&\le |T|_d^{-\frac{1}{2}} \sum_{F \in \mathcal{F}_T} \omega_{TF} |F|_{d-1} |\pi_F^{0,0}(z_g - \pi_T^{0,1} z_g)| \\
&= |T|_d^{-\frac{1}{2}} \sum_{F \in \mathcal{F}_T} \omega_{TF} |F|_{d-1}^{\frac{1}{2}} \|\pi_F^{0,0}(z_g - \pi_T^{0,1} z_g)\|_F,
\end{aligned}$$

where we have used the fact that $i_T(z_g - \pi_T^{0,1} z_g)$ is constant in the first line, and we have applied the definitions of i_T and $\|\cdot\|_F$ in the second and third line, respectively. Invoking then the $L^2(F)$-boundedness of $\pi_F^{0,0}$ and the trace approximation property (1.75) of $\pi_T^{0,1}$ with $p = 2$, $m = 0$ and $s = 2$, we continue with

$$\|i_T(z_g - \pi_T^{0,1} z_g)\|_T \lesssim |T|_d^{-\frac{1}{2}} \sum_{F \in \mathcal{F}_T} \omega_{TF} |F|_{d-1}^{\frac{1}{2}} h_T^{\frac{3}{2}} |z_g|_{H^2(T)} \lesssim h_T^2 |z_g|_{H^2(T)}, \tag{5.54}$$

the second inequality following from the estimate (5.4) on ω_{TF} together with the mesh regularity assumption that ensures $\mathrm{card}(\mathcal{F}_T) \lesssim 1$ and $|F|_{d-1} h_T \lesssim |T|_d$ (by (1.5)–(1.8)).

Inserting $i_T(\pi_T^{0,1} z_g) - \pi_T^{0,0}(\pi_T^{0,1} z_g) = 0$ (by (5.5)), we can then write

$$\begin{aligned}
|\mathfrak{T}_{2,1}'| &\le \sum_{T \in \mathcal{T}_h} \|f\|_T \|i_T(z_g - \pi_T^{0,1} z_g) - \pi_T^{0,0}(z_g - \pi_T^{0,1} z_g)\|_T \\
&\lesssim \sum_{T \in \mathcal{T}_h} \|f\|_T h_T^2 |z_g|_{H^2(T)},
\end{aligned}$$

the first line following from a Cauchy–Schwarz inequality, and the conclusion being obtained invoking a triangle inequality, (5.54), and the $L^2(T)$-boundedness of $\pi_T^{0,0}$ together with the approximation property (1.74) with $X = T$ and $(l, p, m, s) = (1, 2, 0, 2)$ to write

$$\|\pi_T^{0,0}(z_g - \pi_T^{0,1} z_g)\|_T \le \|z_g - \pi_T^{0,1} z_g\|_T \lesssim h_T^2 |z_g|_{H^2(T)}.$$

Using a Cauchy–Schwarz inequality on the sum over $T \in \mathcal{T}_h$ and the elliptic regularity, we infer the bound

$$|\mathfrak{T}'_{2,1}| \lesssim h^2 \|f\| |z_g|_{H^2(\mathcal{T}_h)} \lesssim h^2 \|f\| \|g\|$$

which, plugged into (5.53), yields the estimate of $\mathcal{E}_h(u; \underline{I}_h^{0,-1} z_g)$ required to conclude the proof. □

5.1.8 *Numerical Tests*

We provide here numerical illustrations of the energy and L^2-error estimates for the HHO(k,ℓ) scheme (5.39) contained in Theorems 5.12 and 5.16.

5.1.8.1 Two-Dimensional Test Case

The computational domain is $\Omega = (0, 1)^2$ and the exact solution is

$$u(x_1, x_2) = \sin(\pi x_1) \sin(\pi x_2).$$

The source term is therefore $f(x_1, x_2) = 2\pi^2 \sin(\pi x_1) \sin(\pi x_2)$. We run the simulations for $k \in \{0, 1, 2\}$ and $\ell \in \{k - 1, k, k + 1\}$ on two families of predominantly hexagonal and locally refined Cartesian meshes; see Fig. 5.2 for representatives of these families. Letting $\underline{\hat{u}}_h = \underline{I}_h^{k,\ell} u$ be the interpolate of the

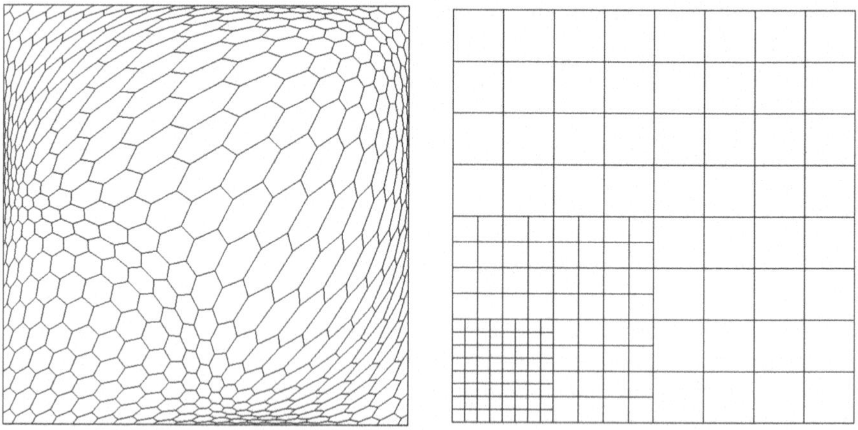

Fig. 5.2 Examples of meshes for the tests of Sect. 5.1.8.1: hexagonal mesh (left); locally refined Cartesian mesh (right)

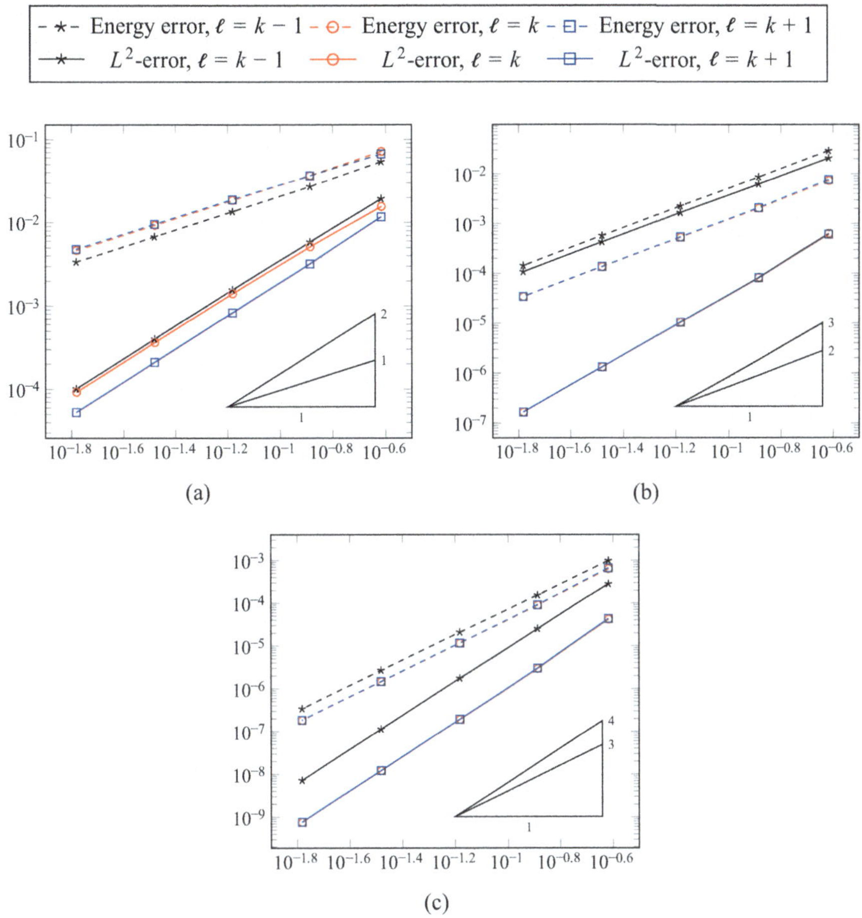

Fig. 5.3 Errors vs. h for the 2D test case in Sect. 5.1.8.1, hexagonal meshes. Energy error refers to $\|\underline{u}_h - \underline{\hat{u}}_h\|_{1,h}$, L^2-error refers to $\|u_h - \hat{u}_h\|$, where $\underline{\hat{u}}_h = \underline{I}_h^{k,\ell} u$. (**a**) $k = 0$. (**b**) $k = 1$. (**c**) $k = 2$

exact solution, we display the errors measured in the energy norm $\|\underline{u}_h - \underline{\hat{u}}_h\|_{1,h}$ (which satisfies the estimate (5.42) since it is uniformly equivalent to $\|\cdot\|_{a,h}$ owing to (2.41)), and in the L^2-norm $\|u_h - \hat{u}_h\|$ (which satisfies (5.47)).

The results on the family of hexagonal meshes, presented in Fig. 5.3, confirm the theoretical $O(h^{k+1})$ convergence in energy norm, and the $O(h^{k+2})$ superconvergence in L^2-norm, except in the case $(k, \ell) = (1, 0)$. As noticed in Remark 5.17, the choice $(k, \ell) = (1, 0)$ is not covered by Theorem 5.16, and the results here show that the rate of convergence in L^2-norm in this case is not better than the $O(h^2)$ rate in energy norm.

Similar conclusions can be drawn from the tests on the locally refined Cartesian meshes, see Fig. 5.4. We however notice, on these meshes, an unexpected supercon-

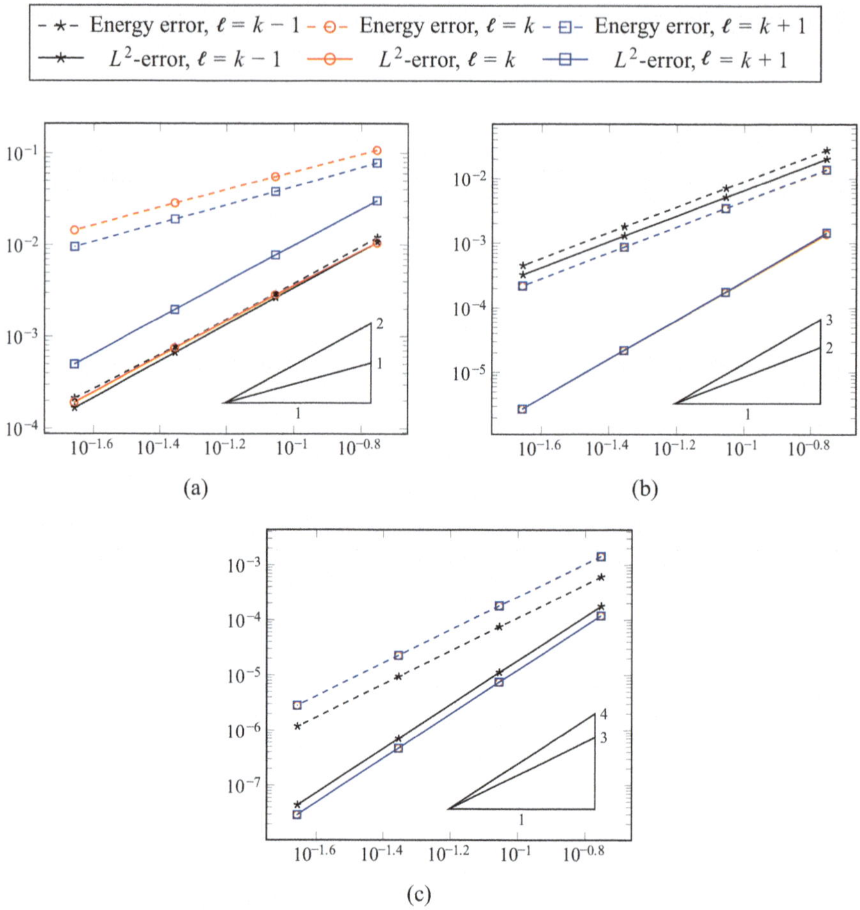

Fig. 5.4 Errors vs. h for the 2D test case in Sect. 5.1.8.1, locally refined Cartesian meshes. Energy error refers to $\|\underline{u}_h - \underline{\hat{u}}_h\|_{1,h}$, L^2-error refers to $\|u_h - \hat{u}_h\|$, where $\underline{\hat{u}}_h = \underline{I}_h^{k,\ell} u$. **(a)** $k = 0$. **(b)** $k = 1$. **(c)** $k = 2$

vergence in energy norm in the case $(k, \ell) = (0, -1)$. This phenomenon is strongly related to the particular mesh we consider here, and does not occur on less structured meshes as seen in Fig. 5.3.

On both families of meshes, we also notice that increasing ℓ does not necessarily lead to better errors—the effect of the choice of ℓ is minimal, and varies with the mesh and polynomial degree k. In passing, when comparing the displayed error measures, one should keep in mind that they have an intrinsic dependence on ℓ via the use of the interpolator $\underline{I}_h^{k,\ell}$.

5.1.8.2 Three-Dimensional Test Case

The 3D computational domain is the unit cube $\Omega = (0, 1)^3$, and the exact solution is

$$u(x_1, x_2, x_3) = \sin(\pi x_1)\sin(\pi x_2)\sin(\pi x_3),$$

corresponding to the source term $f(x_1, x_2, x_3) = 3\pi^2 \sin(\pi x_1)\sin(\pi x_2)\sin(\pi x_3)$. Two families of meshes are considered: a family of matching simplicial meshes, and a family of Voronoi meshes; one example for each family is presented in Fig. 5.5.

Figures 5.6 and 5.7 present the convergence graphs for polynomial degrees $k \in \{0, 1, 2\}$ and $\ell \in \{k-1, k, k+1\}$. As in the 2D cases, we display the errors measured in the energy norm $\|\underline{u}_h - \underline{\hat{u}}_h\|_{1,h}$ and in L^2-norm $\|u_h - \hat{u}_h\|$, with $\underline{\hat{u}}_h = \underline{I}_h^{k,\ell} u$.

The results corresponding to the simplicial meshes, collected in Fig. 5.6, clearly show that the convergence in energy norm is in $O(h^{k+1})$ for all three choices of ℓ. Likewise, except in the case $(k, \ell) = (1, 0)$, a superconvergence in $O(h^{k+2})$ can be observed in the L^2-norm, independently of ℓ.

The results on Voronoi meshes, collected in Fig. 5.7, show the same trend, including the loss of superconvergence in the case $(k, \ell) = (1, 0)$. Some rates of convergence are however sub-optimal. For example, for $(k, \ell) = (1, 1)$, the average slope of the energy error is 1.66 (compared to an expected slope of 2); for $(k, \ell) = (2, 2)$, the average rate of convergence of the L^2 error is 3.53 (for an expected rate of 4), and the average rate of the energy error is 2.57 (for an expected 3). In all the tests, though, the rates of convergence between the last two members of the mesh family get closer to the theoretical rates. The reason for these losses of optimal convergence is to be found in the regularity of the meshes. We present in

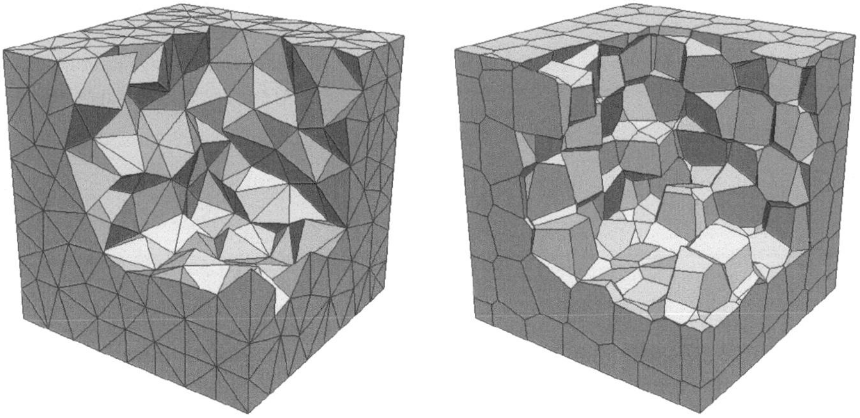

Fig. 5.5 Examples of meshes for the tests of Sect. 5.1.8: matching simplicial mesh (left); Voronoi mesh (right)

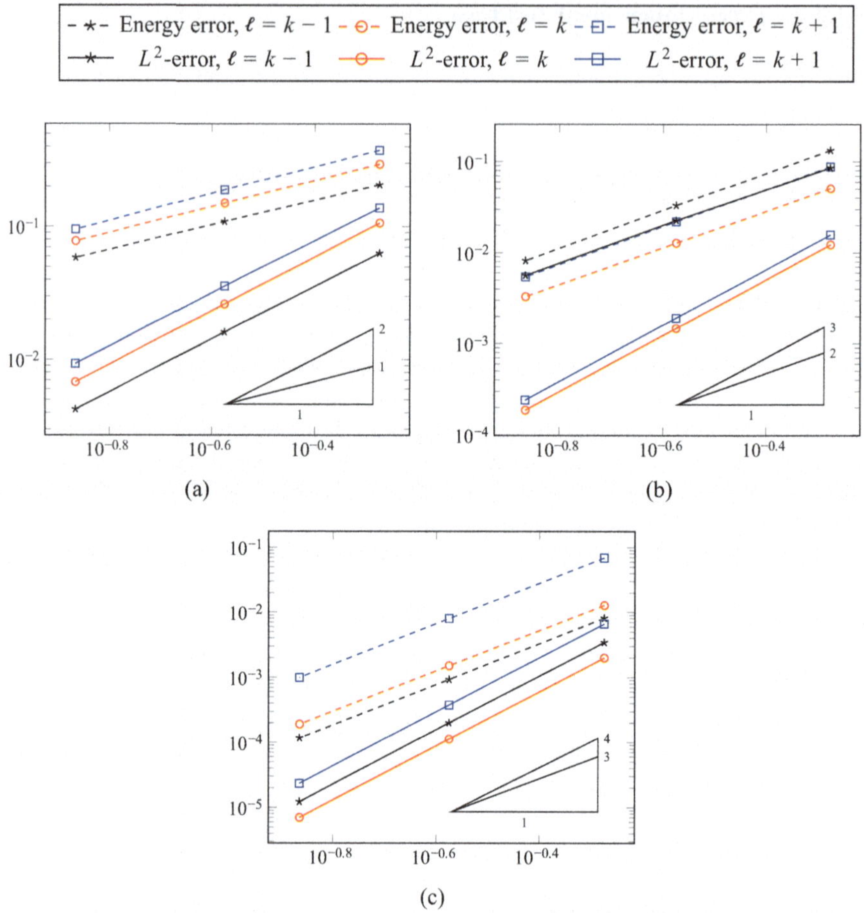

Fig. 5.6 Errors vs. h for the 3D test case in Sect. 5.1.8.2, simplicial meshes. Energy error refers to $\|\underline{u}_h - \underline{\hat{u}}_h\|_{1,h}$, L^2-error refers to $\|u_h - \hat{u}_h\|$, where $\underline{\hat{u}}_h = \underline{I}_h^{k,\ell} u$. **(a)** $k = 0$. **(b)** $k = 1$. **(c)** $k = 2$

Table 5.1 the values of the following mesh regularity parameter:

$$\widetilde{\varrho}_h = \max_{T \in \mathcal{T}_h} \left[\frac{h_T}{|T|_3^{1/3}}, \max_{F \in \mathcal{F}_T} \left(\frac{h_T}{h_F}, \frac{h_F}{|F|_2^{1/2}} \right) \right].$$

This factor measures similar mesh regularity properties as ϱ in Definition 1.9, but is more practical to compute. The table shows that the considered Voronoi mesh family does not form a very regular mesh sequence. It is however interesting to notice that, despite the dramatic loss of regularity with refinement for these meshes, the HHO method still performs relatively well, with only a small reduction of the expected rates of convergence.

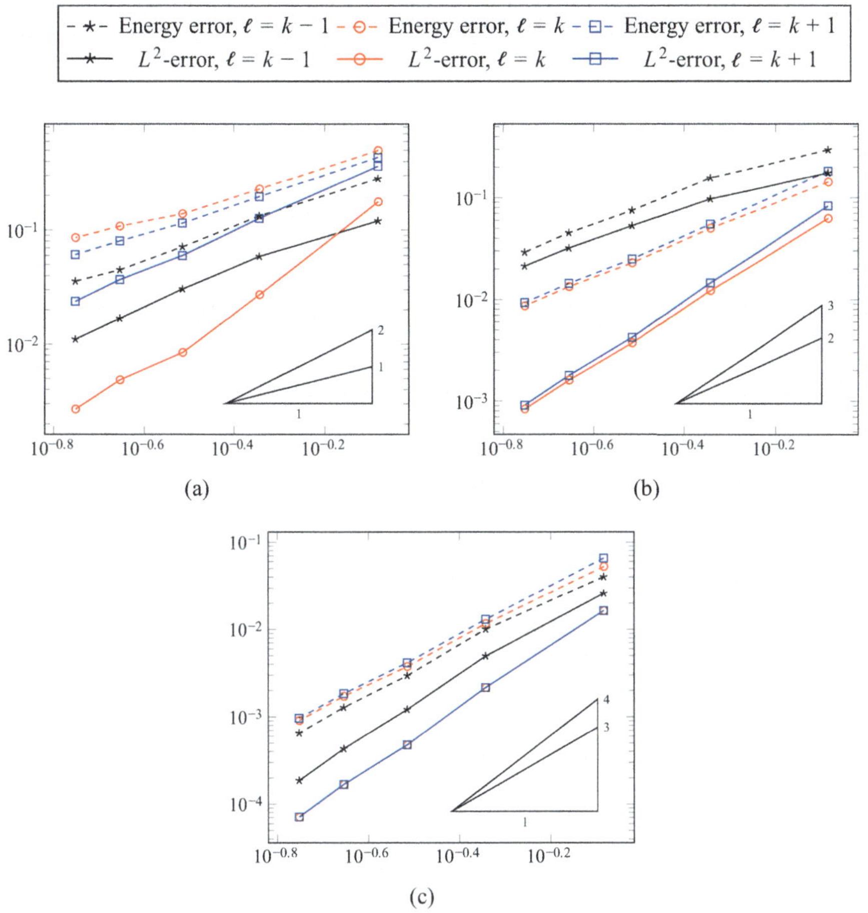

Fig. 5.7 Errors vs. h for the 3D test case in Sect. 5.1.8.2, Voronoi meshes. Energy error refers to $\|\underline{u}_h - \underline{\hat{u}}_h\|_{1,h}$, L^2-error refers to $\|u_h - \hat{u}_h\|$, where $\underline{\hat{u}}_h = \underline{I}_h^{k,\ell} u$. **(a)** $k = 0$. **(b)** $k = 1$. **(c)** $k = 2$

Table 5.1 Mesh regularity parameter for the 3D Voronoi family of meshes used in Sect. 5.1.8.2

Mesh number	Regularity parameter $\widetilde{\varrho}_h$
1	107
2	377
3	1.7e+3
4	2.6e+5
5	1.8e+4

5.2 Nonconforming $\mathbb{P}^1$ Finite Element

We show here that the HHO(0, −1) method of Sect. 5.1 is, on matching simplicial meshes, strongly related to the nonconforming $\mathbb{P}^1$ Finite Element method.

5.2.1 *Presentation of the Nonconforming $\mathbb{P}^1$ Finite Element*

Let $\mathcal{M}_h$ be a matching simplicial mesh in the sense of Definition 1.7. The space of nonconforming $\mathbb{P}^1$ Finite Element functions, with homogeneous Dirichlet boundary conditions, is the space of piecewise linear functions on $\mathcal{T}_h$ whose averages on the faces are continuous across the interfaces and vanish on the boundary faces. In other words, recalling the definition (1.22) of the jumps across interfaces, and setting $[w]_F := (w_{|T})_F$ whenever $F \in \mathcal{F}_h^{\mathrm{b}}$ with $\mathcal{T}_F = \{T\}$, this space is

$$V_{h,0}^{\mathrm{nc}} := \left\{ w \in \mathbb{P}^1(\mathcal{T}_h) \,:\, \pi_F^{0,0}[w]_F = 0 \quad \forall F \in \mathcal{F}_h \right\}.$$

Remark 5.18 (Alternative Continuity Condition) For polynomials of degree 1, the average over a face is equal to the value at the centre of mass of the face. The continuity conditions on a nonconforming $\mathbb{P}^1$ Finite Element function w can therefore be equivalently stated as: w is continuous at the centre of mass of each interface, and vanishes at the centre of mass of each boundary face.

The nonconforming $\mathbb{P}^1$ Finite Element scheme for the Poisson problem (2.1) reads:

$$\text{Find } U_h \in V_{h,0}^{\mathrm{nc}} \text{ such that, for all } w_h \in V_{h,0}^{\mathrm{nc}}, \quad \int_\Omega \nabla_h U_h \cdot \nabla_h w_h = \int_\Omega f w_h, \tag{5.55}$$

where ∇_h is the standard broken gradient (1.21).

5.2.2 *Properties of the Low-Order Potential Reconstruction on Simplices*

To establish a link between the nonconforming $\mathbb{P}^1$ Finite Element scheme and the HHO(0, −1) scheme for the Poisson problem, we first identify properties of the low-order potential reconstruction, specific to simplicial elements and meshes.

Lemma 5.19 (Potential Reconstruction on a Simplex) *Let T be a simplex. For all $\underline{v}_T \in \underline{U}_T^{0,-1}$, $\tilde{\mathrm{p}}_T^1 \underline{v}_T$ defined by (5.26) for $(k, \ell) = (0, -1)$ is the unique element in $\mathbb{P}^1(T)$ that satisfies*

$$\pi_F^{0,0}(\tilde{\mathrm{p}}_T^1 \underline{v}_T) = v_F \qquad \forall F \in \mathcal{F}_T. \tag{5.56}$$

Proof Let $\underline{v}_T \in \underline{U}_T^{0,-1}$ and let us first establish the existence of $q_{\underline{v}_T} \in \mathbb{P}^1(T)$ that satisfies (5.56). Let $(\overline{x}_0, \ldots, \overline{x}_d)$ be the centres of mass of the faces $(F_0, \ldots, F_d)$ of the simplex T. These centres of mass do not lie on the same hyperplane so, setting $z_i := \overline{x}_i - \overline{x}_0$ for all $i \in \{1, \ldots, d\}$, it holds

$$D := \det(z_1, \ldots, z_d) \neq 0.$$

We can then define

$$q_{\underline{v}_T}(\boldsymbol{x}) := v_{F_0} + \sum_{i=1}^{d} (v_{F_i} - v_{F_0}) \frac{\det(z_1, \ldots, z_{i-1}, \boldsymbol{x} - \overline{\boldsymbol{x}}_0, z_{i+1}, \ldots, z_d)}{D}.$$

The linearity of the determinant with respect to each of its columns show that $q_{\underline{v}_T}$ is affine, that is, belongs to $\mathbb{P}^1(T)$. Moreover, for any $j \in \{1, \ldots, d\}$, if $\boldsymbol{x} = \overline{\boldsymbol{x}}_j$ then $\boldsymbol{x} - \overline{\boldsymbol{x}}_0 = z_j$ and all the determinants above except for $j = i$ vanish, since two of their columns are equal. The determinant corresponding to $j = i$ is equal to D, which shows that $q_{\underline{v}_T}(\overline{\boldsymbol{x}}_j) = v_{F_0} + v_{F_j} - v_{F_0} = v_{F_j}$. We also obviously have $q_{\underline{v}_T}(\overline{\boldsymbol{x}}_0) = v_{F_0}$. Hence, $q_{\underline{v}_T} \in \mathbb{P}^1(T)$ takes, for any $F \in \mathcal{F}_T$, the value v_F at the centre of mass of F. By Remark 5.18, we infer that $\pi_F^{0,0} q_{\underline{v}_T} = v_F$ for all $F \in \mathcal{F}_T$.

Let us now prove that $\tilde{\mathrm{p}}_T^1 \underline{v}_T$ has the same property. Applying the definition (5.26a) we have, for any $w \in \mathbb{P}^1(T)$, since $\Delta w = 0$ and $\pi_F^{0,0} q_{\underline{v}_T} = v_F$ for all $F \in \mathcal{F}_T$,

$$\begin{aligned}
(\nabla \tilde{\mathrm{p}}_T^1 \underline{v}_T, \nabla w)_T &= \sum_{F \in \mathcal{F}_T} (v_F, \nabla w \cdot \boldsymbol{n}_{TF})_F \\
&= \sum_{F \in \mathcal{F}_T} (\pi_F^{0,0} q_{\underline{v}_T}, \nabla w \cdot \boldsymbol{n}_{TF})_F \\
&= \sum_{F \in \mathcal{F}_T} (q_{\underline{v}_T}, \nabla w \cdot \boldsymbol{n}_{TF})_F = (\nabla q_{\underline{v}_T}, \nabla w)_T,
\end{aligned}$$

where the removal of the $L^2(F)$-projectors in the third line is justified since $\nabla w \cdot \boldsymbol{n}_{TF} \in \mathbb{P}^0(F)$ for all $F \in \mathcal{F}_T$, and the conclusion follows integrating by parts. Since this relation holds for all $w \in \mathbb{P}^1(T)$, we infer that $\nabla \tilde{\mathrm{p}}_T^1 \underline{v}_T = \nabla q_{\underline{v}_T}$ and thus

$\tilde{\mathrm{p}}_T^1 \underline{v}_T = q_{\underline{v}_T} + C$ for some constant C. Hence, for all $F \in \mathcal{F}_T$,

$$(\tilde{\mathrm{p}}_T^1 \underline{v}_T, 1)_F = (q_{\underline{v}_T} + C, 1)_F = (\pi_F^{0,0} q_{\underline{v}_T} + C, 1)_F = (v_F + C, 1)_F,$$

the introduction of the projector being valid since $1 \in \mathbb{P}^0(F)$. Plugging this relation into the closure equation (5.26d) and using the assumption (5.2b) on the weights yields

$$0 = \sum_{F \in \mathcal{F}_T} \omega_{TF}(C, 1)_F = (C, 1)_T,$$

which proves that $C = 0$. Hence, $\tilde{\mathrm{p}}_T^1 \underline{v}_T = q_{\underline{v}_T}$ and $\tilde{\mathrm{p}}_T^1 \underline{v}_T$ satisfies (5.56).

It remains to show that there can only be one element of $\mathbb{P}^1(T)$ that satisfies this property. By linearity, it suffices to show that if $r \in \mathbb{P}^1(T)$ satisfies $\pi_F^{0,0} r = 0$ for all $F \in \mathcal{F}_T$, then $r = 0$. Integrating by parts and introducing the projectors, we have, by the same arguments as above,

$$(\nabla r, \nabla r)_T = \sum_{F \in \mathcal{F}_T} (r, \nabla r \cdot \boldsymbol{n}_{TF})_F = \sum_{F \in \mathcal{F}_T} (\pi_F^{0,0} r, \nabla r \cdot \boldsymbol{n}_{TF})_F = 0.$$

Hence, r is a constant polynomial and, picking an arbitrary face $F \in \mathcal{F}_T$, we have $r = \pi_F^{0,0} r = 0$, which concludes the proof. □

Let us recall that the patched potential reconstruction $\tilde{\mathrm{p}}_h^1 : \underline{U}_h^{0,-1} \to \mathbb{P}^1(\mathcal{T}_h)$ is defined by: For all $\underline{v}_h \in \underline{U}_h^{0,-1}$,

$$(\tilde{\mathrm{p}}_h^1 \underline{v}_h)_{|T} := \tilde{\mathrm{p}}_T^1 \underline{v}_T \qquad \forall T \in \mathcal{T}_h. \tag{5.57}$$

Lemma 5.20 (Isomorphism Between $\underline{U}_{h,0}^{0,-1}$ and $V_{h,0}^{\mathrm{nc}}$) *The patched potential reconstruction $\tilde{\mathrm{p}}_h^1$ is an isomorphism between $\underline{U}_{h,0}^{0,-1}$ and $V_{h,0}^{\mathrm{nc}}$.*

Proof If $\underline{v}_h \in \underline{U}_{h,0}^{0,-1}$ then, by (5.56), for all $F \in \mathcal{F}_h^{\mathrm{i}}$ with $\mathcal{T}_F = \{T_1, T_2\}$ numbered according to the definition (1.22) of the jump across F, we have

$$\pi_F^{0,0}[\tilde{\mathrm{p}}_h^1 \underline{v}_h]_F = \pi_F^{0,0} \tilde{\mathrm{p}}_{T_1}^1 \underline{v}_{T_1} - \pi_F^{0,0} \tilde{\mathrm{p}}_{T_2}^1 \underline{v}_{T_2} = v_F - v_F = 0.$$

Similarly, if $F \in \mathcal{F}_h^{\mathrm{b}}$ with $\mathcal{T}_F = \{T\}$,

$$\pi_F^{0,0}[\tilde{\mathrm{p}}_h^1 \underline{v}_h]_F = \pi_F^{0,0} \tilde{\mathrm{p}}_T^1 \underline{v}_T = v_F = 0$$

by the boundary condition embedded in $\underline{U}_{h,0}^{0,-1}$. Hence, $\tilde{\mathrm{p}}_h^1 \underline{v}_h \in V_{h,0}^{\mathrm{nc}}$.

Let us show that $\tilde{\mathrm{p}}_h^1$ is one-to-one. If $\underline{v}_h \in \underline{U}_{h,0}^{0,-1}$ is such that $\tilde{\mathrm{p}}_h^1 \underline{v}_h = 0$ then, for all $F \in \mathcal{F}_h$, picking an element $T \in \mathcal{T}_F$, we have $\tilde{\mathrm{p}}_T^1 \underline{v}_T = 0$ and thus, by (5.56),

$$0 = \pi_F^{0,0} \tilde{\mathrm{p}}_T^1 \underline{v}_T = v_F.$$

Hence, all face values of $\underline{v}_h$ vanish. By definition (5.1b) of $\underline{U}_T^{0,-1}$, we deduce that all elements values of $\underline{v}_h$ also vanish, and thus that $\underline{v}_h = \underline{0}$. This proves that $\tilde{\mathrm{p}}_h^1$ is one-to-one.

It remains to prove that $\tilde{\mathrm{p}}_h^1$ is onto. Let $w \in V_{h,0}^{\mathrm{nc}}$. By the continuity condition embedded into $V_{h,0}^{\mathrm{nc}}$, the quantity $v_F = \pi_F^{0,0} w$ is single-valued for any face $F \in \mathcal{F}_h$, and vanishes for boundary faces. These values $(v_F)_{F \in \mathcal{F}_h}$ define a vector $\underline{v}_h \in \underline{U}_{h,0}^{0,-1}$, the element values being reconstructed from the face values. For $T \in \mathcal{T}_h$, by (5.56), $\tilde{\mathrm{p}}_T^1 \underline{v}_T$ and $w_{|T}$ are two polynomials in $\mathbb{P}^1(T)$ whose average value on each $F \in \mathcal{F}_T$ is equal to v_F. Lemma 5.19 thus shows that $\tilde{\mathrm{p}}_T^1 \underline{v}_T = w_{|T}$. Since this is true for all $T \in \mathcal{T}_h$, this proves that $\tilde{\mathrm{p}}_h^1 \underline{v}_h = w$, and thus that $\tilde{\mathrm{p}}_h^1$ is onto. □

5.2.3 Link with HHO(0, −1)

The following theorem establishes a link between the HHO(0, −1) scheme and the nonconforming $\mathbb{P}^1$ Finite Element scheme for the Poisson problem. Specifically, it shows that these two schemes are equivalent, up to a modification of the source term.

Theorem 5.21 (Link Between the HHO(0, −1) Scheme and the Nonconforming $\mathbb{P}^1$ Finite Element Scheme) *Let $\mathcal{M}_h$ be a matching simplicial mesh as in Definition 1.7, and let us consider the following modification of the HHO(0, −1) scheme (5.39), in which only the source term is changed: Find $\underline{u}_h \in \underline{U}_{h,0}^{0,-1}$ such that*

$$\mathrm{a}_h(\underline{u}_h, \underline{v}_h) = (f, \tilde{\mathrm{p}}_h^1 \underline{v}_h) \qquad \forall \underline{v}_h \in \underline{U}_{h,0}^{0,-1}, \tag{5.58}$$

with a_h satisfying (5.39b).

Then, $\underline{u}_h \in \underline{U}_{h,0}^{0,-1}$ is the solution of (5.58) if and only if $\tilde{\mathrm{p}}_h^1 \underline{u}_h$ defined by (5.57) is the solution of the nonconforming $\mathbb{P}^1$ Finite Element scheme (5.55).

Proof We first show that, if T is a simplicial element and $(k, \ell) = (0, -1)$, any local symmetric stabilisation form s_T satisfying (S3) in Assumption 5.9 vanishes.

Following the proof of Lemma 2.11 it is easily established that s_T depends on its arguments only through the difference operators (5.31). It therefore suffices to see that these difference operators are identically zero on $\underline{U}_{h,0}^{0,-1}$. Since $\mathbb{P}^{-1}(T) = \{0\}$ we immediately have $\tilde{\delta}_T^{-1} \equiv 0$. Taking $F \in \mathcal{F}_T$, the relation (5.56) gives $\tilde{\delta}_{TF}^{0}\underline{v}_T = 0$ for all $\underline{v}_T \in \underline{U}_T^{0,-1}$. This concludes the proof that $\mathrm{s}_T \equiv 0$.

On a matching simplicial mesh, the HHO(0, −1) scheme (5.58) therefore does not have any stabilisation term and is written: Find $\underline{u}_h \in \underline{U}_{h,0}^{0,-1}$ such that

$$\sum_{T\in\mathcal{T}_h} (\nabla\tilde{\mathrm{p}}_T^1\underline{u}_T, \nabla\tilde{\mathrm{p}}_T^1\underline{v}_T)_T = (f, \tilde{\mathrm{p}}_h^1\underline{v}_h) \qquad \forall \underline{v}_h \in \underline{U}_{h,0}^{0,-1}.$$

Exploit the surjectivity property of $\tilde{\mathrm{p}}_h^1$ and set $w_h = \tilde{\mathrm{p}}_h^1\underline{v}_h$ to recast this equation into

$$(\nabla_h\tilde{\mathrm{p}}_h^1\underline{u}_h, \nabla_h w_h)_T = (f, w_h) \qquad \forall w_h \in V_{h,0}^{\mathrm{nc}}.$$

This exactly states that $\tilde{\mathrm{p}}_h^1\underline{u}_h \in V_{h,0}^{\mathrm{nc}}$ is the solution to (5.55). □

5.3 Hybrid Mimetic Mixed Method

The Hybrid Mimetic Mixed (HMM) method is a family of schemes, introduced in [176], that encompasses Hybrid Finite Volumes [190], low-order mixed/hybrid Mimetic Finite Differences [82], and Mixed Finite Volumes [172]. We prove in this section that, in most instances, the HMM method is equivalent to the lowest order HHO method (with $k = 0$). In what follows, if $X \subset \mathbb{R}^n$, we identify $\mathbb{P}^0(X)$ with $\mathbb{R}$ so that, in particular,

$$\underline{U}_h^0 = \left\{\underline{v}_h = ((v_T)_{T\in\mathcal{T}_h}, (v_F)_{F\in\mathcal{F}_h}) \ :\ v_T \in \mathbb{R} \quad \forall T \in \mathcal{T}_h \text{ and } v_F \in \mathbb{R} \quad \forall F \in \mathcal{F}_h\right\}.$$

5.3.1 *The HMM Method*

Let $\mathcal{M}_h = (\mathcal{T}_h, \mathcal{F}_h)$ be a polytopal mesh as in Definition 1.4, and let $(\boldsymbol{x}_T)_{T\in\mathcal{T}_h}$ be a family of points such that each T is star-shaped with respect to $\boldsymbol{x}_T$. Let T be a mesh element, and recall the definition (2.55) of the boundary space $\underline{D}_{\partial T}^k$ with $k = 0$, that is to say

$$\underline{D}_{\partial T}^0 := \left\{\underline{\alpha}_{\partial T} = (\alpha_{TF})_{F\in\mathcal{F}_T} \ :\ \alpha_{TF} \in \mathbb{R} \quad \forall F \in \mathcal{F}_T\right\}.$$

We define the operators $\boldsymbol{\nabla}_T^{\rm HMM} : \underline{U}_T^0 \to \mathbb{R}^d$ and $\mathfrak{d}_{\partial T} = (\mathfrak{d}_{TF})_{F\in\mathcal{F}_T} : \underline{U}_T^0 \to \underline{D}_{\partial T}^0$ by: For all $\underline{v}_T \in \underline{U}_T^0$,

$$\boldsymbol{\nabla}_T^{\rm HMM}\underline{v}_T := \frac{1}{|T|_d}\sum_{F\in\mathcal{F}_T}|F|_{d-1}v_F\boldsymbol{n}_{TF} \tag{5.59}$$

and

$$\mathfrak{d}_{TF}\underline{v}_T = v_F - v_T - \boldsymbol{\nabla}_T^{\rm HMM}\underline{v}_T\cdot(\overline{\boldsymbol{x}}_F - \boldsymbol{x}_T), \tag{5.60}$$

where $\overline{\boldsymbol{x}}_F$ is the centre of mass of F.

An HMM scheme for the Poisson problem is built as follows. For each $T \in \mathcal{T}_h$, we take a symmetric positive definite form $\mathrm{s}_T^{\rm HMM}$ on $\underline{D}_{\partial T}^0$, and we look for $\underline{u}_h \in \underline{U}_h^0$ such that

$$\sum_{T\in\mathcal{T}_h}|T|_d\boldsymbol{\nabla}_T^{\rm HMM}\underline{u}_T\cdot\boldsymbol{\nabla}_T^{\rm HMM}\underline{v}_T + \sum_{T\in\mathcal{T}_h}\mathrm{s}_T^{\rm HMM}(\mathfrak{d}_{\partial T}\underline{u}_T, \mathfrak{d}_{\partial T}\underline{v}_T) = \sum_{T\in\mathcal{T}_h}\int_T f v_T, \qquad \forall \underline{v}_h \in \underline{U}_h^0. \tag{5.61}$$

In the design and analysis of the HMM method, it is assumed that each element T is star-shaped with respect to all points in a ball centred at $\boldsymbol{x}_T$ and of radius $\gtrsim h_T$, with hidden constant independent of h and T. Combined with the mesh regularity assumptions in Definition 1.9, the coercivity and boundedness imposed on the bilinear form $\mathrm{s}_T^{\rm HMM}$ are then equivalent to: For all $\underline{v}_T \in \underline{U}_T^0$,

$$\|\boldsymbol{\nabla}_T^{\rm HMM}\underline{v}_T\|_T^2 + \mathrm{s}_T^{\rm HMM}(\mathfrak{d}_{\partial T}\underline{v}_T, \mathfrak{d}_{\partial T}\underline{v}_T) \simeq \|\underline{v}_T\|_{1,T}^2, \tag{5.62}$$

where the hidden constants are independent of h, T and $\underline{v}_T$, and $\|{\cdot}\|_{1,T}$ is defined by (2.7).

5.3.2 *Equivalence Between HMM and HHO with k = 0*

To establish this equivalence, let us first prove a lemma that relates the local HMM operators to the HHO potential reconstruction and difference operators.

Lemma 5.22 (Link Between Local HMM and HHO Operators) *Let $T \in \mathcal{T}_h$ and recall the definition* (2.19) *(with $k = 0$) of the difference operators δ_T^0 and $(\delta_{TF}^0)_{F\in\mathcal{F}_T}$. Assume that $\boldsymbol{x}_T$ is the centre of mass $\overline{\boldsymbol{x}}_T$ of T. Then, for all $\underline{v}_T \in \underline{U}_T^0$, it holds:*

(i) $\boldsymbol{\nabla}_T^{\rm HMM}\underline{v}_T = \boldsymbol{\nabla}\mathrm{p}_T^1\underline{v}_T$,
(ii) $\mathrm{p}_T^1\underline{v}_T(\boldsymbol{x}) = v_T + \boldsymbol{\nabla}_T^{\rm HMM}\underline{v}_T\cdot(\boldsymbol{x} - \overline{\boldsymbol{x}}_T)$ *for all* $\boldsymbol{x} \in T$,
(iii) $\delta_T^0\underline{v}_T = 0$ *and* $\delta_{TF}^0\underline{v}_T = -\mathfrak{d}_{TF}\underline{v}_T$, *for all* $F \in \mathcal{F}_T$.

Proof (i) Let $\boldsymbol{\xi}$ be an arbitrary vector in $\mathbb{R}^d$ and apply the definition (2.11a) of $\nabla \mathrm{p}_T^1 \underline{v}_T$ with $\underline{v}_T \in \underline{U}_T^0$ and $w(\boldsymbol{x}) = \boldsymbol{\xi}\cdot\boldsymbol{x}$, so that $\nabla w = \boldsymbol{\xi}$. This gives

$$(\nabla \mathrm{p}_T^1 \underline{v}_T, \boldsymbol{\xi})_T = \sum_{F\in\mathcal{F}_T} (v_F, \boldsymbol{\xi}\cdot\boldsymbol{n}_{TF})_F.$$

All the functions in the L^2-inner products on T and F are actually constant, and the equation above can thus be written

$$\begin{aligned}|T|_d \nabla \mathrm{p}_T^1 \underline{v}_T\cdot\boldsymbol{\xi} &= \sum_{F\in\mathcal{F}_T} |F|_{d-1} v_F \boldsymbol{\xi}\cdot\boldsymbol{n}_{TF} = \left(\sum_{F\in\mathcal{F}_T} |F|_{d-1} v_F \boldsymbol{n}_{TF}\right)\cdot\boldsymbol{\xi} \\ &= |T|_d \nabla_T^{\mathrm{HMM}} \underline{v}_T\cdot\boldsymbol{\xi},\end{aligned}$$

the conclusion coming from the definition (5.59) of ∇_T^{HMM}. Simplifying by $|T|_d$ and recalling that $\boldsymbol{\xi}$ is arbitrary in $\mathbb{R}^d$, this shows that $\nabla \mathrm{p}_T^1 \underline{v}_T = \nabla_T^{\mathrm{HMM}} \underline{v}_T$ and the proof is complete.

(ii) Since $\nabla \mathrm{p}_T^1 \underline{v}_T = \nabla_T^{\mathrm{HMM}} \underline{v}_T$ and $\mathrm{p}_T^1 \underline{v}_T \in \mathbb{P}^1(T)$, there exists a constant C such that, for all $\boldsymbol{x} \in T$,

$$\mathrm{p}_T^1 \underline{v}_T(\boldsymbol{x}) = C + \nabla_T^{\mathrm{HMM}} \underline{v}_T\cdot\boldsymbol{x}. \tag{5.63}$$

The closure equation (2.11b) on p_T^1 yields $\pi_T^{0,0} \mathrm{p}_T^1 \underline{v}_T = \pi_T^{0,0} v_T = v_T$ and thus

$$v_T = \pi_T^{0,0}(C + \nabla_T^{\mathrm{HMM}} \underline{v}_T\cdot\boldsymbol{x}) = C + \nabla_T^{\mathrm{HMM}} \underline{v}_T\cdot\overline{\boldsymbol{x}}_T,$$

where the conclusion follows using the fact that $\overline{\boldsymbol{x}}_T$ is the centre of mass of T (that is, $\overline{\boldsymbol{x}}_T = \boldsymbol{\pi}_T^{0,0}\boldsymbol{x}$). Hence, $C = v_T - \nabla_T^{\mathrm{HMM}} \underline{v}_T\cdot\overline{\boldsymbol{x}}_T$. Plugged into (5.63) this shows that $\mathrm{p}_T^1 \underline{v}_T(\boldsymbol{x}) = v_T + \nabla_T^{\mathrm{HMM}} \underline{v}_T\cdot(\boldsymbol{x} - \overline{\boldsymbol{x}}_T)$.

(iii) As noticed above, the closure equation (2.11b) gives $\pi_T^{0,0} \mathrm{p}_T^1 \underline{v}_T = \pi_T^{0,0} v_T = v_T$, which readily shows that $\delta_T^0 \underline{v}_T = 0$. We now turn to $\delta_{TF}^0 \underline{v}_T$. Using the result of Point (ii) in the lemma, we have

$$\begin{aligned}\delta_{TF}^0 \underline{v}_T &= \pi_F^{0,0}(\mathrm{p}_T^1 \underline{v}_T - v_F) \\ &= \pi_F^{0,0}(v_T + \nabla_T^{\mathrm{HMM}} \underline{v}_T\cdot(\boldsymbol{x} - \overline{\boldsymbol{x}}_T) - v_F) \\ &= v_T + \nabla_T^{\mathrm{HMM}} \underline{v}_T\cdot(\overline{\boldsymbol{x}}_F - \boldsymbol{x}_T) - v_F = -\mathfrak{d}_{TF} \underline{v}_T,\end{aligned}$$

where the third line follows from $\overline{\boldsymbol{x}}_T = \boldsymbol{x}_T$ and from the definition $\overline{\boldsymbol{x}}_F = \pi_F^{0,0}\boldsymbol{x}$ of the centre of mass of F. □

We can now establish the equivalence theorem between the HMM schemes for the Poisson problem, and the lowest order HHO schemes for this model.

Theorem 5.23 (Equivalence Between HMM and HHO with $k = 0$) *Let $\mathcal{M}_h$ be a polytopal mesh as in Definition 1.4, and assume that the points $(\boldsymbol{x}_T)_{T\in\mathcal{T}_h}$ are the centres of mass of the mesh elements. Then, the HMM scheme (5.61) and the HHO scheme (2.48) with $k = 0$ are equivalent, in the sense that:*

(i) for any choice of HMM stabilisation forms $(\mathrm{s}_T^{\rm HMM})_{T\in\mathcal{T}_h}$, there is a choice of HHO stabilisation forms $(\mathrm{s}_T)_{T\in\mathcal{T}_h}$ such that (5.61) and (2.48) are the same equations, and
(ii) for any choice of HHO stabilisation forms $(\mathrm{s}_T)_{T\in\mathcal{T}_h}$, there is a choice of HMM stabilisation forms $(\mathrm{s}_T^{\rm HMM})_{T\in\mathcal{T}_h}$ such that (5.61) and (2.48) are the same equations.

Proof An inspection of the HHO scheme (2.48) (where a_h is given by (2.39) with each a_T defined by (2.15)) and of the HMM scheme (5.61) shows that the result of the theorem holds if we can prove that, for all $T \in \mathcal{T}_h$ and all $\underline{u}_T, \underline{v}_T \in \underline{U}_T^0$,

$$(\nabla \mathrm{p}_T^1 \underline{u}_T, \nabla \mathrm{p}_T^1 \underline{v}_T)_T + \mathrm{s}_T(\underline{u}_T, \underline{v}_T) = |T|_d \nabla_T^{\rm HMM} \underline{u}_T \cdot \nabla_T^{\rm HMM} \underline{v}_T + \mathrm{s}_T^{\rm HMM}(\mathfrak{d}_{\partial T} \underline{u}_T, \mathfrak{d}_{\partial T} \underline{v}_T). \tag{5.64}$$

By Point (i) in Lemma 5.22 the first terms on each side are identical, and we therefore only have to prove that for any choice of HMM stabilisation form $\mathrm{s}_T^{\rm HMM}$ (resp. any choice of HHO stabilisation form s_T), there is an HHO stabilisation form s_T (resp. an HMM stabilisation form $\mathrm{s}_T^{\rm HMM}$) such that

$$\mathrm{s}_T(\underline{u}_T, \underline{v}_T) = \mathrm{s}_T^{\rm HMM}(\mathfrak{d}_{\partial T} \underline{u}_T, \mathfrak{d}_{\partial T} \underline{v}_T). \tag{5.65}$$

(i) *From HMM to HHO*. Let $\mathrm{s}_T^{\rm HMM}$ be an HMM stabilisation form, and define s_T by (5.65). Then, s_T clearly satisfies (S1) in Assumption 2.4. The property (S2) comes straight from (5.62) since (5.65) ensures (5.64). Finally, (S3) is a consequence of $\mathfrak{d}_{TF} = -\delta_{TF}^0$ (see Point (iii) in Lemma 5.22) and of the polynomial consistency (2.21) of this difference operator.

(ii) *From HHO to HMM*. Take now a stabilisation form s_T that satisfies Assumption 2.4. By Lemma 2.11, it only depends on its arguments through the difference operators (2.19) which means, owing to Point (iii) in Lemma 5.22, that it only depends on its arguments through $\mathfrak{d}_{\partial T} = (\mathfrak{d}_{TF})_{F\in\mathcal{F}_T}$. Thus, there exists a bilinear form $\mathrm{s}_T^{\rm HMM}$ on $\underline{D}_{\partial T}^0$ such that (5.65) holds. The property (S2) in Assumption 2.4 and (5.64) (consequence of (5.65)) then show that $\mathrm{s}_T^{\rm HMM}$ satisfies (5.62). □

Remark 5.24 (Limits of the Equivalence Between HMM and HHO) Although the algebraic description of the HMM method does not require each element T to be star-shaped with respect to $\boldsymbol{x}_T$, its analysis (e.g. in [176] or [178, Chapter 13]) is always performed under this assumption, which is not imposed for the convergence analysis of HHO schemes (see Definition 1.9). The HHO method with $k = 0$ can thus be considered as an extension of the HMM method to meshes made of possibly non-star-shaped elements.

On the other hand, the design and analysis of HMM does not require each $\boldsymbol{x}_T$, for $T \in \mathcal{T}_h$, to be the centre of mass of T. As a consequence of this relative freedom of choice for the element point, on certain meshes the HMM family contains the Two-Point Flux Approximation (TPFA) finite volume scheme [178, Section 13.3], a historical and popular scheme in fluid mechanics; this inclusion of TPFA into HMM enabled the proof of a superconvergence result for the TPFA method [174]. Additionally, the convergence analysis of HMM can be carried out, exploiting the low order of the method, on sequences of meshes that are not regular in the sense of Definition 1.9 (in particular because they have faces that are very small compared to their neighbouring elements); see [178, Chapter 13] and also Remark 1.11 on degenerate faces.

5.4 The Mixed High-Order Method

In this section, we discuss the Mixed High-Order (MHO) method for the Poisson problem originally introduced in [149] and we show that, after hybridisation and local elimination of the flux variables, it coincides with the HHO scheme (2.48). This link between MHO and HHO methods was first highlighted in [9]; see also [56], where an equivalence between mixed and primal formulations is established for a large set of related classical and new generation discretisation methods.

5.4.1 The Poisson Problem in Mixed Formulation

Mixed methods for the Poisson problem use as a starting point a formulation where the flux and the potential appear as separate unknowns. Specifically, a classical mixed formulation of the homogeneous Dirichlet problem (2.1) consists in seeking $\boldsymbol{\sigma} : \Omega \to \mathbb{R}^d$ and $u : \Omega \to \mathbb{R}$ such that

$$\boldsymbol{\sigma} + \nabla u = 0 \qquad \text{in } \Omega, \tag{5.66a}$$

$$\nabla \cdot \boldsymbol{\sigma} = f \qquad \text{in } \Omega. \tag{5.66b}$$

Problem (5.66) admits a straightforward physical interpretation: Eq. (5.66b) represents an infinitesimal balance of fluxes, while Eq. (5.66a) is the linear constitutive law linking the potential and the flux. Defining the spaces

$$\boldsymbol{\Sigma} := \boldsymbol{H}(\mathrm{div};\Omega), \qquad U := L^2(\Omega),$$

a classical weak formulation of problem (5.66) reads: Find $(\boldsymbol{\sigma}, u) \in \boldsymbol{\Sigma} \times U$ such that

$$m(\boldsymbol{\sigma},\boldsymbol{\tau}) + b(\boldsymbol{\tau},u) = 0 \qquad \forall \boldsymbol{\tau} \in \boldsymbol{\Sigma}, \tag{5.67a}$$

$$-b(\boldsymbol{\sigma},v) = (f,v) \qquad \forall v \in U, \tag{5.67b}$$

where the bilinear forms $m : \boldsymbol{\Sigma} \times \boldsymbol{\Sigma} \to \mathbb{R}$ and $b : \boldsymbol{\Sigma} \times U \to \mathbb{R}$ are such that, for all $\boldsymbol{\sigma}, \boldsymbol{\tau} \in \boldsymbol{\Sigma}$ and all $q \in U$,

$$m(\boldsymbol{\sigma},\boldsymbol{\tau}) := (\boldsymbol{\sigma},\boldsymbol{\tau}), \qquad b(\boldsymbol{\tau},q) := -(\nabla\cdot\boldsymbol{\tau},q). \tag{5.68}$$

5.4.2 Local Spaces of Discrete Unknowns

Let a mesh element $T \in \mathcal{T}_h$ be fixed and, for any integer $l \geq 0$ set, for the sake of brevity,

$$\mathbb{G}_T^l := \nabla\mathbb{P}^{l+1}(T) \subset \mathbb{P}^l(T)^d.$$

We define the local space of discrete flux unknowns (see Fig. 5.8):

$$\underline{\boldsymbol{\Sigma}}_T^k := \left\{ \underline{\boldsymbol{\tau}}_T = (\boldsymbol{\tau}_T, (\tau_{TF})_{F\in\mathcal{F}_T}) \;:\; \boldsymbol{\tau}_T \in \mathbb{G}_T^{k-1} \text{ and } \tau_{TF} \in \mathbb{P}^k(F) \quad \forall F \in \mathcal{F}_T \right\}.$$

The corresponding interpolator $\underline{\boldsymbol{I}}_{\boldsymbol{\Sigma},T}^k : H^1(T)^d \to \underline{\boldsymbol{\Sigma}}_T^k$ is such that, for any $\boldsymbol{\tau} \in H^1(T)^d$,

$$\underline{\boldsymbol{I}}_{\boldsymbol{\Sigma},T}^k \boldsymbol{\tau} := (\boldsymbol{\pi}_{\mathbb{G},T}^{k-1}\boldsymbol{\tau}, (\pi_F^{0,k}(\boldsymbol{\tau}\cdot\boldsymbol{n}_{TF}))_{F\in\mathcal{F}_T}), \tag{5.69}$$

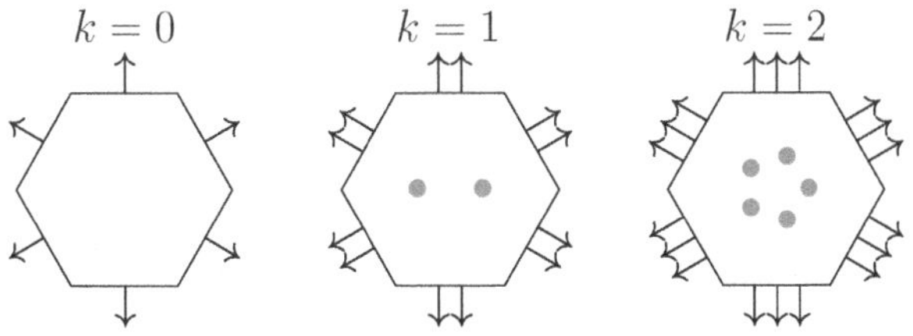

Fig. 5.8 Discrete unknowns in $\underline{\boldsymbol{\Sigma}}_T^k$ for $k \in \{0, 1, 2\}$

where $\boldsymbol{\pi}_{\mathbb{G},T}^{k-1}$ denotes the L^2-orthogonal projector on $\mathbb{G}_T^{k-1}$ such that

$$(\boldsymbol{\pi}_{\mathbb{G},T}^{k-1}\boldsymbol{\tau} - \boldsymbol{\tau}, \boldsymbol{v})_T = 0 \qquad \forall \boldsymbol{v} \in \mathbb{G}_T^{k-1}. \tag{5.70}$$

We equip $\underline{\boldsymbol{\Sigma}}_T^k$ with the following $L^2(T)^d$-like norm: For any $\underline{\boldsymbol{\tau}}_T \in \underline{\boldsymbol{\Sigma}}_T^k$,

$$\|\underline{\boldsymbol{\tau}}_T\|_{\boldsymbol{\Sigma},T} := \left(\|\boldsymbol{\tau}_T\|_T^2 + \sum_{F\in\mathcal{F}_T} h_F \|\tau_{TF}\|_F^2 \right)^{\frac{1}{2}}. \tag{5.71}$$

5.4.3 *Local Divergence and Flux Reconstructions*

The MHO method hinges on local reconstructions of the divergence and of the flux. The divergence reconstruction is inspired by the following integration by parts formula in the spirit of Sect. 2.1.1: For all $\boldsymbol{\tau} \in H^1(T)^d$ and all $v \in C^\infty(\overline{T})$,

$$(\nabla\cdot\boldsymbol{\tau}, v)_T = -(\boldsymbol{\tau}, \nabla v)_T + \sum_{F\in\mathcal{F}_T} (\boldsymbol{\tau}\cdot\boldsymbol{n}_{TF}, v)_F.$$

Specialising this formula to $v \in \mathbb{P}^k(T)$ and using the definition (1.57) of the L^2-orthogonal projectors on $\mathbb{P}^k(T)$ and $\mathbb{P}^k(F)$ along with the definition (5.70) of the L^2-orthogonal projector on $\mathbb{G}_T^{k-1}$ to insert them into the products, we get

$$(\pi_T^{0,k}(\nabla\cdot\boldsymbol{\tau}), v)_T = -(\boldsymbol{\pi}_{\mathbb{G},T}^{k-1}\boldsymbol{\tau}, \nabla v)_T + \sum_{F\in\mathcal{F}_T} (\pi_F^{0,k}(\boldsymbol{\tau}\cdot\boldsymbol{n}_{TF}), v)_F.$$

Inspired by this formula, we define the local divergence operator $\mathrm{D}_T^k : \underline{\boldsymbol{\Sigma}}_T^k \to \mathbb{P}^k(T)$ such that, for any $\underline{\boldsymbol{\tau}}_T \in \underline{\boldsymbol{\Sigma}}_T^k$,

$$(\mathrm{D}_T^k\underline{\boldsymbol{\tau}}_T, v)_T = -(\boldsymbol{\tau}_T, \nabla v)_T + \sum_{F\in\mathcal{F}_T} (\tau_{TF}, v)_F \qquad \forall v \in \mathbb{P}^k(T). \tag{5.72}$$

Existence and uniqueness of $\mathrm{D}_T^k\underline{\boldsymbol{\tau}}_T$ immediately follow from the Riesz representation theorem in $\mathbb{P}^k(T)$ for the $L^2(T)$-inner product. By construction we have, for all $\boldsymbol{\tau} \in H^1(T)^d$,

$$\mathrm{D}_T^k\underline{\boldsymbol{I}}_{\boldsymbol{\Sigma},T}^k\boldsymbol{\tau} = \pi_T^{0,k}(\nabla\cdot\boldsymbol{\tau}). \tag{5.73}$$

This commutation property is illustrated in Fig. 5.9.

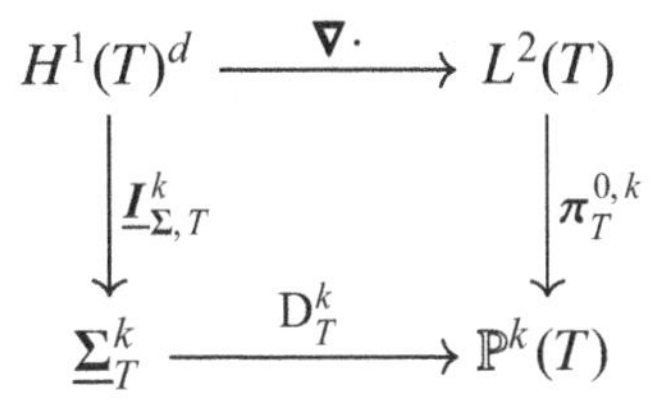

Fig. 5.9 Illustration of the commutation property (5.73) of D^k_T

We next introduce the flux reconstruction operator $\mathbf{F}^k_T : \underline{\Sigma}^k_T \to \mathbb{G}^k_T$ such that, for any $\underline{\tau}_T \in \underline{\Sigma}^k_T$,

$$(\mathbf{F}^k_T\underline{\tau}_T, \nabla w)_T = -(\mathrm{D}^k_T\underline{\tau}_T, w)_T + \sum_{F\in\mathcal{F}_T} (\tau_{TF}, w)_F \qquad \forall w \in \mathbb{P}^{k+1}(T). \tag{5.74}$$

Once again, existence and uniqueness of $\mathbf{F}^k_T\underline{\tau}_T$ are a consequence of the Riesz representation theorem in $\mathbb{G}^k_T$ for the $L^2(T)^d$-inner product. The flux reconstruction is designed to satisfy the following polynomial consistency property: For all $v \in \mathbb{P}^{k+1}(T)$,

$$\mathbf{F}^k_T\underline{I}^k_{\Sigma,T}\nabla v = \nabla v, \tag{5.75}$$

as can be checked writing (5.74) for $\underline{\tau}_T = \underline{I}^k_{\Sigma,T}\nabla v$, observing that $\mathrm{D}^k_T\underline{I}^k_{\Sigma,T}\nabla v = \pi^{0,k}_T(\Delta v) = \Delta v$ owing to (5.73) along with $\Delta v \in \mathbb{P}^{k-1}(T)$, that $\pi^{0,k}_F((\nabla v)_{|T}\cdot\boldsymbol{n}_{TF}) = (\nabla v)_{|T}\cdot\boldsymbol{n}_{TF} \in \mathbb{P}^k(F)$ for all $F \in \mathcal{F}_T$, and integrating by parts the right-hand side.

Remark 5.25 (Commutation Property for $\mathbf{F}^k_T$) Contrary to D^k_T, the operator $\mathbf{F}^k_T$ does not enjoy a general commutation property with the interpolator $\underline{I}^k_{\Sigma,T}$ and the L^2-orthogonal projection on $\mathbb{G}^k_T$. Indeed, if $v \in H^1(T)$ is a general function then (5.74) with $\underline{\tau}_T = \underline{I}^k_{\Sigma,T}\nabla v$ leads to

$$(\mathbf{F}^k_T\underline{I}^k_{\Sigma,T}\nabla v, \nabla w)_T = -(\pi^{0,k}_T(\Delta v), w)_T + \sum_{F\in\mathcal{F}_T} (\pi^{0,k}_F(\nabla v\cdot\boldsymbol{n}_{TF}), w)_F. \tag{5.76}$$

In this equation, however, $w \in \mathbb{P}^{k+1}(T)$ and the orthogonal projectors $\pi^{0,k}_T$ and $\pi^{0,k}_F$ therefore cannot be removed, as would be required for the commutation property to hold. Recalling Proposition 1.35, property (5.75) ensures, however, that $(\mathbf{F}^k_T \circ \underline{I}^k_{\Sigma,T})$ is a projector on $\mathbb{G}^k_T$ (albeit different from the L^2-orthogonal projector).

5.4.4 Local Bilinear Forms

The discrete local versions of the continuous bilinear forms m and b defined by (5.68) are the bilinear forms $\mathrm{m}_T : \underline{\Sigma}^k_T \times \underline{\Sigma}^k_T \to \mathbb{R}$ and $\mathrm{b}_T : \underline{\Sigma}^k_T \times \mathbb{P}^k(T) \to \mathbb{R}$

such that, for any $\underline{\sigma}_T, \underline{\tau}_T \in \underline{\Sigma}_T^k$ and any $v \in \mathbb{P}^k(T)$,

$$\begin{aligned} \mathrm{m}_T(\underline{\sigma}_T, \underline{\tau}_T) &:= (\mathbf{F}_T^k\underline{\sigma}_T, \mathbf{F}_T^k\underline{\tau}_T)_T + \mathrm{s}_{\Sigma,T}(\underline{\sigma}_T, \underline{\tau}_T), \\ \mathrm{b}_T(\underline{\tau}_T, v) &:= -(\mathrm{D}_T^k\underline{\tau}_T, v)_T. \end{aligned} \tag{5.77}$$

The first term in m_T is a consistent contribution mimicking the L^2-product of fluxes, while the second is a stabilisation contribution which satisfies the following assumption, originally proposed in [56].

Assumption 5.26 (MHO Stabilisation Bilinear Form) *The local stabilisation bilinear form* $\mathrm{s}_{\Sigma,T} : \underline{\Sigma}_T^k \times \underline{\Sigma}_T^k \to \mathbb{R}$ *satisfies the following properties:*

(SM1) Symmetry and positivity. $\mathrm{s}_{\Sigma,T}$ *is symmetric and positive semidefinite;*
(SM2) Stability and boundedness. *It holds, for all* $\underline{\tau}_T \in \underline{\Sigma}_T^k$,

$$\mathrm{m}_T(\underline{\tau}_T, \underline{\tau}_T)^{\frac{1}{2}} \simeq \|\underline{\tau}_T\|_{\Sigma,T}, \tag{5.78}$$

with hidden constant independent of h and T.
(SM3) Polynomial consistency. *For all* $w \in \mathbb{P}^{k+1}(T)$ *and all* $\underline{\tau}_T \in \underline{\Sigma}_T^k$, *it holds*

$$\mathrm{s}_{\Sigma,T}(\underline{I}_{\Sigma,T}^k \nabla w, \underline{\tau}_T) = 0. \tag{5.79}$$

The classical MHO stabilisation originally introduced in [149] is obtained setting, for all $\underline{\sigma}_T, \underline{\tau}_T \in \underline{\Sigma}_T^k$,

$$\mathrm{s}_{\Sigma,T}(\underline{\sigma}_T, \underline{\tau}_T) = \sum_{F \in \mathcal{F}_T} h_F(\mathbf{F}_T^k\underline{\sigma}_T \cdot \boldsymbol{n}_{TF} - \sigma_{TF}, \mathbf{F}_T^k\underline{\tau}_T \cdot \boldsymbol{n}_{TF} - \tau_{TF})_F.$$

(SM1) can be checked by inspection, while (SM3) is an immediate consequence of the polynomial consistency (5.75) of the flux reconstruction operator. The proof of (SM2) can be found in [149, Lemma 4].

5.4.5 *Global Spaces of Discrete Unknowns and Discrete Problem*

The global space of discrete flux unknowns is defined as follows:

$$\underline{\Sigma}_h^k := \left\{ \underline{\tau}_h = (\underline{\tau}_T)_{T \in \mathcal{T}_h} \; : \; \underline{\tau}_T \in \underline{\Sigma}_T^k \quad \forall T \in \mathcal{T}_h \text{ and } \sum_{T \in \mathcal{T}_F} \tau_{TF} = 0 \quad \forall F \in \mathcal{F}_h^{\mathrm{i}} \right\},$$

where we remind the reader that, for any $F \in \mathcal{F}_h$, $\mathcal{T}_F$ collects the mesh elements to which F belongs; see (1.2). Notice that the condition on the interface unknowns in $\underline{\boldsymbol{\Sigma}}_T^k$ mimics at the discrete level the continuity of the normal trace of the flux; see the discussion in Sect. 2.2.5 and, in particular, (2.50b). The potential is sought in the space of broken polynomials of total degree k on $\mathcal{T}_h$:

$$U_h^k := \mathbb{P}^k(\mathcal{T}_h).$$

Denote by $\mathrm{m}_h : \underline{\boldsymbol{\Sigma}}_h^k \times \underline{\boldsymbol{\Sigma}}_h^k \to \mathbb{R}$ and $\mathrm{b}_h : \underline{\boldsymbol{\Sigma}}_h^k \times U_h^k \to \mathbb{R}$ the global bilinear forms obtained by element by element assembly of the local contributions defined in (5.77), that is, for all $\underline{\boldsymbol{\sigma}}_h, \underline{\boldsymbol{\tau}}_h \in \underline{\boldsymbol{\Sigma}}_h^k$ and all $v_h \in U_h^k$,

$$\mathrm{m}_h(\underline{\boldsymbol{\sigma}}_h, \underline{\boldsymbol{\tau}}_h) := \sum_{T\in\mathcal{T}_h} \mathrm{m}_T(\underline{\boldsymbol{\sigma}}_T, \underline{\boldsymbol{\tau}}_T), \qquad \mathrm{b}_h(\underline{\boldsymbol{\tau}}_h, v_h) := \sum_{T\in\mathcal{T}_h} \mathrm{b}_T(\underline{\boldsymbol{\tau}}_T, v_T),$$

where, for all $T \in \mathcal{T}_h$, $v_T := (v_h)_{|T}$. The global problem reads: Find $(\underline{\boldsymbol{\sigma}}_h, u_h) \in \underline{\boldsymbol{\Sigma}}_h^k \times U_h^k$ such that

$$\mathrm{m}_h(\underline{\boldsymbol{\sigma}}_h, \underline{\boldsymbol{\tau}}_h) + \mathrm{b}_h(\underline{\boldsymbol{\tau}}_h, u_h) = 0 \qquad \forall \underline{\boldsymbol{\tau}}_h \in \underline{\boldsymbol{\Sigma}}_h^k, \tag{5.80a}$$

$$-\mathrm{b}_h(\underline{\boldsymbol{\sigma}}_h, v_h) = (f, v_h) \qquad \forall v_h \in \mathbb{P}^k(\mathcal{T}_h). \tag{5.80b}$$

5.4.6 Hybridisation and Equivalent Primal Formulation

The MHO method (5.80) can be recast into a primal formulation that fits into the framework of Chap. 2 for a specific choice of the local stabilisation bilinear form. To prove it, we proceed in two steps: first, we perform the so-called *hybridisation* of the method, which consists in enforcing the single-valuedness of interface unknowns via Lagrange multipliers; second, we eliminate the flux unknowns by locally inverting the discrete constitutive law inside each element.

5.4.6.1 Hybridisation

Define the space of fully discontinuous discrete flux unknowns

$$\underline{\check{\boldsymbol{\Sigma}}}_h^k := \left\{ \underline{\boldsymbol{\tau}}_h := (\underline{\boldsymbol{\tau}}_T)_{T\in\mathcal{T}_h} \; : \; \underline{\boldsymbol{\tau}}_T \in \underline{\boldsymbol{\Sigma}}_T^k \quad \forall T \in \mathcal{T}_h \right\},$$

and recall the definition (2.32) of the space $\underline{U}_h^k$ of discrete unknowns for the HHO method, and (2.36) of its subspace $\underline{U}_{h,0}^k$ with strongly enforced homogeneous Dirichlet boundary conditions. We additionally define the bilinear form $\check{\mathrm{b}}_h : \underline{\check{\boldsymbol{\Sigma}}}_h^k \times$

$\underline{U}_h^k \to \mathbb{R}$ such that, for all $\underline{\boldsymbol{\tau}}_h \in \check{\underline{\boldsymbol{\Sigma}}}_h^k$ and all $\underline{v}_h \in \underline{U}_h^k$,

$$\check{\mathrm{b}}_h(\underline{\boldsymbol{\tau}}_h, \underline{v}_h) := \sum_{T \in \mathcal{T}_h} \check{\mathrm{b}}_T(\underline{\boldsymbol{\tau}}_T, \underline{v}_T),$$

where, for all $T \in \mathcal{T}_h$,

$$\check{\mathrm{b}}_T(\underline{\boldsymbol{\tau}}_T, \underline{v}_T) := \mathrm{b}_T(\underline{\boldsymbol{\tau}}_T, v_T) + \sum_{F \in \mathcal{F}_T} (\tau_{TF}, v_F)_F \tag{5.81}$$

$$= (\boldsymbol{\tau}_T, \nabla v_T)_T + \sum_{F \in \mathcal{F}_T} (\tau_{TF}, v_F - v_T)_F, \tag{5.82}$$

where we have expanded first b_T then $\mathrm{D}_T^k \underline{\boldsymbol{\tau}}_T$ according to their respective definitions (5.77) and (5.72). The mixed hybrid reformulation of problem (5.80), obtained using Lagrange multipliers to enforce the continuity of discrete boundary flux unknowns, reads: Find $(\underline{\boldsymbol{\sigma}}_h, \underline{u}_h) \in \check{\underline{\boldsymbol{\Sigma}}}_h^k \times \underline{U}_{h,0}^k$ such that

$$\mathrm{m}_T(\underline{\boldsymbol{\sigma}}_T, \underline{\boldsymbol{\tau}}_T) + \check{\mathrm{b}}_T(\underline{\boldsymbol{\tau}}_T, \underline{u}_T) = 0 \qquad \forall T \in \mathcal{T}_h \qquad \forall \underline{\boldsymbol{\tau}}_T \in \underline{\boldsymbol{\Sigma}}_T^k, \tag{5.83a}$$

$$-\check{\mathrm{b}}_h(\underline{\boldsymbol{\sigma}}_h, \underline{v}_h) = (f, v_h) \qquad \forall \underline{v}_h \in \underline{U}_{h,0}^k. \tag{5.83b}$$

Equation (5.83a) locally enforces the discrete constitutive law inside each element. Equation (5.83b), on the other hand, expresses local balances and a global transmission condition.

5.4.6.2 Potential-to-Flux Operator

The next step in order to derive from (5.83) an equivalent primal formulation analogous to (2.48) is to eliminate the flux unknowns by locally inverting (5.83a). Let an element $T \in \mathcal{T}_h$ be fixed. We define the potential-to-flux operator $\underline{\boldsymbol{\varsigma}}_T^k : \underline{U}_T^k \to \underline{\boldsymbol{\Sigma}}_T^k$ such that, for all $\underline{v}_T \in \underline{U}_T^k$,

$$\mathrm{m}_T(\underline{\boldsymbol{\varsigma}}_T^k \underline{v}_T, \underline{\boldsymbol{\tau}}_T) = -\check{\mathrm{b}}_T(\underline{\boldsymbol{\tau}}_T, \underline{v}_T) \qquad \forall \underline{\boldsymbol{\tau}}_T \in \underline{\boldsymbol{\Sigma}}_T^k. \tag{5.84}$$

Lemma 5.27 (Properties of the Potential-to-Flux Operator) *Let a mesh element $T \in \mathcal{T}_h$ be fixed, and let $\mathrm{s}_{\boldsymbol{\Sigma},T}$ denote a bilinear form satisfying Assumption 5.26. Then, the corresponding potential-to-flux operator $\underline{\boldsymbol{\varsigma}}_T^k$ is well-defined and has the following properties:*

(i) Stability and boundedness. *For all $\underline{v}_T \in \underline{U}_T^k$, it holds*

$$\|\underline{\boldsymbol{\varsigma}}_T^k \underline{v}_T\|_{\boldsymbol{\Sigma},T} \simeq \|\underline{v}_T\|_{1,T}, \tag{5.85}$$

with norm $\|\cdot\|_{\Sigma,T}$ *on* $\underline{\mathbf{\Sigma}}_T^k$ *and seminorm* $\|\cdot\|_{1,T}$ *on* $\underline{U}_T^k$ *defined by* (5.71) *and* (2.7)*, respectively, and hidden constants independent of* h*,* T*, and* $\underline{v}_T$*.*

(ii) Polynomial consistency. *For all* $w \in \mathbb{P}^{k+1}(T)$*, it holds*

$$\underline{\boldsymbol{\varsigma}}_T^k \underline{I}_T^k w = -\underline{\boldsymbol{I}}_{\Sigma,T}^k \nabla w. \tag{5.86}$$

(iii) Link with the potential reconstruction operator. *It holds, for all* $\underline{v}_T \in \underline{U}_T^k$*,*

$$\mathbf{F}_T^k \underline{\boldsymbol{\varsigma}}_T^k \underline{v}_T = -\nabla \mathrm{p}_T^{k+1} \underline{v}_T, \tag{5.87}$$

where the local potential reconstruction operator p_T^{k+1} *is defined by* (2.11) *and the local flux reconstruction operator by* (5.74)*. This commutation property is illustrated in Fig. 5.10.*

Proof Owing to assumptions (SM1) and (SM2), m_T defines a scalar product on $\underline{\mathbf{\Sigma}}_T^k$. Hence, the fact that $\underline{\boldsymbol{\varsigma}}_T^k$ is well-defined is once more an application of the Riesz representation theorem applied to $\underline{\mathbf{\Sigma}}_T^k$ equipped with the scalar product defined by m_T.

(i) *Stability and boundedness.* We start by noticing the following boundedness property for $\check{\mathrm{b}}_T$, obtained from (5.82) applying Cauchy–Schwarz inequalities first on the L^2-inner products then on the sums: For all $\underline{\boldsymbol{\tau}}_T \in \underline{\mathbf{\Sigma}}_T^k$ and all $\underline{v}_T \in \underline{U}_T^k$,

$$\begin{aligned}\left|\check{\mathrm{b}}_T(\underline{\boldsymbol{\tau}}_T, \underline{v}_T)\right| &\le \|\boldsymbol{\tau}_T\|_T \|\nabla v_T\|_T + \sum_{F\in\mathcal{F}_T} h_F^{\frac{1}{2}} \|\tau_{TF}\|_F \, h_F^{-\frac{1}{2}} \|v_F - v_T\|_F \\ &\lesssim \|\underline{\boldsymbol{\tau}}_T\|_{\Sigma,T} \|\underline{v}_T\|_{1,T}.\end{aligned} \tag{5.88}$$

Let now $\underline{v}_T \in \underline{U}_T^k$. Using the local norm equivalence expressed by (5.78) followed by the definition (5.84) of the potential-to-flux operator and the above boundedness property for $\check{\mathrm{b}}_T$, we infer

$$\|\underline{\boldsymbol{\varsigma}}_T^k \underline{v}_T\|_{\Sigma,T}^2 \simeq \mathrm{m}_T(\underline{\boldsymbol{\varsigma}}_T^k \underline{v}_T, \underline{\boldsymbol{\varsigma}}_T^k \underline{v}_T) = -\check{\mathrm{b}}_T(\underline{\boldsymbol{\varsigma}}_T^k \underline{v}_T, \underline{v}_T) \lesssim \|\underline{\boldsymbol{\varsigma}}_T^k \underline{v}_T\|_{\Sigma,T} \|\underline{v}_T\|_{1,T},$$

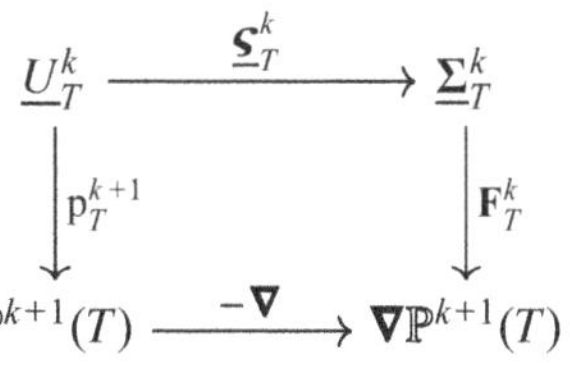

Fig. 5.10 Illustration of the commutation property (5.87)

which yields, after simplification, $\|\underline{\varsigma}_T^k \underline{v}_T\|_{\boldsymbol{\Sigma},T} \lesssim \|\underline{v}_T\|_{1,T}$. To prove the converse inequality, let $\underline{\boldsymbol{\tau}}_T = (\boldsymbol{\tau}_T, (\tau_{TF})_{F\in\mathcal{F}_T})$ be such that $\boldsymbol{\tau}_T = \nabla v_T$ and $\tau_{TF} = h_F^{-1}(v_F - v_T)$ for all $F \in \mathcal{F}_T$, and observe that

$$
\begin{aligned}
\|\underline{v}_T\|_{1,T}^2 &= \check{\mathrm{b}}_T(\underline{\boldsymbol{\tau}}_T, \underline{v}_T)\\
&= -\mathrm{m}_T(\underline{\varsigma}_T^k \underline{v}_T, \underline{\boldsymbol{\tau}}_T)\\
&\lesssim \|\underline{\varsigma}_T^k \underline{v}_T\|_{\boldsymbol{\Sigma},T} \|\underline{\boldsymbol{\tau}}\|_{\boldsymbol{\Sigma},T} = \|\underline{\varsigma}_T^k \underline{v}_T\|_{\boldsymbol{\Sigma},T} \|\underline{v}_T\|_{1,T},
\end{aligned}
$$

where we have used the expression (5.82) of $\check{\mathrm{b}}_T$ along with the definition of $\underline{\boldsymbol{\tau}}_T$ in the first line, the definition (5.84) of $\underline{\varsigma}_T^k$ in the second line, the Cauchy–Schwarz inequality on m_T together with the norm equivalence (5.78) in the third line, and the definition of $\underline{\boldsymbol{\tau}}_T$ and of $\|\cdot\|_{\boldsymbol{\Sigma},T}$ (see (5.71)) to conclude. After simplification, we obtain $\|\underline{v}_T\|_{1,T} \lesssim \|\underline{\varsigma}_T^k \underline{v}_T\|_{\boldsymbol{\Sigma},T}$, which concludes the proof of (5.85).

(ii) *Polynomial consistency.* Let $w \in \mathbb{P}^{k+1}(T)$. Using the definition (5.84) of $\underline{\varsigma}_T^k$ with $\underline{v}_T = \underline{I}_T^k w$, and recalling the definitions (5.81) and (5.77) of $\check{\mathrm{b}}_T$ and b_T, we obtain, for all $\underline{\boldsymbol{\tau}}_T \in \underline{\boldsymbol{\Sigma}}_T^k$,

$$
\begin{aligned}
\mathrm{m}_T(\underline{\varsigma}_T^k \underline{I}_T^k w, \underline{\boldsymbol{\tau}}_T) &= (\pi_T^{0,k} w, \mathrm{D}_T^k \underline{\boldsymbol{\tau}}_T)_T - \sum_{F\in\mathcal{F}_T} (\pi_F^{0,k} w_{|F}, \tau_{TF})_F\\
&= (w, \mathrm{D}_T^k \underline{\boldsymbol{\tau}}_T)_T - \sum_{F\in\mathcal{F}_T} (w, \tau_{TF})_F = -(\nabla w, \mathbf{F}_T^k \underline{\boldsymbol{\tau}}_T)_T,
\end{aligned}
\tag{5.89}
$$

where we have used $\mathrm{D}_T^k \underline{\boldsymbol{\tau}}_T \in \mathbb{P}^k(T)$ and $\tau_{TF} \in \mathbb{P}^k(F)$ for all $F \in \mathcal{F}_T$ along with the definition (1.57) of the projectors to cancel them in the second line, and the definition (5.74) of the flux reconstruction operator to conclude. On the other hand, using the definition (5.77) of m_T followed by the polynomial consistency properties (5.75) of $\mathbf{F}_T^k$ and (5.79) of $\mathrm{s}_{\boldsymbol{\Sigma},T}$, for all $\underline{\boldsymbol{\tau}}_T \in \underline{\boldsymbol{\Sigma}}_T^k$ we have that

$$
\begin{aligned}
\mathrm{m}_T(\underline{\boldsymbol{I}}_{\boldsymbol{\Sigma},T}^k \nabla w, \underline{\boldsymbol{\tau}}_T) &= (\mathbf{F}_T^k \underline{\boldsymbol{I}}_{\boldsymbol{\Sigma},T}^k \nabla w, \mathbf{F}_T^k \underline{\boldsymbol{\tau}}_T)_T + \mathrm{s}_{\boldsymbol{\Sigma},T}(\underline{\boldsymbol{I}}_{\boldsymbol{\Sigma},T}^k \nabla w, \underline{\boldsymbol{\tau}}_T)\\
&= (\nabla w, \mathbf{F}_T^k \underline{\boldsymbol{\tau}}_T)_T.
\end{aligned}
\tag{5.90}
$$

Summing (5.89) and (5.90), we infer that

$$
\mathrm{m}_T(\underline{\varsigma}_T^k \underline{I}_T^k w + \underline{\boldsymbol{I}}_{\boldsymbol{\Sigma},T}^k \nabla w, \underline{\boldsymbol{\tau}}_T) = 0 \qquad \forall \underline{\boldsymbol{\tau}}_T \in \underline{\boldsymbol{\Sigma}}_T^k,
$$

which, since the bilinear form m_T defines an inner product on $\underline{\boldsymbol{\Sigma}}_T^k$, implies (5.86).

(iii) *Link with the potential reconstruction operator.* Let $\underline{v}_T \in \underline{U}_T^k$, $w \in \mathbb{P}^{k+1}(T)$, and set $\underline{\boldsymbol{\tau}}_T := \underline{\boldsymbol{I}}_{\boldsymbol{\Sigma},T}^k \nabla w$. Recalling the definition (5.77) of m_T, and using the polynomial consistency (5.75) of $\mathbf{F}_T^k$ together with (5.79), it is readily inferred that

$$\mathrm{m}_T(\underline{\boldsymbol{\varsigma}}_T^k \underline{v}_T, \underline{\boldsymbol{\tau}}_T) = (\mathbf{F}_T^k \underline{\boldsymbol{\varsigma}}_T^k \underline{v}_T, \nabla w)_T. \tag{5.91}$$

On the other hand, recalling the definition (5.81) of $\check{\mathrm{b}}_T$, we get

$$\begin{aligned}
\check{\mathrm{b}}_T(\underline{\boldsymbol{\tau}}_T, \underline{v}_T) &= -(v_T, \mathrm{D}_T^k \underline{\boldsymbol{\tau}}_T)_T + \sum_{F \in \mathcal{F}_T} (v_F, \tau_{TF})_F \\
&= -(v_T, \pi_T^{0,k}(\Delta w))_T + \sum_{F \in \mathcal{F}_T} (v_F, \pi_F^{0,k}(\nabla w \cdot \boldsymbol{n}_{TF}))_F \\
&= -(v_T, \Delta w)_T + \sum_{F \in \mathcal{F}_T} (v_F, \nabla w \cdot \boldsymbol{n}_{TF})_F = (\nabla \mathrm{p}_T^{k+1} \underline{v}_T, \nabla w)_T,
\end{aligned} \tag{5.92}$$

where we have used the commutation property (5.73) of the discrete divergence operator along with the definition (5.69) of $\underline{\boldsymbol{I}}_{\boldsymbol{\Sigma},T}^k$ in the second line, (1.57) to remove the projectors in the third line, and the definition (2.11) of the local potential reconstruction to conclude. Adding (5.91) to (5.92) and recalling the definition (5.84) of the potential-to-flux operator, we arrive at

$$(\mathbf{F}_T^k \underline{\boldsymbol{\varsigma}}_T^k \underline{v}_T + \nabla \mathrm{p}_T^{k+1} \underline{v}_T, \nabla w)_T = 0 \qquad \forall w \in \mathbb{P}^{k+1}(T),$$

and (5.87) follows after observing that ∇w spans $\mathbb{G}_T^k$ when w spans $\mathbb{P}^{k+1}(T)$. □

5.4.6.3 Equivalence of the Mixed, Mixed Hybrid and Primal Formulations

We next show that the MHO scheme (5.80) is equivalent to the following problem in primal formulation: Find $(\underline{\boldsymbol{\sigma}}_h, \underline{u}_h) \in \check{\underline{\boldsymbol{\Sigma}}}_h^k \times \underline{U}_{h,0}^k$ such that

$$\underline{\boldsymbol{\sigma}}_T = \underline{\boldsymbol{\varsigma}}_T^k \underline{u}_T \qquad \forall T \in \mathcal{T}_h, \tag{5.93a}$$

with $\underline{u}_h \in \underline{U}_{h,0}^k$ solution of

$$\mathrm{a}_h(\underline{u}_h, \underline{v}_h) = (f, v_h) \qquad \forall \underline{v}_h \in \underline{U}_{h,0}^k, \tag{5.93b}$$

where the bilinear form $\mathrm{a}_h : \underline{U}_h^k \times \underline{U}_h^k \to \mathbb{R}$ is such that

$$\mathrm{a}_h(\underline{u}_h, \underline{v}_h) := \sum_{T\in\mathcal{T}_h} \mathrm{a}_T(\underline{u}_T, \underline{v}_T) \text{ with } \mathrm{a}_T(\underline{u}_T, \underline{v}_T) := \mathrm{m}_T(\underline{\varsigma}_T^k \underline{u}_T, \underline{\varsigma}_T^k \underline{v}_T). \tag{5.93c}$$

Theorem 5.28 (Equivalence of the Mixed, Mixed Hybrid and Primal Formulations for the Poisson Problem) *For all $T \in \mathcal{T}_h$, let $\mathrm{s}_{\Sigma,T}$ denote a stabilisation bilinear form matching Assumption 5.26. Let $(\underline{\sigma}_h, \underline{u}_h) \in \check{\underline{\Sigma}}_h^k \times \underline{U}_{h,0}^k$, and let $u_h \in \mathbb{P}^k(\mathcal{T}_h)$ be such that $(u_h)_{|T} = u_T$ for all $T \in \mathcal{T}_h$. Then, the following statements are equivalent:*

(i) $\underline{\sigma}_h \in \underline{\Sigma}_h^k$ and $(\underline{\sigma}_h, u_h)$ solves the mixed problem (5.80)*;*
(ii) $(\underline{\sigma}_h, \underline{u}_h)$ solves the mixed hybrid problem (5.83)*;*
(iii) $(\underline{\sigma}_h, \underline{u}_h)$ solves the primal hybrid problem (5.93).

Proof The equivalence (i) $\Longleftrightarrow$ (ii) classically follows from the theory of Lagrange multipliers. To conclude, it suffices to prove that (ii) $\Longleftrightarrow$ (iii). The equivalence between Eqs. (5.93a) and (5.83a) immediately follows recalling the definition (5.84) of the potential-to-flux operator. As a consequence, it holds for all $T \in \mathcal{T}_h$ and all $\underline{v}_T \in \underline{U}_T^k$,

$$-\check{\mathrm{b}}_T(\underline{\sigma}_T, \underline{v}_T) = -\check{\mathrm{b}}_T(\underline{\varsigma}_T^k \underline{u}_T, \underline{v}_T) = \mathrm{m}_T(\underline{\varsigma}_T^k \underline{u}_T, \underline{\varsigma}_T^k \underline{v}_T) = \mathrm{a}_T(\underline{u}_T, \underline{v}_T),$$

where we have used the definition (5.84) of the potential-to-flux operator together with the symmetry of m_T in the second equality, and the definition (5.93c) of a_T to conclude. Summing this relation over $T \in \mathcal{T}_h$ implies that Eq. (5.93b) is equivalent to (5.83b), thus concluding the proof. □

5.4.7 *Link with the HHO Method*

We are finally ready to show that the primal formulation (5.93) enters the framework of Chap. 2.

Theorem 5.29 (Link Between the Mixed High-Order and Hybrid High-Order Methods) *For all $T \in \mathcal{T}_h$, let $\mathrm{s}_{\Sigma,T}$ denote a bilinear form satisfying Assumption 5.26, and set, for any $\underline{u}_T, \underline{v}_T \in \underline{U}_T^k$,*

$$\mathrm{s}_T(\underline{u}_T, \underline{v}_T) := \mathrm{s}_{\Sigma,T}(\underline{\varsigma}_T^k \underline{u}_T, \underline{\varsigma}_T^k \underline{v}_T), \tag{5.94}$$

(continued)

Theorem 5.29 (continued)
with potential-to-flux operator $\underline{\varsigma}_T^k$ *defined by* (5.84). *Then,*

(i) Properties of s_T. *The stabilisation bilinear forms* s_T, $T \in \mathcal{T}_h$, *satisfy Assumption 2.4.*
(ii) Link with the Hybrid High-Order method. $\underline{u}_h \in \underline{U}_{h,0}^k$ *solves the discrete problem* (2.48) *with stabilisation bilinear forms as in* (5.94) *if and only if* $(\underline{\sigma}_h, \underline{u}_h) \in \check{\underline{\boldsymbol{\Sigma}}}_h^k \times \underline{U}_{h,0}^k$, *with* $\underline{\sigma}_h = (\underline{\varsigma}_T^k \underline{u}_T)_{T \in \mathcal{T}_h}$, *solves the mixed hybrid problem* (5.83).

Proof (i) *Properties of* s_T. Let an element $T \in \mathcal{T}_h$ be fixed. The bilinear form s_T clearly inherits the symmetry and positive semi-definiteness properties of $\mathrm{s}_{\boldsymbol{\Sigma},T}$.
To prove the local seminorm equivalence (2.16) expressing the stability and boundedness of s_T, it suffices to observe that, for all $\underline{v}_T \in \underline{U}_T^k$,

$$\mathrm{a}_T(\underline{v}_T, \underline{v}_T) = \mathrm{m}_T(\underline{\varsigma}_T^k \underline{v}_T, \underline{\varsigma}_T^k \underline{v}_T) \simeq \|\underline{\varsigma}_T^k \underline{v}_T\|_{\boldsymbol{\Sigma},T}^2 \simeq \|\underline{v}_T\|_{1,T}^2,$$

where we have successively used the definition (5.93c) of a_T, (5.78), and the stability and boundedness (5.85) of $\underline{\varsigma}_T^k$.

In order to prove the polynomial consistency property (2.17), we let $w \in \mathbb{P}^{k+1}(T)$ and observe that, for all $\underline{v}_T \in \underline{U}_T^k$, we have

$$\mathrm{s}_T(\underline{I}_T^k w, \underline{v}_T) = \mathrm{s}_{\boldsymbol{\Sigma},T}(\underline{\varsigma}_T^k \underline{I}_T^k w, \underline{\varsigma}_T^k \underline{v}_T) = -\mathrm{s}_{\boldsymbol{\Sigma},T}(\underline{\boldsymbol{I}}_{\boldsymbol{\Sigma},T}^k \nabla w, \underline{\varsigma}_T^k \underline{v}_T) = 0,$$

where we have used the definition (5.94) of s_T and the polynomial consistencies (5.86) and (5.79) of $\underline{\varsigma}_T^k$ and $\mathrm{s}_{\boldsymbol{\Sigma},T}$, respectively.

(ii) *Link with the Hybrid High-Order method.* Compare the primal hybrid scheme (5.93) (recalling the definition (5.77) of m_T and the relation (5.87) between $\mathbf{F}_T^k \underline{\varsigma}_T^k$ and p_T^{k+1}) with the HHO scheme (2.48), and use the equivalence stated in Theorem 5.28 of the primal hybrid scheme with the mixed hybrid problem (5.83). □

5.5 Virtual Elements

In this section, we derive a Virtual Element reformulation of the HHO scheme (2.48) and establish a link with the Nonconforming Virtual Element Method of [27].

We also present the Conforming Virtual Element Method [45, 80], discuss differences with the virtual formulation of the HHO scheme, and develop an analysis in which we establish, in particular, the approximation properties of the relevant projector operator in a non-Hilbertian setting.

In what follows, we denote by $k \geq 0$ a fixed integer corresponding to the polynomial degree of the HHO scheme.

5.5.1 Local Virtual Space

Let an element $T \in \mathcal{T}_h$ be fixed, and define the following space:

$$\mathfrak{U}_T^k := \left\{ \mathfrak{v}_T \in H^1(T) : \Delta \mathfrak{v}_T \in \mathbb{P}^k(T) \text{ and } (\nabla \mathfrak{v}_T)_{|F} \cdot \boldsymbol{n}_{TF} \in \mathbb{P}^k(F) \quad \forall F \in \mathcal{F}_T \right\}. \tag{5.95}$$

It is a simple matter to check that $\mathbb{P}^{k+1}(T) \subset \mathfrak{U}_T^k$. The functions in $\mathfrak{U}_T^k$ are virtual in the sense that, for general polynomial degrees and element shapes, it is not possible (or computationally feasible) to find an explicit expression for use in a Finite Element code. This difficulty is circumvented in the Virtual Element framework by using computable projections to formulate the consistency terms, in conjunction with local stabilisation terms similar to the ones discussed in Sect. 2.1.4 for the HHO method.

Recall the definitions (2.6) of the local space of discrete HHO unknowns, that is,

$$\underline{U}_T^k := \left\{ \underline{v}_T = (v_T, (v_F)_{F \in \mathcal{F}_T}) \ : \ v_T \in \mathbb{P}^k(T) \text{ and } v_F \in \mathbb{P}^k(F) \quad \forall F \in \mathcal{F}_T \right\},$$

and (2.8) of the local interpolator $\underline{I}_T^k : W^{1,1}(T) \to \underline{U}_T^k$ such that, for any $v \in W^{1,1}(T)$,

$$\underline{I}_T^k v := (\pi_T^{0,k} v, (\pi_F^{0,k} v_{|F})_{F \in \mathcal{F}_T}).$$

The following lemma establishes an important link between $\mathfrak{U}_T^k$ and $\underline{U}_T^k$.

Lemma 5.30 (Link Between $\mathfrak{U}_T^k$ and $\underline{U}_T^k$) *The local interpolator $\underline{I}_T^k$ defines a bijective mapping from $\mathfrak{U}_T^k$ to $\underline{U}_T^k$. As a result, the spaces $\mathfrak{U}_T^k$ and $\underline{U}_T^k$ are isomorphic.*

Proof With a small abuse of notation, throughout the proof we still use the symbol $\underline{I}_T^k$ for the linear mapping $\underline{I}_T^k : \mathfrak{U}_T^k \to \underline{U}_T^k$ obtained by restricting the domain of the local interpolator from $W^{1,1}(T)$ to $\mathfrak{U}_T^k \subset W^{1,1}(T)$.

The first part of the proof consists in showing that $\underline{I}_T^k$ is injective so that, by the rank-nullity theorem, we can infer $\dim(\mathfrak{U}_T^k) = \dim(\mathrm{Im}(\underline{I}_T^k)) \leq \dim(\underline{U}_T^k)$. Notice that this implies, in particular, that $\mathfrak{U}_T^k$ has finite dimension. In the second part of the proof, we construct an injective linear mapping $\mathfrak{L}_T^k : \underline{U}_T^k \to \mathfrak{U}_T^k$. Applying again the rank-nullity theorem, this time to $\mathfrak{L}_T^k$, yields $\dim(\underline{U}_T^k) = \dim(\mathrm{Im}(\mathfrak{L}_T^k)) \le \dim(\mathfrak{U}_T^k)$ so that, in conclusion,

$$\dim(\underline{U}_T^k) = \dim(\mathfrak{U}_T^k).$$

Being injective linear mappings between vector spaces with the same finite dimension, both $\underline{I}_T^k$ and $\mathfrak{L}_T^k$ are bijective. Thus, the spaces $\mathfrak{U}_T^k$ and $\underline{U}_T^k$ are isomorphic.

(i) *Injectivity of* $\underline{I}_T^k$. It suffices to prove that $\mathrm{Ker}(\underline{I}_T^k) = \{0\}$. Let $\mathfrak{v}_T \in \mathfrak{U}_T^k$ be such that $\underline{I}_T^k \mathfrak{v}_T = \underline{0}$, that is, $\pi_T^{0,k}\mathfrak{v}_T = 0$ and $\pi_F^{0,k}(\mathfrak{v}_T)_{|F} = 0$ for all $F \in \mathcal{F}_T$. Using an integration by parts, we can write

$$\begin{aligned}
\|\nabla \mathfrak{v}_T\|_T^2 &= -(\mathfrak{v}_T, \Delta\mathfrak{v}_T)_T + \sum_{F\in\mathcal{F}_T} (\mathfrak{v}_T, \nabla\mathfrak{v}_T\cdot\boldsymbol{n}_{TF})_F \\
&= -(\pi_T^{0,k}\mathfrak{v}_T, \Delta\mathfrak{v}_T)_T + \sum_{F\in\mathcal{F}_T} (\pi_F^{0,k}(\mathfrak{v}_T)_{|F}, \nabla\mathfrak{v}_T\cdot\boldsymbol{n}_{TF})_F = 0,
\end{aligned}$$

where we have used the definitions (5.95) of the local virtual space and (1.57) of the L^2-orthogonal projector to insert $\pi_T^{0,k}$ into the first term and $\pi_F^{0,k}$ into the second. As a result, $\mathfrak{v}_T$ is a constant function over T. Consequently, $\mathfrak{v}_T = \pi_T^{0,k}\mathfrak{v}_T = 0$, so that $\mathfrak{v}_T$ is the null function over T. This proves the injectivity of $\underline{I}_T^k : \mathfrak{U}_T^k \to \underline{U}_T^k$.

(ii) *Construction and injectivity of* $\mathfrak{L}_T^k$. Define the linear mapping $\mathfrak{L}_T^k : \underline{U}_T^k \to \mathfrak{U}_T^k$ such that, for any $\underline{v}_T = (v_T, (v_F)_{F\in\mathcal{F}_T}) \in \underline{U}_T^k$, $\mathfrak{L}_T^k\underline{v}_T$ solves the following Neumann problem:

$$-\Delta\mathfrak{L}_T^k\underline{v}_T = v_T - \frac{1}{|T|_d}\left(\int_T v_T - \sum_{F\in\mathcal{F}_T}\int_F v_F\right) \qquad \text{in } T, \tag{5.96a}$$

$$\nabla\mathfrak{L}_T^k\underline{v}_T\cdot\boldsymbol{n}_{TF} = v_F \qquad \text{on all } F \in \mathcal{F}_T, \tag{5.96b}$$

$$\int_T \mathfrak{L}_T^k\underline{v}_T = \int_T v_T. \tag{5.96c}$$

Problem (5.96) defines a unique element $\mathfrak{L}_T^k\underline{v}_T$ of $\mathfrak{U}_T^k$: Eqs. (5.96a) and (5.96b) classically define $\mathfrak{L}_T^k\underline{v}_T$ up to an additive constant since the usual compatibility condition for Neumann problems

$$-\int_T \Delta\mathfrak{L}_T^k\underline{v}_T = \sum_{F\in\mathcal{F}_T}\int_F \nabla\mathfrak{L}_T^k\underline{v}_T\cdot\boldsymbol{n}_{TF}$$

is verified. This constant is then fixed by the closure condition (5.96c).

To prove that $\mathfrak{L}_T^k$ is injective, it suffices to check that $\mathrm{Ker}(\mathfrak{L}_T^k) = \{\underline{0}\}$. Let $\underline{v}_T \in \underline{U}_T^k$ be such that $\mathfrak{L}_T^k \underline{v}_T = 0$. Then, (5.96b) and (5.96c) imply, respectively, $v_F = 0$ for all $F \in \mathcal{F}_T$ and $\int_T v_T = 0$. Plugging these relations into (5.96a) yields $v_T = 0$, so that $\underline{v}_T = \underline{0}$ and the proof of the injectivity of $\mathfrak{L}_T^k$ is concluded. □

5.5.2 Virtual Reformulation of the Local HHO Bilinear Form

Having proved that $\underline{I}_T^k : \mathfrak{U}_T^k \to \underline{U}_T^k$ is bijective, we denote by $(\underline{I}_T^k)^{-1}$ the corresponding inverse mapping. For any $\underline{v}_T \in \underline{U}_T^k$, writing the commutation property (2.14) for $v = \mathfrak{v}_T := (\underline{I}_T^k)^{-1}\underline{v}_T$, we infer that it holds

$$\mathrm{p}_T^{k+1}\underline{v}_T = \pi_T^{1,k+1}\mathfrak{v}_T,$$

which shows, in particular, that the elliptic projection of $\mathfrak{v}_T$ can be computed in terms of the discrete unknowns collected in $\underline{v}_T$. We therefore have the following reformulation for the consistency term in the local HHO bilinear form (2.15): For all $\underline{u}_T, \underline{v}_T \in \underline{U}_T^k$, letting $\mathfrak{u}_T := (\underline{I}_T^k)^{-1}\underline{u}_T$ and $\mathfrak{v}_T := (\underline{I}_T^k)^{-1}\underline{v}_T$,

$$(\nabla \mathrm{p}_T^{k+1}\underline{u}_T, \nabla \mathrm{p}_T^{k+1}\underline{v}_T)_T = (\nabla \pi_T^{1,k+1}\mathfrak{u}_T, \nabla \pi_T^{1,k+1}\mathfrak{v}_T)_T.$$

Let now s_T denote a local stabilisation bilinear form that satisfies Assumption 2.4. By Lemma 2.11, it therefore depends on its arguments only through the difference operators defined by (2.19). We start by noticing that, for any $\underline{v}_T \in \underline{U}_T^k$, letting as before $\mathfrak{v}_T := (\underline{I}_T^k)^{-1}\underline{v}_T$,

$$\delta_T^k \underline{v}_T = \pi_T^{0,k}(\pi_T^{1,k+1}\mathfrak{v}_T - \mathfrak{v}_T), \qquad \delta_{TF}^k \underline{v}_T = \pi_F^{0,k}(\pi_T^{1,k+1}\mathfrak{v}_T - \mathfrak{v}_T) \quad \forall F \in \mathcal{F}_T,$$

where, proceeding as in the proof of Proposition 2.6, we have used the linearity and polynomial invariance (1.56) of $\pi_T^{0,k}$ and $\pi_F^{0,k}$ to cancel them from the second terms in parentheses. Thus, there is a bilinear form $\mathfrak{s}_T : \mathfrak{U}_T^k \times \mathfrak{U}_T^k \to \mathbb{R}$ such that, for any $\underline{u}_T, \underline{v}_T \in \underline{U}_T^k$,

$$\mathrm{s}_T(\underline{u}_T, \underline{v}_T) = \mathfrak{s}_T(\pi_T^{1,k+1}\mathfrak{u}_T - \mathfrak{u}_T, \pi_T^{1,k+1}\mathfrak{v}_T - \mathfrak{v}_T),$$

where again $\mathfrak{u}_T := (\underline{I}_T^k)^{-1}\underline{u}_T$ and $\mathfrak{v}_T := (\underline{I}_T^k)^{-1}\underline{v}_T$.

Define now the discrete virtual bilinear form $\mathfrak{a}_T : \mathfrak{U}_T^k \times \mathfrak{U}_T^k \to \mathbb{R}$ such that, for any $\mathfrak{u}_T, \mathfrak{v}_T \in \mathfrak{U}_T^k$,

$$\mathfrak{a}_T(\mathfrak{u}_T, \mathfrak{v}_T) := (\nabla \pi_T^{1,k+1}\mathfrak{u}_T, \nabla \pi_T^{1,k+1}\mathfrak{v}_T)_T + \mathfrak{s}_T(\pi_T^{1,k+1}\mathfrak{u}_T - \mathfrak{u}_T, \pi_T^{1,k+1}\mathfrak{v}_T - \mathfrak{v}_T).$$

Accounting for the previous remarks, it holds by construction that, for any $\mathfrak{u}_T, \mathfrak{v}_T \in \mathfrak{U}_T^k$,

$$\mathfrak{a}_T(\mathfrak{u}_T, \mathfrak{v}_T) = \mathrm{a}_T(\underline{I}_T^k \mathfrak{u}_T, \underline{I}_T^k \mathfrak{v}_T).$$

5.5.3 Global Virtual Space and Global Bilinear Form

Recalling the jump operator $[\cdot]_F$ defined by (1.22), we can now define the global virtual space

$$\mathfrak{U}_h^k := \left\{ \mathfrak{v}_h \in H^1(\mathcal{T}_h) \ : \ (\mathfrak{v}_h)_{|T} \in \mathfrak{U}_T^k \quad \forall T \in \mathcal{T}_h \text{ and } \pi_F^{0,k}([\mathfrak{v}_h]_F) = 0 \quad \forall F \in \mathcal{F}_h^{\mathrm{i}} \right\},$$

along with its subspace with strongly enforced homogeneous Dirichlet boundary conditions:

$$\mathfrak{U}_{h,0}^k := \left\{ \mathfrak{v}_h \in \mathfrak{U}_h^k \ : \ \pi_F^{0,k}(\mathfrak{v}_h)_{|F} = 0 \quad \forall F \in \mathcal{F}_h^{\mathrm{b}} \right\}.$$

It can be checked that the domain of the global interpolator $\underline{I}_h^k$ defined by (2.34) can be extended to $W^{1,1}(\Omega) + \mathfrak{U}_h^k$ thanks to the continuity of the L^2-projection of degree k of the traces of functions in $\mathfrak{U}_h^k$ across interfaces. Hence, for any $v \in W^{1,1}(\Omega) + \mathfrak{U}_h^k$,

$$\underline{I}_h^k v := ((\pi_T^{0,k} v_{|T})_{T \in \mathcal{T}_h}, (\pi_F^{0,k} v_{|F})_{F \in \mathcal{F}_h}).$$

Moreover, it can easily be inferred from Lemma 5.30 that $\underline{I}_h^k$ defines a bijection from $\mathfrak{U}_h^k$ to the space of global unknowns defined by (2.32) and recalled here for the sake of convenience:

$$\underline{U}_h^k := \Big\{ \underline{v}_h = ((v_T)_{T \in \mathcal{T}_h}, (v_F)_{F \in \mathcal{F}_h}) \ : \\ v_T \in \mathbb{P}^k(T) \quad \forall T \in \mathcal{T}_h \text{ and } v_F \in \mathbb{P}^k(F) \quad \forall F \in \mathcal{F}_h \Big\}.$$

Notice that, when $\underline{v}_h$ belongs to $\underline{U}_{h,0}^k$ (the subspace of $\underline{U}_h^k$ with strongly enforced homogeneous Dirichlet boundary conditions defined by (2.36)), the corresponding virtual function $\mathfrak{v}_h = (\underline{I}_h^k)^{-1} \underline{v}_h$ lies in $\mathfrak{U}_{h,0}^k$.

Define the global virtual bilinear form $\mathfrak{a}_h : \mathfrak{U}_h^k \times \mathfrak{U}_h^k \to \mathbb{R}$ such that

$$\mathfrak{a}_h(\mathfrak{u}_h, \mathfrak{u}_h) := \sum_{T \in \mathcal{T}_h} \mathfrak{a}_T\left((\mathfrak{u}_h)_{|T}, (\mathfrak{v}_h)_{|T}\right).$$

Accounting for the previous remarks and recalling the definition (2.39) of the global HHO bilinear form a_h, it holds, for any $\mathfrak{u}_h, \mathfrak{v}_h \in \mathfrak{U}_h^k$,

$$\mathfrak{a}_h(\mathfrak{u}_h, \mathfrak{v}_h) = \mathrm{a}_h(\underline{I}_h^k \mathfrak{u}_h, \underline{I}_h^k \mathfrak{v}_h). \tag{5.97}$$

5.5.4 Virtual Reformulation of the HHO Scheme

We consider the following Virtual Element scheme for the Poisson problem (2.1): Find $\mathfrak{u}_h \in \mathfrak{U}_{h,0}^k$ such that

$$\mathfrak{a}_h(\mathfrak{u}_h, \mathfrak{v}_h) = (f, \pi_h^{0,k} \mathfrak{v}_h) \qquad \forall \mathfrak{v}_h \in \mathfrak{U}_{h,0}^k, \tag{5.98}$$

where we remind the reader that the global L^2-orthogonal projector is such that, for any $v \in L^1(\Omega)$, $(\pi_h^{0,k} v)_{|T} := \pi_T^{0,k} v_{|T}$ for all $T \in \mathcal{T}_h$; see Definition 1.38. Notice that the linear form in the right-hand side is computable from $\underline{I}_h^k \mathfrak{v}_h$, which contains the L^2-projections of degree k of $\mathfrak{v}_h$ inside each mesh element. The link between the original fully discrete HHO formulation (2.48) and the virtual formulation (5.98) is established in the following theorem.

Theorem 5.31 (Equivalence of the Fully Discrete and Virtual Formulations) *Let $\underline{u}_h \in \underline{U}_{h,0}^k$ and $\mathfrak{u}_h \in \mathfrak{U}_{h,0}^k$ denote the unique solutions to the HHO scheme* (2.48) *and the Virtual Element scheme* (5.98)*, respectively. Then, it holds that*

$$\underline{u}_h = \underline{I}_h^k \mathfrak{u}_h. \tag{5.99}$$

Proof Let $\mathfrak{v}_h \in \mathfrak{U}_{h,0}^k$, and set $\underline{v}_h := \underline{I}_h^k \mathfrak{v}_h$. Using the equivalence (5.97) of the HHO and virtual bilinear forms together with the definition (2.33) of the broken polynomial field $v_h \in \mathbb{P}^k(\mathcal{T}_h)$ (obtained patching element unknowns) to write $v_h = \pi_h^{0,k} \mathfrak{v}_h$, it is inferred from (5.98) and (2.48) that

$$\mathrm{a}_h(\underline{I}_h^k \mathfrak{u}_h, \underline{v}_h) = \mathfrak{a}_h(\mathfrak{u}_h, \mathfrak{v}_h) = (f, \pi_h^{0,k} \mathfrak{v}_h) = (f, v_h) = \mathrm{a}_h(\underline{u}_h, \underline{v}_h).$$

Hence, using the linearity of a_h in its first argument, and observing that $\underline{v}_h$ spans $\underline{U}_{h,0}^k$ when $\mathfrak{v}_h$ spans $\mathfrak{U}_{h,0}^k$, we deduce that

$$\mathrm{a}_h(\underline{I}_h^k \mathfrak{u}_h - \underline{u}_h, \underline{v}_h) = 0 \qquad \forall \underline{v}_h \in \underline{U}_{h,0}^k,$$

which implies (5.99), by coercivity (2.41) of a_h. □

5.5.5 Link with Nonconforming Virtual Elements

We briefly discuss in this section the links between the virtual reformulation of the HHO method and the Nonconforming Virtual Element method of [27]. Let a mesh element $T \in \mathcal{T}_h$ be fixed. The Nonconforming Virtual Element method is based on a variation of the local virtual space (5.95) where element-based unknowns are taken one degree less than face-based unknowns, that is,

$$\mathfrak{U}_T^{k,k-1} := \left\{ \mathfrak{v}_T \in H^1(T) \ : \ \Delta \mathfrak{v}_T \in \mathbb{P}^{k-1}(T) \text{ and } (\nabla \mathfrak{v}_T)_{|F} \cdot \boldsymbol{n}_{TF} \in \mathbb{P}^k(F) \quad \forall F \in \mathcal{F}_T \right\},$$

where we remind the reader that $\mathbb{P}^{-1}(T) := \{0\}$. Recall the definitions (5.1), (5.6) and (5.26) of the space $\underline{U}_T^{k,k-1}$, interpolator $\underline{I}_T^{k,k-1}$ and local potential reconstruction $\tilde{\mathrm{p}}_T^{k+1}$ for the HHO(k, ℓ)-method of Sect. 5.1, with $\ell = k - 1$. We note that if $k \geq 1$ then $\ell \geq 0$ and, formally, $\tilde{\mathrm{p}}_T^{k+1} = \mathrm{p}_T^{k+1}$. In a similar way as in Lemma 5.30, it can be seen that $\underline{I}_T^{k,k-1}$ is an isomorphism between $\mathfrak{U}_T^{k,k-1}$ and $\underline{U}_T^{k,k-1}$.

For a given virtual function $\mathfrak{v}_T \in \mathfrak{U}_T^{k,k-1}$, the commutation properties (5.27) show that

$$\widehat{\pi}_T^{1,k+1} \mathfrak{v}_T = \tilde{\mathrm{p}}_T^{k+1} \underline{I}_T^{k,k-1} \mathfrak{v}_T, \tag{5.100}$$

where $\widehat{\pi}_T^{1,k+1} = \pi_T^{1,k+1}$ if $k \geq 1$, and $\widehat{\pi}_T^{1,1} = \tilde{\pi}_T^{1,1}$ (see (5.14)). The elliptic projection $\widehat{\pi}_T^{1,k+1} \mathfrak{v}_T$ can thus be computed from the discrete unknowns collected in $\underline{I}_T^{k,k-1} \mathfrak{v}_T$. Thus, the local bilinear form for the Nonconforming Virtual Element method is defined as: For all $\mathfrak{u}_T, \mathfrak{v}_T \in \mathfrak{U}_T^{k,k-1}$,

$$\mathfrak{a}_T^{\mathrm{vem}}(\mathfrak{u}_T, \mathfrak{v}_T) := (\nabla \widehat{\pi}_T^{1,k+1} \mathfrak{u}_T, \nabla \widehat{\pi}_T^{1,k+1} \mathfrak{v}_T)_T + \mathfrak{s}_T^{\mathrm{vem}}(\mathfrak{u}_T, \mathfrak{v}_T).$$

Here, $\mathfrak{s}_T^{\mathrm{vem}} : \mathfrak{U}_T^{k,k-1} \times \mathfrak{U}_T^{k,k-1} \to \mathbb{R}$ is a stabilisation bilinear form inspired by Mimetic Finite Difference methods, which satisfies

$$\mathfrak{s}_T^{\mathrm{vem}}(\mathfrak{v}_T, \mathfrak{v}_T)^{\frac{1}{2}} \simeq \|\nabla \mathfrak{v}_T\|_T \qquad \forall \mathfrak{v}_T \in \mathfrak{U}_T^{k,k-1} \text{ such that } \widehat{\pi}_T^{1,k+1} \mathfrak{v}_T = 0,$$

where the hidden constant is independent of both h and T.

The global Nonconforming Virtual Element space is defined by

$$\mathfrak{U}_h^{k,k-1} := \Big\{ \mathfrak{v}_h \in H^1(\mathcal{T}_h) \ : \\ (\mathfrak{v}_h)_{|T} \in \mathfrak{U}_T^{k,k-1} \quad \forall T \in \mathcal{T}_h \text{ and } \pi_F^{0,k}([\mathfrak{v}_h]_F) = 0 \quad \forall F \in \mathcal{F}_h^{\mathrm{i}} \Big\},$$

and its subspace with strongly enforced homogeneous Dirichlet boundary conditions by

$$\mathfrak{U}_{h,0}^{k,k-1} := \left\{ \mathfrak{v}_h \in \mathfrak{U}_h^{k,k-1} \;:\; \pi_F^{0,k}(\mathfrak{v}_h)_{|F} = 0 \quad \forall F \in \mathcal{F}_h^{\mathrm{b}} \right\}.$$

Letting $\mathfrak{a}_h^{\mathrm{vem}} : \mathfrak{U}_{h,0}^{k,k-1} \times \mathfrak{U}_{h,0}^{k,k-1} \to \mathbb{R}$ denote the global bilinear form defined by assembling the elementary contributions $(\mathfrak{a}_T^{\mathrm{vem}})_{T\in\mathcal{T}_h}$, the Nonconforming Virtual Element scheme reads: Find $\mathfrak{u}_h \in \mathfrak{U}_{h,0}^k$ such that, for all $\mathfrak{v}_h \in \mathfrak{U}_{h,0}^k$, setting $\underline{v}_h := \underline{I}_h^{k,k-1}\mathfrak{v}_h$,

$$\mathfrak{a}_h^{\mathrm{vem}}(\mathfrak{u}_h, \mathfrak{v}_h) = \sum_{T\in\mathcal{T}_h} (f, v_T)_T. \tag{5.101}$$

Note that in the case $k \geq 1$, the right-hand side reduces to $(f, \pi_h^{0,k-1}\mathfrak{v}_h)$. In a similar way as in Theorem 5.31, it can be proved that, for a proper choice of the HHO stabilisation bilinear forms $(s_T)_{T\in\mathcal{T}_h}$, this Virtual Element scheme is equivalent to the HHO$(k, k-1)$ scheme (5.39) through the isomorphism $\underline{I}_h^{k,k-1} : \mathfrak{U}_{h,0}^{k,k-1} \to \underline{U}_{h,0}^{k,k-1}$.

Remark 5.32 (Degrees of Freedom for the Nonconforming Virtual Element Method) This isomorphism also shows that natural degrees of freedom for the Nonconforming Virtual Elements Method are the L^2-projections of the virtual functions on $\mathbb{P}^k(F)$, for all $F \in \mathcal{F}_h^{\mathrm{i}}$, and, if $k \geq 1$, the L^2-projections of the virtual functions on $\mathbb{P}^{k-1}(T)$, for all $T \in \mathcal{T}_h$. These degrees of freedom enable the explicit calculation of all the quantities required to implement the scheme (5.101).

5.5.6 The Conforming Virtual Element Method

We present here the Conforming Virtual Element Method [45], considering only, for the sake of simplicity, the case $d = 2$ and the standard (non-enhanced) virtual element space. Some comparisons are drawn with the virtual element formulation of the HHO method that, as seen in Sect. 5.5.5, corresponds to a Nonconforming Virtual Element. We also develop a non-standard analysis of the Conforming Virtual Element Method: instead of endowing the local virtual element space with the H^1-norm of virtual functions (which is not computable), we equip it with a discrete norm that mimics the discrete HHO norm and is fully computable from the degrees of freedom (DOFs) of virtual functions. The main advantage of this approach is that it readily applies to the non-Hilbertian setting, which enables the convergence analysis for fully non-linear models such as the p-Laplace and Leray–Lions models (see Chap. 6 for the analysis of these models in the HHO framework). Usual analyses based on Sobolev norms of virtual functions, such as in [80], would require inverse Lebesgue inequalities in local virtual spaces, which do not seem straightforward to obtain on generic polytopal meshes.

For an alternative HHO-inspired approach to the formulation and analysis of Virtual Element Methods, we refer to [229], which covers both the conforming and nonconforming versions, in two and three space dimensions, albeit in an Hilbertian setting.

5.5.6.1 Local Space and Interpolator

Let $\ell \geq 1$ be a polynomial degree, Ω be an open polygonal set of $\mathbb{R}^2$, and $\mathcal{M}_h = (\mathcal{T}_h, \mathcal{F}_h)$ be a polygonal mesh of Ω from a regular sequence. An element $T \in \mathcal{T}_h$ being given, we let

$$\mathbb{P}^\ell(\partial T) = \left\{ v \in C(\partial T) \, : \, v_{|F} \in \mathbb{P}^\ell(F) \quad \forall F \in \mathcal{F}_T \right\}$$

be the space of continuous, piecewise polynomial functions on the boundary of T. The local Conforming Virtual Elements space is then defined by

$$\mathfrak{U}_T^\ell := \left\{ \mathfrak{v}_T \in H^1(T) : \Delta \mathfrak{v}_T \in \mathbb{P}^{\ell-2}(T), \ (\mathfrak{v}_T)_{|\partial T} \in \mathbb{P}^\ell(\partial T) \right\}. \tag{5.102}$$

Remark 5.33 (Polynomial Degrees) The degree ℓ in Conforming Virtual Elements corresponds, in terms of convergence rates, to the degree $k = \ell - 1$ in the context of HHO methods.

It is a simple matter to notice that $\mathbb{P}^\ell(T) \subset \mathfrak{U}_T^\ell$. To define the DOFs of virtual functions, for each edge F we select $(\ell - 1)$ distinct points $(\boldsymbol{x}_{F,i})_{i=1,\ldots,\ell-1}$ inside the edge (no such point is selected if $\ell = 1$); letting $\boldsymbol{x}_{F,0}$ and $\boldsymbol{x}_{F,\ell}$ be the two endpoints of F, any polynomial in $\mathbb{P}^\ell(F)$ is uniquely determined by its values at $(\boldsymbol{x}_{F,i})_{i=0,\ldots,\ell}$. We assume that these points are chosen to ensure that

$$\|w\|_{L^\infty(F)} \lesssim \max_{i \in \{0,\ldots,\ell\}} |w(\boldsymbol{x}_{F,i})| \qquad \forall w \in \mathbb{P}^\ell(F), \tag{5.103}$$

where the hidden multiplicative constant does not depend on h, F, or w. This bound is satisfied as soon as the points remain well-spaced on the edge, that is, the quantity

$$\frac{\max_{i \in \{0,\ldots,\ell-1\}} |\boldsymbol{x}_{F,i+1} - \boldsymbol{x}_{F,i}|}{\min_{i \in \{0,\ldots,\ell-1\}} |\boldsymbol{x}_{F,i+1} - \boldsymbol{x}_{F,i}|}$$

remains uniformly bounded above as $h \to 0$.

Proposition 5.34 (Local Degrees of Freedom of the Conforming Virtual Element Method) *The following degrees of freedom uniquely determine an element* $\mathfrak{v}_T \in \mathfrak{U}_T^\ell$*:*

(i) The values $(\mathfrak{v}_T(s))_{s \in \mathcal{V}_T}$ *of* $\mathfrak{v}_T$ *at the vertices* $\mathcal{V}_T$ *of* T*,*
(ii) For each edge $F \in \mathcal{F}_T$*, the values* $(\mathfrak{v}_T(\boldsymbol{x}_{F,i}))_{i=1,\ldots,\ell-1}$ *of* $\mathfrak{v}_T$ *at the points* $(\boldsymbol{x}_{F,i})_{i=1,\ldots,\ell-1}$*,*
(iii) The L^2*-projection* $\pi_T^{0,\ell-2}\mathfrak{v}_T$ *of* $\mathfrak{v}_T$ *on* $\mathbb{P}^{\ell-2}(T)$*.*

Remark 5.35 (Lowest-Order Case) In the case $\ell = 1$, given the definition of $\mathbb{P}^{\ell-2} = \mathbb{P}^{-1} = \{0\}$, the last two sets of DOFs do not give any information on the virtual function, and only the values at the vertices are relevant.

Remark 5.36 (Alternate Degrees of Freedom on the Edges) An alternate choice to (ii) of DOFs on the edge F is the L^2-orthogonal projection $\pi_F^{0,\ell-2}\mathfrak{v}_T$ of the virtual function on $\mathbb{P}^{\ell-2}(F)$. Then, (5.103) has to be replaced with the following estimate, which holds with a hidden constant independent of h or F:

$$\|w\|_{L^\infty(F)} \lesssim \max\left(|w(\boldsymbol{x}_{F,0})|, |w(\boldsymbol{x}_{F,\ell})|, \|\pi_F^{0,\ell-2}w\|_{L^\infty(F)}\right) \qquad \forall w \in \mathbb{P}^\ell(F),$$

where we recall that $\boldsymbol{x}_{F,0}$ and $\boldsymbol{x}_{F,\ell}$ are the endpoints of F. The analysis presented hereafter can easily be adapted to this choice of DOFs.

Proof (Proposition 5.34) The mapping

$$\mathfrak{v}_T \in \mathfrak{U}_T^\ell \mapsto ((\mathfrak{v}_T)_{|\partial T}, \Delta\mathfrak{v}_T) \in \mathbb{P}^\ell(\partial T) \times \mathbb{P}^{\ell-2}(T)$$

is an isomorphism, the inverse of which consisting in solving a Dirichlet–Laplace boundary value problem with known boundary conditions and source term. Hence, $\mathfrak{U}_T^\ell$ has the same dimension as $\mathbb{P}^\ell(\partial T) \times \mathbb{P}^{\ell-2}(T)$.

For a given edge $F \in \mathcal{F}_T$, since $(\mathfrak{v}_T)_{|F} \in \mathbb{P}^\ell(F)$, the DOFs (i) and (ii) entirely determine $(\mathfrak{v}_T)_{|F}$. Patching together these polynomial functions determines $(\mathfrak{v}_T)_{|\partial T} \in \mathbb{P}^\ell(\partial T)$, the global continuity being ensured by the DOFs (i). The dimension of $\mathfrak{U}_T^\ell$ is therefore equal to the number of (scalar) DOFs in (i)–(iii), and proving that these degrees entirely determine $\mathfrak{v}_T$ amounts to proving that, if all DOFs vanish, then $\mathfrak{v}_T = 0$.

If all DOFs vanish, the reasoning above shows that $\mathfrak{v}_T = 0$ on ∂T, and thus, integrating by parts, that

$$\int_T |\nabla\mathfrak{v}_T|^2 = -\int_T (\Delta\mathfrak{v}_T)\mathfrak{v}_T = -\int_T (\Delta\mathfrak{v}_T)\pi_T^{0,\ell-2}\mathfrak{v}_T,$$

the introduction of the L^2-orthogonal projector being valid since $\Delta\mathfrak{v}_T \in \mathbb{P}^{\ell-2}(T)$. By assumption, $\pi_T^{0,\ell-2}\mathfrak{v}_T$ also vanishes, which proves that $\nabla\mathfrak{v}_T = 0$, and thus that $\mathfrak{v}_T = 0$ on T since this function vanishes on ∂T. □

Virtual functions are not explicitly known, only their DOFs can be explicitly accessed. As a consequence, it makes sense to carry out an analysis using only these DOFs. Notice that, as seen in the proof above, knowledge of these DOFs give complete explicit knowledge of the virtual function on ∂T.

We next want to define a norm on $\mathfrak{U}_T^\ell$, based on the DOFs and mimicking the discrete norm used in HHO; this requires the knowledge of part of the virtual function inside the element (to penalise the difference between this information on the function in the element and the function on the boundary). Such a knowledge is readily accessible if $\ell \geq 2$, through the degree of freedom $\pi_T^{0,\ell-2}\mathfrak{v}_T$. For $\ell = 1$, we have to reconstruct a quantity representing the average of the function in the element, in a similar way as done in the HHO space $\underline{U}_T^{0,-1}$ defined by (5.1b). We therefore set, for $\mathfrak{v}_T \in \mathfrak{U}_T^\ell$,

$$
\Pi_T^{\ell-2}\mathfrak{v}_T := \begin{cases} \pi_T^{0,\ell-2}\mathfrak{v}_T & \text{if } \ell \geq 2, \\ \dfrac{1}{|T|_d} \displaystyle\sum_{F \in \mathcal{F}_T} \omega_{TF}(\mathfrak{v}_T, 1)_F & \text{if } \ell = 1, \end{cases} \tag{5.104}
$$

where the weights $(\omega_{TF})_{F \in \mathcal{F}_T}$ are chosen to satisfy (5.2).

Even though we will only fully analyse Conforming Virtual Elements for the Poisson problem, as previously mentioned, we aim at developing tools that can serve for the analysis of more complex, non-linear models such as the p-Laplace equation. We therefore endow $\mathfrak{U}_T^\ell$ with a discrete (semi)norm that mimics the $W^{1,p}$-seminorm, as presented in Sect. 6.2.1 for HHO. For a given $p \in (1, +\infty)$, we let

$$
\|\mathfrak{v}_T\|_{\mathrm{cvem},p,T} := \left(\|\nabla \pi_T^{0,\ell-2}\mathfrak{v}_T\|_{L^p(T)^2}^p + \sum_{F \in \mathcal{F}_T} h_F^{1-p} \|\mathfrak{v}_T - \Pi_T^{\ell-2}\mathfrak{v}_T\|_{L^p(F)}^p \right)^{\frac{1}{p}}. \tag{5.105}
$$

The relevant norm for the Poisson problem corresponds to $p = 2$. Finally, the local interpolator $\mathfrak{I}_T^\ell : C(\overline{T}) \to \mathfrak{U}_T^\ell$ is defined as follows: For all $v \in C(\overline{T})$,

$$
\mathfrak{I}_T^\ell v \text{ is the unique element in } \mathfrak{U}_T^\ell \text{ that has the same DOFs as } v. \tag{5.106}
$$

Since $\mathbb{P}^\ell(T) \subset \mathfrak{U}_T^\ell$, we clearly have

$$
\mathfrak{I}_T^\ell w = w \qquad \forall w \in \mathbb{P}^\ell(T). \tag{5.107}
$$

Remark 5.37 (Domain of Interpolator) Using Lemma 5.30, a natural interpolator for the virtual formulation of HHO is $\mathfrak{I}_T^{k,\mathrm{nc}}$ defined as: $\mathfrak{I}_T^{k,\mathrm{nc}}v$ is the unique element in the virtual space (5.95) such that $\underline{I}_T^k \mathfrak{I}_T^{k,\mathrm{nc}} v = \underline{I}_T^k v$. Since $\underline{I}_T^k$ has domain $W^{1,1}(T)$, the interpolator $\mathfrak{I}_T^{k,\mathrm{nc}}$ also has domain $W^{1,1}(T)$.

This highlights a first important difference between the virtual formulation of HHO and Conforming Virtual Elements: their local interpolators have different domains. The Conforming Virtual Elements interpolator has domain $C(\overline{T})$ to ensure that the values of the functions at the vertices of T are well-defined. This has an impact on the analysis which, when based on Sobolev spaces, requires to consider spaces with enough derivatives to ensure, through Sobolev embeddings, the continuity of the considered functions.

We note for future usage the following estimate:

$$\|\Pi_T^{\ell-2}\mathfrak{I}_T^{\ell}v\|_{L^\infty(T)} \lesssim \|v\|_{C(\overline{T})} \qquad \forall v \in C(\overline{T}), \tag{5.108}$$

which follows, if $\ell \geq 2$, from $\pi_T^{0,\ell-2}\mathfrak{I}_T^{\ell}v = \pi_T^{0,\ell-2}v$ together with the boundedness (1.71) of $\pi_T^{0,\ell-2}$ with $p = \infty$ and, if $\ell = 1$, from the estimates (5.4) and (5.103), together with the geometric bounds (1.7)–(1.8) and the definition of $\mathfrak{I}_T^{\ell}$, which imply, for all $F \in \mathcal{F}_T$,

$$\begin{aligned}|\omega_{TF}(\mathfrak{I}_T^{\ell}v, 1)_F| \lesssim h_T|F|_{d-1}\|(\mathfrak{I}_T^{\ell}v)_{|F}\|_{L^\infty(F)} &\lesssim |T|_d \max_{i\in\{0,\ldots,\ell\}} |\mathfrak{I}_T^{\ell}v(\boldsymbol{x}_{F,i})| \\ &= |T|_d \max_{i\in\{0,\ldots,\ell\}} |v(\boldsymbol{x}_{F,i})|.\end{aligned}$$

5.5.6.2 Boundedness of the Interpolator and Approximation Properties of the Projector

As already noticed in the convergence analysis of HHO (see Lemma 2.14), the boundedness of the interpolator is essential to ensure optimal approximation properties of the stabilisation terms. This is also the case in the present context. The following lemma is the counterpart, in the Conforming Virtual Elements framework, of Proposition 2.2.

Lemma 5.38 (Boundedness of $\mathfrak{I}_T^{\ell}$) *Let $T \in \mathcal{T}_h$, $p \in (1, +\infty)$, and $q \in \mathbb{N}$ be such that $qp > 2$. Then, for all $v \in W^{q,p}(T)$,*

$$\|\mathfrak{I}_T^{\ell}v\|_{\mathrm{cvem},p,T} \lesssim \sum_{r=1}^{q} h_T^{r-1}|v|_{W^{r,p}(T)}, \tag{5.109}$$

with hidden constant independent of h, T, and v.

Proof Since we consider in this section the dimension $d = 2$, the condition on p and q ensures that $W^{q,p}(T) \subset C(\overline{T})$ (so that $\mathfrak{I}_T^{\ell}v$ is well defined for $v \in W^{q,p}(T)$), and that, for all $w \in W^{q,p}(T)$,

$$\|w\|_{C(\overline{T})} \lesssim |T|_d^{-\frac{1}{p}} \sum_{r=0}^{q} h_T^r|w|_{W^{r,p}(T)}, \tag{5.110}$$

with hidden constant independent of h, T or w. This estimate is established in [77, Lemma 4.3.4] without the explicit dependencies on $|T|_d$ or h_T, but with a constant that only depends on q, p, and γ, where γ is such that T is star-shaped with respect to all points in a ball of radius γh_T. To obtain the dependencies on $|T|_d$ and h_T as in (5.110), one simply has to scale T to an element of diameter 1, apply [77, Lemma 4.3.4] on that element, and come back to T accounting for the various scaling properties of each derivative in the sum, in a similar way as in the proof of Lemma 1.28. Using the mesh regularity property (Definition 1.9) together with the geometric bounds stated in Lemma 1.12, the inequality (5.110) is proved for T a mesh element by taking the maximum of the same inequalities written on each simplex $\tau \in \mathfrak{T}_T$.

Let us now establish (5.109). If $\ell \leq 2$, the volumetric contribution to $\|\cdot\|_{\mathrm{cvem},p,T}$ vanishes. If $\ell > 2$, by definition of $\mathfrak{I}_T^\ell$, we have $\pi_T^{0,\ell-2}\mathfrak{I}_T^\ell v = \pi_T^{0,\ell-2} v$. Using the boundedness property (1.77) of $\pi_T^{0,\ell-2}$, we infer

$$\|\nabla \pi_T^{0,\ell-2}\mathfrak{I}_T^\ell v\|_{L^p(T)^2} = \|\nabla \pi_T^{0,\ell-2} v\|_{L^p(T)^2} \lesssim |v|_{W^{1,p}(T)}. \tag{5.111}$$

We now consider the boundary contributions to $\|\mathfrak{I}_T^\ell v\|_{\mathrm{cvem},p,T}$. Let $\widetilde{v} := v - \pi_T^{0,0} v$. Using the polynomial invariance (5.107) of $\mathfrak{I}_T^\ell$ we have $\mathfrak{I}_T^\ell \widetilde{v} = \mathfrak{I}_T^\ell v - \pi_T^{0,0} v$ and, by definition (5.104) of $\Pi_T^{\ell-2}$ (and choice (5.2b) of the weights if $\ell = 1$), $\Pi_T^{\ell-2}\pi_T^{0,0} v = \pi_T^{0,0} v$. Hence, $\mathfrak{I}_T^\ell v - \Pi_T^{\ell-2}\mathfrak{I}_T^\ell v = \mathfrak{I}_T^\ell \widetilde{v} - \Pi_T^{\ell-2}\mathfrak{I}_T^\ell \widetilde{v}$. Fixing $F \in \mathcal{F}_T$, we then write

$$\begin{aligned}
\|\mathfrak{I}_T^\ell v - \Pi_T^{\ell-2}\mathfrak{I}_T^\ell v\|_{L^p(F)} &\leq |F|_{d-1}^{\frac{1}{p}} \|\mathfrak{I}_T^\ell \widetilde{v} - \Pi_T^{\ell-2}\mathfrak{I}_T^\ell \widetilde{v}\|_{L^\infty(F)} \\
&\lesssim |F|_{d-1}^{\frac{1}{p}} \max_{i\in\{0,\ldots,\ell\}} |\mathfrak{I}_T^\ell \widetilde{v}(\boldsymbol{x}_{F,i})| + |F|_{d-1}^{\frac{1}{p}} \|\Pi_T^{\ell-2}\mathfrak{I}_T^\ell \widetilde{v}\|_{C(\overline{T})} \\
&\lesssim |F|_{d-1}^{\frac{1}{p}} \|\widetilde{v}\|_{C(\overline{T})} \\
&\lesssim |F|_{d-1}^{\frac{1}{p}} |T|_d^{-\frac{1}{p}} \sum_{r=0}^{q} h_T^r |v - \pi_T^{0,0} v|_{W^{r,p}(T)},
\end{aligned} \tag{5.112}$$

where we have used the triangle inequality and (5.103) with $w = (\mathfrak{I}_T^\ell \widetilde{v})_{|F}$ in the second line, followed by the definition (5.106) of $\mathfrak{I}_T^\ell \widetilde{v}$ (which ensures $\mathfrak{I}_T^\ell \widetilde{v}(\boldsymbol{x}_{F,i}) = \widetilde{v}(\boldsymbol{x}_{F,i})$ for all $i \in \{0,\ldots,\ell\}$) and the estimate (5.108) (with $\widetilde{v}$ instead of v) in the third line, and we have concluded invoking (5.110) with $w = \widetilde{v} = v - \pi_T^{0,0} v$. The approximation property (1.74) of $\pi_T^{0,0}$ gives

$$\|v - \pi_T^{0,0} v\|_{L^p(T)} \lesssim h_T |v|_{W^{1,p}(T)},$$

and, since $\pi_T^{0,0} v$ is constant, we have $|v - \pi_T^{0,0} v|_{W^{r,p}(T)} = |v|_{W^{r,p}(T)}$ whenever $r \geq 1$. Hence, (5.112) yields

$$\|\mathfrak{I}_T^\ell v - \Pi_T^{\ell-2} \mathfrak{I}_T^\ell v\|_{L^p(F)} \lesssim h_T^{-\frac{1}{p}} \sum_{r=1}^{q} h_T^r |v|_{W^{r,p}(T)}, \tag{5.113}$$

where the mesh regularity property (see Lemma 1.12) was used to write $|F|_{d-1}^{\frac{1}{p}} |T|_d^{-\frac{1}{p}} \lesssim h_T^{-\frac{1}{p}}$. Raising (5.113) to the power p, multiplying by $h_F^{1-p} \lesssim h_T^{1-p}$ and summing over $F \in \mathcal{F}_T$ leads to

$$\begin{aligned}\sum_{F \in \mathcal{F}_T} h_F^{1-p} \|\mathfrak{I}_T^\ell v - \Pi_T^{\ell-2} \mathfrak{I}_T^\ell v\|_{L^p(F)}^p &\lesssim h_T^{-p} \left(\sum_{r=1}^{q} h_T^r |v|_{W^{r,p}(T)} \right)^p \\ &= \left(\sum_{r=1}^{q} h_T^{r-1} |v|_{W^{r,p}(T)} \right)^p .\end{aligned}$$

Together with (5.111), this concludes the proof of (5.109). □

As it will become clear in the analysis carried out in Sect. 5.5.6.3, the operator whose approximation properties are essential to the error analysis of Conforming Virtual Elements is

$$\pi_{T,\mathrm{cvem}}^{1,\ell} := \tilde{\pi}_T^{1,\ell} \mathfrak{I}_T^\ell : C(\overline{T}) \to \mathbb{P}^\ell(T), \tag{5.114}$$

where $\tilde{\pi}_T^{1,\ell}$ is the modified elliptic projector defined by (5.14). Note that by (5.107) and the idempotency of $\tilde{\pi}_T^{1,\ell}$, the mapping $\pi_{T,\mathrm{cvem}}^{1,\ell}$ is a polynomial projector. The approximation properties of this projector are established in Theorem 5.40 below. The next lemma states a boundedness result that will be essential to prove this theorem.

Lemma 5.39 (Estimate on the Elliptic Projection of Virtual Functions) *Let $T \in \mathcal{T}_h$ and $p \in (1, +\infty)$. It holds*

$$\|\nabla \tilde{\pi}_T^{1,\ell} \mathfrak{v}_T\|_{L^p(T)^2} \lesssim \|\mathfrak{v}_T\|_{\mathrm{cvem},p,T} \qquad \forall \mathfrak{v}_T \in \mathfrak{U}_T^\ell, \tag{5.115}$$

where the hidden constant is independent of h, T and $\mathfrak{v}_T$, and $\mathfrak{U}_T^\ell$ is the conforming virtual space defined by (5.102).

Proof (i) *The case* $p = 2$. Using the definition (5.14a) of the modified elliptic projector and an integration by parts, we have, for all $w \in \mathbb{P}^\ell(T)$,

$$\begin{aligned}(\nabla\tilde{\pi}_T^{1,\ell}\mathrm{v}_T, \nabla w)_T &= -(\mathrm{v}_T, \Delta w)_T + \sum_{F\in\mathcal{F}_T}(\mathrm{v}_T, \nabla w\cdot\boldsymbol{n}_{TF})_F \\ &= -(\pi_T^{0,\ell-2}\mathrm{v}_T, \Delta w)_T + \sum_{F\in\mathcal{F}_T}(\mathrm{v}_T, \nabla w\cdot\boldsymbol{n}_{TF})_F \\ &= (\nabla\pi_T^{0,\ell-2}\mathrm{v}_T, \nabla w)_T + \sum_{F\in\mathcal{F}_T}(\mathrm{v}_T - \pi_T^{0,\ell-2}\mathrm{v}_T, \nabla w\cdot\boldsymbol{n}_{TF})_F,\end{aligned} \tag{5.116}$$

where the introduction of $\pi_T^{0,\ell-2}$ in the second line is justified by $\Delta w \in \mathbb{P}^{\ell-2}(T)$, and the conclusion follows from another integration by parts. In the case $\ell = 1$, since $w \in \mathbb{P}^1(T)$, we have

$$\sum_{F\in\mathcal{F}_T}(Z, \nabla w\cdot\boldsymbol{n}_{TF})_F = -(Z, \Delta w)_T = 0 \qquad \forall Z \in \mathbb{R}.$$

Applied to $Z = \Pi_T^{\ell-2}\mathrm{v}_T - \pi_T^{0,\ell-2}\mathrm{v}_T$ (constant since $\ell = 1$), this gives

$$\sum_{F\in\mathcal{F}_T}(\mathrm{v}_T - \pi_T^{0,\ell-2}\mathrm{v}_T, \nabla w\cdot\boldsymbol{n}_{TF})_F = \sum_{F\in\mathcal{F}_T}(\mathrm{v}_T - \Pi_T^{\ell-2}\mathrm{v}_T, \nabla w\cdot\boldsymbol{n}_{TF})_F.$$

This relation obviously also holds for $\ell \geq 2$, since $\Pi_T^{\ell-2} = \pi_T^{0,\ell-2}$ in this case. Hence, (5.116) yields, for any $\ell \geq 1$,

$$(\nabla\tilde{\pi}_T^{1,\ell}\mathrm{v}_T, \nabla w)_T = (\nabla\pi_T^{0,\ell-2}\mathrm{v}_T, \nabla w)_T + \sum_{F\in\mathcal{F}_T}(\mathrm{v}_T - \Pi_T^{\ell-2}\mathrm{v}_T, \nabla w\cdot\boldsymbol{n}_{TF})_F. \tag{5.117}$$

Specifying $w = \tilde{\pi}_T^{1,\ell}\mathrm{v}_T$ and using a Cauchy–Schwarz inequality for the volumetric term, generalised Hölder inequalities with exponents $(2, 2, \infty)$ for the boundary terms (together with $\|\boldsymbol{n}_{TF}\|_{L^\infty(F)^2} = 1$), and the discrete trace inequality (1.55) with $p = 2$ on $\nabla\tilde{\pi}_T^{1,\ell}\mathrm{v}_T$, we infer

$$\begin{aligned}\|\nabla\tilde{\pi}_T^{1,\ell}\mathrm{v}_T\|_{L^2(T)^2}^2 &\lesssim \|\nabla\pi_T^{0,\ell-2}\mathrm{v}_T\|_{L^2(T)^2}\|\nabla\tilde{\pi}_T^{1,\ell}\mathrm{v}_T\|_{L^2(T)^2} \\ &\quad + \sum_{F\in\mathcal{F}_T}\|\mathrm{v}_T - \Pi_T^{\ell-2}\mathrm{v}_T\|_{L^2(F)}h_T^{-\frac{1}{2}}\|\nabla\tilde{\pi}_T^{1,\ell}\mathrm{v}_T\|_{L^2(T)^2}.\end{aligned}$$

Simplify by $\|\nabla\widetilde{\pi}_T^{1,\ell}\mathfrak{v}_T\|_{L^2(T)^2}$, use a Cauchy–Schwarz inequality together with the bounds $h_T^{-1} \le h_F^{-1}$ and $\mathrm{card}(\mathcal{F}_T) \lesssim 1$ (see (1.5)), and recall the definition (5.105) of the norm on $\mathfrak{U}_T^\ell$ to obtain (5.115) for $p = 2$.

(ii) *The case $p \in (1, +\infty)$*. The functions $\nabla\widetilde{\pi}_T^{1,\ell}\mathfrak{v}_T$, $\nabla\pi_T^{0,\ell-2}\mathfrak{v}_T$, $(\mathfrak{v}_T - \Pi_T^{\ell-2}\mathfrak{v}_T)_{|F}$ (for $F \in \mathcal{F}_T$) appearing in the left-hand side of (5.115) and in the addends in $\|\mathfrak{v}_T\|_{\mathrm{cvem},p,T}$ in the right-hand side of this equation are all polynomial functions. They are therefore amenable to the inverse Lebesgue inequalities (1.35), and we can write

$$
\begin{aligned}
\|\nabla\widetilde{\pi}_T^{1,\ell}\mathfrak{v}_T\|_{L^p(T)^2} &\lesssim |T|_d^{\frac{1}{p}-\frac{1}{2}}\|\nabla\widetilde{\pi}_T^{1,\ell}\mathfrak{v}_T\|_{L^2(T)^2}\\
&\lesssim |T|_d^{\frac{1}{p}-\frac{1}{2}}\left(\|\nabla\pi_T^{0,\ell-2}\mathfrak{v}_T\|_{L^2(T)^2}^2 + \sum_{F\in\mathcal{F}_T} h_F^{-1}\|\mathfrak{v}_T - \Pi_T^{\ell-2}\mathfrak{v}_T\|_{L^2(F)}^2\right)^{\frac{1}{2}}\\
&\lesssim |T|_d^{\frac{1}{p}-\frac{1}{2}}\left(|T|_d^{1-\frac{2}{p}}\|\nabla\pi_T^{0,\ell-2}\mathfrak{v}_T\|_{L^p(T)^2}^2\right.\\
&\qquad\left. + \sum_{F\in\mathcal{F}_T} h_F^{-1}|F|_{d-1}^{1-\frac{2}{p}}\|\mathfrak{v}_T - \Pi_T^{\ell-2}\mathfrak{v}_T\|_{L^p(F)}^2\right)^{\frac{1}{2}},
\end{aligned}
$$

where the second line corresponds to (5.115) for $p = 2$. By the geometric bounds (1.6), (1.7) and (1.8), it holds $h_F^{-1}|F|_{d-1}^{1-\frac{2}{p}} \lesssim h_F^{\frac{2}{p}-2}|T|_d^{1-\frac{2}{p}}$ and thus, using the uniform bound (1.5) on the number of edges of T,

$$
\begin{aligned}
\|\nabla\widetilde{\pi}_T^{1,\ell}\mathfrak{v}_T\|_{L^p(T)^2} &\lesssim \|\nabla\pi_T^{0,\ell-2}\mathfrak{v}_T\|_{L^p(T)^2} + \sum_{F\in\mathcal{F}_T} h_F^{\frac{1}{p}-1}\|\mathfrak{v}_T - \Pi_T^{\ell-2}\mathfrak{v}_T\|_{L^p(F)}\\
&\lesssim \left(\|\nabla\pi_T^{0,\ell-2}\mathfrak{v}_T\|_{L^p(T)^2}^p\right.\\
&\qquad\left. + \sum_{F\in\mathcal{F}_T} h_F^{1-p}\|\mathfrak{v}_T - \Pi_T^{\ell-2}\mathfrak{v}_T\|_{L^p(F)}^p\right)^{\frac{1}{p}},
\end{aligned}
$$

which concludes the proof of (5.115). □

We now establish the approximation properties of the projector $\pi_{T,\mathrm{cvem}}^{1,\ell}$ defined by (5.114).

Theorem 5.40 (Approximation Properties of the Projector $\pi^{1,\ell}_{T,\mathrm{cvem}}$) *Let $(\mathcal{M}_h)_{h\in\mathcal{H}} = (\mathcal{T}_h, \mathcal{F}_h)_{h\in\mathcal{H}}$ be a regular mesh sequence in the sense of Definition 1.9. Let a real number $p \in (1, \infty)$ and a natural number $q \geq 1$ be chosen such that $qp > 2$. Let a polynomial degree $\ell \geq \max(1, q-1)$ and an integer $s \in \{q, \ldots, \ell+1\}$ be given. Then, for all $T \in \mathcal{T}_h$, all $v \in W^{s,p}(T)$, and all $m \in \{0, \ldots, s\}$,*

$$|v - \pi^{1,\ell}_{T,\mathrm{cvem}} v|_{W^{m,p}(T)} \lesssim h_T^{s-m} |v|_{W^{s,p}(T)}. \tag{5.118}$$

Moreover, for all $F \in \mathcal{F}_T$ and all $m \in \{0, \ldots, s-1\}$, it holds that

$$h_T^{\frac{1}{p}} |v - \pi^{1,\ell}_{T,\mathrm{cvem}} v|_{W^{m,p}(F)} \lesssim h_T^{s-m} |v|_{W^{s,p}(T)}. \tag{5.119}$$

In the above estimates, the hidden constants depend only on d, ϱ, p, q, ℓ, and s.

Remark 5.41 (Commutation of Interpolator and Elliptic Projection: HHO and Conforming Virtual Elements) We defined in Remark 5.37 the interpolator $\mathfrak{I}^{\ell,\mathrm{nc}}_T$ for the virtual element interpretation of HHO by: $\underline{I}^k_T \mathfrak{I}^{k,\mathrm{nc}}_T v = \underline{I}^k_T v$. Taking $v \in W^{1,1}(T)$ and applying p^{k+1}_T to this relation, the commutation property (2.14) shows that $\pi^{1,k+1}_T \mathfrak{I}^{k,\mathrm{nc}}_T v = \pi^{1,k+1}_T v$. In other words, for the virtual HHO reformulation, a function and its interpolate on the virtual space have the same elliptic projection. The equivalent of Theorem 5.40 in that case is nothing but Theorem 1.48, which states the approximation properties of $\pi^{1,k+1}_T$.

On the contrary, for Conforming Virtual Elements, we have, in general, $\tilde{\pi}^{1,\ell}_T \mathfrak{I}^\ell_T \neq \tilde{\pi}^{1,\ell}_T$ (that is, $\pi^{1,\ell}_{T,\mathrm{cvem}} \neq \tilde{\pi}^{1,\ell}_T$). This is due to the fact that a function and its virtual interpolate do not have the same moments up to degree ℓ on each edge. As a consequence, the approximation properties of $\pi^{1,\ell}_{T,\mathrm{cvem}}$ have to be established separately.

Proof (Theorem 5.40) By choice of q and p, $W^{q,p}(T)$ is contained in $C(\overline{T})$ and $\pi^{1,\ell}_{T,\mathrm{cvem}}$ is thus well defined on $W^{q,p}(T)$. We already noticed that this mapping is a polynomial projector, so its approximation properties (5.118) follow from Lemma 1.43, provided we establish (1.63). The trace approximation properties (5.119) are then a consequence of (5.118), using the same argument as in the proof of Theorem 1.45. We now focus on proving (1.63).

(i) *Case $m = 1$*. Recall the definition (5.114) of $\pi^{1,\ell}_{T,\mathrm{cvem}}$ and combine (5.115) and (5.109) to write

$$\|\nabla \pi^{1,\ell}_{T,\mathrm{cvem}} v\|_{L^p(T)^2} \lesssim \|\mathfrak{I}^\ell_T v\|_{\mathrm{cvem},p,T} \lesssim \sum_{r=1}^{q} h_T^{r-1} |v|_{W^{r,p}(T)}, \tag{5.120}$$

which is exactly the relevant estimate (1.63a) since $m = 1 \le q$ (note that, in the case $q = m$, (1.63a) and (1.63b) are identical).

(ii) *Case* $m = 0$. We reason as in the proof of Theorem 5.7 (with $l = \ell$). Introducing $\pm\pi_T^{0,0}(\pi_{T,\text{cvem}}^{1,\ell}v)$, using a triangle inequality, and applying the approximation property (1.74) of $\pi_T^{0,0}$ to $\pi_{T,\text{cvem}}^{1,\ell}v$ instead of v, we have

$$\begin{aligned}\|\pi_{T,\text{cvem}}^{1,\ell}v\|_{L^p(T)} &\le \|\pi_{T,\text{cvem}}^{1,\ell}v - \pi_T^{0,0}(\pi_{T,\text{cvem}}^{1,\ell}v)\|_{L^p(T)} + \|\pi_T^{0,0}(\pi_{T,\text{cvem}}^{1,\ell}v)\|_{L^p(T)}\\ &\lesssim h_T\|\nabla\pi_{T,\text{cvem}}^{1,\ell}v\|_{L^p(T)^2} + \|\pi_T^{0,0}(\pi_{T,\text{cvem}}^{1,\ell}v)\|_{L^p(T)}.\end{aligned} \tag{5.121}$$

Applying (5.23) to $\mathfrak{I}_T^{\ell}v$ instead of v, we find

$$|T|_d\left|\pi_T^{0,0}(\pi_{T,\text{cvem}}^{1,\ell}v)\right| \lesssim h_T|T|_d^{\frac{1}{p'}}\|\nabla\tilde{\pi}_T^{1,\ell}\mathfrak{I}_T^{\ell}v\|_{L^p(T)^2} + \left|\sum_{F\in\mathcal{F}_T}\omega_{TF}(\mathfrak{I}_T^{\ell}v,1)_F\right|.$$

Combined with $\|\pi_T^{0,0}(\pi_{T,\text{cvem}}^{1,\ell}v)\|_{L^p(T)} = |T|_d^{\frac{1}{p}}|\pi_T^{0,0}(\pi_{T,\text{cvem}}^{1,\ell}v)|$, (5.121) and (5.120), this leads to

$$\|\pi_{T,\text{cvem}}^{1,\ell}v\|_{L^p(T)} \lesssim \sum_{r=1}^{q} h_T^r|v|_{W^{r,p}(T)} + |T|_d^{\frac{1}{p}-1}\left|\sum_{F\in\mathcal{F}_T}\omega_{TF}(\mathfrak{I}_T^{\ell}v,1)_F\right|. \tag{5.122}$$

To estimate the last addend, we write

$$\begin{aligned}\left|\sum_{F\in\mathcal{F}_T}\omega_{TF}(\mathfrak{I}_T^{\ell}v,1)_F\right| &\le \sum_{F\in\mathcal{F}_T}|\omega_{TF}||F|_{d-1}\|\mathfrak{I}_T^{\ell}v\|_{L^\infty(F)}\\ &\lesssim \sum_{F\in\mathcal{F}_T}h_T|F|_{d-1}\|v\|_{C(\overline{T})}\\ &\lesssim |T|_d|T|_d^{-\frac{1}{p}}\sum_{r=0}^{q}h_T^r|v|_{W^{r,p}(T)},\end{aligned}$$

where the second line follows from (5.4) and (5.103) applied to $w = (\mathfrak{I}_T^{\ell}v)_{|F}$ for each $F \in \mathcal{F}_T$ (recall that $\mathfrak{I}_T^{\ell}v(\boldsymbol{x}_{F,i}) = v(\boldsymbol{x}_{F,i})$ for all $i \in \{0,\ldots,\ell\}$), and the conclusion is obtained using the geometric bounds stated in Lemma 1.12

together with the estimate (5.110). Plugged into (5.122), this shows that

$$\|\pi_{T,\mathrm{cvem}}^{1,\ell} v\|_{L^p(T)} \lesssim \sum_{r=0}^{q} h_T^r |v|_{W^{r,p}(T)},$$

which is exactly (1.63a) in the case $m = 0$.

(iii) *Case $m \geq 2$ and $m \leq q$*. Using the linearity and idempotency of the projector $\pi_{T,\mathrm{cvem}}^{1,\ell}$, we have

$$\pi_{T,\mathrm{cvem}}^{1,\ell} v = \pi_{T,\mathrm{cvem}}^{1,\ell}(v - \pi_T^{0,\ell} v) + \pi_T^{0,\ell} v.$$

Taking the $W^{m,p}(T)$-seminorm and using the inverse inequality (1.46) on the polynomial function $\pi_{T,\mathrm{cvem}}^{1,\ell}(v - \pi_T^{0,\ell} v)$, together with the boundedness property (1.77) of $\pi_T^{0,\ell}$ with m instead of s, we infer

$$|\pi_{T,\mathrm{cvem}}^{1,\ell} v|_{W^{m,p}(T)} \lesssim h_T^{-(m-1)} |\pi_{T,\mathrm{cvem}}^{1,\ell}(v - \pi_T^{0,\ell} v)|_{W^{1,p}(T)} + |v|_{W^{m,p}(T)}. \tag{5.123}$$

Apply (5.120) with $v - \pi_T^{0,\ell} v$ instead of v to get

$$\begin{aligned}
|\pi_{T,\mathrm{cvem}}^{1,\ell}(v - \pi_T^{0,\ell} v)|_{W^{1,p}(T)} &\lesssim \sum_{r=1}^{q} h_T^{r-1} |v - \pi_T^{0,\ell} v|_{W^{r,p}(T)} \\
&\lesssim \sum_{r=1}^{m} h_T^{r-1} h_T^{m-r} |v|_{W^{m,p}(T)} + \sum_{r=m+1}^{q} h_T^{r-1} |v|_{W^{r,p}(T)} \\
&\lesssim h_T^{m-1} |v|_{W^{m,p}(T)} + \sum_{r=m+1}^{q} h_T^{r-1} |v|_{W^{r,p}(T)},
\end{aligned}$$

where the second line follows using, for each $r \leq m$, the approximation property (1.74) of $\pi_T^{0,\ell}$ with (r, m) in lieu of (m, s) (note that $m \leq \ell + 1$) and, for each $r \geq m + 1$, the boundedness (1.77) of $\pi_T^{0,\ell}$ with r instead of s. Plugged into (5.123), this estimate yields

$$\begin{aligned}
|\pi_{T,\mathrm{cvem}}^{1,\ell} v|_{W^{m,p}(T)} &\lesssim |v|_{W^{m,p}(T)} + \sum_{r=m+1}^{q} h_T^{r-m} |v|_{W^{r,p}(T)} \\
&= \sum_{r=m}^{q} h_T^{r-m} |v|_{W^{r,p}(T)},
\end{aligned} \tag{5.124}$$

which establishes (1.63a).

(iv) *Case $m \geq q$*. Making $m = q$ in (5.124) proves (1.63b). □

5.5.6.3 A Conforming Virtual Elements scheme for the Poisson Problem

In this section, we present and analyse the Conforming Virtual Element Method for the Poisson problem (2.1).

The relation (5.116) and the closure equation (5.14b) show that, for any $\mathfrak{v}_T \in \mathfrak{U}_T^\ell$, $\tilde{\pi}_T^{1,\ell}\mathfrak{v}_T$ is computable from the DOFs of $\mathfrak{v}_T$. This justifies the following definition of the local bilinear form $\mathfrak{a}_T^{\rm cvem} : \mathfrak{U}_T^\ell \times \mathfrak{U}_T^\ell \to \mathbb{R}$, mimicking $(\nabla u, \nabla v)_T$, with consistent contribution based on the modified elliptic projector: For all $\mathfrak{u}_T, \mathfrak{v}_T \in \mathfrak{U}_T^\ell$,

$$\mathfrak{a}_T^{\rm cvem}(\mathfrak{u}_T, \mathfrak{v}_T) := (\nabla\tilde{\pi}_T^{1,\ell}\mathfrak{u}_T, \nabla\tilde{\pi}_T^{1,\ell}\mathfrak{v}_T)_T + \mathfrak{s}_T^{\rm cvem}(\mathfrak{u}_T, \mathfrak{v}_T),$$

where $\mathfrak{s}_T^{\rm cvem} : \mathfrak{U}_T^\ell \times \mathfrak{U}_T^\ell \to \mathbb{R}$ is a symmetric positive semidefinite stabilisation bilinear form, computable from the DOFs and which satisfies the following polynomial consistency and stability properties:

$$\mathfrak{s}_T^{\rm cvem}(w, \mathfrak{v}_T) = 0 \qquad \forall (w, \mathfrak{v}_T) \in \mathbb{P}^\ell(T) \times \mathfrak{U}_T^\ell, \tag{5.125}$$

$$\|\nabla\tilde{\pi}_T^{1,\ell}\mathfrak{v}_T\|_T^2 + \mathfrak{s}_T^{\rm cvem}(\mathfrak{v}_T, \mathfrak{v}_T) \simeq \|\mathfrak{v}_T\|_{{\rm cvem},2,T}^2 \qquad \forall \mathfrak{v}_T \in \mathfrak{U}_T^\ell, \tag{5.126}$$

where the hidden constants are independent of both h and T.

Remark 5.42 (Alternative for the Consistent Term) In the case $\ell \geq 2$, since the average on a mesh element of a virtual function is computable from the DOFs, the standard elliptic projector $\pi_T^{1,\ell}$ can be used instead of the modified elliptic projector $\tilde{\pi}_T^{1,\ell}$.

As usual, the global space is defined patching together the local spaces and strongly enforcing the homogeneous Dirichlet boundary conditions:

$$\mathfrak{U}_{h,0}^\ell := \left\{\mathfrak{v}_h \in H_0^1(\Omega) \,:\, \mathfrak{v}_T := (\mathfrak{v}_h)_{|T} \in \mathfrak{U}_T^\ell \quad \forall T \in \mathcal{T}_h\right\}.$$

This space is endowed with the norm

$$\|\mathfrak{v}_h\|_{{\rm cvem},2,h} := \left(\sum_{T\in\mathcal{T}_h} \|\mathfrak{v}_T\|_{{\rm cvem},2,T}^2\right)^{\frac{1}{2}} \qquad \forall \mathfrak{v}_h \in \mathfrak{U}_{h,0}^\ell. \tag{5.127}$$

Since each local virtual element space is made of functions that are continuous (on the boundary as well as inside the elements), a function in $\mathfrak{U}_{h,0}^\ell$ is actually continuous over $\overline{\Omega}$. The global DOFs are also obtained patching together the local DOFs, and are therefore made of: the values of the virtual functions at the mesh vertices, their values at each $(\boldsymbol{x}_{F,i})_{i=1,\ldots,\ell-1}$ for all $F \in \mathcal{F}_h^{\rm i}$, and their local L^2-orthogonal projections on $\mathbb{P}^{\ell-2}(T)$ for all $T \in \mathcal{T}_h$. The global interpolator

$\mathfrak{I}_h^\ell : C(\overline{\Omega}) \cap H_0^1(\Omega) \to \mathfrak{U}_{h,0}^\ell$ is defined such that, for all $v \in C(\overline{\Omega}) \cap H_0^1(\Omega)$, $\mathfrak{I}_h^\ell v$ is the unique element in $\mathfrak{U}_{h,0}^\ell$ that has the same global DOFs as v.

Finally, we assemble the global bilinear form $\mathfrak{a}_h^{\text{cvem}} : \mathfrak{U}_{h,0}^\ell \times \mathfrak{U}_{h,0}^\ell \to \mathbb{R}$ from the local contributions:

$$\mathfrak{a}_h^{\text{cvem}}(\mathfrak{u}_h, \mathfrak{v}_h) := \sum_{T \in \mathcal{T}_h} \mathfrak{a}_T^{\text{cvem}}(\mathfrak{u}_T, \mathfrak{v}_T) \qquad \forall \mathfrak{u}_h, \mathfrak{v}_h \in \mathfrak{U}_{h,0}^\ell.$$

The energy norm is then defined by

$$\|\mathfrak{v}_h\|_{\text{cvem},\mathfrak{a},h} := \mathfrak{a}_h^{\text{cvem}}(\mathfrak{v}_h, \mathfrak{v}_h)^{\frac{1}{2}} \qquad \forall \mathfrak{v}_h \in \mathfrak{U}_{h,0}^\ell.$$

The property (5.126) ensures the following norm equivalence, in which the hidden constant does not depend on h:

$$\|\mathfrak{v}_h\|_{\text{cvem},\mathfrak{a},h} \simeq \|\mathfrak{v}_h\|_{\text{cvem},2,h} \qquad \forall \mathfrak{v}_h \in \mathfrak{U}_{h,0}^\ell. \tag{5.128}$$

The Conforming Virtual Elements scheme for the Poisson problem then reads: Find $\mathfrak{u}_h \in \mathfrak{U}_{h,0}^\ell$ such that

$$\mathfrak{a}_h^{\text{cvem}}(\mathfrak{u}_h, \mathfrak{v}_h) = (f, \Pi_h^{\ell-2}\mathfrak{v}_h) \qquad \forall \mathfrak{v}_h \in \mathfrak{U}_{h,0}^\ell, \tag{5.129}$$

where $(\Pi_h^{\ell-2}\mathfrak{v}_h)_{|T} = \Pi_T^{\ell-2}\mathfrak{v}_T$ for all $T \in \mathcal{T}_h$. Existence and uniqueness of the solution to this scheme follows from the Lax-Milgram lemma (Lemma 2.20). An error estimate in discrete energy norm is provided in the following theorem.

Theorem 5.43 (Discrete Energy Error Estimate for Conforming Virtual Elements) *Let $(\mathcal{M}_h)_{h\in\mathcal{H}}$ denote a regular mesh sequence in the sense of Definition 1.9. Let a polynomial degree $\ell \geq 1$ be fixed. Let $u \in H_0^1(\Omega)$ denote the unique solution to (2.2), for which we assume the additional regularity $u \in H^{r+2}(\mathcal{T}_h)$ for some $r \in \{0, \ldots, \ell-1\}$. For all $h \in \mathcal{H}$, let $\mathfrak{u}_h \in \mathfrak{U}_{h,0}^\ell$ denote the unique solution to (5.129). Then,*

$$\|\mathfrak{u}_h - \mathfrak{I}_h^\ell u\|_{\text{cvem},\mathfrak{a},h} \lesssim h^{r+1}|u|_{H^{r+2}(\mathcal{T}_h)}, \tag{5.130}$$

where the hidden constant is independent of both h and u.

Remark 5.44 (L^2-Error Estimates) In the case $\ell \geq 3$, an optimal error estimate in L^2-norm can also be obtained using the Aubin–Nitsche trick (Theorem A.10). For $\ell = 1$ this requires, as for the HHO(0, −1) method (see Theorem 5.16), to choose weights that satisfy the quadrature rule (5.5) of order one. For $\ell = 2$, the discretisation of the source term must be modified in a similar way as for the HHO(1, 0) scheme, see Remark 5.17: $(f, \Pi_h^0 \mathfrak{v}_h)$ is replaced with

$$\sum_{T\in\mathcal{T}_h} (f, \widetilde{\mathfrak{v}}_T)_T$$

where, for all $T \in \mathcal{T}_h$ and all $\boldsymbol{x} \in \overline{T}$, $\widetilde{\mathfrak{v}}_T(\boldsymbol{x}) = \pi_T^{0,0}\mathfrak{v}_T + \pi_T^{0,0}(\nabla \mathfrak{v}_T)\cdot(\boldsymbol{x} - \overline{\boldsymbol{x}}_T)$, with $\overline{\boldsymbol{x}}_T = \pi_T^{0,0}(\boldsymbol{x})$ the centre of mass of T; see [46] for details.

Remark 5.45 (Small Edges) The analysis of the Conforming Virtual Element Method is carried out here under our standard mesh regularity assumption, which allows for non-star-shaped elements but imposes each edge diameter to be comparable to the diameters of the elements it belongs to. For linear problems, an analysis of Conforming Virtual Elements can be carried out without the latter restriction [54, 79], assuming that each element T is star-shaped with respect to a ball of diameter comparable to h_T. Understanding if the analysis tools developed for non-linear problems in Sect. 5.5.6.2 can be adapted to meshes with small edges is currently an open question.

Proof (Theorem 5.43) We use the Third Strang Lemma A.7, with $\mathrm{U}_h = \mathfrak{U}_h^\ell$ endowed with the norm $\|\cdot\|_{\mathrm{cvem},\mathfrak{a},h}$ (for which $\mathfrak{a}_h^{\mathrm{cvem}}$ is obviously coercive with constant 1), $\mathrm{a}_h = \mathfrak{a}_h^{\mathrm{cvem}}$, $\mathrm{l}_h(\mathfrak{v}_h) = (f, \Pi_h^{\ell-2}\mathfrak{v}_h)$, and $\mathrm{I}_h u = \mathfrak{I}_h^\ell u$. We only have to show that the dual norm of the consistency error is bounded above by the right-hand side in (5.130).

Recalling that $f = -\Delta u$ and the definition (5.114) of $\pi_{T,\mathrm{cvem}}^{1,\ell}$, this consistency error is

$$\begin{aligned}\mathcal{E}_h(u;\mathfrak{v}_h) = &\sum_{T\in\mathcal{T}_h} -(\Delta u, \Pi_T^{\ell-2}\mathfrak{v}_T)_T - \sum_{T\in\mathcal{T}_h} (\nabla \pi_{T,\mathrm{cvem}}^{1,\ell} u, \nabla \widetilde{\pi}_T^{1,\ell}\mathfrak{v}_T)_T \\ &- \sum_{T\in\mathcal{T}_h} \mathfrak{s}_T^{\mathrm{cvem}}(\mathfrak{I}_T^\ell u, \mathfrak{v}_T) =: \mathfrak{T}_1 + \mathfrak{T}_2 + \mathfrak{T}_3.\end{aligned} \tag{5.131}$$

Let us first consider the stabilisation terms. Using the polynomial consistency (5.125) of $\mathfrak{s}_T^{\mathrm{cvem}}$ together with the polynomial invariance (5.107) of $\mathfrak{I}_T^\ell$, we have

$$\begin{aligned}\mathfrak{s}_T^{\mathrm{cvem}}(\mathfrak{I}_T^\ell u, \mathfrak{I}_T^\ell u)^{\frac12} &= \mathfrak{s}_T^{\mathrm{cvem}}(\mathfrak{I}_T^\ell(u - \pi_T^{0,\ell}u), \mathfrak{I}_T^\ell(u - \pi_T^{0,\ell}u))^{\frac12} \\ &\lesssim \|\mathfrak{I}_T^\ell(u - \pi_T^{0,\ell}u)\|_{\mathrm{cvem},2,T} \\ &\lesssim |u - \pi_T^{0,\ell}u|_{H^1(T)} + h_T|u - \pi_T^{0,\ell}u|_{H^2(T)} \\ &\lesssim h_T^{r+1}|u|_{H^{r+2}(T)},\end{aligned}$$

where the second inequality follows from the boundedness (5.126) of $\mathfrak{s}_T^{\text{cvem}}$, the third inequality from (5.109) with $q = p = 2$ and $v = u - \pi_T^{0,\ell} u$, and the conclusion from the approximation properties (1.74) of $\pi_T^{0,\ell} u$. Using Cauchy–Schwarz inequalities on each $\mathfrak{s}_T^{\text{cvem}}$ and on the sum, together with the definition of $\|\cdot\|_{\text{cvem},\mathfrak{a},h}$, we infer

$$
\begin{aligned}
|\mathfrak{T}_3| &\le \sum_{T\in\mathcal{T}_h} \mathfrak{s}_T^{\text{cvem}}(\mathfrak{I}_T^\ell u, \mathfrak{I}_T^\ell u)^{\frac{1}{2}} \mathfrak{s}_T^{\text{cvem}}(\mathfrak{v}_T, \mathfrak{v}_T)^{\frac{1}{2}} \\
&\lesssim h^{r+1} |u|_{H^{r+2}(\mathcal{T}_h)} \|\mathfrak{v}_h\|_{\text{cvem},\mathfrak{a},h}.
\end{aligned}
\tag{5.132}
$$

We now turn to the first two terms in (5.131). Integrating by parts and noticing that, for any value of $\ell \ge 1$, the definition (5.104) of $\Pi_T^{\ell-2}$ gives $\nabla \Pi_T^{\ell-2} = \nabla \pi_T^{0,\ell-2}$, we find

$$
\mathfrak{T}_1 = \sum_{T\in\mathcal{T}_h} (\nabla u, \nabla \pi_T^{0,\ell-2} \mathfrak{v}_T)_T - \sum_{T\in\mathcal{T}_h} \sum_{F\in\mathcal{F}_T} (\nabla u \cdot \boldsymbol{n}_{TF}, \Pi_T^{\ell-2} \mathfrak{v}_T - \mathfrak{v}_T)_F.
$$

where the introduction of $\mathfrak{v}_T = (\mathfrak{v}_h)_{|T}$ into the face integrals is justified by Corollary 1.19 after recalling that $\mathfrak{v}_h$ is continuous. On the other hand, applying (5.117) to $w = \pi_{T,\text{cvem}}^{1,\ell} u$ for each $T \in \mathcal{T}_h$, we have

$$
\begin{aligned}
&\mathfrak{T}_2 = \\
&- \sum_{T\in\mathcal{T}_h} (\nabla \pi_{T,\text{cvem}}^{1,\ell} u, \nabla \pi_T^{0,\ell-2} \mathfrak{v}_T)_T - \sum_{T\in\mathcal{T}_h} \sum_{F\in\mathcal{F}_T} ((\nabla \pi_{T,\text{cvem}}^{1,\ell} u) \cdot \boldsymbol{n}_{TF}, \mathfrak{v}_T - \Pi_T^{\ell-2} \mathfrak{v}_T)_F.
\end{aligned}
$$

We infer that

$$
\begin{aligned}
|\mathfrak{T}_1 + \mathfrak{T}_2| &\le \Bigg| \sum_{T\in\mathcal{T}_h} (\nabla (u - \pi_{T,\text{cvem}}^{1,\ell} u), \nabla \pi_T^{0,\ell-2} \mathfrak{v}_T)_T \\
&\qquad + \sum_{T\in\mathcal{T}_h} \sum_{F\in\mathcal{F}_T} (\nabla (u - \pi_{T,\text{cvem}}^{1,\ell} u) \cdot \boldsymbol{n}_{TF}, \mathfrak{v}_T - \Pi_T^{\ell-2} \mathfrak{v}_T)_F \Bigg| \\
&\le \sum_{T\in\mathcal{T}_h} \|\nabla (u - \pi_{T,\text{cvem}}^{1,\ell} u)\|_T \|\nabla \pi_T^{0,\ell-2} \mathfrak{v}_T\|_T \\
&\qquad + \sum_{T\in\mathcal{T}_h} \sum_{F\in\mathcal{F}_T} h_F^{\frac{1}{2}} \|\nabla (u - \pi_{T,\text{cvem}}^{1,\ell} u)\|_F \, h_F^{-\frac{1}{2}} \|\mathfrak{v}_T - \Pi_T^{\ell-2} \mathfrak{v}_T\|_F \\
&\lesssim h^{r+1} |u|_{H^{r+2}(\mathcal{T}_h)} \|\mathfrak{v}_h\|_{\text{cvem},\mathfrak{a},h},
\end{aligned}
$$

where the second bound follows using Cauchy–Schwarz inequalities on the volumetric terms and generalised Hölder inequalities with exponents $(2, 2, \infty)$ on the boundary terms along with $\|\boldsymbol{n}_{TF}\|_{L^\infty(F)^2} = 1$, and the conclusion is obtained using the approximation properties of $\pi_{T,\mathrm{cvem}}^{1,\ell}$ stated in Theorem 5.40 together with discrete Cauchy–Schwarz inequalities, the definition (5.127) (see also (5.105)) of the norm $\|\cdot\|_{\mathrm{cvem},2,h}$, and the norm equivalence (5.128).

Combined with (5.132) and (5.131), this bound on $|\mathfrak{T}_1 + \mathfrak{T}_2|$ shows that

$$\sup_{\mathfrak{v}_h \in \mathfrak{U}_{h,0}^{\ell} \setminus \{0\}} \frac{|\mathcal{E}_h(u; \mathfrak{v}_h)|}{\|\mathfrak{v}_h\|_{\mathrm{cvem},\mathfrak{a},h}} \lesssim h^{r+1} |u|_{H^{r+2}(\mathcal{T}_h)},$$

which concludes the proof. □

5.6 Gradient Discretisation Method

The Gradient Discretisation Method (GDM) is a generic framework for the design and analysis of schemes for linear and nonlinear diffusion problems [178]. It covers many classical methods, such as Finite Elements (conforming, nonconforming and mixed), Finite Volumes, Discontinuous Galerkin, etc. The principle of the GDM is to replace, in the weak formulation of the PDE, the continuous space and operators by discrete counterparts; each choice of discrete space and operator corresponds to a different numerical method, that is usable on a range of different models (the same discrete space and operators can be re-used for different PDEs).

Our purpose here is to construct a Gradient Discretisation Method inspired by the HHO method. Because the GDM is designed to tackle fully anisotropic and heterogeneous (possibly nonlinear) diffusion models, it makes more sense to base our construction on the HHO method designed in Sect. 4.2 for the locally variable diffusion model (4.33).

The material presented here is adapted from [163], which also covers GDMs designed from the HHO method of Chap. 2 and from the HHO(k, ℓ) method of Sect. 5.1.

5.6.1 *The Gradient Discretisation Method*

The GDM consists in replacing, in the continuous weak formulation of the problem, the infinite-dimensional space, functions, and gradients with a finite dimensional space and reconstructions of functions and gradients. The set of these discrete elements (space, reconstruction of functions, reconstruction of gradients) is called a Gradient Discretisation (GD).

Definition 5.46 (Gradient Discretisation) A Gradient Discretisation (for homogeneous Dirichlet boundary conditions) is a triplet $\mathcal{D} = (X_{\mathcal{D},0}, \Pi_{\mathcal{D}}, \nabla_{\mathcal{D}})$ where

- $X_{\mathcal{D},0}$ is a finite dimensional space (the space of discrete unknowns of the method),
- $\Pi_{\mathcal{D}} : X_{\mathcal{D},0} \to L^2(\Omega)$ is a linear operator that reconstructs functions from vectors of discrete unknowns,
- $\nabla_{\mathcal{D}} : X_{\mathcal{D},0} \to L^2(\Omega)^d$ is a linear operator that reconstructs "gradients" from vectors of discrete unknowns; it must be chosen such that $v \mapsto \|\nabla_{\mathcal{D}} v\|$ is a norm on $X_{\mathcal{D},0}$.

A Gradient Discretisation $\mathcal{D}$ having been chosen, using its discrete elements in lieu of their continuous counterparts in the weak formulation of the problem leads to a Gradient Scheme (GS) for that problem. Let us illustrate this principle on the locally variable diffusion model of Sect. 4.2. We recall that $\boldsymbol{K} : \Omega \to \mathbb{R}^{d\times d}_{\text{sym}}$ is a bounded uniformly coercive diffusion tensor (see (4.34)), and that the weak formulation of the problem is (see (4.35))

$$\text{Find } u \in H^1_0(\Omega) \text{ such that } \int_\Omega \boldsymbol{K}\nabla u\cdot\nabla v = \int_\Omega f v \qquad \forall v \in H^1_0(\Omega). \tag{5.133}$$

The Gradient Scheme for this problem is then: Find $u_{\mathcal{D}} \in X_{\mathcal{D},0}$ such that

$$\int_\Omega \boldsymbol{K}\nabla_{\mathcal{D}} u_{\mathcal{D}}\cdot\nabla_{\mathcal{D}} v_{\mathcal{D}} = \int_\Omega f \Pi_{\mathcal{D}} v \qquad \forall v_{\mathcal{D}} \in X_{\mathcal{D},0}. \tag{5.134}$$

The existence and uniqueness of a solution to problem (5.134) follows from the Lax–Milgram lemma (Lemma 2.20). Specific choices of GDs lead to GSs that correspond to known schemes for (5.133), see [178, Chapters 8–14]. In Sect. 5.6.2 we construct a Gradient Discretisation such that (5.134) is the HHO discretisation of the locally variable diffusion problem.

Remark 5.47 (Usage of Integral Signs) Above and throughout the rest of this section, we write integrals instead of L^2-inner products. The reason is twofold: first, because this is the way GDM has been historically presented; second, to emphasise the ability of the GDM to generate schemes for nonlinear problems, possibly posed in a non-Hilbertian setting, such as the ones considered in Chap. 6.

The accuracy of a Gradient Discretisation is measured through three quantities. The first one is the *discrete Poincaré constant*

$$C_{\mathcal{D}} = \max_{v \in X_{\mathcal{D},0}\backslash\{0\}} \frac{\|\Pi_{\mathcal{D}} v\|}{\|\nabla_{\mathcal{D}} v\|}. \tag{5.135}$$

The second is the *interpolation error* $S_{\mathcal{D}} : H^1_0(\Omega) \to \mathbb{R}$ defined by

$$S_{\mathcal{D}}(\varphi) = \min_{v \in X_{\mathcal{D},0}} \left(\|\Pi_{\mathcal{D}} v - \varphi\| + \|\nabla_{\mathcal{D}} v - \nabla\varphi\|\right) \qquad \forall \varphi \in H^1_0(\Omega). \tag{5.136}$$

The third and last measure of accuracy associated with $\mathcal{D}$ is $W_{\mathcal{D}} : \boldsymbol{H}(\text{div}; \Omega) \to \mathbb{R}$ that measures the *defect of conformity* of the method (how well a discrete Stokes formula holds):

$$W_{\mathcal{D}}(\boldsymbol{\psi}) = \max_{v \in X_{\mathcal{D},0} \setminus \{0\}} \frac{1}{\|\nabla_{\mathcal{D}} v\|} \left| \int_{\Omega} (\nabla_{\mathcal{D}} v \cdot \boldsymbol{\psi} + \Pi_{\mathcal{D}} v \, \nabla \cdot \boldsymbol{\psi}) \right| \qquad \forall \boldsymbol{\psi} \in \boldsymbol{H}(\text{div}; \Omega). \tag{5.137}$$

Based on these quantities, an error estimate for the GS can be established.

Theorem 5.48 (Error Estimate for the Gradient Scheme) *Let u be the weak solution to the locally variable diffusion problem* (5.133)*, and let $u_{\mathcal{D}}$ be the solution to the Gradient Scheme* (5.134)*. Then,*

$$\|\nabla u - \nabla_{\mathcal{D}} u_{\mathcal{D}}\| \le \underline{K}^{-1} \left[W_{\mathcal{D}}(\boldsymbol{K} \nabla u) + (\underline{K} + \overline{K}) S_{\mathcal{D}}(u) \right] \tag{5.138}$$

and

$$\|u - \Pi_{\mathcal{D}} u_{\mathcal{D}}\| \le \underline{K}^{-1} \left[C_{\mathcal{D}} W_{\mathcal{D}}(\boldsymbol{K} \nabla u) + (C_{\mathcal{D}} \overline{K} + \underline{K}) S_{\mathcal{D}}(u) \right], \tag{5.139}$$

where $\underline{K}$ and $\overline{K}$ are as in (4.34)*.*

Proof A standalone proof can be found in [178, Theorem 2.28]. For the sake of completeness, we show here how these error estimates can be deduced from the abstract framework of Appendix A. The setting we consider is: $\mathtt{U} = H_0^1(\Omega)$, $\mathtt{a}(u, v) = (\boldsymbol{K} \nabla u, \nabla v)$, $\mathtt{l}(v) = (f, v)$, $\mathtt{U}_h = X_{\mathcal{D},0}$ endowed with the norm $\|\nabla_{\mathcal{D}} \cdot\|$, $\mathtt{a}_h(u_{\mathcal{D}}, v_{\mathcal{D}}) = (\boldsymbol{K} \nabla_{\mathcal{D}} u_{\mathcal{D}}, \nabla_{\mathcal{D}} v_{\mathcal{D}})$, and $\mathtt{l}_h(v_{\mathcal{D}}) = (f, \Pi_{\mathcal{D}} v_{\mathcal{D}})$. The interpolate $\mathtt{I}_h u = \mathtt{I}_{\mathcal{D}} u$ is defined as the element of $X_{\mathcal{D},0}$ that realises the minimum defining $S_{\mathcal{D}}(u)$, so that

$$S_{\mathcal{D}}(u) = \|\Pi_{\mathcal{D}} \mathtt{I}_{\mathcal{D}} u - u\| + \|\nabla_{\mathcal{D}} \mathtt{I}_{\mathcal{D}} u - \nabla u\|. \tag{5.140}$$

We note that the index "h" was kept in $\mathtt{U}_h$, $\mathtt{a}_h$, etc. to match the notations in Appendix A, but that in the GDM framework such an index is rarely used (as GDMs need not be mesh-based), and replaced with the index $\mathcal{D}$.

Inserting $\pm \nabla_{\mathcal{D}} \mathtt{I}_{\mathcal{D}} u$ into the norm and using the triangle inequality together with (5.140), we have

$$\begin{aligned} \|\nabla u - \nabla_{\mathcal{D}} u_{\mathcal{D}}\| &\le \|\nabla u - \nabla_{\mathcal{D}} \mathtt{I}_{\mathcal{D}} u\| + \|\nabla_{\mathcal{D}} \mathtt{I}_{\mathcal{D}} u - \nabla_{\mathcal{D}} u_{\mathcal{D}}\| \\ &\le S_{\mathcal{D}}(u) + \|\nabla_{\mathcal{D}}(\mathtt{I}_{\mathcal{D}} u - u_{\mathcal{D}})\|. \end{aligned}$$

Similarly, inserting $\pm\Pi_{\mathcal{D}}\mathrm{I}_{\mathcal{D}}u$, using the triangle inequality, (5.140) and the definition (5.135) of $C_{\mathcal{D}}$, we have

$$\begin{aligned}\|u - \Pi_{\mathcal{D}}u_{\mathcal{D}}\| &\le \|u - \Pi_{\mathcal{D}}\mathrm{I}_{\mathcal{D}}u\| + \|\Pi_{\mathcal{D}}(\mathrm{I}_{\mathcal{D}}u - u_{\mathcal{D}})\| \\ &\le S_{\mathcal{D}}(u) + C_{\mathcal{D}}\|\nabla_{\mathcal{D}}(\mathrm{I}_{\mathcal{D}}u - u_{\mathcal{D}})\|.\end{aligned}$$

These estimates prove (5.138)–(5.139) provided we can establish that

$$\|\nabla_{\mathcal{D}}(\mathrm{I}_{\mathcal{D}}u - u_{\mathcal{D}})\| \le \underline{K}^{-1}\left(W_{\mathcal{D}}(\boldsymbol{K}\nabla u) + \overline{K}S_{\mathcal{D}}(u)\right). \tag{5.141}$$

This bound will follow from the Third Strang lemma (Lemma A.7). By assumption (4.34) on $\boldsymbol{K}$, the bilinear form a_h is coercive with respect to the norm on $X_{\mathcal{D},0}$, with constant $\underline{K}$. We now estimate the consistency error

$$\mathcal{E}_h(u; v_{\mathcal{D}}) := \int_\Omega f\,\Pi_{\mathcal{D}}v_{\mathcal{D}} - \int_\Omega \boldsymbol{K}\nabla_{\mathcal{D}}\mathrm{I}_{\mathcal{D}}u\cdot\nabla_{\mathcal{D}}v_{\mathcal{D}}.$$

Recalling that $f = -\nabla\cdot(\boldsymbol{K}\nabla u)$ and inserting $\pm\int_\Omega \boldsymbol{K}\nabla u\cdot\nabla_{\mathcal{D}}v_{\mathcal{D}}$, we write

$$\begin{aligned}&|\mathcal{E}_h(u; v_{\mathcal{D}})| \\ &\quad= \left|-\int_\Omega (\nabla\cdot(\boldsymbol{K}\nabla u)\Pi_{\mathcal{D}}v_{\mathcal{D}} + \boldsymbol{K}\nabla u\cdot\nabla_{\mathcal{D}}v_{\mathcal{D}}) + \int_\Omega \boldsymbol{K}\left(\nabla u - \nabla_{\mathcal{D}}\mathrm{I}_{\mathcal{D}}u\right)\cdot\nabla_{\mathcal{D}}v_{\mathcal{D}}\right| \\ &\quad\le W_{\mathcal{D}}(\boldsymbol{K}\nabla u)\|\nabla_{\mathcal{D}}v_{\mathcal{D}}\| + \left|\int_\Omega \boldsymbol{K}(\nabla u - \nabla_{\mathcal{D}}\mathrm{I}_{\mathcal{D}}u)\cdot\nabla_{\mathcal{D}}v_{\mathcal{D}}\right| \\ &\quad\le W_{\mathcal{D}}(\boldsymbol{K}\nabla u)\|\nabla_{\mathcal{D}}v_{\mathcal{D}}\| + \overline{K}S_{\mathcal{D}}(u)\|\nabla_{\mathcal{D}}v_{\mathcal{D}}\|,\end{aligned}$$

where we have used the triangle inequality and the definition (5.137) of $W_{\mathcal{D}}(\boldsymbol{K}\nabla u)$ in the first inequality, and invoked the Cauchy–Schwarz inequality together with (5.140) and (4.34) to conclude. Recalling that $\|\nabla_{\mathcal{D}}v_{\mathcal{D}}\|$ is the norm of $v_{\mathcal{D}}$ in $X_{\mathcal{D},0}$, this estimate on $\mathcal{E}_h(u; v_{\mathcal{D}})$ readily gives

$$\|\mathcal{E}_h(u;\cdot)\|_{X^\star_{\mathcal{D},0}} \le W_{\mathcal{D}}(\boldsymbol{K}\nabla u) + \overline{K}S_{\mathcal{D}}(u)$$

which, according to (A.6) and the coercivity of a_h, yields (5.141). □

5.6.2 Discontinuous Skeletal Gradient Discretisations

In this section, we construct a GD inspired by the HHO method designed for the locally variable diffusion problem. We show that, in case of piecewise-constant diffusion tensor, the Gradient Scheme corresponding to this Gradient Discretisation is exactly the HHO scheme (4.63) for (5.133). This Gradient Discretisation is called Discontinuous Skeletal Gradient Discretisation (DSGD) to recall one of the main features of HHO schemes, that will carry to the GS corresponding to this GD: the main unknowns of the scheme, after static condensation (see Sect. B.3.2), are polynomial functions on the mesh faces (the skeleton of the mesh), without continuity conditions at the vertices (if $d = 2$) or at the edges (if $d = 3$).

Let $\mathcal{M}_h$ be a polytopal mesh and define $\mathcal{D}_h = (X_{\mathcal{D}_h,0}, \Pi_{\mathcal{D}_h}, \nabla_{\mathcal{D}_h})$ by setting:

$$X_{\mathcal{D}_h,0} := \underline{U}_{h,0}^k, \tag{5.142a}$$

$$\Pi_{\mathcal{D}_h}\underline{v}_h := v_h \qquad \forall \underline{v}_h \in \underline{U}_{h,0}^k = X_{\mathcal{D}_h,0}, \tag{5.142b}$$

$$\nabla_{\mathcal{D}_h}\underline{v}_h := \mathbf{G}_h^k\underline{v}_h + \mathbf{S}_h\underline{v}_h \qquad \forall \underline{v}_h \in \underline{U}_{h,0}^k = X_{\mathcal{D}_h,0}, \tag{5.142c}$$

where $\underline{U}_{h,0}^k$ is defined by (2.36), v_h by (2.33), $\mathbf{G}_h^k$ is the global gradient reconstruction (4.74) (with $\mathbf{G}_T^k$ defined by (4.37)), and the stabilisation term $\mathbf{S}_h : \underline{U}_h^k \to L^2(\Omega)^d$ is given by

$$(\mathbf{S}_h\underline{v}_h)_{|T} := \mathbf{S}_T\underline{v}_T \qquad \forall T \in \mathcal{T}_h\,,\ \forall \underline{v}_h \in \underline{U}_h^k, \tag{5.142d}$$

with local stabilisation terms $(\mathbf{S}_T)_{T\in\mathcal{T}_h}$ satisfying the design properties in Assumption 5.49 below. To state these properties, for a given $T \in \mathcal{T}_h$ we introduce the difference seminorm $|\cdot|_{\delta,\partial T}$ on $\underline{U}_T^k$ such that

$$|\underline{v}_T|_{\delta,\partial T}^2 := \sum_{F\in\mathcal{F}_T} h_F^{-1}\|(\delta_{TF}^k - \delta_T^k)\underline{v}_T\|_F^2 \qquad \forall \underline{v}_T \in \underline{U}_T^k, \tag{5.143}$$

where the difference operators δ_T^k and $(\delta_{TF}^k)_{F\in\mathcal{F}_T}$ are defined by (2.19).

Assumption 5.49 (DSGD Local Stabilisation Term $\mathbf{S}_T$) *The local stabilisation term $\mathbf{S}_T : \underline{U}_T^k \to L^2(T)^d$ is a linear map that satisfies the following conditions:*

(GDM-S1) Stability and boundedness. *For all $\underline{v}_T \in \underline{U}_T^k$, it holds that*

$$\|\mathbf{S}_T\underline{v}_T\|_T^2 \simeq |\underline{v}_T|_{\delta,\partial T}^2, \tag{5.144}$$

where the hidden constants are independent of h, T and $\underline{v}_T$ (but may depend on $\boldsymbol{K}$).

(GDM-S2) Orthogonality. *For all $\underline{v}_T \in \underline{U}_T^k$ and all $\boldsymbol{\phi} \in \mathbb{P}^k(T)^d$, it holds*

$$(\mathbf{S}_T\underline{v}_T, \boldsymbol{\phi})_T = 0. \tag{5.145}$$

(GDM-S3) Polynomial consistency. *For all $w \in \mathbb{P}^{k+1}(T)$, it holds $\mathbf{S}_T\underline{I}_T^k w = 0$.*

An example of such a stabilisation term is constructed in Sect. 5.6.3.

Remark 5.50 (Non-Hilbertian Setting) To use the DSGD in non-Hilbertian setting (e.g., for the p-Laplacian equation, see Chap. 6), we additionally need to assume that the range of $\mathbf{S}_T$ is contained in a space of piecewise polynomials on T; see [163] for details.

The following theorem makes explicit the link between $\mathcal{D}_h$ constructed above and the HHO scheme for the locally variable diffusion problem (5.133).

Theorem 5.51 (HHO is a GDM) *Let $\mathcal{M}_h$ be a polytopal mesh, and assume that $\boldsymbol{K}$ is piecewise constant on $\mathcal{T}_h$. Then, there exists a choice of stabilisation terms $(\mathbf{S}_T)_{T\in\mathcal{T}_h}$, satisfying Assumption 5.49, such that the Gradient Scheme* (5.134) *with $\mathcal{D} = \mathcal{D}_h$ given by* (5.142) *is the HHO scheme* (4.63) *for* (5.133).

Remark 5.52 (Locally Variable Diffusion Tensor) An inspection of the proof below shows that the result extends to the case where $\boldsymbol{K}$ varies inside each cell, provided that the orthogonality condition (5.145) holds for all $\boldsymbol{\phi} = \boldsymbol{K}_{|T}\boldsymbol{\psi}$ with $\boldsymbol{\psi} \in \mathbb{P}^k(T)^d$.

Proof The definition (5.142b) of $\Pi_{\mathcal{D}_h}$ shows that the right-hand sides of (4.63) and (5.134) are identical. We therefore only have to prove that, for all $\underline{u}_h, \underline{v}_h \in \underline{U}_{h,0}^k$, recalling the definition (4.45) of $\mathrm{a}_{\boldsymbol{K},h}$, it holds

$$\mathrm{a}_{\boldsymbol{K},h}(\underline{u}_h, \underline{v}_h) = \int_\Omega \boldsymbol{K}\nabla_{\mathcal{D}_h}\underline{u}_h\cdot\nabla_{\mathcal{D}_h}\underline{v}_h.$$

Developing the right-hand side according to (5.142c), this amounts to showing that

$$\begin{aligned}
\mathrm{a}_{\boldsymbol{K},h}(\underline{u}_h, \underline{v}_h) &= (\boldsymbol{K}\mathbf{G}_h^k\underline{u}_h, \mathbf{G}_h^k\underline{v}_h) + (\boldsymbol{K}\mathbf{G}_h^k\underline{u}_h, \mathbf{S}_h\underline{v}_h) \\
&\quad + (\mathbf{S}_h\underline{u}_h, \boldsymbol{K}\mathbf{G}_h^k\underline{v}_h) + (\boldsymbol{K}\mathbf{S}_h\underline{u}_h, \mathbf{S}_h\underline{v}_h) \\
&= (\boldsymbol{K}\mathbf{G}_h^k\underline{u}_h, \mathbf{G}_h^k\underline{v}_h) + (\boldsymbol{K}\mathbf{S}_h\underline{u}_h, \mathbf{S}_h\underline{v}_h),
\end{aligned} \tag{5.146}$$

where the cross terms have been eliminated in the third line using (GDM-S2) in Assumption 5.49 which yields, since $\boldsymbol{K}_{|T}\mathbf{G}_T^k\underline{u}_T \in \mathbb{P}^k(T)^d$ (because $\boldsymbol{K}_{|T}$ is constant) for all $T \in \mathcal{T}_h$,

$$(\boldsymbol{K}\mathbf{G}_h^k\underline{u}_h, \mathbf{S}_h\underline{v}_h) = \sum_{T\in\mathcal{T}_h}(\boldsymbol{K}_{|T}\mathbf{G}_T^k\underline{u}_T, \mathbf{S}_T\underline{v}_T)_T = 0$$

(and similarly with $\underline{u}_h$ and $\underline{v}_h$ swapped). By definition (4.47) of $\mathrm{a}_{\boldsymbol{K},h}$ (see also (4.45)), a sufficient condition for (5.146) to hold is that, for all $T \in \mathcal{T}_h$,

$$(\boldsymbol{K}_{|T}\mathbf{S}_T\underline{u}_T, \mathbf{S}_T\underline{v}_T)_T = \mathrm{s}_{\boldsymbol{K},T}(\underline{u}_T, \underline{v}_T), \tag{5.147}$$

where $\mathrm{s}_{\boldsymbol{K},T}$ is given by (4.46).

Let $V_T := \underline{U}_T^k/[\underline{I}_T^k\mathbb{P}^{k+1}(T)]$ be the quotient space of $\underline{U}_T^k$ for the relation $\sim$ defined by: $\underline{u}_T \sim \underline{v}_T$ if and only if $\underline{u}_T - \underline{v}_T \in \underline{I}_T^k\mathbb{P}^{k+1}(T)$. Let $P : \underline{U}_T^k \to V_T$ be the canonical projection. If $\underline{u}_T, \underline{v}_T, \underline{w}_T \in \underline{U}_T^k$ with $\underline{u}_T \sim \underline{w}_T$, then the polynomial consistency (2.21) of the difference operators and the definition of $\mathrm{s}_{\boldsymbol{K},T}$ yield $\mathrm{s}_{\boldsymbol{K},T}(\underline{u}_T, \underline{v}_T) = \mathrm{s}_{\boldsymbol{K},T}(\underline{w}_T, \underline{v}_T)$. Hence, $\mathrm{s}_{\boldsymbol{K},T}$ defines a symmetric bilinear form $\tilde{\mathrm{s}}_{\boldsymbol{K},T}$ on $V_T \times V_T$ such that

$$\tilde{\mathrm{s}}_{\boldsymbol{K},T}(P\underline{u}_T, P\underline{v}_T) := \mathrm{s}_{\boldsymbol{K},T}(\underline{u}_T, \underline{v}_T) \qquad \forall \underline{u}_T, \underline{v}_T \in \underline{U}_T^k. \tag{5.148}$$

Let us show that $\tilde{\mathrm{s}}_{\boldsymbol{K},T}$ is an inner product on V_T. It is clearly positive semidefinite, so we only have to show that, if $\tilde{\mathrm{s}}_{\boldsymbol{K},T}(P\underline{v}_T, P\underline{v}_T) = 0$, then $P\underline{v}_T = 0$. Assuming the former relation, we have $\mathrm{s}_{\boldsymbol{K},T}(\underline{v}_T, \underline{v}_T) = 0$ and thus, using the polynomial consistency (2.21) of the difference operators and the definition (4.46) of $\mathrm{s}_{\boldsymbol{K},T}$,

$$\mathrm{s}_{\boldsymbol{K},T}(\underline{v}_T - \underline{I}_T^k\mathrm{p}_T^{k+1}\underline{v}_T, \underline{v}_T - \underline{I}_T^k\mathrm{p}_T^{k+1}\underline{v}_T) = 0.$$

Let $\underline{z}_T := \underline{v}_T - \underline{I}_T^k\mathrm{p}_T^{k+1}\underline{v}_T$, so that the relation above becomes $\mathrm{s}_{\boldsymbol{K},T}(\underline{z}_T, \underline{z}_T) = 0$. Using then (4.53) with $\underline{z}_T$ instead of $\underline{v}_T$, we infer

$$\sum_{F\in\mathcal{F}_T} \frac{K_{TF}}{h_F}\|z_F - z_T\|_F^2 \lesssim \overline{K}_T\|\nabla\mathrm{p}_T^{k+1}\underline{z}_T\|_T^2, \tag{5.149}$$

where the hidden constant is independent of h, T, and $\underline{z}_T$. The commutation property (2.14) and the polynomial invariance of the elliptic projector yield

$$\begin{aligned}\mathrm{p}_T^{k+1}\underline{z}_T = \mathrm{p}_T^{k+1}\underline{v}_T - \mathrm{p}_T^{k+1}\underline{I}_T^k(\mathrm{p}_T^{k+1}\underline{v}_T) &= \mathrm{p}_T^{k+1}\underline{v}_T - \pi_T^{1,k+1}\mathrm{p}_T^{k+1}\underline{v}_T \\ &= \mathrm{p}_T^{k+1}\underline{v}_T - \mathrm{p}_T^{k+1}\underline{v}_T = 0.\end{aligned} \tag{5.150}$$

Hence, (5.149) shows that $z_F = (z_T)_{|F}$ for all $F \in \mathcal{F}_T$. Plugged together with (5.150) into the definition (2.12) of $\mathrm{p}_T^{k+1}\underline{z}_T$, this yields $(\nabla z_T, \nabla w)_T = 0$ for all $w \in \mathbb{P}^{k+1}(T)$, and thus $\nabla z_T = 0$. As a consequence, z_T is constant equal to some $c \in \mathbb{R}$, and thus $z_F = z_T = c$ for all $F \in \mathcal{F}_T$. We therefore have $\underline{z}_T = \underline{I}_T^k c$, from which we infer $\underline{v}_T = \underline{I}_T^k(\mathrm{p}_T^{k+1}\underline{v}_T + c) \in \underline{I}_T^k\mathbb{P}^{k+1}(T)$. Hence, $P\underline{v}_T = 0$ and $\tilde{\mathrm{s}}_{\boldsymbol{K},T}$ is indeed an inner product on V_T.

Let now $\boldsymbol{E} : V_T \to [\mathbb{P}^k(T)^d]^\perp \subset L^2(T)^d$ be an embedding, where $[\mathbb{P}^k(T)^d]^\perp$ denotes the orthogonal complement of $\mathbb{P}^k(T)^d$ in $L^2(T)^d$. Such an embedding can be constructed, for example, taking an arbitrary basis $(\boldsymbol{v}_i)_{i\in I}$ on V_T, a linearly-independent family $(\boldsymbol{\phi}_i)_{i\in I}$ in $[\mathbb{P}^k(T)^d]^\perp$ (this space is infinite-dimensional), and setting $\boldsymbol{E}(\sum_{i\in I}\alpha_i\boldsymbol{v}_i) = \sum_{i\in I}\alpha_i\boldsymbol{\phi}_i$ for any real numbers $(\alpha_i)_{i\in I}$. The mapping $\mathrm{Im}(\boldsymbol{E}) \times \mathrm{Im}(\boldsymbol{E}) \ni (\boldsymbol{\phi}, \boldsymbol{\psi}) \mapsto \tilde{\mathrm{s}}_{\boldsymbol{K},T}(\boldsymbol{E}^{-1}(\boldsymbol{\phi}), \boldsymbol{E}^{-1}(\boldsymbol{\psi})) \in \mathbb{R}$ is an inner product on $\mathrm{Im}(\boldsymbol{E})$. The bilinear form $(\boldsymbol{\phi}, \boldsymbol{\psi}) \mapsto (\boldsymbol{K}_{|T}\boldsymbol{\phi}, \boldsymbol{\psi})_T$ is another inner product on $\mathrm{Im}(\boldsymbol{E})$. Applying Lemma 5.53 below on $\mathrm{Im}(\boldsymbol{E})$ endowed with these inner products yields an isomorphism $\mathcal{L}$ of $\mathrm{Im}(\boldsymbol{E})$ such that

$$\tilde{\mathrm{s}}_{\boldsymbol{K},T}(\boldsymbol{E}^{-1}(\boldsymbol{\phi}), \boldsymbol{E}^{-1}(\boldsymbol{\psi})) = (\boldsymbol{K}_{|T}\mathcal{L}\boldsymbol{\phi}, \mathcal{L}\boldsymbol{\psi})_T \qquad \forall \boldsymbol{\phi}, \boldsymbol{\psi} \in \mathrm{Im}(\boldsymbol{E}).$$

Applying this to $\boldsymbol{\phi} = \boldsymbol{E}(P\underline{u}_T)$ and $\boldsymbol{\psi} = \boldsymbol{E}(P\underline{v}_T)$, for arbitrary $\underline{u}_T, \underline{v}_T \in \underline{U}_T^k$, we obtain

$$\tilde{\mathrm{s}}_{\boldsymbol{K},T}(P\underline{u}_T, P\underline{v}_T) = (\boldsymbol{K}_{|T}\mathcal{L}\boldsymbol{E}(P\underline{u}_T), \mathcal{L}\boldsymbol{E}(P\underline{v}_T))_T.$$

Recalling (5.148) then yields (5.147) with $\mathbf{S}_T = \mathcal{L}\boldsymbol{E}P$. The coercivity and boundedness (GDM-S1) for $\mathbf{S}_T$ follows from (5.147) and from the definition (4.46) of $\mathrm{s}_{\boldsymbol{K},T}$. The polynomial consistency (GDM-S3) is also a consequence of (5.147) (using $\underline{u}_T = \underline{v}_T = \underline{I}_T^k w$), and of the polynomial consistency (2.21) of the difference operators which yields $\mathrm{s}_{\boldsymbol{K},T}(\underline{I}_T^k w, \underline{I}_T^k w) = 0$ whenever $w \in \mathbb{P}^{k+1}(T)$. Finally, the range of $\mathbf{S}_T$ is contained $\mathrm{Im}(\mathcal{L}) = \mathrm{Im}(\boldsymbol{E}) \subset [\mathbb{P}^k(T)^d]^\perp$, and (GDM-S2) thus holds. □

The following lemma, used in the proof above, corresponds to [177, Lemma 5.2].

Lemma 5.53 *Let V be a finite-dimensional vector space, and let $\langle\cdot,\cdot\rangle_{V,1}$ and $\langle\cdot,\cdot\rangle_{V,2}$ be two inner products on V. Then, there exists an isomorphism $\mathcal{L} : V \to V$ such that, for all $(x, y) \in V^2$,*

$$\langle x, y\rangle_{V,1} = \langle \mathcal{L}x, \mathcal{L}y\rangle_{V,2}.$$

5.6.3 Construction of a Stabilisation Term Satisfying the Design Conditions

The proof of Theorem 5.51 provides a way to construct suitable stabilisation terms $(\mathbf{S}_T)_{T\in\mathcal{T}_h}$ that satisfy Assumption 5.49. This construction however appears cumbersome. We present another, more practical, way to construct stabilisation terms, using a Raviart–Thomas–Nédélec space on a simplicial subdivision of the elements.

Let $\mathcal{M}_h = (\mathcal{T}_h, \mathcal{F}_h)$ be a mesh from a regular sequence, and let $T \in \mathcal{T}_h$. In this section, the hidden constants in $\lesssim$ and $\gtrsim$ may depend on the regularity parameter ϱ of this sequence and on the polynomial degree k, but not on h or T.

5.6.3.1 A Key Property and a Lifting of Face Differences

We start with a few preliminary comments and results. Let $\boldsymbol{\delta}_{\nabla,T}^k : \underline{U}_T^k \to \mathbb{P}^k(T)^d$ be the operator defined by

$$\boldsymbol{\delta}_{\nabla,T}^k := \nabla \mathrm{p}_T^{k+1} - \mathbf{G}_T^k. \tag{5.151}$$

For all $\underline{v}_T \in \underline{U}_T^k$ and $\boldsymbol{\phi} \in \mathbb{P}^k(T)^d$, it holds

$$\begin{aligned}
(\boldsymbol{\delta}_{\nabla,T}^k \underline{v}_T, \boldsymbol{\phi})_T &= (\nabla \mathrm{p}_T^{k+1} \underline{v}_T, \boldsymbol{\phi})_T - (\mathbf{G}_T^k \underline{v}_T, \boldsymbol{\phi})_T \\
&= -(\mathrm{p}_T^{k+1} \underline{v}_T, \nabla \cdot \boldsymbol{\phi})_T + \sum_{F \in \mathcal{F}_T} (\mathrm{p}_T^{k+1} \underline{v}_T, \boldsymbol{\phi} \cdot \boldsymbol{n}_{TF})_F \\
&\quad + (v_T, \nabla \cdot \boldsymbol{\phi})_T - \sum_{F \in \mathcal{F}_T} (v_F, \boldsymbol{\phi} \cdot \boldsymbol{n}_{TF})_F \\
&= (v_T - \mathrm{p}_T^{k+1} \underline{v}_T, \nabla \cdot \boldsymbol{\phi})_T + \sum_{F \in \mathcal{F}_T} (\mathrm{p}_T^{k+1} \underline{v}_T - v_F, \boldsymbol{\phi} \cdot \boldsymbol{n}_{TF})_F,
\end{aligned}$$

where we have used the definition of $\boldsymbol{\delta}_{\nabla,T}^k$ in the first line, an integration by parts together with the definition (4.37) of $\mathbf{G}_T^k$ in the second line, and we have gathered the volumetric and boundary contributions in the third line. Since $\nabla \cdot \boldsymbol{\phi} \in \mathbb{P}^{k-1}(T) \subset \mathbb{P}^k(T)$ and $\boldsymbol{\phi}_{|F} \cdot \boldsymbol{n}_{TF} \in \mathbb{P}^k(F)$, we can introduce the L^2-orthogonal projectors $\pi_T^{0,k}$ and $\pi_F^{0,k}$ and continue with

$$\begin{aligned}
&(\boldsymbol{\delta}_{\nabla,T}^k \underline{v}_T, \boldsymbol{\phi})_T \\
&\quad = (\pi_T^{0,k}(v_T - \mathrm{p}_T^{k+1} \underline{v}_T), \nabla \cdot \boldsymbol{\phi})_T + \sum_{F \in \mathcal{F}_T} (\pi_F^{0,k}(\mathrm{p}_T^{k+1} \underline{v}_T - v_F), \boldsymbol{\phi} \cdot \boldsymbol{n}_{TF})_F \\
&\quad = -(\delta_T^k \underline{v}_T, \nabla \cdot \boldsymbol{\phi})_T + \sum_{F \in \mathcal{F}_T} (\delta_{TF}^k \underline{v}_T, \boldsymbol{\phi} \cdot \boldsymbol{n}_{TF})_F \\
&\quad = (\nabla \delta_T^k \underline{v}_T, \boldsymbol{\phi})_T + \sum_{F \in \mathcal{F}_T} ((\delta_{TF}^k - \delta_T^k) \underline{v}_T, \boldsymbol{\phi} \cdot \boldsymbol{n}_{TF})_F,
\end{aligned}$$

where the second equality follows from the definitions (2.19) of the difference operators, and the conclusion is obtained integrating by parts the volumetric term. Rearranging the terms, we arrive at

$$\begin{aligned}
-((\boldsymbol{\delta}_{\nabla,T}^k - \nabla \delta_T^k) \underline{v}_T, \boldsymbol{\phi})_T & \\
+ \sum_{F \in \mathcal{F}_T} ((\delta_{TF}^k - \delta_T^k) \underline{v}_T, \boldsymbol{\phi} \cdot \boldsymbol{n}_{TF})_F &= 0 \qquad \forall \boldsymbol{\phi} \in \mathbb{P}^k(T)^d.
\end{aligned} \tag{5.152}$$

When evaluated with a function $\boldsymbol{\phi}$ from a space larger than $\mathbb{P}^k(T)^d$, the left-hand side no longer vanishes in general. It then defines a residual that is linear with respect to $\boldsymbol{\phi}$, and can thus be lifted as a function over T using the Riesz representation theorem.

More precisely, let $\mathfrak{S}_T$ be a subspace of $L^2(T)^d$ such that, for all $\boldsymbol{\phi} \in \mathfrak{S}_T$ and all $F \in \mathcal{F}_T$, it holds $\boldsymbol{\phi}_{|F} \cdot \boldsymbol{n}_{TF} \in L^2(F)$. For a given $\underline{v}_T \in \underline{U}_T^k$, the left-hand side of (5.152) is linear on $\mathfrak{S}_T$ and therefore has a Riesz representation, i.e., there is a unique $\mathbf{S}_T \underline{v}_T \in \mathfrak{S}_T$ such that

$$\begin{aligned}(\mathbf{S}_T \underline{v}_T, \boldsymbol{\phi})_T = &-((\boldsymbol{\delta}_{\nabla,T}^k - \nabla \delta_T^k)\underline{v}_T, \boldsymbol{\phi})_T \\ &+ \sum_{F \in \mathcal{F}_T} ((\delta_{TF}^k - \delta_T^k)\underline{v}_T, \boldsymbol{\phi} \cdot \boldsymbol{n}_{TF})_F \qquad \forall \boldsymbol{\phi} \in \mathfrak{S}_T.\end{aligned} \tag{5.153}$$

The mapping $\underline{U}_T^k \ni \underline{v}_T \mapsto \mathbf{S}_T \underline{v}_T \in \mathfrak{S}_T$ thus defined satisfies two of the three properties in Assumption 5.49.

Lemma 5.54 (Orthogonality and Polynomial Consistency of $\mathbf{S}_T$) *The mapping $\mathbf{S}_T$ defined by* (5.153) *is linear and satisfies (GDM-S3). If $\mathbb{P}^k(T)^d \subset \mathfrak{S}_T$, then $\mathbf{S}_T$ also satisfies (GDM-S2).*

Proof The linearity easily follows from the uniqueness of the Riesz representation, and the fact that the right-hand side of (5.153) is linear with respect to $\underline{v}_T$.

By definition (5.151) of $\boldsymbol{\delta}_{\nabla,T}^k$ we have, for all $w \in \mathbb{P}^{k+1}(T)$,

$$\begin{aligned}\boldsymbol{\delta}_{\nabla,T}^k \underline{I}_T^k w &= \nabla \mathrm{p}_T^{k+1} \underline{I}_T^k w - \mathbf{G}_T^k \underline{I}_T^k w \\ &= \nabla \pi_T^{1,k+1} w - \boldsymbol{\pi}_T^{0,k}(\nabla w) \\ &= \nabla w - \nabla w = 0,\end{aligned} \tag{5.154}$$

where we have used, to pass to the second line, the commutation properties (2.14) and (4.40) of p_T^{k+1} and $\mathbf{G}_T^k$, respectively, and the conclusion follows from the polynomial invariance of the elliptic and orthogonal projectors. Plugged into (5.153) together with the polynomial consistency (2.21) of the difference operators, (5.154) shows that $(\mathbf{S}_T \underline{I}_T^k w, \boldsymbol{\phi})_T = 0$ for all $w \in \mathbb{P}^{k+1}(T)$ and all $\boldsymbol{\phi} \in \mathfrak{S}_T$. Hence, $\mathbf{S}_T \underline{I}_T^k w = 0$ and $\mathbf{S}_T$ satisfies (GDM-S3) in Assumption 5.49.

We now turn to (GDM-S2), assuming that $\mathbb{P}^k(T)^d \subset \mathfrak{S}_T$. Equation (5.153) can thus be applied to any $\boldsymbol{\phi} \in \mathbb{P}^k(T)^d$ and, using (5.152), leads to $(\mathbf{S}_T \underline{v}_T, \boldsymbol{\phi})_T = 0$, which precisely proves (GDM-S2). □

Remark 5.55 (Lifting of Face Differences) Equation (5.153) can be recast as

$$(\mathbf{S}_T \underline{v}_T + (\boldsymbol{\delta}_{\nabla,T}^k - \nabla \delta_T^k)\underline{v}_T, \boldsymbol{\phi}) = \sum_{F \in \mathcal{F}_T} ((\delta_{TF}^k - \delta_T^k)\underline{v}_T, \boldsymbol{\phi} \cdot \boldsymbol{n}_{TF})_F,$$

showing that the function $\mathbf{S}_T \underline{v}_T + (\boldsymbol{\delta}_{\nabla,T}^k - \nabla \delta_T^k)\underline{v}_T \in L^2(T)^d$ is indeed a lifting on T of terms based on the face differences on ∂T.

5.6.3.2 A Stabilisation Term Based on a Local Raviart–Thomas–Nédélec Space

We now have to find a finite-dimensional space $\mathfrak{S}_T$ which is "rich" enough so that $\mathbf{S}_T$ satisfies the stability assumption (GDM-S1). This space will be the Raviart–Thomas–Nédélec space on the simplicial subdivision of T.

Let $\mathfrak{M}_h = (\mathfrak{T}_h, \mathfrak{F}_h)$ be a matching simplicial submesh given by the definition of regular mesh sequence (see Definition 1.9). We recall that $\mathfrak{T}_T$ is the set of simplices $\tau \in \mathfrak{T}_h$ contained in T. For $F \in \mathcal{F}_T$, we denote by $\mathfrak{F}_F$ the set of simplicial faces $\sigma \in \mathfrak{F}_h$ contained in F. Finally, $\mathfrak{F}_T^{\mathrm{i}}$ is the set of simplicial faces of $\mathfrak{F}_h$ contained in T but not in any of its faces (and, as usual, if $\sigma \in \mathfrak{F}_T^{\mathrm{i}}$, then $\mathfrak{T}_\sigma$ is the set of two simplices in $\mathfrak{T}_T$ that share σ as a face).

Let us recall a few facts on the Raviart–Thomas–Nédélec (RTN) spaces, for a proof of which we refer, e.g., to [196]. For a simplex $\tau \in \mathfrak{T}_h$, the RTN space of degree $(k+1)$ on τ is $\mathbb{RTN}^{k+1}(\tau) := \mathbb{P}^{k+1}(\tau)^d + \boldsymbol{x}\mathbb{P}^{k+1}(\tau)^d$. If $\boldsymbol{\eta} \in \mathbb{RTN}^{k+1}(\tau)$ then, for all $\sigma \in \mathfrak{F}_\tau$ with unit normal $\boldsymbol{n}_{\tau\sigma}$ pointing out of τ, $\boldsymbol{\eta}_{|\sigma}\cdot\boldsymbol{n}_{\tau\sigma} \in \mathbb{P}^{k+1}(\sigma)$, and $\boldsymbol{\eta}$ is entirely determined by its L^2-orthogonal projection on $\mathbb{P}^k(\tau)^d$ and its normal traces on $\partial\tau$. More precisely, for all $\boldsymbol{q} \in \mathbb{P}^k(\tau)^d$ and $(q_\sigma)_{\sigma\in\mathfrak{F}_\tau} \in (\mathbb{P}^{k+1}(\sigma))_{\sigma\in\mathfrak{F}_\tau}$, there exists a unique $\boldsymbol{\eta} \in \mathbb{RTN}^{k+1}(\tau)$ such that $\boldsymbol{\pi}_\tau^{0,k}\boldsymbol{\eta} = \boldsymbol{q}$ and $\boldsymbol{\eta}_{|\sigma}\cdot\boldsymbol{n}_{\tau\sigma} = q_\sigma$ for all $\sigma \in \mathfrak{T}_\tau$. Moreover,

$$\|\boldsymbol{\eta}\|_\tau^2 \simeq \|\boldsymbol{\pi}_\tau^{0,k}\boldsymbol{\eta}\|_\tau^2 + \sum_{\sigma\in\mathfrak{F}_\tau} h_\sigma \|\boldsymbol{\eta}_{|\sigma}\cdot\boldsymbol{n}_{\tau\sigma}\|_\sigma^2. \tag{5.155}$$

In what follows, we consider the local RTN space on the subdivision $\mathfrak{T}_T$ of T, obtained by patching the RTN spaces on each simplex and imposing the continuity of the normal traces:

$$\begin{aligned}\mathbb{RTN}^{k+1}(\mathfrak{T}_T) := \big\{\boldsymbol{\eta} \in L^2(T)^d \ :\ &\boldsymbol{\eta}_{|\tau} \in \mathbb{RTN}^{k+1}(\tau) \quad \forall\tau \in \mathfrak{T}_T \text{ and} \\ &\boldsymbol{\eta}_{|\tau_1}\cdot\boldsymbol{n}_{\tau_1\sigma} + \boldsymbol{\eta}_{|\tau_2}\cdot\boldsymbol{n}_{\tau_2\sigma} = 0 \quad \forall\sigma \in \mathfrak{F}_T^{\mathrm{i}}, \text{ with } \{\tau_1,\tau_2\} = \mathfrak{T}_\sigma\big\}.\end{aligned} \tag{5.156}$$

We will prove that $\mathbb{RTN}^{k+1}(\mathfrak{T}_T)$ is a proper choice for $\mathfrak{S}_T$. To do so, we need the following lemma.

Lemma 5.56 (Control of the Element-Based Difference Through Face-Based Differences) *Recalling the definition* (5.143) *of the difference seminorm, it holds: For all* $\underline{v}_T \in \underline{U}_T^k$,

$$\|(\boldsymbol{\delta}_{\nabla,T}^k - \nabla\delta_T^k)\underline{v}_T\|_T \lesssim |\underline{v}_T|_{\delta,\partial T}. \tag{5.157}$$

Proof Since $(\boldsymbol{\delta}^k_{\nabla,T} - \nabla\delta^k_T)\underline{v}_T \in \mathbb{P}^k(T)^d$, we have

$$\begin{aligned}
\|(\boldsymbol{\delta}^k_{\nabla,T} - \nabla\delta^k_T)\underline{v}_T\|_T &= \sup_{\boldsymbol{\phi}\in\mathbb{P}^k(T)^d,\|\boldsymbol{\phi}\|_T=1} ((\boldsymbol{\delta}^k_{\nabla,T} - \nabla\delta^k_T)\underline{v}_T, \boldsymbol{\phi})_T \\
&= \sup_{\boldsymbol{\phi}\in\mathbb{P}^k(T)^d,\|\boldsymbol{\phi}\|_T=1} \sum_{F\in\mathcal{F}_T} ((\delta^k_{TF} - \delta^k_T)\underline{v}_T, \boldsymbol{\phi}\cdot\boldsymbol{n}_{TF})_F \\
&\le \sup_{\boldsymbol{\phi}\in\mathbb{P}^k(T)^d,\|\boldsymbol{\phi}\|_T=1} \sum_{F\in\mathcal{F}_T} h_F^{-\frac{1}{2}} \|(\delta^k_{TF} - \delta^k_T)\underline{v}_T\|_F \, h_T^{\frac{1}{2}} \|\boldsymbol{\phi}\|_F \\
&\le \sup_{\boldsymbol{\phi}\in\mathbb{P}^k(T)^d,\|\boldsymbol{\phi}\|_T=1} |\underline{v}_T|_{\delta,\partial T} \, h_T^{\frac{1}{2}} \|\boldsymbol{\phi}\|_{\partial T},
\end{aligned}$$

where we have used (5.152) in the second line, a generalised Hölder inequality with exponents $(2, 2, \infty)$ on the integrals over F together with $\|\boldsymbol{n}_{TF}\|_{L^\infty(F)^d} = 1$ and $h_F \le h_T$ in the third line, and a Cauchy–Schwarz inequality on the sum to conclude. Invoking the discrete trace inequality (1.55) with $p = 2$ and $v =$ components of $\boldsymbol{\phi}$, we deduce

$$\|(\boldsymbol{\delta}^k_{\nabla,T} - \nabla\delta^k_T)\underline{v}_T\|_T \lesssim \sup_{\boldsymbol{\phi}\in\mathbb{P}^k(T)^d,\|\boldsymbol{\phi}\|_T=1} |\underline{v}_T|_{\delta,\partial T} \|\boldsymbol{\phi}\|_T = |\underline{v}_T|_{\delta,\partial T},$$

and the proof is complete. □

We can now conclude the construction of an explicit and computable stabilisation term for the DSGD.

Theorem 5.57 (DSGD Stabilisation Based on $\mathbb{RTN}^{k+1}$) *If* $\mathfrak{S}_T = \mathbb{RTN}^{k+1}(\mathfrak{T}_T)$, *given by* (5.156)*, then* $\mathbf{S}_T$ *defined by* (5.153) *satisfies Assumption 5.49.*

Proof The property (GDM-S3) has already been established in Lemma 5.54.

If $\boldsymbol{\eta} \in \mathbb{P}^k(T)^d$, then $\boldsymbol{\eta}_{|\tau} \in \mathbb{P}^k(\tau)^d \subset \mathbb{RTN}^{k+1}(\tau)$ for all $\tau \in \mathfrak{T}_T$, and $\boldsymbol{\eta}$ is continuous over T so its (normal) traces across the faces in $\mathfrak{F}^{\rm i}_T$ are continuous. Hence, $\boldsymbol{\eta} \in \mathbb{RTN}^{k+1}(\mathfrak{T}_T)$. This proves that $\mathbb{P}^k(T)^d \subset \mathbb{RTN}^{k+1}(\mathfrak{T}_T)$ and, by Lemma 5.54, that $\mathbf{S}_T$ satisfies (GDM-S2).

It remains to prove (GDM-S1), which we do by establishing two inequalities. Making $\boldsymbol{\phi} = \mathbf{S}_T\underline{v}_T$ in (5.153), we have

$$\begin{aligned}
\|\mathbf{S}_T\underline{v}_T\|_T^2 &\le \|(\boldsymbol{\delta}^k_{\nabla,T} - \nabla\delta^k_T)\underline{v}_T\|_T \|\mathbf{S}_T\underline{v}_T\|_T + \sum_{F\in\mathcal{F}_T} \|(\delta^k_{TF} - \delta^k_T)\underline{v}_T\|_F \|\mathbf{S}_T\underline{v}_T\|_F \\
&\lesssim |\underline{v}_T|_{\delta,\partial T} \|\mathbf{S}_T\underline{v}_T\|_T + \sum_{F\in\mathcal{F}_T} h_F^{-\frac{1}{2}} \|(\delta^k_{TF} - \delta^k_T)\underline{v}_T\|_F \|\mathbf{S}_T\underline{v}_T\|_T \\
&\lesssim |\underline{v}_T|_{\delta,\partial T} \|\mathbf{S}_T\underline{v}_T\|_T,
\end{aligned}$$

where we have used in the first line the Cauchy–Schwarz and generalised Hölder inequalities (with exponents $(2, 2, \infty)$ for the second one) along with $\|\boldsymbol{n}_{TF}\|_{L^\infty(F)^d} = 1$, followed in the second line by (5.157) and the discrete trace inequality (1.55) with $v =$ components of $\mathbf{S}_T\underline{v}_T$ (valid by the mesh regularity property, since this function is polynomial on each simplex in $\mathfrak{T}_T$), and we have concluded invoking the Cauchy–Schwarz inequality on the sum and the definition (5.143) of the difference seminorm. Simplifying leads to

$$\|\mathbf{S}_T\underline{v}_T\|_T \lesssim |\underline{v}_T|_{\delta,\partial T}. \tag{5.158}$$

To prove the converse inequality, let $\boldsymbol{\eta} \in \mathbb{RTN}^{k+1}(\mathfrak{T}_T)$ be defined by:

$$\boldsymbol{\pi}_\tau^{0,k}\boldsymbol{\eta} = 0 \qquad \forall \tau \in \mathfrak{T}_T, \tag{5.159a}$$

$$\boldsymbol{\eta}_{|\sigma}\cdot\boldsymbol{n}_{\tau\sigma} = 0 \qquad \forall \sigma \in \mathfrak{F}_T^{\mathrm{i}}, \tag{5.159b}$$

$$\boldsymbol{\eta}_{|\sigma}\cdot\boldsymbol{n}_{TF} = h_F^{-1}(\delta_{TF}^k - \delta_T^k)\underline{v}_T \qquad \forall F \in \mathcal{F}_T\,, \forall \sigma \in \mathfrak{F}_F. \tag{5.159c}$$

These equations properly define $\boldsymbol{\eta}$ since, in (5.159c), $\boldsymbol{n}_{TF} = \boldsymbol{n}_{\tau\sigma}$ and $(\delta_{TF}^k - \delta_T^k)\underline{v}_T \in \mathbb{P}^k(\sigma)$. Moreover, summing (5.155) over $\tau \in \mathfrak{T}_T$, we have

$$\begin{aligned}
\|\boldsymbol{\eta}\|_T^2 &\simeq \sum_{\tau\in\mathfrak{T}_T}\sum_{\sigma\in\mathfrak{F}_\tau\setminus\mathfrak{F}_T^{\mathrm{i}}} h_\sigma h_F^{-2}\|(\delta_{TF}^k - \delta_T^k)\underline{v}_T\|_\sigma^2 \\
&\simeq \sum_{F\in\mathcal{F}_T} h_F^{-1}\sum_{\sigma\in\mathfrak{F}_F}\|(\delta_{TF}^k - \delta_T^k)\underline{v}_T\|_\sigma^2 = |\underline{v}_T|_{\delta,\partial T}^2,
\end{aligned} \tag{5.160}$$

where we have gathered, in the second line, the sum by faces of T (noting that all simplicial faces σ in the first line lie on ∂T), and we have used $h_\sigma \le h_F$ (for $\sigma \in \mathfrak{F}_F$) in the second line. Making $\boldsymbol{\phi} = \boldsymbol{\eta}$ in (5.153), we obtain

$$\begin{aligned}
(\mathbf{S}_T\underline{v}_T, \boldsymbol{\eta})_T &= -((\boldsymbol{\delta}_{\nabla,T}^k - \nabla\delta_T^k)\underline{v}_T, \boldsymbol{\eta})_T + \sum_{F\in\mathcal{F}_T}((\delta_{TF}^k - \delta_T^k)\underline{v}_T, \boldsymbol{\eta}\cdot\boldsymbol{n}_{TF})_F \\
&= -\sum_{\tau\in\mathfrak{T}_T}\cancel{((\boldsymbol{\delta}_{\nabla,T}^k - \nabla\delta_T^k)\underline{v}_T, \boldsymbol{\eta})_\tau} + \sum_{F\in\mathcal{F}_T}\sum_{\sigma\in\mathfrak{F}_F}((\delta_{TF}^k - \delta_T^k)\underline{v}_T, \boldsymbol{\eta}\cdot\boldsymbol{n}_{TF})_\sigma \\
&= \sum_{F\in\mathcal{F}_T} h_F^{-1}\sum_{\sigma\in\mathfrak{F}_F}\|(\delta_{TF}^k - \delta_T^k)\underline{v}_T\|_\sigma^2 = |\underline{v}_T|_{\delta,\partial T}^2,
\end{aligned}$$

where the cancellation in the second line is justified by (5.159a) along with the fact that $((\boldsymbol{\delta}_{\nabla,T}^k - \nabla\delta_T^k)\underline{v}_T)_{|\tau} \in \mathbb{P}^k(\tau)^d$ (see (5.151)), and the conclusion follows from (5.159c). Using the Cauchy–Schwarz inequality in the left-hand side together with (5.160), we infer

$$\|\mathbf{S}_T\underline{v}_T\|_T|\underline{v}_T|_{\delta,\partial T} \gtrsim |\underline{v}_T|_{\delta,\partial T}^2$$

which, after simplification, yields $\|\mathbf{S}_T \underline{v}_T\|_T \gtrsim |\underline{v}_T|_{\delta,\partial T}$. Combined with (5.158), this proves (GDM-S1). □

5.6.4 Properties of Discontinuous Skeletal Gradient Discretisations

We prove here the following theorem, which gives the expected estimates on the measures $C_{\mathcal{D}}$, $S_{\mathcal{D}}$ and $W_{\mathcal{D}}$ for the DSGD.

Theorem 5.58 (Properties of the DSGD) *Let $(\mathcal{M}_h)_{h\in\mathcal{H}}$ be a regular mesh sequence in the sense of Definition 1.9 and, for each $h \in \mathcal{H}$, let $\mathcal{D}_h$ be a Discontinuous Skeletal Gradient Discretisation defined by (5.142) with stabilisation terms satisfying Assumption 5.49. Then,*

$$C_{\mathcal{D}_h} \lesssim 1 \tag{5.161}$$

and, for all $r \in \{0,\ldots,k\}$,

$$S_{\mathcal{D}_h}(\varphi) \lesssim h^{r+1}\|\varphi\|_{H^{r+2}(\mathcal{T}_h)} \qquad \forall\varphi \in H_0^1(\Omega)\cap H^{r+2}(\mathcal{T}_h), \tag{5.162}$$

$$W_{\mathcal{D}_h}(\boldsymbol{\psi}) \lesssim h^{r+1}\|\boldsymbol{\psi}\|_{H^{r+1}(\mathcal{T}_h)^d} \qquad \forall\boldsymbol{\psi} \in \boldsymbol{H}(\mathrm{div};\Omega)\cap H^{r+1}(\mathcal{T}_h)^d, \tag{5.163}$$

where $C_{\mathcal{D}_h}$, $S_{\mathcal{D}_h}$, and $W_{\mathcal{D}_h}$ are respectively defined by (5.135), (5.136), (5.137), while the hidden constants depend only on Ω, d, k, and the mesh regularity parameter ϱ.

Proof (i) *Estimate on $C_{\mathcal{D}_h}$*. Let $\underline{v}_h \in \underline{U}_{h,0}^k\backslash\{\underline{0}\}$. By (GDM-S2), for all $T \in \mathcal{T}_h$, the functions $\mathbf{G}_T^k \underline{v}_T$ and $\mathbf{S}_T \underline{v}_T$ are $L^2(T)^d$-orthogonal. The Pythagorean theorem together with the definition (5.142c) of $\nabla_{\mathcal{D}_h}\underline{v}_h$ and (GDM-S1) thus gives

$$\|\nabla_{\mathcal{D}_h}\underline{v}_h\|_T^2 = \|\mathbf{G}_T^k\underline{v}_T\|_T^2 + \|\mathbf{S}_T\underline{v}_T\|_T^2 \simeq \|\mathbf{G}_T^k\underline{v}_T\|_T^2 + |\underline{v}_T|_{\delta,\partial T}^2. \tag{5.164}$$

Using the seminorm equivalence (6.19) proved in Chap. 6 below with $p = 2$, we infer $\|\underline{v}_T\|_{1,T}^2 \lesssim \|\nabla_{\mathcal{D}_h}\underline{v}_h\|_T^2$. Summing these estimates over $T \in \mathcal{T}_h$ and taking the square root gives

$$\|\underline{v}_h\|_{1,h} \lesssim \|\nabla_{\mathcal{D}_h}\underline{v}_h\|. \tag{5.165}$$

The discrete Poincaré inequality (2.37) together with the definition (5.142b) of $\Pi_{\mathcal{D}_h}\underline{v}_h$ then yield

$$\|\Pi_{\mathcal{D}_h}\underline{v}_h\| = \|v_h\| \lesssim \|\underline{v}_h\|_{1,h} \lesssim \|\nabla_{\mathcal{D}_h}\underline{v}_h\|.$$

Dividing by $\|\nabla_{\mathcal{D}_h}\underline{v}_h\|$ and taking the supremum over $\underline{v}_h \in \underline{U}^k_{h,0}\backslash\{\underline{0}\}$ yields $C_{\mathcal{D}_h} \lesssim 1$.

(ii) *Estimate on* $S_{\mathcal{D}_h}$. Let $\varphi \in H^1_0(\Omega) \cap H^{r+2}(\mathcal{T}_h)$ and set $\underline{v}_h := \underline{I}^k_h\varphi \in \underline{U}^k_{h,0}$. By definitions (5.136) of $S_{\mathcal{D}_h}$, (5.142b) of $\Pi_{\mathcal{D}_h}$ and (5.142c) of $\nabla_{\mathcal{D}_h}$,

$$\begin{aligned}
S_{\mathcal{D}_h}(\varphi) &\le \|\Pi_{\mathcal{D}_h}\underline{v}_h - \varphi\| + \|\nabla_{\mathcal{D}_h}\underline{v}_h - \nabla\varphi\| \\
&\le \|\pi^{0,k}_h\varphi - \varphi\| + \|\mathbf{G}^k_h\underline{I}^k_h\varphi - \nabla\varphi\| + \|\mathbf{S}_h\underline{I}^k_h\varphi\| \\
&= \|\pi^{0,k}_h\varphi - \varphi\| + \|\boldsymbol{\pi}^{0,k}_h(\nabla\varphi) - \nabla\varphi\| + \|\mathbf{S}_h\underline{I}^k_h\varphi\|,
\end{aligned}$$

where the global projector $\pi^{0,k}_h$ is defined by (1.59), and we have used the commutation property (4.40). Invoking the approximation property (1.74) of the L^2-projector with $(l, s, m, p) = (k, r+1, 0, 2)$ and $v = \varphi$ or $v =$ components of $\nabla\varphi$, together with (GDM-S1), we infer

$$S_{\mathcal{D}_h}(\varphi) \lesssim h^{r+1}|\varphi|_{H^{r+1}(\mathcal{T}_h)} + h^{r+1}|\nabla\varphi|_{H^{r+1}(\mathcal{T}_h)^d} + \left(\sum_{T\in\mathcal{T}_h} |\underline{I}^k_T\varphi|^2_{\delta,\partial T}\right)^{\frac{1}{2}}. \tag{5.166}$$

The consistency property (2.31) applied to the HHO stabilisation bilinear form (2.22) yields $|\underline{I}^k_T\varphi|_{\delta,\partial T} \lesssim h^{r+1}_T|\varphi|_{H^{r+2}(T)}$. Plugged into (5.166), this proves (5.162).

(iii) *Estimate on* $W_{\mathcal{D}_h}$. Let $\boldsymbol{\psi} \in \boldsymbol{H}(\mathrm{div};\Omega) \cap H^{r+1}(\mathcal{T}_h)^d$, and take an arbitrary $\underline{v}_h \in \underline{U}^k_{h,0}\backslash\{\underline{0}\}$. Recalling the definition (5.142c) of $\nabla_{\mathcal{D}_h}\underline{v}_h$, we write

$$\begin{aligned}
\int_\Omega \nabla_{\mathcal{D}_h}\underline{v}_h\cdot\boldsymbol{\psi} &= \sum_{T\in\mathcal{T}_h} (\mathbf{G}^k_T\underline{v}_T, \boldsymbol{\psi})_T + \sum_{T\in\mathcal{T}_h} (\mathbf{S}_T\underline{v}_T, \boldsymbol{\psi})_T \\
&= \sum_{T\in\mathcal{T}_h} (\nabla v_T, \boldsymbol{\psi})_T + \sum_{T\in\mathcal{T}_h}\sum_{F\in\mathcal{F}_T} (v_F - v_T, (\boldsymbol{\pi}^{0,k}_T\boldsymbol{\psi})\cdot\boldsymbol{n}_{TF})_F \\
&\quad + \sum_{T\in\mathcal{T}_h} (\mathbf{S}_T\underline{v}_T, \boldsymbol{\psi} - \boldsymbol{\pi}^{0,k}_T\boldsymbol{\psi})_T \\
&= \sum_{T\in\mathcal{T}_h} -(v_T, \nabla\cdot\boldsymbol{\psi})_T + \sum_{T\in\mathcal{T}_h}\sum_{F\in\mathcal{F}_T} (v_F - v_T, [(\boldsymbol{\pi}^{0,k}_T\boldsymbol{\psi}) - \boldsymbol{\psi}]\cdot\boldsymbol{n}_{TF})_F \\
&\quad + \sum_{T\in\mathcal{T}_h} (\mathbf{S}_T\underline{v}_T, \boldsymbol{\psi} - \boldsymbol{\pi}^{0,k}_T\boldsymbol{\psi})_T,
\end{aligned}$$

where the second line follows from the property (4.41) of $\mathbf{G}_T^k \underline{v}_T$ with $\boldsymbol{\tau} = \boldsymbol{\psi}$ and from the orthogonality property (GDM-S2), and we have concluded using element-wise integration by parts and (1.28) with $(\boldsymbol{\tau}, (\varphi_F)_{F\in\mathcal{F}_h}) = (\boldsymbol{\psi}, (v_F)_{F\in\mathcal{F}_h})$ to write

$$\sum_{T\in\mathcal{T}_h}\sum_{F\in\mathcal{F}_T}(v_F, \boldsymbol{\psi}\cdot\boldsymbol{n}_{TF})_F = 0.$$

Recalling the definition (5.142b) of $\Pi_{\mathcal{D}_h}\underline{v}_h$, we infer

$$\begin{aligned}\int_\Omega (\nabla_{\mathcal{D}_h}\underline{v}_h\cdot\boldsymbol{\psi} + \Pi_{\mathcal{D}_h}\underline{v}_h\nabla\cdot\boldsymbol{\psi}) &= \sum_{T\in\mathcal{T}_h}\sum_{F\in\mathcal{F}_T}(v_F - v_T, [(\boldsymbol{\pi}_T^{0,k}\boldsymbol{\psi}) - \boldsymbol{\psi}]\cdot\boldsymbol{n}_{TF})_F \\ &\quad + \sum_{T\in\mathcal{T}_h}(\mathbf{S}_T\underline{v}_T, \boldsymbol{\psi} - \boldsymbol{\pi}_T^{0,k}\boldsymbol{\psi})_T \\ &=: \mathfrak{T}_1 + \mathfrak{T}_2.\end{aligned} \tag{5.167}$$

Using the generalised Hölder inequality with exponents $(2, 2, \infty)$ along with $\|\boldsymbol{n}_{TF}\|_{L^\infty(F)^d} = 1$, we write

$$\begin{aligned}|\mathfrak{T}_1| &\le \sum_{T\in\mathcal{T}_h}\sum_{F\in\mathcal{F}_T} h_F^{-\frac{1}{2}}\|v_F - v_T\|_F\, h_F^{\frac{1}{2}}\|(\boldsymbol{\pi}_T^{0,k}\boldsymbol{\psi}) - \boldsymbol{\psi}\|_F \\ &\lesssim \left(\sum_{T\in\mathcal{T}_h}\sum_{F\in\mathcal{F}_T} h_F^{-1}\|v_F - v_T\|_F^2\right)^{\frac{1}{2}}\left(\sum_{T\in\mathcal{T}_h}\sum_{F\in\mathcal{F}_T} h_F h_T^{2r+1}|\boldsymbol{\psi}|_{H^{r+1}(T)^d}^2\right)^{\frac{1}{2}} \\ &\lesssim \|\underline{v}_h\|_{1,h} h^{r+1}|\boldsymbol{\psi}|_{H^{r+1}(\mathcal{T}_h)^d},\end{aligned}$$

where we have used the trace approximation property (1.75) of the local L^2-projector with $(l, s, m, p) = (k, r+1, 0, 2)$, and we have concluded using the definition (2.35) of $\|\cdot\|_{1,h}$ (see also (2.7)) and $h_F \le h_T$. Using (5.165), we deduce

$$|\mathfrak{T}_1| \lesssim \|\nabla_{\mathcal{D}_h}\underline{v}_h\| h^{r+1}|\boldsymbol{\psi}|_{H^{r+1}(\mathcal{T}_h)^d}. \tag{5.168}$$

To estimate $\mathfrak{T}_2$, we write

$$
\begin{aligned}
|\mathfrak{T}_2| &\le \sum_{T\in\mathcal{T}_h} \|\mathbf{S}_T \underline{v}_T\|_T \|\boldsymbol{\psi} - \boldsymbol{\pi}_T^{0,k}\boldsymbol{\psi}\|_T \\
&\lesssim \sum_{T\in\mathcal{T}_h} \|\nabla_{\mathcal{D}_h}\underline{v}_h\|_T h_T^{r+1} |\boldsymbol{\psi}|_{H^{r+1}(T)^d} \\
&\lesssim \|\nabla_{\mathcal{D}_h}\underline{v}_h\| h^{r+1} |\boldsymbol{\psi}|_{H^{r+1}(\mathcal{T}_h)^d},
\end{aligned}
\tag{5.169}
$$

where we have used Cauchy–Schwarz inequalities on the integrals over the mesh elements T in the first line, (5.164) and the approximation property (1.74) with $(l, s, m, p) = (k, r+1, 0, 2)$ in the second line, and we have concluded using a Cauchy–Schwarz inequality on the sum. Plugging (5.168) and (5.169) into (5.167) yields

$$
\left| \int_\Omega \left(\nabla_{\mathcal{D}_h}\underline{v}_h \cdot \boldsymbol{\psi} + \Pi_{\mathcal{D}_h}\underline{v}_h \nabla\cdot\boldsymbol{\psi} \right) \right| \lesssim \|\nabla_{\mathcal{D}_h}\underline{v}_h\| h^{r+1} |\boldsymbol{\psi}|_{H^{r+1}(\mathcal{T}_h)^d}.
$$

Dividing by $\|\nabla_{\mathcal{D}_h}\underline{v}_h\|$ and taking the supremum over $\underline{v}_h \in \underline{U}_{h,0}^k \backslash \{\underline{0}\}$ concludes the proof of (5.163). □

Part II
Applications to Advanced Models

Chapter 6
p-Laplacian and Leray–Lions

We consider in this chapter an extension of the HHO method to fully nonlinear elliptic equations of Leray–Lions kind [230]. This class of problems contains as a special case the p-Laplace equation, which appears in the modelling of glacier motion [201], of incompressible turbulent flows in porous media [164], in airfoil design [200], and can be regarded as a simplified version of the viscous terms in power-law fluids. The pure diffusion linear problems treated in Chap. 2, Sects. 3.1, and 4.2 can also be recovered as special cases of the framework developed here.

Several novelties are present with respect to the previous chapters. The first obvious difference is that the continuous problem (and, therefore, its HHO approximation) are possibly nonlinear. This will give us the opportunity to introduce general techniques for the discretisation and analysis of nonlinear problems, as well as a set of functional analysis results of independent interest. A second difference, related to the first, is that Leray–Lions problems are naturally posed in a non-Hilbertian setting. This will require to emulate a Sobolev structure at the discrete level, which we do by extending the discrete norms of Chap. 2 and associated results to the $W^{1,p}$-setting. Finally, unlike previous chapters, we consider non-homogeneous Neumann boundary conditions to illustrate how the HHO method is constructed and analysed in this case.

The material is organised as follows. In Sect. 6.1 we state the general Leray–Lions problem and formulate the assumptions on the flux function. Section 6.2 focuses on the HHO discretisation. We first introduce the general setting required to deal with problems posed in a non-Hilbertian setting. As for the locally variable diffusion model studied in Sect. 4.2, the gradient of the potential reconstruction (2.11) (see also (3.22)) is not a valid choice to discretise fully nonlinear problems, and the HHO scheme for such problems is rather based on the gradient $\mathbf{G}_T^k$ reconstructed in the full space $\mathbb{P}^k(T)^d$, see (4.37). Section 6.3 covers the special case of the p-Laplace equation mentioned in Remark 6.2 below. In this case, the flux function enjoys stronger monotonicity and continuity properties than

D. A. Di Pietro, J. Droniou, *The Hybrid High-Order Method for Polytopal Meshes*,
MS&A 19, https://doi.org/10.1007/978-3-030-37203-3_6

general Leray–Lions flux functions, which permit to establish error estimates. Numerical results are provided to illustrate the practical behaviour of the HHO scheme for this nonlinear equation. Finally, in Sect. 6.4, we go back to generic Leray–Lions equations and prove the convergence of the HHO method using compactness arguments. The analysis follows, in this case, a well-established pattern [171, Section 1.2]: first, an a priori estimate, uniform in h, is established on the discrete solution; second, compactness properties are inferred for a sequence of discrete solutions on refined meshes; finally, the limit of such a sequence (up to the extraction of a subsequence) is shown to solve the PDE model.

The analysis done in this chapter follows ideas originally developed in [141, 142], where the homogeneous Dirichlet case is considered.

6.1 Model

The Leray–Lions problem reads: Find $u : \Omega \to \mathbb{R}$ such that

$$-\nabla\cdot\boldsymbol{\sigma}(\boldsymbol{x}, u, \nabla u) = f \qquad \text{in } \Omega, \tag{6.1a}$$

$$\boldsymbol{\sigma}(\boldsymbol{x}, u, \nabla u)\cdot\boldsymbol{n}_\Omega = g \qquad \text{on } \partial\Omega, \tag{6.1b}$$

where $\boldsymbol{\sigma}$ is a possibly nonlinear flux function, f is a volumetric source term, g is the non-homogeneous Neumann boundary condition, and $\boldsymbol{n}_\Omega$ denotes the outer unit normal to Ω on $\partial\Omega$. The precise assumptions on the problem data are discussed in what follows. Let $p \in (1, \infty)$ be fixed, and denote its conjugate exponent by

$$p' := \frac{p}{p-1}.$$

Concerning f and g, we assume that

$$f \in L^{p'}(\Omega),\ g \in L^{p'}(\partial\Omega), \text{ and } \int_\Omega f + \int_{\partial\Omega} g = 0. \tag{6.2}$$

The third relation above is a compatibility condition, obtained by integrating (6.1a) over the domain Ω and using the divergence theorem and the Neumann boundary condition (6.1b) to write

$$\int_\Omega f = -\int_\Omega \nabla\cdot\boldsymbol{\sigma}(\boldsymbol{x}, u, \nabla u) = -\int_{\partial\Omega} \boldsymbol{\sigma}(\boldsymbol{x}, u, \nabla u)\cdot\boldsymbol{n}_\Omega = -\int_{\partial\Omega} g.$$

The requirements on the flux function are gathered in the following standard assumption on Leary–Lions operators.

Assumption 6.1 (Leray–Lions Flux function) *The following holds:*

(i) Carathéodory function. *The Leray–Lions flux function*

$$\boldsymbol{\sigma} : \Omega \times \mathbb{R} \times \mathbb{R}^d \to \mathbb{R}^d \text{ is a Carathéodory function,} \tag{6.3a}$$

that is, $\boldsymbol{\sigma}(\cdot, s, \boldsymbol{\xi})$ *is measurable for all* $(s, \boldsymbol{\xi}) \in \mathbb{R} \times \mathbb{R}^d$ *and* $\boldsymbol{\sigma}(\boldsymbol{x}, \cdot, \cdot)$ *is continuous for a.e.* $\boldsymbol{x} \in \Omega$*;*

(ii) Growth. *Setting* $\hat{p} := \frac{dp}{d-p}$ *if* $p < d$*, or* $\hat{p} := \infty$ *if* $p \geq d$*, there exists a function* $\overline{\sigma} \in L^{p'}(\Omega)$ *along with real numbers* $\beta_\sigma \in (0, \infty)$ *and* $0 \leq t < \frac{\hat{p}}{p'}$ *such that, for a.e.* $\boldsymbol{x} \in \Omega$ *and all* $(s, \boldsymbol{\xi}) \in \mathbb{R} \times \mathbb{R}^d$*,*

$$|\boldsymbol{\sigma}(\boldsymbol{x}, s, \boldsymbol{\xi})| \leq \overline{\sigma}(\boldsymbol{x}) + \beta_\sigma |s|^t + \beta_\sigma |\boldsymbol{\xi}|^{p-1}; \tag{6.3b}$$

(iii) Monotonicity. *It holds, for a.e.* $\boldsymbol{x} \in \Omega$ *and all* $(s, \boldsymbol{\xi}, \boldsymbol{\eta}) \in \mathbb{R} \times \mathbb{R}^d \times \mathbb{R}^d$*,*

$$[\boldsymbol{\sigma}(\boldsymbol{x}, s, \boldsymbol{\xi}) - \boldsymbol{\sigma}(\boldsymbol{x}, s, \boldsymbol{\eta})] \cdot (\boldsymbol{\xi} - \boldsymbol{\eta}) \geq 0; \tag{6.3c}$$

(iv) Coercivity. *There is a real number* $\lambda_\sigma \in (0, \infty)$ *such that, for a.e.* $\boldsymbol{x} \in \Omega$ *and all* $(s, \boldsymbol{\xi}) \in \mathbb{R} \times \mathbb{R}^d$*,*

$$\boldsymbol{\sigma}(\boldsymbol{x}, s, \boldsymbol{\xi}) \cdot \boldsymbol{\xi} \geq \lambda_\sigma |\boldsymbol{\xi}|^p. \tag{6.3d}$$

Remark 6.2 (p-Laplace Equation) A classical example of a Leray–Lions flux function is given by

$$\boldsymbol{\sigma}(\boldsymbol{x}, s, \boldsymbol{\xi}) := |\boldsymbol{\xi}|^{p-2}\boldsymbol{\xi}. \tag{6.4}$$

The corresponding problem (6.1a) is then called the *p-Laplace equation*, and it generalises the Poisson problem considered in Chap. 2 (recovered taking $p = 2$). Notice that, in this case, the flux function does not actually depend on its first two arguments.

Under Assumptions (6.2) and (6.3), and setting

$$W^{1,p}_\star(\Omega) := \left\{ u \in W^{1,p}(\Omega) \, : \, \int_\Omega u = 0 \right\}, \tag{6.5}$$

the weak formulation of problem (6.1) reads: Find $u \in W^{1,p}_\star(\Omega)$ such that, for all $v \in W^{1,p}_\star(\Omega)$,

$$A(u; v) = \int_\Omega f v + \int_{\partial\Omega} g v_{|\partial\Omega} \,, \tag{6.6}$$

where the function $A : W^{1,p}(\Omega) \times W^{1,p}(\Omega) \to \mathbb{R}$ is such that

$$A(u; v) := \int_\Omega \boldsymbol{\sigma}(u, \nabla u) \cdot \nabla v. \tag{6.7}$$

Remark 6.3 (Notation) Unlike the models encountered so far, problem (6.6) is possibly nonlinear and posed in a non-Hilbertian setting. For this reason, a few changes in the notation are due.

First, in order to distinguish the function A (which is only linear in its second argument) from the bilinear form a used in previous chapters for linear problems, we use a capital letter and a semi-colon, instead of a colon, to separate its arguments. A similar notation is adopted for the functions A_h and S_T, respectively defined by (6.27) and (6.28) below.

Second, the L^2-product notation introduced in Remark 1.14 is systematically dropped in favour of integrals. In order to alleviate the notations, in (6.7) and in the integrals that follow we do not explicitly indicate the dependence of $\boldsymbol{\sigma}$, u and other functions on $\boldsymbol{x}$, and we also omit the measure, which can be unequivocally inferred from the context. Thus, in (6.6), the last integral is to be understood for the $(d-1)$-dimensional measure on $\partial\Omega$, and $v_{|\partial\Omega}$ is taken in the sense of the trace of a $W^{1,p}(\Omega)$ function. Similar considerations hold for the other boundary integrals appearing in the rest of this chapter.

Remark 6.4 (Zero Average Condition) The zero average condition in $W^{1,p}_\star(\Omega)$ is used to ensure that a priori estimates on the solution to (6.6) can be obtained using the Poincaré–Wirtinger inequality valid in this space (see, e.g., [81, Comments on Chapter 9]). In the case of linear equations with Neumann boundary conditions, this zero average condition also ensures the uniqueness of the solution (since any two solutions only differ by an additive constant). This is not necessarily the case for nonlinear models such as (6.1); see Remark 6.16.

6.2 Discrete Problem

In this section, after introducing a global discrete HHO space that incorporates in a suitable way the zero-average condition, we equip it with a norm $\|\cdot\|_{1,p,h}$ that generalises to the $W^{1,p}_\star$-setting the one defined by (2.35). Three key discrete functional analysis results are then stated, and two additional reconstruction-based norms are introduced and shown to be equivalent to $\|\cdot\|_{1,p,h}$, uniformly in h. Finally, we state the discrete problem and, based on the previous tools along with standard results from nonlinear analysis, study the existence of a solution.

6.2.1 Discrete $W_\star^{1,p}$ Space and Discrete Functional Analysis

In Chap. 2, the discrete space $\underline{U}_{h,0}^k$ (see (2.36)) and the norm $\|\cdot\|_{1,h}$ (see (2.35) and (2.7)) played the role of the Hilbert space $H_0^1(\Omega)$ and of the seminorm $|\cdot|_{H^1(\Omega)}$, respectively (notice that $|\cdot|_{H^1(\Omega)}$ is a norm on $H_0^1(\Omega)$ by virtue of the continuous Poincaré inequality). For the Leray–Lions equation (6.1) with Neumann boundary conditions, the discretisation space must replace at the discrete level the Sobolev space $W_\star^{1,p}(\Omega)$ (cf. (6.5)) with its associated norm. A natural candidate for this discrete space is

$$\underline{U}_{h,\star}^k := \left\{ \underline{v}_h \in \underline{U}_h^k \;:\; \int_\Omega v_h = 0 \right\}, \tag{6.8}$$

where we remind the reader that, for any $\underline{v}_h \in \underline{U}_h^k$, the broken polynomial function $v_h \in \mathbb{P}^k(\mathcal{T}_h)$ is given by (2.33), that is, $(v_h)_{|T} = v_T$ for all $T \in \mathcal{T}_h$. The counterpart, on $\underline{U}_{h,\star}^k$, of the seminorm $|\cdot|_{W^{1,p}(\Omega)}$ is the following map $\|\cdot\|_{1,p,h}$ defined by generalising (2.7) and (2.35) to the non-Hilbertian setting: For all $\underline{v}_h \in \underline{U}_{h,\star}^k$,

$$\begin{aligned}
\|\underline{v}_h\|_{1,p,h} &:= \left(\sum_{T\in\mathcal{T}_h} \|\underline{v}_T\|_{1,p,T}^p \right)^{\frac{1}{p}}, \\
\|\underline{v}_T\|_{1,p,T} &:= \left(\|\nabla v_T\|_{L^p(T)^d}^p + \sum_{F\in\mathcal{F}_T} h_F^{1-p} \|v_F - v_T\|_{L^p(F)}^p \right)^{\frac{1}{p}} \qquad \forall T \in \mathcal{T}_h.
\end{aligned} \tag{6.9}$$

The power of h_F in the second term ensures that both contributions have the same scaling. Note that it holds $\|\cdot\|_{1,2,h} = \|\cdot\|_{1,h}$ (cf. (2.35) and (2.7)).

Because we are dealing with non-homogeneous Neumann boundary conditions, we will also need the discrete trace operator $\gamma_h : \underline{U}_h^k \to L^p(\partial\Omega)$ such that, for all $\underline{v}_h \in \underline{U}_h^k$,

$$(\gamma_h \underline{v}_h)_{|F} := v_F \qquad \forall F \in \mathcal{F}_h^{\mathrm{b}}. \tag{6.10}$$

Three theorems on $\underline{U}_{h,\star}^k$ will be required for the analysis of the HHO method for (6.1): a discrete Sobolev–Poincaré–Wirtinger inequality, a trace inequality, and a compactness property. These theorems mimic, at the discrete level, known functional analysis results on $W_\star^{1,p}(\Omega)$. Following [178, Appendix B], we therefore refer to them as *discrete functional analysis* results. Their proofs are postponed to Sect. 6.5. The discrete Sobolev–Poincaré–Wirtinger and trace inequalities will be instrumental in establishing a priori estimates. The compactness theorem will be essential to prove the convergence of the HHO solution in cases where error estimates cannot be obtained, due to the lack of an appropriate structure of $\boldsymbol{\sigma}$.

Theorem 6.5 (Discrete Sobolev–Poincaré–Wirtinger Inequality) *Let a polynomial degree $k \geq 0$ and an index $p \in (1, \infty)$ be fixed. Let $(\mathcal{M}_h)_{h \in \mathcal{H}}$ denote a regular sequence of meshes in the sense of Definition 1.9. Let $1 \leq q \leq \frac{dp}{d-p}$ if $1 \leq p < d$, and $1 \leq q < \infty$ if $p \geq d$. Then, for all $\underline{v}_h \in \underline{U}_{h,\star}^k$,*

$$\|v_h\|_{L^q(\Omega)} \lesssim \|\underline{v}_h\|_{1,p,h}, \tag{6.11}$$

where the hidden multiplicative constant depends only on Ω, ϱ, k, p, and q.

Proof See Sect. 6.5. □

Remark 6.6 (Discrete Sobolev–Poincaré–Wirtinger Inequalities on Broken Spaces) Discrete Sobolev embeddings for the HHO space (2.36) strongly incorporating the homogeneous Dirichlet boundary conditions can be found in [142, Proposition 5.4]. The proof provided therein hinges on similar results for broken polynomial spaces; see Remark 2.17 for further details. In the Finite Elements literature, discrete counterparts of Sobolev embeddings for broken spaces are proved, e.g., in [225] for $p = 2$ and in [88] for generic exponents p. In both cases, stronger assumptions on the mesh are made, namely the fact that every mesh element is the image through an affine map of one polyhedron out of a finite set of reference polyhedra.

Theorem 6.7 (Global Discrete Trace Inequality in $\underline{U}_{h,\star}^k$) *Let a polynomial degree $k \geq 0$ and an index $p \in (1, \infty)$ be fixed. Let $(\mathcal{M}_h)_{h \in \mathcal{H}}$ denote a regular sequence of meshes in the sense of Definition 1.9. Then, for all $\underline{v}_h \in \underline{U}_{h,\star}^k$,*

$$\|\gamma_h \underline{v}_h\|_{L^p(\partial\Omega)} \lesssim \|\underline{v}_h\|_{1,p,h}, \tag{6.12}$$

where the hidden multiplicative constant depends only on Ω, ϱ, k, and p.

Proof See Sect. 6.5. □

To state the discrete compactness result, we need the global discrete gradient operator $\mathbf{G}_h^k : \underline{U}_h^k \to \mathbb{P}^k(\mathcal{T}_h)^d$ defined by (4.74) and such that, for all $\underline{v}_h \in \underline{U}_{h,\star}^k$,

$$(\mathbf{G}_h^k \underline{v}_h)_{|T} := \mathbf{G}_T^k \underline{v}_T \qquad \forall T \in \mathcal{T}_h,$$

where, for any $T \in \mathcal{T}_h$, the local gradient reconstruction $\mathbf{G}_T^k : \underline{U}_T^k \to \mathbb{P}^k(T)^d$, given by (4.37), is such that, for all $\underline{v}_T \in \underline{U}_T^k$ and all $\boldsymbol{\tau} \in \mathbb{P}^k(T)^d$,

$$\int_T \mathbf{G}_T^k \underline{v}_T \cdot \boldsymbol{\tau} = -\int_T v_T (\boldsymbol{\nabla} \cdot \boldsymbol{\tau}) + \sum_{F \in \mathcal{F}_T} \int_F v_F (\boldsymbol{\tau} \cdot \boldsymbol{n}_{TF}) \tag{6.13}$$

$$= \int_T \boldsymbol{\nabla} v_T \cdot \boldsymbol{\tau} + \sum_{F \in \mathcal{F}_T} \int_F (v_F - v_T)(\boldsymbol{\tau} \cdot \boldsymbol{n}_{TF}). \tag{6.14}$$

We also need the potential reconstruction defined by (2.11). To avoid confusion with the exponent p, throughout this chapter we rename this reconstruction $\mathrm{r}_T^{k+1} : \underline{U}_T^k \to \mathbb{P}^{k+1}(T)$. For the sake of convenience, its definition is recalled hereafter:

$$\begin{aligned} \int_T \boldsymbol{\nabla} \mathrm{r}_T^{k+1} \underline{v}_T \cdot \boldsymbol{\nabla} w &= -\int_T v_T \Delta w + \sum_{F \in \mathcal{F}_T} \int_F v_F (\boldsymbol{\nabla} w \cdot \boldsymbol{n}_{TF}) \quad \forall w \in \mathbb{P}^{k+1}(T), \\ \int_T \mathrm{r}_T^{k+1} \underline{v}_T &= \int_T v_T. \end{aligned} \tag{6.15}$$

We finally recall the definition of the global potential reconstruction $\mathrm{r}_h^{k+1} : \underline{U}_h^k \to \mathbb{P}^{k+1}(\mathcal{T}_h)$ such that, for all $\underline{v}_h \in \underline{U}_h^k$,

$$(\mathrm{r}_h^{k+1} \underline{v}_h)_{|T} := \mathrm{r}_T^{k+1} \underline{v}_T \qquad \forall T \in \mathcal{T}_h.$$

Theorem 6.8 (Discrete Compactness) *Let a polynomial degree $k \geq 0$ and an index $p \in (1, \infty)$ be fixed. Let $(\mathcal{M}_h)_{h \in \mathcal{H}}$ denote a regular sequence of meshes in the sense of Definition 1.9. Let $(\underline{v}_h)_{h \in \mathcal{H}} \in (\underline{U}_{h,\star}^k)_{h \in \mathcal{H}}$ be a sequence for which there exists a real number $C > 0$ independent of h such that*

$$\|\underline{v}_h\|_{1,p,h} \leq C \qquad \forall h \in \mathcal{H}.$$

Then, there exists $v \in W_\star^{1,p}(\Omega)$ such that, up to a subsequence as $h \to 0$,

(i) $v_h \to v$ and $\mathrm{r}_h^{k+1} \underline{v}_h \to v$ strongly in $L^q(\Omega)$ for all $1 \leq q < \frac{dp}{d-p}$ if $1 \leq p < d$, and $1 \leq q < \infty$ if $p \geq d$;
(ii) $\gamma_h \underline{v}_h \to v_{|\partial\Omega}$ strongly in $L^p(\partial\Omega)$;
(iii) $\mathbf{G}_h^k \underline{v}_h \rightharpoonup \boldsymbol{\nabla} v$ weakly in $L^p(\Omega)^d$.

Proof See Sect. 6.5. □

Remark 6.9 (Role of Discrete Sobolev–Poincaré–Wirtinger Embeddings) The discrete embeddings (6.11) are essential to obtain the convergence of v_h in $L^q(\Omega)$, and thus to deal with the dependency of $\boldsymbol{\sigma}$ with respect to u (in particular through the growth condition involving t in (6.3b)).

6.2.2 Reconstruction-Based Discrete $W^{1,p}$-Norms

The norm (6.9) on $\underline{U}_{h,\star}^k$ mimics the $W^{1,p}$-norm and is used to establish properties on sequences in this space (e.g., boundedness of reconstructed functions in Lebesgue norms through Theorem 6.5, or compactness through Theorem 6.8). For the convergence analysis, two additional norms based on local reconstructions that we introduce hereafter will be useful.

6.2.2.1 Gradient Reconstruction-Based $W^{1,p}$-norm

The first norm, based on the discrete gradient operator, is given by

$$\begin{aligned}
\|\underline{v}_h\|_{\mathbf{G},p,h} &:= \left(\sum_{T\in\mathcal{T}_h} \|\underline{v}_T\|_{\mathbf{G},p,T}^p\right)^{\frac{1}{p}}, \\
\|\underline{v}_T\|_{\mathbf{G},p,T} &:= \left(\|\mathbf{G}_T^k\underline{v}_T\|_{L^p(T)^d}^p + |\underline{v}_T|_{\delta,p,T}^p\right)^{\frac{1}{p}} \qquad \forall T\in\mathcal{T}_h,
\end{aligned} \tag{6.16}$$

where, recalling the difference operators δ_T^k and δ_{TF}^k defined by (2.19), the difference seminorm $|\cdot|_{\delta,p,T}$ is such that

$$|\underline{v}_T|_{\delta,p,T} := \left(\sum_{F\in\mathcal{F}_T} h_F^{1-p}\|\delta_{TF}^k\underline{v}_T - \delta_T^k\underline{v}_T\|_{L^p(F)}^p\right)^{\frac{1}{p}}. \tag{6.17}$$

Remark 6.10 (Choice of the Consistent Gradient) The locally variable diffusion case covered in Sect. 4.2 is a special case of (6.1a), obtained by selecting $\boldsymbol{\sigma}(\boldsymbol{x}, u, \nabla u) = \boldsymbol{K}(\boldsymbol{x})\nabla u$. As a consequence, it is clear from the discussion at the beginning of Sect. 4.2.1 that the gradient of the potential reconstruction is not a good choice for the consistent contribution in the discretisation of (6.7), and that the gradient reconstruction $\mathbf{G}_T^k$ should be used instead. The norm $\|\cdot\|_{\mathbf{G},p,h}$ will therefore appear as the natural norm to establish a priori estimates on the solution to the HHO scheme for (6.1).

6.2.2.2 Potential Reconstruction-Based $W^{1,p}$-Norm

The second norm hinges on the potential reconstruction defined by (6.15):

$$\|\underline{v}_h\|_{\nabla r,p,h} := \left(\sum_{T\in\mathcal{T}_h} \|\underline{v}_T\|_{\nabla r,p,T}^p\right)^{\frac{1}{p}},$$
$$\|\underline{v}_T\|_{\nabla r,p,T}^p := \left(\|\nabla r_T^{k+1}\underline{v}_T\|_{L^p(T)^d}^p + |\underline{v}_T|_{\delta,p,T}^p\right)^{\frac{1}{p}} \qquad \forall T\in\mathcal{T}_h, \tag{6.18}$$

where $|\cdot|_{\delta,p,T}$ is given by (6.17). This norm generalises to the non-Hilbertian setting the norm $\|\cdot\|_{a,h}$ defined by (2.41) with the choice (2.22). It will enable the proof of convergence of the reconstructed potential.

6.2.2.3 Equivalence of Reconstruction-Based $W^{1,p}$-norms

The following lemma establishes the equivalence, uniform in h, of the three norms constructed on $\underline{U}_{h,\star}^k$.

Lemma 6.11 (Equivalence of Norms) *It holds, with hidden constants depending only on k, p and* ϱ:

$$\|\underline{v}_T\|_{1,p,T} \simeq \|\underline{v}_T\|_{\nabla r,p,T} \simeq \|\underline{v}_T\|_{\mathbf{G},p,T} \quad \forall T\in\mathcal{T}_h\,,\ \forall \underline{v}_T\in\underline{U}_T^k. \tag{6.19}$$

As a consequence,

$$\|\underline{v}_h\|_{1,p,h} \simeq \|\underline{v}_h\|_{\nabla r,p,h} \simeq \|\underline{v}_h\|_{\mathbf{G},p,h} \quad \forall \underline{v}_h\in\underline{U}_h^k. \tag{6.20}$$

Remark 6.12 Note that $\|\cdot\|_{1,p,h}$, $\|\cdot\|_{\nabla r,p,h}$ and $\|\cdot\|_{\mathbf{G},p,h}$ are only seminorms on $\underline{U}_h^k$, but genuine norms on $\underline{U}_{h,\star}^k$. Indeed, if $\underline{v}_h\in\underline{U}_{h,\star}^k$ and $\|\underline{v}_h\|_{1,p,h}=0$, then reasoning as in the proof of Corollary 2.16 shows that all element and face components in $\underline{v}_h$ are constant and equal, and by the Poincaré–Sobolev–Wirtinger inequality (6.11) that this constant is zero.

Proof The global equivalence (6.20) is obtained by raising the local equivalences (6.19) to the power p, summing over $T\in\mathcal{T}_h$, and taking the power $1/p$ of the resulting relation. We therefore only have to prove the local equivalences.

(i) *The case* $p=2$. For $p=2$, we have $\|\cdot\|_{1,2,T}=\|\cdot\|_{1,T}$ (defined by (2.7)), and $\|\cdot\|_{\nabla r,2,T}=a_T(\cdot,\cdot)^{\frac{1}{2}}$, where a_T is given by (2.15) with the stabilisation term (2.22). Proposition 2.13 shows that $\|\cdot\|_{1,2,T}\simeq\|\cdot\|_{\nabla r,2,T}$, thus proving the first equivalence in (6.19). To conclude the proof in the case $p=2$, it only remains to establish the following two estimates:

$$\|\cdot\|_{\nabla\mathrm{r},2,T} \lesssim \|\cdot\|_{\mathbf{G},2,T} \tag{6.21}$$

and

$$\|\cdot\|_{\mathbf{G},2,T} \lesssim \|\cdot\|_{1,2,T}. \tag{6.22}$$

Let $\underline{v}_T \in \underline{U}_T^k$. By Remark 4.9, $\nabla \mathrm{r}_T^{k+1}\underline{v}_T$ is the $L^2(T)^d$-orthogonal projection of $\mathbf{G}_T^k\underline{v}_T$ onto $\nabla\mathbb{P}^{k+1}(T)$. Hence, $\|\nabla \mathrm{r}_T^{k+1}\underline{v}_T\|_{L^2(T)^d} \le \|\mathbf{G}_T^k\underline{v}_T\|_{L^2(T)^d}$, which readily proves (6.21) (with actually $\le$ instead of the less precise $\lesssim$).

We now turn to (6.22). Since $\|\underline{v}_T\|_{\nabla\mathrm{r},2,T} \lesssim \|\underline{v}_T\|_{1,2,T}$, we have $|\underline{v}_T|_{\delta,2,T} \lesssim \|\underline{v}_T\|_{1,2,T}$, and proving (6.22) thus reduces to showing that $\|\mathbf{G}_T^k\underline{v}_T\|_{L^2(T)^d} \lesssim \|\underline{v}_T\|_{1,2,T}$. To establish this, take $\boldsymbol{\tau} = \mathbf{G}_T^k\underline{v}_T$ in (6.14) and write

$$\begin{aligned}
\|\mathbf{G}_T^k\underline{v}_T\|_{L^2(T)^d}^2 &= \int_T \nabla v_T\cdot\mathbf{G}_T^k\underline{v}_T + \sum_{F\in\mathcal{F}_T}\int_F (v_F - v_T)(\mathbf{G}_T^k\underline{v}_T\cdot\boldsymbol{n}_{TF})\\
&\le \|\nabla v_T\|_{L^2(T)^d}\|\mathbf{G}_T^k\underline{v}_T\|_{L^2(T)^d} + \sum_{F\in\mathcal{F}_T}\|v_F - v_T\|_{L^2(F)}\|\mathbf{G}_T^k\underline{v}_T\|_{L^2(F)}\\
&\lesssim \|\nabla v_T\|_{L^2(T)^d}\|\mathbf{G}_T^k\underline{v}_T\|_{L^2(T)^d} + \sum_{F\in\mathcal{F}_T} h_F^{-\frac{1}{2}}\|v_F - v_T\|_{L^2(F)}\|\mathbf{G}_T^k\underline{v}_T\|_{L^2(T)^d}\\
&\lesssim \|\underline{v}_T\|_{1,2,T}\|\mathbf{G}_T^k\underline{v}_T\|_{L^2(T)^d},
\end{aligned} \tag{6.23}$$

where we have used a Cauchy–Schwarz inequality and generalised Hölder inequalities with exponents $(2, 2, \infty)$ along with $\|\boldsymbol{n}_{TF}\|_{L^\infty(F)^d} = 1$ to pass to the second line, the discrete trace inequality (1.55) applied with $p = 2$ to the components of $\mathbf{G}_T^k\underline{v}_T$ along with $h_F \le h_T$ to pass to the third line, and the definition (6.9) of the $\|\cdot\|_{1,2,T}$-norm to conclude. After simplification, (6.23) shows that $\|\mathbf{G}_T^k\underline{v}_T\|_{L^2(T)^d} \lesssim \|\underline{v}_T\|_{1,2,T}$, which concludes the proof of (6.22).

(ii) *The case* $p \neq 2$. The idea is to leverage the equivalence proved in the case $p = 2$ by using the direct and inverse Lebesgue inequalities (1.35) on mesh elements and faces; see Remark 1.27. These inequalities show that, for all $\boldsymbol{\varphi}_T \in \mathbb{P}^k(T)^d$,

$$\|\boldsymbol{\varphi}_T\|_{L^p(T)^d}^p \simeq |T|_d^{1-\frac{p}{2}}\|\boldsymbol{\varphi}_T\|_{L^2(T)^d}^p = |T|_d^{1-\frac{p}{2}}\left(\|\boldsymbol{\varphi}_T\|_{L^2(T)^d}^2\right)^{\frac{p}{2}} \tag{6.24}$$

and, for all $F \in \mathcal{F}_T$ and $\varphi_F \in \mathbb{P}^k(F)$,

$$\begin{aligned}
h_F^{1-p}\|\varphi_F\|_{L^p(F)}^p &\simeq h_F^{1-p}|F|_{d-1}^{1-\frac{p}{2}}\|\varphi_F\|_{L^2(F)}^p \\
&= h_F^{-\frac{p}{2}}(h_F|F|_{d-1})^{1-\frac{p}{2}}\|\varphi_F\|_{L^2(F)}^p \\
&\simeq |T|_d^{1-\frac{p}{2}}\left(h_F^{-1}\|\varphi_F\|_{L^2(F)}^2\right)^{\frac{p}{2}},
\end{aligned} \tag{6.25}$$

where we have used the mesh regularity property to write $h_F|F|_{d-1} \simeq |T|_d$, owing to (1.6), (1.7) and (1.8). We also notice that, for any finite family $(a_i)_{i\in I}$ of non-negative numbers and any exponent $\alpha > 0$,

$$\sum_{i\in I} a_i^\alpha \le \mathrm{card}(I)\left(\sum_{i\in I} a_i\right)^\alpha \le \mathrm{card}(I)^{1+\alpha}\sum_{i\in I} a_i^\alpha. \tag{6.26}$$

The first inequality is obtained writing $a_i \le \sum_{j\in I} a_j$ for all $i \in I$, while the second follows from $(\sum_{i\in I} a_i)^\alpha \le (\mathrm{card}(I)\max_{i\in I} a_i)^\alpha = \mathrm{card}(I)^\alpha \max_{i\in I} a_i^\alpha \le \mathrm{card}(I)^\alpha \sum_{i\in I} a_i^\alpha$. Summing (6.25) over $F \in \mathcal{F}_T$, adding (6.24) and using (6.26) (with $I = \{T\} \cup \mathcal{F}_T$, $a_T = \|\boldsymbol{\varphi}_T\|_{L^2(T)^d}^2$, $a_F = h_F^{-1}\|\varphi_F\|_{L^2(F)}^2$ for all $F \in \mathcal{F}_T$, and $\alpha = p/2$), we have

$$\|\boldsymbol{\varphi}_T\|_{L^p(T)^d}^p + \sum_{F\in\mathcal{F}_T} h_F^{1-p}\|\varphi_F\|_{L^p(F)}^p \simeq |T|_d^{1-\frac{p}{2}}\left(\|\boldsymbol{\varphi}_T\|_{L^2(T)^d}^2 + \sum_{F\in\mathcal{F}_T} h_F^{-1}\|\varphi_F\|_{L^2(F)}^2\right)^{\frac{p}{2}}.$$

Apply this relation to

$$\begin{aligned}
(\boldsymbol{\varphi}_T, (\varphi_F)_{F\in\mathcal{F}_T}) &= (\nabla v_T, (v_F - v_T)_{F\in\mathcal{F}_T}), \\
(\boldsymbol{\varphi}_T, (\varphi_F)_{F\in\mathcal{F}_T}) &= (\nabla \mathrm{r}_T^{k+1}\underline{v}_T, (\delta_{TF}^k\underline{v}_T - \delta_T^k\underline{v}_T)_{F\in\mathcal{F}_T}), \\
(\boldsymbol{\varphi}_T, (\varphi_F)_{F\in\mathcal{F}_T}) &= (\mathbf{G}_T^k\underline{v}_T, (\delta_{TF}^k\underline{v}_T - \delta_T^k\underline{v}_T)_{F\in\mathcal{F}_T})
\end{aligned}$$

to obtain, respectively,

$$\begin{aligned}
\|\underline{v}_T\|_{1,p,T}^p &\simeq |T|_d^{1-\frac{p}{2}}\|\underline{v}_T\|_{1,2,T}^{\frac{p}{2}}, \\
\|\underline{v}_T\|_{\nabla\mathrm{r},p,T}^p &\simeq |T|_d^{1-\frac{p}{2}}\|\underline{v}_T\|_{\nabla\mathrm{r},2,T}^{\frac{p}{2}}, \\
\|\underline{v}_T\|_{\mathbf{G},p,T}^p &\simeq |T|_d^{1-\frac{p}{2}}\|\underline{v}_T\|_{\mathbf{G},2,T}^{\frac{p}{2}}.
\end{aligned}$$

The equivalences (6.19) then follow from the case $p = 2$ covered in Point (i). □

6.2.3 Discrete Problem and Well-Posedness

The discrete counterpart of the function A defined by (6.7) is the function $\mathrm{A}_h : \underline{U}_h^k \times \underline{U}_h^k \to \mathbb{R}$ such that, for all $\underline{u}_h, \underline{v}_h \in \underline{U}_h^k$,

$$\mathrm{A}_h(\underline{u}_h; \underline{v}_h) := \int_\Omega \boldsymbol{\sigma}(u_h, \mathbf{G}_h^k \underline{u}_h) \cdot \mathbf{G}_h^k \underline{v}_h + \sum_{T \in \mathcal{T}_h} \mathrm{S}_T(\underline{u}_T; \underline{v}_T). \tag{6.27}$$

Here, for all $T \in \mathcal{T}_h$, $\mathrm{S}_T : \underline{U}_T^k \times \underline{U}_T^k \to \mathbb{R}$ is a local stabilisation function which can be obtained, e.g., by generalising (2.22) to the non-Hilbertian setting:

$$\mathrm{S}_T(\underline{u}_T; \underline{v}_T) :=$$

$$\sum_{F \in \mathcal{F}_T} h_F^{1-p} \int_F |\delta_{TF}^k \underline{u}_T - \delta_T^k \underline{u}_T|^{p-2} (\delta_{TF}^k \underline{u}_T - \delta_T^k \underline{u}_T)(\delta_{TF}^k \underline{v}_T - \delta_T^k \underline{v}_T), \tag{6.28}$$

where we remind the reader that the difference operators δ_T^k and δ_{TF}^k are given by (2.19).

Remark 6.13 (Scaling of S_T*)* As in Sect. 4.2.2, a scaling factor can be introduced in the stabilisation term S_T to match the (local) magnitude of $\boldsymbol{\sigma}$, when such a thing can be defined. See, e.g., the discussion in [73, Remark 7], where nonlinear elasticity problems with $p = 2$ are considered.

Recalling the definition (6.10) of γ_h, the discrete problem reads: Find $\underline{u}_h \in \underline{U}_{h,\star}^k$ such that

$$\mathrm{A}_h(\underline{u}_h; \underline{v}_h) = \int_\Omega f v_h + \int_{\partial\Omega} g\, \gamma_h \underline{v}_h \qquad \forall \underline{v}_h \in \underline{U}_{h,\star}^k. \tag{6.29}$$

The existence of a solution to (6.29) can be established using results from the topological degree theory.

Lemma 6.14 (Existence of a Solution and a Priori Bound) *Problem* (6.29) *admits at least one solution, and any solution* $\underline{u}_h$ *to this problem satisfies*

$$\|\underline{u}_h\|_{1,p,h} \lesssim \left(\|f\|_{L^{p'}(\Omega)} + \|g\|_{L^{p'}(\partial\Omega)} \right)^{\frac{1}{p-1}}, \tag{6.30}$$

where the hidden constant is independent of h.

Proof Endow the space $\underline{U}_{h,\star}^k$ with an inner product $\langle\cdot,\cdot\rangle$ and associated norm $|\cdot|$. For all $\underline{u}_h \in \underline{U}_{h,\star}^k$, the mapping $\mathrm{A}_h(\underline{u}_h; \cdot)$ is linear on $\underline{U}_{h,\star}^k$, and the Riesz representation theorem thus gives a unique $\Phi(\underline{u}_h) \in \underline{U}_{h,\star}^k$ such that

$$\mathrm{A}_h(\underline{u}_h; \underline{v}_h) = \langle \Phi(\underline{u}_h), \underline{v}_h \rangle \qquad \forall \underline{v}_h \in \underline{U}_{h,\star}^k.$$

Likewise, there is a unique $\underline{w}_h \in \underline{U}_{h,\star}^k$ such that $\int_\Omega f v_h + \int_{\partial\Omega} g\, \gamma_h \underline{v}_h = \langle \underline{w}_h, \underline{v}_h \rangle$ for all $\underline{v}_h \in \underline{U}_{h,\star}^k$. We note that $\underline{u}_h$ is a solution to (6.29) if and only if $\Phi(\underline{u}_h) = \underline{w}_h$.

Using the assumptions (6.3a) and (6.3b) on $\boldsymbol{\sigma}$, it can easily be checked that the mapping $\Phi : \underline{U}_{h,\star}^k \to \underline{U}_{h,\star}^k$ is continuous since, the space $\underline{U}_{h,\star}^k$ being finite-dimensional, convergence of a sequence in this space for the norm $|\cdot|$ is equivalent to the convergence of all its components on a fixed basis of the space. Moreover, by the coercivity condition (6.3d) and the definition of S_T, for all $\underline{u}_h \in \underline{U}_{h,\star}^k$,

$$\begin{aligned}
\langle \Phi(\underline{u}_h), \underline{u}_h \rangle &= \mathrm{A}_h(\underline{u}_h; \underline{u}_h) \\
&\ge \lambda_\sigma \|\mathbf{G}_h^k \underline{u}_h\|_{L^p(\Omega)^d}^p + \sum_{T\in\mathcal{T}_h} \sum_{F\in\mathcal{F}_T} h_F^{1-p} \|\delta_{TF}^k \underline{u}_T - \delta_T^k \underline{u}_T\|_{L^p(F)}^p \\
&\gtrsim \|\underline{u}_h\|_{\mathbf{G},p,h}^p \qquad (6.31)\\
&\ge c_h |\underline{u}_h|^p,
\end{aligned}$$

where we have used, in the conclusion, the equivalence of $\|\cdot\|_{\mathbf{G},p,h}$ and $|\cdot|$ on the finite-dimensional space $\underline{U}_{h,\star}^k$; here, $c_h > 0$ possibly depends on h but not on $\underline{u}_h$. This estimate shows that $\frac{\langle \Phi(\underline{u}_h), \underline{u}_h \rangle}{|\underline{u}_h|} \to \infty$ as $|\underline{u}_h| \to \infty$. Using Theorem 6.15 below, we infer the existence of $\underline{u}_h \in \underline{U}_{h,\star}^k$ such that $\Phi(\underline{u}_h) = \underline{w}_h$, which precisely means that $\underline{u}_h$ is a solution to (6.29).

To establish (6.30), we write

$$\begin{aligned}
\|\underline{u}_h\|_{1,p,h}^p &\lesssim \|\underline{u}_h\|_{\mathbf{G},p,h}^p \\
&\lesssim \mathrm{A}_h(\underline{u}_h; \underline{u}_h) = \int_\Omega f u_h + \int_{\partial\Omega} g\, \gamma_h \underline{u}_h \\
&\le \|f\|_{L^{p'}(\Omega)} \|u_h\|_{L^p(\Omega)} + \|g\|_{L^{p'}(\partial\Omega)} \|\gamma_h \underline{u}_h\|_{L^p(\partial\Omega)} \\
&\lesssim \left(\|f\|_{L^{p'}(\Omega)} + \|g\|_{L^{p'}(\partial\Omega)} \right) \|\underline{u}_h\|_{1,p,h},
\end{aligned}$$

where the first line follows from the equivalence (6.20), the second line is a consequence of (6.31) and of the fact that $\underline{u}_h$ solves (6.29), the third line is obtained using Hölder's inequalities, and the conclusion follows from the Sobolev–Poincaré–Wirtinger Theorem 6.5 with $q = p$ and from the trace inequality (6.12). Simplify by $\|\underline{u}_h\|_{1,p,h}$ on each side and take the power $\frac{1}{p-1}$ to get (6.30). □

The following theorem, used in the proof above, corresponds to [139, Theorem 3.3]

Theorem 6.15 (Surjectivity of Nonlinear Coercive Mappings) *Let E be a Euclidean space with inner product $\langle\cdot,\cdot\rangle$ and norm $|\cdot|$. If $f : E \to E$ is a continuous function that satisfies*

$$\lim_{|x|\to\infty} \frac{\langle f(x), x\rangle}{|x|} = \infty,$$

then f is surjective, that is, for any $y \in E$ there exists at least one $x \in E$ such that $f(x) = y$.

Remark 6.16 (Uniqueness) The uniqueness of the continuous and discrete solutions cannot be established under Assumption 6.1; see [177, Remark 3.4] in the case of homogeneous Dirichlet boundary conditions. Uniqueness can be proved under the additional assumption (6.82) of strict monotonicity of the flux function; see Theorem 6.19 (and its proof in Sect. 6.3.4) for the p-Laplace equation. For more general Leray–Lions models, we refer to [230] for the continuous model and to [169] for a Mixed Finite Volumes approximation (close to an HHO method of degree $k = 0$). We also mention [68, Theorem 7] concerning nonlinear elasticity problems.

6.2.4 Flux Formulation

We state here an equivalent formulation of the discrete problem (6.29) using numerical fluxes. We start with some preliminaries, for an arbitrary mesh element $T \in \mathcal{T}_h$, on the stabilisation function S_T defined by (6.28). This function only depends on its arguments through the difference operators $(\delta_T^k, (\delta_{TF}^k)_{F\in\mathcal{F}_T})$. Following the proof of Proposition 2.24, we can therefore write

$$\mathrm{S}_T(\underline{u}_T; \underline{v}_T) = \mathrm{S}_T\left((0, \underline{\Delta}_{\partial T}^k \underline{u}_T); (0, \underline{\Delta}_{\partial T}^k \underline{v}_T)\right), \tag{6.32}$$

where $\underline{\Delta}_{\partial T}^k : \underline{U}_T^k \to \underline{D}_{\partial T}^k$ is given by (2.56). We then define, in a similar way as in (2.59), a (nonlinear) boundary residual operator $\underline{R}_{\partial T}^k : \underline{U}_T^k \to \underline{D}_{\partial T}^k$ such that, for all $\underline{u}_T \in \underline{U}_T^k$, $\underline{R}_{\partial T}^k(\underline{u}_T) = (R_{TF}^k(\underline{u}_T))_{F\in\mathcal{F}_T}$ with, for all $\underline{\alpha}_{\partial T} = (\alpha_{TF})_{F\in\mathcal{F}_T} \in \underline{D}_{\partial T}^k$,

$$-\sum_{F\in\mathcal{F}_T} \int_F R_{TF}^k(\underline{u}_T)\alpha_{TF} = \mathrm{S}_T((0, \underline{\Delta}_{\partial T}^k \underline{u}_T); (0, \underline{\alpha}_{\partial T})). \tag{6.33}$$

Lemma 6.17 (Flux Formulation for the HHO Approximation of the Leray–Lions Problem) *Let $\mathcal{M}_h$ denote a polytopal mesh in the sense of Definition 1.4. Let $\underline{u}_h \in \underline{U}_{h,\star}^k$ and, for all $T \in \mathcal{T}_h$ and all $F \in \mathcal{F}_T$, define the numerical normal trace of the flux*

$$\Phi_{TF}(\underline{u}_T) := -\boldsymbol{\pi}_T^{0,k}\left(\boldsymbol{\sigma}(u_T, \mathbf{G}_T^k \underline{u}_T)\right)\cdot\boldsymbol{n}_{TF} + R_{TF}^k(\underline{u}_T)$$

with R_{TF}^k given by (6.33).

Then, $\underline{u}_h$ is a solution of problem (6.29) *if and only if the following three properties hold:*

(i) Local balance. *For all $T \in \mathcal{T}_h$ and all $v_T \in \mathbb{P}^k(T)$, it holds*

$$\int_T \boldsymbol{\sigma}(u_T, \mathbf{G}_T^k \underline{u}_T)\cdot\nabla v_T + \sum_{F\in\mathcal{F}_T}\int_F \Phi_{TF}(\underline{u}_T) v_T = \int_T f v_T. \tag{6.34a}$$

(ii) Continuity of the numerical normal traces of the flux. *For any interface $F \in \mathcal{F}_h^{\rm i}$ such that $F \subset \partial T_1 \cap \partial T_2$ with distinct mesh elements $T_1, T_2 \in \mathcal{T}_h$, it holds*

$$\Phi_{T_1F}(\underline{u}_{T_1}) + \Phi_{T_2F}(\underline{u}_{T_2}) = 0. \tag{6.34b}$$

(iii) Values of boundary fluxes. *For any boundary face $F \in \mathcal{F}_h^{\rm b}$, letting $T \in \mathcal{T}_h$ be such that $F \in \mathcal{F}_T$, it holds*

$$\Phi_{TF}(\underline{u}_T) = -\pi_F^{0,k} g. \tag{6.34c}$$

Remark 6.18 (Neumann Boundary Conditions) Comparing with Lemma 2.25, we notice a third property involving the boundary fluxes. This additional property corresponds to the Neumann boundary condition considered here (see (6.1b)).

Proof The proof follows the ideas developed in Sect. 2.2.5, with adaptations to deal with non-homogeneous Neumann boundary conditions.

Let $\underline{v}_h \in \underline{U}_h^k$. Using the relation (4.41) with $\boldsymbol{\tau} = \boldsymbol{\sigma}(u_T, \mathbf{G}_T^k \underline{u}_T)$, we have

$$\int_T \boldsymbol{\sigma}(u_T, \mathbf{G}_T^k \underline{u}_T)\cdot\mathbf{G}_T^k \underline{v}_T = \int_T \boldsymbol{\sigma}(u_T, \mathbf{G}_T^k \underline{u}_T)\cdot\nabla v_T \\ + \sum_{F\in\mathcal{F}_T}\int_F \boldsymbol{\pi}_T^{0,k}(\boldsymbol{\sigma}(u_T, \mathbf{G}_T^k \underline{u}_T))\cdot\boldsymbol{n}_{TF}(v_F - v_T).$$

The relations (6.32) and (6.33) yield

$$\mathrm{S}_T(\underline{u}_T;\underline{v}_T) = -\sum_{F\in\mathcal{F}_T}\int_F R^k_{TF}(\underline{u}_T)(v_F - v_T).$$

Plugging these two equations into (6.27), we see that the scheme (6.29) can be recast as:

$$\begin{aligned}\sum_{T\in\mathcal{T}_h}\Big(\int_T \boldsymbol{\sigma}(u_T,\mathbf{G}^k_T\underline{u}_T)\cdot\nabla v_T - \sum_{F\in\mathcal{F}_T}\int_F \Phi_{TF}(\underline{u}_T)(v_F - v_T)\Big)\\ = \int_\Omega f v_h + \int_{\partial\Omega} g\,\gamma_h\underline{v}_h \qquad \forall \underline{v}_h \in \underline{U}^k_{h,\star}.\end{aligned} \tag{6.35}$$

Define $\underline{\mathbf{1}}_h \in \underline{U}^k_h$ such that $\mathbf{1}_T = 1$ for all $T \in \mathcal{T}_h$ and $\mathbf{1}_F = 1$ for all $F \in \mathcal{F}_h$. If $\underline{v}_h \in \underline{U}^k_h$ and $m = \frac{1}{|\Omega|}\int_\Omega v_h$, then $\underline{v}_h - m\underline{\mathbf{1}}_h \in \underline{U}^k_{h,\star}$ can be used in (6.35). A simple inspection shows that the terms involving $\underline{\mathbf{1}}_h$ in the left-hand side then vanish, while the compatibility condition (6.2) ensures that the corresponding terms in the right-hand side cancel out. This proves that (6.35) also holds for $\underline{v}_h \in \underline{U}^k_h$.

The conclusion of the proof is then similar to the proof of Lemma 2.21, selecting for $\underline{v}_h$ elements that span a basis of $\underline{U}^k_h$, that is, the same elements as in the proof of Lemma 2.21 but including also non-zero polynomials on boundary faces. The cell basis functions yield (6.34a) and, noticing that the normal trace of the numerical flux $\Phi_{TF}(\underline{u}_T)$ is a polynomial of degree k on F, the internal face basis functions yield (6.34b), while the boundary face basis functions correspond to (6.34c). □

6.3 Error Estimates for the *p*-Laplacian

When dealing with high-order methods, it is important to determine the convergence rates attained when the solution is regular enough. General Leray–Lions operators lack the structure that enables the proof of theoretical rates of convergence. It is however possible, for certain operators, to establish error estimates and deduce from them convergence rates. The goal of this section is precisely to prove such estimates for the p-Laplace equation, that is, (6.1) with flux function given by (6.4). For this choice, stronger monotonicity and continuity properties hold than the ones in Assumption 6.1, which are crucial to our goal.

The material is organised as follows: in Sect. 6.3.1 we state and comment the main result, Theorem 6.19, and give a general overview of its proof; Sects. 6.3.2 and 6.3.3 contain preliminary results, namely a study of the consistency properties of the stabilisation function applied to interpolates of smooth functions, and a proof of the monotonicity and continuity properties of the p-Laplace flux function; finally, Sect. 6.3.4 contains the proof of Theorem 6.19.

6.3.1 Statement of the Error Estimates

The main result is the following estimate in a discrete energy norm, where the discrete solution is compared to the interpolate of the continuous one.

Theorem 6.19 (Error Estimate in Discrete Energy Norm) *Let a polynomial degree $k \geq 0$ and an index $p \in (1, \infty)$ be fixed, and assume that*

$$\sigma(\boldsymbol{x}, s, \boldsymbol{\xi}) = |\boldsymbol{\xi}|^{p-2}\boldsymbol{\xi} \quad \text{for a.e. } \boldsymbol{x} \in \Omega,\ \forall s \in \mathbb{R},\ \forall \boldsymbol{\xi} \in \mathbb{R}^d.$$

Let $(\mathcal{M}_h)_{h\in\mathcal{H}}$ denote a regular mesh sequence in the sense of Definition 1.9. Let $h \in \mathcal{H}$ and assume that the solution to (6.6) satisfies $u \in W^{r+2,p}(\mathcal{T}_h)$ and $\sigma(\nabla u) \in W^{r+1,p'}(\mathcal{T}_h)^d$ for some $r \in \{0, \ldots, k\}$. Define $E_h(u) \in \mathbb{R}^+$ the following way:

(i) If $p \geq 2$,

$$E_h(u) := h^{r+1}|u|_{W^{r+2,p}(\mathcal{T}_h)} + h^{\frac{r+1}{p-1}}\left(|u|^{\frac{1}{p-1}}_{W^{r+2,p}(\mathcal{T}_h)} + |\sigma(\nabla u)|^{\frac{1}{p-1}}_{W^{r+1,p'}(\mathcal{T}_h)^d}\right);$$

(ii) If $p < 2$,

$$E_h(u) := h^{(r+1)(p-1)}|u|^{p-1}_{W^{r+2,p}(\mathcal{T}_h)} + h^{r+1}|\sigma(\nabla u)|_{W^{r+1,p'}(\mathcal{T}_h)^d}.$$

Then, there exists a unique $\underline{u}_h \in \underline{U}^k_{h,\star}$ solution to (6.29), and it satisfies

$$\|\underline{u}_h - \underline{I}^k_h u\|_{\mathbf{G},p,h} \lesssim E_h(u) \tag{6.36}$$

with hidden constant independent of h and u, but possibly depending on Ω, p, ϱ, and on an upper bound of $\|f\|_{L^{p'}(\Omega)} + \|g\|_{L^{p'}(\partial\Omega)}$.

Proof See Sect. 6.3.4, based on the results of Sects. 6.3.2 and 6.3.3. □

From this estimate in a discrete energy norm, we can derive a convergence result for the error measured as the difference between the discrete and continuous gradient.

Corollary 6.20 (Error Estimate for Reconstructed Gradient) *Under the assumptions and notations of Theorem 6.19, it holds*

$$\|\mathbf{G}^k_h\underline{u}_h - \nabla u\|_{L^p(\Omega)^d} + |\underline{u}_h|_{\delta,p,h} \lesssim E_h(u) + h^{r+1}|u|_{W^{r+2,p}(\mathcal{T}_h)}, \tag{6.37}$$

where we have introduced the seminorm $|\cdot|_{\delta,p,h}$ *on* $\underline{U}_h^k$ *such that, for all* $\underline{v}_h \in \underline{U}_h^k$,

$$|\underline{v}_h|_{\delta,p,h}^p := \sum_{T\in\mathcal{T}_h} |\underline{v}_T|_{\delta,p,T}^p = \sum_{T\in\mathcal{T}_h} \mathrm{S}_T(\underline{v}_T;\underline{v}_T).$$

A few remarks are in order.

Remark 6.21 (Order of Convergence) The asymptotic scaling for the approximation error in the left-hand sides of (6.36) and (6.37) is determined by the leading terms in the right-hand side $\mathcal{E}_h(u)$, namely:

$$\mathcal{E}_h(u) \lesssim \begin{cases} h^{\frac{r+1}{p-1}} & \text{if } p \ge 2, \\ h^{(r+1)(p-1)} & \text{if } p < 2. \end{cases} \tag{6.38}$$

Remark 6.22 (Error Estimate in Other Norms) The norm equivalence (6.20) shows that the error estimate (6.36) also holds for $\|\underline{u}_h - \underline{I}_h^k u\|_{1,p,h}$ and for $\|\underline{u}_h - \underline{I}_h^k u\|_{\nabla\mathrm{r},p,h}$. As a consequence, the error estimate (6.37) could also be stated with $\nabla_h \mathrm{r}_h^{k+1}\underline{u}_h$ instead of $\mathbf{G}_h^k \underline{u}_h$, see [141, Theorem 3.2 and Corollary 3.1].

Remark 6.23 (Error Estimates for the Dirichlet Problem) Error estimates analogous to the ones of Theorem 6.19 and Corollary 6.20 for the Dirichlet problem have been proved in [141], to which we refer for further details.

Proof (Corollary 6.20) Let $T \in \mathcal{T}_h$. Inserting $\mathbf{G}_T^k \underline{I}_T^k u - \boldsymbol{\pi}_T^{0,k}(\nabla u) = 0$ (see (4.40)) into the norm and invoking the approximation property (1.74) of the L^2-orthogonal projector with $X = T$, $l = k$, $m = 0$, and $s = r + 1$ and $v =$ components of ∇u, we write

$$\begin{aligned}\|\mathbf{G}_T^k \underline{u}_T - \nabla u\|_{L^p(T)^d} &\le \|\mathbf{G}_T^k(\underline{u}_T - \underline{I}_T^k u)\|_{L^p(T)^d} + \|\boldsymbol{\pi}_T^{0,k}(\nabla u) - \nabla u\|_{L^p(T)^d} \\ &\lesssim \|\mathbf{G}_T^k(\underline{u}_T - \underline{I}_T^k u)\|_{L^p(T)^d} + h_T^{r+1}|\nabla u|_{W^{r+1,p}(T)^d}.\end{aligned} \tag{6.39}$$

Since $|\cdot|_{\delta,p,h}$ is a seminorm, we can invoke the triangle inequality to write

$$|\underline{u}_h|_{\delta,p,h} \le |\underline{u}_h - \underline{I}_h^k u|_{\delta,p,h} + |\underline{I}_h^k u|_{\delta,p,h}. \tag{6.40}$$

Apply (6.48) below to $\phi = u$, raise to the power p, and sum over $T \in \mathcal{T}_h$ to see that

$$|\underline{I}_h^k u|_{\delta,p,h}^p \lesssim \sum_{T\in\mathcal{T}_h} h_T^{(r+1)p} |u|_{W^{r+2,p}(T)}^p \le h^{(r+1)p} |u|_{W^{r+2,p}(\mathcal{T}_h)}^p. \tag{6.41}$$

Raising (6.39) to the power p, summing over $T \in \mathcal{T}_h$, adding the power p of (6.40), recalling the definition (6.16) of $\|\cdot\|_{\mathbf{G},p,h}$, and invoking (6.41), we find

$$\|\mathbf{G}_h^k \underline{u}_h - \nabla u\|_{L^p(\Omega)^d}^p + |\underline{u}_h|_{\delta,p,h}^p \lesssim \|\underline{u}_h - \underline{I}_h^k u\|_{\mathbf{G},p,h}^p + h^{(r+1)p} |u|_{W^{r+2,p}(\mathcal{T}_h)}^p.$$

The proof of (6.37) is completed by taking the power $1/p$ and using (6.36). □

Let us now briefly discuss the approach to proving Theorem 6.19. As the continuous equation (6.6), the HHO scheme (6.29) is of course nonlinear in general. The theory of Appendix A is therefore not directly applicable. That being said, some similarities with this linear theory can be found and exploited. Specifically, the HHO scheme (6.29) shows that, for all $\underline{v}_h \in \underline{U}_{h,\star}^k$,

$$\mathrm{A}_h(\underline{u}_h; \underline{v}_h) - \mathrm{A}_h(\underline{I}_h^k u; \underline{v}_h) = \ell_h(\underline{v}_h) - \mathrm{A}_h(\underline{I}_h^k u; \underline{v}_h), \tag{6.42}$$

where

$$\ell_h(\underline{v}_h) := \int_\Omega f v_h + \int_{\partial\Omega} g\, \gamma_h \underline{v}_h.$$

The right-hand side in equation (6.42) is a linear map in $\underline{v}_h$ that defines a consistency error, identical to (A.5) in the linear setting. Since the left-hand side of this equation involves $\underline{u}_h$ and $\underline{I}_h^k u$, (6.42) can be considered as a sort of error equation except that, contrary to the linear case (see (A.7)), this left-hand side is not an expression of the difference $\underline{u}_h - \underline{I}_h^k u$. We can nonetheless follow the principles of the analysis in Appendix A, namely:

1. estimate the consistency error in an appropriate dual norm, and
2. establish a stability property of the left-hand side in the corresponding primal norm.

The consistency error will be estimated, as for linear equations, by expressing ℓ_h in terms of u through the relations $f = -\nabla\cdot(\boldsymbol{\sigma}(\nabla u))$ and $g = \boldsymbol{\sigma}(\nabla u)\cdot\boldsymbol{n}_\Omega$, and by invoking optimal approximation properties of the reconstructions composed with the interpolator. The stability property, on the other hand, will not directly rely on an inf–sup or coercivity condition, but rather on strong monotonicity properties of A_h (stemming, in turn, from the monotonicity properties of $\boldsymbol{\sigma}$).

6.3.2 *Consistency of the Stabilisation Function*

The first preliminary result, required to estimate the consistency error, concerns the consistency properties of the global stabilisation term when one of its arguments is the interpolate of a smooth function. As in the Hilbertian setting of Chap. 2, these consistency properties follow from the boundedness of the local interpolator $\underline{I}_T^k$ and from the polynomial consistency of the local stabilisation terms S_T.

Proposition 6.24 (Boundedness of the Local Interpolator, $W^{1,p}$-Setting) *For all $T \in \mathcal{T}_h$ and all $v \in W^{1,p}(T)$,*

$$\|\underline{I}_T^k v\|_{1,p,T} \lesssim |v|_{W^{1,p}(T)}, \tag{6.43}$$

where the hidden constant depends only on d, p, ϱ, and k.

Proof The proof is similar to that of Proposition 2.2, with L^2-norms replaced with L^p-norms. Using the definitions (2.8) and (6.9) of $\underline{I}_T^k$ and $\|\cdot\|_{1,p,T}$, we write

$$\begin{aligned}\|\underline{I}_T^k v\|_{1,p,T}^p &= \|\nabla \pi_T^{0,k} v\|_{L^p(T)^d}^p + \sum_{F\in\mathcal{F}_T} h_F^{1-p}\|\pi_F^{0,k} v - \pi_T^{0,k} v\|_{L^p(F)}^p \\ &\lesssim \|\nabla v\|_{L^p(T)^d}^p + \sum_{F\in\mathcal{F}_T} h_F^{1-p}\|v - \pi_T^{0,k} v\|_{L^p(F)}^p,\end{aligned}$$

where we have used, to pass to the second line, the boundedness (1.77) of $\pi_T^{0,k}$ with $s = 1$ together with the idempotency, linearity, and boundedness (1.77) of $\pi_F^{0,k}$ with $s = 0$ to write

$$\|\pi_F^{0,k} v - \pi_T^{0,k} v\|_{L^p(F)} = \|\pi_F^{0,k}(v - \pi_T^{0,k} v)\|_{L^p(F)} \lesssim \|v - \pi_T^{0,k} v\|_{L^p(F)}.$$

We then continue invoking the trace approximation property (1.75) of the L^2-orthogonal projector with $l = k$, $s = 1$, and $m = 0$:

$$\|\underline{I}_T^k v\|_{1,p,T}^p \lesssim \|\nabla v\|_{L^p(T)^d}^p + \sum_{F\in\mathcal{F}_T} h_F^{1-p} h_T^{p-1}\|\nabla v\|_{L^p(T)^d}^p.$$

The estimate (6.43) follows using the mesh regularity assumption to write $h_T^{p-1} \lesssim h_F^{p-1}$ and $\mathrm{card}(\mathcal{F}_T) \lesssim 1$; see Lemma 1.12. □

Proposition 6.25 (Consistency of S_h) *For all $r \in \{0, \ldots, k\}$, all $\phi \in W^{r+2,p}(\mathcal{T}_h)$, and all $\underline{v}_h \in \underline{U}_h^k$, it holds*

$$|\mathrm{S}_h(\underline{v}_h; \underline{I}_h^k \phi)| \lesssim \|\underline{v}_h\|_{\mathbf{G},p,h}^{p-1} \left(\sum_{T\in\mathcal{T}_h} h_T^{p(r+1)} |\phi|_{W^{r+2,p}(T)}^p\right)^{\frac{1}{p}} \tag{6.44}$$

and

$$|\mathrm{S}_h(\underline{I}_h^k\phi;\underline{v}_h)| \lesssim \|\underline{v}_h\|_{\mathbf{G},p,h}\left(\sum_{T\in\mathcal{T}_h} h_T^{p(r+1)}|\phi|_{W^{r+2,p}(T)}^p\right)^{\frac{p-1}{p}}, \tag{6.45}$$

where the hidden constants are independent of h, ϕ and $\underline{v}_h$, and we have set, for all $\underline{w}_h, \underline{z}_h \in \underline{U}_h^k$,

$$\mathrm{S}_h(\underline{w}_h;\underline{z}_h) := \sum_{T\in\mathcal{T}_h}\mathrm{S}_T(\underline{w}_T;\underline{z}_T). \tag{6.46}$$

Proof Let $\underline{w}_h, \underline{z}_h \in \underline{U}_h^k$. Using the Hölder inequality with exponents p and $p' = \frac{p}{p-1}$ we have, for any $T\in\mathcal{T}_h$,

$$|\mathrm{S}_T(\underline{w}_T;\underline{z}_T)| \le \sum_{F\in\mathcal{F}_T} h_F^{1-p}\|\delta_{TF}^k\underline{w}_T - \delta_T^k\underline{w}_T\|_{L^p(F)}^{p-1}\|\delta_{TF}^k\underline{z}_T - \delta_T^k\underline{z}_T\|_{L^p(F)}.$$

Hence, writing $h_F^{1-p} = h_F^{(1-p)\frac{1}{p'}}h_F^{(1-p)\frac{1}{p}}$ and using a discrete Hölder inequality on the summations, with the same exponents as before, we infer

$$\begin{aligned}
|\mathrm{S}_h(\underline{w}_h;\underline{z}_h)| &\le \left(\sum_{T\in\mathcal{T}_h}\sum_{F\in\mathcal{F}_T} h_F^{1-p}\|\delta_{TF}^k\underline{w}_T - \delta_T^k\underline{w}_T\|_{L^p(F)}^p\right)^{\frac{p-1}{p}}\\
&\quad\times\left(\sum_{T\in\mathcal{T}_h}\sum_{F\in\mathcal{F}_T} h_F^{1-p}\|\delta_{TF}^k\underline{z}_T - \delta_T^k\underline{z}_T\|_{L^p(F)}^p\right)^{\frac{1}{p}}\\
&= \left(\sum_{T\in\mathcal{T}_h}|\underline{w}_T|_{\delta,p,T}^p\right)^{\frac{p-1}{p}}\left(\sum_{T\in\mathcal{T}_h}|\underline{z}_T|_{\delta,p,T}^p\right)^{\frac{1}{p}}.
\end{aligned} \tag{6.47}$$

Estimates (6.44) and (6.45) follow applying (6.47) to, respectively, $(\underline{w}_h, \underline{z}_h) = (\underline{v}_h, \underline{I}_h^k\phi)$ and $(\underline{w}_h, \underline{z}_h) = (\underline{I}_h^k\phi, \underline{v}_h)$, using $|\underline{v}_T|_{\delta,p,T}^p \le \|\underline{v}_T\|_{\mathbf{G},p,T}^p$, and invoking the following estimate: For all $T\in\mathcal{T}_h$ and all $\phi\in W^{r+2,p}(T)$,

$$|\underline{I}_T^k\phi|_{\delta,p,T} \lesssim h_T^{r+1}|\phi|_{W^{r+2,p}(T)}, \tag{6.48}$$

where the hidden constant is additionally independent of T.

Let us prove (6.48). The polynomial consistency property (2.21) of the difference operators shows that $\delta_T^k\underline{I}_T^k\phi = \delta_T^k\underline{I}_T^k(\phi - \pi_T^{0,k+1}\phi)$ and $\delta_{TF}^k\underline{I}_T^k\phi =$

$\delta^k_{TF}\underline{I}^k_T(\phi - \pi^{0,k+1}_T\phi)$ for all $F \in \mathcal{F}_T$. Hence,

$$\begin{aligned}|\underline{I}^k_T\phi|_{\delta,p,T} &= |\underline{I}^k_T(\phi - \pi^{0,k+1}_T\phi)|_{\delta,p,T}\\ &\le \|\underline{I}^k_T(\phi - \pi^{0,k+1}_T\phi)\|_{\mathbf{G},p,T}\\ &\lesssim |\phi - \pi^{0,k+1}_T\phi|_{W^{1,p}(T)},\end{aligned}$$

where the first inequality follows from the definition (6.16) of the $\|\cdot\|_{\mathbf{G},p,T}$-seminorm, while the second is a consequence of the seminorm equivalence (6.19) and of the boundedness property (6.43) of $\underline{I}^k_T$. The approximation property (1.74) of $\pi^{0,k+1}_T$ with $s = r + 2$ and $m = 1$ then concludes the proof of (6.48). □

6.3.3 *Strong Monotonicity and Continuity of the p-Laplace Flux Function*

The second preliminary result, contained in this section, concerns the strong monotonicity and continuity properties of the p-Laplace flux function $\boldsymbol{\sigma}$ defined by (6.4). As discussed at the end of Sect. 6.3.1, these properties are stronger than the ones listed in Assumption 6.1, and are instrumental to the consistency and strong monotonicity analysis carried out in the next section. Similar results can be found in [35, Lemma 2.1] and in [178, Lemma 2.40], with different proofs.

Lemma 6.26 (Strong Monotonicity of the p-Laplace Flux Function) *Let $p \in (1,\infty)$ and $\boldsymbol{\sigma}(\boldsymbol{\tau}) = |\boldsymbol{\tau}|^{p-2}\boldsymbol{\tau}$ for all $\boldsymbol{\tau} \in \mathbb{R}^d$. Set $C(p) = 2^{1-p}$ if $p \ge 2$, and $C(p) = p-1$ if $p < 2$. Then, for any $\boldsymbol{\xi},\boldsymbol{\eta} \in \mathbb{R}^d$,*

$$(\boldsymbol{\sigma}(\boldsymbol{\xi}) - \boldsymbol{\sigma}(\boldsymbol{\eta}))\cdot(\boldsymbol{\xi} - \boldsymbol{\eta}) \ge C(p)\,(|\boldsymbol{\xi}| + |\boldsymbol{\eta}|)^{p-2}\,|\boldsymbol{\xi} - \boldsymbol{\eta}|^2. \tag{6.49}$$

As a consequence, for any $\boldsymbol{\xi},\boldsymbol{\eta} \in \mathbb{R}^d$,

(i) If $p \ge 2$,

$$|\boldsymbol{\xi} - \boldsymbol{\eta}|^p \le 2^{p-1}\,(\boldsymbol{\sigma}(\boldsymbol{\xi}) - \boldsymbol{\sigma}(\boldsymbol{\eta}))\cdot(\boldsymbol{\xi} - \boldsymbol{\eta}). \tag{6.50}$$

(ii) If $p < 2$,

$$|\boldsymbol{\xi} - \boldsymbol{\eta}|^p \le (p-1)^{-\frac{p}{2}}2^{(p-1)\frac{2-p}{2}}\Big[(\boldsymbol{\sigma}(\boldsymbol{\xi}) - \boldsymbol{\sigma}(\boldsymbol{\eta}))\cdot(\boldsymbol{\xi} - \boldsymbol{\eta})\Big]^{\frac{p}{2}}\Big[|\boldsymbol{\xi}|^p + |\boldsymbol{\eta}|^p\Big]^{\frac{2-p}{2}}. \tag{6.51}$$

Remark 6.27 (Case $d = 1$) The case $d = 1$ will also be of interest to us, as we will see in the proof of Theorem 6.19 that it is related to monotonicity properties of the stabilisation term (6.28). For future use, we make here explicit the estimates (6.50) and (6.51) for $d = 1$: For all $s, t \in \mathbb{R}$,

(i) If $p \geq 2$,

$$|s - t|^p \leq 2^{p-1} \left(|s|^{p-2}s - |t|^{p-2}t \right) (s - t). \tag{6.52}$$

(ii) If $p < 2$,

$$|s - t|^p \\ \leq (p-1)^{-\frac{p}{2}} 2^{(p-1)\frac{2-p}{p}} \left[\left(|s|^{p-2}s - |t|^{p-2}t \right) (s - t) \right]^{\frac{p}{2}} \left[|s|^p + |t|^p \right]^{\frac{2-p}{2}}. \tag{6.53}$$

Proof (i) *Proof of* (6.49). Let us first consider the case $p \geq 2$. Writing

$$|\xi|^{p-2} = \frac{|\xi|^{p-2} + |\eta|^{p-2}}{2} + \frac{|\xi|^{p-2} - |\eta|^{p-2}}{2}$$

and

$$|\eta|^{p-2} = \frac{|\xi|^{p-2} + |\eta|^{p-2}}{2} - \frac{|\xi|^{p-2} - |\eta|^{p-2}}{2},$$

we have

$$\begin{aligned}
(\sigma(\xi) - \sigma(\eta)) \cdot (\xi - \eta) \\
&= \frac{|\xi|^{p-2} + |\eta|^{p-2}}{2} (\xi - \eta) \cdot (\xi - \eta) + \frac{|\xi|^{p-2} - |\eta|^{p-2}}{2} (\xi + \eta) \cdot (\xi - \eta) \\
&= \frac{|\xi|^{p-2} + |\eta|^{p-2}}{2} |\xi - \eta|^2 + \frac{|\xi|^{p-2} - |\eta|^{p-2}}{2} (|\xi|^2 - |\eta|^2).
\end{aligned}$$

The last term in the above expression is nonnegative. To see this, notice that, since $p \geq 2$, the mappings $s \mapsto s^{p-2}$ and $s \mapsto s^2$ are both increasing, so the quantities $|\xi|^{p-2} - |\eta|^{p-2}$ and $|\xi|^2 - |\eta|^2$ have the same sign. As a result,

$$(\sigma(\xi) - \sigma(\eta)) \cdot (\xi - \eta) \geq \frac{|\xi|^{p-2} + |\eta|^{p-2}}{2} |\xi - \eta|^2.$$

The proof of (6.49) is completed using the estimate

$$(|\boldsymbol{\xi}| + |\boldsymbol{\eta}|)^{p-2} \le (2\max(|\boldsymbol{\xi}|, |\boldsymbol{\eta}|))^{p-2} \le 2^{p-2}(|\boldsymbol{\xi}|^{p-2} + |\boldsymbol{\eta}|^{p-2}).$$

We now deal with the case $p < 2$. Outside $\mathbf{0}$, $\boldsymbol{\sigma}$ is differentiable with derivative $D\boldsymbol{\sigma}(z)\boldsymbol{h} = |z|^{p-2}\boldsymbol{h} + (p-2)|z|^{p-4}(z\cdot\boldsymbol{h})z$ for all $\boldsymbol{h} \in \mathbb{R}^d$. Since $p - 2 < 0$, the Cauchy–Schwarz inequality shows that

$$\begin{aligned} D\boldsymbol{\sigma}(z)\boldsymbol{h}\cdot\boldsymbol{h} &= |z|^{p-2}|\boldsymbol{h}|^2 + (p-2)|z|^{p-4}(z\cdot\boldsymbol{h})^2 \\ &\ge |z|^{p-2}|\boldsymbol{h}|^2 + (p-2)|z|^{p-4}|z|^2|\boldsymbol{h}|^2 \\ &= (p-1)|z|^{p-2}|\boldsymbol{h}|^2. \end{aligned}$$

Hence, if the segment $[\boldsymbol{\xi}, \boldsymbol{\eta}]$ does not contain $\mathbf{0}$, a Taylor expansion of $\boldsymbol{\sigma}(\boldsymbol{\xi}) - \boldsymbol{\sigma}(\boldsymbol{\eta})$ yields

$$\begin{aligned} (\boldsymbol{\sigma}(\boldsymbol{\xi}) - \boldsymbol{\sigma}(\boldsymbol{\eta}))\cdot(\boldsymbol{\xi} - \boldsymbol{\eta}) &= \int_0^1 D\boldsymbol{\sigma}(s\boldsymbol{\xi} + (1-s)\boldsymbol{\eta})(\boldsymbol{\xi} - \boldsymbol{\eta})\cdot(\boldsymbol{\xi} - \boldsymbol{\eta})\,\mathrm{d}s \\ &\ge (p-1)|\boldsymbol{\xi} - \boldsymbol{\eta}|^2 \int_0^1 |s\boldsymbol{\xi} + (1-s)\boldsymbol{\eta}|^{p-2}\,\mathrm{d}s. \end{aligned} \tag{6.54}$$

We have $|s\boldsymbol{\xi} + (1-s)\boldsymbol{\eta}| \le s|\boldsymbol{\xi}| + (1-s)|\boldsymbol{\eta}| \le |\boldsymbol{\xi}| + |\boldsymbol{\eta}|$ and thus, since $p - 2 < 0$,

$$|s\boldsymbol{\xi} + (1-s)\boldsymbol{\eta}|^{p-2} \ge (|\boldsymbol{\xi}| + |\boldsymbol{\eta}|)^{p-2}.$$

Plugging this estimate into (6.54) concludes the proof of (6.49) for $p < 2$, in the case where $\mathbf{0} \notin [\boldsymbol{\xi}, \boldsymbol{\eta}]$. If $\mathbf{0}$ is on that segment, then we approximate $\boldsymbol{\xi}, \boldsymbol{\eta}$ by vectors $(\boldsymbol{\xi}_i)_{i\in\mathbb{N}}$ and $(\boldsymbol{\eta}_i)_{i\in\mathbb{N}}$ such that $\mathbf{0} \notin [\boldsymbol{\xi}_i, \boldsymbol{\eta}_i]$ (such vectors always exist in dimension $d \ge 2$), and pass to the limit $i \to \infty$ in (6.49) applied to $\boldsymbol{\xi}_i, \boldsymbol{\eta}_i$. The estimate (6.49) for $d = 1$, namely

$$(|s|^{p-2}s - |t|^{p-2}t)(s-t) \ge C(p)(|s| + |t|)^{p-2}|s-t|^2 \qquad \forall s, t \in \mathbb{R},$$

follows by applying (6.49) for $d = 2$ to the vectors $\boldsymbol{\xi} = s\boldsymbol{e}$ and $\boldsymbol{\eta} = t\boldsymbol{e}$, where $\boldsymbol{e}$ is a fixed unit vector in $\mathbb{R}^2$.

(ii) *Proof of* (6.50) *and* (6.51). If $p \ge 2$, we write

$$|\boldsymbol{\xi} - \boldsymbol{\eta}|^p = |\boldsymbol{\xi} - \boldsymbol{\eta}|^2|\boldsymbol{\xi} - \boldsymbol{\eta}|^{p-2} \le |\boldsymbol{\xi} - \boldsymbol{\eta}|^2\,(|\boldsymbol{\xi}| + |\boldsymbol{\eta}|)^{p-2}$$

and we invoke (6.49), recalling that $C(p) = 2^{1-p}$ in this case.

If $p < 2$, assuming without loss of generality that $|\boldsymbol{\xi}| + |\boldsymbol{\eta}| \neq 0$, raise (6.49) to the power $p/2$ and multiply by $C(p)^{-\frac{p}{2}}(|\boldsymbol{\xi}| + |\boldsymbol{\eta}|)^{(2-p)\frac{p}{2}}$ to find

$$|\boldsymbol{\xi} - \boldsymbol{\eta}|^p \leq C(p)^{-\frac{p}{2}} \Big[(\boldsymbol{\sigma}(\boldsymbol{\xi}) - \boldsymbol{\sigma}(\boldsymbol{\eta})) \cdot (\boldsymbol{\xi} - \boldsymbol{\eta}) \Big]^{\frac{p}{2}} \Big[|\boldsymbol{\xi}| + |\boldsymbol{\eta}| \Big]^{p\frac{2-p}{2}}.$$

The conclusion follows recalling that $C(p) = p - 1$ and writing $(|\boldsymbol{\xi}| + |\boldsymbol{\eta}|)^p \leq 2^{p-1}(|\boldsymbol{\xi}|^p + |\boldsymbol{\eta}|^p)$, by convexity of $s \mapsto s^p$. □

Lemma 6.28 (Continuity of the p-Laplace Flux Function) *Let $p \in (1, \infty)$ and $\boldsymbol{\sigma}(\boldsymbol{\tau}) = |\boldsymbol{\tau}|^{p-2}\boldsymbol{\tau}$ for all $\boldsymbol{\tau} \in \mathbb{R}^d$. Then, for all $\boldsymbol{\xi}, \boldsymbol{\eta} \in \mathbb{R}^d$,*

(i) If $p \geq 2$,

$$|\boldsymbol{\sigma}(\boldsymbol{\xi}) - \boldsymbol{\sigma}(\boldsymbol{\eta})| \leq (p-1)|\boldsymbol{\xi} - \boldsymbol{\eta}| \Big(|\boldsymbol{\xi}|^{p-2} + |\boldsymbol{\eta}|^{p-2} \Big). \tag{6.55}$$

(ii) If $p < 2$,

$$|\boldsymbol{\sigma}(\boldsymbol{\xi}) - \boldsymbol{\sigma}(\boldsymbol{\eta})| \leq 2^{2-p}|\boldsymbol{\xi} - \boldsymbol{\eta}|^{p-1}. \tag{6.56}$$

Proof We first notice that we can always assume that $\mathbf{0} \notin [\boldsymbol{\xi}, \boldsymbol{\eta}]$. Otherwise, we reason as in Point (i) of the proof of Lemma 6.26, taking $(\boldsymbol{\xi}_i)_{i\in\mathbb{N}}$ and $(\boldsymbol{\eta}_i)_{i\in\mathbb{N}}$ that respectively converge to $\boldsymbol{\xi}$ and $\boldsymbol{\eta}$ and such that $\mathbf{0} \notin [\boldsymbol{\xi}_i, \boldsymbol{\eta}_i]$ for any $i \in \mathbb{N}$, and then passing to the limit in the corresponding inequalities applied to $\boldsymbol{\xi}_i, \boldsymbol{\eta}_i$.

Let us first consider $p \geq 2$. As seen in the proof of Lemma 6.26, for $z \neq 0$ we have $D\boldsymbol{\sigma}(z)\boldsymbol{h} = |z|^{p-2}\boldsymbol{h} + (p-2)|z|^{p-4}(z \cdot \boldsymbol{h})z$, and thus, for the induced norm,

$$|D\boldsymbol{\sigma}(z)| \leq |z|^{p-2} + (p-2)|z|^{p-4}|z|\,|z| = (p-1)|z|^{p-2}.$$

The mean value theorem then yields

$$|\boldsymbol{\sigma}(\boldsymbol{\xi}) - \boldsymbol{\sigma}(\boldsymbol{\eta})| \leq |\boldsymbol{\xi} - \boldsymbol{\eta}| \sup_{z\in[\boldsymbol{\xi},\boldsymbol{\eta}]} \Big[(p-1)|z|^{p-2} \Big].$$

Since $p \geq 2$, it holds $|z|^{p-2} \leq \max(|\boldsymbol{\xi}|^{p-2}, |\boldsymbol{\eta}|^{p-2}) \leq |\boldsymbol{\xi}|^{p-2} + |\boldsymbol{\eta}|^{p-2}$ whenever $z \in [\boldsymbol{\xi}, \boldsymbol{\eta}]$, which concludes the proof of (6.55).

We now deal with the case $p < 2$. We still have $|D\boldsymbol{\sigma}(z)| \leq (p-1)|z|^{p-2}$ so

$$|\boldsymbol{\sigma}(\boldsymbol{\xi}) - \boldsymbol{\sigma}(\boldsymbol{\eta})| \leq (p-1) \int_{[\boldsymbol{\xi},\boldsymbol{\eta}]} |z|^{p-2}\, \mathrm{d}l,$$

where dl is the integration element over the line $(\boldsymbol{\xi}, \boldsymbol{\eta})$. Let $\widetilde{\mathbf{0}}$ be the orthogonal projection of $\mathbf{0}$ on this line. The line $(\boldsymbol{\xi}, \boldsymbol{\eta})$ is equipped with an arbitrary orientation

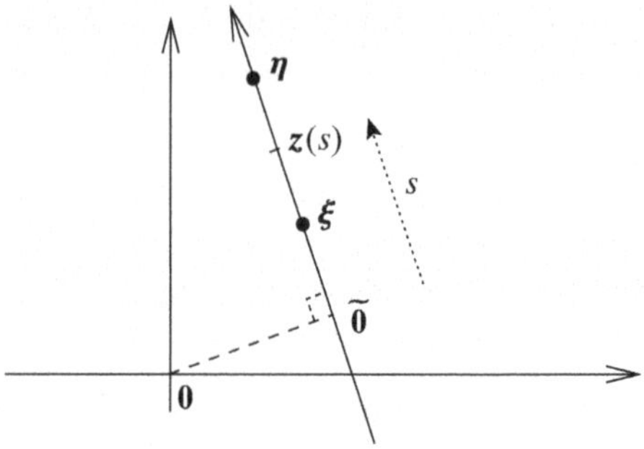

Fig. 6.1 Illustration of the case $p < 2$ in the proof of Lemma 6.28

and a parametrisation $z = z(s)$ of its points using their signed distance s from $\widetilde{\mathbf{0}}$ (see Fig. 6.1). Then,

$$|\boldsymbol{\sigma}(\boldsymbol{\xi}) - \boldsymbol{\sigma}(\boldsymbol{\eta})| \leq (p-1) \int_a^{a+L} |z(s)|^{p-2} \mathrm{d}s,$$

where a is the smallest between the abscissa of $\boldsymbol{\xi}$ and $\boldsymbol{\eta}$ on this line, and $L := |\boldsymbol{\xi} - \boldsymbol{\eta}|$. By definition of the orthogonal projection, $|s| = |z(s) - \widetilde{\mathbf{0}}| \leq |z(s) - \mathbf{0}| = |z(s)|$ and thus, since $p - 2 < 0$,

$$|\boldsymbol{\sigma}(\boldsymbol{\xi}) - \boldsymbol{\sigma}(\boldsymbol{\eta})| \leq (p-1) \int_a^{a+L} |s|^{p-2} \mathrm{d}s.$$

The shape of the function $s \mapsto |s|^{p-2}$ shows that, for a fixed L, the maximum of $\mathbb{R} \ni b \mapsto \int_b^{b+L} |s|^{p-2} \mathrm{d}s$ is reached for $b = -L/2$ (a straightforward analysis of the variations of this function also leads to the same conclusion). Hence,

$$|\boldsymbol{\sigma}(\boldsymbol{\xi}) - \boldsymbol{\sigma}(\boldsymbol{\eta})| \leq (p-1) \int_{-L/2}^{L/2} |s|^{p-2} \mathrm{d}s = 2(p-1) \int_0^{L/2} |s|^{p-2} \mathrm{d}s = 2^{2-p} L^{p-1}.$$

Recalling that $L = |\boldsymbol{\xi} - \boldsymbol{\eta}|$ concludes the proof of (6.56). □

6.3.4 *Proof of the Error Estimates*

We are now ready to prove Theorem 6.19.

Proof (Theorem 6.19) The existence of a solution $\underline{u}_h$ to (6.29) is established in Lemma 6.14. To prove its uniqueness when $\boldsymbol{\sigma}(\boldsymbol{x}, s, \boldsymbol{\xi}) = \boldsymbol{\sigma}(\boldsymbol{\xi}) = |\boldsymbol{\xi}|^{p-2}\boldsymbol{\xi}$, assume that $\underline{u}_h$ and $\underline{w}_h$ are two such solutions, subtract the corresponding equations and take $\underline{v}_h = \underline{u}_h - \underline{w}_h$ as a test function to obtain

$$\int_\Omega \left[\boldsymbol{\sigma}(\mathbf{G}_h^k \underline{u}_h) - \boldsymbol{\sigma}(\mathbf{G}_h^k \underline{w}_h)\right] \cdot \left[\mathbf{G}_h^k \underline{u}_h - \mathbf{G}_h^k \underline{w}_h\right] + \sum_{T \in \mathcal{T}_h} \left(\mathrm{S}_T(\underline{u}_h; \underline{u}_h - \underline{w}_h) - \mathrm{S}_T(\underline{w}_h; \underline{u}_h - \underline{w}_h)\right) = 0.$$

The monotonicity properties (6.49) (with $\boldsymbol{\xi} = \mathbf{G}_h^k \underline{u}_h$ and $\boldsymbol{\eta} = \mathbf{G}_h^k \underline{w}_h$) and (6.52)–(6.53) (with $s = \delta_{TF}^k \underline{u}_T - \delta_T^k \underline{u}_T$ and $t = \delta_{TF}^k \underline{w}_T - \delta_T^k \underline{w}_T$, for $T \in \mathcal{T}_h$ and $F \in \mathcal{F}_T$) show that all addends in the above equation are positive, and must thus vanish. Hence, the same monotonicity properties ensure that $\mathbf{G}_h^k \underline{u}_h = \mathbf{G}_h^k \underline{w}_h$ and $\delta_{TF}^k \underline{u}_T - \delta_T^k \underline{u}_T = \delta_{TF}^k \underline{w}_T - \delta_T^k \underline{w}_T$ for all $T \in \mathcal{T}_h$ and $F \in \mathcal{F}_T$. This proves that $\|\underline{u}_h - \underline{w}_h\|_{\mathbf{G},p,h} = 0$, and thus that $\underline{u}_h = \underline{w}_h$ by virtue of Remark 6.12.

We now turn to the error estimate (6.36). Following the discussion at the end of Sect. 6.3.1, we have to estimate the consistency error in the right-hand side of (6.42), and establish stability properties for the left-hand side.

(i) *Estimate of the consistency error.* We have $-\nabla\cdot(\boldsymbol{\sigma}(\nabla u)) = f$ in Ω and, by the assumed regularity on $\boldsymbol{\sigma}(\nabla u)$, it holds $\boldsymbol{\sigma}(\nabla u)\cdot\boldsymbol{n}_\Omega = g$ on $\partial\Omega$. Hence, for any $\underline{v}_h \in \underline{U}_{h,\star}^k$, we write

$$\begin{aligned}
\ell_h(\underline{v}_h) &= \int_\Omega f v_h + \int_{\partial\Omega} g\, \gamma_h \underline{v}_h \\
&= -\sum_{T\in\mathcal{T}_h} \int_T \nabla\cdot\boldsymbol{\sigma}(\nabla u) v_T + \sum_{F\in\mathcal{F}_h^{\mathrm{b}},\, \mathcal{T}_F=\{T\}} \int_F \left(\boldsymbol{\sigma}(\nabla u)_{|T}\cdot\boldsymbol{n}_{TF}\right) v_F \\
&= \sum_{T\in\mathcal{T}_h} \int_T \boldsymbol{\sigma}(\nabla u)\cdot\nabla v_T - \sum_{T\in\mathcal{T}_h}\sum_{F\in\mathcal{F}_T} \int_F \left(\boldsymbol{\sigma}(\nabla u)_{|T}\cdot\boldsymbol{n}_{TF}\right) v_T \\
&\quad + \sum_{F\in\mathcal{F}_h^{\mathrm{b}},\, \mathcal{T}_F=\{T\}} \int_F \left(\boldsymbol{\sigma}(\nabla u)_{|T}\cdot\boldsymbol{n}_{TF}\right) v_F \\
&= \sum_{T\in\mathcal{T}_h} \int_T \boldsymbol{\sigma}(\nabla u)\cdot\nabla v_T + \sum_{T\in\mathcal{T}_h}\sum_{F\in\mathcal{F}_T} \int_F \left(\boldsymbol{\sigma}(\nabla u)_{|T}\cdot\boldsymbol{n}_{TF}\right)(v_F - v_T),
\end{aligned}$$

where the third equality follows from an element-wise integration by parts, and the conclusion is a consequence of Corollary 1.19 with $\boldsymbol{\tau} = \boldsymbol{\sigma}(\nabla u) \in \boldsymbol{W}^{p'}(\mathrm{div};\Omega) \cap W^{1,p'}(\mathcal{T}_h)^d$ and $(\varphi_F)_{F\in\mathcal{F}_h} = (v_F)_{F\in\mathcal{F}_h}$. Invoking then (4.41) with $\boldsymbol{\tau} = \boldsymbol{\sigma}(\nabla u)$ to substitute the volumetric terms involving ∇v_T, we obtain

$$\ell_h(v_h) = \sum_{T\in\mathcal{T}_h} \int_T \boldsymbol{\sigma}(\nabla u)\cdot \mathbf{G}_T^k \underline{v}_T + \sum_{T\in\mathcal{T}_h} \sum_{F\in\mathcal{F}_T} \int_F \left[\boldsymbol{\sigma}(\nabla u)_{|T} - \boldsymbol{\pi}_T^{0,k}(\boldsymbol{\sigma}(\nabla u))\right] \cdot \boldsymbol{n}_{TF}\,(v_F - v_T).$$

Recalling the definitions (6.27) of A_h and (6.46) of S_h, and since $\mathbf{G}_T^k \underline{I}_T^k u = \boldsymbol{\pi}_T^{0,k}(\nabla u)$ by virtue of (4.40), the consistency error can be recast as

$$\ell_h(\underline{v}_h) - \mathrm{A}_h(\underline{I}_h^k u; \underline{v}_h) = \underbrace{\sum_{T\in\mathcal{T}_h} \int_T \left[\boldsymbol{\sigma}(\nabla u) - \boldsymbol{\sigma}(\boldsymbol{\pi}_T^{0,k}(\nabla u))\right] \cdot \mathbf{G}_T^k \underline{v}_T}_{\mathfrak{T}_1} - \underbrace{\mathrm{S}_h(\underline{I}_h^k u; \underline{v}_h)}_{\mathfrak{T}_2} + \underbrace{\sum_{T\in\mathcal{T}_h} \sum_{F\in\mathcal{F}_T} \int_F \left[\boldsymbol{\sigma}(\nabla u)_{|T} - \boldsymbol{\pi}_T^{0,k}(\boldsymbol{\sigma}(\nabla u))\right] \cdot \boldsymbol{n}_{TF}(v_F - v_T)}_{\mathfrak{T}_3}.$$

The term $\mathfrak{T}_2$ has already been estimated in Proposition 6.25 (see (6.45)):

$$|\mathfrak{T}_2| = |\mathrm{S}_h(\underline{I}_h^k u; \underline{v}_h)| \lesssim h^{(r+1)(p-1)} |u|_{W^{r+2,p}(\mathcal{T}_h)}^{p-1} \|\underline{v}_h\|_{\mathbf{G},p,h}. \tag{6.57}$$

For $\mathfrak{T}_3$, we write

$$\begin{aligned}
|\mathfrak{T}_3| &\le \sum_{T\in\mathcal{T}_h} \sum_{F\in\mathcal{F}_T} h_F^{\frac{1}{p'}} \|\boldsymbol{\sigma}(\nabla u)_{|T} - \boldsymbol{\pi}_T^{0,k}(\boldsymbol{\sigma}(\nabla u))\|_{L^{p'}(F)^d}\, h_F^{\frac{1}{p}-1} \|v_F - v_T\|_{L^p(F)} \\
&\le \left(\sum_{T\in\mathcal{T}_h} \sum_{F\in\mathcal{F}_T} h_T \|\boldsymbol{\sigma}(\nabla u)_{|T} - \boldsymbol{\pi}_T^{0,k}(\boldsymbol{\sigma}(\nabla u))\|_{L^{p'}(F)^d}^{p'}\right)^{\frac{1}{p'}} \\
&\quad \times \left(\sum_{T\in\mathcal{T}_h} \sum_{F\in\mathcal{F}_T} h_F^{1-p} \|v_F - v_T\|_{L^p(F)}^p\right)^{\frac{1}{p}} \\
&\lesssim \left(\sum_{T\in\mathcal{T}_h} h_T^{(r+1)p'} |\boldsymbol{\sigma}(\nabla u)|_{W^{r+1,p'}(T)^d}^{p'}\right)^{\frac{1}{p'}} \|\underline{v}_h\|_{1,p,h} \\
&\lesssim h^{r+1} |\boldsymbol{\sigma}(\nabla u)|_{W^{r+1,p'}(\mathcal{T}_h)^d} \|\underline{v}_h\|_{\mathbf{G},p,h},
\end{aligned} \tag{6.58}$$

where we obtained the first inequality using a generalised Hölder inequality (on the integrals over F) with exponents (p', ∞, p) along with $\|\boldsymbol{n}_{TF}\|_{L^\infty(F)^d} = 1$ and $\frac{1}{p'} + \frac{1}{p} - 1 = 0$, we invoked the Hölder inequality on the sums with exponents (p', p) together with $h_F \le h_T$ to write the second inequality, we applied the trace approximation property (1.75) of the $L^2(T)$-projector (with $l = k$, $m = 0$, $s = r + 1$, $v =$ components of $\boldsymbol{\sigma}(\nabla u)$, and p' instead of p) together with the definition (6.9) of $\|\cdot\|_{1,p,h}$ to pass to the penultimate line, and we concluded by the norm equivalence (6.20).

To estimate $\mathfrak{T}_1$, we first use a Hölder inequality (on the integrals and the sum) with exponents (p', p) and then recall the definition (6.16) of $\|\cdot\|_{\mathbf{G},p,h}$:

$$
\begin{aligned}
|\mathfrak{T}_1| &\le \left(\sum_{T\in\mathcal{T}_h} \|\boldsymbol{\sigma}(\nabla u) - \boldsymbol{\sigma}(\boldsymbol{\pi}_T^{0,k}(\nabla u))\|_{L^{p'}(T)^d}^{p'} \right)^{\frac{1}{p'}} \left(\sum_{T\in\mathcal{T}_h} \|\mathbf{G}_T^k \underline{v}_T\|_{L^p(T)^d}^p \right)^{\frac{1}{p}} \\
&\lesssim \left(\sum_{T\in\mathcal{T}_h} \int_T |\boldsymbol{\sigma}(\nabla u) - \boldsymbol{\sigma}(\boldsymbol{\pi}_T^{0,k}(\nabla u))|^{p'} \right)^{\frac{1}{p'}} \|\underline{v}_h\|_{\mathbf{G},p,h}.
\end{aligned} \tag{6.59}
$$

Continuing further requires the continuity property of $\boldsymbol{\sigma}$ (see Lemma 6.28), and therefore a separation of the cases $p \ge 2$ and $p < 2$.

(i.A) *Case* $p \ge 2$. The continuity property (6.55) and a Hölder inequality with exponents $(p-1, \frac{p-1}{p-2})$ yield

$$
\begin{aligned}
&\int_T |\boldsymbol{\sigma}(\nabla u) - \boldsymbol{\sigma}(\boldsymbol{\pi}_T^{0,k}(\nabla u))|^{p'} \\
&\qquad \lesssim \int_T |\nabla u - \boldsymbol{\pi}_T^{0,k}(\nabla u)|^{p'} \left(|\nabla u|^{p-2} + |\boldsymbol{\pi}_T^{0,k}(\nabla u)|^{p-2} \right)^{p'} \\
&\qquad \lesssim \|\nabla u - \boldsymbol{\pi}_T^{0,k}(\nabla u)\|_{L^p(T)^d}^{p'} \left(\|\nabla u\|_{L^p(T)^d} + \|\boldsymbol{\pi}_T^{0,k}(\nabla u)\|_{L^p(T)^d} \right)^{p'(p-2)},
\end{aligned}
$$

the conclusion following from the relation $(p-2)p'\frac{p-1}{p-2} = p$ and multiple usages of the inequality $(a+b)^\alpha \le 2^\alpha(a^\alpha + b^\alpha)$, valid for any non-negative numbers a, b, α. Invoking the approximation property (1.74) (with $X = T$, $l = k$, $m = 0$, $s = r + 1$ and $v =$ components of ∇u) and the boundedness property (1.77) (with the same choices except $s = 0$) of the local L^2-projector, we infer

$$
\int_T |\boldsymbol{\sigma}(\nabla u) - \boldsymbol{\sigma}(\boldsymbol{\pi}_T^{0,k}(\nabla u))|^{p'} \lesssim h_T^{p'(r+1)} |\nabla u|_{W^{r+1,p}(T)^d}^{p'} \|\nabla u\|_{L^p(T)^d}^{p'(p-2)}. \tag{6.60}
$$

The following estimate is obtained choosing $v = u$ in (6.6) and using the coercivity (6.3d) of $\boldsymbol{\sigma}$ together with a Poincaré and trace inequality in $W^{1,p}_{\star}(\Omega)$:

$$\|\nabla u\|_{L^p(\Omega)^d}^{p-1} \lesssim \|f\|_{L^{p'}(\Omega)} + \|g\|_{L^{p'}(\partial\Omega)} \lesssim 1. \tag{6.61}$$

Plugging (6.60) into (6.59), using the Hölder inequality on the sum with exponents $\left(p-1, \frac{p-1}{p-2}\right)$, and invoking (6.61), we obtain

$$|\mathfrak{T}_1| \le h^{r+1}|u|_{W^{r+2,p}(\mathcal{T}_h)}\|\underline{v}_h\|_{\mathbf{G},p,h}. \tag{6.62}$$

(i.B) *Case* $p < 2$. Invoking (6.56) in (6.59) gives

$$\begin{aligned}
|\mathfrak{T}_1| &\lesssim \left(\sum_{T\in\mathcal{T}_h} \|\nabla u - \boldsymbol{\pi}_T^{0,k}(\nabla u)\|_{L^p(T)^d}^p\right)^{\frac{1}{p'}} \|\underline{v}_h\|_{\mathbf{G},p,h} \\
&\lesssim \left(\sum_{T\in\mathcal{T}_h} h_T^{p(r+1)}|\nabla u|_{W^{r+1,p}(T)^d}^p\right)^{\frac{1}{p'}} \|\underline{v}_h\|_{\mathbf{G},p,h} \\
&\lesssim h^{(r+1)(p-1)}|u|_{W^{r+2,p}(\mathcal{T}_h)}^{p-1}\|\underline{v}_h\|_{\mathbf{G},p,h},
\end{aligned} \tag{6.63}$$

where we have used in the second line the approximation property (1.74) of the $L^2(T)$-projector with $l = k$, $m = 0$, $s = r+1$ and $v =$ components of ∇u.

Gathering (6.57), (6.58) and either (6.62) or (6.63), we arrive at the consistency estimate

$$\begin{aligned}
&\ell_h(\underline{v}_h) - \mathrm{A}_h(\underline{I}_h^k u; \underline{v}_h) \\
&\quad\lesssim \|\underline{v}_h\|_{\mathbf{G},p,h}\Bigg[h^{(r+1)(p-1)}|u|_{W^{r+2,p}(\mathcal{T}_h)}^{p-1} + h^{r+1}|\boldsymbol{\sigma}(\nabla u)|_{W^{r+1,p'}(\mathcal{T}_h)^d} \\
&\qquad\qquad + \begin{cases} h^{r+1}|u|_{W^{r+2,p}(\mathcal{T}_h)} & \text{if } p \ge 2 \\ 0 & \text{if } p < 2 \end{cases}\Bigg].
\end{aligned} \tag{6.64}$$

(ii) *Stability property.* The stability property of A_h hinges on the strong monotonicity of $\boldsymbol{\sigma}$ expressed in Lemma 6.26. As a consequence, the cases $p \ge 2$ and $p < 2$ have to be handled separately. Before dealing with each case, we write the stabilisation terms in a more condensed form: For $T \in \mathcal{T}_h$ and $F \in \mathcal{F}_T$, we set $d_{TF}^k := \delta_{TF}^k - \delta_T^k$ so that, recalling the definition (6.28) of S_T, for all

$\underline{w}_T, \underline{z}_T \in \underline{U}_T^k$,

$$\mathrm{S}_T(\underline{w}_T; \underline{z}_T) = \sum_{F \in \mathcal{F}_T} h_F^{1-p} \int_F |d_{TF}^k \underline{w}_T|^{p-2} d_{TF}^k \underline{w}_T \, d_{TF}^k \underline{z}_T. \tag{6.65}$$

(ii.A) *Case* $p \geq 2$. For any $T \in \mathcal{T}_h$, by (6.50) we have

$$\int_T \left(\boldsymbol{\sigma}(\mathbf{G}_T^k \underline{u}_T) - \boldsymbol{\sigma}(\mathbf{G}_T^k \underline{I}_T^k u) \right) \cdot (\mathbf{G}_T^k \underline{u}_T - \mathbf{G}_T^k \underline{I}_T^k u) \gtrsim \|\mathbf{G}_T^k (\underline{u}_T - \underline{I}_T^k u)\|_{L^p(T)^d}^p. \tag{6.66}$$

Using (6.65) with $(\underline{w}_T, \underline{z}_T) = (\underline{u}_T, \underline{u}_T - \underline{I}_T^k u)$ and $(\underline{w}_T, \underline{z}_T) = (\underline{I}_T^k u, \underline{u}_T - \underline{I}_T^k u)$, and invoking (6.52) with $s = d_{TF}^k \underline{u}_T$ and $t = d_{TF}^k \underline{I}_T^k u$, we can write

$$\begin{aligned}
&\mathrm{S}_T(\underline{u}_T; \underline{u}_T - \underline{I}_T^k u) - \mathrm{S}_T(\underline{I}_T^k u; \underline{u}_T - \underline{I}_T^k u) \\
&= \sum_{F \in \mathcal{F}_T} h_F^{1-p} \int_F \left(|d_{TF}^k \underline{u}_T|^{p-2} d_{TF}^k \underline{u}_T - |d_{TF}^k \underline{I}_T^k u|^{p-2} d_{TF}^k \underline{I}_T^k u \right) \\
&\qquad \times d_{TF}^k (\underline{u}_T - \underline{I}_T^k u) \\
&\gtrsim \sum_{F \in \mathcal{F}_T} h_F^{1-p} \int_F |d_{TF}^k (\underline{u}_T - \underline{I}_T^k u)|^p.
\end{aligned} \tag{6.67}$$

Adding together (6.66) and (6.67), summing over $T \in \mathcal{T}_h$, and recalling the definitions (6.27) of A_h and (6.16) of $\|\cdot\|_{\mathbf{G},p,h}$, we arrive at

$$\mathrm{A}_h(\underline{u}_h; \underline{u}_h - \underline{I}_h^k u) - \mathrm{A}_h(\underline{I}_h^k u; \underline{u}_h - \underline{I}_h^k u) \gtrsim \|\underline{u}_h - \underline{I}_h^k u\|_{\mathbf{G},p,h}^p. \tag{6.68}$$

(ii.B) *Case* $p < 2$. Let $T \in \mathcal{T}_h$. Starting from (6.51) with $\boldsymbol{\xi} = \mathbf{G}_T^k \underline{u}_T$ and $\boldsymbol{\eta} = \mathbf{G}_T^k \underline{I}_T^k u$, integrating over T and using a Hölder inequality with exponents $\left(\frac{2}{p}, \frac{2}{2-p}\right)$, we have

$$\begin{aligned}
\|\mathbf{G}_T^k (\underline{u}_T - \underline{I}_T^k u)\|_{L^p(T)^d}^p &\lesssim \left(\int_T \left(\boldsymbol{\sigma}(\mathbf{G}_T^k \underline{u}_T) - \boldsymbol{\sigma}(\mathbf{G}_T^k \underline{I}_T^k u) \right) \cdot \mathbf{G}_T^k (\underline{u}_T - \underline{I}_T^k u) \right)^{\frac{p}{2}} \\
&\quad \times \left(\|\mathbf{G}_T^k \underline{u}_T\|_{L^p(T)^d}^p + \|\mathbf{G}_T^k \underline{I}_T^k u\|_{L^p(T)^d}^p \right)^{\frac{2-p}{2}}.
\end{aligned}$$

Summing over $T \in \mathcal{T}_h$ and using a discrete Hölder inequality with the same exponents as above yields

$$\sum_{T\in\mathcal{T}_h} \|\mathbf{G}_T^k(\underline{u}_T - \underline{I}_T^k u)\|_{L^p(T)^d}^p \lesssim \left(\sum_{T\in\mathcal{T}_h}\int_T \left(\boldsymbol{\sigma}(\mathbf{G}_T^k\underline{u}_T) - \boldsymbol{\sigma}(\mathbf{G}_T^k\underline{I}_T^k u)\right)\cdot\mathbf{G}_T^k(\underline{u}_T - \underline{I}_T^k u)\right)^{\frac{p}{2}}$$
$$\times\left(\sum_{T\in\mathcal{T}_h}\|\mathbf{G}_T^k\underline{u}_T\|_{L^p(T)^d}^p + \sum_{T\in\mathcal{T}_h}\|\mathbf{G}_T^k\underline{I}_T^k u\|_{L^p(T)^d}^p\right)^{\frac{2-p}{2}}. \tag{6.69}$$

We estimate the second factor in the right-hand side as

$$\begin{aligned}
\sum_{T\in\mathcal{T}_h}\|\mathbf{G}_T^k\underline{u}_T\|_{L^p(T)^d}^p + \sum_{T\in\mathcal{T}_h}\|\mathbf{G}_T^k\underline{I}_T^k u\|_{L^p(T)^d}^p &\le \|\underline{u}_h\|_{\mathbf{G},p,h}^p + \sum_{T\in\mathcal{T}_h}\|\boldsymbol{\pi}_T^{0,k}(\nabla u)\|_{L^p(T)^d}^p \\
&\lesssim \|\underline{u}_h\|_{1,p,h}^p + \|\nabla u\|_{L^p(\Omega)^d}^p \\
&\lesssim 1,
\end{aligned}$$

where we have used in the first line the definition (6.16) of $\|\cdot\|_{\mathbf{G},p,h}$ together with the commutation property $\mathbf{G}_T^k\underline{I}_T^k u = \boldsymbol{\pi}_T^{0,k}(\nabla u)$ (see (4.40)), passed to the second line invoking the norm equivalence (6.20) for the first term and, for all $T \in \mathcal{T}_h$, the boundedness property (1.77) of $\pi_T^{0,k}$ with $s = 0$ and $v =$ components of ∇u, and concluded using the discrete and continuous energy estimates (6.30) and (6.61). Plugged into (6.69), this leads to

$$\sum_{T\in\mathcal{T}_h}\|\mathbf{G}_T^k(\underline{u}_T - \underline{I}_T^k u)\|_{L^p(T)^d}^p$$
$$\lesssim \left(\sum_{T\in\mathcal{T}_h}\int_T\left(\boldsymbol{\sigma}(\mathbf{G}_T^k\underline{u}_T) - \boldsymbol{\sigma}(\mathbf{G}_T^k\underline{I}_T^k u)\right)\cdot\mathbf{G}_T^k(\underline{u}_T - \underline{I}_T^k u)\right)^{\frac{p}{2}}. \tag{6.70}$$

Let us now turn to the stabilisation term. Starting from its representation (6.65), the monotonicity property (6.53) and the arguments that led to (6.69) give

$$\sum_{T\in\mathcal{T}_h}\sum_{F\in\mathcal{F}_T} h_F^{1-p}\|d_{TF}^k(\underline{u}_T - \underline{I}_T^k u)\|_{L^p(F)}^p$$
$$\lesssim\left(\sum_{T\in\mathcal{T}_h}\mathrm{S}_T(\underline{u}_T;\underline{u}_T - \underline{I}_T^k u) - \mathrm{S}_T(\underline{I}_T^k u;\underline{u}_T - \underline{I}_T^k u)\right)^{\frac{p}{2}}$$

$$\times\left(\sum_{T\in\mathcal{T}_h}\sum_{F\in\mathcal{F}_T} h_F^{1-p}\|d_{TF}^k(\underline{u}_T)\|_{L^p(F)}^p + \sum_{T\in\mathcal{T}_h}\sum_{F\in\mathcal{F}_T} h_F^{1-p}\|d_{TF}^k(\underline{I}_T^k u)\|_{L^p(F)}^p\right)^{\frac{2-p}{2}}$$

$$\lesssim\left(\sum_{T\in\mathcal{T}_h}\mathrm{S}_T(\underline{u}_T;\underline{u}_T-\underline{I}_T^k u)-\mathrm{S}_T(\underline{I}_T^k u;\underline{u}_T-\underline{I}_T^k u)\right)^{\frac{p}{2}}$$

$$\times\left(\sum_{T\in\mathcal{T}_h}|\underline{u}_T|_{\delta,p,T}^p+\sum_{T\in\mathcal{T}_h}|\underline{I}_T^k u|_{\delta,p,T}^p\right)^{\frac{2-p}{2}}, \tag{6.71}$$

where the conclusion follows from the definitions of d_{TF}^k and $|\cdot|_{\delta,p,T}$ (see (6.17)). For the first term in the last bracket, we have

$$\sum_{T\in\mathcal{T}_h}|\underline{u}_T|_{\delta,p,T}^p\le\|\underline{u}_h\|_{\mathbf{G},p,h}^p\lesssim\|\underline{u}_h\|_{1,p,h}^p\lesssim 1, \tag{6.72}$$

where the first inequality is deduced from the definition (6.16) of $\|\cdot\|_{\mathbf{G},p,h}$, the second inequality follows from the norm equivalence (6.20), and the third inequality is obtained invoking the discrete a priori estimate (6.30). The second term in the last bracket of (6.71) is bounded invoking the local seminorm equivalence (6.19) together with the boundedness property (6.43) of $\underline{I}_T^k$ to write

$$\begin{aligned}\sum_{T\in\mathcal{T}_h}|\underline{I}_T^k u|_{\delta,p,T}^p\le\sum_{T\in\mathcal{T}_h}\|\underline{I}_T^k u\|_{\mathbf{G},p,T}^p&\lesssim\sum_{T\in\mathcal{T}_h}\|\underline{I}_T^k u\|_{1,p,T}^p\\&\lesssim\sum_{T\in\mathcal{T}_h}|u|_{W^{1,p}(T)}^p=|u|_{W^{1,p}(\Omega)}^p\lesssim 1,\end{aligned} \tag{6.73}$$

the last inequality being a consequence of the continuous a priori estimate (6.61). Plugging (6.72) and (6.73) into (6.71), we infer

$$\begin{aligned}&\sum_{T\in\mathcal{T}_h}\sum_{F\in\mathcal{F}_T}h_F^{1-p}\|d_{TF}^k(\underline{u}_T-\underline{I}_T^k u)\|_{L^p(F)}^p\\&\qquad\lesssim\left(\sum_{T\in\mathcal{T}_h}\mathrm{S}_T(\underline{u}_T;\underline{u}_T-\underline{I}_T^k u)-\mathrm{S}_T(\underline{I}_T^k u;\underline{u}_T-\underline{I}_T^k u)\right)^{\frac{p}{2}}.\end{aligned} \tag{6.74}$$

Adding together (6.70) and (6.74), using the inequality $a^{\frac{p}{2}} + b^{\frac{p}{2}} \leq 2(a+b)^{\frac{p}{2}}$ (valid for any non-negative numbers a, b), and recalling the definition (6.27) of A_h yields

$$\|\underline{u}_h - \underline{I}_h^k u\|_{\mathbf{G},p,h}^p \lesssim \left(\mathrm{A}_h(\underline{u}_h; \underline{u}_h - \underline{I}_h^k u) - \mathrm{A}_h(\underline{I}_h^k u; \underline{u}_h - \underline{I}_h^k u)\right)^{\frac{p}{2}}. \tag{6.75}$$

To summarise (6.68) and (6.75),

$$\mathrm{A}_h(\underline{u}_h; \underline{u}_h - \underline{I}_h^k u) - \mathrm{A}_h(\underline{I}_h^k u; \underline{u}_h - \underline{I}_h^k u) \gtrsim \begin{cases} \|\underline{u}_h - \underline{I}_h^k u\|_{\mathbf{G},p,h}^p & \text{if } p \geq 2, \\ \|\underline{u}_h - \underline{I}_h^k u\|_{\mathbf{G},p,h}^2 & \text{if } p < 2. \end{cases} \tag{6.76}$$

(iii) *Conclusion.* The error bound (6.36) is obtained plugging the consistency estimate (6.64) with $\underline{v}_h = \underline{u}_h - \underline{I}_h^k u$ and the stability property (6.76) into the error equation (6.42), and simplifying by $\|\underline{u}_h - \underline{I}_h^k u\|_{\mathbf{G},p,h}$. □

6.3.5 Numerical Example

To illustrate the performance of the HHO method, we solve the p-Laplacian version of the problem considered in Sect. 2.5.1, for $p \in \{2, 3, 4\}$. This test is taken from [141, Section 3.5], where Dirichlet boundary conditions are considered; see Remark 6.23. The domain is again the unit square $\Omega = (0, 1)^2$, the exact solution is given by (2.89), and the volumetric source term f is inferred from (6.1). The convergence results for the same triangular and polygonal mesh families of Sect. 2.5.1 (see Fig. 1.1a, c) are displayed in Fig. 6.2. Here, the error is measured by the quantity $\|\underline{I}_h^k u - \underline{u}_h\|_{1,p,h}$, for which analogous estimates as those in Theorem 6.19 hold by Remark 6.22. For $p = 2$, we recover results coherent with those expected for the Poisson problem (2.2). For $p \in \{3, 4\}$, better orders of convergence than the asymptotic ones in (6.38) are observed in some cases. One possible explanation is that the lowest-order terms in $\mathcal{E}_h(u)$ are not yet dominant for the specific problem data and mesh. Another possibility is that compensations occur among terms that are separately estimated in the proof of Theorem 6.19.

6.4 Convergence by Compactness for General Leray–Lions Operators

We now come back to the generic model (6.1) under Assumption 6.1. In this situation, establishing an error estimate is in general impossible (because such an error estimate would impose the uniqueness of the solution to the continuous model, which sometimes fails; see Remark 6.16). The convergence analysis of the HHO

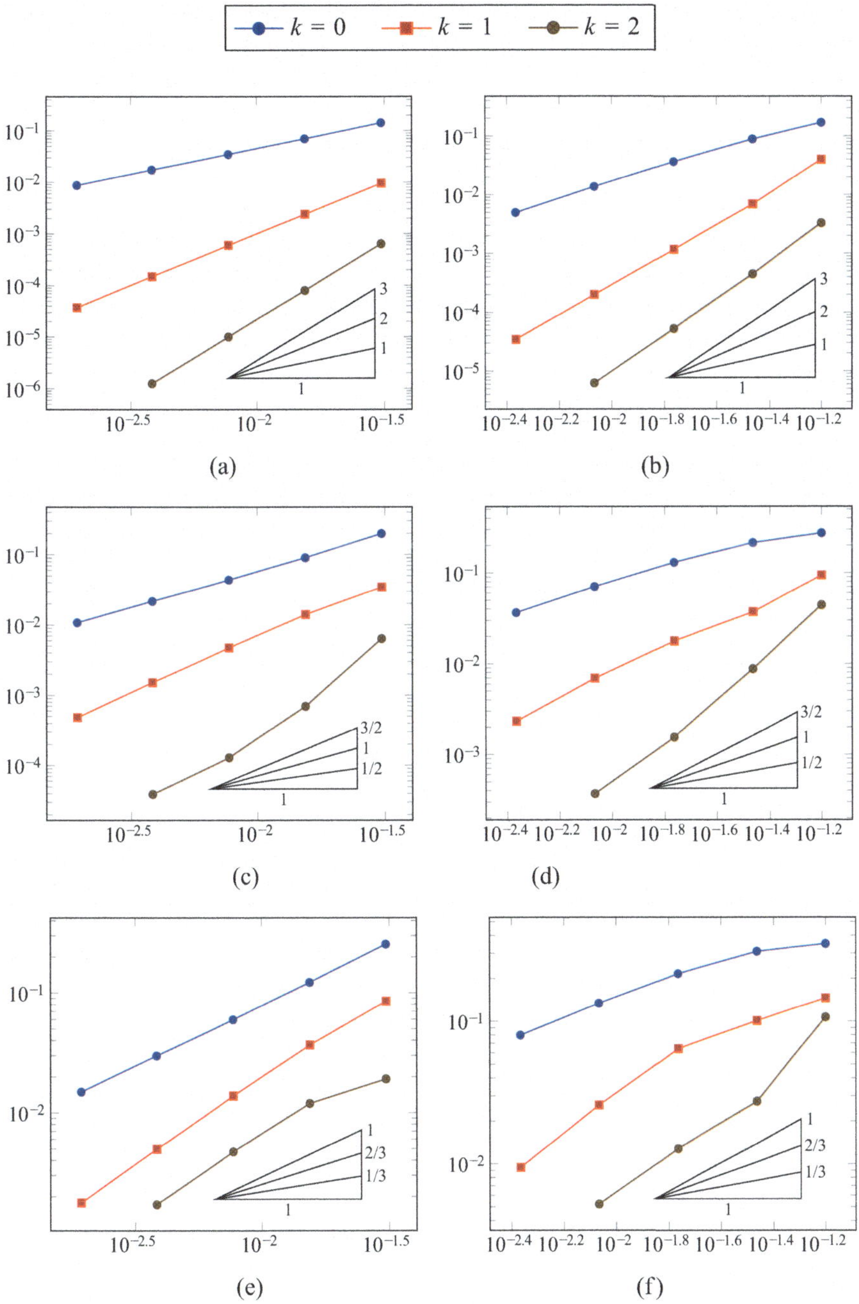

Fig. 6.2 $\|\underline{I}_h^k u - \underline{u}_h\|_{1,p,h}$ vs. h for the test case of Sect. 6.3.5. (**a**) Triangular mesh, $p = 2$. (**b**) Polygonal mesh, $p = 2$. (**c**) Triangular mesh, $p = 3$. (**d**) Polygonal mesh, $p = 3$. (**e**) Triangular mesh, $p = 4$. (**f**) Polygonal mesh, $p = 4$

scheme is then performed using compactness techniques, following the general process described in [171, Section 1.2]. The first step of this process consists in establishing a priori estimates on the solution to the scheme. This was done in Lemma 6.14. The second step, consisting in showing that sequences that satisfy such estimates enjoy compactness properties, is covered by Theorem 6.8. The last step is to show that limits of such sequences solve the continuous PDE.

The convergence of solutions to problem (6.29) is stated in the following theorem. Notice that this convergence is proved for exact solutions that display only the minimal regularity property $u \in W^{1,p}_\star(\Omega)$ required by the weak formulation (6.6). This is an important point when dealing with nonlinear problems, for which further regularity is sometimes unknown, or possibly requires assumptions on the data that are too strong to be matched in practical situations.

Theorem 6.29 (Convergence of the HHO Scheme for the Leray–Lions Problem) *Let a polynomial degree $k \ge 0$ and an index $p \in (1, \infty)$ be fixed. Assume that $\boldsymbol{\sigma}$ satisfies Assumption 6.1, and that f and g satisfy (6.2). Let $(\mathcal{M}_h)_{h\in\mathcal{H}}$ denote a regular mesh sequence in the sense of Definition 1.9 and, for each $h \in \mathcal{H}$, denote by $\underline{u}_h \in \underline{U}^k_{h,\star}$ a solution to (6.29) on $\mathcal{M}_h$. Then, there exists a solution $u \in W^{1,p}_\star(\Omega)$ to (6.6) such that, along a subsequence as $h \to 0$, it holds:*

(i) $u_h \to u$ and $\mathrm{r}_h^{k+1}\underline{u}_h \to u$ strongly in $L^q(\Omega)$, for all $1 \le q < \frac{dp}{d-p}$ if $1 \le p < d$ and all $1 \le q < \infty$ if $p \ge d$;
(ii) $\gamma_h\underline{u}_h \to u_{|\partial\Omega}$ strongly in $L^p(\partial\Omega)$;
(iii) $\mathbf{G}_h^k\underline{u}_h \rightharpoonup \nabla u$ weakly in $L^p(\Omega)^d$.

Proof Combining the discrete a priori estimate (6.30) and Theorem 6.8, we obtain the existence of $u \in W^{1,p}_\star(\Omega)$ such that, along a subsequence, the convergences (i), (ii) and (iii) stated in the theorem hold. It remains to prove that u satisfies (6.6).

Using the growth condition (6.3b) on $\boldsymbol{\sigma}$ we have

$$\|\boldsymbol{\sigma}(u_h, \mathbf{G}_h^k\underline{u}_h)\|_{L^{p'}(\Omega)^d} \le \|\overline{\sigma}\|_{L^{p'}(\Omega)} + \beta_\sigma \left(\int_\Omega |u_h|^{p't}\right)^{\frac{1}{p'}} + \beta_\sigma \|\mathbf{G}_h^k\underline{u}_h\|^{p-1}_{L^p(\Omega)^d}.$$

Since $q := p't < \hat{p}$ and $\hat{p} = \frac{dp}{d-p}$ if $p < d$, $\hat{p} = \infty$ if $p \ge d$, the convergence $u_h \to u$ in $L^q(\Omega)$ shows that $\int_\Omega |u_h|^{p't}$ is bounded as $h \to 0$. Similarly, the weak convergence $\mathbf{G}_h^k\underline{u}_h \rightharpoonup \nabla u$ in $L^p(\Omega)^d$ shows that $\|\mathbf{G}_h^k\underline{u}_h\|_{L^p(\Omega)^d}$ is bounded. Hence, $\boldsymbol{\sigma}(u_h, \mathbf{G}_h^k\underline{u}_h)$ is bounded in $L^{p'}(\Omega)^d$ and, up to a subsequence as $h \to 0$, converges weakly in this space to some $\boldsymbol{\chi}$.

Let $\phi \in W_\star^{1,p}(\Omega) \cap W^{2,p}(\Omega)$ and use the scheme (6.29) with $\underline{v}_h = \underline{I}_h^k \phi$:

$$\int_\Omega \boldsymbol{\sigma}(u_h, \mathbf{G}_h^k \underline{u}_h) \cdot \mathbf{G}_h^k \underline{I}_h^k \phi = \int_\Omega f \pi_h^{0,k} \phi + \int_{\partial\Omega} g \pi_{h,\partial}^{0,k} \phi - \sum_{T \in \mathcal{T}_h} \mathrm{S}_T(\underline{u}_T; \underline{I}_T^k \phi), \tag{6.77}$$

where $\pi_{h,\partial}^{0,k}$ is the patched projector on the boundary, defined in a similar way as $\pi_h^{0,k}$ (see Definition 1.38), that is, for all $F \in \mathcal{F}_h^{\mathrm{b}}$, $(\pi_{h,\partial}^{0,k}\phi)_{|F} := \pi_F^{0,k}\phi_{|F}$. The commutation property (4.40), the $W^{2,p}$-regularity of ϕ, and the approximation properties (1.74) of $\pi_T^{0,k}$ (with $s = 1$, $m = 0$, and $v = \phi$ or $v =$ components of $\nabla\phi$) show that, as $h \to 0$, $\mathbf{G}_h^k \underline{I}_h^k \phi = \boldsymbol{\pi}_h^{0,k}(\nabla\phi) \to \nabla\phi$ in $L^p(\Omega)^d$ and $\pi_h^{0,k}\phi \to \phi$ in $L^p(\Omega)$. Since $\phi_{|\partial\Omega} \in W^{1,p}(\partial\Omega)$, the same approximation properties (1.74) but for $\pi_F^{0,k}$ show that $\pi_{h,\partial}^{0,k}\phi \to \phi$ in $L^p(\partial\Omega)$. Additionally, by the consistency property (6.44) with $r = 0$, the norm equivalence (6.20) and the estimate (6.30),

$$\left| \sum_{T \in \mathcal{T}_h} \mathrm{S}_T(\underline{u}_T; \underline{I}_T^k \phi) \right| \lesssim h \|\underline{u}_h\|_{1,p,h}^{p-1} |\phi|_{W^{2,p}(\Omega)} \to 0 \text{ as } h \to 0.$$

Gathering all these convergences in (6.77) leads to

$$\int_\Omega \boldsymbol{\chi} \cdot \nabla\phi = \int_\Omega f\phi + \int_{\partial\Omega} g\phi. \tag{6.78}$$

By density of $W_\star^{1,p}(\Omega) \cap W^{2,p}(\Omega)$ in $W_\star^{1,p}(\Omega)$, this relation also holds if ϕ is merely in $W_\star^{1,p}(\Omega)$.

We now conclude the proof using the Minty trick (cf. [238]). Take $\boldsymbol{\Lambda} \in L^p(\Omega)^d$ and write, using the monotonicity (6.3c) of $\boldsymbol{\sigma}$,

$$\int_\Omega [\boldsymbol{\sigma}(u_h, \mathbf{G}_h^k \underline{u}_h) - \boldsymbol{\sigma}(u_h, \boldsymbol{\Lambda})] \cdot [\mathbf{G}_h^k \underline{u}_h - \boldsymbol{\Lambda}] \geq 0. \tag{6.79}$$

The scheme (6.29) with $\underline{v}_h = \underline{u}_h$ shows that

$$\begin{aligned} \int_\Omega \boldsymbol{\sigma}(u_h, \mathbf{G}_h^k \underline{u}_h) \cdot \mathbf{G}_h^k \underline{u}_h &= \int_\Omega f u_h + \int_{\partial\Omega} g\, \gamma_h \underline{u}_h - \sum_{T \in \mathcal{T}_h} \mathrm{S}_T(\underline{u}_T; \underline{u}_T) \\ &\leq \int_\Omega f u_h + \int_{\partial\Omega} g\, \gamma_h \underline{u}_h. \end{aligned} \tag{6.80}$$

Developing (6.79) and using this relation gives

$$\int_\Omega f u_h + \int_{\partial\Omega} g\, \gamma_h \underline{u}_h - \int_\Omega \boldsymbol{\sigma}(u_h, \mathbf{G}_h^k \underline{u}_h)\cdot\boldsymbol{\Lambda} - \int_\Omega \boldsymbol{\sigma}(u_h, \boldsymbol{\Lambda})\cdot[\mathbf{G}_h^k \underline{u}_h - \boldsymbol{\Lambda}] \ge 0. \tag{6.81}$$

Using the convergences $u_h \to u$ in $L^p(\Omega)$ and $\gamma_h \underline{u}_h \to u_{|\partial\Omega}$ in $L^p(\partial\Omega)$, we have

$$\int_\Omega f u_h + \int_{\partial\Omega} g\, \gamma_h \underline{u}_h \to \int_\Omega f u + \int_{\partial\Omega} g u_{|\partial\Omega} \quad \text{as } h \to 0.$$

The weak convergence of $\boldsymbol{\sigma}(u_h, \mathbf{G}_h^k \underline{u}_h)$ towards $\boldsymbol{\chi}$ shows that

$$\int_\Omega \boldsymbol{\sigma}(u_h, \mathbf{G}_h^k \underline{u}_h)\cdot\boldsymbol{\Lambda} \to \int_\Omega \boldsymbol{\chi}\cdot\boldsymbol{\Lambda} \quad \text{as } h \to 0.$$

Since $u_h \to u$ in $L^q(\Omega)$ for all $q < \hat{p}$, the continuity and growth properties (6.3a) and (6.3b) of $\boldsymbol{\sigma}$ show that $\boldsymbol{\sigma}(u_h, \boldsymbol{\Lambda}) \to \boldsymbol{\sigma}(u, \boldsymbol{\Lambda})$ strongly in $L^{p'}(\Omega)$ (see, e.g., [175, Lemma A.1]). Hence, using the weak convergence $\mathbf{G}_h^k \underline{u}_h \rightharpoonup \nabla u$ in $L^p(\Omega)^d$,

$$\int_\Omega \boldsymbol{\sigma}(u_h, \boldsymbol{\Lambda})\cdot[\mathbf{G}_h^k \underline{u}_h - \boldsymbol{\Lambda}] \to \int_\Omega \boldsymbol{\sigma}(u, \boldsymbol{\Lambda})\cdot[\nabla u - \boldsymbol{\Lambda}] \quad \text{as } h \to 0.$$

Gathering all these convergences, we pass to the limit in (6.81) and find

$$\int_\Omega f u + \int_{\partial\Omega} g u_{|\partial\Omega} - \int_\Omega \boldsymbol{\chi}\cdot\boldsymbol{\Lambda} \ge \int_\Omega \boldsymbol{\sigma}(u, \boldsymbol{\Lambda})\cdot[\nabla u - \boldsymbol{\Lambda}].$$

Take $v \in W_\star^{1,p}(\Omega)$ and apply this relation to $\boldsymbol{\Lambda} = \nabla u \pm \rho \nabla v$, with $\rho > 0$:

$$\int_\Omega f u + \int_{\partial\Omega} g u_{|\partial\Omega} - \int_\Omega \boldsymbol{\chi}\cdot\nabla(u \pm \rho v) \ge \mp\rho \int_\Omega \boldsymbol{\sigma}(u, \nabla u \pm \rho\nabla v)\cdot\nabla v.$$

Invoking (6.78) with $\phi = u \pm \rho v$ leads to the simplification

$$\mp\rho\left(\int_\Omega f v + \int_{\partial\Omega} g v_{|\partial\Omega}\right) \ge \mp\rho \int_\Omega \boldsymbol{\sigma}(u, \nabla u \pm \rho\nabla v)\cdot\nabla v.$$

Divide by ρ and let $\rho \to 0$. The continuity of $\boldsymbol{\sigma}(u, \cdot)$ and the growth condition (6.3b) show, by dominated convergence theorem, that $\boldsymbol{\sigma}(u, \nabla u \pm \rho\nabla v) \to \boldsymbol{\sigma}(u, \nabla u)$ strongly in $L^{p'}(\Omega)^d$, and thus

$$\mp\left(\int_\Omega f v + \int_{\partial\Omega} g v_{|\partial\Omega}\right) \ge \mp \int_\Omega \boldsymbol{\sigma}(u, \nabla u)\cdot\nabla v.$$

Considering separately the cases $+$ and $-$ in $\mp$ leads to $\int_\Omega f v + \int_{\partial\Omega} g v_{|\partial\Omega} = \int_\Omega \boldsymbol{\sigma}(u, \nabla u)\cdot\nabla v$, which shows that u is indeed a solution to (6.6). □

We conclude this chapter with an improved convergence result on the reconstructed gradients, in the case where $\boldsymbol{\sigma}$ is strictly monotonic, that is, satisfies (6.3c) with a strict inequality when $\boldsymbol{\xi} \neq \boldsymbol{\eta}$.

Theorem 6.30 (Strong Convergence of the Reconstructed Gradients) *Under the assumptions of Theorem 6.29, suppose additionally that, for a.e. $\boldsymbol{x} \in \Omega$ and all $(s, \boldsymbol{\xi}, \boldsymbol{\eta}) \in \mathbb{R} \times \mathbb{R}^d \times \mathbb{R}^d$ with $\boldsymbol{\xi} \neq \boldsymbol{\eta}$,*

$$[\boldsymbol{\sigma}(\boldsymbol{x}, s, \boldsymbol{\xi}) - \boldsymbol{\sigma}(\boldsymbol{x}, s, \boldsymbol{\eta})]\cdot[\boldsymbol{\xi} - \boldsymbol{\eta}] > 0. \tag{6.82}$$

Then, along the same subsequence as in Theorem 6.29, $\mathbf{G}_h^k \underline{u}_h \to \nabla u$ strongly in $L^p(\Omega)^d$.

Proof The proof follows classical arguments, see e.g. [15, 169]. Define $F_h : \Omega \to \mathbb{R}$ by

$$F_h := [\boldsymbol{\sigma}(u_h, \mathbf{G}_h^k \underline{u}_h) - \boldsymbol{\sigma}(u_h, \nabla u)]\cdot[\mathbf{G}_h^k \underline{u}_h - \nabla u].$$

The function F_h is integrable, non-negative, and its integral is equal to the left-hand side of (6.79) with $\boldsymbol{\Lambda} = \nabla u$. Hence, considering (6.81) with this choice of $\boldsymbol{\Lambda}$,

$$\int_\Omega F_h = \int_\Omega f u_h + \int_{\partial\Omega} g\, \gamma_h \underline{u}_h - \int_\Omega \boldsymbol{\sigma}(u_h, \mathbf{G}_h^k \underline{u}_h)\cdot\nabla u - \int_\Omega \boldsymbol{\sigma}(u_h, \nabla u)\cdot[\mathbf{G}_h^k \underline{u}_h - \nabla u].$$

We saw in the proof of Theorem 6.29 that, along a subsequence, $u_h \to u$ in $L^p(\Omega)$, $\gamma_h \underline{u}_h \to u_{|\partial\Omega}$ in $L^p(\partial\Omega)$, $\boldsymbol{\sigma}(u_h, \mathbf{G}_h^k \underline{u}_h) \rightharpoonup \boldsymbol{\chi}$ weakly in $L^{p'}(\Omega)^d$ (where $\boldsymbol{\chi}$ satisfies (6.78)), $\boldsymbol{\sigma}(u_h, \nabla u) \to \boldsymbol{\sigma}(u, \nabla u)$ strongly in $L^{p'}(\Omega)^d$, and $\mathbf{G}_h^k \underline{u}_h \rightharpoonup \nabla u$ weakly in $L^p(\Omega)^d$. Hence, as $h \to 0$ along the same subsequence,

$$\int_\Omega F_h \to \int_\Omega f u + \int_{\partial\Omega} g u_{|\partial\Omega} - \int_\Omega \boldsymbol{\chi}\cdot\nabla u = 0.$$

This proves that $F_h \to 0$ in $L^1(\Omega)$. Along a subsequence, we can assume that the convergence holds a.e. on Ω. By strong convergence in $L^p(\Omega)$ we can also assume that u_h converges a.e. on Ω as $h \to 0$. The strict monotonicity of $\boldsymbol{\sigma}$ and Lemma 6.31 below then show that $\mathbf{G}_h^k \underline{u}_h \to \nabla u$ a.e. on Ω.

By the continuity property (6.3a) of $\boldsymbol{\sigma}(\boldsymbol{x}, \cdot, \cdot)$, we infer that $\boldsymbol{\sigma}(u_h, \mathbf{G}_h^k \underline{u}_h)\cdot\mathbf{G}_h^k \underline{u}_h \to \boldsymbol{\sigma}(u, \nabla u)\cdot\nabla u$ a.e. on Ω. Taking the superior limit of (6.80) and using the fact that u is a solution to (6.6), we also have

$$\limsup_{h\to 0}\int_\Omega \boldsymbol{\sigma}(u_h,\mathbf{G}_h^k\underline{u}_h)\cdot\mathbf{G}_h^k\underline{u}_h \le \int_\Omega fu + \int_{\partial\Omega} gu_{|\partial\Omega} = \int_\Omega \boldsymbol{\sigma}(u,\nabla u)\cdot\nabla u.$$

Together with the a.e. convergence of $\boldsymbol{\sigma}(u_h,\mathbf{G}_h^k\underline{u}_h)\cdot\mathbf{G}_h^k\underline{u}_h$ and Lemma 6.32 below, this proves that $\boldsymbol{\sigma}(u_h,\mathbf{G}_h^k\underline{u}_h)\cdot\mathbf{G}_h^k\underline{u}_h$ converges strongly in $L^1(\Omega)$ to $\boldsymbol{\sigma}(u,\nabla u)\cdot\nabla u$. In particular, $\left(\boldsymbol{\sigma}(u_h,\mathbf{G}_h^k\underline{u}_h)\cdot\mathbf{G}_h^k\underline{u}_h\right)_{h\to 0}$ is equi-integrable in $L^1(\Omega)$, from which we deduce, using the coercivity property (6.3d), that $\left(\mathbf{G}_h^k\underline{u}_h\right)_{h\to 0}$ is equi-integrable in $L^p(\Omega)^d$. Since this sequence converges a.e. to ∇u, the Vitali theorem concludes the proof that $\mathbf{G}_h^k\underline{u}_h\to\nabla u$ strongly in $L^p(\Omega)^d$. □

The following lemma, used in the proof above, is a particular case of [178, Lemma 2.47].

Lemma 6.31 *Let $\boldsymbol{\beta}:\mathbb{R}\times\mathbb{R}^d\to\mathbb{R}^d$ be a continuous function such that*

$$(\boldsymbol{\beta}(s,\boldsymbol{\xi})-\boldsymbol{\beta}(s,\boldsymbol{\eta}))\cdot(\boldsymbol{\xi}-\boldsymbol{\eta})>0,\quad \forall(s,\boldsymbol{\xi},\boldsymbol{\eta})\in\mathbb{R}\times\mathbb{R}^d\times\mathbb{R}^d \textit{ with } \boldsymbol{\xi}\ne\boldsymbol{\eta}.$$

Let $(s_m,\boldsymbol{\xi}_m)_{n\in\mathbb{N}}$ be a sequence in $\mathbb{R}\times\mathbb{R}^d$ and $(s,\boldsymbol{\xi})\in\mathbb{R}\times\mathbb{R}^d$ be such that

$$\left(\boldsymbol{\beta}(s_m,\boldsymbol{\xi}_m)-\boldsymbol{\beta}(s_m,\boldsymbol{\xi})\right)\cdot(\boldsymbol{\xi}_m-\boldsymbol{\xi})\to 0 \textit{ and } s_m\to s \textit{ as } m\to\infty.$$

Then, $\boldsymbol{\xi}_m\to\boldsymbol{\xi}$ as $m\to\infty$.

The lemma below was also used in the proof of Theorem 6.30. Its proof is sketched in [178, Lemma 2.48]; we provide it here in full for the sake of completeness.

Lemma 6.32 *Let X be a measured space, $f\in L^1(X)$ and $(f_m)_{m\in\mathbb{N}}$ be non-negative functions in $L^1(X)$. Assume that $f_m\to f$ a.e. on X as $m\to\infty$, and that $\limsup_{m\to\infty}\int_X f_m\le\int_X f$. Then, $f_m\to f$ in $L^1(X)$ as $m\to\infty$.*

Proof Since $f_m\ge 0$, f is also positive and we have $(f-f_m)^+\le f$ where we recall that, for any $\alpha\in\mathbb{R}$, $\alpha^+:=\frac{1}{2}(|\alpha|+\alpha)=\max(\alpha,0)$ is the positive part of α. Moreover, $(f-f_m)^+\to 0$ a.e. on X. The dominated convergence theorem then shows that $\int_X(f-f_m)^+\to 0$ as $m\to\infty$. Writing $|f-f_m|=2(f-f_m)^++(f_m-f)$, we infer that

$$\begin{aligned}\limsup_{m\to\infty}\int_X|f-f_m| &= \limsup_{m\to\infty}\left(2\int_X(f-f_m)^+ + \int_X(f_m-f)\right)\\ &= \limsup_{m\to\infty}\int_X f_m - \int_X f\le 0,\end{aligned}$$

which proves that $\int_X|f-f_m|\to 0$ as $m\to\infty$. □

6.5 Proofs of the Discrete Functional Analysis Results

The proofs of the discrete Sobolev–Poincaré–Wirtinger inequality (Theorem 6.5), the discrete trace inequality (Theorem 6.7), and the discrete compactness property (Theorem 6.8) hinge on results established in [178, Section B.3] for the lowest-order case $k = 0$ and under a star-shapedness assumptions on the mesh elements. Our approach to leverage these results to high-order and non-star-shaped elements is to project unknowns in $\underline{U}_{h,\star}^k$ onto the lowest-order unknowns on a matching simplicial submesh, and to apply the results of [178] on that simplicial submesh.

6.5.1 *Mapping High-Order Unknowns to Lowest-Order Unknowns on Simplicial Submeshes*

We recall that, for each polytopal mesh $\mathcal{M}_h = (\mathcal{T}_h, \mathcal{F}_h)$ in a regular sequence $(\mathcal{M}_h)_{h\in\mathcal{H}}$, there exists a matching simplicial submesh $\mathfrak{M}_h = (\mathfrak{T}_h, \mathfrak{F}_h)$ of $\mathcal{M}_h$ and that, by Definition 1.9, the sequence of simplicial meshes $(\mathfrak{M}_h)_{h\in\mathcal{H}}$ is itself regular. This means that, for each $\tau \in \mathfrak{T}_h$, the ratio of the diameter of τ over the diameter of the largest ball contained in τ is bounded above independently of h.

Let us fix $h \in \mathcal{H}$, and let $\underline{\mathfrak{U}}_h^0$ be the lowest-order space of unknowns on $\mathfrak{M}_h$, that is

$$\underline{\mathfrak{U}}_h^0 := \Big\{ \underline{\mathfrak{w}}_h = ((\mathfrak{w}_\tau)_{\tau\in\mathfrak{T}_h}, (\mathfrak{w}_\sigma)_{\sigma\in\mathfrak{F}_h}) \; : \; \mathfrak{w}_\tau \in \mathbb{R} \quad \forall \tau \in \mathfrak{T}_h \text{ and } \mathfrak{w}_\sigma \in \mathbb{R} \quad \forall \sigma \in \mathfrak{F}_h \Big\}. \tag{6.83}$$

For $p \in [1, \infty)$, this space is endowed with the (semi)-norm $|||{\cdot}|||_{1,p,h}$ defined by: For all $\underline{\mathfrak{w}}_h \in \underline{\mathfrak{U}}_h^0$,

$$|||\underline{\mathfrak{w}}_h|||_{1,p,h} := \left(\sum_{\tau\in\mathfrak{T}_h} \sum_{\sigma\in\mathfrak{F}_\tau} |\sigma| d_{\tau\sigma} \left| \frac{\mathfrak{w}_\sigma - \mathfrak{w}_\tau}{d_{\tau\sigma}} \right|^p \right)^{\frac{1}{p}}, \tag{6.84}$$

where $\mathfrak{F}_\tau$ is the set of faces of τ and $d_{\tau\sigma}$ is the orthogonal distance between σ and the centre of the largest ball contained in τ.

For each $\tau \in \mathfrak{T}_h$, there is a unique $T \in \mathcal{T}_h$ such that $\tau \subset T$; we denote this element T by $T(\tau)$. Let $\mathfrak{F}_{h,\mathrm{sk}}$ be the set of simplicial faces in $\mathfrak{F}_h$ that lie on the skeleton $\mathcal{F}_h$ of $\mathcal{M}_h$ and, for each $\sigma \in \mathfrak{F}_{h,\mathrm{sk}}$, denote by $F(\sigma)$ the unique face $F \in \mathcal{F}_h$ such that $\sigma \subset F$. If $\sigma \in \mathfrak{F}_h \backslash \mathfrak{F}_{h,\mathrm{sk}}$ is an "internal" simplicial face, i.e., it completely lies inside one mesh element $T \in \mathcal{T}_h$, we set $T(\sigma) := T$. The function $\underline{\Pi}_h : \underline{U}_h^k \to \underline{\mathfrak{U}}_h^0$ that maps high-order unknowns on $\mathcal{M}_h$ to low-order unknowns on $\mathfrak{M}_h$ is then defined by: For all $\underline{v}_h \in \underline{U}_h^k$,

$$\underline{\Pi}_h \underline{v}_h = \underline{\mathfrak{w}}_h \quad \text{with} \quad \begin{cases} \mathfrak{w}_\tau = \pi_\tau^{0,0} v_{T(\tau)} & \forall \tau \in \mathfrak{T}_h, \\ \mathfrak{w}_\sigma = \pi_\sigma^{0,0} v_{F(\sigma)} & \forall \sigma \in \mathfrak{F}_{h,\text{sk}}, \\ \mathfrak{w}_\sigma = \pi_\sigma^{0,0} v_{T(\sigma)} & \forall \sigma \in \mathfrak{F}_h \backslash \mathfrak{F}_{h,\text{sk}}. \end{cases} \tag{6.85}$$

We remark that $\underline{\Pi}_h$ sends $\underline{U}_{h,\star}^k$ into

$$\underline{\mathfrak{U}}_{h,\star}^0 = \left\{ \underline{\mathfrak{w}}_h \in \underline{\mathfrak{U}}_h^0 \, : \, \int_\Omega \mathfrak{w}_h = 0 \right\},$$

where we have denoted by $\mathfrak{w}_h$ the function in $\mathbb{P}^0(\mathfrak{T}_h)$ such that $(\mathfrak{w}_h)_{|\tau} = \mathfrak{w}_\tau$ for all $\tau \in \mathfrak{T}_h$. To prove this, it suffices to notice that, if $\underline{\mathfrak{w}}_h = \underline{\Pi}_h \underline{v}_h$ for some $\underline{v}_h \in \underline{U}_{h,\star}^k$,

$$\begin{aligned} \int_\Omega \mathfrak{w}_h = \sum_{T \in \mathcal{T}_h} \sum_{\tau \in \mathfrak{T}_h,\, \tau \subset T} \int_\tau \mathfrak{w}_\tau = \sum_{T \in \mathcal{T}_h} \sum_{\tau \in \mathfrak{T}_h,\, \tau \subset T} \int_\tau \pi_\tau^{0,0} v_T \\ = \sum_{T \in \mathcal{T}_h} \sum_{\tau \in \mathfrak{T}_h,\, \tau \subset T} \int_\tau v_T = \sum_{T \in \mathcal{T}_h} \int_T v_T = \int_\Omega v_h = 0. \end{aligned}$$

We now prove a boundedness property on $\underline{\Pi}_h$.

Lemma 6.33 (Boundedness of $\underline{\Pi}_h$) *We have*

$$|||\underline{\Pi}_h \underline{v}_h|||_{1,p,h} \lesssim \|\underline{v}_h\|_{1,p,h} \qquad \forall \underline{v}_h \in \underline{U}_h^k, \tag{6.86}$$

with hidden constant depending only on p, k, and ϱ.

Proof Let $\underline{v}_h \in \underline{U}_h^k$ and set $\underline{\mathfrak{w}}_h := \underline{\Pi}_h \underline{v}_h$. Take $\tau \in \mathfrak{T}_h$ and $\sigma \in \mathfrak{F}_\tau$. By the mesh regularity assumption, it holds $d_{\tau\sigma} \simeq h_\tau \simeq h_\sigma$, and thus

$$|\sigma| d_{\tau\sigma} \left| \frac{\mathfrak{w}_\sigma - \mathfrak{w}_\tau}{d_{\tau\sigma}} \right|^p = d_{\tau\sigma}^{1-p} |\sigma| \, |\mathfrak{w}_\sigma - \mathfrak{w}_\tau|^p \simeq h_\sigma^{1-p} \|\mathfrak{w}_\sigma - \mathfrak{w}_\tau\|_{L^p(\sigma)}^p. \tag{6.87}$$

Let $X(\sigma) = F(\sigma)$ if $\sigma \in \mathfrak{F}_{h,\text{sk}}$ and $X(\sigma) = T(\sigma)$ if $\sigma \notin \mathfrak{F}_{h,\text{sk}}$. The linearity and idempotency of $\pi_\sigma^{0,0}$ yield

$$\mathfrak{w}_\sigma - \mathfrak{w}_\tau = \pi_\sigma^{0,0} v_{X(\sigma)} - \pi_\tau^{0,0} v_{T(\tau)} = \pi_\sigma^{0,0} \left(v_{X(\sigma)} - \pi_\tau^{0,0} v_{T(\tau)} \right).$$

Plugging this into (6.87) and using the L^p-stability of $\pi_\sigma^{0,0}$ (see Lemma 1.44), we infer

$$|\sigma| d_{\tau\sigma} \left| \frac{\mathfrak{w}_\sigma - \mathfrak{w}_\tau}{d_{\tau\sigma}} \right|^p \lesssim h_\sigma^{1-p} \| v_{X(\sigma)} - \pi_\tau^{0,0} v_{T(\tau)} \|_{L^p(\sigma)}^p. \tag{6.88}$$

We now separate the cases $\sigma \notin \mathfrak{F}_{h,\text{sk}}$ and $\sigma \in \mathfrak{F}_{h,\text{sk}}$. In the former case, $X(\sigma) = T(\sigma) = T(\tau)$ and thus, by the trace approximation property (1.75) of $\pi_\tau^{0,0}$ with τ instead of T and $(l, s, m) = (0, 1, 0)$,

$$\| v_{X(\sigma)} - \pi_\tau^{0,0} v_{T(\tau)} \|_{L^p(\sigma)}^p \lesssim h_\tau^{p-1} \| \nabla v_{T(\tau)} \|_{L^p(\tau)^d}^p \lesssim h_\sigma^{p-1} \| \nabla v_{T(\tau)} \|_{L^p(\tau)^d}^p, \tag{6.89}$$

where we have additionally used the fact that $h_\tau \lesssim h_\sigma$ by mesh regularity. Plugged into (6.88), this yields

$$|\sigma| d_{\tau\sigma} \left| \frac{\mathfrak{w}_\sigma - \mathfrak{w}_\tau}{d_{\tau\sigma}} \right|^p \lesssim \| \nabla v_{T(\tau)} \|_{L^p(\tau)^d}^p$$

and, after summing over $\sigma \in \mathfrak{F}_\tau \backslash \mathfrak{F}_{h,\text{sk}}$ (there are at most $d+1$ such faces since τ is a simplex) and $\tau \in \mathfrak{T}_h$,

$$\sum_{\tau \in \mathfrak{T}_h} \sum_{\sigma \in \mathfrak{F}_\tau \backslash \mathfrak{F}_{h,\text{sk}}} |\sigma| d_{\tau\sigma} \left| \frac{\mathfrak{w}_\sigma - \mathfrak{w}_\tau}{d_{\tau\sigma}} \right|^p \lesssim \sum_{\tau \in \mathfrak{T}_h} \| \nabla v_{T(\tau)} \|_{L^p(\tau)^d}^p = \sum_{T \in \mathcal{T}_h} \| \nabla v_T \|_{L^p(T)^d}^p, \tag{6.90}$$

the conclusion following from $\sum_{\tau \in \mathfrak{T}_h} \bullet = \sum_{T \in \mathcal{T}_h} \sum_{\tau \in \mathfrak{T}_h,\, \tau \subset T} \bullet$.

Consider now the case $\sigma \in \mathfrak{F}_{h,\text{sk}}$. In this case, $X(\sigma) = F(\sigma)$ and (6.88) gives

$$\begin{aligned} |\sigma| d_{\tau\sigma} \left| \frac{\mathfrak{w}_\sigma - \mathfrak{w}_\tau}{d_{\tau\sigma}} \right|^p &\lesssim h_\sigma^{1-p} \| v_{F(\sigma)} - \pi_\tau^{0,0} v_{T(\tau)} \|_{L^p(\sigma)}^p \\ &\lesssim h_\sigma^{1-p} \| v_{F(\sigma)} - v_{T(\tau)} \|_{L^p(\sigma)}^p + h_\sigma^{1-p} \| v_{T(\tau)} - \pi_\tau^{0,0} v_{T(\tau)} \|_{L^p(\sigma)}^p \\ &\lesssim h_{F(\sigma)}^{1-p} \| v_{F(\sigma)} - v_{T(\tau)} \|_{L^p(\sigma)}^p + \| \nabla v_{T(\tau)} \|_{L^p(\tau)^d}^p, \end{aligned} \tag{6.91}$$

where the second inequality is obtained inserting $\pm v_{T(\tau)}$ into the norm and using the triangle inequality, while the conclusion follows from $h_\sigma \simeq h_{F(\sigma)}$ and (6.89) (in which we recall that $X(\sigma) = T(\sigma) = T(\tau)$). Summing (6.91) over $\sigma \in \mathfrak{F}_\tau \cap \mathfrak{F}_{h,\text{sk}}$ and $\tau \in \mathfrak{T}_h$, rearranging the sums over $T \in \mathcal{T}_h$ and $F \in \mathcal{F}_T$, and noticing that, for all $F \in \mathcal{F}_h$,

$$\sum_{\sigma \in \mathfrak{F}_h,\, \sigma \subset F} h_{F(\sigma)}^{1-p} \| v_{F(\sigma)} - v_T \|_{L^p(\sigma)}^p = h_F^{1-p} \| v_F - v_T \|_{L^p(F)}^p,$$

we infer

$$\sum_{\tau\in\mathfrak{T}_h}\sum_{\sigma\in\mathfrak{F}_\tau\cap\mathfrak{F}_{h,\rm sk}}|\sigma|d_{\tau\sigma}\left|\frac{\mathfrak{w}_\sigma-\mathfrak{w}_\tau}{d_{\tau\sigma}}\right|^p$$

$$\lesssim\sum_{T\in\mathcal{T}_h}\sum_{F\in\mathcal{F}_T}h_F^{1-p}\|v_F-v_T\|_{L^p(F)}^p+\sum_{T\in\mathcal{T}_h}\|\nabla v_T\|_{L^p(T)^d}^p. \tag{6.92}$$

Adding this estimate to (6.90) and recalling the definition (6.9) of $\|{\cdot}\|_{1,p,h}$ yields (6.86). □

Remark 6.34 (Equivalence of Norms on $\underline{\mathfrak{U}}_h^0$) Letting $\|{\cdot}\|_{1,p,\mathfrak{U},h}$ be the standard seminorm $\|{\cdot}\|_{1,p,h}$ on $\underline{\mathfrak{U}}_h^0$ (that is, (6.9) with $k=0$ and $\mathfrak{M}_h$ instead of $\mathcal{M}_h$) and summing (6.87) over $\sigma\in\mathfrak{F}_\tau$ and $\tau\in\mathfrak{T}_h$ gives the uniform equivalence

$$|||\underline{\mathfrak{w}}_h|||_{1,p,h}\simeq\|\underline{\mathfrak{w}}_h\|_{1,p,\mathfrak{U},h}\qquad\forall\underline{\mathfrak{w}}_h\in\underline{\mathfrak{U}}_h^0. \tag{6.93}$$

The following lemma will be the key to comparing $\underline{v}_h$ and $\underline{\Pi}_h\underline{v}_h$, in order to lift discrete functional analysis properties of [178] from $\underline{\mathfrak{U}}_{h,\star}^0$ to $\underline{U}_{h,\star}^k$. It will be used both with the original mesh $\mathcal{M}_h$, and with $\mathcal{M}_h$ replaced by $\mathfrak{M}_h$.

Lemma 6.35 (Poincaré–Wirtinger–Sobolev Inequality for Broken Polynomial Functions with Local Zero Average) *Let $(\mathcal{M}_h)_{h\in\mathcal{H}}$ be a regular mesh sequence, and take a real number r such that*

$$p\le r\le\begin{cases}\frac{dp}{d-p} & \text{if } p<d,\\ \infty & \text{if } p\ge d.\end{cases}$$

If $w\in\mathbb{P}^k(\mathcal{T}_h)$ satisfies $\int_T w=0$ for all $T\in\mathcal{T}_h$, then

$$\|w\|_{L^r(\Omega)}\lesssim h^{1+\frac{d}{r}-\frac{d}{p}}\|\nabla_h w\|_{L^p(\Omega)^d}, \tag{6.94}$$

with hidden multiplicative constant depending on Ω, ϱ, k, r, and p.

Remark 6.36 (Power of h in (6.94)) The proof shows that (6.94) is actually valid for any $r\ge p$, without upper bound restriction. However, in the case $p<d$, if $r>\frac{dp}{d-p}$ then the exponent of h in (6.94) is negative.

Proof For all $T\in\mathcal{T}_h$, we have $\pi_T^{0,0}w=0$ and thus, by the optimal approximation property (1.74) of $\pi_T^{0,0}$ (with $s=1$, $m=0$, and r instead of p),

$$\begin{aligned}\|w\|_{L^r(T)}=\|w-\pi_T^{0,0}w\|_{L^r(T)}&\lesssim h_T\|\nabla w\|_{L^r(T)^d}\\&\lesssim h_T|T|_d^{\frac{1}{r}-\frac{1}{p}}\|\nabla w\|_{L^p(T)^d}\\&\lesssim h_T^{1+\frac{d}{r}-\frac{d}{p}}\|\nabla w\|_{L^p(T)^d},\end{aligned} \tag{6.95}$$

where the second line follows from the inverse Lebesgue inequality (1.35), and the third line is a consequence of (1.7).

If r is finite, take the power r of (6.95), sum over $T \in \mathcal{T}_h$, and use $\|\nabla w\|_{L^p(T)^d}^{r-p} \le \|\nabla_h w\|_{L^p(\Omega)^d}^{r-p}$ (since $r \ge p$) to infer

$$\begin{aligned}
\|w\|_{L^r(\Omega)}^r &\lesssim h^{r+d-\frac{dr}{p}} \sum_{T\in\mathcal{T}_h} \|\nabla w\|_{L^p(T)^d}^r \\
&\le h^{r+d-\frac{dr}{p}} \|\nabla_h w\|_{L^p(\Omega)^d}^{r-p} \sum_{T\in\mathcal{T}_h} \|\nabla w\|_{L^p(T)^d}^p \\
&= h^{r+d-\frac{dr}{p}} \|\nabla_h w\|_{L^p(\Omega)^d}^{r-p} \|\nabla_h w\|_{L^p(\Omega)^d}^p \\
&= h^{r+d-\frac{dr}{p}} \|\nabla_h w\|_{L^p(\Omega)^d}^r.
\end{aligned}$$

Taking the power $1/r$ of this inequality concludes the proof of (6.94).

If $r = \infty$, apply (6.95) to $T \in \mathcal{T}_h$ such that $\|w\|_{L^\infty(T)} = \|w\|_{L^\infty(\Omega)}$ to obtain $\|w\|_{L^\infty(\Omega)} \lesssim h^{1-\frac{d}{p}} \|\nabla w\|_{L^p(T)^d} \le h^{1-\frac{d}{p}} \|\nabla_h w\|_{L^p(\Omega)^d}$. □

6.5.2 Discrete Sobolev–Poincaré–Wirtinger Embeddings

Proof (Theorem 6.5) Let $\underline{v}_h \in \underline{U}_{h,\star}^k$ and q as in the theorem. Without restriction, we can assume that $q \ge p$. Otherwise, we use the Hölder inequality to write $\|v_h\|_{L^q(\Omega)} \lesssim \|v_h\|_{L^p(\Omega)}$ and (6.11) follows from the same estimate with p instead of q.

Set $\mathfrak{w}_h := \underline{\Pi}_h \underline{v}_h \in \underline{\mathfrak{U}}_{h,\star}^0$, and denote by $\pi_{\mathfrak{T}_h}^{0,0}$ the L^2-orthogonal projection on piecewise constant functions on $\mathfrak{T}_h$, that is, $(\pi_{\mathfrak{T}_h}^{0,0}\phi)_{|\tau} := \pi_\tau^{0,0}\phi$ for all $\tau \in \mathfrak{T}_h$ and all $\phi \in L^1(\Omega)$. The definition (6.85) of $\underline{\Pi}_h$ shows that $\pi_{\mathfrak{T}_h}^{0,0} v_h = \mathfrak{w}_h$, and thus $v_h = v_h - \pi_{\mathfrak{T}_h}^{0,0} v_h + \mathfrak{w}_h$. Moreover, by definition of $\pi_{\mathfrak{T}_h}^{0,0}$, it holds $\int_\tau (v_h - \pi_{\mathfrak{T}_h}^{0,0} v_h) = 0$ for all $\tau \in \mathfrak{T}_h$. Hence, letting $\nabla_{\mathfrak{T}_h}$ denote the broken gradient on $\mathfrak{T}_h$, the estimate (6.11) follows from

$$\begin{aligned}
\|v_h\|_{L^q(\Omega)} &\le \|v_h - \pi_{\mathfrak{T}_h}^{0,0} v_h\|_{L^q(\Omega)} + \|\mathfrak{w}_h\|_{L^q(\Omega)} \\
&\lesssim h^{1+\frac{d}{q}-\frac{d}{p}} \|\nabla_{\mathfrak{T}_h} v_h\|_{L^p(\Omega)^d} + |||\underline{\mathfrak{w}}_h|||_{1,p,h} \\
&\lesssim \|\underline{v}_h\|_{1,p,h},
\end{aligned} \tag{6.96}$$

where, to pass to the second line, we have used (6.94) (with $r = q$, $w = v_h - \pi_{\mathfrak{T}_h}^{0,0} v_h$, and $\mathfrak{T}_h$ instead of $\mathcal{T}_h$) together with Lemma 6.37 below, and the conclusion is obtained recalling the definition (6.9) of $\|\cdot\|_{1,p,h}$, the property $1+\frac{d}{q}-\frac{d}{p}\geq 0$ (by choice of q) together with $h\lesssim 1$, and the boundedness property (6.86) of $\underline{\Pi}_h$. □

The following lemma, used in the proof above, is a special case of [178, Lemma B.25].

Lemma 6.37 (Discrete Sobolev Embedding for $\underline{\mathfrak{U}}_{h,\star}^0$) *Let $(\mathfrak{M}_h)_{h\in\mathcal{H}}$ be a regular sequence of matching simplicial meshes, and let $p\in[1,\infty)$. Take $q\in\left[1,\frac{dp}{d-p}\right]$ if $p<d$, and $q\in[1,\infty)$ if $p\geq d$. Then,*

$$\|\mathfrak{w}_h\|_{L^q(\Omega)}\lesssim \|\!|\!|\underline{\mathfrak{w}}_h|\!|\!|_{1,p,h}\qquad \forall\underline{\mathfrak{w}}_h\in\underline{\mathfrak{U}}_{h,\star}^0,$$

with hidden constant depending only on Ω, ϱ, p, and q.

6.5.3 Discrete Trace Inequality

Proof (Theorem 6.7) Let $\underline{v}_h\in\underline{U}_{h,\star}^k$. We naturally map $\underline{v}_h$ to an element $\underline{\widetilde{v}}_h\in\underline{\mathfrak{U}}_{h,\star}^k$ (the space of high-order unknowns on $\mathfrak{M}_h$) by setting, in a similar way as in (6.85),

$$\begin{aligned}\widetilde{v}_\tau &= v_{T(\tau)} &&\forall\tau\in\mathfrak{T}_h,\\ \widetilde{v}_\sigma &= v_{F(\sigma)} &&\forall\sigma\in\mathfrak{F}_{h,\mathrm{sk}},\\ \widetilde{v}_\sigma &= v_{T(\sigma)} &&\forall\sigma\in\mathfrak{F}_h\backslash\mathfrak{F}_{h,\mathrm{sk}}.\end{aligned}\tag{6.97}$$

Recalling that $\|\cdot\|_{1,p,\mathfrak{U},h}$ is the norm $\|\cdot\|_{1,p,h}$ on $\mathfrak{M}_h$, it can easily be checked that

$$\|\underline{\widetilde{v}}_h\|_{1,p,\mathfrak{U},h}\simeq\|\underline{v}_h\|_{1,p,h}\tag{6.98}$$

with hidden constant depending only on d, ϱ, and p. Set $\underline{z}_h:=\underline{\widetilde{v}}_h-\underline{\Pi}_h\underline{v}_h\in\underline{\mathfrak{U}}_{h,\star}^k$. Let $\sigma\in\mathfrak{F}_h^{\mathrm{b}}$, the set of simplicial faces in $\mathfrak{F}_h$ that lie on $\partial\Omega$, and take $\tau_\sigma\in\mathfrak{T}_h$ such that σ is a face of τ_σ. A triangle inequality yields

$$\begin{aligned}\|z_\sigma\|_{L^p(\sigma)}^p&\lesssim\|z_\sigma-z_{\tau_\sigma}\|_{L^p(\sigma)}^p+\|z_{\tau_\sigma}\|_{L^p(\sigma)}^p\\ &\lesssim\|z_\sigma-z_{\tau_\sigma}\|_{L^p(\sigma)}^p+h_{\tau_\sigma}^{p-1}\|\nabla v_{T(\tau_\sigma)}\|_{L^p(\tau_\sigma)^d}^p,\end{aligned}\tag{6.99}$$

the conclusion following from $z_{\tau_\sigma}=v_{T(\tau_\sigma)}-\pi_{\tau_\sigma}^{0,0}v_{T(\tau_\sigma)}$ (see (6.97) and (6.85)) and from the trace approximation property (1.75) of the L^2-orthogonal projector with $X=\tau_\sigma$, $l=0$, $s=1$, and $m=0$. Let $\gamma_{\mathfrak{M}_h}:\underline{\mathfrak{U}}_h^k\to L^p(\partial\Omega)$ be the trace defined by:

$$(\gamma_{\mathfrak{M}_h}\underline{\mathfrak{w}}_h)_{|\sigma}:=\mathfrak{w}_\sigma\qquad\forall\sigma\in\mathfrak{F}_h^{\mathrm{b}},\ \forall\underline{\mathfrak{w}}_h\in\underline{\mathfrak{U}}_h^k.\tag{6.100}$$

Summing (6.99) over $\sigma \in \mathfrak{F}_h^{\mathrm{b}}$, we obtain

$$\begin{aligned}
&\|\gamma_{\mathfrak{M}_h}\underline{z}_h\|_{L^p(\partial\Omega)}^p \\
&\lesssim \sum_{\sigma\in\mathfrak{F}_h^{\mathrm{b}}} \|z_\sigma - z_{\tau_\sigma}\|_{L^p(\sigma)}^p + \sum_{\sigma\in\mathfrak{F}_h^{\mathrm{b}}} h_{\tau_\sigma}^{p-1}\|\nabla v_{T(\tau_\sigma)}\|_{L^p(\tau_\sigma)^d}^p \\
&\lesssim h^{p-1}\sum_{\tau\in\mathfrak{T}_h^{\mathrm{b}}}\sum_{\sigma\in\mathfrak{F}_\tau\cap\mathfrak{F}_h^{\mathrm{b}}} h_\sigma^{1-p}\|z_\sigma - z_\tau\|_{L^p(\sigma)}^p + h^{p-1}\sum_{\tau\in\mathfrak{T}_h^{\mathrm{b}}}\|\nabla v_{T(\tau)}\|_{L^p(\tau)^d}^p,
\end{aligned} \tag{6.101}$$

where $\mathfrak{T}_h^{\mathrm{b}}$ is the set of elements $\tau \in \mathfrak{T}_h$ that have at least one face on $\partial\Omega$, and the second inequality follows from $1 = h_\sigma^{p-1}h_\sigma^{1-p} \le h^{p-1}h_\sigma^{1-p}$ and from

$$\sum_{\sigma\in\mathfrak{F}_h^{\mathrm{b}}} \bullet = \sum_{\tau\in\mathfrak{T}_h^{\mathrm{b}}}\sum_{\sigma\in\mathfrak{F}_\tau\cap\mathfrak{F}_h^{\mathrm{b}}} \bullet.$$

The definition (6.97) shows that $\gamma_{\mathfrak{M}_h}\underline{\widetilde{v}}_h = \gamma_h\underline{v}_h$, and thus that $\gamma_{\mathfrak{M}_h}\underline{z}_h = \gamma_h\underline{v}_h - \gamma_{\mathfrak{M}_h}\underline{\Pi}_h\underline{v}_h$. Hence, recalling the definitions of $\|\cdot\|_{1,p,\mathfrak{U},h}$ (on $\mathfrak{M}_h$) and $\|\cdot\|_{1,p,h}$ (on $\mathcal{M}_h$), we infer from (6.101) that

$$\begin{aligned}
\|\gamma_h\underline{v}_h - \gamma_{\mathfrak{M}_h}\underline{\Pi}_h\underline{v}_h\|_{L^p(\partial\Omega)}^p &\lesssim h^{p-1}\|\underline{z}_h\|_{1,p,\mathfrak{U},h}^p + h^{p-1}\|\underline{v}_h\|_{1,p,h}^p \\
&\lesssim h^{p-1}\|\underline{v}_h\|_{1,p,h}^p,
\end{aligned} \tag{6.102}$$

where the conclusion follows from $\underline{z}_h = \underline{\widetilde{v}}_h - \underline{\Pi}_h\underline{v}_h$, the relation (6.98), the norm equivalence (6.93), and the boundedness property (6.86) of $\underline{\Pi}_h$.

Invoking Lemma 6.38 below, we can write $\|\gamma_{\mathfrak{M}_h}\underline{\Pi}_h\underline{v}_h\|_{L^p(\partial\Omega)} \lesssim |||\underline{\Pi}_h\underline{v}_h|||_{1,p,h}$, and the proof is completed using the triangle inequality, (6.102), the property (6.86), and $h \lesssim 1$ to write

$$\|\gamma_h\underline{v}_h\|_{L^p(\partial\Omega)} \le \|\gamma_h\underline{v}_h - \gamma_{\mathfrak{M}_h}\underline{\Pi}_h\underline{v}_h\|_{L^p(\partial\Omega)} + \|\gamma_{\mathfrak{M}_h}\underline{\Pi}_h\underline{v}_h\|_{L^p(\partial\Omega)} \lesssim \|\underline{v}_h\|_{1,p,h}.$$

□

The following lemma is a straightforward consequence of [178, Eq. (B.58) and Lemma B.24].

Lemma 6.38 (Discrete Trace Inequality in $\underline{\mathfrak{U}}_{h,\star}^0$) *Let $(\mathfrak{M}_h)_{h\in\mathcal{H}}$ be a regular family of matching simplicial meshes, and let $p \in [1,\infty)$. Then, defining the trace $\gamma_{\mathfrak{M}_h} : \underline{\mathfrak{U}}_h^0 \to L^p(\partial\Omega)$ by (6.100) with $k = 0$, it holds*

$$\|\gamma_{\mathfrak{M}_h}\underline{\mathfrak{w}}_h\|_{L^p(\partial\Omega)} \lesssim |||\underline{\mathfrak{w}}_h|||_{1,p,h} \qquad \forall \underline{\mathfrak{w}}_h \in \underline{\mathfrak{U}}_{h,\star}^0,$$

with hidden constant depending only on Ω, ϱ and p.

6.5.4 Discrete Compactness

Proof (Theorem 6.8) Set $\underline{\mathfrak{w}}_h := \underline{\Pi}_h \underline{v}_h \in \underline{\mathfrak{U}}^0_{h,\star}$. The boundedness property (6.86) of $\underline{\Pi}_h$ shows that $(|||\underline{\mathfrak{w}}_h|||_{1,p,h})_{h\in\mathcal{H}}$ is bounded. By Lemma 6.39 below, we get $v \in L^p(\Omega)$ and $\omega \in L^p(\partial\Omega)$ such that, up to a subsequence as $h \to 0$, $\mathfrak{w}_h \to v$ in $L^p(\Omega)$ and $\gamma_{\mathfrak{M}_h}\underline{\mathfrak{w}}_h \to \omega$ in $L^p(\partial\Omega)$. Since $\int_\Omega \mathfrak{w}_h = 0$ for all $h \in \mathcal{H}$, we have $\int_\Omega v = 0$.

As in the proof of Theorem 6.5, we have $v_h = v_h - \pi^{0,0}_{\mathfrak{T}_h} v_h + \mathfrak{w}_h$. Hence, using (6.94) with $r = p$, $w = v_h - \pi^{0,0}_{\mathfrak{T}_h} v_h$ and $\mathfrak{T}_h$ instead of $\mathcal{T}_h$, and recalling that

$$\|\nabla_{\mathfrak{T}_h} v_h\|_{L^p(\Omega)^d} \le \|\underline{v}_h\|_{1,p,h},$$

we find

$$\begin{aligned}\|v_h - v\|_{L^p(\Omega)} &\le \|v_h - \pi^{0,0}_{\mathfrak{T}_h} v_h\|_{L^p(\Omega)} + \|\mathfrak{w}_h - v\|_{L^p(\Omega)}\\ &\lesssim h\|\underline{v}_h\|_{1,p,h} + \|\mathfrak{w}_h - v\|_{L^p(\Omega)}\end{aligned}$$

(here and in the rest of the proof, the hidden constants in $\lesssim$ do not depend on h). This shows that, as $h \to 0$, $v_h \to v$ in $L^p(\Omega)$, and thus also in $L^q(\Omega)$ if $q \le p$. Take now $q > p$ satisfying the condition in Theorem 6.8. We can find $r > q$ such that $r \le \frac{dp}{d-p}$ if $p < d$, and $r < \infty$ if $p \ge d$. The Sobolev–Poincaré–Wirtinger inequality (Theorem 6.5) then shows that $(v_h)_{h\in\mathcal{H}}$ is bounded in $L^r(\Omega)$. Moreover, letting $\theta \in (0,1)$ be such that $\frac{1}{q} = \frac{\theta}{r} + \frac{1-\theta}{p}$, by Hölder's inequality on $|v_h - v|^q = |v_h - v|^{\theta q}|v_h - v|^{(1-\theta)q}$ with exponents $\frac{r}{\theta q}$ and $\frac{p}{(1-\theta)q}$, we have

$$\|v_h - v\|_{L^q(\Omega)} \le \|v_h - v\|^{\theta}_{L^r(\Omega)}\|v_h - v\|^{1-\theta}_{L^p(\Omega)}.$$

Together with the convergence of v_h to v in $L^p(\Omega)$ and its boundedness in $L^r(\Omega)$, this estimate shows that $v_h \to v$ in $L^q(\Omega)$ as $h \to 0$.

Using (6.102) and recalling that $\gamma_{\mathfrak{M}_h}\underline{\mathfrak{w}}_h = \gamma_{\mathfrak{M}_h}\underline{\Pi}_h\underline{v}_h$ converges to ω in $L^p(\partial\Omega)$, we also have $\gamma_h\underline{v}_h \to \omega$ in $L^p(\partial\Omega)$ as $h \to 0$, along the same subsequence as before.

We now turn to the convergence of $\mathrm{r}_h^{k+1}\underline{v}_h$ still assuming, without loss of generality, that $q \ge p$ (the convergence in $L^q(\Omega)$ for $q < p$ follows from the convergence in $L^p(\Omega)$). Since $\int_T (v_h - \mathrm{r}_h^{k+1}\underline{v}_h) = 0$ for all $T \in \mathcal{T}_h$, estimate (6.94) applied to $w = v_h - \mathrm{r}_h^{k+1}\underline{v}_h$ with $r = q$ shows that

$$\begin{aligned}\|v_h - \mathrm{r}_h^{k+1}\underline{v}_h\|_{L^q(\Omega)} &\lesssim h^{1+\frac{d}{q}-\frac{d}{p}}\|\nabla_h v_h - \nabla_h \mathrm{r}_h^{k+1}\underline{v}_h\|_{L^p(\Omega)^d}\\ &\lesssim h^{1+\frac{d}{q}-\frac{d}{p}}\left(\|\underline{v}_h\|_{1,p,h} + \|\underline{v}_h\|_{\nabla\mathrm{r},p,h}\right)\\ &\lesssim h^{1+\frac{d}{q}-\frac{d}{p}}\|\underline{v}_h\|_{1,p,h},\end{aligned}$$

where we have used the definitions (6.9) and (6.18) of the norms $\|\cdot\|_{1,p,h}$ and $\|\cdot\|_{\nabla\mathrm{r},p,h}$ in the second line, and the norm equivalence (6.20) in the third line. The choice of q ensures that $1+\frac{d}{q}-\frac{d}{p}>0$, from which we deduce that $v_h - \mathrm{r}_h^{k+1}\underline{v}_h \to 0$ in $L^q(\Omega)$. This proves that $\mathrm{r}_h^{k+1}\underline{v}_h \to v$ in $L^q(\Omega)$ as $h \to 0$.

It remains to show that $v \in W^{1,p}(\Omega)$, that $\omega = v_{|\partial\Omega}$, and that $\mathbf{G}_h^k\underline{v}_h \rightharpoonup \nabla v$ weakly in $L^p(\Omega)^d$ as $h \to 0$. The norm equivalence (6.20), the boundedness of $\left(\|\underline{v}_h\|_{1,p,h}\right)_{h\in\mathcal{H}}$, and the definition (6.16) of $\|\cdot\|_{\mathbf{G},p,h}$ show that $\mathbf{G}_h^k\underline{v}_h$ is bounded in $L^p(\Omega)^d$. Hence, up to a subsequence, it converges weakly in this space to some $\boldsymbol{G}$. The conclusion follows if we prove that $\boldsymbol{G} = \nabla v$ in the sense of distributions on Ω, and that $\omega = v_{|\partial\Omega}$.

Take $\boldsymbol{\tau} \in C^\infty(\overline{\Omega})^d$ and write

$$
\begin{aligned}
\int_\Omega \mathbf{G}_h^k\underline{v}_h\cdot\boldsymbol{\tau} &= \sum_{T\in\mathcal{T}_h}\int_T \mathbf{G}_T^k\underline{v}_T\cdot\boldsymbol{\tau} \\
&= \sum_{T\in\mathcal{T}_h}\int_T \nabla v_T\cdot\boldsymbol{\tau} + \sum_{T\in\mathcal{T}_h}\sum_{F\in\mathcal{F}_T}\int_F (v_F - v_T)\,(\boldsymbol{\pi}_T^{0,k}\boldsymbol{\tau}\cdot\boldsymbol{n}_{TF}) \\
&= \sum_{T\in\mathcal{T}_h}\int_T \nabla v_T\cdot\boldsymbol{\tau} + \sum_{T\in\mathcal{T}_h}\sum_{F\in\mathcal{F}_T}\int_F (v_F - v_T)\,(\boldsymbol{\tau}\cdot\boldsymbol{n}_{TF}) \\
&\quad + \underbrace{\sum_{T\in\mathcal{T}_h}\sum_{F\in\mathcal{F}_T}\int_F (v_F - v_T)\,(\boldsymbol{\pi}_T^{0,k}\boldsymbol{\tau} - \boldsymbol{\tau})\cdot\boldsymbol{n}_{TF}}_{\mathfrak{T}_{1,h}}
\end{aligned}
$$

where the second line is a consequence of the property (4.41) of $\mathbf{G}_T^k$, and the third equality is obtained introducing $\pm\boldsymbol{\tau}$ in the boundary terms. Continuing with element-wise integration by parts, we find

$$
\begin{aligned}
\int_\Omega \mathbf{G}_h^k\underline{v}_h\cdot\boldsymbol{\tau} &= -\sum_{T\in\mathcal{T}_h}\int_T v_T\,(\nabla\cdot\boldsymbol{\tau}) + \sum_{T\in\mathcal{T}_h}\sum_{F\in\mathcal{F}_T}\int_F v_F\,(\boldsymbol{\tau}\cdot\boldsymbol{n}_{TF}) + \mathfrak{T}_{1,h} \\
&= -\int_\Omega v_h\,(\nabla\cdot\boldsymbol{\tau}) + \int_{\partial\Omega}\gamma_h\underline{v}_h\,(\boldsymbol{\tau}\cdot\boldsymbol{n}_{\partial\Omega}) + \mathfrak{T}_{1,h},
\end{aligned}
\tag{6.103}
$$

the conclusion being a consequence of formula (1.27) in Corollary 1.19 with $(\varphi_F)_{F\in\mathcal{F}_h} = (v_F)_{F\in\mathcal{F}_h}$, together with, by definition (6.10) of γ_h,

$$
\sum_{F\in\mathcal{F}_h^{\mathrm{b}}}\int_F (\boldsymbol{\tau}\cdot\boldsymbol{n}_F)v_F = \int_{\partial\Omega}\gamma_h\underline{v}_h\,(\boldsymbol{\tau}\cdot\boldsymbol{n}_{\partial\Omega}).
$$

We now deal with $\mathfrak{T}_{1,h}$. Using the generalised Hölder inequality with exponents (p, p', ∞) on the integrals over F together with $\|\boldsymbol{n}_{TF}\|_{L^\infty(F)^d} = 1$ and the trace approximation property (1.75) (with $(l, s, m) = (k, k+1, 0)$ and p' instead of p), we write

$$
\begin{aligned}
|\mathfrak{T}_{1,h}| &\lesssim \sum_{T\in\mathcal{T}_h}\sum_{F\in\mathcal{F}_T} \|v_F - v_T\|_{L^p(F)} h_T^{k+1-\frac{1}{p'}} \|\boldsymbol{\tau}\|_{W^{k+1,p'}(T)^d} \\
&\lesssim \sum_{T\in\mathcal{T}_h}\sum_{F\in\mathcal{F}_T} h_F^{\frac{1}{p}-1} \|v_F - v_T\|_{L^p(F)} h_T^{k+1} \|\boldsymbol{\tau}\|_{W^{k+1,p'}(T)^d} \\
&\lesssim h^{k+1}\left(\sum_{T\in\mathcal{T}_h}\sum_{F\in\mathcal{F}_T} h_F^{1-p}\|v_F - v_T\|_{L^p(F)}^p\right)^{\frac{1}{p}} \|\boldsymbol{\tau}\|_{W^{k+1,p'}(\Omega)^d} \\
&\lesssim h^{k+1}\|\underline{v}_h\|_{1,p,h}\|\boldsymbol{\tau}\|_{W^{k+1,p'}(\Omega)^d},
\end{aligned}
$$

where the second line follows from $-\frac{1}{p'} = \frac{1}{p} - 1$ and the uniform equivalence (1.6) of face and element diameters, the third line is obtained applying a Hölder inequality on the sum, and we have used the definition (6.9) of $\|\cdot\|_{1,p,h}$ to conclude. This estimate proves that $\mathfrak{T}_{1,h} \to 0$.

Coming back to (6.103) and using $v_h \to v$ in $L^p(\Omega)$, $\gamma_h \underline{v}_h \to \omega$ in $L^p(\partial\Omega)$ and $\mathbf{G}_h^k \underline{v}_h \rightharpoonup \boldsymbol{G}$ weakly in $L^p(\Omega)^d$, we deduce that

$$
\int_\Omega \boldsymbol{G}\cdot\boldsymbol{\tau} = -\int_\Omega v\,(\nabla\cdot\boldsymbol{\tau}) + \int_{\partial\Omega} \omega\,(\boldsymbol{\tau}\cdot\boldsymbol{n}_{\partial\Omega}).
$$

Specifying $\boldsymbol{\tau} \in C_c^\infty(\Omega)^d$ in this formula, the boundary term disappears and the equation shows that $\boldsymbol{G} = \nabla v$ in the sense of distributions. Taking then any $\boldsymbol{\tau} \in C^\infty(\overline{\Omega})$, we obtain

$$
\begin{aligned}
\int_{\partial\Omega} \omega\,(\boldsymbol{\tau}\cdot\boldsymbol{n}_{\partial\Omega}) &= \int_\Omega \boldsymbol{G}\cdot\boldsymbol{\tau} + \int_\Omega v\,(\nabla\cdot\boldsymbol{\tau}) \\
&= \int_\Omega \nabla v\cdot\boldsymbol{\tau} + \int_\Omega v\,(\nabla\cdot\boldsymbol{\tau}) = \int_{\partial\Omega} v_{|\partial\Omega}\,(\boldsymbol{\tau}\cdot\boldsymbol{n}_{\partial\Omega}),
\end{aligned}
$$

where the conclusion follows from an integration by parts. The generic nature of $\boldsymbol{\tau}$ enables us to conclude that $\omega = v_{|\partial\Omega}$, and the proof is complete. □

The following compactness result is a special case of [178, Lemma B.27].

Lemma 6.39 (Discrete Compactness in $\underline{\mathfrak{U}}_{h,\star}^0$) *Let $(\mathfrak{M}_h)_{h\in\mathcal{H}}$ be a regular family of matching simplicial meshes and let $p \in (1,\infty)$. Let $(\underline{\mathfrak{w}}_h)_{h\in\mathcal{H}} \in (\underline{\mathfrak{U}}_{h,\star}^0)_{h\in\mathcal{H}}$ and assume that $\left(|||\underline{\mathfrak{w}}_h|||_{1,p,h}\right)_{h\in\mathcal{H}}$ is bounded. Then, $(\mathfrak{w}_h)_{h\in\mathcal{H}}$ is relatively compact in $L^p(\Omega)$ and, defining the trace operator $\gamma_{\mathfrak{M}_h} : \underline{\mathfrak{U}}_h^0 \to L^p(\partial\Omega)$ by (6.100) with $k=0$, $(\gamma_{\mathfrak{M}_h}\underline{\mathfrak{w}}_h)_{h\in\mathcal{H}}$ is relatively compact in $L^p(\partial\Omega)$.*

6.6 Discrete Functional Analysis for Homogeneous Dirichlet Boundary Conditions

To adapt the convergence analysis of Sect. 6.4 to homogeneous Dirichlet boundary conditions, Theorems 6.5 and 6.8 have to be modified as described in this section. The results stated in the present section will be useful in Chap. 9 on Navier–Stokes equations.

Theorem 6.40 (Discrete Sobolev–Poincaré Inequality) *Let a polynomial degree $k \geq 0$ and an index $p \in (1, \infty)$ be fixed. Let $(\mathcal{M}_h)_{h\in\mathcal{H}}$ denote a regular sequence of meshes in the sense of Definition 1.9. Let $1 \leq q \leq \frac{dp}{d-p}$ if $1 \leq p < d$, and $1 \leq q < \infty$ if $p \geq d$. Then, for all $\underline{v}_h \in \underline{U}^k_{h,0}$, with $\underline{U}^k_{h,0}$ defined by* (2.36),

$$\|v_h\|_{L^q(\Omega)} \lesssim \|\underline{v}_h\|_{1,p,h},$$

where the hidden multiplicative constant depends only on Ω, ϱ, k, p, and q.

Theorem 6.41 (Discrete Compactness for Homogeneous Dirichlet Boundary Conditions) *Let a polynomial degree $k \geq 0$ and an index $p \in (1, \infty)$ be fixed. Let $(\mathcal{M}_h)_{h\in\mathcal{H}}$ denote a regular sequence of meshes in the sense of Definition 1.9. Let $(\underline{v}_h)_{h\in\mathcal{H}} \in \left(\underline{U}^k_{h,0}\right)_{h\in\mathcal{H}}$ (with $\underline{U}^k_{h,0}$ defined by* (2.36)*) be a sequence for which there exists a real number $C > 0$ independent of h such that*

$$\|\underline{v}_h\|_{1,p,h} \leq C \qquad \forall h \in \mathcal{H}.$$

Then, there exists $v \in W^{1,p}_0(\Omega)$ such that, up to a subsequence as $h \to 0$,

(i) $v_h \to v$ and $\mathrm{r}^{k+1}_h \underline{v}_h \to v$ strongly in $L^q(\Omega)$ for all $1 \leq q < \frac{dp}{d-p}$ if $1 \leq p < d$, and $1 \leq q < \infty$ if $p \geq d$;
(ii) $\mathbf{G}^k_h \underline{v}_h \rightharpoonup \nabla v$ weakly in $L^p(\Omega)^d$.

These theorems can be established following the same ideas as in Sect. 6.5, using the operator $\underline{\Pi}_h$ that sends $\underline{U}^k_{h,0}$ into

$$\underline{\mathfrak{U}}^0_{h,0} := \left\{\underline{\mathfrak{w}}_h \in \underline{\mathfrak{U}}^0_h \,:\, \mathfrak{w}_\sigma = 0 \quad \forall \sigma \in \mathfrak{F}^{\mathrm{b}}_h\right\}$$

and invoking the lemmas below, special cases of [178, Lemmas B.15 and B.19]. We let the reader go over the details of these proofs.

Lemma 6.42 (Discrete Sobolev Embedding for $\underline{\mathfrak{U}}_{h,0}^0$) *Let $(\mathfrak{M}_h)_{h\in\mathcal{H}}$ be a regular sequence of matching simplicial meshes, and let $p \in [1,\infty)$. Then, for all $1 \le q \le \frac{dp}{d-p}$ if $p < d$, and $1 \le q < \infty$ if $p \ge d$, it holds*

$$\|\mathfrak{w}_h\|_{L^q(\Omega)} \lesssim |||\underline{\mathfrak{w}}_h|||_{1,p,h} \qquad \forall \underline{\mathfrak{w}}_h \in \underline{\mathfrak{U}}_{h,0}^0,$$

with hidden constant depending only on Ω, ϱ, p, and q.

Lemma 6.43 (Discrete Compactness in $\underline{\mathfrak{U}}_{h,0}^0$) *Let $(\mathfrak{M}_h)_{h\in\mathcal{H}}$ be a regular family of matching simplicial meshes and let $p \in (1,\infty)$. Let $(\underline{\mathfrak{w}}_h)_{h\in\mathcal{H}} \in \left(\underline{\mathfrak{U}}_{h,0}^0\right)_{h\in\mathcal{H}}$ and assume that $\left(|||\underline{\mathfrak{w}}_h|||_{1,p,h}\right)_{h\in\mathcal{H}}$ is bounded. Then, $(\mathfrak{w}_h)_{h\in\mathcal{H}}$ is relatively compact in $L^p(\Omega)$.*

Chapter 7
Linear Elasticity

In this chapter, we discuss HHO discretisations of linear elasticity. This problem, central in solid mechanics, is encountered when modelling the (small) deformations of a body under a volumetric load. From the mathematical point of view, there are relevant differences with respect to the Poisson problem discussed in Chap. 2, both at the continuous and at the discrete level. The first, obvious, difference is that, in this case, the unknown is vector-valued. The second, far-reaching, difference is that the key differential operator is the symmetric part of the gradient which, applied to the displacement field, yields the infinitesimal strain tensor. As a consequence, well-posedness for the continuous problem hinges on the Korn inequality, which states that, for homogeneous Dirichlet boundary conditions, the L^2-norm of the gradient is controlled by the L^2-norm of its symmetric part.

In Sect. 7.1, we discuss the model. After introducing some tensor-related notations, defining the symmetric and skew-symmetric parts of the gradient of a vector field, and discussing rigid-body motions, we state the linear elasticity problem along with its weak formulation. To study the well-posedness of the continuous weak problem, we recall and prove the first Korn inequality, a result which will be later mimicked at the discrete level in Lemmas 7.23 and 7.24.

In Sect. 7.2, we discuss the local construction. We start by introducing the strain projector which, for polynomial degrees >1, differs from applying the elliptic projector of Definition 1.39 component-wise in that the symmetric part of the gradient replaces the gradient. Adapting the theory of Sect. 1.3.2, we show that this projector has optimal approximation properties. The proof hinges on uniform local Korn inequalities inside mesh elements. We next identify inspiring relations which serve as a starting point for the choice of the discrete unknowns in the HHO method and for the definition of two local reconstruction operators, one for the gradient of the displacement and another for the displacement itself. These ingredients are combined to formulate a local contribution composed, as usual, of consistency and stabilisation terms. The approximation properties of the strain projector are used here to design a particular stabilisation term that satisfies the required assumptions.

D. A. Di Pietro, J. Droniou, *The Hybrid High-Order Method for Polytopal Meshes*,
MS&A 19, https://doi.org/10.1007/978-3-030-37203-3_7

A crucial difference with respect to the Poisson problem is that, in order to match the stability assumption, one needs to use polynomials of degree $k \geq 1$, that is, the lowest-order case $k = 0$ requires a specific treatment.

In Sect. 7.3 we formulate the HHO scheme. More specifically, after defining the global space of discrete unknowns with single-valued interface values, we prove discrete global Korn inequalities in broken polynomial and HHO spaces, define the global bilinear form, state the discrete problem, and prove its well-posedness. The error analysis is carried out in Sect. 7.4 based on the abstract analysis framework of Appendix A. We prove error estimates for both the energy- and L^2-norms of the displacement. To close Sect. 7.4, we discuss the robustness of our estimates in the case when Lamé's first coefficient takes large values, corresponding to quasi-incompressible bodies which deform at constant volume. In this case, it is well-known that lowest-order conforming Finite Elements approximations do not deliver satisfactory approximations owing to their inability to represent non-trivial divergence-free fields; see, e.g., [31]. This phenomenon is often referred to as *numerical locking* in the literature. From a mathematical point of view, this can be avoided by making sure that the error estimates are uniform in Lamé's first coefficient, which we show to be the case for the HHO method studied here.

In Sect. 7.6 we hack the HHO scheme to cover the case $k = 0$, which can be useful to reduce the number of unknowns in large three-dimensional simulations, or whenever the solution is not expected to be smooth. As previously observed, it is not possible in this case to attain stability by means of a local stabilisation term devised inside each mesh element. The solution proposed here consists in adding a jump penalisation term inspired by the discrete Korn inequality in broken polynomial spaces proved in Lemma 7.23, which restores coercivity. The price to pay is that, owing to the presence of the jump penalisation term, additional links are introduced among element-based unknowns, and static condensation is no longer an interesting option. Notice that, despite this fact, a significant cost reduction is achieved taking $k = 0$ in three-dimensional cases when compared to $k = 1$.

While most of the chapter focuses on (homogeneous) Dirichlet boundary conditions corresponding to a clamped boundary, other boundary conditions are briefly discussed in Sect. 7.5.

Finally, in Sect. 7.7 we provide a proof of the uniform local second Korn inequality stated in Lemma 7.7. This inequality plays a key role in proving optimal approximation properties for the strain projector as well as the coercivity of the discrete bilinear form; see Remark 7.10.

7.1 Model

In this section we discuss the continuous setting.

7.1.1 Notations and Concepts Related to Tensors

The tensor product of two vectors $\boldsymbol{a} = (a_i)_{1\le i\le d}$ and $\boldsymbol{b} = (b_i)_{1\le i\le d}$ in $\mathbb{R}^d$ is the matrix $\boldsymbol{a}\otimes\boldsymbol{b} \in \mathbb{R}^{d\times d}$ given by

$$\boldsymbol{a}\otimes\boldsymbol{b} := (a_i b_j)_{1\le i,j\le d}. \tag{7.1}$$

The Frobenius inner product in $\mathbb{R}^{d\times d}$ is defined by: For all $\boldsymbol{\sigma} = (\sigma_{ij})_{1\le i,j\le d} \in \mathbb{R}^{d\times d}$ and all $\boldsymbol{\tau} = (\tau_{ij})_{1\le i,j\le d} \in \mathbb{R}^{d\times d}$,

$$\boldsymbol{\sigma}:\boldsymbol{\tau} := \sum_{i,j=1}^{d} \sigma_{ij}\tau_{ij}. \tag{7.2}$$

The associated norm is obtained setting, for all $\boldsymbol{\tau} \in \mathbb{R}^{d\times d}$, $|\boldsymbol{\tau}| := (\boldsymbol{\tau}:\boldsymbol{\tau})^{\frac{1}{2}}$. This inner product naturally carries out to the space $L^2(\Omega)^{d\times d}$ of square-integrable matrix-valued functions by: For $\boldsymbol{\sigma}$ and $\boldsymbol{\tau}$ in $L^2(\Omega)^{d\times d}$,

$$(\boldsymbol{\sigma},\boldsymbol{\tau}) := \int_\Omega \boldsymbol{\sigma}(\boldsymbol{x}):\boldsymbol{\tau}(\boldsymbol{x})\,\mathrm{d}\boldsymbol{x}.$$

The corresponding norm is obtained setting, for all $\boldsymbol{\tau} \in L^2(\Omega)^{d\times d}$, $\|\boldsymbol{\tau}\| := (\boldsymbol{\tau},\boldsymbol{\tau})^{\frac{1}{2}}$. Coherently with Remark 1.14, these notations are extended as $(\cdot,\cdot)_X$ and $\|\cdot\|_X$ when the domain is no longer Ω but a measurable set X.

The set of symmetric real-valued matrices of size d is denoted by $\mathbb{R}^{d\times d}_{\mathrm{sym}}$, and we denote by $L^2(\Omega;\mathbb{R}^{d\times d}_{\mathrm{sym}})$ the space of square-integrable functions that take values in $\mathbb{R}^{d\times d}_{\mathrm{sym}}$.

Finally, the divergence of a sufficiently smooth tensor-valued function $\boldsymbol{\sigma}:\Omega\to L^2(\Omega)^{d\times d}$ is taken row-wise: If $\boldsymbol{\sigma} = (\sigma_{ij})_{1\le i,j\le d}$, then

$$\nabla\cdot\boldsymbol{\sigma} = ((\nabla\cdot\boldsymbol{\sigma})_i)_{1\le i\le d} \text{ with } (\nabla\cdot\boldsymbol{\sigma})_i = \nabla\cdot(\boldsymbol{\sigma}_{i\bullet}) = \sum_{j=1}^{d} \partial_j\sigma_{ij}, \tag{7.3}$$

with $\boldsymbol{\sigma}_{i\bullet}$ denoting the ith row of $\boldsymbol{\sigma}$.

7.1.2 Symmetric and Skew-Symmetric Gradients, Rigid-Body Motions

Let $d \in \{2,3\}$, and denote by $X \subset \mathbb{R}^d$ a polytopal set in the sense of Definition 1.1. We define the tensor-valued gradient operator $\nabla : H^1(X)^d \to L^2(X)^{d\times d}$ acting on vector-valued functions as follows: For any $\boldsymbol{v} = (v_1, \ldots, v_d) \in H^1(X)^d$,

$$\nabla \boldsymbol{v} := (\partial_j v_i)_{1 \le i,j \le d} = \begin{pmatrix} \partial_1 v_1 & \ldots & \partial_d v_1 \\ \vdots & \ddots & \vdots \\ \partial_1 v_d & \ldots & \partial_d v_d \end{pmatrix},$$

where ∂_j denotes the weak derivative with respect to the jth variable. The symmetric and skew-symmetric parts of the gradient (in short, symmetric and skew-symmetric gradients, respectively) are such that, for all $\boldsymbol{v} \in H^1(X)^d$,

$$\nabla_{\mathrm{s}} \boldsymbol{v} := \frac{1}{2}\left(\nabla \boldsymbol{v} + (\nabla \boldsymbol{v})^\top\right), \qquad \nabla_{\mathrm{ss}} \boldsymbol{v} := \frac{1}{2}\left(\nabla \boldsymbol{v} - (\nabla \boldsymbol{v})^\top\right), \tag{7.4}$$

where $(\nabla \boldsymbol{v})^\top$ denotes the transpose of the matrix-valued function $\nabla \boldsymbol{v}$. We notice for future usage that, if $\boldsymbol{v} \in H^1(X)^d$ and $\boldsymbol{\tau} \in L^2(X, \mathbb{R}^{d\times d}_{\mathrm{sym}})$, then

$$(\nabla \boldsymbol{v}, \boldsymbol{\tau})_X = (\nabla_{\mathrm{s}} \boldsymbol{v}, \boldsymbol{\tau})_X. \tag{7.5}$$

This follows from the relation, valid for a.e. $\boldsymbol{x} \in X$:

$$\begin{aligned}
\nabla \boldsymbol{v}(\boldsymbol{x}) : \boldsymbol{\tau}(\boldsymbol{x}) &= \sum_{i,j=1}^{d} \partial_j v_i(\boldsymbol{x}) \tau_{ij}(\boldsymbol{x}) \\
&= \frac{1}{2} \sum_{i,j=1}^{d} \partial_j v_i(\boldsymbol{x}) \tau_{ij}(\boldsymbol{x}) + \frac{1}{2} \sum_{i,j=1}^{d} \partial_j v_i(\boldsymbol{x}) \tau_{ji}(\boldsymbol{x}) \\
&= \frac{1}{2} \sum_{i,j=1}^{d} \partial_j v_i(\boldsymbol{x}) \tau_{ij}(\boldsymbol{x}) + \frac{1}{2} \sum_{i,j=1}^{d} \partial_i v_j(\boldsymbol{x}) \tau_{ij}(\boldsymbol{x}) \\
&= \sum_{i,j=1}^{d} \frac{1}{2} \left(\partial_j v_i(\boldsymbol{x}) + \partial_i v_j(\boldsymbol{x})\right) \tau_{ij}(\boldsymbol{x}) = \nabla_{\mathrm{s}} \boldsymbol{v}(\boldsymbol{x}) : \boldsymbol{\tau}(\boldsymbol{x}),
\end{aligned}$$

where the symmetry of $\boldsymbol{\tau}$ was used in the second line, and the exchange of indices $i \leftrightarrow j$ was performed in the second sum in the third line. The kernel of the symmetric gradient is the space of *rigid-body motions*:

$$\mathbb{RM}_d(X) := \ker(\nabla_{\mathrm{s}}) = \left\{ \boldsymbol{v} \in H^1(X)^d \;:\; \nabla_{\mathrm{s}} \boldsymbol{v} = \boldsymbol{0} \right\}.$$

It can be checked that $\boldsymbol{v} \in \mathbb{RM}_d(X)$ if and only if there exists a vector $\boldsymbol{t}_{\boldsymbol{v}} \in \mathbb{R}^d$ and a skew-symmetric matrix $\boldsymbol{R}_{\boldsymbol{v}} \in \mathbb{R}^{d\times d}$ such that, for almost every $\boldsymbol{x} \in X$,

$$\boldsymbol{v}(\boldsymbol{x}) = \boldsymbol{t}_{\boldsymbol{v}} + \boldsymbol{R}_{\boldsymbol{v}} \boldsymbol{x}. \tag{7.6}$$

If $\boldsymbol{v}$ is a displacement field, the first term in this expression represents a translation. In dimension $d = 2$, the second term is a rotation of angle $\pi/2$ around the origin composed by a (positive or negative) dilatation. In dimension $d = 3$, it can be seen that there exists $\boldsymbol{\omega}_{\boldsymbol{v}} \in \mathbb{R}^3$ such that $\boldsymbol{R}_{\boldsymbol{v}}\boldsymbol{x} = \boldsymbol{\omega}_{\boldsymbol{v}} \times \boldsymbol{x}$, where $\times$ denotes the cross-product of two vectors in $\mathbb{R}^3$. An immediate consequence of this characterisation of rigid-body motions is that $\mathbb{RM}_d(X)$ is a vector space of dimension 3 when $d = 2$ and 6 when $d = 3$ and, in both cases, it holds that $\mathbb{RM}_d(X) \subset \mathbb{P}^1(X)^d$.

7.1.3 The Elasticity Problem

Consider a body which, in its reference configuration, occupies a region of space corresponding to the domain $\Omega \subset \mathbb{R}^d$, $d \in \{2, 3\}$. We are interested in finding the displacement $\boldsymbol{u} : \Omega \to \mathbb{R}^d$ of the body when it is subjected to a given force per unit volume $\boldsymbol{f} : \Omega \to \mathbb{R}^d$. We work, in what follows, under the small deformations assumption which implies, in particular, that the strain tensor is given by

$$\boldsymbol{\varepsilon} = \boldsymbol{\nabla}_{\mathrm{s}}\boldsymbol{u}.$$

We further assume, for the sake of simplicity, that the body is clamped along its boundary, so that the displacement is zero on $\partial\Omega$. Other boundary conditions are briefly discussed in Sect. 7.5. The displacement field is then obtained solving the following *linear elasticity problem*, which expresses the equilibrium between internal stresses and external loads: Find $\boldsymbol{u} : \Omega \to \mathbb{R}^d$ such that

$$-\boldsymbol{\nabla}{\cdot}(\boldsymbol{\sigma}(\boldsymbol{\nabla}_{\mathrm{s}}\boldsymbol{u})) = \boldsymbol{f} \quad \text{in } \Omega, \tag{7.7a}$$

$$\boldsymbol{u} = \boldsymbol{0} \quad \text{on } \partial\Omega, \tag{7.7b}$$

where the linear mapping $\boldsymbol{\sigma} : \mathbb{R}^{d\times d}_{\mathrm{sym}} \to \mathbb{R}^{d\times d}_{\mathrm{sym}}$ represents the strain-stress law. For isotropic but not necessarily homogeneous media, the strain-stress law is such that, for any $\boldsymbol{\tau} \in \mathbb{R}^{d\times d}_{\mathrm{sym}}$,

$$\boldsymbol{\sigma}(\boldsymbol{\tau}) = 2\mu\boldsymbol{\tau} + \lambda\,\mathrm{tr}(\boldsymbol{\tau})\boldsymbol{I}_d, \tag{7.8}$$

where $\mathrm{tr}(\boldsymbol{\tau}) := \sum_{i=1}^{d} \tau_{ii}$ is the trace of $\boldsymbol{\tau}$ and $\boldsymbol{I}_d$ is the $d \times d$ identity matrix. The two real-valued *Lamé coefficients* $\lambda : \Omega \to \mathbb{R}$ and $\mu : \Omega \to \mathbb{R}$ are such that, for given real numbers $0 < \underline{\mu} \le \overline{\mu}$, $\alpha > 0$, and $\overline{\lambda} \ge 0$ it holds, for almost every $\boldsymbol{x} \in \Omega$,

$$\underline{\mu} \le \mu(\boldsymbol{x}) \le \overline{\mu}, \quad \alpha \le 2\mu(\boldsymbol{x}) - d\lambda^-(\boldsymbol{x}) \quad \text{and} \quad |\lambda(\boldsymbol{x})| \le \overline{\lambda}, \tag{7.9}$$

where $\lambda^- := \frac{1}{2}(|\lambda| - \lambda) = \max(-\lambda, 0)$ denotes the negative part of λ. Clearly, if $\lambda \ge 0$, then the second condition in (7.9) is a consequence of the first one with $\alpha = 2\underline{\mu}$.

In what follows, the Lamé coefficients, the related bounds (7.9), and the body force per unit volume f will be collectively referred to as the *problem data*. When relevant, we will track the dependencies on the problem data of the multiplicative constants in error estimates in order to highlight the robustness properties of the HHO method.

Throughout the rest of this chapter, we make the following additional assumption concerning the spatial dependence of the Lamé coefficients:

Assumption 7.1 (Piecewise Constant Lamé Coefficients) *The Lamé coefficients are piecewise constant on a finite collection $P_\Omega := \{\Omega_i\}_{i\in I}$, $I \subset \mathbb{N}$, of disjoint polytopes Ω_i such that $\overline{\Omega} = \bigcup_{i\in I} \overline{\Omega_i}$, i.e.,*

$$\mu_{|\Omega_i} \in \mathbb{P}^0(\Omega_i) \text{ and } \lambda_{|\Omega_i} \in \mathbb{P}^0(\Omega_i) \text{ for all } i \in I.$$

Assumption 7.1 is often verified in practice, and corresponds to the case of a heterogeneous body composed of homogeneous materials. The extension to the more general case of Lamé coefficients that are piecewise smooth is possible in the spirit of Sect. 4.2.

Remark 7.2 (Two-dimensional Elasticity) The mechanical theory of linear elasticity is inherently three-dimensional. As a result, the interpretation of problem (7.7) for $d = 2$ requires some care. Denote by $\boldsymbol{\sigma} = (\sigma_{ij})_{1\le i,j\le d}$ the value of the stress tensor at a point, and let us consider, for instance, the plane stress problem, for which $\sigma_{13} = \sigma_{23} = \sigma_{33} = 0$, and the non-zero components of $\boldsymbol{\sigma}$ depend only on x_1 and x_2. Further assume, for the sake of simplicity, that both λ and μ are constant over Ω. Enforcing $\sigma_{33} = 0$ in (7.8) with $\boldsymbol{\tau} = \boldsymbol{\varepsilon}$ reveals that the component ε_{33} of the strain tensor cannot be zero in general, but is instead equal to $-\frac{\lambda}{2\mu+\lambda}(\varepsilon_{11} + \varepsilon_{22})$. Plugging this relation into the constitutive law leads us again to a problem of the form (7.7) with $d = 2$, but where the quantity $\lambda\frac{2\mu}{2\mu+\lambda}$ replaces Lamé's first coefficient λ in the second term of (7.8). Further developments lead to altogether different mathematical models, such as those encountered in the theory of thin plates and shells. Since our book focuses on the numerical approximation, we will not develop further this topic, and refer the interested reader to textbooks in mechanics such as, e.g., [204]. An HHO method for the approximation of Kirchhoff–Love plates is discussed in [60]; see also [19] for related developments in the context of nonconforming Virtual Elements.

7.1.4 Weak Formulation and Well-Posedness

Assume $f \in L^2(\Omega)^d$. Classically, a weak formulation of problem (7.7) reads: Find $\boldsymbol{u} \in H_0^1(\Omega)^d$ such that

$$a(\boldsymbol{u}, \boldsymbol{v}) = (\boldsymbol{f}, \boldsymbol{v}), \tag{7.10}$$

with bilinear form $a : H^1(\Omega)^d \times H^1(\Omega)^d \to \mathbb{R}$ defined by

$$a(\boldsymbol{u}, \boldsymbol{v}) := (\boldsymbol{\sigma}(\nabla_{\mathrm{s}}\boldsymbol{u}), \nabla_{\mathrm{s}}\boldsymbol{v}) = (2\mu\nabla_{\mathrm{s}}\boldsymbol{u}, \nabla_{\mathrm{s}}\boldsymbol{v}) + (\lambda\nabla{\cdot}\boldsymbol{u}, \nabla{\cdot}\boldsymbol{v}), \tag{7.11}$$

where we have expanded the stress tensor according to its definition (7.8) to conclude. The well-posedness of problem (7.10) hinges on the first Korn inequality, which states that the L^2-norm of the symmetric part of the gradient controls the L^2-norm of the gradient for functions in $H_0^1(\Omega)^d$.

Proposition 7.3 (First Korn Inequality) *For all* $\boldsymbol{v} \in H_0^1(\Omega)^d$,

$$\|\nabla\boldsymbol{v}\| \le \sqrt{2}\|\nabla_{\mathrm{s}}\boldsymbol{v}\|. \tag{7.12}$$

Proof For any $\boldsymbol{v} \in C_{\mathrm{c}}^{\infty}(\Omega)^d$, expanding the symmetric gradient according to its definition (7.4), it is inferred that

$$2\int_\Omega |\nabla_{\mathrm{s}}\boldsymbol{v}|^2 = \int_\Omega |\nabla\boldsymbol{v}|^2 + \int_\Omega \nabla\boldsymbol{v}{:}(\nabla\boldsymbol{v})^{\intercal}.$$

Using an integration by parts and the fact that $\boldsymbol{v} = \boldsymbol{0}$ on $\partial\Omega$, for all $1 \le i, j \le d$, we have

$$\int_\Omega \partial_j v_i\ \partial_i v_j = -\int_\Omega v_i\ \partial_j\partial_i v_j = -\int_\Omega v_i\ \partial_i\partial_j v_j = \int_\Omega \partial_i v_i\ \partial_j v_j.$$

Hence, since $\nabla\boldsymbol{v}{:}(\nabla\boldsymbol{v})^{\intercal} = \sum_{i,j=1}^d \partial_j v_i\ \partial_i v_j$,

$$2\int_\Omega |\nabla_{\mathrm{s}}\boldsymbol{v}|^2 = \int_\Omega |\nabla\boldsymbol{v}|^2 + \int_\Omega |\nabla{\cdot}\boldsymbol{v}|^2 \ge \int_\Omega |\nabla\boldsymbol{v}|^2.$$

The conclusion follows taking the square root of this inequality and invoking the density of $C_{\mathrm{c}}^{\infty}(\Omega)^d$ in $H_0^1(\Omega)^d$. □

The first Korn inequality implies, in particular, that the strain seminorm $\|{\cdot}\|_{\boldsymbol{\varepsilon}}$ such that

$$\|\boldsymbol{v}\|_{\boldsymbol{\varepsilon}} := \|\nabla_{\mathrm{s}}\boldsymbol{v}\| \qquad \forall \boldsymbol{v} \in H_0^1(\Omega)^d$$

is a norm on $H_0^1(\Omega)^d$. We notice, in passing, that the extension of this result to more general boundary conditions is not trivial; see, e.g., the discussion in [211].

Proposition 7.4 (Well-Posedness of Problem (7.10)**)** *Problem* (7.10) *is well-posed with a priori estimate*

$$\|\nabla_{\mathrm{s}}\boldsymbol{u}\| \le \frac{\sqrt{2}C_\Omega}{\alpha}\|\boldsymbol{f}\|,$$

where α is given in (7.9) *and* $C_\Omega > 0$ *is the constant of the continuous Poincaré inequality* $\|v\| \le C_\Omega \|\nabla v\|$ *valid for all* $v \in H_0^1(\Omega)$.

Proof We check the assumptions of the Lax–Milgram Lemma 2.20 with $\mathtt{U} = H_0^1(\Omega)^d$, $\mathtt{a} = a$, and $\langle \mathtt{f}, \boldsymbol{v}\rangle_{\mathtt{U}^\star,\mathtt{U}} = (\boldsymbol{f}, \boldsymbol{v})$. It can be easily verified that $H_0^1(\Omega)^d$ equipped with the inner product norm $\|\cdot\|_{\boldsymbol{\varepsilon}}$ is a Hilbert space. Let us check that the bilinear form a is $\mathtt{U}$-coercive and continuous. We first notice that, for almost every $\boldsymbol{x} \in \Omega$, it holds $\nabla\cdot\boldsymbol{v}(\boldsymbol{x}) = \mathrm{tr}(\nabla\boldsymbol{v}(\boldsymbol{x})) = \mathrm{tr}(\nabla_{\mathrm{s}}\boldsymbol{v}(\boldsymbol{x}))$, so that, by the Cauchy–Schwarz inequality on the sum defining the trace,

$$|\nabla\cdot\boldsymbol{v}(\boldsymbol{x})|^2 \le d|\nabla_{\mathrm{s}}\boldsymbol{v}(\boldsymbol{x})|^2. \tag{7.13}$$

As a consequence, continuity holds with constant $(2\overline{\mu} + d|\overline{\lambda}|)$. For the coercivity, using the definition (7.11) of the bilinear form a, we infer

$$\begin{aligned}
a(\boldsymbol{v}, \boldsymbol{v}) &= (2\mu\nabla_{\mathrm{s}}\boldsymbol{v}, \nabla_{\mathrm{s}}\boldsymbol{v}) + (\lambda\nabla\cdot\boldsymbol{v}, \nabla\cdot\boldsymbol{v}) \\
&\ge \int_\Omega 2\mu|\nabla_{\mathrm{s}}\boldsymbol{v}|^2 - \lambda^-|\nabla\cdot\boldsymbol{v}|^2 \\
&\ge \int_\Omega (2\mu - d\lambda^-)|\nabla_{\mathrm{s}}\boldsymbol{v}|^2 \\
&\ge \alpha\|\nabla_{\mathrm{s}}\boldsymbol{v}\|^2,
\end{aligned}$$

where we have used (7.13) to pass to the third line and the definition (7.9) of α to conclude. This relation proves that a is $\mathtt{U}$-coercive with coercivity constant α. To conclude the proof, it suffices to observe that, owing to the continuous Poincaré and Korn (7.12) inequalities, it holds, for any $\boldsymbol{v} \in H_0^1(\Omega)^d$,

$$|(\boldsymbol{f}, \boldsymbol{v})| \le \|\boldsymbol{f}\|\,\|\boldsymbol{v}\| \le C_\Omega\|\boldsymbol{f}\|\,\|\nabla\boldsymbol{v}\| \le \sqrt{2}C_\Omega\|\boldsymbol{f}\|\,\|\nabla_{\mathrm{s}}\boldsymbol{v}\|,$$

which implies that the dual norm of the linear form $\mathtt{f} : \boldsymbol{v} \mapsto (\boldsymbol{f}, \boldsymbol{v})$ is bounded above by $\sqrt{2}C_\Omega\|\boldsymbol{f}\|$. □

7.2 Local Construction

In this section we state a uniform local second Korn inequality, introduce a novel local projector geared towards vector problems, and present the local ingredients underlying the HHO construction: discrete unknowns, reconstruction operators, and local approximation of the global continuous bilinear form defined by (7.11).

Throughout the rest of this chapter, we let $(\mathcal{M}_h)_{h\in\mathcal{H}}$ denote a regular mesh sequence in the sense of Definition 1.9. We additionally make the following assumption.

Assumption 7.5 (Compliant Mesh Sequence) *For all $h \in \mathcal{H}$, we assume that $\mathcal{M}_h$ is* compliant *with the partition P_Ω introduced in Assumption 7.1 in the sense that, for all $T \in \mathcal{T}_h$, there exists a unique index $i \in I$ such that $T \subset \Omega_i$.*

Assumption 7.5 is typically satisfied in the context of computer-assisted modelling, where the different parts corresponding to each subdomain Ω_i are first separately created, and then assembled together to form the complete model. Combining Assumptions 7.1 and 7.5 we have that

$$\mu_{|T} \in \mathbb{P}^0(T) \text{ and } \lambda_{|T} \in \mathbb{P}^0(T) \text{ for all } T \in \mathcal{T}_h. \tag{7.14}$$

For any mesh element $T \in \mathcal{T}_h$, we let $\mu_T := \mu_{|T}$ and $\lambda_T := \lambda_{|T}$ denote the constant values of μ and λ inside T, respectively.

7.2.1 Regular Mesh Sequence with Star-Shaped Elements

Some results established hereafter require the following assumption on the mesh elements, which is slightly stronger than the one introduced in Chap. 1.

Assumption 7.6 (Regular Mesh Sequence with Star-Shaped Elements) *Let $(\mathcal{M}_h)_{h\in\mathcal{H}}$ denote a sequence of polygonal (if $d = 2$) or polyhedral (if $d = 3$) meshes in the sense of Definition 1.4. We assume that there exists a real number $\varrho > 0$ such that $(\mathcal{M}_h)_{h\in\mathcal{H}}$ is regular in the sense of Definition 1.9 with parameter ϱ and, for all $h \in \mathcal{H}$, every mesh element $T \in \mathcal{T}_h$ is* star-shaped *(see Definition 1.40) with respect to every point of a ball of radius ϱh_T.*

Assumption 7.6 is verified, in particular, by regular mesh sequences, in the sense of Definition 1.9, made of convex elements. The following uniform second Korn inequality inside mesh elements plays a key role in the proof of optimal approximation properties for the strain projector (see Theorem 7.9 below), and of the local norm equivalence (7.46) for the HHO stabilisation bilinear form (7.54); see Proposition 7.21 below.

Lemma 7.7 (Uniform Local Second Korn Inequality) *Denoting by $(\mathcal{M}_h)_{h\in\mathcal{H}}$ a mesh sequence satisfying Assumption 7.6 it holds, for all $h \in \mathcal{H}$ and all $T \in \mathcal{T}_h$,*

$$\|\nabla \boldsymbol{u} - \boldsymbol{\pi}_T^{0,0}(\nabla_{\rm ss}\boldsymbol{u})\|_T \lesssim \|\nabla_{\rm s}\boldsymbol{u}\|_T \qquad \forall \boldsymbol{u} \in H^1(T)^d, \tag{7.15}$$

where the symmetric and skew-symmetric parts of the gradient are defined by (7.4) and the hidden constant depends only on d and ϱ.

Proof See Sect. 7.7. □

7.2.2 The Strain Projector

Throughout the rest of this section we work on a fixed mesh element $T \in \mathcal{T}_h$. We study a novel projector for vector-valued functions that will be used to formulate a stabilisation term in the HHO discretisation of problem (7.10). Specifically, for a given integer $l \geq 1$, the *strain projector* $\boldsymbol{\pi}_T^{\boldsymbol{\varepsilon},l} : H^1(T)^d \to \mathbb{P}^l(T)^d$ is such that, for any $\boldsymbol{v} \in H^1(T)^d$,

$$(\boldsymbol{\nabla}_{\mathrm{s}}(\boldsymbol{\pi}_T^{\boldsymbol{\varepsilon},l}\boldsymbol{v} - \boldsymbol{v}), \boldsymbol{\nabla}_{\mathrm{s}}\boldsymbol{w})_T = 0 \qquad \forall \boldsymbol{w} \in \mathbb{P}^l(T)^d. \tag{7.16a}$$

By the Riesz representation theorem in $\boldsymbol{\nabla}_{\mathrm{s}}\mathbb{P}^l(T)^d$ for the inner product of $L^2(T)^{d\times d}$, relation (7.16a) defines a unique element $\boldsymbol{\nabla}_{\mathrm{s}}\boldsymbol{\pi}_T^{\boldsymbol{\varepsilon},l}\boldsymbol{v} \in \boldsymbol{\nabla}_{\mathrm{s}}\mathbb{P}^l(T)^d$, and thus a unique polynomial $\boldsymbol{\pi}_T^{\boldsymbol{\varepsilon},l}\boldsymbol{v}$ up to a rigid-body motion. The latter can be fixed by further prescribing that

$$\int_T \boldsymbol{\pi}_T^{\boldsymbol{\varepsilon},l}\boldsymbol{v} = \int_T \boldsymbol{v}, \qquad \int_T \boldsymbol{\nabla}_{\mathrm{ss}}\boldsymbol{\pi}_T^{\boldsymbol{\varepsilon},l}\boldsymbol{v} = \int_T \boldsymbol{\nabla}_{\mathrm{ss}}\boldsymbol{v}. \tag{7.16b}$$

To see why these relations fix $\boldsymbol{\pi}_T^{\boldsymbol{\varepsilon},l}\boldsymbol{v}$, let us write $\boldsymbol{\pi}_T^{\boldsymbol{\varepsilon},l}\boldsymbol{v} = \boldsymbol{q} + \boldsymbol{w}$ where $\boldsymbol{q}$ is a fixed polynomial in $\mathbb{P}^l(T)^d$ that satisfies (7.16a), and $\boldsymbol{w}$ is an unknown rigid-body motion. The representation (7.6) of $\boldsymbol{w}$ can be recast under the form

$$\boldsymbol{w}(\boldsymbol{x}) = (\boldsymbol{t}_{\boldsymbol{w}} + \boldsymbol{R}_{\boldsymbol{w}}\overline{\boldsymbol{x}}_T) + \boldsymbol{R}_{\boldsymbol{w}}(\boldsymbol{x} - \overline{\boldsymbol{x}}_T),$$

with $\overline{\boldsymbol{x}}_T = \pi_T^{0,0}\boldsymbol{x}$ the centre of mass of T, and the conditions (7.16b) admit straightforward interpretations: the second one prescribes the matrix $\boldsymbol{R}_{\boldsymbol{w}}$ (since $\boldsymbol{R}_{\boldsymbol{w}} = \boldsymbol{\nabla}_{\mathrm{ss}}\boldsymbol{w}$), whereas the first one prescribes $\boldsymbol{t}_{\boldsymbol{w}} + \boldsymbol{R}_{\boldsymbol{w}}\overline{\boldsymbol{x}}_T$. The following characterisation holds:

$$\boldsymbol{\pi}_T^{\boldsymbol{\varepsilon},l}\boldsymbol{v} = \underset{\boldsymbol{w}\in \boldsymbol{V}^l(T)}{\operatorname{argmin}} \, \|\boldsymbol{\nabla}_{\mathrm{s}}(\boldsymbol{w} - \boldsymbol{v})\|_T^2, \tag{7.17}$$

where $\boldsymbol{V}^l(T) := \left\{\boldsymbol{w} \in \mathbb{P}^l(T)^d : \int_T \boldsymbol{w} = \int_T \boldsymbol{v} \text{ and } \int_T \boldsymbol{\nabla}_{\mathrm{ss}}\boldsymbol{w} = \int_T \boldsymbol{\nabla}_{\mathrm{ss}}\boldsymbol{v}\right\}$. As a matter of fact, (7.16a) is nothing but the Euler equation for the minimisation problem (7.17). To check that $\boldsymbol{\pi}_T^{\boldsymbol{\varepsilon},l}$ satisfies the polynomial invariance requirement (1.56) (and thus, by Proposition 1.35, that it meets the conditions of Definition 1.34) it suffices to observe that, if $\boldsymbol{v} \in \mathbb{P}^l(T)^d$, then making $\boldsymbol{w} = \boldsymbol{\pi}_T^{\boldsymbol{\varepsilon},l}\boldsymbol{v} - \boldsymbol{v}$ in (7.16a) implies $\boldsymbol{\nabla}_{\mathrm{s}}(\boldsymbol{\pi}_T^{\boldsymbol{\varepsilon},l}\boldsymbol{v} - \boldsymbol{v}) = \boldsymbol{0}$; hence $(\boldsymbol{\pi}_T^{\boldsymbol{\varepsilon},l}\boldsymbol{v} - \boldsymbol{v})$ is a rigid-body motion inside T, from which we deduce, using (7.16b), that $\boldsymbol{\pi}_T^{\boldsymbol{\varepsilon},l}\boldsymbol{v} - \boldsymbol{v} = \boldsymbol{0}$.

Remark 7.8 (Link with the Elliptic Projector for $l = 1$) When $l = 1$, combining the second condition in (7.16b) and (7.16a) with $\boldsymbol{\nabla}_{\mathrm{s}}\boldsymbol{w}$ replaced by $\boldsymbol{\nabla}\boldsymbol{w}$ (this is possible by virtue of (7.5)), and observing that the latter spans $\mathbb{P}^0(T)^{d\times d}$ (since $\boldsymbol{\nabla}\mathbb{P}^1(T)^d$ coincides with this space), we infer, for any $\boldsymbol{v} \in H^1(T)^d$,

$$(\boldsymbol{\nabla}(\boldsymbol{\pi}_T^{\boldsymbol{\varepsilon},1}\boldsymbol{v} - \boldsymbol{v}), \boldsymbol{\nabla}\boldsymbol{w})_T = 0 \qquad \forall \boldsymbol{w} \in \mathbb{P}^1(T)^d.$$

Combining this result and the first condition in (7.16b), and comparing with the definition (1.60) of the elliptic projector, we conclude that

$$\boldsymbol{\pi}_T^{\boldsymbol{\varepsilon},1} = \boldsymbol{\pi}_T^{1,1}, \tag{7.18}$$

where $\boldsymbol{\pi}_T^{1,1}$ is obtained applying $\pi_T^{1,1}$ component-wise.

We next study the approximation properties of the strain projector, focusing on the Hilbertian case relevant for the developments considered in this chapter. These results are instrumental to proving optimal consistency properties for the local stabilisation bilinear form; see Remark 7.20.

Theorem 7.9 (Approximation Properties of the Strain Projector) *Let a polynomial degree $l \geq 1$ be fixed, denote by $(\mathcal{M}_h)_{h\in\mathcal{H}} = (\mathcal{T}_h, \mathcal{F}_h)_{h\in\mathcal{H}}$ a regular mesh sequence in the sense of Definition 1.9 and, if $l > 1$, further assume that $(\mathcal{M}_h)_{h\in\mathcal{H}}$ satisfies Assumption 7.6. Let an integer $s \in \{1, \ldots, l+1\}$ be given. Then, for all $T \in \mathcal{T}_h$, all $\boldsymbol{v} \in H^s(T)^d$, and all $m \in \{0, \ldots, s\}$,*

$$|\boldsymbol{v} - \boldsymbol{\pi}_T^{\boldsymbol{\varepsilon},l}\boldsymbol{v}|_{H^m(T)^d} \lesssim h_T^{s-m}|\boldsymbol{v}|_{H^s(T)^d}. \tag{7.19}$$

Moreover, if $m \leq s-1$, then, for all $F \in \mathcal{F}_T$,

$$h_T^{\frac{1}{2}}|\boldsymbol{v} - \boldsymbol{\pi}_T^{\boldsymbol{\varepsilon},l}\boldsymbol{v}|_{H^m(F)^d} \lesssim h_T^{s-m}|\boldsymbol{v}|_{H^s(T)^d}. \tag{7.20}$$

The hidden constants in (7.19) and (7.20) depend only on d, ϱ, l, s, and m.

Proof We start by noticing that (7.20) can be deduced from (7.19) proceeding as in Theorem 1.45. It therefore suffices to prove (7.19). If $l = 1$, accounting for (7.18), (7.19) follows from Theorem 1.48, which does not require Assumption 7.6. If $l > 1$, we can apply the abstract results of Lemma 1.43, which readily extend to the vector case. This requires to prove that it holds, for all $\boldsymbol{v} \in H^1(T)^d$,

$$|\boldsymbol{\pi}_T^{\boldsymbol{\varepsilon},l}\boldsymbol{v}|_{H^1(T)^d} \lesssim |\boldsymbol{v}|_{H^1(T)^d}, \qquad \text{if } m \geq 1, \tag{7.21a}$$

$$\|\boldsymbol{\pi}_T^{\boldsymbol{\varepsilon},l}\boldsymbol{v}\|_T \lesssim \|\boldsymbol{v}\|_T + h_T|\boldsymbol{v}|_{H^1(T)^d} \qquad \text{if } m = 0. \tag{7.21b}$$

(i) *The case $m \geq 1$.* We start by observing that Eq. (7.16a) implies

$$\|\boldsymbol{\nabla}_{\rm s}\boldsymbol{\pi}_T^{\boldsymbol{\varepsilon},l}\boldsymbol{v}\|_T \leq \|\boldsymbol{\nabla}_{\rm s}\boldsymbol{v}\|_T, \tag{7.22}$$

as can be easily checked letting $\boldsymbol{w} = \boldsymbol{\pi}_T^{\varepsilon,l}\boldsymbol{v}$ and using a Cauchy–Schwarz inequality. We can now write

$$\begin{aligned}
|\boldsymbol{\pi}_T^{\varepsilon,l}\boldsymbol{v}|_{H^1(T)^d} &\lesssim \|\nabla\boldsymbol{\pi}_T^{\varepsilon,l}\boldsymbol{v}\|_T \\
&\le \|\nabla\boldsymbol{\pi}_T^{\varepsilon,l}\boldsymbol{v} - \boldsymbol{\pi}_T^{0,0}(\nabla_{\rm ss}\boldsymbol{\pi}_T^{\varepsilon,l}\boldsymbol{v})\|_T + \|\boldsymbol{\pi}_T^{0,0}(\nabla_{\rm ss}\boldsymbol{v})\|_T \\
&\lesssim \|\nabla_{\rm s}\boldsymbol{\pi}_T^{\varepsilon,l}\boldsymbol{v}\|_T + \|\boldsymbol{\pi}_T^{0,0}(\nabla_{\rm ss}\boldsymbol{v})\|_T \\
&\lesssim \|\nabla_{\rm s}\boldsymbol{v}\|_T + \|\nabla_{\rm ss}\boldsymbol{v}\|_T \lesssim |\boldsymbol{v}|_{H^1(T)^d},
\end{aligned} \tag{7.23}$$

where we have inserted $\mathbf{0} = \boldsymbol{\pi}_T^{0,0}(\nabla_{\rm ss}\boldsymbol{v}) - \boldsymbol{\pi}_T^{0,0}(\nabla_{\rm ss}\boldsymbol{\pi}_T^{\varepsilon,l}\boldsymbol{v})$ (see (7.16b)) into the norm and used the triangle inequality to pass to the second line, we have used the uniform local Korn inequality (7.15) to pass to the third line, and we have invoked (7.22) together with the boundedness of $\boldsymbol{\pi}_T^{0,0}$ expressed by (1.77) with $l = 0$, $X = T$, $s = 0$, and $p = 2$ to conclude. This proves (7.21a).

(ii) *The case* $m = 0$. We can write

$$\|\boldsymbol{\pi}_T^{\varepsilon,l}\boldsymbol{v}\|_T \le \|\boldsymbol{v}\|_T + \|\boldsymbol{\pi}_T^{\varepsilon,l}\boldsymbol{v} - \boldsymbol{v}\|_T \lesssim \|\boldsymbol{v}\|_T + h_T\|\nabla(\boldsymbol{\pi}_T^{\varepsilon,l}\boldsymbol{v} - \boldsymbol{v})\|_T \lesssim \|\boldsymbol{v}\|_T + h_T|\boldsymbol{v}|_{H^1(T)^d},$$

where we have inserted $\pm\boldsymbol{v}$ into the norm and used the triangle inequality in the first bound, the local Poincaré–Wirtinger inequality (1.76) for the zero-average function $(\boldsymbol{\pi}_T^{\varepsilon,l}\boldsymbol{v} - \boldsymbol{v})$ in the second bound, and we have concluded using the triangle inequality together with (7.23) to write $\|\nabla(\boldsymbol{\pi}_T^{\varepsilon,l}\boldsymbol{v} - \boldsymbol{v})\|_T \le \|\nabla\boldsymbol{\pi}_T^{\varepsilon,l}\boldsymbol{v}\|_T + \|\nabla\boldsymbol{v}\|_T \lesssim |\boldsymbol{v}|_{H^1(T)^d}$. This proves (7.21b). □

Remark 7.10 (Uniform Local Korn Inequalities on Mesh Elements) In the above proof, the uniform local Korn inequality (7.15) is invoked in the case $l > 1$ to pass to the third line in (7.23), for which a crucial requirement is that the hidden constant is independent of the element shape.

7.2.3 *Two Inspiring Relations*

Let a polynomial degree $k \ge 0$ be fixed. In a similar way as in Sect. 2.1.1, we derive two relations that will inspire the choice of the discrete unknowns and the definitions of the local reconstructions.

The starting point in both cases is the following integration by parts formula: For all $\boldsymbol{v} \in H^1(T)^d$ and all $\boldsymbol{\tau} \in C^\infty(\overline{T})^{d\times d}$,

$$(\nabla\boldsymbol{v}, \boldsymbol{\tau})_T = -(\boldsymbol{v}, \nabla\cdot\boldsymbol{\tau})_T + \sum_{F\in\mathcal{F}_T}(\boldsymbol{v}, \boldsymbol{\tau}\boldsymbol{n}_{TF})_F. \tag{7.24}$$

Specialising this formula to $\boldsymbol{\tau} \in \mathbb{P}^k(T)^{d\times d}$, we obtain the first inspiring result:

$$(\boldsymbol{\pi}_T^{0,k}(\nabla \boldsymbol{v}), \boldsymbol{\tau})_T = -(\boldsymbol{\pi}_T^{0,k}\boldsymbol{v}, \nabla\cdot\boldsymbol{\tau})_T + \sum_{F\in\mathcal{F}_T} (\boldsymbol{\pi}_F^{0,k}\boldsymbol{v}, \boldsymbol{\tau}\boldsymbol{n}_{TF})_F, \tag{7.25}$$

where, to insert the L^2-orthogonal projectors, we have used their definition (1.57) after observing that $\nabla\cdot\boldsymbol{\tau} \in \mathbb{P}^{k-1}(T)^d \subset \mathbb{P}^k(T)^d$ and $\boldsymbol{\tau}_{|F}\boldsymbol{n}_{TF} \in \mathbb{P}^k(F)^d$ for all $F \in \mathcal{F}_T$. Relation (7.25) shows that the L^2-orthogonal projection $\boldsymbol{\pi}_T^{0,k}(\nabla\boldsymbol{v})$ of the gradient of $\boldsymbol{v}$ can be computed from the L^2-projections $\boldsymbol{\pi}_T^{0,k}\boldsymbol{v}$ and $(\boldsymbol{\pi}_F^{0,k}\boldsymbol{v})_{F\in\mathcal{F}_T}$. This is a straightforward extension to the vector case of the discussion in Sect. 4.2.1.

Letting now $\boldsymbol{\tau} = \nabla_{\mathrm{s}}\boldsymbol{w}$ with $\boldsymbol{w} \in C^\infty(\overline{T})^d$ in (7.24) and using (7.5) to write $(\nabla\boldsymbol{v}, \nabla_{\mathrm{s}}\boldsymbol{w})_T = (\nabla_{\mathrm{s}}\boldsymbol{v}, \nabla_{\mathrm{s}}\boldsymbol{w})_T$ in the left-hand side, we obtain

$$(\nabla_{\mathrm{s}}\boldsymbol{v}, \nabla_{\mathrm{s}}\boldsymbol{w})_T = -(\boldsymbol{v}, \nabla\cdot\nabla_{\mathrm{s}}\boldsymbol{w})_T + \sum_{F\in\mathcal{F}_T} (\boldsymbol{v}, \nabla_{\mathrm{s}}\boldsymbol{w}\boldsymbol{n}_{TF})_F. \tag{7.26}$$

Specialising this formula to $\boldsymbol{w} \in \mathbb{P}^{k+1}(T)^d$ gives

$$(\nabla_{\mathrm{s}}\boldsymbol{\pi}_T^{\boldsymbol{\varepsilon},k+1}\boldsymbol{v}, \nabla_{\mathrm{s}}\boldsymbol{w})_T = -(\boldsymbol{\pi}_T^{0,k}\boldsymbol{v}, \nabla\cdot\nabla_{\mathrm{s}}\boldsymbol{w})_T + \sum_{F\in\mathcal{F}_T} (\boldsymbol{\pi}_F^{0,k}\boldsymbol{v}, \nabla_{\mathrm{s}}\boldsymbol{w}\boldsymbol{n}_{TF})_F, \tag{7.27a}$$

where we have used (7.16a) to insert the strain projector $\boldsymbol{\pi}_T^{\boldsymbol{\varepsilon},k+1}$ into the left-hand side and (1.57) to insert the L^2-orthogonal projectors $\boldsymbol{\pi}_T^{0,k}$ and $\boldsymbol{\pi}_F^{0,k}$ into the right-hand side. Moreover, using again the definition (1.57) of the L^2-orthogonal projectors over T and its faces, we infer that

$$\int_T \boldsymbol{\pi}_T^{0,k}\boldsymbol{v} = \int_T \boldsymbol{v} \tag{7.27b}$$

and

$$\begin{aligned}
&\frac{1}{2}\sum_{F\in\mathcal{F}_T}\int_F \left(\boldsymbol{\pi}_F^{0,k}\boldsymbol{v}\otimes\boldsymbol{n}_{TF} - \boldsymbol{n}_{TF}\otimes\boldsymbol{\pi}_F^{0,k}\boldsymbol{v}\right)\\
&\qquad = \frac{1}{2}\sum_{F\in\mathcal{F}_T}\int_F (\boldsymbol{v}\otimes\boldsymbol{n}_{TF} - \boldsymbol{n}_{TF}\otimes\boldsymbol{v})\\
&\qquad = \frac{1}{2}\int_T \left(\nabla\boldsymbol{v} - (\nabla\boldsymbol{v})^\intercal\right) = \int_T \nabla_{\mathrm{ss}}\boldsymbol{v},
\end{aligned} \tag{7.27c}$$

where we have used an integration by parts in the second equality and (7.4) to conclude. The relations (7.27) show that the strain projection $\boldsymbol{\pi}_T^{\boldsymbol{\varepsilon},k+1}\boldsymbol{v}$ can also be computed from the L^2-projections $\boldsymbol{\pi}_T^{0,k}\boldsymbol{v}$ and $(\boldsymbol{\pi}_F^{0,k}\boldsymbol{v})_{F\in\mathcal{F}_T}$.

Remark 7.11 (Degree of Orthogonal Projection in the Cell) In the spirit of Remark 2.1, we could have replaced $\boldsymbol{\pi}_T^{0,k}v$ by $\boldsymbol{\pi}_T^{0,k-1}v$ in both (7.25) and (7.27a). This would have required a separate treatment for the case $k = 0$ in order to express the condition (7.27b) in terms of face unknowns only; see Sect. 5.1.

7.2.4 Local Space of Discrete Unknowns

The discussion in the previous section motivates the introduction of the following local space of discrete unknowns:

$$\underline{\boldsymbol{U}}_T^k := \left\{ \underline{\boldsymbol{v}}_T = (\boldsymbol{v}_T, (\boldsymbol{v}_F)_{F\in\mathcal{F}_T}) \ : \ \boldsymbol{v}_T \in \mathbb{P}^k(T)^d \text{ and } \boldsymbol{v}_F \in \mathbb{P}^k(F)^d \quad \forall F \in \mathcal{F}_T \right\}.$$

We define the local interpolator $\underline{\boldsymbol{I}}_T^k : H^1(T)^d \to \underline{\boldsymbol{U}}_T^k$ such that

$$\underline{\boldsymbol{I}}_T^k \boldsymbol{v} := (\boldsymbol{\pi}_T^{0,k}\boldsymbol{v}, (\boldsymbol{\pi}_F^{0,k}\boldsymbol{v})_{F\in\mathcal{F}_T}) \qquad \forall \boldsymbol{v} \in H^1(T)^d,$$

as well as the local strain seminorm $\|{\cdot}\|_{\boldsymbol{\varepsilon},T}$ such that, for all $\underline{\boldsymbol{v}}_T \in \underline{\boldsymbol{U}}_T^k$,

$$\|\underline{\boldsymbol{v}}_T\|_{\boldsymbol{\varepsilon},T} := \left(\|\nabla_{\mathrm{s}}\boldsymbol{v}_T\|_T^2 + |\underline{\boldsymbol{v}}_T|_{1,\partial T}^2 \right)^{\frac{1}{2}} \text{ with } |\underline{\boldsymbol{v}}_T|_{1,\partial T} := \left(\sum_{F\in\mathcal{F}_T} h_F^{-1} \|\boldsymbol{v}_F - \boldsymbol{v}_T\|_F^2 \right)^{\frac{1}{2}}, \tag{7.28}$$

where the negative power of the diameter of F in the boundary seminorm ensures that both contributions have the same scaling.

In analogy with (2.7), for $\underline{\boldsymbol{v}}_T \in \underline{\boldsymbol{U}}_T^k$ we set

$$\|\underline{\boldsymbol{v}}_T\|_{1,T} := \left(\|\nabla \boldsymbol{v}_T\|_T^2 + |\underline{\boldsymbol{v}}_T|_{1,\partial T}^2 \right)^{\frac{1}{2}}. \tag{7.29}$$

Applying Proposition 2.2 component-wise yields

$$\|\underline{\boldsymbol{I}}_T^k \boldsymbol{v}\|_{1,T} \lesssim \|\nabla \boldsymbol{v}\|_T \qquad \forall \boldsymbol{v} \in H^1(T)^d,$$

with hidden constant depending only on d, ϱ and k. Since $\|{\cdot}\|_{\boldsymbol{\varepsilon},T} \le \|{\cdot}\|_{1,T}$, we infer that

$$\|\underline{\boldsymbol{I}}_T^k \boldsymbol{v}\|_{\boldsymbol{\varepsilon},T} \lesssim \|\nabla \boldsymbol{v}\|_T \qquad \forall \boldsymbol{v} \in H^1(T)^d. \tag{7.30}$$

In the case $k \geq 1$, this result can be improved using only the symmetric gradient of $\boldsymbol{v}$, as seen in the following lemma.

Lemma 7.12 (Boundedness of $\underline{\boldsymbol{I}}_T^k$, Case $k \geq 1$) *Assume that $k \geq 1$ and that Assumption 7.6 holds. Then, for all $T \in \mathcal{T}_h$ and $\boldsymbol{v} \in H^1(T)^d$, it holds, with hidden constant depending only on d, ϱ and k,*

$$\|\underline{\boldsymbol{I}}_T^k \boldsymbol{v}\|_{\boldsymbol{\varepsilon},T} \lesssim \|\boldsymbol{\nabla}_{\mathrm{s}} \boldsymbol{v}\|_T. \tag{7.31}$$

Proof Let $\boldsymbol{v} \in H^1(T)^d$ and $\boldsymbol{v}_{\mathrm{rm},T} \in \mathbb{RM}_d(T)$ be the rigid-body motion such that $\boldsymbol{v}_{\mathrm{rm},T}(\boldsymbol{x}) = \boldsymbol{R}\boldsymbol{x}$ with $\boldsymbol{R} = \boldsymbol{\pi}_T^{0,0}(\boldsymbol{\nabla}_{\mathrm{ss}}\boldsymbol{v})$. Then, $\boldsymbol{\nabla}_{\mathrm{s}}\boldsymbol{v}_{\mathrm{rm},T} = \boldsymbol{0}$ and $\boldsymbol{\nabla}_{\mathrm{ss}}\boldsymbol{v}_{\mathrm{rm},T} = \boldsymbol{\pi}_T^{0,0}(\boldsymbol{\nabla}_{\mathrm{ss}}\boldsymbol{v})$. Since $\boldsymbol{v}_{\mathrm{rm},T} \in \mathbb{P}^1(T)^d \subset \mathbb{P}^k(T)^d$ (as $k \geq 1$) we also have, by idempotency of $\boldsymbol{\pi}_T^{0,k}$ and $\boldsymbol{\pi}_F^{0,k}$,

$$\boldsymbol{\pi}_T^{0,k}\boldsymbol{v}_{\mathrm{rm},T} = \boldsymbol{v}_{\mathrm{rm},T}, \quad \boldsymbol{\pi}_F^{0,k}\boldsymbol{v}_{\mathrm{rm},T} = (\boldsymbol{v}_{\mathrm{rm},T})_{|F} \quad \forall F \in \mathcal{F}_T.$$

Hence,

$$\begin{aligned}
\|\underline{\boldsymbol{I}}_T^k \boldsymbol{v}_{\mathrm{rm},T}\|_{\boldsymbol{\varepsilon},T}^2 &= \|\boldsymbol{\nabla}_{\mathrm{s}}\boldsymbol{\pi}_T^{0,k}\boldsymbol{v}_{\mathrm{rm},T}\|_T^2 + \sum_{F\in\mathcal{F}_T} h_F^{-1}\|\boldsymbol{\pi}_F^{0,k}\boldsymbol{v}_{\mathrm{rm},T} - \boldsymbol{\pi}_T^{0,k}\boldsymbol{v}_{\mathrm{rm},T}\|_F^2 \\
&= \|\boldsymbol{\nabla}_{\mathrm{s}}\boldsymbol{v}_{\mathrm{rm},T}\|_T^2 + \sum_{F\in\mathcal{F}_T} h_F^{-1}\|\boldsymbol{v}_{\mathrm{rm},T} - \boldsymbol{v}_{\mathrm{rm},T}\|_F^2 = 0.
\end{aligned} \tag{7.32}$$

We then write

$$\begin{aligned}
\|\underline{\boldsymbol{I}}_T^k \boldsymbol{v}\|_{\boldsymbol{\varepsilon},T} &\leq \|\underline{\boldsymbol{I}}_T^k(\boldsymbol{v} - \boldsymbol{v}_{\mathrm{rm},T})\|_{\boldsymbol{\varepsilon},T} + \|\underline{\boldsymbol{I}}_T^k\boldsymbol{v}_{\mathrm{rm},T}\|_{\boldsymbol{\varepsilon},T} \\
&\lesssim \|\boldsymbol{\nabla}(\boldsymbol{v} - \boldsymbol{v}_{\mathrm{rm},T})\|_T \\
&\lesssim \|\boldsymbol{\nabla}_{\mathrm{s}}\boldsymbol{v}\|_T,
\end{aligned}$$

where we have introduced $\pm\boldsymbol{v}_{\mathrm{rm},T}$ and used the triangle inequality in the first line, we have invoked (7.30) and (7.32) to pass to the second line, and the conclusion follows using $\boldsymbol{\nabla}\boldsymbol{v}_{\mathrm{rm},T} = \boldsymbol{\nabla}_{\mathrm{ss}}\boldsymbol{v}_{\mathrm{rm},T} = \boldsymbol{\pi}_T^{0,0}(\boldsymbol{\nabla}_{\mathrm{ss}}\boldsymbol{v})$ and the local second Korn inequality (7.15). □

Remark 7.13 (Assumption $k \geq 1$) The assumption $k \geq 1$ ensures that $\mathbb{RM}_d(T) \subset \mathbb{P}^k(T)^d$, which is not the case for $k = 0$; see Sect. 7.1.2 on this point.

7.2.5 Local Reconstructions

Inspired by (7.25), we introduce the gradient reconstruction operator $\mathbf{G}_T^k : \underline{\boldsymbol{U}}_T^k \to \mathbb{P}^k(T)^{d\times d}$ such that, for all $\underline{\boldsymbol{v}}_T \in \underline{\boldsymbol{U}}_T^k$ and all $\boldsymbol{\tau} \in \mathbb{P}^k(T)^{d\times d}$,

$$(\mathbf{G}_T^k \underline{\boldsymbol{v}}_T, \boldsymbol{\tau})_T = -(\boldsymbol{v}_T, \nabla\cdot\boldsymbol{\tau})_T + \sum_{F\in\mathcal{F}_T} (\boldsymbol{v}_F, \boldsymbol{\tau}\boldsymbol{n}_{TF})_F \tag{7.33}$$

$$= (\nabla \boldsymbol{v}_T, \boldsymbol{\tau})_T + \sum_{F\in\mathcal{F}_T} (\boldsymbol{v}_F - \boldsymbol{v}_T, \boldsymbol{\tau}\boldsymbol{n}_{TF})_F, \tag{7.34}$$

where we have integrated by parts the first term in the right-hand side to pass to the second line. From $\mathbf{G}_T^k$, we can define the local symmetric gradient reconstruction operator $\mathbf{G}_{\mathrm{s},T}^k : \underline{\boldsymbol{U}}_T^k \to \mathbb{P}^k(T;\mathbb{R}^{d\times d}_{\mathrm{sym}})$ and the local divergence reconstruction operator $\mathrm{D}_T^k : \underline{\boldsymbol{U}}_T^k \to \mathbb{P}^k(T)$ setting, for all $\underline{\boldsymbol{v}}_T \in \underline{\boldsymbol{U}}_T^k$,

$$\mathbf{G}_{\mathrm{s},T}^k \underline{\boldsymbol{v}}_T := \frac{1}{2}\left(\mathbf{G}_T^k \underline{\boldsymbol{v}}_T + (\mathbf{G}_T^k \underline{\boldsymbol{v}}_T)^\intercal\right), \quad \mathrm{D}_T^k \underline{\boldsymbol{v}}_T := \mathrm{tr}(\mathbf{G}_T^k \underline{\boldsymbol{v}}_T) = \mathrm{tr}(\mathbf{G}_{\mathrm{s},T}^k \underline{\boldsymbol{v}}_T). \tag{7.35}$$

Remark 7.14 (Characterisation of $\mathbf{G}_{\mathrm{s},T}^k$) Combining (7.35) with (7.34), and recalling (7.5), it can be checked that $\mathbf{G}_{\mathrm{s},T}^k$ satisfies, for all $\boldsymbol{\tau} \in \mathbb{P}^k(T;\mathbb{R}^{d\times d}_{\mathrm{sym}})$,

$$(\mathbf{G}_{\mathrm{s},T}^k \underline{\boldsymbol{v}}_T, \boldsymbol{\tau})_T = (\nabla_{\mathrm{s}} \boldsymbol{v}_T, \boldsymbol{\tau})_T + \sum_{F\in\mathcal{F}_T} (\boldsymbol{v}_F - \boldsymbol{v}_T, \boldsymbol{\tau}\boldsymbol{n}_{TF})_F. \tag{7.36}$$

This formula can be used in place of (7.34) to actually compute $\mathbf{G}_{\mathrm{s},T}^k$.

By construction it holds, for all $\boldsymbol{v} \in H^1(T)^d$,

$$(\mathbf{G}_T^k \circ \underline{\boldsymbol{I}}_T^k)\boldsymbol{v} = \boldsymbol{\pi}_T^{0,k}(\nabla \boldsymbol{v}) \tag{7.37}$$

and, as a result,

$$(\mathbf{G}_{\mathrm{s},T}^k \circ \underline{\boldsymbol{I}}_T^k)\boldsymbol{v} = \boldsymbol{\pi}_T^{0,k}(\nabla_{\mathrm{s}} \boldsymbol{v}), \qquad (\mathrm{D}_T^k \circ \underline{\boldsymbol{I}}_T^k)\boldsymbol{v} = \pi_T^{0,k}(\nabla\cdot\boldsymbol{v}). \tag{7.38}$$

These commutation properties are illustrated in Fig. 7.1.

Bearing in mind (7.27a), we also introduce the displacement reconstruction operator $\mathbf{p}_T^{k+1} : \underline{\boldsymbol{U}}_T^k \to \mathbb{P}^{k+1}(T)^d$ such that, for all $\underline{\boldsymbol{v}}_T \in \underline{\boldsymbol{U}}_T^k$ and all $\boldsymbol{w} \in \mathbb{P}^{k+1}(T)^d$,

$$(\nabla_{\mathrm{s}} \mathbf{p}_T^{k+1} \underline{\boldsymbol{v}}_T, \nabla_{\mathrm{s}} \boldsymbol{w})_T = -(\boldsymbol{v}_T, \nabla\cdot\nabla_{\mathrm{s}} \boldsymbol{w})_T + \sum_{F\in\mathcal{F}_T} (\boldsymbol{v}_F, \nabla_{\mathrm{s}} \boldsymbol{w}\boldsymbol{n}_{TF})_F \tag{7.39a}$$

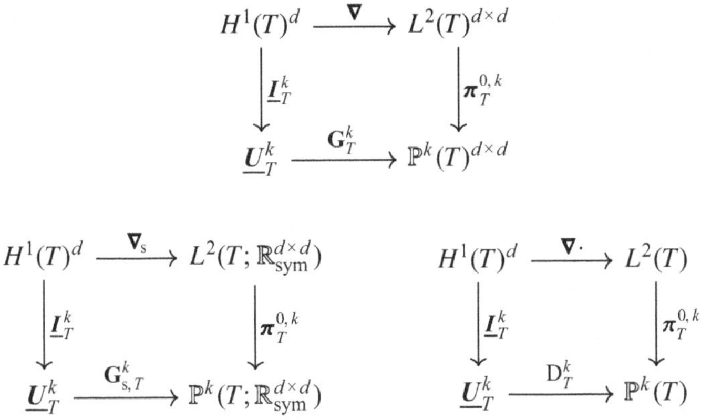

Fig. 7.1 Illustration of the commutation properties (7.37) of $\mathbf{G}_T^k$ (above) and (7.38) of $\mathbf{G}_{{\rm s},T}^k$ and D_T^k (below)

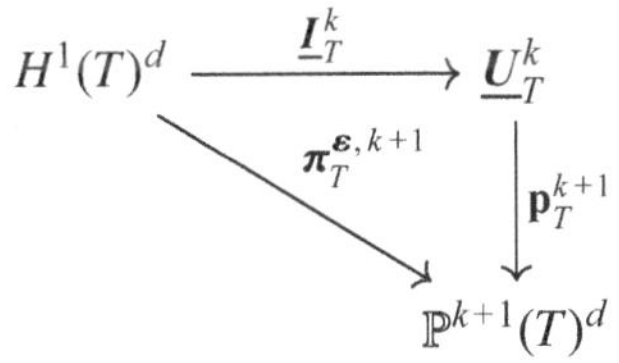

Fig. 7.2 Illustration of the commutation property (7.40) of $\mathbf{p}_T^{k+1}$

and, in accordance with (7.27b) and (7.27c),

$$\int_T \mathbf{p}_T^{k+1}\underline{\boldsymbol{v}}_T = \int_T \boldsymbol{v}_T,$$
$$\int_T \nabla_{\rm ss}\mathbf{p}_T^{k+1}\underline{\boldsymbol{v}}_T = \frac{1}{2}\sum_{F\in\mathcal{F}_T}\int_F (\boldsymbol{v}_F\otimes\boldsymbol{n}_{TF} - \boldsymbol{n}_{TF}\otimes\boldsymbol{v}_F). \tag{7.39b}$$

By construction, it holds, for all $\boldsymbol{v}\in H^1(T)^d$,

$$(\mathbf{p}_T^{k+1}\circ\underline{\boldsymbol{I}}_T^k)\boldsymbol{v} = \boldsymbol{\pi}_T^{\varepsilon,k+1}\boldsymbol{v}. \tag{7.40}$$

This commutation property is illustrated in Fig. 7.2. For future use, we record the following equivalent reformulation of (7.39a), obtained integrating by parts the first term in the right-hand side:

$$(\nabla_{\rm s}\mathbf{p}_T^{k+1}\underline{\boldsymbol{v}}_T, \nabla_{\rm s}\boldsymbol{w})_T = (\nabla_{\rm s}\boldsymbol{v}_T, \nabla_{\rm s}\boldsymbol{w})_T + \sum_{F\in\mathcal{F}_T}(\boldsymbol{v}_F - \boldsymbol{v}_T, \nabla_{\rm s}\boldsymbol{w}\boldsymbol{n}_{TF})_F. \tag{7.41}$$

To close this section, we highlight a link between $\mathbf{G}_{\mathrm{s},T}^k$ and $\nabla_{\mathrm{s}}\mathbf{p}_T^{k+1}$. Taking $\boldsymbol{w} \in \mathbb{P}^{k+1}(T)^d$, making $\boldsymbol{\tau} = \nabla_{\mathrm{s}}\boldsymbol{w}$ in (7.36), and using (7.41) shows that

$$(\mathbf{G}_{\mathrm{s},T}^k \underline{\boldsymbol{v}}_T - \nabla_{\mathrm{s}}\mathbf{p}_T^{k+1}\underline{\boldsymbol{v}}_T, \nabla_{\mathrm{s}}\boldsymbol{w})_T = 0 \qquad \forall \boldsymbol{w} \in \mathbb{P}^{k+1}(T)^d.$$

Since $\nabla_{\mathrm{s}}\mathbf{p}_T^{k+1}\underline{\boldsymbol{v}}_T \in \nabla_{\mathrm{s}}\mathbb{P}^{k+1}(T)^d$, this shows that

$$\nabla_{\mathrm{s}}\mathbf{p}_T^{k+1}\underline{\boldsymbol{v}}_T \text{ is the } L^2\text{-orthogonal projection of } \mathbf{G}_{\mathrm{s},T}^k\underline{\boldsymbol{v}}_T \text{ on } \nabla_{\mathrm{s}}\mathbb{P}^{k+1}(T)^d. \tag{7.42}$$

As a consequence, we have the following estimate:

$$\|\nabla_{\mathrm{s}}\mathbf{p}_T^{k+1}\underline{\boldsymbol{v}}_T\|_T \le \|\mathbf{G}_{\mathrm{s},T}^k\underline{\boldsymbol{v}}_T\|_T \qquad \forall \underline{\boldsymbol{v}}_T \in \underline{\boldsymbol{U}}_T^k. \tag{7.43}$$

Remark 7.15 (Equivalent Definition of $\mathbf{p}_T^{k+1}$*)* Let α_T and β_T be two non-zero real numbers. In the spirit of Remark 2.3, the definition (7.39) of $\mathbf{p}_T^{k+1}$ is equivalent to: For all $\underline{\boldsymbol{v}}_T \in \underline{\boldsymbol{U}}_T^k$ and all $\boldsymbol{w} \in \mathbb{P}^{k+1}(T)^d$,

$$\begin{aligned}
&(\nabla_{\mathrm{s}}\mathbf{p}_T^{k+1}\underline{\boldsymbol{v}}_T, \nabla_{\mathrm{s}}\boldsymbol{w})_T + \alpha_T\left(\int_T \mathbf{p}_T^{k+1}\underline{\boldsymbol{v}}_T\right)\cdot\left(\int_T \boldsymbol{w}\right) + \beta_T\left(\int_T \nabla_{\mathrm{ss}}\mathbf{p}_T^{k+1}\underline{\boldsymbol{v}}_T\right):\left(\int_T \nabla_{\mathrm{ss}}\boldsymbol{w}\right)\\
&= (\nabla_{\mathrm{s}}\boldsymbol{v}_T, \nabla_{\mathrm{s}}\boldsymbol{w})_T + \sum_{F\in\mathcal{F}_T}(\boldsymbol{v}_F - \boldsymbol{v}_T, \nabla_{\mathrm{s}}\boldsymbol{w}\boldsymbol{n}_{TF})_F + \alpha_T\left(\int_T \boldsymbol{v}_T\right)\cdot\left(\int_T \boldsymbol{w}\right)\\
&\quad + \beta_T\left(\frac{1}{2}\sum_{F\in\mathcal{F}_T}\int_F(\boldsymbol{v}_F\otimes\boldsymbol{n}_{TF} - \boldsymbol{n}_{TF}\otimes\boldsymbol{v}_F)\right):\left(\int_T \nabla_{\mathrm{ss}}\boldsymbol{w}\right).
\end{aligned} \tag{7.44}$$

This relation is obtained from (7.39) taking the scalar product of each equation in (7.39b) with $\alpha_T \int_T \boldsymbol{w}$ and $\beta_T \int_T \nabla_{\mathrm{ss}}\boldsymbol{w}$, respectively, and by adding the two resulting equations to (7.41) (equivalent to (7.39a)). Conversely, taking $\boldsymbol{w}$ a generic constant vector (resp. $\boldsymbol{w}(\boldsymbol{x}) = \boldsymbol{R}(\boldsymbol{x} - \overline{\boldsymbol{x}}_T)$ with $\boldsymbol{R}$ a generic skew-symmetric matrix and $\overline{\boldsymbol{x}}_T = \pi_T^{0,0}\boldsymbol{x})$ in (7.44) show that the first (resp. second) equation in (7.39b) hold, and thus also that (7.44) boils down to (7.41).

The characterisation (7.44) of $\mathbf{p}_T^{k+1}$ is particularly useful for computing, in an implementation of the HHO method, the matrix corresponding to this operator; see Sect. B.2.1 for the scalar case.

7.2.6 Local Contribution

Inside T, we approximate the continuous bilinear form a defined by (7.11) by the discrete bilinear form $\mathrm{a}_T : \underline{\boldsymbol{U}}_T^k \times \underline{\boldsymbol{U}}_T^k \to \mathbb{R}$ such that, for all $\underline{\boldsymbol{u}}_T, \underline{\boldsymbol{v}}_T \in \underline{\boldsymbol{U}}_T^k$,

$$\mathrm{a}_T(\underline{\boldsymbol{u}}_T, \underline{\boldsymbol{v}}_T) := (\boldsymbol{\sigma}(\mathbf{G}_{\mathrm{s},T}^k\underline{\boldsymbol{u}}_T), \mathbf{G}_{\mathrm{s},T}^k\underline{\boldsymbol{v}}_T)_T + (2\mu_T)\mathrm{s}_T(\underline{\boldsymbol{u}}_T, \underline{\boldsymbol{v}}_T), \tag{7.45}$$

where the first term is the usual Galerkin contribution in charge of consistency, while $\mathrm{s}_T : \underline{\boldsymbol{U}}_T^k \times \underline{\boldsymbol{U}}_T^k \to \mathbb{R}$ is a local stabilisation bilinear form on which we make the following assumptions.

Assumption 7.16 (Local Stabilisation Bilinear form s_T) *The local stabilisation bilinear form* $\mathrm{s}_T : \underline{\boldsymbol{U}}_T^k \times \underline{\boldsymbol{U}}_T^k \to \mathbb{R}$ *has the following properties:*

(SE1) Symmetry and positivity. s_T *is symmetric and positive semidefinite;*
(SE2) Stability and boundedness. *It holds, for all* $T \in \mathcal{T}_h$ *and all* $\underline{\boldsymbol{v}}_T \in \underline{\boldsymbol{U}}_T^k$,

$$\|\mathbf{G}_{\mathrm{s},T}^k \underline{\boldsymbol{v}}_T\|_T^2 + \mathrm{s}_T(\underline{\boldsymbol{v}}_T, \underline{\boldsymbol{v}}_T) \simeq \|\underline{\boldsymbol{v}}_T\|_{\boldsymbol{\varepsilon},T}^2 \tag{7.46}$$

with local discrete strain seminorm defined by (7.28) *and hidden constants independent of* h, T, *and of the problem data;*
(SE3) Polynomial consistency. *For all* $\boldsymbol{w} \in \mathbb{P}^{k+1}(T)^d$ *and all* $\underline{\boldsymbol{v}}_T \in \underline{\boldsymbol{U}}_T^k$, *it holds*

$$\mathrm{s}_T(\underline{\boldsymbol{I}}_T^k \boldsymbol{w}, \underline{\boldsymbol{v}}_T) = 0.$$

Some remarks are in order.

Remark 7.17 (Reformulation of the Consistency Term) Recalling the expression (7.8) of the strain-stress law, we have the following reformulation of the local bilinear form:

$$\begin{aligned} \mathrm{a}_T(\underline{\boldsymbol{u}}_T, \underline{\boldsymbol{v}}_T) = (2\mu_T)\Big((\mathbf{G}_{\mathrm{s},T}^k \underline{\boldsymbol{u}}_T, \mathbf{G}_{\mathrm{s},T}^k \underline{\boldsymbol{v}}_T)_T + \mathrm{s}_T(\underline{\boldsymbol{u}}_T, \underline{\boldsymbol{v}}_T)\Big) \\ + \lambda_T(\mathrm{D}_T^k \underline{\boldsymbol{u}}_T, \mathrm{D}_T^k \underline{\boldsymbol{v}}_T)_T. \end{aligned} \tag{7.47}$$

Remark 7.18 (Coercivity and Boundedness of a_T*)* Observing that, owing to (7.35),

$$\|\mathrm{D}_T^k \underline{\boldsymbol{v}}_T\|_T^2 \le d \|\mathbf{G}_{\mathrm{s},T}^k \underline{\boldsymbol{v}}_T\|_T^2 \qquad \forall \underline{\boldsymbol{v}}_T \in \underline{\boldsymbol{U}}_T^k, \tag{7.48}$$

we have the local semi-norm equivalence: For all $\underline{\boldsymbol{v}}_T \in \underline{\boldsymbol{U}}_T^k$,

$$\begin{aligned} &(2\mu_T - d\lambda_T^-)\|\mathbf{G}_{\mathrm{s},T}^k \underline{\boldsymbol{v}}_T\|^2 + (2\mu_T)\mathrm{s}_T(\underline{\boldsymbol{v}}_T, \underline{\boldsymbol{v}}_T) \\ &\quad \lesssim \mathrm{a}_T(\underline{\boldsymbol{v}}_T, \underline{\boldsymbol{v}}_T) \lesssim (2\mu_T + d|\lambda_T|)\|\mathbf{G}_{\mathrm{s},T}^k \underline{\boldsymbol{v}}_T\|^2 + (2\mu_T)\mathrm{s}_T(\underline{\boldsymbol{v}}_T, \underline{\boldsymbol{v}}_T). \end{aligned} \tag{7.49}$$

Combined with (7.46), this also yields

$$(2\mu_T - d\lambda_T^-)\|\underline{\boldsymbol{v}}_T\|_{\boldsymbol{\varepsilon},T}^2 \lesssim \mathrm{a}_T(\underline{\boldsymbol{v}}_T, \underline{\boldsymbol{v}}_T) \lesssim (2\mu_T + d|\lambda_T|)\|\underline{\boldsymbol{v}}_T\|_{\boldsymbol{\varepsilon},T}^2. \tag{7.50}$$

In these estimates, $\lambda_T^- := \frac{1}{2}(|\lambda_T| - \lambda_T)$ denotes the negative part of λ_T, and the hidden constants are independent of h, T, $\underline{\boldsymbol{v}}_T$, and of the problem data.

Remark 7.19 (Comparison with the Original HHO Bilinear Form) In the original work [146] on HHO methods for linear elasticity, a different local bilinear form is considered, corresponding to (7.47) with $\mathbf{G}^k_{\mathrm{s},T}$ replaced by $\boldsymbol{\nabla}_{\mathrm{s}}\mathbf{p}_T^{k+1}$ in the first term; see Eqs. (24) and (17) therein. Coercivity for this alternative bilinear form requires the stronger condition $\lambda \geq 0$ on the first Lamé coefficient (compare with (7.9)) since the bound (7.48) no longer holds in general if we replace $\mathbf{G}^k_{\mathrm{s},T}$ with $\boldsymbol{\nabla}_{\mathrm{s}}\mathbf{p}_T^{k+1}$. While the condition $\lambda \geq 0$ is often verified in practice, we have preferred here the expression (7.47) for a_T which covers the entire range of physical values for the first Lamé coefficient.

Remark 7.20 (Consistency of s_T) As in Proposition 2.14, using Assumption 7.16 and the boundedness (7.30) of $\underline{\boldsymbol{I}}^k_T$, we can show that, for all $r \in \{0, \ldots, k\}$ and all $\boldsymbol{v} \in H^{r+2}(T)^d$,

$$\mathrm{s}_T(\underline{\boldsymbol{I}}^k_T \boldsymbol{v}, \underline{\boldsymbol{I}}^k_T \boldsymbol{v})^{\frac{1}{2}} \lesssim h_T^{r+1} |\boldsymbol{v}|_{H^{r+2}(T)^d}, \tag{7.51}$$

with hidden constant independent of h, T, $\boldsymbol{v}$, and of the problem data.

As for scalar problems, the stabilisation bilinear form s_T can be obtained penalising, in a least square sense, differences that vanish for interpolates of polynomial functions in $\mathbb{P}^{k+1}(T)^d$. Specifically, reasoning as in Lemma 2.11, one can prove that a symmetric bilinear form s_T satisfies (SE3) if and only if it depends on its arguments through the vector difference operators $\boldsymbol{\delta}^k_T : \underline{\boldsymbol{U}}^k_T \to \mathbb{P}^k(T)^d$ and, for all $F \in \mathcal{F}_T$, $\boldsymbol{\delta}^k_{TF} : \underline{\boldsymbol{U}}^k_T \to \mathbb{P}^k(F)^d$ such that, for all $\underline{\boldsymbol{v}}_T \in \underline{\boldsymbol{U}}^k_T$,

$$\boldsymbol{\delta}^k_T \underline{\boldsymbol{v}}_T := \boldsymbol{\pi}^{0,k}_T (\mathbf{p}^{k+1}_T \underline{\boldsymbol{v}}_T - \boldsymbol{v}_T), \qquad \boldsymbol{\delta}^k_{TF} \underline{\boldsymbol{v}}_T := \boldsymbol{\pi}^{0,k}_F (\mathbf{p}^{k+1}_T \underline{\boldsymbol{v}}_T - \boldsymbol{v}_F) \quad \forall F \in \mathcal{F}_T. \tag{7.52}$$

Proceeding as in the proof of Proposition 2.6, it is readily inferred that it holds, for all $\boldsymbol{w} \in \mathbb{P}^{k+1}(T)^d$,

$$\boldsymbol{\delta}^k_T \underline{\boldsymbol{I}}^k_T \boldsymbol{w} = \boldsymbol{0}, \qquad \boldsymbol{\delta}^k_{TF} \underline{\boldsymbol{I}}^k_T \boldsymbol{w} = \boldsymbol{0} \quad \forall F \in \mathcal{F}_T. \tag{7.53}$$

In the following proposition we study the original HHO stabilisation of [146], which generalises (2.22) to the vector case.

Proposition 7.21 (Original HHO Stabilisation) *Assume that $k \geq 1$ and that Assumption 7.6 holds. Then, the stabilisation bilinear form $\mathrm{s}_T : \underline{\boldsymbol{U}}^k_T \times \underline{\boldsymbol{U}}^k_T \to \mathbb{R}$ such that, for all $\underline{\boldsymbol{u}}_T, \underline{\boldsymbol{v}}_T \in \underline{\boldsymbol{U}}^k_T$,*

$$\mathrm{s}_T(\underline{\boldsymbol{u}}_T, \underline{\boldsymbol{v}}_T) := \sum_{F \in \mathcal{F}_T} h_F^{-1} ((\boldsymbol{\delta}^k_{TF} - \boldsymbol{\delta}^k_T)\underline{\boldsymbol{u}}_T, (\boldsymbol{\delta}^k_{TF} - \boldsymbol{\delta}^k_T)\underline{\boldsymbol{v}}_T)_F \tag{7.54}$$

satisfies Assumption 7.16.

Proof The bilinear form s_T is clearly symmetric and positive semidefinite, so that property (SE1) holds. Property (SE3), on the other hand, is a consequence of the fact that s_T depends on its arguments only via the difference operators (7.52), and that the latter vanish when their arguments are of the form $\underline{\boldsymbol{I}}_T^k \boldsymbol{w}$ with $\boldsymbol{w} \in \mathbb{P}^{k+1}(T)^d$, see (7.53). It only remains to prove (SE2). The rest of the proof follows closely the arguments in the proof of Proposition 2.13, using the boundedness property (7.31) of $\underline{\boldsymbol{I}}_T^k$ (valid here since we assume $k \geq 1$ and work under Assumption 7.6). We denote by $\underline{\boldsymbol{v}}_T$ a generic element of $\underline{\boldsymbol{U}}_T^k$ and, for the sake of brevity, set

$$\check{\boldsymbol{v}}_T := \mathbf{p}_T^{k+1} \underline{\boldsymbol{v}}_T.$$

The notation $A \lesssim B$ is understood with hidden constant independent of h, T, $\underline{\boldsymbol{v}}_T$, and of the problem data.

(i) *Estimates on the volumetric components.* We prove in this step that

$$\|\boldsymbol{\nabla}_{\rm s} \boldsymbol{v}_T\|_T^2 \lesssim \|\mathbf{G}_{{\rm s},T}^k \underline{\boldsymbol{v}}_T\|_T^2 + |\underline{\boldsymbol{v}}_T|_{1,\partial T}^2 \tag{7.55}$$

and

$$\|\mathbf{G}_{{\rm s},T}^k \underline{\boldsymbol{v}}_T\|_T^2 \lesssim \|\underline{\boldsymbol{v}}_T\|_{\boldsymbol{\varepsilon},T}^2. \tag{7.56}$$

We first consider (7.55). Let $\boldsymbol{\tau} = \boldsymbol{\nabla}_{\rm s} \boldsymbol{v}_T$ in (7.36) and rearrange the terms to write

$$\begin{aligned}
\|\boldsymbol{\nabla}_{\rm s} \boldsymbol{v}_T\|_T^2 &= (\mathbf{G}_{{\rm s},T}^k \underline{\boldsymbol{v}}_T, \boldsymbol{\nabla}_{\rm s} \boldsymbol{v}_T)_T - \sum_{F \in \mathcal{F}_T} (\boldsymbol{v}_F - \boldsymbol{v}_T, \boldsymbol{\nabla}_{\rm s} \boldsymbol{v}_T \boldsymbol{n}_{TF})_F \\
&\leq \|\mathbf{G}_{{\rm s},T}^k \underline{\boldsymbol{v}}_T\|_T \|\boldsymbol{\nabla}_{\rm s} \boldsymbol{v}_T\|_T + \left(\sum_{F \in \mathcal{F}_T} h_F^{-1} \|\boldsymbol{v}_F - \boldsymbol{v}_T\|_F^2 \right)^{\frac{1}{2}} \left(\sum_{F \in \mathcal{F}_T} h_F \|\boldsymbol{\nabla}_{\rm s} \boldsymbol{v}_T\|_F^2 \right)^{\frac{1}{2}} \\
&\lesssim \left(\|\mathbf{G}_{{\rm s},T}^k \underline{\boldsymbol{v}}_T\|_T + |\underline{\boldsymbol{v}}_T|_{1,\partial T} \right) \|\boldsymbol{\nabla}_{\rm s} \boldsymbol{v}_T\|_T,
\end{aligned}$$

where, to pass to the second line, we have used a Cauchy–Schwarz inequality for the volumetric term along with generalised Hölder inequalities with exponents $(2, 2, \infty)$ and $\|\boldsymbol{n}_{TF}\|_{L^\infty(F)^d} = 1$ for the boundary terms, while, to pass to the third line, we have invoked the discrete trace inequality (1.55) with $p = 2$ together with the uniform bound (1.5) on the number of faces of T. Simplifying, we get $\|\boldsymbol{\nabla}_{\rm s} \boldsymbol{v}_T\|_T \lesssim \|\mathbf{G}_{{\rm s},T}^k \underline{\boldsymbol{v}}_T\|_T + |\underline{\boldsymbol{v}}_T|_{1,\partial T}$. Squaring this inequality and using the fact that $(a + b)^2 \leq 2a^2 + 2b^2$ for any $a, b \in \mathbb{R}$, we arrive at (7.55).

We now estimate $\|\mathbf{G}_{{\rm s},T}^k \underline{\boldsymbol{v}}_T\|_T$. Using the characterisation (7.36) of $\mathbf{G}_{{\rm s},T}^k$ with $\boldsymbol{\tau} = \mathbf{G}_{{\rm s},T}^k \underline{\boldsymbol{v}}_T$ and proceeding as above with Cauchy–Schwarz, generalised Hölder, and discrete trace inequalities, we have that

$$\begin{aligned}\|\mathbf{G}^k_{\mathrm{s},T}\underline{\boldsymbol{v}}_T\|_T^2 &= (\boldsymbol{\nabla}_{\mathrm{s}}\boldsymbol{v}_T, \mathbf{G}^k_{\mathrm{s},T}\underline{\boldsymbol{v}}_T) + \sum_{F\in\mathcal{F}_T}(\boldsymbol{v}_F - \boldsymbol{v}_T, \mathbf{G}^k_{\mathrm{s},T}\underline{\boldsymbol{v}}_T\boldsymbol{n}_{TF})_F \\ &\le \|\boldsymbol{\nabla}_{\mathrm{s}}\boldsymbol{v}_T\|_T\|\mathbf{G}^k_{\mathrm{s},T}\underline{\boldsymbol{v}}_T\|_T + |\underline{\boldsymbol{v}}_T|_{1,\partial T}\left(\sum_{F\in\mathcal{F}_T}h_F\|\mathbf{G}^k_{\mathrm{s},T}\underline{\boldsymbol{v}}_T\|_F^2\right)^{\frac{1}{2}} \\ &\lesssim \|\underline{\boldsymbol{v}}_T\|_{\boldsymbol{\varepsilon},T}\|\mathbf{G}^k_{\mathrm{s},T}\underline{\boldsymbol{v}}_T\|_T.\end{aligned}$$

Simplifying the above inequality and squaring yields (7.56).

(ii) *Proof of (SE2)*. Set

$$\underline{\boldsymbol{z}}_T := \underline{\boldsymbol{I}}^k_T\check{\boldsymbol{v}}_T - \underline{\boldsymbol{v}}_T = (\boldsymbol{\delta}^k_T\underline{\boldsymbol{v}}_T, (\boldsymbol{\delta}^k_{TF}\underline{\boldsymbol{v}}_T)_{F\in\mathcal{F}_T})$$

and notice that

$$\mathrm{s}_T(\underline{\boldsymbol{v}}_T, \underline{\boldsymbol{v}}_T) = |\underline{\boldsymbol{z}}_T|_{1,\partial T}^2. \tag{7.57}$$

We also remark that, by the boundedness (7.31) of $\underline{\boldsymbol{I}}^k_T$ and the estimate (7.43),

$$|\underline{\boldsymbol{I}}^k_T\check{\boldsymbol{v}}_T|_{1,\partial T} \lesssim \|\boldsymbol{\nabla}_{\mathrm{s}}\check{\boldsymbol{v}}_T\|_T \lesssim \|\mathbf{G}^k_{\mathrm{s},T}\underline{\boldsymbol{v}}_T\|_T. \tag{7.58}$$

Let us now prove (7.46). We first write

$$\begin{aligned}\|\underline{\boldsymbol{v}}_T\|_{\boldsymbol{\varepsilon},T}^2 &= \|\boldsymbol{\nabla}_{\mathrm{s}}\boldsymbol{v}_T\|_T^2 + |\underline{\boldsymbol{v}}_T|_{1,\partial T}^2 \\ &\lesssim \|\mathbf{G}^k_{\mathrm{s},T}\underline{\boldsymbol{v}}_T\|_T^2 + |\underline{\boldsymbol{I}}^k_T\check{\boldsymbol{v}}_T - \underline{\boldsymbol{z}}_T|_{1,\partial T}^2 \\ &\lesssim \|\mathbf{G}^k_{\mathrm{s},T}\underline{\boldsymbol{v}}_T\|_T^2 + \mathrm{s}_T(\underline{\boldsymbol{v}}_T, \underline{\boldsymbol{v}}_T),\end{aligned}$$

where we have used the definition (7.28) of $\|\cdot\|_{\boldsymbol{\varepsilon},T}$ in the first line, the estimate (7.55) together with $\underline{\boldsymbol{v}}_T = \underline{\boldsymbol{I}}^k_T\check{\boldsymbol{v}}_T - \underline{\boldsymbol{z}}_T$ to pass to the second line, and we have concluded using the triangle inequality and invoking (7.58) and (7.57). This establishes $\gtrsim$ in (7.46).

To establish the converse inequality, we start from (7.57) and substitute the definition of $\underline{\boldsymbol{z}}_T$ to get

$$\mathrm{s}_T(\underline{\boldsymbol{v}}_T, \underline{\boldsymbol{v}}_T) \le 2|\underline{\boldsymbol{I}}^k_T\check{\boldsymbol{v}}_T|_{1,\partial T}^2 + 2|\underline{\boldsymbol{v}}_T|_{1,\partial T}^2 \lesssim \|\mathbf{G}^k_{\mathrm{s},T}\underline{\boldsymbol{v}}_T\|_T^2 + |\underline{\boldsymbol{v}}_T|_{1,\partial T}^2,$$

where the second inequality is a consequence of (7.58). Adding $\|\mathbf{G}^k_{\mathrm{s},T}\underline{\boldsymbol{v}}_T\|_T^2$ to both sides of this relation and invoking (7.56), the proof of $\lesssim$ in (7.46) is complete. □

Remark 7.22 (Case $k = 0$) Assumption 7.6 along with $k \geq 1$ are essential to the proof of Proposition 7.21. Their role is to ensure that the boundedness property (7.31) of $\underline{\boldsymbol{I}}_T^k$ holds, and thus that (7.58) is valid.

Actually, in the case $k = 0$, (SE2) and (SE3) are incompatible. To see this, assume (SE3), consider a rigid-body motion $\boldsymbol{v}_{\mathrm{rm},T}$ and take $\underline{\boldsymbol{v}}_T = \underline{\boldsymbol{I}}_T^0 \boldsymbol{v}_{\mathrm{rm},T}$. Since $\boldsymbol{v}_{\mathrm{rm},T} \in \mathbb{P}^1(T)^d$, (7.37) shows that $\mathbf{G}_T^0 \underline{\boldsymbol{v}}_T = \boldsymbol{\pi}_T^{0,0}(\nabla \boldsymbol{v}_{\mathrm{rm},T}) = \nabla \boldsymbol{v}_{\mathrm{rm},T} = \nabla_{\mathrm{ss}} \boldsymbol{v}_{\mathrm{rm},T}$ so that, in particular, $\mathbf{G}_T^0 \underline{\boldsymbol{v}}_T$ is skew-symmetric. Hence, $\mathbf{G}_{\mathrm{s},T}^0 \underline{\boldsymbol{v}}_T = \mathbf{0}$. Moreover, by (SE3), $\mathrm{s}_T(\underline{\boldsymbol{v}}_T, \underline{\boldsymbol{v}}_T) = \mathrm{s}_T(\underline{\boldsymbol{I}}_T^0 \boldsymbol{v}_{\mathrm{rm},T}, \underline{\boldsymbol{v}}_T) = 0$. Hence, the left-hand side of (7.46) vanishes for all $\underline{\boldsymbol{v}}_T = \underline{\boldsymbol{I}}_T^0 \boldsymbol{v}_{\mathrm{rm},T}$. It is however easy to construct $\boldsymbol{v}_{\mathrm{rm},T}$ such that $|\underline{\boldsymbol{v}}_T|_{1,\partial T} \neq 0$, which shows that the right-hand side of (7.46) does not vanish and thus that (SE2) cannot hold.

The consequence is that, for the lowest-order method, a stabilisation term cannot be constructed such that Assumption 7.16 holds. An alternative approach must be considered, that we present in Sect. 7.6.

7.3 Discrete Problem

In this section we formulate the discrete problem, based on the local contributions introduced in the previous section, and study its well-posedness.

7.3.1 Global Space of Discrete Unknowns

The global space of discrete unknowns is defined as

$$\underline{\boldsymbol{U}}_h^k := \Big\{ \underline{\boldsymbol{v}}_h = ((\boldsymbol{v}_T)_{T \in \mathcal{T}_h}, (\boldsymbol{v}_F)_{F \in \mathcal{F}_h}) \; : \\ \boldsymbol{v}_T \in \mathbb{P}^k(T)^d \quad \forall T \in \mathcal{T}_h \text{ and } \boldsymbol{v}_F \in \mathbb{P}^k(F)^d \quad \forall F \in \mathcal{F}_h \Big\}.$$

Given $\underline{\boldsymbol{v}}_h \in \underline{\boldsymbol{U}}_h^k$, for all $T \in \mathcal{T}_h$ we denote by $\underline{\boldsymbol{v}}_T := (\boldsymbol{v}_T, (\boldsymbol{v}_F)_{F \in \mathcal{F}_T}) \in \underline{\boldsymbol{U}}_T^k$ its restriction to T. We also define the broken polynomial function $\boldsymbol{v}_h \in \mathbb{P}^k(\mathcal{T}_h)^d$ such that

$$(\boldsymbol{v}_h)_{|T} := \boldsymbol{v}_T \qquad \forall T \in \mathcal{T}_h. \tag{7.59}$$

The discrete unknowns corresponding to a function $\boldsymbol{v} \in H^1(\Omega)^d$ are obtained via the global interpolator $\underline{\boldsymbol{I}}_h^k : H^1(\Omega)^d \to \underline{\boldsymbol{U}}_h^k$ such that

$$\underline{\boldsymbol{I}}_h^k \boldsymbol{v} := ((\boldsymbol{\pi}_T^{0,k} \boldsymbol{v})_{T \in \mathcal{T}_h}, (\boldsymbol{\pi}_F^{0,k} \boldsymbol{v})_{F \in \mathcal{F}_h}).$$

We define on $\underline{U}_h^k$ the global strain seminorm $\|\cdot\|_{\varepsilon,h}$ such that, for all $\underline{v}_h \in \underline{U}_h^k$,

$$\|\underline{v}_h\|_{\varepsilon,h} := \left(\sum_{T\in\mathcal{T}_h} \|\underline{v}_T\|_{\varepsilon,T}^2\right)^{\frac{1}{2}}, \tag{7.60}$$

with local seminorm $\|\cdot\|_{\varepsilon,T}$ defined by (7.28). For future use, we also denote by $|\cdot|_{s,h}$ the global seminorm associated with the stabilisation bilinear forms: For all $\underline{v}_h \in \underline{U}_h^k$,

$$|\underline{v}_h|_{s,h} := \left(\sum_{T\in\mathcal{T}_h} s_T(\underline{v}_T, \underline{v}_T)\right)^{\frac{1}{2}}. \tag{7.61}$$

Finally, to account for the homogeneous Dirichlet boundary condition (7.7b) in a strong manner, we introduce the subspace

$$\underline{U}_{h,0}^k := \left\{\underline{v}_h \in \underline{U}_h^k \ :\ v_F = \mathbf{0} \quad \forall F \in \mathcal{F}_h^{\mathrm{b}}\right\}. \tag{7.62}$$

7.3.2 Global Discrete Korn Inequalities in Broken Polynomial and HHO Spaces

We next prove discrete counterparts of the Korn inequality (7.12), first for broken polynomial spaces, then for HHO spaces. In the former case, the proof hinges on the node-averaging operator on the submesh introduced in Remark 4.4, whose definition and properties are briefly summarised here for the sake of convenience. Let $\mathcal{M}_h = (\mathcal{T}_h, \mathcal{F}_h)$ denote a polytopal mesh in the sense of Definition 1.4, and let $\mathfrak{M}_h = (\mathfrak{T}_h, \mathfrak{F}_h)$ denote a matching simplicial submesh of $\mathcal{M}_h$ in the sense of Definition 1.8. Given an integer $l \ge 1$, the node-averaging operator $\mathcal{I}_{\mathrm{av},h}^l : \mathbb{P}^l(\mathcal{T}_h) \to \mathbb{P}^l(\mathfrak{T}_h) \cap H_0^1(\Omega)$ is such that, for any function $v_h \in \mathbb{P}^l(\mathcal{T}_h)$ and any Lagrange node $\boldsymbol{N}$ of $\mathfrak{T}_h$ (see, e.g., [183, Section 1.2.3] or [77, Section 3.2] for a precise definition of Lagrange nodes),

$$(\mathcal{I}_{\mathrm{av},h}^l v_h)(\boldsymbol{N}) := \begin{cases} \dfrac{1}{\mathrm{card}(\mathfrak{T}_{\boldsymbol{N}})} \displaystyle\sum_{\tau\in\mathfrak{T}_{\boldsymbol{N}}} (v_h)_{|\tau}(\boldsymbol{N}) & \text{if } \boldsymbol{N} \in \Omega, \\ 0 & \text{if } \boldsymbol{N} \in \partial\Omega, \end{cases}$$

where $\mathfrak{T}_{\boldsymbol{N}} \subset \mathfrak{T}_h$ collects the simplices to which $\boldsymbol{N}$ belongs. The vector-version, denoted by $\boldsymbol{\mathcal{I}}_{\mathrm{av},h}^l$, acts component-wise. Reasoning as in [145, Section 5.5.2] (where the original results of [216] on simplicial meshes are extended to polytopal meshes), we infer that it holds, for all $T \in \mathcal{T}_h$,

$$\|v_h - \mathcal{I}_{\mathrm{av},h}^l v_h\|_T^2 \lesssim \sum_{F \in \mathcal{F}_{N,T}} h_F \|[v_h]_F\|_F^2, \tag{7.63}$$

where $\mathcal{F}_{N,T} \subset \mathcal{F}_h$, defined by (4.22), denotes the set of faces whose closure has nonempty intersection with ∂T. In (7.63), the hidden constant is independent of h, T, and v_h, but possibly depends on d, l, and ϱ. Combining this result with the inverse inequality (1.46) (see also Remark 1.33 concerning its validity for $\mathcal{I}_{\mathrm{av},h}^l v_h$) for $p = 2$ we obtain, with hidden constants as before,

$$\begin{aligned}
|v_h - \mathcal{I}_{\mathrm{av},h}^l v_h|_{H^1(\mathcal{T}_h)}^2 &= \sum_{T \in \mathcal{T}_h} \|\nabla(v_h - \mathcal{I}_{\mathrm{av},h}^l v_h)\|_T^2 \\
&\lesssim \sum_{T \in \mathcal{T}_h} h_T^{-2} \|v_h - \mathcal{I}_{\mathrm{av},h}^l v_h\|_T^2 \\
&\lesssim \sum_{T \in \mathcal{T}_h} h_T^{-2} \sum_{F \in \mathcal{F}_{N,T}} h_F \|[v_h]_F\|_F^2 \\
&\lesssim \sum_{F \in \mathcal{F}_h} \sum_{T \in \mathcal{T}_{N,F}} h_F^{-1} \|[v_h]_F\|_F^2,
\end{aligned}$$

where we have used (7.63) to pass to the third line, while the conclusion was obtained invoking the geometric bound (4.24) to write $h_F h_T^{-2} \lesssim h_F^{-1}$ for all $F \in \mathcal{F}_{N,T}$ and exchanging the order of the sums after introducing the notation $\mathcal{T}_{N,F} := \{T \in \mathcal{T}_h : \overline{F} \cap \partial T \neq \emptyset\}$ for the set of mesh elements whose boundary intersects the closure of F. Using the geometric bound (4.23) to infer that $\mathrm{card}(\mathcal{T}_{N,F}) \leq \sum_{T \in \mathcal{T}_F} \mathrm{card}(\mathcal{T}_{N,T}) \lesssim 1$, we arrive at

$$|v_h - \mathcal{I}_{\mathrm{av},h}^l v_h|_{H^1(\mathcal{T}_h)}^2 \lesssim \sum_{F \in \mathcal{F}_h} h_F^{-1} \|[v_h]_F\|_F^2. \tag{7.64}$$

We are now ready to prove the discrete Korn inequality in broken polynomial spaces.

Lemma 7.23 (Discrete Korn Inequality in Broken Polynomial Spaces) *Let an integer $l \geq 0$ be fixed and set, for all $\boldsymbol{v}_h \in \mathbb{P}^l(\mathcal{T}_h)^d$,*

$$\|\boldsymbol{v}_h\|_{\boldsymbol{\varepsilon},\mathrm{j},h} := \left(\|\boldsymbol{\nabla}_{\mathrm{s},h}\boldsymbol{v}_h\|^2 + |\boldsymbol{v}_h|_{\mathrm{j},h}^2\right)^{\frac{1}{2}} \text{ with } |\boldsymbol{v}_h|_{\mathrm{j},h} := \left(\sum_{F \in \mathcal{F}_h} h_F^{-1} \|[\boldsymbol{v}_h]_F\|_F^2\right)^{\frac{1}{2}}. \tag{7.65}$$

Above, the broken symmetric gradient $\nabla_{s,h} : H^1(\mathcal{T}_h)^d \to L^2(\Omega; \mathbb{R}^{d\times d}_{\text{sym}})$ *is such that, for all* $\boldsymbol{v} \in H^1(\mathcal{T}_h)^d$,

$$(\nabla_{s,h}\boldsymbol{v})_{|T} := \nabla_s \boldsymbol{v}_{|T} \qquad \forall T \in \mathcal{T}_h,$$

while the jump operator is defined by (1.22) *on internal faces and extended to boundary faces* $F \in \mathcal{F}_h^{\text{b}}$ *by setting* $[\boldsymbol{v}_h]_F := (\boldsymbol{v}_h)_{|F}$. *Then, it holds, for all* $\boldsymbol{v}_h \in \mathbb{P}^l(\mathcal{T}_h)^d$, *with hidden constant depending only on* Ω, d, l, *and* ϱ:

$$|\boldsymbol{v}_h|_{H^1(\mathcal{T}_h)^d} \lesssim \|\boldsymbol{v}_h\|_{\boldsymbol{\varepsilon},\text{j},h}. \tag{7.66}$$

Proof The case $l = 0$ is trivial, so we consider hereafter $l \geq 1$. The proof adapts the arguments of [76, Lemma 2.2]. We write

$$\begin{aligned}
|\boldsymbol{v}_h|^2_{H^1(\mathcal{T}_h)^d} &\lesssim |\boldsymbol{I}^l_{\text{av},h}\boldsymbol{v}_h|^2_{H^1(\Omega)^d} + |\boldsymbol{v}_h - \boldsymbol{I}^l_{\text{av},h}\boldsymbol{v}_h|^2_{H^1(\mathcal{T}_h)^d} \\
&\lesssim \|\nabla_s \boldsymbol{I}^l_{\text{av},h}\boldsymbol{v}_h\|^2 + |\boldsymbol{v}_h|^2_{\text{j},h} \\
&\lesssim \|\nabla_{s,h}\boldsymbol{v}_h\|^2 + \|\nabla_{s,h}(\boldsymbol{I}^l_{\text{av},h}\boldsymbol{v}_h - \boldsymbol{v}_h)\|^2 + |\boldsymbol{v}_h|^2_{\text{j},h} \\
&\lesssim \|\nabla_{s,h}\boldsymbol{v}_h\|^2 + |\boldsymbol{v}_h|^2_{\text{j},h} = \|\boldsymbol{v}_h\|^2_{\boldsymbol{\varepsilon},\text{j},h},
\end{aligned}$$

where we have inserted $\pm\boldsymbol{I}^l_{\text{av},h}\boldsymbol{v}_h$ into the seminorm and used a triangle inequality in the first line, we have applied the continuous Korn inequality (7.12) to the first term and invoked (7.64) and the definition (7.65) of the jump seminorm for the second term, we have inserted $\pm\nabla_{s,h}\boldsymbol{v}_h$ and used a triangle inequality to pass to the third line, we have invoked again (7.64) to estimate the second term in the right-hand side and pass to the fourth line, and we have used the definition (7.65) of the $\|\cdot\|_{\boldsymbol{\varepsilon},\text{j},h}$-norm to conclude. □

Based on the result of Lemma 7.23, we next establish a Korn inequality in HHO spaces.

Lemma 7.24 (Discrete Korn Inequality in HHO Spaces) *Assume* $k \geq 1$. *Then it holds, for all* $\underline{\boldsymbol{v}}_h \in \underline{\boldsymbol{U}}^k_{h,0}$,

$$\|\underline{\boldsymbol{v}}_h\|_{1,h} \lesssim \|\underline{\boldsymbol{v}}_h\|_{\boldsymbol{\varepsilon},h}, \tag{7.67}$$

where the hidden constant depends only on Ω, d, k *and* ϱ, *and, in analogy with* (2.35) *and recalling the definition* (7.29) *of the* $\|\cdot\|_{1,T}$*-seminorm, we have set*

$$\|\underline{\boldsymbol{v}}_h\|_{1,h} := \left(\sum_{T\in\mathcal{T}_h} \|\underline{\boldsymbol{v}}_T\|^2_{1,T}\right)^{\frac{1}{2}} = \left(\sum_{T\in\mathcal{T}_h} \|\nabla \boldsymbol{v}_T\|^2_T + |\underline{\boldsymbol{v}}_T|^2_{1,\partial T}\right)^{\frac{1}{2}}. \tag{7.68}$$

Proof Let $\underline{v}_h \in \underline{U}_{h,0}^k$ and recall the definition (7.59) of v_h. By (7.66) we have that

$$\|\nabla_h v_h\|^2 \lesssim \|\nabla_{s,h} v_h\|^2 + |v_h|_{j,h}^2 =: \mathfrak{T}_1 + \mathfrak{T}_2. \tag{7.69}$$

Recalling the definitions (7.60) and (7.28) of the global and local strain norms on HHO spaces, it is readily inferred that

$$\mathfrak{T}_1 \lesssim \|\underline{v}_h\|_{\varepsilon,h}^2. \tag{7.70}$$

For the second term, on the other hand, we can write

$$\begin{aligned}
\mathfrak{T}_2 &= \sum_{F\in\mathcal{F}_h} h_F^{-1}\|[v_h]_F\|_F^2 \\
&\lesssim \sum_{F\in\mathcal{F}_h}\sum_{T\in\mathcal{T}_F} h_F^{-1}\|v_T - v_F\|_F^2 \\
&= \sum_{T\in\mathcal{T}_h}\sum_{F\in\mathcal{F}_T} h_F^{-1}\|v_T - v_F\|_F^2 = \sum_{T\in\mathcal{T}_h} |\underline{v}_T|_{1,\partial T}^2 \lesssim \|\underline{v}_h\|_{\varepsilon,h}^2,
\end{aligned} \tag{7.71}$$

where we have expanded the jump seminorm according to its definition (7.65) in the first line, we have used in the second line the definition (1.22) of the jump operator followed by a triangle inequality after inserting $\pm v_F$ inside the norm for all $F \in \mathcal{F}_h^{\rm i}$ and used the fact that $v_F = \mathbf{0}$ for all $F \in \mathcal{F}_h^{\rm b}$, we have exchanged the order of the summations according to (1.25) to pass to the third line, and we have used the definitions (7.60) and (7.28) of the global and local strain (semi)norms to conclude. Plugging (7.70) and (7.71) into (7.69) yields

$$\sum_{T\in\mathcal{T}_h} \|\nabla v_T\|_T^2 \lesssim \|\underline{v}_h\|_{\varepsilon,h}^2.$$

The proof of (7.67) is completed by recalling that $\sum_{T\in\mathcal{T}_h} |\underline{v}_T|_{1,\partial T}^2 \le \|\underline{v}_h\|_{\varepsilon,h}^2$. □

Remark 7.25 (Discrete Strain Norm) Recalling Corollary 2.16, an immediate consequence of (7.67) is that $\|{\cdot}\|_{\varepsilon,h}$ defines a norm on $\underline{U}_{h,0}^k$.

Remark 7.26 (Discrete Korn–Poincaré Inequalities) Combining the discrete Poincaré inequality resulting from [144, Theorem 6.1] (see also [145, Theorem 5.3 and Corollary 5.4]) with (7.66), we infer that it holds, for all $v_h \in \mathbb{P}^l(\mathcal{T}_h)^d$,

$$\|v_h\| \lesssim \|v_h\|_{\varepsilon,j,h}, \tag{7.72}$$

with hidden constant independent of h and v_h. Similarly, combining the discrete Poincaré inequality (2.37) with (7.67) we infer, for all $\underline{v}_h \in \underline{U}_{h,0}^k$, with hidden constant as before,

$$\|v_h\| \lesssim \|\underline{v}_h\|_{\varepsilon,h}. \tag{7.73}$$

7.3.3 Global Bilinear Form

We define the global bilinear form $\mathrm{a}_h : \underline{U}_h^k \times \underline{U}_h^k \to \mathbb{R}$ by element by element assembly setting, for all $\underline{u}_h, \underline{v}_h \in \underline{U}_h^k$,

$$\mathrm{a}_h(\underline{u}_h, \underline{v}_h) := \sum_{T \in \mathcal{T}_h} \mathrm{a}_T(\underline{u}_T, \underline{v}_T). \tag{7.74}$$

Lemma 7.27 (Properties of a_h) *Let a polynomial degree $k \geq 1$ be fixed. The bilinear form a_h enjoys the following properties:*

(i) Stability and boundedness. *For all $\underline{v}_h \in \underline{U}_h^k$ it holds*

$$\alpha\|\underline{v}_h\|_{\varepsilon,h}^2 \lesssim \mathrm{a}_h(\underline{v}_h, \underline{v}_h) \lesssim (2\overline{\mu} + d\overline{\lambda})\|\underline{v}_h\|_{\varepsilon,h}^2, \tag{7.75}$$

where the hidden constant is independent of h, $\underline{v}_h$, and of the problem data.

(ii) Consistency. *It holds for all $r \in \{0, \ldots, k\}$ and all $w \in H_0^1(\Omega)^d \cap H^{r+2}(\mathcal{T}_h)^d$ such that $\nabla \cdot \sigma(\nabla_s w) \in L^2(\Omega)^d$,*

$$\sup_{\underline{v}_h \in \underline{U}_{h,0}^k, \|\underline{v}_h\|_{\varepsilon,h}=1} |\mathcal{E}_h(w; \underline{v}_h)| \lesssim h^{r+1}\left[|\sigma(\nabla_s w)|_{H^{r+1}(\mathcal{T}_h)^{d\times d}} + (2\overline{\mu})|w|_{H^{r+2}(\mathcal{T}_h)^d}\right], \tag{7.76}$$

where the hidden constant is independent of w, h, and of the problem data, and the linear form $\mathcal{E}_h(w; \cdot) : \underline{U}_{h,0}^k \to \mathbb{R}$ representing the consistency error is such that, for all $\underline{v}_h \in \underline{U}_{h,0}^k$,

$$\mathcal{E}_h(w; \underline{v}_h) := -(\nabla \cdot \sigma(\nabla_s w), v_h) - \mathrm{a}_h(\underline{I}_h^k w, \underline{v}_h). \tag{7.77}$$

Remark 7.28 (Regularity of $\sigma(\nabla_s w)$) By property (7.14) on the Lamé coefficients, if $w \in H^{r+2}(\mathcal{T}_h)^d$, then $\sigma(\nabla_s w) \in H^{r+1}(\mathcal{T}_h)^{d\times d}$. The right-hand side of (7.76) is thus well-defined.

Proof (i) *Stability and boundedness.* Summing (7.50) over $T \in \mathcal{T}_h$ and accounting for the bounds (7.9) yields (7.75).

(ii) *Consistency.* Let $\underline{v}_h \in \underline{U}_{h,0}^k$. For the sake of brevity we let, for all $T \in \mathcal{T}_h$,

$$\check{\sigma}_T := \sigma(\mathbf{G}_{s,T}^k \underline{I}_T^k w) = \sigma(\pi_T^{0,k}(\nabla_s w)) = \pi_T^{0,k}(\sigma(\nabla_s w)) \in \mathbb{P}^k(T; \mathbb{R}_{\mathrm{sym}}^{d\times d}), \tag{7.78}$$

where we have used the commutation property (7.38) to replace $\mathbf{G}^k_{\mathrm{s},T}\underline{\boldsymbol{I}}^k_T\boldsymbol{w}$ with $\boldsymbol{\pi}^{0,k}_T(\boldsymbol{\nabla}_{\mathrm{s}}\boldsymbol{w})$ in the first passage, and the fact that the Lamé coefficients are constant over T (cf. (7.14)) together with the linearity of the L^2-orthogonal projector to conclude. Integrating by parts element by element, we write

$$
\begin{aligned}
&-(\boldsymbol{\nabla}\cdot\boldsymbol{\sigma}(\boldsymbol{\nabla}_{\mathrm{s}}\boldsymbol{w}),\boldsymbol{v}_h)\\
&\quad=\sum_{T\in\mathcal{T}_h}\left((\boldsymbol{\sigma}(\boldsymbol{\nabla}_{\mathrm{s}}\boldsymbol{w}),\boldsymbol{\nabla}_{\mathrm{s}}\boldsymbol{v}_T)_T+\sum_{F\in\mathcal{F}_T}(\boldsymbol{\sigma}(\boldsymbol{\nabla}_{\mathrm{s}}\boldsymbol{w})\boldsymbol{n}_{TF},\boldsymbol{v}_F-\boldsymbol{v}_T)_F\right),
\end{aligned}
\tag{7.79}
$$

where we have replaced, in the volumetric terms, $\boldsymbol{\nabla}\boldsymbol{v}_T$ with $\boldsymbol{\nabla}_{\mathrm{s}}\boldsymbol{v}_T$ using (7.5) with $\boldsymbol{\tau}=\boldsymbol{\sigma}(\boldsymbol{\nabla}_{\mathrm{s}}\boldsymbol{w})$ and, to insert $\boldsymbol{v}_F$ into the second term in parentheses, we have used Corollary 1.19 with, for any $1\le i\le d$, $\boldsymbol{\tau}$ equal to the ith line of $\boldsymbol{\sigma}(\boldsymbol{\nabla}_{\mathrm{s}}\boldsymbol{w})$ and $(\varphi_F)_{F\in\mathcal{F}_h}=(v_{F,i})_{F\in\mathcal{F}_h}$, $v_{F,i}$ being the ith component of $\boldsymbol{v}_F$. On the other hand, plugging the definition (7.45) of a_T into (7.74), and expanding $\mathbf{G}^k_{\mathrm{s},T}\underline{\boldsymbol{v}}_T$ according to (7.36) with $\boldsymbol{\tau}=\check{\boldsymbol{\sigma}}_T$, it is inferred that

$$
\begin{aligned}
&\mathrm{a}_h(\underline{\boldsymbol{I}}^k_h\boldsymbol{w},\underline{\boldsymbol{v}}_h)\\
&\quad=\sum_{T\in\mathcal{T}_h}\left((\check{\boldsymbol{\sigma}}_T,\boldsymbol{\nabla}_{\mathrm{s}}\boldsymbol{v}_T)_T+\sum_{F\in\mathcal{F}_T}(\check{\boldsymbol{\sigma}}_T\boldsymbol{n}_{TF},\boldsymbol{v}_F-\boldsymbol{v}_T)_F+(2\mu_T)\mathrm{s}(\underline{\boldsymbol{I}}^k_T\boldsymbol{w},\underline{\boldsymbol{v}}_T)\right).
\end{aligned}
\tag{7.80}
$$

Subtracting (7.80) from (7.79), taking absolute values, invoking (7.78) followed by the definition (1.57) of $\boldsymbol{\pi}^{0,k}_T$ to cancel the first terms in the corresponding summations, and using the triangle inequality we get

$$
\begin{aligned}
&|\mathcal{E}_h(\boldsymbol{w};\underline{\boldsymbol{v}}_h)|\\
&\quad\le\left|\sum_{T\in\mathcal{T}_h}\sum_{F\in\mathcal{F}_T}((\boldsymbol{\sigma}(\boldsymbol{\nabla}_{\mathrm{s}}\boldsymbol{w})-\check{\boldsymbol{\sigma}}_T)\boldsymbol{n}_{TF},\boldsymbol{v}_F-\boldsymbol{v}_T)_F\right|+\sum_{T\in\mathcal{T}_h}(2\mu_T)|\mathrm{s}_T(\underline{\boldsymbol{I}}^k_T\boldsymbol{w},\underline{\boldsymbol{v}}_T)|\\
&\quad\le\left(\sum_{T\in\mathcal{T}_h}h_T\|\boldsymbol{\sigma}(\boldsymbol{\nabla}_{\mathrm{s}}\boldsymbol{w})-\check{\boldsymbol{\sigma}}_T\|^2_{\partial T}\right)^{\frac12}\left(\sum_{T\in\mathcal{T}_h}|\underline{\boldsymbol{v}}_T|^2_{1,\partial T}\right)^{\frac12}+(2\overline{\mu})|\underline{\boldsymbol{I}}^k_h\boldsymbol{w}|_{\mathrm{s},h}|\underline{\boldsymbol{v}}_h|_{\mathrm{s},h},
\end{aligned}
\tag{7.81}
$$

where the conclusion follows using generalised Hölder inequalities with exponents $(2,\infty,2)$ together with $\|\boldsymbol{n}_{TF}\|_{L^\infty(F)^d}=1$ for the first term, continuous Cauchy–Schwarz inequalities on the positive semidefinite bilinear forms s_T, and discrete Cauchy–Schwarz inequalities on the sums over $T\in\mathcal{T}_h$. Recalling (7.78) and invoking the trace approximation properties (1.75) of the L^2-orthogonal projector with $l = k$, $s = r + 1$, and $m = 0$, it is inferred that

$$h_T^{\frac{1}{2}} \|\boldsymbol{\sigma}(\boldsymbol{\nabla}_{\mathrm{s}}\boldsymbol{w}) - \check{\boldsymbol{\sigma}}_T\|_{\partial T} \lesssim h_T^{r+1} |\boldsymbol{\sigma}(\boldsymbol{\nabla}_{\mathrm{s}}\boldsymbol{w})|_{H^{r+1}(T)^{d\times d}}. \tag{7.82}$$

Plugging this bound into (7.81) and using the approximation properties (7.51) of the stabilisation bilinear form for the first factor in the second term, we get

$$|\mathcal{E}_h(\boldsymbol{w}; \underline{\boldsymbol{v}}_h)| \lesssim h^{r+1} \left(|\boldsymbol{\sigma}(\boldsymbol{\nabla}_{\mathrm{s}}\boldsymbol{w})|_{H^{r+1}(\mathcal{T}_h)^{d\times d}} + (2\overline{\mu}) |\boldsymbol{w}|_{H^{r+2}(\mathcal{T}_h)^d} \right) \\ \times \left[\left(\sum_{T\in\mathcal{T}_h} |\underline{\boldsymbol{v}}_T|_{1,\partial T}^2 \right)^{\frac{1}{2}} + |\underline{\boldsymbol{v}}_h|_{\mathrm{s},h} \right]. \tag{7.83}$$

Summing the local seminorm equivalence (7.46) over $T \in \mathcal{T}_h$ and taking the square root of the resulting inequality to infer the estimate $|\underline{\boldsymbol{v}}_h|_{\mathrm{s},h} \lesssim \|\underline{\boldsymbol{v}}_h\|_{\boldsymbol{\varepsilon},h}$, then passing to the supremum over $\left\{\underline{\boldsymbol{v}}_h \in \underline{\boldsymbol{U}}_{h,0}^k \,:\, \|\underline{\boldsymbol{v}}_h\|_{\boldsymbol{\varepsilon},h} = 1\right\}$, (7.76) follows. □

7.3.4 *Discrete Problem and Well-Posedness*

The HHO scheme for the approximation of problem (7.10) reads: Find $\underline{\boldsymbol{u}}_h \in \underline{\boldsymbol{U}}_{h,0}^k$ such that

$$\mathrm{a}_h(\underline{\boldsymbol{u}}_h, \underline{\boldsymbol{v}}_h) = (\boldsymbol{f}, \boldsymbol{v}_h) \qquad \forall \underline{\boldsymbol{v}}_h \in \underline{\boldsymbol{U}}_{h,0}^k. \tag{7.84}$$

Remark 7.29 (Static Condensation for Problem (7.84)*)* Following the ideas exposed in Sect. B.3.2, an efficient resolution algorithm for problem (7.84) consists in statically condensing the element-based discrete unknowns first, and then solving a reduced global system of size $d\,\mathrm{card}(\mathcal{F}_h^{\mathrm{i}})\binom{k+d-1}{d-1}$ where only face-based unknowns are present.

Lemma 7.30 (Well-Posedness of Problem (7.84)**)** *Let a polynomial degree $k \geq 1$ be fixed. Problem* (7.84) *is well-posed, and we have the following a priori bound for the unique discrete solution $\underline{\boldsymbol{u}}_h \in \underline{\boldsymbol{U}}_{h,0}^k$:*

$$\|\underline{\boldsymbol{u}}_h\|_{\boldsymbol{\varepsilon},h} \lesssim \alpha^{-\frac{1}{2}} \|\boldsymbol{f}\|, \tag{7.85}$$

where the hidden constant is independent of both h and of the problem data.

Proof We check the assumptions of the Lax–Milgram Lemma 2.20 with $\mathrm{U} = \underline{\boldsymbol{U}}_{h,0}^k$, $\mathrm{a} = \mathrm{a}_h$, and $\langle \mathrm{f}, \underline{\boldsymbol{v}}_h\rangle_{\mathrm{U}^\star,\mathrm{U}} = (\boldsymbol{f}, \boldsymbol{v}_h)$. Clearly, $\underline{\boldsymbol{U}}_{h,0}^k$ equipped with the norm $\|{\cdot}\|_{\boldsymbol{\varepsilon},h}$ is a Hilbert space. By (7.75), the bilinear form a_h is U-coercive with coercivity constant $\gtrsim \alpha$. To conclude, observe that, owing to the discrete Korn–Poincaré inequality (7.73), we have $|(\boldsymbol{f}, \boldsymbol{v}_h)| \le \|\boldsymbol{f}\| \, \|\boldsymbol{v}_h\| \lesssim \|\boldsymbol{f}\| \, \|\underline{\boldsymbol{v}}_h\|_{\boldsymbol{\varepsilon},h}$ so that, in particular, the dual norm of the linear form $\mathtt{f} : \underline{\boldsymbol{v}}_h \mapsto (\boldsymbol{f}, \boldsymbol{v}_h)$ is $\lesssim \|\boldsymbol{f}\|$. □

7.3.5 Flux Formulation

The following lemma identifies a flux formulation for the HHO scheme (7.84). Its proof is a straightforward adaptation of that of Lemma 2.25, and is left as an exercise to the reader.

Lemma 7.31 (Flux Formulation) *Let a polynomial degree $k \ge 1$ be fixed. Let $\mathcal{M}_h$ denote a polytopal mesh in the sense of Definition 1.4. Let, for any $T \in \mathcal{T}_h$, s_T be a stabilisation bilinear form that satisfies Assumption 7.16. Set $\underline{\boldsymbol{D}}_{\partial T}^k := \{\underline{\boldsymbol{\alpha}}_{\partial T} := (\boldsymbol{\alpha}_F)_{F \in \mathcal{F}_T} : \boldsymbol{\alpha}_F \in \mathbb{P}^k(F)^d \quad \forall F \in \mathcal{F}_T\}$ and define the boundary residual operator $\underline{\boldsymbol{R}}_{\partial T}^k := (\boldsymbol{R}_{TF}^k)_{F \in \mathcal{F}_T} : \underline{\boldsymbol{U}}_T^k \to \underline{\boldsymbol{D}}_{\partial T}^k$ such that, for all $\underline{\boldsymbol{v}}_T \in \underline{\boldsymbol{U}}_T^k$,*

$$-\sum_{F \in \mathcal{F}_T} (\boldsymbol{R}_{TF}^k \underline{\boldsymbol{v}}_T, \boldsymbol{\alpha}_F)_F = \mathrm{s}_T(\underline{\boldsymbol{v}}_T, (\boldsymbol{0}, \underline{\boldsymbol{\alpha}}_{\partial T})) \qquad \forall \underline{\boldsymbol{\alpha}}_{\partial T} \in \underline{\boldsymbol{D}}_{\partial T}^k. \tag{7.86}$$

Let $\underline{\boldsymbol{u}}_h \in \underline{\boldsymbol{U}}_{h,0}^k$ and, for all $T \in \mathcal{T}_h$ and all $F \in \mathcal{F}_T$, define the numerical normal trace of the flux

$$\boldsymbol{\Phi}_{TF}(\underline{\boldsymbol{u}}_T) := -\boldsymbol{\sigma}_{|T}(\mathbf{G}_{\mathrm{s},T}^k \underline{\boldsymbol{u}}_T)\boldsymbol{n}_{TF} + (2\mu_T)\boldsymbol{R}_{TF}^k \underline{\boldsymbol{u}}_T.$$

Then $\underline{\boldsymbol{u}}_h$ is the unique solution of problem (7.84) if and only if the following two properties hold:

(i) Local balance. *For all $T \in \mathcal{T}_h$ and all $\boldsymbol{v}_T \in \mathbb{P}^k(T)^d$, it holds*

$$(\boldsymbol{\sigma}(\mathbf{G}_{\mathrm{s},T}^k \underline{\boldsymbol{u}}_T), \nabla_{\mathrm{s}} \boldsymbol{v}_T)_T + \sum_{F \in \mathcal{F}_T} (\boldsymbol{\Phi}_{TF}(\underline{\boldsymbol{u}}_T), \boldsymbol{v}_T)_F = (\boldsymbol{f}, \boldsymbol{v}_T)_T. \tag{7.87a}$$

(ii) Continuity of the numerical normal traces of the flux. *For any interface $F \in \mathcal{F}_h^{\mathrm{i}}$ such that $F \subset \partial T_1 \cap \partial T_2$ with distinct mesh elements $T_1, T_2 \in \mathcal{T}_h$,*

$$\boldsymbol{\Phi}_{T_1F}(\underline{\boldsymbol{u}}_{T_1}) + \boldsymbol{\Phi}_{T_2F}(\underline{\boldsymbol{u}}_{T_2}) = \boldsymbol{0}. \tag{7.87b}$$

Remark 7.32 (Mechanical Interpretation) In the context of the linear elasticity problem (7.7), for all $T \in \mathcal{T}_h$ and all $F \in \mathcal{F}_T$, the quantity $\boldsymbol{\sigma}_{|T}(\nabla_{\mathrm{s}} \boldsymbol{u})\boldsymbol{n}_{TF}$ can be interpreted as a traction acting on F. We have favoured here the terminology "flux formulation" over "traction formulation" to emphasise the conceptual link with the other problems considered throughout the book.

7.4 Error Analysis

This section contains the error analysis of the method: we start by deriving an estimate for the discrete error measured in the $\|\cdot\|_{\varepsilon,h}$-norm; from this estimate, an error on strains is inferred. We then derive an improved L^2-error estimate on the displacement under the usual elliptic regularity assumption; finally, we discuss the robustness of our error estimates in the quasi-incompressible limit.

7.4.1 Energy Error Estimate

We start by deriving a convergence result in the energy-norm, using the interpolate of the solution to the continuous problem. This result is a direct application of the generic analysis framework presented in Appendix A.

Theorem 7.33 (Discrete Energy Error Estimate) *Let $(\mathcal{M}_h)_{h\in\mathcal{H}}$ denote a regular mesh sequence in the sense of Definition 1.9. Let a polynomial degree $k \geq 1$ be fixed. Let $\boldsymbol{u} \in H_0^1(\Omega)^d$ denote the unique solution to (7.10), for which we assume the additional regularity $\boldsymbol{u} \in H^{r+2}(\mathcal{T}_h)^d$ for some $r \in \{0, \ldots, k\}$. For all $h \in \mathcal{H}$, let $\underline{\boldsymbol{u}}_h \in \underline{\boldsymbol{U}}_{h,0}^k$ denote the unique solution to (7.84) with stabilisation bilinear forms s_T, $T \in \mathcal{T}_h$, in (7.45) satisfying Assumption 7.16. Then,*

$$\|\underline{\boldsymbol{u}}_h - \underline{\boldsymbol{I}}_h^k \boldsymbol{u}\|_{\varepsilon,h} \lesssim \alpha^{-1} h^{r+1} \left(|\boldsymbol{\sigma}(\boldsymbol{\nabla}_{\mathrm{s}} \boldsymbol{u})|_{H^{r+1}(\mathcal{T}_h)^{d\times d}} + (2\overline{\mu}) |\boldsymbol{u}|_{H^{r+2}(\mathcal{T}_h)^d} \right), \tag{7.88}$$

where the norm $\|\cdot\|_{\varepsilon,h}$ is defined in (7.60) and the hidden constant is independent of h, $\boldsymbol{u}$, and of the problem data.

Proof We invoke the Third Strang Lemma A.7 with $\mathrm{U} = H_0^1(\Omega)^d$, $\mathrm{a} = a$ defined by (7.11), $\mathrm{l}(v) = (\boldsymbol{f}, \boldsymbol{v})$, $\mathrm{U}_h = \underline{\boldsymbol{U}}_{h,0}^k$ endowed with the norm $\|\cdot\|_{\varepsilon,h}$, $\mathrm{a}_h = \mathrm{a}_h$, $\mathrm{l}_h(\underline{v}_h) = (\boldsymbol{f}, \boldsymbol{v}_h)$, and $\mathrm{I}_h \boldsymbol{u} = \underline{\boldsymbol{I}}_h^k \boldsymbol{u}$. By (7.75), a_h is coercive for $\|\cdot\|_{\varepsilon,h}$ with constant $\gtrsim \alpha$ and, since $-\boldsymbol{\nabla}\cdot\boldsymbol{\sigma}(\boldsymbol{\nabla}_{\mathrm{s}}\boldsymbol{u}) = \boldsymbol{f}$, the consistency error (A.5) is exactly (7.77) with $\boldsymbol{w} = \boldsymbol{u}$. Hence, (7.88) follows plugging (7.76) into (A.6). □

Remark 7.34 (Discrete Error Estimate in the Norm Induced by a_h*)* An estimate can also be obtained for the error measured in the norm induced by the discrete bilinear form a_h such that, for all $\underline{\boldsymbol{v}}_h \in \underline{\boldsymbol{U}}_{h,0}^k$,

$$\|\underline{\boldsymbol{v}}_h\|_{\mathrm{a},h} := \mathrm{a}_h(\underline{\boldsymbol{v}}_h, \underline{\boldsymbol{v}}_h)^{\frac{1}{2}}.$$

Specifically, we obtain in this case, under the assumptions of Theorem 7.33,

$$\|\underline{\boldsymbol{u}}_h - \underline{\boldsymbol{I}}_h^k \boldsymbol{u}\|_{\mathrm{a},h} \lesssim h^{r+1} \left(\alpha^{-\frac{1}{2}} |\boldsymbol{\sigma}(\boldsymbol{\nabla}_{\mathrm{s}} \boldsymbol{u})|_{H^{r+1}(\mathcal{T}_h)^{d\times d}} + (2\overline{\mu})^{\frac{1}{2}} |\boldsymbol{u}|_{H^{r+2}(\mathcal{T}_h)^d} \right),$$

with hidden constant independent of h, $\boldsymbol{u}$, and of the problem data.

From the error estimate (7.88) in the discrete strain norm, we prove an estimate for the error measured with respect to the continuous solution. To this end, we define the global symmetric gradient reconstruction $\mathbf{G}_{\mathrm{s},h}^k : \underline{\boldsymbol{U}}_h^k \to \mathbb{P}^k(\mathcal{T}_h; \mathbb{R}_{\mathrm{sym}}^{d\times d})$ such that, for all $\underline{\boldsymbol{v}}_h \in \underline{\boldsymbol{U}}_h^k$,

$$(\mathbf{G}_{\mathrm{s},h}^k \underline{\boldsymbol{v}}_h)_{|T} := \mathbf{G}_{\mathrm{s},T}^k \underline{\boldsymbol{v}}_T \qquad \forall T \in \mathcal{T}_h.$$

Theorem 7.35 (Energy Error Estimate for the Reconstructed Approximate Solution) *Under the assumptions of Theorem 7.33, it holds*

$$\begin{aligned} &\|\mathbf{G}_{\mathrm{s},h}^k \underline{\boldsymbol{u}}_h - \boldsymbol{\nabla}_{\mathrm{s}} \boldsymbol{u}\| + |\underline{\boldsymbol{u}}_h|_{\mathrm{s},h} \\ &\lesssim h_T^{r+1} \left(\alpha^{-1} |\boldsymbol{\sigma}(\boldsymbol{\nabla}_{\mathrm{s}} \boldsymbol{u})|_{H^{r+1}(\mathcal{T}_h)^{d\times d}} + (1 + 2\overline{\mu}\alpha^{-1}) |\boldsymbol{u}|_{H^{r+2}(\mathcal{T}_h)^d} \right), \end{aligned} \tag{7.89}$$

where the hidden constant is independent of h, $\boldsymbol{u}$*, and of the problem data.*

Proof We insert $\boldsymbol{\pi}_h^{0,k}(\boldsymbol{\nabla}_{\mathrm{s}} \boldsymbol{u}) - \mathbf{G}_{\mathrm{s},h}^k \underline{\boldsymbol{I}}_h^k \boldsymbol{u} = \mathbf{0}$ (see (7.38)) into the first norm in the left-hand side of (7.89), we add and subtract $\underline{\boldsymbol{I}}_h^k \boldsymbol{u}$ into the second seminorm, and we apply triangle inequalities to infer

$$\begin{aligned} \|\mathbf{G}_{\mathrm{s},h}^k \underline{\boldsymbol{u}}_h - \boldsymbol{\nabla}_{\mathrm{s}} \boldsymbol{u}\| + |\underline{\boldsymbol{u}}_h|_{\mathrm{s},h} &\lesssim \underbrace{\|\mathbf{G}_{\mathrm{s},h}^k (\underline{\boldsymbol{u}}_h - \underline{\boldsymbol{I}}_h^k \boldsymbol{u})\| + |\underline{\boldsymbol{u}}_h - \underline{\boldsymbol{I}}_h^k \boldsymbol{u}|_{\mathrm{s},h}}_{\mathfrak{T}_1} \\ &\quad + \underbrace{\|\boldsymbol{\pi}_h^{0,k}(\boldsymbol{\nabla}_{\mathrm{s}} \boldsymbol{u}) - \boldsymbol{\nabla}_{\mathrm{s}} \boldsymbol{u}\|}_{\mathfrak{T}_2} + \underbrace{|\underline{\boldsymbol{I}}_h^k \boldsymbol{u}|_{\mathrm{s},h}}_{\mathfrak{T}_3}. \end{aligned}$$

Summing (7.46) over $T \in \mathcal{T}_h$ and passing to the square root yields the estimate $\mathfrak{T}_1 \lesssim \|\underline{\boldsymbol{u}}_h - \underline{\boldsymbol{I}}_h^k \boldsymbol{u}\|_{\boldsymbol{\varepsilon},h}$. The conclusion follows invoking the error estimate (7.88) to estimate $\mathfrak{T}_1$, using the approximation properties (1.74) of the L^2-orthogonal projector with X successively equal to the elements of $\mathcal{T}_h$, $l = k$, $p = 2$, $m = 0$, and $s = r + 1$ to bound $\mathfrak{T}_2$, and the consistency properties (7.51) of s_T to bound $\mathfrak{T}_3$. □

7.4.2 L^2-Error Estimate

Combining the discrete Korn–Poincaré inequality (7.73) with the energy estimate (7.88) gives, under the regularity assumptions of Theorem 7.33, an estimate in h^{r+1} for the L^2-norm of the error. As for the Poisson problem (see Sect. 2.3.3), however, an improved L^2-error estimate can be derived also for the linear elasticity problem. Throughout this section, we will make the following assumption, which is justified by the elliptic regularity requirement; see Remark 3.21 for a discussion in the context of variable diffusion problems.

Assumption 7.36 (Constant Normalised Lamé Coefficients) *The Lamé coefficients λ and μ are constant over Ω. Without loss of generality, we take $2\mu = 1$ (when $2\mu \neq 1$, it suffices to divide equation* (7.7a) *by 2μ and to replace λ with $\lambda/(2\mu)$ and $\boldsymbol{f}$ with $\boldsymbol{f}/(2\mu)$).*

In the context of linear elasticity problems, elliptic regularity holds if, for all $\boldsymbol{g} \in L^2(\Omega)^d$, the unique solution $\boldsymbol{z}_{\boldsymbol{g}} \in H^1_0(\Omega)^d$ of the dual problem

$$a(\boldsymbol{v}, \boldsymbol{z}_{\boldsymbol{g}}) = (\boldsymbol{g}, \boldsymbol{v}) \qquad \forall \boldsymbol{v} \in H^1_0(\Omega)^d \tag{7.90}$$

satisfies the a priori estimate

$$\|\boldsymbol{z}_{\boldsymbol{g}}\|_{H^2(\Omega)^d} \le C\|\boldsymbol{g}\|. \tag{7.91}$$

This regularity property holds when Ω is convex. Notice that the dual problem (7.90) coincides with the primal problem since a is symmetric. A stronger a priori estimate valid for $\lambda \ge 0$ is given in Lemma 7.39 below.

Theorem 7.37 (Superconvergence of Element Unknowns) *Let Assumption 7.36 hold true. Let $(\mathcal{M}_h)_{h\in\mathcal{H}}$ denote a regular mesh sequence in the sense of Definition 1.9. Denote by $\boldsymbol{u} \in H^1_0(\Omega)^d$ the unique solution of* (7.10)*, for which we assume the additional regularity $\boldsymbol{u} \in H^{r+2}(\mathcal{T}_h)^d$ for some $r \in \{0, \ldots, k\}$. Let a polynomial degree $k \ge 1$ be fixed and, for all $h \in \mathcal{H}$, let $\underline{\boldsymbol{u}}_h \in \underline{\boldsymbol{U}}^k_{h,0}$ denote the unique solution to* (7.84) *with stabilisation bilinear forms s_T, $T \in \mathcal{T}_h$, in* (7.45) *satisfying Assumption 7.16. Further assuming elliptic regularity, it holds, for all $h \in \mathcal{H}$,*

(continued)

Theorem 7.37 (continued)

$$\|u_h - \pi_h^{0,k} u\| \lesssim h^{r+2} |u|_{H^{r+2}(\mathcal{T}_h)^d}, \tag{7.92}$$

with hidden constant independent of both h and $\boldsymbol{u}$*, but possibly depending on* Ω*, d,* ϱ*, k, r, and* λ*.*

Remark 7.38 (L^2-Error Estimate) Following similar arguments as in Sect. 2.3.3, one can prove from Theorem 7.37 that the displacement reconstruction also converges in h^{r+2}. The details are left as an exercise to the reader.

Proof The result follows from the Aubin–Nitsche Lemma A.10 in Appendix A, with the same setting as in Theorem 7.33, that is: $\mathtt{U} = H_0^1(\Omega)^d$, $\mathtt{a} = a$ defined by (7.11), $\mathtt{l}(\boldsymbol{v}) = (\boldsymbol{f}, \boldsymbol{v})$, $\mathtt{U}_h = \underline{\boldsymbol{U}}_{h,0}^k$, $\|\cdot\|_{\mathtt{U}_h} = \|\cdot\|_{\boldsymbol{\varepsilon},h}$, $\mathtt{a}_h = \mathrm{a}_h$, $\mathtt{l}_h(\underline{\boldsymbol{v}}_h) = (\boldsymbol{f}, \boldsymbol{v}_h)$ and $\mathtt{I}_h \boldsymbol{u} = \underline{\boldsymbol{I}}_h^k \boldsymbol{u}$. Additionally, we take $\mathtt{L} = L^2(\Omega)^d$ and $\mathtt{r}_h : \underline{\boldsymbol{U}}_{h,0}^k \to L^2(\Omega)^d$ defined by $\mathtt{r}_h \underline{\boldsymbol{v}}_h = \boldsymbol{v}_h$. In what follows, the hidden constants in the inequalities $A \lesssim B$ do not depend on h, $\boldsymbol{f}$, $\boldsymbol{u}$ or $\boldsymbol{g}$ in the dual problem (7.90) (but possibly depend on λ).

Since $\mathtt{a}$ is symmetric, the dual problem (A.10) coincides with (7.90) and, by choice of $\mathtt{r}_h$, the dual consistency error $\mathcal{E}_h^{\mathrm{d}}(\boldsymbol{z}_{\boldsymbol{g}}; \cdot)$ is equal to the primal consistency error $\mathcal{E}_h(\boldsymbol{z}_{\boldsymbol{g}}; \cdot)$ defined by (7.77). By definition of $\mathtt{r}_h$ and $\underline{\boldsymbol{I}}_h^k$, the Aubin–Nitsche Lemma A.10 therefore shows that

$$\begin{aligned} \|\boldsymbol{u}_h - \boldsymbol{\pi}_h^{0,k}\boldsymbol{u}\| \le{} & \underbrace{\|\underline{\boldsymbol{u}}_h - \underline{\boldsymbol{I}}_h^k \boldsymbol{u}\|_{\boldsymbol{\varepsilon},h} \sup_{\boldsymbol{g}\in L^2(\Omega)^d,\ \|\boldsymbol{g}\|\le 1} \|\mathcal{E}_h(\boldsymbol{z}_{\boldsymbol{g}}; \cdot)\|_{\boldsymbol{\varepsilon},h,\star}}_{\mathfrak{T}_1} \\ & + \underbrace{\sup_{\boldsymbol{g}\in L^2(\Omega)^d,\ \|\boldsymbol{g}\|\le 1} |\mathcal{E}_h(\boldsymbol{u}; \underline{\boldsymbol{I}}_h^k \boldsymbol{z}_{\boldsymbol{g}})|}_{\mathfrak{T}_2}, \end{aligned} \tag{7.93}$$

where $\|\cdot\|_{\boldsymbol{\varepsilon},h,\star}$ denotes the norm dual to $\|\cdot\|_{\boldsymbol{\varepsilon},h}$.

(i) *Estimate of* $\mathfrak{T}_1$. Since $\boldsymbol{z}_{\boldsymbol{g}} \in H_0^1(\Omega)^d \cap H^2(\Omega)^d$, the estimate (7.76) with $r = 0$ yields

$$\|\mathcal{E}_h(\boldsymbol{z}_{\boldsymbol{g}}; \cdot)\|_{\boldsymbol{\varepsilon},h,\star} \lesssim h \left(|\boldsymbol{\sigma}(\boldsymbol{\nabla}_{\mathrm{s}} \boldsymbol{z}_{\boldsymbol{g}})|_{H^1(\Omega)^{d\times d}} + |\boldsymbol{z}_{\boldsymbol{g}}|_{H^2(\Omega)^d} \right) \lesssim h \|\boldsymbol{g}\|,$$

where we have invoked the elliptic regularity estimate (7.91) to conclude. Combining this bound with (7.88), the first term in the right-hand side of (7.93) is estimated as

$$\mathfrak{T}_1 \lesssim h^{r+2} |\boldsymbol{u}|_{H^{r+2}(\mathcal{T}_h)^d}. \tag{7.94}$$

(ii) *Estimate of* $\mathfrak{T}_2$. Apply (7.83) to $\boldsymbol{w} = \boldsymbol{u}$ and $\underline{\boldsymbol{v}}_h = \underline{\boldsymbol{I}}_h^k \boldsymbol{z_g}$. After accounting for Assumption 7.36 to estimate the first factor, we get

$$|\mathcal{E}_h(\boldsymbol{u}; \underline{\boldsymbol{I}}_h^k \boldsymbol{z_g})| \lesssim h^{r+1} |\boldsymbol{u}|_{H^{r+2}(\mathcal{T}_h)^d} \left[\left(\sum_{T \in \mathcal{T}_h} |\underline{\boldsymbol{I}}_T^k \boldsymbol{z_g}|_{1,\partial T}^2 \right)^{\frac{1}{2}} + |\underline{\boldsymbol{I}}_h^k \boldsymbol{z_g}|_{\mathrm{s},h} \right]. \tag{7.95}$$

Using (7.51) with $r = 0$ for all $T \in \mathcal{T}_h$, we see that

$$|\underline{\boldsymbol{I}}_h^k \boldsymbol{z_g}|_{\mathrm{s},h} \lesssim h |\boldsymbol{z_g}|_{H^2(\Omega)^d}.$$

On the other hand, recalling the definition (7.28) of $|\cdot|_{1,\partial T}$ we write, for any $T \in \mathcal{T}_h$,

$$\begin{aligned}
|\underline{\boldsymbol{I}}_h^k \boldsymbol{z_g}|_{1,\partial T}^2 &= \sum_{F \in \mathcal{F}_T} h_F^{-1} \|\boldsymbol{\pi}_F^{0,k} \boldsymbol{z_g} - \boldsymbol{\pi}_T^{0,k} \boldsymbol{z_g}\|_F^2 \\
&= \sum_{F \in \mathcal{F}_T} h_F^{-1} \|\boldsymbol{\pi}_F^{0,k} (\boldsymbol{z_g} - \boldsymbol{\pi}_T^{0,k} \boldsymbol{z_g})\|_F^2 \\
&\le \sum_{F \in \mathcal{F}_T} h_F^{-1} \|\boldsymbol{z_g} - \boldsymbol{\pi}_T^{0,k} \boldsymbol{z_g}\|_F^2 \lesssim h_T^2 |\boldsymbol{z_g}|_{H^2(T)^d}^2,
\end{aligned} \tag{7.96}$$

where we have used the definition of $\underline{\boldsymbol{I}}_h^k \boldsymbol{z_g}$ in the first equality followed by the linearity and polynomial invariance (1.56) of $\boldsymbol{\pi}_F^{0,k}$ in the second equality, the L^2-boundedness of $\boldsymbol{\pi}_F^{0,k}$ in the third line, and concluded by the trace approximation property (1.75) with $l = k$, $m = 0$ and $s = 2$ (we have $s \le l + 1$ since, here, $k \ge 1$) along with the uniform equivalence of face and element diameters (1.6). Plugging the above bounds into (7.95) and recalling the elliptic regularity estimate (7.91), we infer that $|\mathcal{E}_h(\boldsymbol{u}; \underline{\boldsymbol{I}}_h^k \boldsymbol{z_g})| \lesssim h^{r+2} |\boldsymbol{u}|_{H^{r+2}(\mathcal{T}_h)^d} \|\boldsymbol{g}\|$, hence

$$\mathfrak{T}_2 \lesssim h^{r+2} |\boldsymbol{u}|_{H^{r+2}(\mathcal{T}_h)^d}. \tag{7.97}$$

Plugging (7.94) and (7.97) into (7.93), the conclusion follows. □

7.4.3 *Robustness in the Quasi-Incompressible Limit*

To simplify the discussion, throughout this section we work under Assumption 7.36. We are interested in the case $\lambda \gg 1$ (which implies, in particular, $\lambda \ge 0$). This situation corresponds to the quasi-incompressible limit, for which the displacement field is nearly divergence-free. It is well known that the accuracy of $\mathbb{P}^1$-conforming Finite Elements approximations deteriorates in this case, a phenomenon often referred to as *numerical locking*; see, e.g., [30]. The underlying reason is that this Finite Elements space is unable to accurately represent divergence-free displacement fields other than the constant ones, so that the numerical solution is "locked" to a constant value. In the corresponding error estimates, this translates into the fact that the factor that multiplies the meshsize explodes with λ. Robustness for $\lambda \gg 1$ is achieved, on the other hand, when this factor is independent of λ, and the corresponding error estimate (which is therefore uniform in λ) is referred to as *locking-free*. We show here that the error estimates for the HHO method (7.84) are indeed locking-free. In order to derive from Theorem 7.35 a locking-free error estimate, we must show that the right-hand side of (7.89) can be estimated uniformly in λ. The key lies in the following regularity result.

Lemma 7.39 (A Priori Bound on the Exact Solution) *Assume $d = 2$. Let Ω denote a convex polygonal set, and let Assumption 7.36 hold true along with $\lambda \geq \lambda_0$, where $\lambda_0 > 0$ denotes a sufficiently large real number. Then, problem (7.10) has a unique solution $\boldsymbol{u} \in H_0^1(\Omega)^d \cap H^2(\Omega)^d$, and it holds that*

$$\|\boldsymbol{\sigma}(\boldsymbol{\nabla}_{\rm s}\boldsymbol{u})\|_{H^1(\Omega)^{d\times d}} \lesssim \|\boldsymbol{u}\|_{H^2(\Omega)^d} + \lambda\|\boldsymbol{\nabla}\cdot\boldsymbol{u}\|_{H^1(\Omega)} \lesssim \|\boldsymbol{f}\|, \tag{7.98}$$

where the hidden constants depend only on Ω.

Proof The first inequality is an immediate consequence of the expression (7.8) of the stress tensor together with Assumption 7.36 on the Lamé coefficients. The proof of the second inequality can be obtained reasoning as in [78, Lemma 2.2], leveraging the regularity estimates of [217] for the two-dimensional Stokes problem (analogous estimates for the three-dimensional case are derived in [138]). Notice that regularity estimates for the planar elasticity problem in convex domains can also be found in [33]. □

Combining (7.98) with (7.89) written for $r = 0$ and observing that, having assumed $\lambda \geq 0$ and $2\mu = 1$, we can take $\alpha = 1$ in (7.9), we infer that

$$\|\mathbf{G}_{{\rm s},h}^k\underline{\boldsymbol{u}}_h - \boldsymbol{\nabla}_{\rm s}\boldsymbol{u}\| + |\underline{\boldsymbol{u}}_h|_{{\rm s},h} \lesssim h\|\boldsymbol{f}\|,$$

where the hidden constant is independent of h, $\boldsymbol{u}$, and of the problem data. Clearly, this error estimate is locking-free since it is uniform in λ. The crucial point that makes this possible is the commutation property (7.38), which yields the estimate (7.82) on the stress in terms of its H^1-seminorm, the latter being in turn bounded by the L^2-norm of the volumetric force term (see (7.98)). We close this section with two remarks.

Remark 7.40 (High-Order Estimates) Higher-order estimates can be proved proceeding as above whenever the following regularity shift hold for the continuous problem: For any $\boldsymbol{f} \in H^r(\Omega)^d$, it holds $\boldsymbol{u} \in H^{r+2}(\Omega)^d$ and, with hidden constant depending only on Ω,

$$\|\boldsymbol{u}\|_{H^{r+2}(\Omega)^d} + \lambda\|\boldsymbol{\nabla}\cdot\boldsymbol{u}\|_{H^{r+1}(\Omega)} \lesssim \|\boldsymbol{f}\|_{H^r(\Omega)^d}.$$

This requires, in general, further regularity on the domain; see, e.g., [13], where the corresponding results for the Stokes problem are detailed.

Remark 7.41 (Locking-Free L^2-Error Estimates) Minor modifications of the proof of Theorem 7.37 accounting for the a priori estimate (7.98) reveal that a locking-free error estimate can also be obtained for the L^2-norm of the displacement; see also [146, Theorem 11] on this subject.

7.4.4 Numerical Examples

To illustrate the above results, we consider a test case inspired by [75]: we solve on the unit square $\Omega = (0,1)^2$ the Dirichlet problem corresponding to the exact solution such that

$$\boldsymbol{u}(\boldsymbol{x}) = \begin{pmatrix} (\cos(2\pi x_1) - 1)\sin(2\pi x_2) + \frac{1}{1+\lambda}\sin(\pi x_1)\sin(\pi x_2) \\ (1 - \cos(2\pi x_2))\sin(2\pi x_1) + \frac{1}{1+\lambda}\sin(\pi x_1)\sin(\pi x_2) \end{pmatrix}.$$

The corresponding forcing term is

$$\begin{aligned} \boldsymbol{f}(\boldsymbol{x}) = -\mu &\begin{pmatrix} 4\sin(2\pi x_2)\,(1 - 2\cos(2\pi x_1)) - \frac{2}{1+\lambda}\sin(\pi x_1)\sin(\pi x_2) \\ 4\sin(2\pi x_1)\,(2\cos(2\pi x_2) - 1) - \frac{2}{1+\lambda}\sin(\pi x_1)\sin(\pi x_2) \end{pmatrix} \\ &- \frac{\lambda + \mu}{1 + \lambda}\begin{pmatrix} \cos(\pi(x_1 + x_2)) \\ \cos(\pi(x_1 + x_2)) \end{pmatrix}. \end{aligned}$$

We take $\mu = 1$ and, in order to assess the robustness of the method in the quasi-incompressible limit, we let λ vary in $\{1, 10^3, 10^6\}$. For the numerical resolution, we consider a family of deformed quadrangular meshes.

The numerical results are collected in Tables 7.1 and 7.2, where the following quantities are monitored: $N_{\mathrm{dof},h}$, the number of degrees of freedom; $N_{\mathrm{nz},h}$, the number of non-zero entries in the problem matrix; $\|\underline{\boldsymbol{u}}_h - \underline{\boldsymbol{I}}_h^k \boldsymbol{u}\|_{\mathrm{a},h}$, the energy-norm of the error (see Remark 7.34); and $\|\boldsymbol{u}_h - \boldsymbol{\pi}_h^{0,k}\boldsymbol{u}\|$, the L^2-norm of the error estimated in Theorem 7.37. We additionally display the Estimated Order of Convergence (EOC) which, denoting by e_i an error measure on the ith mesh refinement with meshsize h_i, is computed as

$$\mathrm{EOC} = \frac{\log e_i - \log e_{i+1}}{\log h_i - \log h_{i+1}}.$$

In all the cases, the asymptotic EOC match the ones predicted by the theory, that is, $(k+1)$ for the energy-norm and $(k+2)$ for the L^2-norm. The results additionally highlight the robustness of the method in the quasi-incompressible limit, showing errors of comparable magnitude irrespective of the value of λ.

Table 7.1 Numerical results for the test of Sect. 7.4.4, distorted quadrangular mesh family, $k = 1$

$N_{\mathrm{dof},h}$	$N_{\mathrm{nz},h}$	$\|\underline{u}_h - \underline{I}_h^k u\|_{\mathrm{a},h}$	EOC	$\|u_h - \pi_h^{0,k} u\|$	EOC
$\lambda = 1$					
96	2048	1.83e+00	–	1.02e−01	–
448	11,136	5.26e−01	1.80	1.58e−02	2.70
1920	50,816	1.32e−01	2.00	1.94e−03	3.02
7936	216,192	3.36e−02	1.97	2.46e−04	2.98
32256	891,008	8.48e−03	1.98	3.10e−05	2.99
130048	3,616,896	2.12e−03	2.00	3.86e−06	3.00
$\lambda = 10^3$					
96	2048	1.81e+00	–	1.02e−01	–
448	11,136	5.22e−01	1.79	1.57e−02	2.69
1920	50,816	1.31e−01	2.00	1.93e−03	3.03
7936	216,192	3.34e−02	1.97	2.45e−04	2.98
32256	891,008	8.43e−03	1.98	3.09e−05	2.99
130048	3,616,896	2.11e−03	2.00	3.85e−06	3.00
$\lambda = 10^6$					
96	2048	2.23e+00	–	1.02e−01	–
448	11,136	5.39e−01	2.05	1.57e−02	2.69
1920	50,816	1.31e−01	2.04	1.93e−03	3.03
7936	216,192	3.34e−02	1.97	2.45e−04	2.98
32256	891,008	8.43e−03	1.98	3.09e−05	2.99
130048	3,616,896	2.11e−03	2.00	3.85e−06	3.00

7.5 Other Boundary Conditions

We hint in this section at the treatment of more general boundary conditions. Specifically, we consider the case where the displacement is prescribed on a portion of the boundary Γ_{D} with non-zero $(d-1)$-dimensional Hausdorff measure, while tractions are prescribed on the remaining portion $\Gamma_{\mathrm{N}} := \partial\Omega \setminus \Gamma_{\mathrm{D}}$. The extension to the pure traction case is also possible up to minor modifications. Let $\boldsymbol{g}_{\mathrm{D}} := (\boldsymbol{u}_{\mathrm{D}})_{|\Gamma_{\mathrm{D}}}$ with $\boldsymbol{u}_{\mathrm{D}} \in H^1(\Omega)^d$, $\boldsymbol{g}_{\mathrm{N}} \in L^2(\Gamma_{\mathrm{N}})^d$, and consider the problem: Find $\boldsymbol{u} : \Omega \to \mathbb{R}^d$ such that

$$\begin{aligned} -\nabla\cdot\boldsymbol{\sigma}(\nabla_{\mathrm{s}}\boldsymbol{u}) &= \boldsymbol{f} && \text{in } \Omega, \\ \boldsymbol{u} &= \boldsymbol{g}_{\mathrm{D}} && \text{on } \Gamma_{\mathrm{D}}, \\ \boldsymbol{\sigma}(\nabla_{\mathrm{s}}\boldsymbol{u})\boldsymbol{n}_\Omega &= \boldsymbol{g}_{\mathrm{N}} && \text{on } \Gamma_{\mathrm{N}}, \end{aligned} \tag{7.99}$$

where $\boldsymbol{n}_\Omega$ denotes the outer unit normal to Ω on $\partial\Omega$. Denote by $H^1_{\mathrm{D}}(\Omega)$ the space of functions in $H^1(\Omega)$ which vanish (in the sense of traces) on Γ_{D}. Classically, a weak

Table 7.2 Numerical results for the test of Sect. 7.4.4, distorted quadrangular mesh family, $k = 2$

$N_{\mathrm{dof},h}$	$N_{\mathrm{nz},h}$	$\|\underline{\boldsymbol{u}}_h - \underline{\boldsymbol{I}}_h^k \boldsymbol{u}\|_{\mathrm{a},h}$	EOC	$\|\boldsymbol{u}_h - \boldsymbol{\pi}_h^{0,k} \boldsymbol{u}\|$	EOC
$\lambda = 1$					
144	4608	5.45e−01	–	2.80e−02	–
672	25,056	7.56e−02	2.85	1.98e−03	3.82
2880	114,336	1.05e−02	2.85	1.35e−04	3.87
11,904	486,432	1.32e−03	2.99	8.44e−06	4.00
48,384	2,004,768	1.65e−04	2.99	5.28e−07	4.00
195,072	8,138,016	2.06e−05	3.00	3.29e−08	4.01
$\lambda = 10^3$					
144	4608	5.42e−01	–	2.78e−02	–
672	25,056	7.54e−02	2.85	1.98e−03	3.81
2880	114,336	1.04e−02	2.85	1.35e−04	3.87
11,904	486,432	1.31e−03	2.99	8.43e−06	4.00
48,384	2,004,768	1.65e−04	2.99	5.28e−07	4.00
195,072	8,138,016	2.06e−05	3.00	3.29e−08	4.01
$\lambda = 10^6$					
144	4608	5.59e−01	–	2.78e−02	–
672	25,056	7.56e−02	2.89	1.98e−03	3.81
2880	114,336	1.04e−02	2.86	1.35e−04	3.87
11,904	486,432	1.31e−03	2.99	8.43e−06	4.00
48,384	2,004,768	1.65e−04	2.99	5.28e−07	4.00
195,072	8,138,016	2.06e−05	3.00	3.95e−08	3.74

solution can be obtained as $\boldsymbol{u} = \boldsymbol{u}_0 + \boldsymbol{u}_{\mathrm{D}}$ where $\boldsymbol{u}_0 \in H^1_{\mathrm{D}}(\Omega)^d$ is such that

$$(\boldsymbol{\sigma}(\nabla_{\mathrm{s}}\boldsymbol{u}_0), \nabla_{\mathrm{s}}\boldsymbol{v}) = (\boldsymbol{f}, \boldsymbol{v}) - (\boldsymbol{\sigma}(\nabla_{\mathrm{s}}\boldsymbol{u}_{\mathrm{D}}), \nabla_{\mathrm{s}}\boldsymbol{v}) + (\boldsymbol{g}_{\mathrm{N}}, \boldsymbol{v})_{\Gamma_{\mathrm{N}}} \qquad \forall \boldsymbol{v} \in H^1_{\mathrm{D}}(\Omega)^d. \tag{7.100}$$

In order to write the HHO discretisation of problem (7.100), we consider a polygonal (if $d = 2$) or polyhedral (if $d = 3$) mesh $\mathcal{M}_h$ that is boundary-datum compliant (cf. Assumption 2.34). For a fixed polynomial degree $k \geq 1$, we also introduce the space

$$\underline{\boldsymbol{U}}_{h,\mathrm{D}}^k := \left\{ \underline{\boldsymbol{v}}_h \in \underline{\boldsymbol{U}}_h^k \ : \ \boldsymbol{u}_F = \boldsymbol{0} \quad \forall F \in \mathcal{F}_h^D \right\},$$

and we let $\underline{\boldsymbol{u}}_{h,\mathrm{D}} \in \underline{\boldsymbol{U}}_h^k$ be such that

$$\boldsymbol{u}_{T,\mathrm{D}} = \boldsymbol{0} \quad \forall T \in \mathcal{T}_h, \quad \boldsymbol{u}_{F,\mathrm{D}} = \boldsymbol{\pi}_F^{0,k} \boldsymbol{g}_{\mathrm{D}} \quad \forall F \in \mathcal{F}_h^D, \quad \boldsymbol{u}_{F,\mathrm{D}} = \boldsymbol{0} \quad \forall F \in \mathcal{F}_h^{\not D},$$

where $\mathcal{F}_h^{\not D}$ is the set of non-Dirichlet faces defined by (2.87). Then, the HHO solution $\underline{\boldsymbol{u}}_h \in \underline{\boldsymbol{U}}_h^k$ is obtained as $\underline{\boldsymbol{u}}_h = \underline{\boldsymbol{u}}_{h,0} + \underline{\boldsymbol{u}}_{h,\mathrm{D}}$ with $\underline{\boldsymbol{u}}_{h,0} \in \underline{\boldsymbol{U}}_{h,\mathrm{D}}^k$ such that

$$\mathrm{a}_h(\underline{\boldsymbol{u}}_{h,0}, \underline{\boldsymbol{v}}_h) = (\boldsymbol{f}, \boldsymbol{v}_h) - \mathrm{a}_h(\underline{\boldsymbol{u}}_{h,\mathrm{D}}, \underline{\boldsymbol{v}}_h) + \sum_{F \in \mathcal{F}_h^{\mathrm{N}}} (\boldsymbol{g}_{\mathrm{N}}, \boldsymbol{v}_F)_F \qquad \forall \underline{\boldsymbol{v}}_h \in \underline{\boldsymbol{U}}_{h,\mathrm{D}}^k. \tag{7.101}$$

7.6 The Lowest-Order Case

As seen in Remark 7.22, in the lowest-order case corresponding to $k = 0$ assumptions (SE2) and (SE3) are incompatible, and one therefore cannot design a proper *local* stabilisation term. In this section we show, following [71], that a stable and convergent method for $k = 0$ can be recovered adding a jump penalisation term inspired by the discrete Korn inequality (7.66) in broken polynomial spaces. This comes at the price of introducing additional links among element-based unknowns. Throughout the rest of this section we work, for the sake of simplicity, under Assumption 7.36 (that is, $2\mu = 1$ and λ is constant).

7.6.1 A Global Discrete Strain Norm Including Jumps

Let the global displacement reconstruction $\mathbf{p}_h^1 : \underline{\boldsymbol{U}}_h^0 \to \mathbb{P}^1(\mathcal{T}_h)^d$ be such that, for all $\underline{\boldsymbol{v}}_h \in \underline{\boldsymbol{U}}_h^0$,

$$(\mathbf{p}_h^1 \underline{\boldsymbol{v}}_h)_{|T} := \mathbf{p}_T^1 \underline{\boldsymbol{v}}_T \qquad \forall T \in \mathcal{T}_h.$$

We define the map $|||{\cdot}|||_{\boldsymbol{\varepsilon},h} : \underline{\boldsymbol{U}}_h^0 \to \mathbb{R}^+$ setting, for any $\underline{\boldsymbol{v}}_h \in \underline{\boldsymbol{U}}_h^0$,

$$|||\underline{\boldsymbol{v}}_h|||_{\boldsymbol{\varepsilon},h} := \left(\|\mathbf{p}_h^1 \underline{\boldsymbol{v}}_h\|_{\boldsymbol{\varepsilon},\mathrm{j},h}^2 + |\underline{\boldsymbol{v}}_h|_{\mathrm{s},h}^2 \right)^{\frac{1}{2}} \tag{7.102}$$

with $\|{\cdot}\|_{\boldsymbol{\varepsilon},\mathrm{j},h}$-norm defined by (7.65) and $|{\cdot}|_{\mathrm{s},h}$-seminorm defined by (7.61) from the local stabilisation bilinear forms given, for any $T \in \mathcal{T}_h$, by (7.54) with $k = 0$. We have the following norm equivalence, upon which rests the stability of the HHO method studied in this section.

Lemma 7.42 (Global Stability and Boundedness) *For all $\underline{\boldsymbol{v}}_h \in \underline{\boldsymbol{U}}_{h,0}^0$, it holds*

$$\|\boldsymbol{\nabla}_{\mathrm{s},h} \mathbf{p}_h^1 \underline{\boldsymbol{v}}_h\|^2 + |\underline{\boldsymbol{v}}_h|_{\mathrm{s},h}^2 \lesssim \|\underline{\boldsymbol{v}}_h\|_{1,h}^2 \lesssim |||\underline{\boldsymbol{v}}_h|||_{\boldsymbol{\varepsilon},h}^2, \tag{7.103}$$

with $\|\cdot\|_{1,h}$*-norm defined by* (7.68) *and hidden constants independent of both h and* $\underline{v}_h$*, but possibly depending on* Ω*, d, and* ϱ*. As a consequence, the map* $|||\cdot|||_{\varepsilon,h}$ *defines a norm on* $\underline{U}_{h,0}^0$*.*

Proof By Remark 7.8, the local displacement reconstruction $\mathbf{p}_T^1$ coincides with the component-wise application of the reconstruction operator p_T^1 defined by (2.11). Hence, it follows from (2.41) that

$$\|\nabla_h \mathbf{p}_h^1 \underline{v}_h\|^2 + |\underline{v}_h|_{\mathrm{s},h}^2 \simeq \|\underline{v}_h\|_{1,h}^2. \tag{7.104}$$

On the other hand, using the definition of the symmetric gradient for the first bound and the global discrete Korn inequality (7.66) for the second, we can write

$$\|\nabla_{\mathrm{s},h} \mathbf{p}_h^1 \underline{v}_h\|^2 \lesssim \|\nabla_h \mathbf{p}_h^1 \underline{v}_h\|^2 \lesssim \|\mathbf{p}_h^1 \underline{v}_h\|_{\varepsilon,\mathrm{j},h}^2. \tag{7.105}$$

Combining (7.105) with (7.104) yields (7.103). The fact that $|||\cdot|||_{\varepsilon,h}$ defines a norm on $\underline{U}_{h,0}^0$ follows observing that $\|\cdot\|_{1,h}$ is itself a norm on this space (see Corollary 2.16). □

7.6.2 A Global Bilinear Form with Jump Penalisation

We consider the bilinear form $\mathrm{a}_h^{\mathrm{lo}} : \underline{U}_h^0 \times \underline{U}_h^0 \to \mathbb{R}$ such that, for all $\underline{u}_h, \underline{v}_h \in \underline{U}_h^0$,

$$\mathrm{a}_h^{\mathrm{lo}}(\underline{u}_h, \underline{v}_h) := \sum_{T\in\mathcal{T}_h} \mathrm{a}_T(\underline{u}_T, \underline{v}_T) + \mathrm{j}_h(\mathbf{p}_h^1 \underline{u}_h, \mathbf{p}_h^1 \underline{v}_h), \tag{7.106}$$

where, for all $T \in \mathcal{T}_h$, the local bilinear form a_T is given by (7.45) with local stabilisation bilinear form s_T as in (7.54) with $k = 0$, while the jump penalisation bilinear form $\mathrm{j}_h : H^1(\mathcal{T}_h)^d \times H^1(\mathcal{T}_h)^d \to \mathbb{R}$ is such that, for all $\boldsymbol{u}, \boldsymbol{v} \in H^1(\mathcal{T}_h)^d$,

$$\mathrm{j}_h(\boldsymbol{u}, \boldsymbol{v}) := \sum_{F\in\mathcal{F}_h} h_F^{-1}([\boldsymbol{u}]_F, [\boldsymbol{v}]_F)_F.$$

We next reformulate $\mathrm{a}_h^{\mathrm{lo}}$ in a more convenient way. Observing that it holds $\nabla_{\mathrm{s}}\mathbb{P}^1(T)^d = \mathbb{P}^0(T; \mathbb{R}_{\mathrm{sym}}^{d\times d})$ for all $T \in \mathcal{T}_h$ and using (7.42), we infer that

$$(\nabla_{\mathrm{s},h} \mathbf{p}_h^1 \underline{v}_h)_{|T} = \mathbf{G}_{\mathrm{s},T}^0 \underline{v}_T \qquad \forall T \in \mathcal{T}_h. \tag{7.107}$$

Expanding a_T according to its definition (7.45) and using (7.107), we arrive at the following equivalent expression for $\mathrm{a}_h^{\mathrm{lo}}$: For all $\underline{u}_h, \underline{v}_h \in \underline{U}_{h,0}^0$,

$$\mathrm{a}_h^{\mathrm{lo}}(\underline{u}_h, \underline{v}_h) := (\boldsymbol{\sigma}(\nabla_{\mathrm{s},h} \mathbf{p}_h^1 \underline{u}_h), \nabla_{\mathrm{s},h} \mathbf{p}_h^1 \underline{v}_h) + \mathrm{j}_h(\mathbf{p}_h^1 \underline{u}_h, \mathbf{p}_h^1 \underline{v}_h) + \mathrm{s}_h(\underline{u}_h, \underline{v}_h), \tag{7.108}$$

where $\mathrm{s}_h(\underline{\boldsymbol{u}}_h, \underline{\boldsymbol{v}}_h) := \sum_{T\in\mathcal{T}_h} \mathrm{s}_T(\underline{\boldsymbol{u}}_T, \underline{\boldsymbol{v}}_T)$. Inside the proof of the next lemma, we will need the global strain projector $\boldsymbol{\pi}_h^{\boldsymbol{\varepsilon},1} : H^1(\Omega)^d \to \mathbb{P}^1(\mathcal{T}_h)^d$ such that

$$(\boldsymbol{\pi}_h^{\boldsymbol{\varepsilon},1}\boldsymbol{v})_{|T} := \boldsymbol{\pi}_T^{\boldsymbol{\varepsilon},1}\boldsymbol{v}_{|T}.$$

Lemma 7.43 (Properties of $\mathrm{a}_h^{\mathrm{lo}}$) *The bilinear form $\mathrm{a}_h^{\mathrm{lo}}$ enjoys the following properties:*

(i) Stability and boundedness. *For all $\underline{\boldsymbol{v}}_h \in \underline{\boldsymbol{U}}_h^0$, it holds*

$$\alpha|||\underline{\boldsymbol{v}}_h|||_{\boldsymbol{\varepsilon},h}^2 \lesssim \mathrm{a}_h^{\mathrm{lo}}(\underline{\boldsymbol{v}}_h, \underline{\boldsymbol{v}}_h) \lesssim (1 + d|\lambda|)\, |||\underline{\boldsymbol{v}}_h|||_{\boldsymbol{\varepsilon},h}^2, \tag{7.109}$$

where the hidden constant is independent of h, $\underline{\boldsymbol{v}}_h$ and of the problem data, and the triple-bar strain norm $|||\cdot|||_{\boldsymbol{\varepsilon},h}$ is defined by (7.102).

(ii) Consistency. *It holds, for all $\boldsymbol{w} \in H_0^1(\Omega)^d \cap H^2(\mathcal{T}_h)^d$ such that $\boldsymbol{\nabla}\cdot\boldsymbol{\sigma}(\boldsymbol{\nabla}_{\mathrm{s}}\boldsymbol{w}) \in L^2(\Omega)^d$,*

$$\sup_{\underline{\boldsymbol{v}}_h\in\underline{\boldsymbol{U}}_{h,0}^0,\, |||\underline{\boldsymbol{v}}_h|||_{\boldsymbol{\varepsilon},h}=1} \left|\mathcal{E}_h^{\mathrm{lo}}(\boldsymbol{w}; \underline{\boldsymbol{v}}_h)\right| \lesssim h\left(|\boldsymbol{\sigma}(\boldsymbol{\nabla}_{\mathrm{s}}\boldsymbol{w})|_{H^1(\mathcal{T}_h)^{d\times d}} + |\boldsymbol{w}|_{H^2(\mathcal{T}_h)^d}\right), \tag{7.110}$$

where the hidden constant is independent of $\boldsymbol{w}$, h and of the problem data, and the linear form $\mathcal{E}_h^{\mathrm{lo}}(\boldsymbol{w};\cdot) : \underline{\boldsymbol{U}}_{h,0}^0 \to \mathbb{R}$ representing the consistency error is such that, for all $\underline{\boldsymbol{v}}_h \in \underline{\boldsymbol{U}}_{h,0}^0$,

$$\mathcal{E}_h^{\mathrm{lo}}(\boldsymbol{w}; \underline{\boldsymbol{v}}_h) := -(\boldsymbol{\nabla}\cdot\boldsymbol{\sigma}(\boldsymbol{\nabla}_{\mathrm{s}}\boldsymbol{w}), \boldsymbol{v}_h) - \mathrm{a}_h^{\mathrm{lo}}(\underline{\boldsymbol{I}}_h^0\boldsymbol{w}, \underline{\boldsymbol{v}}_h). \tag{7.111}$$

Proof (i) *Stability and boundedness.* Accounting for Assumption 7.36, the norm equivalence in (7.109) is an immediate consequence of the definition (7.102) of the triple-bar strain norm $|||\cdot|||_{\boldsymbol{\varepsilon},h}$ and of the reformulation (7.108) of the bilinear form $\mathrm{a}_h^{\mathrm{lo}}$.

(ii) *Consistency.* Let $\underline{\boldsymbol{v}}_h \in \underline{\boldsymbol{U}}_{h,0}^0$. Proceeding as in the proof of the second point of Lemma 7.27, we deduce that

$$\left|\mathcal{E}_h^{\mathrm{lo}}(\boldsymbol{w}; \underline{\boldsymbol{v}}_h)\right| = \underbrace{\left(\sum_{T\in\mathcal{T}_h} h_T\|\boldsymbol{\sigma}(\boldsymbol{\nabla}_{\mathrm{s}}\boldsymbol{w})_{|T} - \boldsymbol{\pi}_T^{0,0}(\boldsymbol{\sigma}(\boldsymbol{\nabla}_{\mathrm{s}}\boldsymbol{w}))\|_{\partial T}^2\right)^{\frac{1}{2}}}_{\mathfrak{T}_1} \left(\sum_{T\in\mathcal{T}_h} |\underline{\boldsymbol{v}}_T|_{1,\partial T}^2\right)^{\frac{1}{2}}$$
$$+ \underbrace{|\mathbf{p}_h^1\underline{\boldsymbol{I}}_h^0\boldsymbol{w}|_{\mathrm{j},h}}_{\mathfrak{T}_2}\, |\mathbf{p}_h^1\underline{\boldsymbol{v}}_h|_{\mathrm{j},h} + \underbrace{|\underline{\boldsymbol{I}}_h^0\boldsymbol{w}|_{\mathrm{s},h}}_{\mathfrak{T}_3}\, |\underline{\boldsymbol{v}}_h|_{\mathrm{s},h}.$$

Using the second inequality in (7.103) and recalling the definition (7.102) of the triple-bar strain norm, we arrive at

$$\left|\mathcal{E}_h^{\text{lo}}(\boldsymbol{w}; \underline{\boldsymbol{v}}_h)\right| \lesssim (\mathfrak{T}_1 + \mathfrak{T}_2 + \mathfrak{T}_3) \, |||\underline{\boldsymbol{v}}_h|||_{\boldsymbol{\varepsilon},h}. \tag{7.112}$$

Let us estimate the terms in parentheses. Using the trace approximation properties (1.75) of the L^2-orthogonal projector with $l = 0$, $m = 0$, $p = 2$, and $s = 1$ we get, for any $T \in \mathcal{T}_h$, $h_T^{\frac{1}{2}} \|\boldsymbol{\sigma}(\boldsymbol{\nabla}_{\text{s}}\boldsymbol{w})_{|T} - \boldsymbol{\pi}_T^{0,0}(\boldsymbol{\sigma}(\boldsymbol{\nabla}_{\text{s}}\boldsymbol{w}))\|_{\partial T} \le h_T |\boldsymbol{\sigma}(\boldsymbol{\nabla}_{\text{s}}\boldsymbol{w})|_{H^1(T)^{d\times d}}$, which gives for the first term

$$\mathfrak{T}_1 \lesssim h |\boldsymbol{\sigma}(\boldsymbol{\nabla}_{\text{s}}\boldsymbol{w})|_{H^1(\mathcal{T}_h)^{d\times d}}. \tag{7.113}$$

Moving to the second term, we can write

$$\begin{aligned}
\mathfrak{T}_2^2 &= \sum_{F\in\mathcal{F}_h} h_F^{-1} \|[\boldsymbol{\pi}_h^{\boldsymbol{\varepsilon},1}\boldsymbol{w}]_F\|_F^2 \\
&= \sum_{F\in\mathcal{F}_h} h_F^{-1} \|[\boldsymbol{\pi}_h^{\boldsymbol{\varepsilon},1}\boldsymbol{w} - \boldsymbol{w}]_F\|_F^2 \\
&\lesssim \sum_{F\in\mathcal{F}_h} \sum_{T\in\mathcal{T}_F} h_F^{-1} \|\boldsymbol{\pi}_T^{\boldsymbol{\varepsilon},1}\boldsymbol{w} - \boldsymbol{w}\|_F^2 \\
&\lesssim \sum_{T\in\mathcal{T}_h} h_T^{-1} \|\boldsymbol{\pi}_T^{\boldsymbol{\varepsilon},1}\boldsymbol{w} - \boldsymbol{w}\|_{\partial T}^2,
\end{aligned}$$

where we have used (7.40) to replace $\mathbf{p}_h^1\underline{\boldsymbol{I}}_h^0$ by $\boldsymbol{\pi}_h^{\boldsymbol{\varepsilon},1}$ along with the definition (7.65) of the jump seminorm in the first line, the fact that the jumps of $\boldsymbol{w}$ vanish across any $F \in \mathcal{F}_h$ (a consequence of the assumed regularity $\boldsymbol{w} \in H_0^1(\Omega)^d$ together with Lemma 1.21 with $p = 2$) to insert them into the second line, the definition (1.22) of the jump operator together with a triangle inequality in the third line, and we have exchanged the order of the summations over faces and elements according to (1.25) and used the mesh regularity to write $h_F^{-1} \lesssim h_T^{-1}$ and conclude. Hence, using the trace approximation properties (7.20) of the strain projector and taking the square root, we arrive at

$$\mathfrak{T}_2 \lesssim h |\boldsymbol{w}|_{H^2(\mathcal{T}_h)^d}. \tag{7.114}$$

For the third term, invoking the consistency property (7.51), which also holds for $k = 0$, of s_T with $r = 0$ for all $T \in \mathcal{T}_h$ readily gives

$$\mathfrak{T}_3 \lesssim h |\boldsymbol{w}|_{H^2(\mathcal{T}_h)^d}. \tag{7.115}$$

Plugging (7.113), (7.114), and (7.115) into (7.112) and passing to the supremum over $\underline{v}_h \in \underline{U}_{h,0}^0$ such that $|||\underline{v}_h|||_{\varepsilon,h} = 1$ yields (7.110). □

7.6.3 *Discrete Problem and Energy Error Estimate*

The lowest-order HHO scheme for the approximation of problem (7.10) reads: Find $\underline{u}_h \in \underline{U}_{h,0}^0$ such that

$$\mathrm{a}_h^{\mathrm{lo}}(\underline{u}_h, \underline{v}_h) = (f, v_h) \qquad \forall \underline{v}_h \in \underline{U}_{h,0}^0. \tag{7.116}$$

The well-posedness of problem (7.116), with corresponding a priori bounds on the discrete solution, can be proved adapting the arguments of Lemma 7.30 and using (7.109); the details are left to the reader.

Remark 7.44 (Static Condensation for Problem (7.116)*)* The jump stabilisation introduces a direct link among element-based discrete unknowns of neighbouring mesh elements. As a result, static condensation (see in Sect. B.3.2) is no longer an interesting option.

A convergence result is stated in the following theorem.

Theorem 7.45 (Discrete Energy Error Estimate for the Lowest-Order HHO Scheme) *Suppose that Assumption 7.36 holds, and let $(\mathcal{M}_h)_{h\in\mathcal{H}} = (\mathcal{T}_h, \mathcal{F}_h)_{h\in\mathcal{H}}$ be a regular mesh sequence in the sense of Definition 1.9. Let $u \in H_0^1(\Omega)^d$ denote the unique solution to* (7.10)*, for which we assume the additional regularity $u \in H^2(\mathcal{T}_h)^d$. For all $h \in \mathcal{H}$, let $\underline{u}_h \in \underline{U}_{h,0}^0$ denote the unique solution to* (7.116)*. Then,*

$$|||\underline{u}_h - \underline{I}_h^0 u|||_{\varepsilon,h} \lesssim h\alpha^{-1}\left(|\sigma(\nabla_s u)|_{H^1(\mathcal{T}_h)^{d\times d}} + |u|_{H^2(\mathcal{T}_h)^d}\right), \tag{7.117}$$

where, according to (7.9)*, $\alpha = 1 - d\lambda^-$, the norm $|||\cdot|||_{\varepsilon,h}$ is defined in* (7.102)*, and the hidden constant is independent of h, u, and of the problem data.*

Proof We invoke the Third Strang Lemma A.7 with $\mathrm{U} = H_0^1(\Omega)^d$, $\mathrm{a} = a$ defined by (7.11), $\mathrm{l}(v) = (f, v)$, $\mathrm{U}_h = \underline{U}_{h,0}^0$ endowed with the norm $|||\cdot|||_{\varepsilon,h}$, $\mathrm{a}_h = \mathrm{a}_h^{\mathrm{lo}}$, $\mathrm{l}_h(\underline{v}_h) = (f, v_h)$ and $\mathrm{I}_h u = \underline{I}_h^0 u$. By (7.109), a_h is coercive for $|||\cdot|||_{\varepsilon,h}$ with constant $\gtrsim \alpha$ and, since $-\nabla\cdot\sigma(\nabla_s u) = f$, the consistency error (A.5) is exactly (7.111) with $w = u$. Hence, (7.117) follows plugging (7.110) into (A.6). □

A few remarks are in order.

Remark 7.46 (Discrete Error Estimate in the Norm Induced by $\mathrm{a}_h^{\mathrm{lo}}$*)* In the spirit of Remark 7.34, quasi-optimal error estimates can also be derived in the norm induced by the bilinear form $\mathrm{a}_h^{\mathrm{lo}}$ and such that, for all $\underline{\boldsymbol{v}}_h \in \underline{\boldsymbol{U}}_{h,0}^0$,

$$|||\underline{\boldsymbol{v}}_h|||_{\mathrm{a},h} := \mathrm{a}_h^{\mathrm{lo}}(\underline{\boldsymbol{v}}_h, \underline{\boldsymbol{v}}_h)^{\frac{1}{2}}. \tag{7.118}$$

Specifically, under the assumptions of Theorem 7.45, it holds that

$$|||\underline{\boldsymbol{u}}_h - \underline{\boldsymbol{I}}_h^k \boldsymbol{u}|||_{\mathrm{a},h} \lesssim h\left(\alpha^{-\frac{1}{2}}|\boldsymbol{\sigma}(\boldsymbol{\nabla}_{\mathrm{s}}\boldsymbol{u})|_{H^1(\mathcal{T}_h)^{d\times d}} + |\boldsymbol{u}|_{H^2(\mathcal{T}_h)^d}\right).$$

Remark 7.47 (L^2-Error Estimate) An estimate in h^2 for the L^2-norm of the error can be derived under the elliptic regularity assumption. The interested reader can find the details in [71, Theorem 12].

Remark 7.48 (Robustness in the Quasi-Incompressible Limit) In the framework of Sect. 7.4.3, combining (7.117) with the a priori bound (7.98) yields the following locking-free energy error estimate:

$$|||\underline{\boldsymbol{u}}_h - \underline{\boldsymbol{I}}_h^0 \boldsymbol{u}|||_{\boldsymbol{\varepsilon},h} \lesssim h\|\boldsymbol{f}\|.$$

Under the same assumptions, it can be proved that the L^2-norm error estimate discussed in the previous remark is also uniform in λ.

7.6.4 Numerical Examples

We close this section with some numerical tests.

7.6.4.1 Quasi-Incompressible Test Case

To illustrate the performance of the lowest-order scheme, we run the test case of Sect. 7.4.4 for the same values of the Lamé coefficients. The results collected in Table 7.3 show that the expected EOC of 1 for the energy-norm and 2 for the L^2-norm are attained.

Table 7.3 Numerical results for the test of Sect. 7.6.4.1, distorted quadrangular mesh family

$N_{\mathrm{dof},h}$	$N_{\mathrm{nz},h}$	$\|\|\|\underline{\boldsymbol{u}}_h - \underline{\boldsymbol{I}}_h^0\boldsymbol{u}\|\|\|_{\mathrm{a},h}$	EOC	$\|\boldsymbol{u}_h - \boldsymbol{\pi}_h^{0,0}\boldsymbol{u}\|$	EOC
$\lambda = 1$					
80	2768	3.51e+00	–	1.89e−01	–
352	15,856	1.91e+00	0.88	5.45e−02	1.79
1472	73,904	1.08e+00	0.82	1.34e−02	2.03
6016	317,488	5.83e−01	0.89	3.52e−03	1.93
24,320	1,314,608	2.97e−01	0.97	9.18e−04	1.94
97,792	5,348,656	1.49e−01	0.99	2.33e−04	1.98
$\lambda = 10^3$					
80	2768	3.44e+00	–	1.96e−01	–
352	15,856	1.87e+00	0.88	5.89e−02	1.73
1472	73,904	1.07e+00	0.81	1.63e−02	1.85
6016	317,488	5.74e−01	0.89	4.48e−03	1.86
24,320	1,314,608	2.92e−01	0.97	1.18e−03	1.93
97,792	5,348,656	1.47e−01	0.99	3.00e−04	1.97
$\lambda = 10^6$					
80	2768	9.12e+00	–	1.96e−01	–
352	15,856	2.27e+00	2.00	5.89e−02	1.73
1472	73,904	1.08e+00	1.07	1.63e−02	1.85
6016	317,488	5.74e−01	0.91	4.48e−03	1.86
24,320	1,314,608	2.92e−01	0.97	1.18e−03	1.93
97,792	5,348,656	1.47e−01	0.99	3.00e−04	1.97

7.6.4.2 Singular Test Case

We next consider the solution of [11, Section 5.1] which, in polar coordinates (r, θ), reads

$$\boldsymbol{u}(r,\theta) = \frac{1}{2G} r^L \begin{pmatrix} (\kappa - Q(L+1))\cos(L\theta) - L\cos((L-2)\theta) \\ (\kappa + Q(L+1))\sin(L\theta) + L\sin((L-2)\theta) \end{pmatrix},$$

where the various parameters take the following numerical values: $\mu = 0.65$, $\lambda = 0.98$, $G = \frac{5}{13}$, $\kappa = \frac{9}{5}$, $L = 0.5444837367825$, $Q = 0.5430755788367$. The forcing term in this case is equal to zero, while the Dirichlet boundary condition is inferred from the exact solution. The domain Ω is illustrated in Fig. 7.3, while the solution on the finest computational mesh considered here is depicted in Fig. 7.4. This test case is representative of real-life situations corresponding to a mode 1 fracture in a plane strain problem. The solution exhibits a singularity at the origin, which prevents the method from attaining the full orders of convergence predicted for smooth solutions on uniformly refined mesh sequences.

For the numerical resolution, we consider a sequence of refined structured quadrangular meshes. The numerical results collected in the top half of Table 7.4

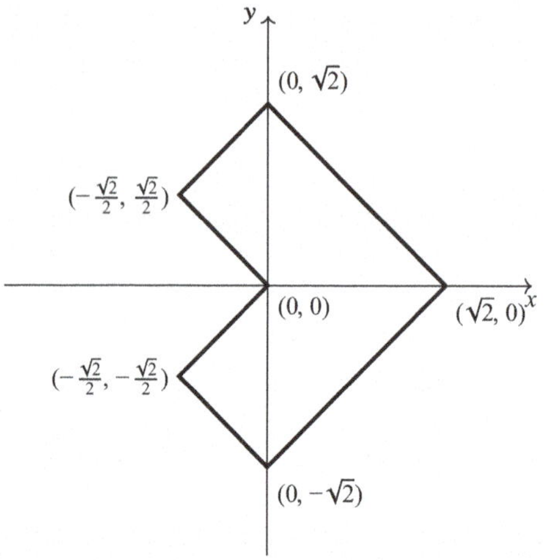

Fig. 7.3 Domain for the test case of Sect. 7.6.4.2

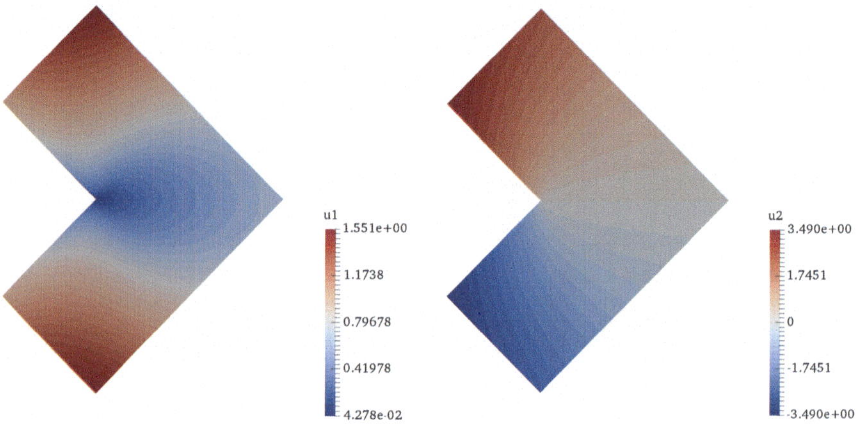

(a) (b)

Fig. 7.4 Numerical solution for the test of Sect. 7.6.4.2. (**a**) u_1, (**b**) u_2

show an asymptotic EOC in the energy-norm of about 0.54, while the asymptotic EOC in the L^2-norm is about 1.31. For the sake of completeness, we show, in the bottom half of Table 7.4, a comparison with the HHO method (7.84) with $k = 1$. The EOC are also limited by the regularity of the solution, and coincide with those observed for the lowest-order method (7.116). As expected, the number of unknowns on a given mesh is larger for the method with $k = 1$ compared to the method with $k = 0$, despite the fact that static condensation is applied in the former case. It has to be noticed, however, that the reduction in the number of unknowns is balanced by the increased number of non-zero entries in the matrix, due to both the

Table 7.4 Numerical results for the test of Sect. 7.6.4.2 and comparison with the high-order method (7.84) with $k = 1$. For the latter, the energy norm is the one associated to the corresponding bilinear form without jump stabilisation

$N_{\mathrm{dof},h}$	$N_{\mathrm{nz},h}$	$\|\underline{\boldsymbol{u}}_h - \underline{\boldsymbol{I}}_h^k \boldsymbol{u}\|_{\mathrm{a},h}$	EOC	$\|\boldsymbol{u}_h - \boldsymbol{\pi}_h^{0,k} \boldsymbol{u}\|$	EOC
Lowest-order HHO method (7.116)					
256	10,616	7.65e−01	–	7.51e−02	–
1088	52,728	5.63e−01	0.44	3.34e−02	1.17
4480	232,568	3.97e−01	0.50	1.40e−02	1.25
18,176	974,712	2.76e−01	0.53	5.72e−03	1.29
73,216	3,988,856	1.90e−01	0.54	2.31e−03	1.31
293,888	16,136,568	1.31e−01	0.54	9.29e−04	1.31
HHO method (7.84) *with* $k = 1$					
320	7584	1.07e−01	–	9.40e−03	–
1408	36,512	7.32e−02	0.55	3.64e−03	1.37
5888	158,880	5.01e−02	0.55	1.41e−03	1.36
24,064	661,664	3.43e−02	0.55	5.52e−04	1.36
97,280	2,699,424	2.35e−02	0.54	2.17e−04	1.35
391,168	10,903,712	1.61e−02	0.54	8.57e−05	1.34

absence of static condensation and the presence of the jump penalisation term. This phenomenon is specific to the two-dimensional case: in dimension $d = 3$, the matrix corresponding to (7.84) with $k = 1$ is in general more dense; see [71, Section 6.3] for an example. The errors in the energy norm appear to be smaller for the method with $k = 1$, but this is in part due to the fact that the natural energy norm associated with the corresponding bilinear form does not contain the norm of the jumps.

7.7 Proof of the Uniform Local Second Korn Inequality

In this section we prove Lemma 7.7. A preliminary step consists in proving the following lemma, which gives the existence of a uniform atlas for all mesh elements that are star-shaped with respect to every point in a ball of radius comparable to their diameter. For a given positive number ζ, recall that $\mathcal{B}_d(\mathbf{0}, \zeta)$ denotes the open ball in $\mathbb{R}^d$ centred at the origin and of radius ζ. For a given unit vector $\boldsymbol{r}$, we define the semi-infinite cylinder

$$M(\boldsymbol{r}, \zeta) := \{\boldsymbol{x} := \boldsymbol{x}^\perp + z\boldsymbol{r} \ :\ \boldsymbol{x}^\perp \in \mathcal{B}_d(\mathbf{0}, \zeta) \text{ is orthogonal to } \boldsymbol{r} \text{ and } z \in [0, \infty)\}.$$

Figure 7.5 provides an illustration of this definition, along with other notations used in the proof of the lemma. In what follows, for the open unit ball centred at the origin, we use the abridged notation $\mathcal{B}_d := \mathcal{B}_d(\mathbf{0}, 1)$.

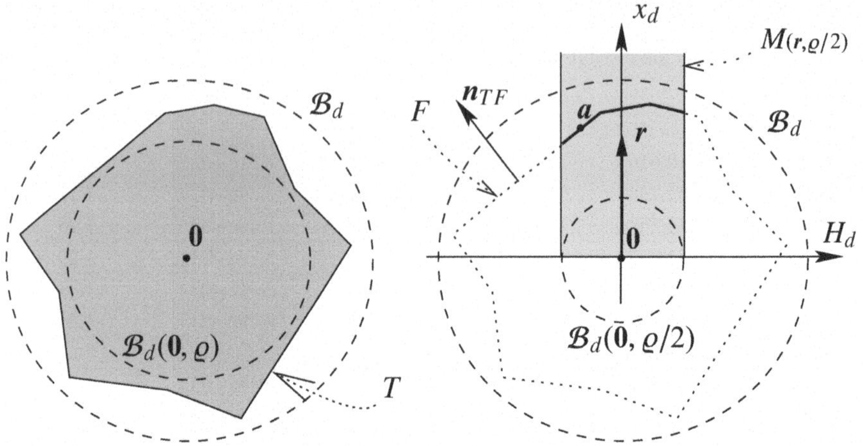

Fig. 7.5 Illustration of the proof of Lemma 7.49

Lemma 7.49 (Uniform Atlas for Star-Shaped Elements) *Let $\varrho > 0$. There exists a finite number $m \in \mathbb{N}$ of unit vectors $\boldsymbol{r}_1, \ldots, \boldsymbol{r}_m \in \mathbb{R}^d$ and a real number $L > 0$, all depending only on d and ϱ, such that $\mathcal{B}_d \subset \bigcup_{l=1}^m M(\boldsymbol{r}_l, \varrho/2)$ and, if T is a polytope of $\mathbb{R}^d$ contained in $\mathcal{B}_d$ and star-shaped with respect to every point in $\mathcal{B}_d(\mathbf{0}, \varrho)$, for any $l = 1, \ldots, m$,*

$$T \cap M(\boldsymbol{r}_l, \varrho/2) = \{\boldsymbol{x} = (x_1, \ldots, x_d) \in M(\boldsymbol{r}_l, \varrho/2) \,:\, x_d \le \varphi_l(x_1, \ldots, x_{d-1})\},$$

where the system of orthonormal coordinates $(x_1, \ldots, x_d)$ is chosen such that x_d is the coordinate along $\boldsymbol{r}_l$, $H_d := \{\boldsymbol{x} \in \mathbb{R}^d \,:\, x_d = 0\}$ is the horizontal hyperplane in this system of coordinates, and $\varphi_l : \mathcal{B}_d(\mathbf{0}, \varrho/2) \cap H_d \to \mathbb{R}$ is a Lipschitz-continuous function with Lipschitz constant bounded by L.

Proof In the following, $a \lesssim b$ means that $a \le Cb$ with C depending only on d and ϱ. We first notice that, since $\mathcal{B}_d$ is determined by d, there is a fixed number m of unit vectors $(\boldsymbol{r}_1, \ldots, \boldsymbol{r}_m)$, depending only on d and ϱ, such that $\mathcal{B}_d \subset \bigcup_{l=1}^m M(\boldsymbol{r}_l, \varrho/2)$. The proof is completed by showing that, in each $M(\boldsymbol{r}_l, \varrho/2)$ and in the coordinates associated with $\boldsymbol{r}_l$ as in the lemma, T is the hypograph of a Lipschitz function φ_l, with a controlled Lipschitz constant. From this point on, we drop the index l for legibility.

The existence of a continuous function φ defined on $M(\boldsymbol{r}, \varrho/2) \cap H_d$ and such that $T \cap M(\boldsymbol{r}, \varrho/2)$ is the hypograph of φ results from the fact that T is star-shaped with respect to all points in $\mathcal{B}_d(\mathbf{0}, \varrho/2)$. Since the boundary of T is made of the faces $F \in \mathcal{F}_T$, this function φ is piecewise affine, and it holds that

$$\boldsymbol{n}_{TF} = \frac{(-\boldsymbol{\nabla}_{d-1}\varphi, 1)}{\left(1 + |\boldsymbol{\nabla}_{d-1}\varphi|^2\right)^{\frac{1}{2}}},$$

where $\nabla_{d-1}\varphi$ is the gradient in H_d of φ with respect to its $(d-1)$ variables. Hence, since $\boldsymbol{r} = (0, 1)$ in the local system of coordinates, it holds $\boldsymbol{r}\cdot\boldsymbol{n}_{TF} = (1 + |\nabla_{d-1}\varphi|^2)^{-\frac{1}{2}}$. If we can prove that

$$1 \lesssim \boldsymbol{n}_{TF}\cdot\boldsymbol{r} \qquad \forall F \in \mathcal{F}_T \text{ such that } F \cap M(\boldsymbol{r}, \varrho/2) \neq \emptyset, \tag{7.119}$$

then we will have $|\nabla_{d-1}\varphi| \lesssim 1$, which yields a uniform control of the Lipschitz constant of φ.

To prove (7.119), let $F \in \mathcal{F}_T$ and $\boldsymbol{a} \in F \cap M(\boldsymbol{r}, \varrho/2)$, and let us translate the fact that T is star-shaped with respect to every point in $\mathcal{B}_d(\boldsymbol{0}, \varrho)$. Working as in the proof of [178, Lemma B.1], we see that this assumption forces $\mathcal{B}_d(\boldsymbol{0}, \varrho)$ to be fully on one side of the hyperplane spanned by F, which translates into

$$(\boldsymbol{a} - \boldsymbol{x})\cdot\boldsymbol{n}_{TF} \geq 0 \qquad \forall \boldsymbol{x} \in \mathcal{B}_d(\boldsymbol{0}, \varrho). \tag{7.120}$$

On the other hand, since $\boldsymbol{a} \in M(\boldsymbol{r}, \varrho/2)$, we have $\boldsymbol{a} = \boldsymbol{a}^\perp + z\boldsymbol{r}$ with $z > 0$ and $\boldsymbol{a}^\perp \in \mathcal{B}_d(\boldsymbol{0}, \varrho/2)$ orthogonal to $\boldsymbol{r}$. Apply (7.120) to $\boldsymbol{x} = \boldsymbol{a}^\perp + (\varrho/2)\boldsymbol{n}_{TF}$, which belongs to $\mathcal{B}_d(\boldsymbol{0}, \varrho)$ since $|\boldsymbol{a}^\perp| \leq \varrho/2$. Noticing that $\boldsymbol{a} - \boldsymbol{x} = z\boldsymbol{r} - (\varrho/2)\boldsymbol{n}_{TF}$, this yields

$$z\boldsymbol{r}\cdot\boldsymbol{n}_{TF} - \frac{\varrho}{2} \geq 0. \tag{7.121}$$

Since $\boldsymbol{a} \in \mathcal{B}_d(\boldsymbol{0}, 1)$ and $\boldsymbol{r}$ is a unit vector, we have $z = \boldsymbol{a}\cdot\boldsymbol{r} \in (0, 1]$ and (7.121) therefore gives $\boldsymbol{r}\cdot\boldsymbol{n}_{TF} = z^{-1}(z\boldsymbol{r}\cdot\boldsymbol{n}_{TF}) \geq z^{-1}\varrho/2 \geq \varrho/2$. The proof of (7.119) is complete. □

We are now in a position to prove the local second Korn inequality.

Proof (Lemma 7.7) The reasoning of [270] shows that, if the following Nečas inequality

$$\|v - \pi_T^{0,0} v\|_T \leq C\|\nabla v\|_{H^{-1}(T)^d} \qquad \forall v \in L^2(T), \tag{7.122}$$

holds with a certain C, then the second Korn inequality (7.15) holds with the constant $\sqrt{1 + 2C^2}$. Hence, we only have to prove that the mesh elements T considered in the proposition satisfy (7.122) with a constant C that depends only on d and ϱ. This is achieved proceeding in two steps: first, we scale the problem in order to reduce the proof to the case of a polytopal set contained in the unit ball and star-shaped with respect to $\mathcal{B}_d(\boldsymbol{0}, \varrho)$; second, we prove the sought result in this scaled case.

(i) *Scaling.* Since the inequality is obviously invariant by translation, we can assume that T is star-shaped with respect to every point in $\mathcal{B}_d(\boldsymbol{0}, \varrho h_T)$. We then scale T so that its diameter is equal to 1. Precisely, define $\widehat{T} = T/h_T$ and, for $f \in L^2(T)$, set $\widehat{f} \in L^2(\widehat{T})$ such that $\widehat{f}(\widehat{\boldsymbol{x}}) = f(h_T\widehat{\boldsymbol{x}})$ for all $\widehat{\boldsymbol{x}} \in \widehat{T}$. Then $h_{\widehat{T}} = 1$ and $\widehat{T}$ is star-shaped with respect to every point in $\mathcal{B}_d(\boldsymbol{0}, \varrho)$. Moreover, by the change of variable

$\widehat{T} \ni \widehat{\boldsymbol{x}} \mapsto \boldsymbol{x} = h_T \widehat{\boldsymbol{x}} \in T$, it holds that

$$\int_T f = h_T^d \int_{\widehat{T}} \widehat{f} \tag{7.123}$$

and, if $f \in H^1(T)$,

$$\widehat{\nabla} \widehat{f} = h_T \widehat{\nabla f}, \tag{7.124}$$

where $\widehat{\nabla}$ is the gradient in the coordinates $\widehat{\boldsymbol{x}} \in \widehat{T}$. These properties show that, for any $v \in L^2(T)$,

$$\|v - \pi_T^{0,0} v\|_T = h_T^{d/2} \|\widehat{v} - \pi_{\widehat{T}}^{0,0} \widehat{v}\|_{\widehat{T}}, \tag{7.125}$$

and that, furnishing $H_0^1(T)^d$ with the norm $\|\cdot\|_{H_0^1(T)^d} := \|\nabla \cdot\|_T$,

$$\begin{aligned}
\|\nabla v\|_{H^{-1}(T)^d} &= \sup_{\boldsymbol{\psi} \in H_0^1(T)^d} \frac{\langle \nabla v, \boldsymbol{\psi} \rangle_{H^{-1}(T)^d, H_0^1(T)^d}}{\|\boldsymbol{\psi}\|_{H_0^1(T)^d}} \\
&= \sup_{\boldsymbol{\psi} \in H_0^1(T)^d} \frac{-\int_T v \nabla \cdot \boldsymbol{\psi}}{\|\boldsymbol{\psi}\|_{H_0^1(T)^d}} \\
&= \sup_{\boldsymbol{\psi} \in H_0^1(T)^d} \frac{-h_T^d \int_{\widehat{T}} \widehat{v}\, \widehat{\nabla \cdot \boldsymbol{\psi}}}{\|\nabla \boldsymbol{\psi}\|_T} \\
&= \sup_{\widehat{\boldsymbol{\psi}} \in H_0^1(\widehat{T})^d} \frac{-h_T^d \int_{\widehat{T}} \widehat{v} h_T^{-1} \widehat{\nabla} \cdot \widehat{\boldsymbol{\psi}}}{h_T^{d/2} \|\widehat{\nabla \boldsymbol{\psi}}\|_{\widehat{T}}} \\
&= \sup_{\widehat{\boldsymbol{\psi}} \in H_0^1(\widehat{T})^d} \frac{-h_T^d h_T^{-1} \int_{\widehat{T}} \widehat{v}\, \widehat{\nabla} \cdot \widehat{\boldsymbol{\psi}}}{h_T^{d/2} h_T^{-1} \|\widehat{\nabla} \widehat{\boldsymbol{\psi}}\|_{\widehat{T}}} = h_T^{d/2} \|\widehat{\nabla} \widehat{v}\|_{H^{-1}(\widehat{T})^d},
\end{aligned} \tag{7.126}$$

where we have used the definition of the norm in $H^{-1}(T)^d$ in the first line, the definition of the weak gradient of v in the second line, the change of variable (7.123) with $f = v \nabla \cdot \boldsymbol{\psi}$ in the third line, (7.124) with $f =$ components of $\boldsymbol{\psi}$ and the change of variables (7.123) with $f = |\nabla \boldsymbol{\psi}|$ in the fourth line, once again the relation (7.124) with $f =$ components of $\boldsymbol{\psi}$ to pass to the fifth line, and the definition of $\|\widehat{\nabla} \widehat{v}\|_{H^{-1}(\widehat{T})^d}$ to conclude. If we prove (7.122), with C depending only on d and ϱ, for all polytopal sets $\widehat{T}$ of diameter 1 and star-shaped with respect to every point in $\mathcal{B}_d(\mathbf{0}, \varrho)$, the relations (7.125)–(7.126) show that (7.122) also holds for T with the same constant. To simplify the notations, in the following we drop the hat symbol and we simply write T and v for $\widehat{T}$ and $\widehat{v}$. In other words, we reduced the proof to the case where T is a polytopal set contained in $\mathcal{B}_d$ and star-shaped with respect to every point in $\mathcal{B}_d(\mathbf{0}, \varrho)$.

(ii) *Proof of* (7.122) *in the scaled case.* [74, Theorem IV.1.1] establishes the existence of C_T such that

$$\|w\|_T \leq C_T \left(\|w\|_{H^{-1}(T)} + \|\nabla w\|_{H^{-1}(T)^d}\right) \qquad \forall w \in L^2(T). \tag{7.127}$$

The proof [74, Theorem IV.1.1] gives a clear dependency on the constant C_T in terms of an atlas of ∂T. Lemma 7.49 provides an atlas, whose elements (open coverings, domains, upper bound of the Lipschitz constants of the maps) depend only on d and ϱ, for all T contained in $\mathcal{B}_d$ and star-shaped with respect to every point in $\mathcal{B}_d(\mathbf{0}, \varrho)$. Using this atlas in the proof of [74, Theorem IV.1.1], we see that (7.127) holds with $C_T = C_0$ depending only on d and ϱ. Applying this inequality to $w = v - \pi_T^{0,0} v$ (that has a zero integral over T), the Nečas estimate (7.122) follows if we show that

$$\|w\|_{H^{-1}(T)} \leq C_1 \|\nabla w\|_{H^{-1}(T)^d} \text{ for all } w \in L^2(T) \text{ such that } \int_T w = 0, \tag{7.128}$$

with C_1 depending only on d and ϱ. This estimate is established in [74, Proposition IV.1.7], but with a proof by contradiction that does not directly provide the independence of C_1 with respect to the domain T. We adapt here this proof to show that (7.128) holds with a constant that is uniform with respect to the mesh element T.

The proof proceeds by contradiction. Assume that (7.128) does not hold uniformly with respect to T. Then, there is a sequence $(T_n, w_n)_{n\in\mathbb{N}}$ such that T_n is contained in $\mathcal{B}_d$ and is star-shaped with respect to every point in $\mathcal{B}_d(\mathbf{0}, \varrho)$, $w_n \in L^2(T_n)$ has a zero average over T_n, and

$$\|w_n\|_{H^{-1}(T_n)} > n\|\nabla w_n\|_{H^{-1}(T_n)^d}. \tag{7.129}$$

Replacing w_n with $w_n/\|w_n\|_{H^{-1}(T_n)}$, we can also assume that

$$\|w_n\|_{H^{-1}(T_n)} = 1. \tag{7.130}$$

Let $\widetilde{w}_n$ be the extension of w_n to $\mathcal{B}_d$ by 0 outside T_n. By (7.127), (7.129) and (7.130), $\|w_n\|_{T_n}$ is bounded, and so $\widetilde{w}_n$ is bounded in $L^2(\mathcal{B}_d)$. Hence, $L^2(\mathcal{B}_d)$ being compactly embedded in $H^{-1}(\mathcal{B}_d)$, we find $w \in L^2(\mathcal{B}_d)$ such that, upon extracting a subsequence,

$$\widetilde{w}_n \to w \text{ weakly in } L^2(\mathcal{B}_d) \text{ and strongly in } H^{-1}(\mathcal{B}_d) \text{ as } n \to \infty. \tag{7.131}$$

The weak convergence in $L^2(\mathcal{B}_d)$ together with the relation $0 = \int_{T_n} w_n = \int_{\mathcal{B}_d} \widetilde{w}_n$ shows that

$$\int_{\mathcal{B}_d} w = 0. \tag{7.132}$$

Considering the uniform atlas of ∂T_n given by Lemma 7.49 (whose covering and domains of mappings are independent of n), we see that the corresponding maps $(\varphi_{l,n})_{l=1,\ldots,m}$ are uniformly Lipschitz, with a constant not depending on n. Hence, upon extracting another subsequence, we can assume that these maps converge uniformly as $n \to \infty$ to some Lipschitz functions $(\varphi_l)_{l=1,\ldots,m}$. The hypographs of these Lipschitz functions define a Lipschitz open set U and, by uniform convergence of the maps, the following two properties hold:

1. the characteristic function χ_{T_n} of T_n converges strongly in $L^2(\mathcal{B}_d)$ towards the characteristic function χ_U of U, and
2. for any $\boldsymbol{\psi} \in C_c^\infty(U)^d$ there is an $N(\boldsymbol{\psi}) \in \mathbb{N}$ such that $\mathrm{supp}(\boldsymbol{\psi}) \subset T_n$ for all $n \geq N(\boldsymbol{\psi})$.

We exploit Property 1 by writing $\widetilde{w}_n = \chi_{T_n}\widetilde{w}_n$ (since $\widetilde{w}_n$ is equal to zero outside T_n), and by passing to the L^2-weak limit in the left-hand side and the weak/strong distributional limit in the right-hand side, to see that $w = \chi_U w$. In particular, this shows that $w = 0$ outside U and, together with (7.132), that

$$\int_U w = 0. \tag{7.133}$$

Consider now Property 2 of $(T_n)_{n\in\mathbb{N}}$. Fixing $\boldsymbol{\psi} \in C_c^\infty(U)^d$, for any $n \geq N(\boldsymbol{\psi})$ we can write

$$\begin{aligned}\left|\int_{\mathcal{B}_d} \widetilde{w}_n \nabla\cdot\boldsymbol{\psi}\right| &= \left|\int_{T_n} w_n \nabla\cdot\boldsymbol{\psi}\right| = \left|-\langle \nabla w_n, \boldsymbol{\psi}\rangle_{H^{-1}(T_n)^d, H_0^1(T_n)^d}\right| \\ &\leq \|\nabla w_n\|_{H^{-1}(T_n)^d}\|\boldsymbol{\psi}\|_{H_0^1(T_n)^d} \leq \frac{1}{n}\|\boldsymbol{\psi}\|_{H_0^1(U)^d}\end{aligned}$$

where the first line follows from the definitions of $\widetilde{w}_n$ and ∇w_n together with the fact that $\boldsymbol{\psi} \in C_c^\infty(T_n)^d$ (since $\mathrm{supp}(\boldsymbol{\psi}) \subset T_n$), and the second line is a consequence of (7.129)–(7.130) and of the fact that $\boldsymbol{\psi}$ has a compact support in U. Combined with the weak convergence in (7.131) this shows that

$$\int_{\mathcal{B}_d} w\nabla\cdot\boldsymbol{\psi} = 0.$$

Since it is true for any $\boldsymbol{\psi} \in C_c^\infty(U)^d$, this property proves that $\nabla w = 0$ in $\mathcal{D}'(U)^d$. By construction, the open set U is connected, and thus w is constant over U. Invoking (7.133), we deduce that $w = 0$ on U and thus, since $w = 0$ outside

U, that $w = 0$ on $\mathcal{B}_d$. The strong convergence in (7.131) therefore shows that

$$\widetilde{w}_n \to 0 \text{ strongly in } H^{-1}(\mathcal{B}_d) \text{ as } n \to \infty. \tag{7.134}$$

To conclude the proof, recall (7.130) and notice that any function $\varphi \in H_0^1(T_n)$ can be considered, after extension by 0 outside T_n, as a function in $H_0^1(\mathcal{B}_d)$ with $\|\varphi\|_{H_0^1(T_n)} = \|\varphi\|_{H_0^1(\mathcal{B}_d)}$. Hence, by definition of the norms in $H^{-1}(\mathcal{B}_d)$ and $H^{-1}(T_n)$,

$$\|\widetilde{w}_n\|_{H^{-1}(\mathcal{B}_d)} = \sup_{\varphi \in H_0^1(\mathcal{B}_d)} \frac{\int_{\mathcal{B}_d} \widetilde{w}_n \varphi}{\|\varphi\|_{H_0^1(\mathcal{B}_d)}} \geq \sup_{\varphi \in H_0^1(T_n)} \frac{\int_{T_n} w_n \varphi}{\|\varphi\|_{H_0^1(T_n)}} = \|w_n\|_{H^{-1}(T_n)} = 1.$$

However, (7.134) shows that the left-hand side goes to 0 as $n \to \infty$, which establishes the contradiction. □

Remark 7.50 (Second Korn Inequality in L^q) Following [74, Remark IV.1.1], we could as well establish a uniform local second Korn inequality in L^q spaces, with $1 < q < \infty$, rather than in the L^2 space.

Chapter 8
Stokes

In this chapter, we apply the HHO method to the discretisation of the steady Stokes problem, which models fluid flows where convective inertial forces are small compared to viscous forces. From a physical point of view, the Stokes problem is obtained writing momentum and mass balance equations. In the case of a uniform density fluid, the mass balance translates into a zero-divergence constraint on the velocity, enabling an interpretation as a constrained minimisation (saddle-point) problem with the pressure acting as the Lagrange multiplier; see Remark 8.7. As a consequence, the well-posedness of the Stokes problem hinges on an inf–sup rather than a coercivity condition. This property has to be reproduced at the discrete level, which requires to select the discrete spaces for the velocity and pressure so that the discrete divergence operator from the former to the latter is surjective. In the context of Finite Element Methods, a large effort has been devoted to devising inf–sup stable space couples; see, e.g., the discussions in [254, Chapter 9], [183, Chapter 4], and [57, Chapter 8]. As we will see, in the framework of HHO methods inf–sup stability can be achieved via a suitably designed divergence reconstruction operator.

The chapter is organised as follows. In Sect. 8.1 we establish the continuous setting for the model, state the weak formulation, discuss the continuous inf–sup condition and, for the sake of completeness, sketch its proof.

In Sect. 8.2 we describe local constructions required to design the HHO scheme. After introducing the local space of discrete velocity unknowns, we define two local reconstruction operators, one for the velocity and one for its divergence. The velocity reconstruction is designed so that its composition with the local interpolator coincides with the elliptic projector. The divergence reconstruction, on the other hand, yields the L^2-orthogonal projection of the continuous divergence when composed with the interpolator. This commutation property plays a crucial role in the proof of the discrete inf–sup condition.

In Sect. 8.3 we discuss the discrete problem. After introducing the global spaces of discrete unknowns for the velocity and the pressure, we prove a second key property for the discrete inf–sup condition, namely the uniform H^1-boundedness of

D. A. Di Pietro, J. Droniou, *The Hybrid High-Order Method for Polytopal Meshes*,
MS&A 19, https://doi.org/10.1007/978-3-030-37203-3_8

the velocity interpolator. Together with the commutation property for the divergence reconstruction, this expresses the fact that the latter is a Fortin interpolator. We next discuss the discretisation of the various terms in the Stokes equations. The viscous term is discretised by adapting the bilinear form for the Poisson problem to the vector case. The pressure–velocity coupling term, on the other hand, hinges on the discrete divergence reconstruction introduced in the previous section. A central result of this section is the proof of consistency and inf–sup stability for the discrete pressure–velocity coupling bilinear form. In Sect. 8.4 we derive a reformulation of the HHO method in terms of numerical fluxes, which shows that both momentum and mass are conserved inside each element, and that the corresponding fluxes are continuous across interfaces.

In Sect. 8.5, we carry out an error analysis. We first derive energy error estimates showing, for smooth enough solutions, convergence in h^{k+1} (with h and $k \geq 0$ denoting, as usual, the meshsize and polynomial degree, respectively) for an H^1-like norm of the velocity error and the L^2-norm of the pressure error. We next derive an improved L^2-error estimate in h^{k+2} for the velocity under an elliptic regularity assumption. These theoretical estimates are illustrated by a numerical example.

Finally, in Sect. 8.6 we discuss a variation of the method which delivers an error estimate for the velocity independent of both the pressure and the viscosity. In practice, this property is relevant when dealing with large irrotational body forces, as it typically delivers a better approximation of the velocity.

8.1 Model

In this section we discuss the continuous setting for the model.

8.1.1 *The Stokes Problem*

Let $d \in \{2, 3\}$ and take $\Omega \subset \mathbb{R}^d$ that satisfies Assumption 1.3. We additionally assume that Ω has a Lipschitz continuous boundary, that is, for any $\boldsymbol{x} \in \partial\Omega$ there is a neighbourhood $O_{\boldsymbol{x}}$ of $\boldsymbol{x}$ in $\mathbb{R}^d$ such that $\Omega \cap O_{\boldsymbol{x}}$ is, in a suitable set of Cartesian coordinates, the epigraph of a Lipschitz-continuous function.

Let $\nu > 0$ denote a real number representing the kinematic viscosity, and let $\boldsymbol{f} : \Omega \to \mathbb{R}^d$ denote a body force. The steady Stokes problem for a uniform density, Newtonian fluid consists in finding the velocity $\boldsymbol{u} : \Omega \to \mathbb{R}^d$ and the pressure $p : \Omega \to \mathbb{R}$ such that

$$-\nu\Delta\boldsymbol{u} + \nabla p = \boldsymbol{f} \qquad \text{in } \Omega, \tag{8.1a}$$

$$\nabla\cdot\boldsymbol{u} = 0 \qquad \text{in } \Omega, \tag{8.1b}$$

$$\boldsymbol{u} = \boldsymbol{0} \qquad \text{on } \partial\Omega, \tag{8.1c}$$

$$\int_\Omega p = 0. \tag{8.1d}$$

Equation (8.1a) expresses the momentum balance. The first term in (8.1a) accounts for viscous phenomena, and its expression is specialised to the case of uniform kinematic viscosity: for variable viscosity, one should replace $-\nu\Delta\boldsymbol{u}$ by $-\boldsymbol{\nabla}\cdot(\nu\boldsymbol{\nabla}_{\mathrm{s}}\boldsymbol{u})$, with $\boldsymbol{\nabla}_{\mathrm{s}}$ denoting the symmetric gradient defined by (7.4). The second term in (8.1a) represents an isotropic contribution to internal stresses which, as we will see, is intimately linked to the enforcement of the incompressibility constraint (8.1b); see Remark 8.7. To close problem (8.1a), we have considered, for the sake of simplicity, the so-called *wall* (homogeneous Dirichlet) boundary condition (8.1c); the discussion extends without difficulties to other standard boundary conditions. Finally, condition (8.1d) is introduced to uniquely identify the pressure, which would otherwise be defined only up to an additive constant.

Remark 8.1 (Constant Kinematic Viscosity) As we have assumed constant kinematic viscosity, dividing (8.1a) by ν and replacing $p \leftarrow \nu^{-1}p$ and $\boldsymbol{f} \leftarrow \nu^{-1}\boldsymbol{f}$ shows that we could in fact have simply taken $\nu = 1$. This argument, however, does not apply to the full Navier–Stokes problem owing to the presence of an additional nonlinear term in (8.1a). Thus, in view of Chap. 9, we will keep the kinematic viscosity throughout this chapter, with the exception of Sect. 8.5.2, which contains material that will not be further developed in the context of the Navier–Stokes problem.

8.1.2 Weak Formulation

We next discuss a standard weak formulation of the Stokes problem. For the sake of uniformity with the sibling Chap. 9, the L^2-product notation introduced in Remark 1.14 will often be dropped in favour of integrals. As a consequence, when they are used, L^2-norms and inner products are explicitly identified for the sake of coherence.

Assume $\boldsymbol{f} \in L^2(\Omega)^d$ and define the following spaces for the velocity and the pressure:

$$\boldsymbol{U} := H_0^1(\Omega)^d, \qquad P := \left\{q \in L^2(\Omega) \,:\, \int_\Omega q = 0\right\}. \tag{8.2}$$

A classical weak formulation of problem (8.1) reads: Find $(\boldsymbol{u}, p) \in \boldsymbol{U} \times P$ such that

$$\nu a(\boldsymbol{u}, \boldsymbol{v}) + b(\boldsymbol{v}, p) = \int_\Omega \boldsymbol{f}\cdot\boldsymbol{v} \qquad \forall \boldsymbol{v} \in \boldsymbol{U}, \tag{8.3a}$$

$$-b(\boldsymbol{u}, q) = 0 \qquad \forall q \in L^2(\Omega), \tag{8.3b}$$

with bilinear forms $a : \boldsymbol{U} \times \boldsymbol{U} \to \mathbb{R}$ and $b : \boldsymbol{U} \times L^2(\Omega) \to \mathbb{R}$ defined by

$$a(\boldsymbol{w}, \boldsymbol{v}) := \int_\Omega \boldsymbol{\nabla} \boldsymbol{w}:\boldsymbol{\nabla} \boldsymbol{v}, \quad b(\boldsymbol{v}, q) := -\int_\Omega (\boldsymbol{\nabla}\cdot\boldsymbol{v})\, q, \tag{8.4}$$

where we remind the reader that the Frobenius product is such that, for all $\boldsymbol{\sigma}, \boldsymbol{\tau} \in \mathbb{R}^{d\times d}$, $\boldsymbol{\sigma}:\boldsymbol{\tau} := \sum_{i=1}^d \sum_{j=1}^d \sigma_{ij}\tau_{ij}$.

Remark 8.2 (Test Space in (8.3b)*)* In (8.3b), it is possible to take $L^2(\Omega)$ instead of P as a test space because the following compatibility condition is verified:

$$-b(\boldsymbol{u}, 1) = \int_\Omega \boldsymbol{\nabla}\cdot\boldsymbol{u} = \int_{\partial\Omega} \boldsymbol{u}\cdot\boldsymbol{n} = 0, \tag{8.5}$$

where we have integrated by parts in the second passage and used the wall boundary condition strongly incorporated in $\boldsymbol{U}$ to conclude.

An equivalent formulation of problem (8.3) is obtained the following way. Set

$$X := \boldsymbol{U} \times P,$$

and introduce the global bilinear form $\mathcal{A} : X \times X \to \mathbb{R}$ such that, for all $(\boldsymbol{w}, r), (\boldsymbol{v}, q) \in X$,

$$\mathcal{A}((\boldsymbol{w}, r), (\boldsymbol{v}, q)) := \nu a(\boldsymbol{w}, \boldsymbol{v}) + b(\boldsymbol{v}, r) - b(\boldsymbol{w}, q). \tag{8.6}$$

Then, (8.3) is equivalent to: Find $(\boldsymbol{u}, p) \in X$ such that

$$\mathcal{A}((\boldsymbol{u}, p), (\boldsymbol{v}, q)) = \int_\Omega \boldsymbol{f}\cdot\boldsymbol{v} \qquad \forall(\boldsymbol{v}, q) \in X. \tag{8.7}$$

The well-posedness of problem (8.3) (or, equivalently, (8.7)) hinges on two key properties: (1) the coercivity of the bilinear form a, resulting from the Poincaré inequality, and (2) the inf–sup stability of the bilinear form b which, as we will see, corresponds to the surjectivity of the divergence operator in the pressure space.

8.1.3 Inf–Sup Stability of the Pressure–Velocity Coupling

We introduce the pressure–velocity coupling operator such that

$$B : \boldsymbol{U} \ni \boldsymbol{v} \longmapsto B\boldsymbol{v} := -\boldsymbol{\nabla}\cdot\boldsymbol{v} \in P. \tag{8.8}$$

To check that, for any $\boldsymbol{v} \in \boldsymbol{U}$, $B\boldsymbol{v}$ has zero mean-value on Ω (hence it belongs to P), it suffices to proceed as in (8.5). Additionally, it is a simple matter to see that $B\boldsymbol{v}$ is the Riesz representation of the linear form $b(\boldsymbol{v}, \cdot)$ in $L^2(\Omega)$ equipped with the usual inner product $(\cdot, \cdot)_{L^2(\Omega)}$, that is

$$(B\boldsymbol{v}, q)_{L^2(\Omega)} = b(\boldsymbol{v}, q) \qquad \forall q \in L^2(\Omega). \tag{8.9}$$

Lemma 8.3 (Continuous inf–sup Condition) *There exists a real number $\beta > 0$ depending only on Ω such that*

$$\forall q \in P, \exists \boldsymbol{v}_q \in \boldsymbol{U} \text{ such that}$$
$$q = B\boldsymbol{v}_q = -\nabla\cdot\boldsymbol{v}_q \text{ and } \beta\|\boldsymbol{v}_q\|_{H^1(\Omega)^d} \le \|q\|_{L^2(\Omega)}. \tag{8.10}$$

Moreover, property (8.10) *is equivalent to the following inf–sup condition:*

$$\forall q \in P, \qquad \beta\|q\|_{L^2(\Omega)} \le \sup_{\boldsymbol{v}\in\boldsymbol{U}\setminus\{\boldsymbol{0}\}} \frac{b(\boldsymbol{v}, q)}{\|\boldsymbol{v}\|_{H^1(\Omega)^d}}. \tag{8.11}$$

Remark 8.4 (Properties (8.10)*)* The first condition in (8.10) expresses the fact that B is surjective. The second condition in (8.10) is inferred from the Open Mapping Theorem (see, e.g., [81, Theorem 2.6]), whose statement is recalled next.

Theorem 8.5 (Open Mapping) *Let E and F be two Banach spaces, and let L be a continuous linear operator from E to F that is surjective. Then, there is a constant $c > 0$ such that $\mathcal{B}_F(0, c) \subset L(\mathcal{B}_E(0, 1))$ where, with X Banach space and $r > 0$, $\mathcal{B}_X(0, r) := \{x \in X : \|x\|_X < r\}$ denotes the open ball in X centred at 0 and of radius r.*

Apply this result to $E = \boldsymbol{U}$, $F = P$, and $L = B$, assuming that the surjectivity of the latter operator has been established. Then, for all $q \in P\setminus\{0\}$, $\frac{c}{2\|q\|_{L^2(\Omega)}} q \in \mathcal{B}_P(0, c)$ and thus there exists $\hat{\boldsymbol{v}}_q \in \mathcal{B}_U(0, 1)$ such that $B\hat{\boldsymbol{v}}_q = \frac{c}{2\|q\|_{L^2(\Omega)}} q$. Setting $\boldsymbol{v}_q := \frac{2\|q\|_{L^2(\Omega)}}{c}\hat{\boldsymbol{v}}_q$, we have $B\boldsymbol{v}_q = q$ and $\|\boldsymbol{v}_q\|_{H^1(\Omega)^d} \le \frac{2\|q\|_{L^2(\Omega)}}{c}$, which shows that (8.10) holds with $\beta = c/2$. This constant β depends only on Ω since c, provided by the Open Mapping Theorem, depends only on B, which is fully determined by Ω.

Proof (Lemma 8.3) The proof is classical, and can be found in several textbooks; see, e.g. [59, 180, 199, 259]. It is summarised here to make the exposition self-contained. We proceed in two steps: first, we prove that conditions (8.10) and (8.11) are equivalent; then we show that the inf–sup condition (8.11) holds.

(i) *Proof of the Equivalence* (8.10) $\Longleftrightarrow$ (8.11). Let us assume that (8.10) holds. Then, by definition of $\boldsymbol{v}_q$ and (8.9),

$$\sup_{\boldsymbol{v}\in\boldsymbol{U}\setminus\{\boldsymbol{0}\}} \frac{b(\boldsymbol{v}, q)}{\|\boldsymbol{v}\|_{H^1(\Omega)^d}} \ge \frac{b(\boldsymbol{v}_q, q)}{\|\boldsymbol{v}_q\|_{H^1(\Omega)^d}} = \frac{\|q\|^2_{L^2(\Omega)}}{\|\boldsymbol{v}_q\|_{H^1(\Omega)^d}} \ge \beta\|q\|_{L^2(\Omega)},$$

which is precisely (8.11).

We now assume that (8.11) holds. We denote by $\boldsymbol{U}^\star := [H^{-1}(\Omega)]^d$ the dual space of $\boldsymbol{U}$, and we identify P with its dual space. Let $B^\star : P \to \boldsymbol{U}^\star$ be the adjoint operator of B such that

$$\forall q \in P, \qquad \langle B^\star q, \boldsymbol{v}\rangle_{\boldsymbol{U}^\star,\boldsymbol{U}} = (B\boldsymbol{v}, q)_{L^2(\Omega)} \qquad \forall \boldsymbol{v} \in \boldsymbol{U},$$

where $\langle\cdot,\cdot\rangle_{\boldsymbol{U}^\star,\boldsymbol{U}}$ denotes the duality pairing between $\boldsymbol{U}^\star$ and $\boldsymbol{U}$. Using the characterisation (8.9) of B followed by the above equation, we have that

$$\sup_{\boldsymbol{v}\in\boldsymbol{U}\setminus\{\boldsymbol{0}\}} \frac{b(\boldsymbol{v}, q)}{\|\boldsymbol{v}\|_{H^1(\Omega)^d}} = \sup_{\boldsymbol{v}\in\boldsymbol{U}\setminus\{\boldsymbol{0}\}} \frac{(B\boldsymbol{v}, q)_{L^2(\Omega)}}{\|\boldsymbol{v}\|_{H^1(\Omega)^d}} = \sup_{\boldsymbol{v}\in\boldsymbol{U}\setminus\{\boldsymbol{0}\}} \frac{\langle B^\star q, \boldsymbol{v}\rangle_{\boldsymbol{U}^\star,\boldsymbol{U}}}{\|\boldsymbol{v}\|_{H^1(\Omega)^d}} = \|B^\star q\|_{\boldsymbol{U}^\star},$$

where the conclusion follows from the standard definition of the dual norm. Hence, condition (8.11) can be reformulated as follows:

$$\forall q \in P, \qquad \beta\|q\|_{L^2(\Omega)} \le \|B^\star q\|_{\boldsymbol{U}^\star}. \tag{8.12}$$

Condition (8.12) implies that

$$\mathrm{Ker}(B^\star) = \{0\},$$

as can be checked observing that, for any $q \in P$, if $B^\star q = 0$ then $\|q\|_{L^2(\Omega)} \le \beta^{-1}\|B^\star q\|_{\boldsymbol{U}^\star} = 0$, i.e., $q = 0$ since $\|\cdot\|_{L^2(\Omega)}$ is a norm on P.

Let us now prove that $\mathrm{Im}(B^\star)$ is closed in $\boldsymbol{U}^\star$. Let $\{B^\star q_n\}_{n\in\mathbb{N}}$ be a converging (and thus Cauchy) sequence in $\boldsymbol{U}^\star$, for some sequence $\{q_n\}_{n\in\mathbb{N}}$ in P. Then, by (8.12), we have

$$\|q_n - q_m\|_{L^2(\Omega)} \le \beta^{-1}\|B^\star q_n - B^\star q_m\|_{\boldsymbol{U}^\star},$$

which shows that $\{q_n\}_{n\in\mathbb{N}}$ is a Cauchy sequence in P and thus converges toward some q in this space. The continuity of $B^\star$ then ensures that $B^\star q_n \to B^\star q$, which proves that $\mathrm{Im}(B^\star)$ is closed in $\boldsymbol{U}^\star$.

We next apply the Closed Range Theorem (see, e.g., [183, Theorem A.34]), whose statement is recalled next.

Theorem 8.6 (Closed Range) *Let E and F be two real Banach spaces, and let L be a continuous linear operator from E to F. Then, denoting by $L^\star$ the adjoint operator of L, the following statements are equivalent: (i)* $\mathrm{Im}(L)$ *is closed in F; (ii)* $\mathrm{Im}(L^\star)$ *is closed in $E^\star$; (iii)* $\mathrm{Im}(L) = (\mathrm{Ker}(L^\star))^\perp$*; (iv)* $\mathrm{Im}(L^\star) = (\mathrm{Ker}(L))^\perp$.

By virtue of this theorem, having proved that $\mathrm{Im}(B^\star)$ is closed in $\boldsymbol{U}^\star$, it follows that $\mathrm{Im}(B) = (\mathrm{Ker}(B^\star))^\perp = \{0\}^\perp = P$, i.e., B is surjective. The argument in Remark 8.4 then concludes the proof of (8.10).

(ii) *Proof of* (8.11). We start with the following Nečas inequality proved in [243] under the Lipschitz assumption on Ω: There exists $C > 0$ such that

$$\forall q \in L^2(\Omega), \qquad C\|q\|_{L^2(\Omega)} \le \|q\|_{H^{-1}(\Omega)} + \|\nabla q\|_{H^{-1}(\Omega)^d}. \tag{8.13}$$

Reproducing the argument by contradiction used in Step (ii) of the proof of Lemma 7.7 to establish (7.128) with $T = T_n = \Omega$ (fixing the domain eliminates the need to consider star-shaped regions) shows the existence of $C' > 0$ depending only on Ω such that

$$\forall q \in P, \qquad \|q\|_{H^{-1}(\Omega)} \le C'\|\nabla q\|_{H^{-1}(\Omega)^d}.$$

Combining this estimate with (8.13) shows that $\|q\|_{L^2(\Omega)} \le C^{-1}(C' + 1)\|\nabla q\|_{H^{-1}(\Omega)^d}$, which is precisely (8.12) (with $\beta = C(C'+1)^{-1}$) since $B^\star$ is in fact the distributional gradient from P to $\boldsymbol{U}^\star$. The equivalence, established in Point (i) above, of (8.12) and (8.11) concludes the proof. □

Remark 8.7 (Variational Interpretation and Role of the Pressure) Problem (8.3) is equivalent to the Lagrange multiplier formulation of the minimisation problem

$$\min_{\boldsymbol{v}\in\boldsymbol{U},\nabla\cdot\boldsymbol{v}=0} \left(\frac{\nu}{2}\|\nabla\boldsymbol{v}\|^2_{L^2(\Omega)^{d\times d}} - \int_\Omega \boldsymbol{f}\cdot\boldsymbol{v}\right),$$

where the pressure acts as the Lagrange multiplier for the zero-divergence constraint. The condition (8.11) ensures that $\nabla\cdot\boldsymbol{U} = P$.

8.2 Local Construction

We discuss here the local construction underlying the HHO discretisation of problem (8.3). Throughout this section, we work on a fixed mesh element $T \in \mathcal{T}_h$.

8.2.1 Local Space of Discrete Velocity Unknowns

The viscous term in (8.3a) is nothing but the vector version of the pure diffusion operator studied in Chap. 2 obtained applying the latter component-wise. Thus, its HHO discretisation hinges on the following local space of discrete velocity unknowns, which is the natural adaptation to the vector case of the space defined by (2.6):

$$\underline{\boldsymbol{U}}_T^k := \left\{\underline{\boldsymbol{v}}_T = (\boldsymbol{v}_T, (\boldsymbol{v}_F)_{F\in\mathcal{F}_T}) \ :\ \boldsymbol{v}_T \in \mathbb{P}^k(T)^d \text{ and } \boldsymbol{v}_F \in \mathbb{P}^k(F)^d \quad \forall F \in \mathcal{F}_T\right\}.$$

The local interpolator $\underline{I}_T^k : H^1(T)^d \to \underline{U}_T^k$ is such that, for all $v \in H^1(T)^d$,

$$\underline{I}_T^k v := (\pi_T^{0,k} v, (\pi_F^{0,k} v)_{F \in \mathcal{F}_T}). \tag{8.14}$$

We define on $\underline{U}_T^k$ the local H^1-like seminorm $\|\cdot\|_{1,T}$ such that, for all $\underline{v}_T \in \underline{U}_T^k$,

$$\begin{aligned} \|\underline{v}_T\|_{1,T} &:= \left(\|\nabla v_T\|^2_{L^2(T)^{d\times d}} + |\underline{v}_T|^2_{1,\partial T}\right)^{\frac{1}{2}} \\ |\underline{v}_T|_{1,\partial T} &:= \left(\sum_{F\in\mathcal{F}_T} h_F^{-1} \|v_F - v_T\|^2_{L^2(F)^d}\right)^{\frac{1}{2}}, \end{aligned} \tag{8.15}$$

where, as usual, the negative power of the diameter of F in the boundary seminorm ensures that both contributions have the same scaling.

8.2.2 Velocity and Divergence Reconstructions

We next introduce the local reconstructions on which the HHO method hinges: (1) a velocity reconstruction for use in the discretisation of the viscous term obtained by adapting (2.11) to the vector case, and (2) a divergence reconstruction for use in the pressure–velocity coupling term.

Recalling the discussion of Sect. 2.1.3 for the scalar case, the local velocity reconstruction $\mathbf{r}_T^{k+1} : \underline{U}_T^k \to \mathbb{P}^{k+1}(T)^d$ is defined such that, for all $\underline{v}_T \in \underline{U}_T^k$ and all $w \in \mathbb{P}^{k+1}(T)^d$,

$$\int_T \nabla \mathbf{r}_T^{k+1} \underline{v}_T : \nabla w = -\int_T v_T \cdot \Delta w + \sum_{F\in\mathcal{F}_T} \int_F v_F \cdot (\nabla w n_{TF}) \tag{8.16a}$$

and

$$\int_T (\mathbf{r}_T^{k+1} \underline{v}_T - v_T) = \mathbf{0}. \tag{8.16b}$$

Notice that we have used the notation $\mathbf{r}_T^{k+1}$ instead of $\mathbf{p}_T^{k+1}$ to avoid confusion with the displacement reconstruction (7.39) introduced in Chap. 7, as well as with pressure unknowns. Using the exact same arguments that lead to (2.14), we infer that, for any $v \in H^1(T)^d$,

$$\mathbf{r}_T^{k+1} \underline{I}_T^k v = \pi_T^{1,k+1} v, \tag{8.17}$$

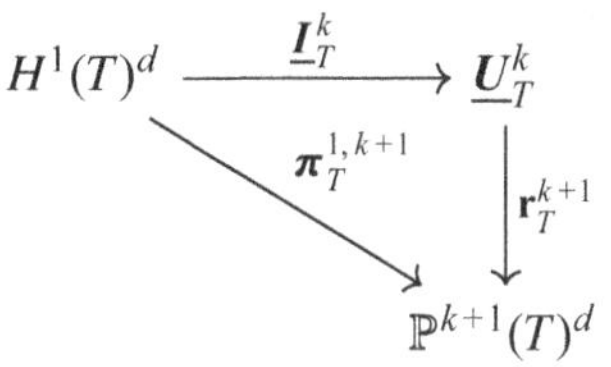

Fig. 8.1 Illustration of the commutation property (8.17) of $\mathbf{r}_T^{k+1}$

where $\boldsymbol{\pi}_T^{1,k+1}$ denotes the vector version of the elliptic projector obtained applying component-wise the scalar counterpart introduced in Definition 1.39. This commutation property is illustrated in Fig. 8.1.

To define the divergence reconstruction, let us start with an inspiring remark, in the spirit of Sects. 2.1.1, 3.1.3.1, and 4.2.1. Let $\boldsymbol{v} \in H^1(T)^d$. We have, for all $q \in \mathbb{P}^k(T)$,

$$\begin{aligned}\int_T (\nabla\cdot\boldsymbol{v})\, q &= -\int_T \boldsymbol{v}\cdot\nabla q + \sum_{F\in\mathcal{F}_T}\int_F (\boldsymbol{v}\cdot\boldsymbol{n}_{TF})\, q \\ &= -\int_T \boldsymbol{\pi}_T^{0,k}\boldsymbol{v}\cdot\nabla q + \sum_{F\in\mathcal{F}_T}\int_F (\boldsymbol{\pi}_F^{0,k}\boldsymbol{v}\cdot\boldsymbol{n}_{TF})\, q, \end{aligned} \tag{8.18}$$

where we have used an integration by parts, and (1.57) to insert the projectors in the second line after observing that $\nabla q \in \mathbb{P}^{k-1}(T)^d \subset \mathbb{P}^k(T)^d$ and $q_{|F}\boldsymbol{n}_{TF} \in \mathbb{P}^k(F)^d$ for all $F \in \mathcal{F}_T$. This formula shows that the L^2-orthogonal projection of $\nabla\cdot\boldsymbol{v}$ on $\mathbb{P}^k(T)$ can be computed using the projections of $\boldsymbol{v}$ on $\mathbb{P}^k(T)^d$ and on $\mathbb{P}^k(F)^d$ for all $F \in \mathcal{F}_T$. It justifies the following definition of the divergence reconstruction $\mathrm{D}_T^k : \underline{\boldsymbol{U}}_T^k \to \mathbb{P}^k(T)$: For all $\underline{\boldsymbol{v}}_T \in \underline{\boldsymbol{U}}_T^k$ and all $q \in \mathbb{P}^k(T)$,

$$\int_T \mathrm{D}_T^k\underline{\boldsymbol{v}}_T\, q = -\int_T \boldsymbol{v}_T\cdot\nabla q + \sum_{F\in\mathcal{F}_T}\int_F (\boldsymbol{v}_F\cdot\boldsymbol{n}_{TF})\, q \tag{8.19}$$

$$= \int_T (\nabla\cdot\boldsymbol{v}_T) q + \sum_{F\in\mathcal{F}_T}\int_F (\boldsymbol{v}_F - \boldsymbol{v}_T)\cdot\boldsymbol{n}_{TF}\, q, \tag{8.20}$$

where the second equation follows from the first one by an integration by parts. By the Riesz representation theorem in $\mathbb{P}^k(T)$ endowed with the $L^2(T)$-inner product, $\mathrm{D}_T^k\underline{\boldsymbol{v}}_T$ is well-defined. As a consequence of (8.18), and recalling the definition (8.14) of the local interpolator, the following commutation property holds (see illustration in Fig. 8.2): For all $\boldsymbol{v} \in H^1(T)^d$,

$$\mathrm{D}_T^k\underline{\boldsymbol{I}}_T^k\boldsymbol{v} = \pi_T^{0,k}(\nabla\cdot\boldsymbol{v}). \tag{8.21}$$

Remark 8.8 (Local Divergence Operator) We notice, in passing, that this local divergence operator D_T^k is the same as the one defined in (7.35): formula (8.19) can indeed be recovered writing (7.33) with test function $\boldsymbol{\tau} = q\boldsymbol{I}_d$ and q spanning $\mathbb{P}^k(T)$.

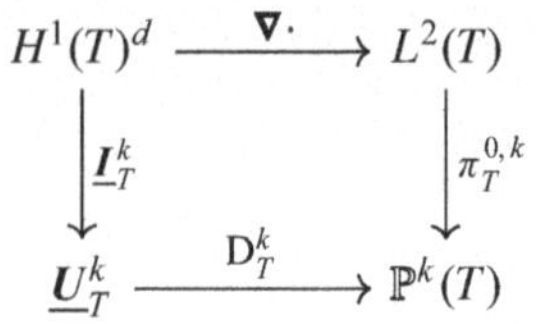

Fig. 8.2 Illustration of the commutation property (8.21) of D_T^k

8.3 Discrete Problem

In this section we define the global spaces of discrete velocity and pressure unknowns, formulate the discrete counterparts of the viscous and pressure–velocity coupling terms, and state the discrete problem.

8.3.1 Global Spaces of Discrete Unknowns

The global space of discrete unknowns is defined as

$$\underline{U}_h^k := \Big\{ \underline{v}_h = ((v_T)_{T\in\mathcal{T}_h}, (v_F)_{F\in\mathcal{F}_h}) \; : \\ v_T \in \mathbb{P}^k(T)^d \quad \forall T \in \mathcal{T}_h \text{ and } v_F \in \mathbb{P}^k(F)^d \quad \forall F \in \mathcal{F}_h \Big\}. \tag{8.22}$$

Given $\underline{v}_h \in \underline{U}_h^k$, for all $T \in \mathcal{T}_h$ we denote by $\underline{v}_T := (v_T, (v_F)_{F\in\mathcal{F}_T}) \in \underline{U}_T^k$ its restriction to T. We also define the broken polynomial function $v_h \in \mathbb{P}^k(\mathcal{T}_h)^d$ such that

$$(v_h)_{|T} := v_T \qquad \forall T \in \mathcal{T}_h.$$

The discrete unknowns corresponding to a smooth function $v \in H^1(\Omega)^d$ are obtained via the global interpolator $\underline{I}_h^k : H^1(\Omega)^d \to \underline{U}_h^k$ such that

$$\underline{I}_h^k v := ((\pi_T^{0,k} v)_{T\in\mathcal{T}_h}, (\pi_F^{0,k} v)_{F\in\mathcal{F}_h}). \tag{8.23}$$

Finally, we define on $\underline{U}_h^k$ the global H^1-like seminorm $\|\cdot\|_{1,h}$ such that, for all $\underline{v}_h \in \underline{U}_h^k$,

$$\|\underline{v}_h\|_{1,h} := \left(\sum_{T\in\mathcal{T}_h} \|\underline{v}_T\|_{1,T}^2 \right)^{\frac{1}{2}}, \tag{8.24}$$

with local seminorm $\|\cdot\|_{1,T}$ given by (8.15). The following uniform boundedness property of the global interpolator will be crucial to prove the discrete inf–sup condition (8.36).

Proposition 8.9 (Boundedness of the Global Interpolator) *There exists a real number $C_I > 0$ independent of h, but possibly depending on d, ϱ, and k, such that, for all $\boldsymbol{v} \in H^1(\Omega)^d$,*

$$\|\underline{\boldsymbol{I}}_h^k \boldsymbol{v}\|_{1,h} \le C_I |\boldsymbol{v}|_{H^1(\Omega)^d}. \tag{8.25}$$

Proof Square the local boundedness property (2.9) applied to each component v_i of $\boldsymbol{v} = (v_i)_{1 \le i \le d}$, sum over $T \in \mathcal{T}_h$ and over $1 \le i \le d$, and take the square root of the resulting inequality. □

With the above definitions, the global spaces of discrete unknowns for the velocity and the pressure, respectively accounting for the wall boundary condition (8.1c) and the zero-average condition (8.1d), are

$$\begin{aligned}\underline{\boldsymbol{U}}_{h,0}^k &:= \left\{\underline{\boldsymbol{v}}_h \in \underline{\boldsymbol{U}}_h^k \;:\; \boldsymbol{v}_F = \boldsymbol{0} \quad \forall F \in \mathcal{F}_h^{\rm b}\right\},\\ P_h^k &:= \mathbb{P}^k(\mathcal{T}_h) \cap P = \left\{q_h \in \mathbb{P}^k(\mathcal{T}_h) \;:\; \int_\Omega q_h = 0\right\}.\end{aligned} \tag{8.26}$$

8.3.2 Viscous Term

The viscous term is discretised by means of the bilinear form $\mathrm{a}_h : \underline{\boldsymbol{U}}_h^k \times \underline{\boldsymbol{U}}_h^k \to \mathbb{R}$ such that, for all $\underline{\boldsymbol{w}}_h, \underline{\boldsymbol{v}}_h \in \underline{\boldsymbol{U}}_h^k$,

$$\mathrm{a}_h(\underline{\boldsymbol{w}}_h, \underline{\boldsymbol{v}}_h) := \sum_{T \in \mathcal{T}_h} \mathrm{a}_T(\underline{\boldsymbol{w}}_T, \underline{\boldsymbol{v}}_T), \tag{8.27}$$

where, in analogy with (2.15), the local contribution is such that

$$\mathrm{a}_T(\underline{\boldsymbol{w}}_T, \underline{\boldsymbol{v}}_T) := \int_\Omega \nabla \mathbf{r}_T^{k+1} \underline{\boldsymbol{w}}_T : \nabla \mathbf{r}_T^{k+1} \underline{\boldsymbol{v}}_T + \mathrm{s}_T(\underline{\boldsymbol{w}}_T, \underline{\boldsymbol{v}}_T). \tag{8.28}$$

In the above expression, the first term in the right-hand side is the usual Galerkin contribution responsible for consistency, while $\mathrm{s}_T : \underline{\boldsymbol{U}}_T^k \times \underline{\boldsymbol{U}}_T^k \to \mathbb{R}$ is a local stabilisation bilinear form that satisfies the following conditions, adapting those in Assumption 2.4 to the vector case.

Assumption 8.10 (Local Stabilisation Bilinear Form s_T) *The local stabilisation bilinear form $\mathrm{s}_T : \underline{\boldsymbol{U}}_T^k \times \underline{\boldsymbol{U}}_T^k \to \mathbb{R}$ satisfies the following properties:*

(S1) Symmetry and positivity. *s_T is symmetric and positive semidefinite;*

(S2) Stability and boundedness. *There is a real number $C_{\mathrm{a}} > 0$ independent of h and of T such that, for all $\underline{\boldsymbol{v}}_T \in \underline{\boldsymbol{U}}_T^k$,*

$$C_{\mathrm{a}}^{-1} \|\underline{\boldsymbol{v}}_T\|_{1,T}^2 \leq \mathrm{a}_T(\underline{\boldsymbol{v}}_T, \underline{\boldsymbol{v}}_T) \leq C_{\mathrm{a}} \|\underline{\boldsymbol{v}}_T\|_{1,T}^2; \tag{8.29}$$

(S3) Polynomial consistency. *For all $\boldsymbol{w} \in \mathbb{P}^{k+1}(T)^d$ and all $\underline{\boldsymbol{v}}_T \in \underline{\boldsymbol{U}}_T^k$, it holds*

$$\mathrm{s}_T(\underline{\boldsymbol{I}}_T^k \boldsymbol{w}, \underline{\boldsymbol{v}}_T) = 0.$$

As for the Poisson problem, viable local stabilisation bilinear forms can be obtained penalising, in a least square sense, the high-order differences obtained through the operators $\boldsymbol{\delta}_T^k : \underline{\boldsymbol{U}}_T^k \to \mathbb{P}^k(T)^d$ and, for all $F \in \mathcal{F}_T$, $\boldsymbol{\delta}_{TF}^k : \underline{\boldsymbol{U}}_T^k \to \mathbb{P}^k(F)^d$ such that, for all $\underline{\boldsymbol{v}}_T \in \underline{\boldsymbol{U}}_T^k$,

$$\boldsymbol{\delta}_T^k \underline{\boldsymbol{v}}_T := \boldsymbol{\pi}_T^{0,k}(\mathbf{r}_T^{k+1} \underline{\boldsymbol{v}}_T - \boldsymbol{v}_T), \qquad \boldsymbol{\delta}_{TF}^k \underline{\boldsymbol{v}}_T := \boldsymbol{\pi}_F^{0,k}(\mathbf{r}_T^{k+1} \underline{\boldsymbol{v}}_T - \boldsymbol{v}_F) \quad \forall F \in \mathcal{F}_T. \tag{8.30}$$

Specifically, the vector versions of the stabilisation bilinear forms discussed in Examples 2.7 and 2.8 are, respectively,

$$\mathrm{s}_T(\underline{\boldsymbol{u}}_T, \underline{\boldsymbol{v}}_T) = \sum_{F \in \mathcal{F}_T} h_F^{-1} \int_F (\boldsymbol{\delta}_{TF}^k \underline{\boldsymbol{u}}_T - \boldsymbol{\delta}_T^k \underline{\boldsymbol{u}}_T) \cdot (\boldsymbol{\delta}_{TF}^k \underline{\boldsymbol{v}}_T - \boldsymbol{\delta}_T^k \underline{\boldsymbol{v}}_T)$$

and

$$\mathrm{s}_T(\underline{\boldsymbol{u}}_T, \underline{\boldsymbol{v}}_T) = h_T^{-2} \int_T \boldsymbol{\delta}_T^k \underline{\boldsymbol{u}}_T \cdot \boldsymbol{\delta}_T^k \underline{\boldsymbol{v}}_T + \sum_{F \in \mathcal{F}_T} h_F^{-1} \int_F \boldsymbol{\delta}_{TF}^k \underline{\boldsymbol{u}}_T \cdot \boldsymbol{\delta}_{TF}^k \underline{\boldsymbol{v}}_T.$$

Lemma 8.11 (Properties of a_h) *The bilinear form a_h enjoys the following properties:*

(i) Stability and boundedness. *For all $\underline{\boldsymbol{v}}_h \in \underline{\boldsymbol{U}}_{h,0}^k$, it holds with C_{a} as in* (8.29) *that*

$$C_{\mathrm{a}}^{-1} \|\underline{\boldsymbol{v}}_h\|_{1,h}^2 \leq \|\underline{\boldsymbol{v}}_h\|_{\mathrm{a},h}^2 \leq C_{\mathrm{a}} \|\underline{\boldsymbol{v}}_h\|_{1,h}^2 \text{ where } \|\underline{\boldsymbol{v}}_h\|_{\mathrm{a},h} := \mathrm{a}_h(\underline{\boldsymbol{v}}_h, \underline{\boldsymbol{v}}_h)^{\frac{1}{2}}. \tag{8.31}$$

(ii) Consistency. *It holds, for all $r \in \{0, \ldots, k\}$ and all $\boldsymbol{w} \in H_0^1(\Omega)^d \cap H^{r+2}(\mathcal{T}_h)^d$ such that $\Delta \boldsymbol{w} \in L^2(\Omega)^d$,*

$$\sup_{\underline{\boldsymbol{v}}_h \in \underline{\boldsymbol{U}}_{h,0}^k, \|\underline{\boldsymbol{v}}_h\|_{1,h}=1} |\mathcal{E}_{\mathrm{a},h}(\boldsymbol{w}; \underline{\boldsymbol{v}}_h)| \lesssim h^{r+1} |\boldsymbol{w}|_{H^{r+2}(\mathcal{T}_h)^d}, \tag{8.32}$$

where the hidden constant is independent of $\boldsymbol{w}$ *and* h, *and the linear form* $\mathcal{E}_{\mathrm{a},h}(\boldsymbol{w};\cdot) : \underline{\boldsymbol{U}}_{h,0}^k \to \mathbb{R}$ *representing the consistency error is such that, for all* $\underline{\boldsymbol{v}}_h \in \underline{\boldsymbol{U}}_{h,0}^k$,

$$\mathcal{E}_{\mathrm{a},h}(\boldsymbol{w};\underline{\boldsymbol{v}}_h) := -\int_\Omega \Delta\boldsymbol{w}\cdot\boldsymbol{v}_h - \mathrm{a}_h(\underline{\boldsymbol{I}}_h^k\boldsymbol{w},\underline{\boldsymbol{v}}_h). \tag{8.33}$$

Proof Property (8.31) readily follows summing (8.29) over $T \in \mathcal{T}_h$. The proof of property (8.32) is obtained repeating the arguments in Point (ii) of Lemma 2.18, with the sole difference that the second factor in (2.47) is bounded by $\|\underline{\boldsymbol{v}}_h\|_{1,h}$ instead of $\|\underline{\boldsymbol{v}}_h\|_{\mathrm{a},h}$ (this is possible thanks to the norm equivalence proved in Point (i)). □

8.3.3 Pressure–Velocity Coupling

The pressure–velocity coupling is realised through the bilinear form $\mathrm{b}_h : \underline{\boldsymbol{U}}_h^k \times \mathbb{P}^k(\mathcal{T}_h) \to \mathbb{R}$ such that, for all $\underline{\boldsymbol{v}}_h \in \underline{\boldsymbol{U}}_h^k$ and all $q_h \in \mathbb{P}^k(\mathcal{T}_h)$,

$$\mathrm{b}_h(\underline{\boldsymbol{v}}_h, q_h) := -\sum_{T\in\mathcal{T}_h}\int_T \mathrm{D}_T^k\underline{\boldsymbol{v}}_T\, q_T, \tag{8.34}$$

where we have set, for all $T \in \mathcal{T}_h$, $q_T := (q_h)_{|T}$.

Lemma 8.12 (Properties of b_h**)** *The bilinear form* b_h *enjoys the following properties:*

(i) Consistency/1. *For all* $\boldsymbol{v} \in H^1(\Omega)^d$, *it holds that*

$$\mathrm{b}_h(\underline{\boldsymbol{I}}_h^k\boldsymbol{v}, q_h) = b(\boldsymbol{v}, q_h) \qquad \forall q_h \in \mathbb{P}^k(\mathcal{T}_h). \tag{8.35}$$

(ii) Inf–sup stability. *There is a real number* $C_{\mathrm{b}} > 0$ *independent of* h, *but possibly depending on* Ω, d, ϱ, *and* k, *such that*

$$\forall q_h \in P_h^k, \qquad C_{\mathrm{b}}\|q_h\|_{L^2(\Omega)} \le \sup_{\underline{\boldsymbol{v}}_h\in\underline{\boldsymbol{U}}_{h,0}^k\setminus\{\underline{\boldsymbol{0}}\}} \frac{\mathrm{b}_h(\underline{\boldsymbol{v}}_h, q_h)}{\|\underline{\boldsymbol{v}}_h\|_{1,h}}. \tag{8.36}$$

(iii) Consistency/2. *It holds, for all* $r \in \{0,\ldots k\}$ *and all* $q \in H^1(\Omega)\cap H^{r+1}(\mathcal{T}_h)$,

$$\sup_{\underline{\boldsymbol{v}}_h\in\underline{\boldsymbol{U}}_{h,0}^k,\,\|\underline{\boldsymbol{v}}_h\|_{1,h}=1} |\mathcal{E}_{\mathrm{b},h}(q;\underline{\boldsymbol{v}}_h)| \lesssim h^{r+1}|q|_{H^{r+1}(\mathcal{T}_h)}, \tag{8.37}$$

where the hidden constant is independent of q and h, and the linear form $\mathcal{E}_{\mathrm{b},h}(q;\cdot) : \underline{U}_h^k \to \mathbb{R}$ representing the consistency error is such that, for all $\underline{v}_h \in \underline{U}_h^k$,

$$\mathcal{E}_{\mathrm{b},h}(q;\underline{v}_h) := \int_\Omega \nabla q \cdot \boldsymbol{v}_h - \mathrm{b}_h(\underline{\boldsymbol{v}}_h, \pi_h^{0,k} q). \tag{8.38}$$

Proof (i) *Consistency/1.* It holds

$$\begin{aligned}
\mathrm{b}_h(\underline{\boldsymbol{I}}_h^k \boldsymbol{v}, q_h) &= -\sum_{T\in\mathcal{T}_h} \int_T \mathrm{D}_T^k \underline{\boldsymbol{I}}_T^k \boldsymbol{v}\, q_T \\
&= -\sum_{T\in\mathcal{T}_h} \int_T \pi_T^{0,k}(\nabla\cdot\boldsymbol{v})\, q_T \\
&= -\sum_{T\in\mathcal{T}_h} \int_T (\nabla\cdot\boldsymbol{v})\, q_T = b(\boldsymbol{v}, q_h),
\end{aligned}$$

where we have used the definition (8.34) of the bilinear form b_h in the first line, the commutation property (8.21) of the local discrete divergence operator in the second line, the definition (1.57) of $\pi_T^{0,k}$ together with the fact that $q_T \in \mathbb{P}^k(T)$ for all $T \in \mathcal{T}_h$ to pass to the third line, and the definition (8.4) of the bilinear form b to conclude.

(ii) *Inf–sup stability.* We start by noticing that the properties (8.25) and (8.35) express the fact that $\underline{\boldsymbol{I}}_h^k$ is a Fortin (or B-compatible) operator; see [194] and also [57, Section 5.4]. Let now $q_h \in P_h^k$. From Lemma 8.3, we infer the existence of $\boldsymbol{v}_{q_h} \in H_0^1(\Omega)^d$ such that $-\nabla\cdot\boldsymbol{v}_{q_h} = q_h$ and $\beta\|\boldsymbol{v}_{q_h}\|_{H^1(\Omega)^d} \le \|q_h\|_{L^2(\Omega)}$, with constant β depending only on Ω. Using the above fact, we get

$$\|q_h\|_{L^2(\Omega)}^2 = -\int_\Omega (\nabla\cdot\boldsymbol{v}_{q_h})\, q_h = b(\boldsymbol{v}_{q_h}, q_h) = \mathrm{b}_h(\underline{\boldsymbol{I}}_h^k \boldsymbol{v}_{q_h}, q_h),$$

where we have used the definition (8.4) of the continuous pressure–velocity coupling bilinear form b and the consistency property (8.35) of its discrete counterpart b_h with $\boldsymbol{v} = \boldsymbol{v}_{q_h}$ to conclude. Hence, denoting by $\mathcal{S}_h$ the supremum in the right-hand side of (8.36) and using (8.25), we can write

$$\|q_h\|_{L^2(\Omega)}^2 \le \mathcal{S}_h \|\underline{\boldsymbol{I}}_h^k \boldsymbol{v}_{q_h}\|_{1,h} \lesssim \mathcal{S}_h \|\boldsymbol{v}_{q_h}\|_{H^1(\Omega)^d} \lesssim \mathcal{S}_h \|q_h\|_{L^2(\Omega)}.$$

Simplifying yields (8.36).

(iii) *Consistency/2.* Integrating by parts element by element, and using, in a similar way as in the proof of Corollary 1.19, the fact that the jumps of $q \in H^1(\Omega)$ vanish across interfaces (use Lemma 1.21 with $p = 2$) and that $\boldsymbol{v}_F$ is single-valued for all

$F \in \mathcal{F}_h^{\mathrm{i}}$ while it vanishes on any $F \in \mathcal{F}_h^{\mathrm{b}}$ to insert it into the second term, we can write

$$\int_\Omega \nabla q \cdot \boldsymbol{v}_h = -\sum_{T \in \mathcal{T}_h} \left(\int_T q\, (\nabla \cdot \boldsymbol{v}_T) + \sum_{F \in \mathcal{F}_T} \int_F q\, (\boldsymbol{v}_F - \boldsymbol{v}_T) \cdot \boldsymbol{n}_{TF} \right). \tag{8.39}$$

On the other hand, recalling the definition (8.34) of b_h and expanding, for all $T \in \mathcal{T}_h$, $\mathrm{D}_T^k \underline{\boldsymbol{v}}_T$ according to (8.20) with $\pi_T^{0,k} q$ instead of q, we have that

$$\mathrm{b}_h(\underline{\boldsymbol{v}}_h, \pi_h^{0,k} q) = -\sum_{T \in \mathcal{T}_h} \left(\int_T q\, (\nabla \cdot \boldsymbol{v}_T) + \sum_{F \in \mathcal{F}_T} \int_F \pi_T^{0,k} q\, (\boldsymbol{v}_F - \boldsymbol{v}_T) \cdot \boldsymbol{n}_{TF} \right), \tag{8.40}$$

where we have used the definition (1.57) of $\pi_T^{0,k}$ together with the fact that $(\nabla \cdot \boldsymbol{v}_T) \in \mathbb{P}^{k-1}(T) \subset \mathbb{P}^k(T)$ to remove the projector from the first term inside the summation over $T \in \mathcal{T}_h$. Subtracting (8.40) from (8.39), taking absolute values, and using generalised Hölder inequalities with exponents $(2, 2, \infty)$ followed by $\|\boldsymbol{n}_{TF}\|_{L^\infty(F)^d} = 1$ and $h_F \le h_T$, we get

$$\begin{aligned}
&\left| \int_\Omega \nabla q \cdot \boldsymbol{v}_h - \mathrm{b}_h(\underline{\boldsymbol{v}}_h, \pi_h^{0,k} q) \right| \\
&\quad \le \sum_{T \in \mathcal{T}_h} \sum_{F \in \mathcal{F}_T} h_F^{\frac{1}{2}} \|q - \pi_T^{0,k} q\|_{L^2(F)}\, h_F^{-\frac{1}{2}} \|\boldsymbol{v}_F - \boldsymbol{v}_T\|_{L^2(F)^d}\, \|\boldsymbol{n}_{TF}\|_{L^\infty(F)^d} \\
&\quad \le \sum_{T \in \mathcal{T}_h} \sum_{F \in \mathcal{F}_T} h_T^{\frac{1}{2}} \|q - \pi_T^{0,k} q\|_{L^2(F)}\, h_F^{-\frac{1}{2}} \|\boldsymbol{v}_F - \boldsymbol{v}_T\|_{L^2(F)^d} \\
&\quad \lesssim \sum_{T \in \mathcal{T}_h} h_T^{r+1} |q|_{H^{r+1}(T)} |\underline{\boldsymbol{v}}_T|_{1,\partial T},
\end{aligned} \tag{8.41}$$

where we have concluded using the trace approximation properties (1.75) of the L^2-orthogonal projector with $l = k$, $p = 2$, $s = r + 1$, and $m = 0$. Using a discrete Cauchy–Schwarz inequality on the sum over $T \in \mathcal{T}_h$, noticing that $h_T \le h$ for all $T \in \mathcal{T}_h$, and recalling the definition (8.24) of the norm $\|\cdot\|_{1,h}$ (see also (8.15)) gives

$$\left| \int_\Omega \nabla q \cdot \boldsymbol{v}_h - \mathrm{b}_h(\underline{\boldsymbol{v}}_h, \pi_h^{0,k} q) \right| \lesssim h^{r+1} |q|_{H^{r+1}(\mathcal{T}_h)} \|\underline{\boldsymbol{v}}_h\|_{1,h}.$$

Passing to the supremum over $\left\{ \underline{\boldsymbol{v}}_h \in \underline{\boldsymbol{U}}_{h,0}^k \,:\, \|\underline{\boldsymbol{v}}_h\|_{1,h} = 1 \right\}$ yields (8.37). □

8.3.4 Discrete Problem and Well-Posedness

The HHO scheme for the approximation of problem (8.3) reads: Find $(\underline{\boldsymbol{u}}_h, p_h) \in \underline{\boldsymbol{U}}_{h,0}^k \times P_h^k$ such that

$$\nu \mathrm{a}_h(\underline{\boldsymbol{u}}_h, \underline{\boldsymbol{v}}_h) + \mathrm{b}_h(\underline{\boldsymbol{v}}_h, p_h) = \int_\Omega \boldsymbol{f}\cdot\boldsymbol{v}_h \qquad \forall \underline{\boldsymbol{v}}_h \in \underline{\boldsymbol{U}}_{h,0}^k, \tag{8.42a}$$

$$-\mathrm{b}_h(\underline{\boldsymbol{u}}_h, q_h) = 0 \qquad \forall q_h \in \mathbb{P}^k(\mathcal{T}_h). \tag{8.42b}$$

Remark 8.13 (Test Space in (8.42b)*)* Similarly to the continuous problem (see Remark 8.2), it holds that $-\mathrm{b}_h(\underline{\boldsymbol{v}}_h, 1) = 0$ for all $\underline{\boldsymbol{v}}_h \in \underline{\boldsymbol{U}}_{h,0}^k$, which allows one to take the full broken polynomial space $\mathbb{P}^k(\mathcal{T}_h)$ instead of its zero-average subspace P_h^k as a test space in (8.42b). To check this property, let $\underline{\boldsymbol{v}}_h \in \underline{\boldsymbol{U}}_{h,0}^k$ and use the definitions (8.34) of the bilinear form b_h and (8.19) of the discrete divergence operator to write:

$$\begin{aligned}
\mathrm{b}_h(\underline{\boldsymbol{v}}_h, 1) &= -\sum_{T\in\mathcal{T}_h} \int_T \mathrm{D}_T^k \underline{\boldsymbol{v}}_T \\
&= -\sum_{T\in\mathcal{T}_h} \sum_{F\in\mathcal{F}_T} \int_F \boldsymbol{v}_F\cdot\boldsymbol{n}_{TF} \\
&= -\sum_{F\in\mathcal{F}_h} \sum_{T\in\mathcal{T}_F} \int_F \boldsymbol{v}_F\cdot\boldsymbol{n}_{TF} = 0,
\end{aligned}$$

where the conclusion follows using the fact that $\boldsymbol{v}_F$ is single-valued on every interface $F \in \mathcal{F}_h^{\mathrm{i}}$ (so that, in particular, $\sum_{T\in\mathcal{T}_F} \int_F \boldsymbol{v}_F\cdot\boldsymbol{n}_{TF} = \int_F \boldsymbol{v}_F\cdot(\boldsymbol{n}_{T_1F} + \boldsymbol{n}_{T_2F}) = 0$, with T_1, T_2 the two elements on each side of F), while it vanishes on every boundary face $F \in \mathcal{F}_h^{\mathrm{b}}$.

Remark 8.14 (Efficient Implementation) As originally noticed in [161, Section 6.2], when solving the system of algebraic equations corresponding to (8.42), all element-based velocity unknowns and all but one pressure unknowns per element can be locally eliminated by computing the corresponding Schur complement element-wise. Since all the computations are local, this static condensation procedure is a trivially parallel task which can fully benefit from multi-thread and multi-processor architectures. As a result, after eliminating the velocity unknowns corresponding to Dirichlet boundary conditions, we end up solving a linear system of size

$$N_{\mathrm{dof},h} := d\,\mathrm{card}(\mathcal{F}_h^{\mathrm{i}})\binom{k+d-1}{d-1} + \mathrm{card}(\mathcal{T}_h). \tag{8.43}$$

As for the continuous problem, an equivalent reformulation can be obtained setting

$$X_h^k := \underline{U}_{h,0}^k \times P_h^k, \tag{8.44}$$

introducing the global bilinear form $\mathcal{A}_h : X_h^k \times X_h^k \to \mathbb{R}$ such that, for all $(\underline{w}_h, r_h), (\underline{v}_h, q_h) \in X_h^k$,

$$\mathcal{A}_h((\underline{w}_h, r_h), (\underline{v}_h, q_h)) := \nu \mathrm{a}_h(\underline{w}_h, \underline{v}_h) + \mathrm{b}_h(\underline{v}_h, r_h) - \mathrm{b}_h(\underline{w}_h, q_h), \tag{8.45}$$

and writing: Find $(\underline{u}_h, p_h) \in X_h^k$ such that

$$\mathcal{A}_h((\underline{u}_h, p_h), (\underline{v}_h, q_h)) = \int_\Omega \boldsymbol{f} \cdot \boldsymbol{v}_h \qquad \forall (\underline{v}_h, q_h) \in X_h^k. \tag{8.46}$$

To study the well-posedness of the discrete problem, we need the following discrete Poincaré inequality, which immediately follows from the scalar case proved in Lemma 2.15: For all $\underline{v}_h \in \underline{U}_{h,0}^k$,

$$\|\boldsymbol{v}_h\|_{L^2(\Omega)^d} \le C_{\mathrm{P}} \|\underline{v}_h\|_{1,h}. \tag{8.47}$$

Lemma 8.15 (Well-Posedness of Problem (8.42)**)** *Problem* (8.42) *(or, equivalently,* (8.46)*) is well-posed, and we have the following a priori bounds for the unique discrete solution* $(\underline{u}_h, p_h) \in \underline{U}_{h,0}^k \times P_h^k$*:*

$$\nu \|\underline{u}_h\|_{1,h} \le C_{\mathrm{a}} C_{\mathrm{P}} \|\boldsymbol{f}\|_{L^2(\Omega)^d}, \qquad \|p_h\|_{L^2(\Omega)} \le C_{\mathrm{b}}^{-1}\left(1 + C_{\mathrm{a}}^2\right) C_{\mathrm{P}} \|\boldsymbol{f}\|_{L^2(\Omega)^d}, \tag{8.48}$$

with C_{a} *as in* (8.29)*,* C_{b} *as in* (8.36)*, and* C_{P} *as in* (8.47).

The proof hinges on the following result, which constitutes a special case of [183, Theorem 2.34].

Lemma 8.16 (Well-Posedness of Saddle Point Problems) *Let* $\mathtt{U}$ *and* $\mathtt{P}$ *be two reflexive Banach spaces, let* $\mathtt{a} : \mathtt{U} \times \mathtt{U} \to \mathbb{R}$ *and* $\mathtt{b} : \mathtt{U} \times \mathtt{P} \to \mathbb{R}$ *denote two bounded bilinear forms, and let* $\mathtt{f} \in \mathtt{U}^\star$ *and* $\mathtt{g} \in \mathtt{P}^\star$*, with* $\mathtt{U}^\star$ *and* $\mathtt{P}^\star$ *denoting the dual spaces of* $\mathtt{U}$ *and* $\mathtt{P}$*, respectively. Further assume that:*

(i) The bilinear form $\mathtt{a}$ *is* $\mathtt{U}$*-coercive, i.e., there exists a real number* $\alpha > 0$ *such that, for all* $v \in \mathtt{U}$*,*

$$\alpha \|v\|_{\mathtt{U}}^2 \le \mathtt{a}(v, v);$$

(ii) The bilinear form $\mathtt{b}$ *is inf–sup stable, i.e., there exists a real number* $\beta > 0$ *such that, for all* $q \in \mathtt{P}$*,*

$$\beta \|q\|_{\mathtt{P}} \le \sup_{v \in \mathtt{U} \setminus \{0\}} \frac{\mathtt{b}(v, q)}{\|v\|_{\mathtt{U}}}.$$

Then, the problem: Find $(u, p) \in \mathtt{U} \times \mathtt{P}$ such that

$$\begin{aligned} \mathtt{a}(u, v) + \mathtt{b}(v, p) &= \langle \mathtt{f}, v\rangle_{\mathtt{U}^\star,\mathtt{U}} \qquad \forall v \in \mathtt{U}, \\ -\mathtt{b}(u, q) &= \langle \mathtt{g}, q\rangle_{\mathtt{P}^\star,\mathtt{P}} \qquad \forall q \in \mathtt{P}, \end{aligned}$$

is well-posed, i.e., it admits a unique solution for which the following a priori bounds hold:

$$\|u\|_{\mathtt{U}} \le C_1\|\mathtt{f}\|_{\mathtt{U}^\star} + C_2\|\mathtt{g}\|_{\mathtt{P}^\star}, \qquad \|p\|_{\mathtt{P}} \le C_2\|\mathtt{f}\|_{\mathtt{U}^\star} + C_3\|\mathtt{g}\|_{\mathtt{P}^\star}, \tag{8.49}$$

where $C_1 := \frac{1}{\alpha}$, $C_2 := \frac{1}{\beta}\left(1 + \frac{\|\mathtt{a}\|_{\mathtt{U}\times\mathtt{U}}}{\alpha}\right)$, *and* $C_3 := \frac{\|\mathtt{a}\|_{\mathtt{U}\times\mathtt{U}}}{\beta^2}\left(1 + \frac{\|\mathtt{a}\|_{\mathtt{U}\times\mathtt{U}}}{\alpha}\right)$.

Proof (Lemma 8.15) We apply Lemma 8.16 with $\mathtt{U} = \underline{\boldsymbol{U}}_{h,0}^k$ equipped with the norm $\nu^{\frac{1}{2}}\|\cdot\|_{1,h}$, $\mathtt{P} = P_h^k$ equipped with the norm $\nu^{-\frac{1}{2}}\|\cdot\|_{L^2(\Omega)}$, $\mathtt{a} = \nu \mathrm{a}_h$ (so that, by (8.31), the coercivity constant is $\alpha = C_{\mathrm{a}}^{-1}$ and it holds $\|\mathtt{a}\|_{\mathtt{U}\times\mathtt{U}} \le C_{\mathrm{a}}$), $\mathtt{b} = \mathrm{b}_h$ (so that, by (8.36), the inf–sup constant is $\beta = C_{\mathrm{b}}$), $\langle \mathtt{f}, v\rangle_{\mathtt{U}^\star,\mathtt{U}} = \int_\Omega \boldsymbol{f}\cdot\boldsymbol{v}$, and $\langle \mathtt{g}, q\rangle_{\mathtt{P}^\star,\mathtt{P}} = 0$. The a priori bounds (8.48) follow from (8.49) after estimating

$$|\langle \mathtt{f}, v\rangle_{\mathtt{U}^\star,\mathtt{U}}| = \left|\int_\Omega \boldsymbol{f}\cdot\boldsymbol{v}\right| \le \|\boldsymbol{f}\|_{L^2(\Omega)^d}\,\|\boldsymbol{v}_h\|_{L^2(\Omega)^d} \le \nu^{-\frac{1}{2}} C_{\mathrm{P}}\|\boldsymbol{f}\|_{L^2(\Omega)^d}\,\nu^{\frac{1}{2}}\|\underline{\boldsymbol{v}}_h\|_{1,h},$$

so that, in particular, $\|\mathtt{f}\|_{\mathtt{U}^\star} \le \nu^{-\frac{1}{2}} C_{\mathrm{P}}\|\boldsymbol{f}\|_{L^2(\Omega)^d}$. □

8.4 Flux Formulation

Denote by $(\boldsymbol{u}, p) \in \boldsymbol{U} \times P$ the unique solution to (8.3) and assume, for the sake of simplicity, that $\boldsymbol{u} \in H^2(\mathcal{T}_h)^d$ and $p \in H^1(\mathcal{T}_h)$. At the continuous level, we have the following local momentum and mass balances: For all $T \in \mathcal{T}_h$ and all $(\boldsymbol{v}_T, q_T) \in \mathbb{P}^k(T)^d \times \mathbb{P}^k(T)$,

$$\int_T \nu\boldsymbol{\nabla}\boldsymbol{u}{:}\boldsymbol{\nabla}\boldsymbol{v}_T - \int_T p\,(\boldsymbol{\nabla}\cdot\boldsymbol{v}_T) + \sum_{F\in\mathcal{F}_T}\int_F (-\nu\boldsymbol{\nabla}\boldsymbol{u} + p\boldsymbol{I}_d)_{|T}\,\boldsymbol{n}_{TF}\cdot\boldsymbol{v}_T = \int_T \boldsymbol{f}\cdot\boldsymbol{v}_T \tag{8.50a}$$

$$\int_T \boldsymbol{u}\cdot\boldsymbol{\nabla} q_T - \int_F (\boldsymbol{u}_{|T}\cdot\boldsymbol{n}_{TF})\,q_T = 0. \tag{8.50b}$$

Here, for any $F \in \mathcal{F}_T$, the quantities $(-\nu\boldsymbol{\nabla}\boldsymbol{u} + p\boldsymbol{I}_d)_{|T}\,\boldsymbol{n}_{TF}$ and $-\boldsymbol{u}_{|T}\cdot\boldsymbol{n}_{TF}$ can be interpreted, respectively, as the *momentum* and *mass fluxes* exiting T through F. Using Lemma 1.17 with $\boldsymbol{\tau}$ successively equal to the rows of $(-\nu\boldsymbol{\nabla}\boldsymbol{u} + p\boldsymbol{I}_d)$ for the momentum flux and $\boldsymbol{\tau} = \boldsymbol{u}$ for the mass flux, we infer that their normal traces are continuous across interfaces: For all $F \in \mathcal{F}_h^{\rm i}$ such that $F \subset \partial T_1 \cap \partial T_2$ with distinct mesh elements $T_1, T_2 \in \mathcal{T}_h$, it holds that

$$(-\nu\nabla\boldsymbol{u} + p\boldsymbol{I}_d)_{|T_1}\boldsymbol{n}_{T_1F} + (-\nu\nabla\boldsymbol{u} + p\boldsymbol{I}_d)_{|T_2}\boldsymbol{n}_{T_2F} = \boldsymbol{0}, \tag{8.51a}$$

$$\boldsymbol{u}_{|T_1}\cdot\boldsymbol{n}_{T_1F} + \boldsymbol{u}_{|T_2}\cdot\boldsymbol{n}_{T_2F} = 0. \tag{8.51b}$$

We notice, in passing, that the stronger condition $\boldsymbol{u}_{|T_1} = \boldsymbol{u}_{|T_2}$ holds for the traces of the velocity on either side of F as a result of the regularity $\boldsymbol{u} \in \boldsymbol{U}$ together with Lemma 1.21 applied component-wise with $p = 2$. The conservation property (8.51) can be formulated with weaker regularity, but we do not further develop this point here as it is not relevant to our purpose.

The goal of this section is to show that discrete counterparts of (8.50) and (8.51) hold for the discrete solution. Following the discussion in Sect. 2.2.5, to identify the viscous contribution to the momentum flux, we introduce the boundary difference space

$$\underline{\boldsymbol{D}}_{\partial T}^k := \left\{\underline{\boldsymbol{\alpha}}_{\partial T} := (\boldsymbol{\alpha}_F)_{F\in\mathcal{F}_T} \;:\; \boldsymbol{\alpha}_F \in \mathbb{P}^k(F)^d \quad \forall F \in \mathcal{F}_T\right\},$$

along with the boundary residual operator $\underline{\boldsymbol{R}}_{\partial T}^k := (\boldsymbol{R}_{TF}^k)_{F\in\mathcal{F}_T} : \underline{\boldsymbol{U}}_T^k \to \underline{\boldsymbol{D}}_{\partial T}^k$ such that, for all $\underline{\boldsymbol{v}}_T \in \underline{\boldsymbol{U}}_T^k$,

$$-\sum_{F\in\mathcal{F}_T}\int_F \boldsymbol{R}_{TF}^k\underline{\boldsymbol{v}}_T\cdot\boldsymbol{\alpha}_F = \mathrm{s}_T(\underline{\boldsymbol{v}}_T, (\boldsymbol{0}, \underline{\boldsymbol{\alpha}}_{\partial T})) \qquad \forall \underline{\boldsymbol{\alpha}}_{\partial T} \in \underline{\boldsymbol{D}}_{\partial T}^k. \tag{8.52}$$

Lemma 8.17 (Flux Formulation) *Let $\mathcal{M}_h$ denote a polytopal mesh in the sense of Definition 1.4. Assume that, for any $T \in \mathcal{T}_h$, the stabilisation bilinear form s_T satisfies Assumption 8.10. Let $(\underline{\boldsymbol{u}}_h, p_h) \in \underline{\boldsymbol{U}}_{h,0}^k \times P_h^k$ and, for all $T \in \mathcal{T}_h$ and all $F \in \mathcal{F}_T$, define the numerical normal trace of the viscous momentum flux as follows:*

$$\boldsymbol{\Phi}_{TF}^{\rm visc}(\underline{\boldsymbol{u}}_T) := \nu\left(-\nabla\mathbf{r}_T^{k+1}\underline{\boldsymbol{u}}_T\boldsymbol{n}_{TF} + \boldsymbol{R}_{TF}^k\underline{\boldsymbol{u}}_T\right),$$

with $\boldsymbol{R}_{TF}^k$ defined by (8.52).

Then, $(\underline{\boldsymbol{u}}_h, p_h)$ is the unique solution to the discrete problem (8.42) *(or, equivalently,* (8.46)*) if and only if the following two properties hold:*

(i) Local momentum and mass balance. *For all $T \in \mathcal{T}_h$ and all $(\boldsymbol{v}_T, q_T) \in \mathbb{P}^k(T)^d \times \mathbb{P}^k(T)$,*

(continued)

Lemma 8.17 (continued)

$$\int_T \nu \nabla \mathbf{r}_T^{k+1} \underline{\boldsymbol{u}}_T : \nabla \boldsymbol{v}_T - \int_T p_T \, (\nabla \cdot \boldsymbol{v}_T)$$

$$+ \sum_{F \in \mathcal{F}_T} \int_F \left(\boldsymbol{\Phi}_{TF}^{\mathrm{visc}}(\underline{\boldsymbol{u}}_T) + p_T \boldsymbol{n}_{TF} \right) \cdot \boldsymbol{v}_T = \int_T \boldsymbol{f} \cdot \boldsymbol{v}_T, \tag{8.53a}$$

$$\int_T \boldsymbol{u}_T \cdot \nabla q_T - \sum_{F \in \mathcal{F}_T} \int_F (\boldsymbol{u}_F \cdot \boldsymbol{n}_{TF}) q_T = 0. \tag{8.53b}$$

(ii) Continuity of the numerical normal traces of the momentum and mass fluxes. *For any interface $F \in \mathcal{F}_h^{\mathrm{i}}$ such that $F \subset \partial T_1 \cap \partial T_2$ for distinct mesh elements $T_1, T_2 \in \mathcal{T}_h$, it holds*

$$\left(\boldsymbol{\Phi}_{T_1F}^{\mathrm{visc}}(\underline{\boldsymbol{u}}_{T_1}) + p_{T_1} \boldsymbol{n}_{T_1F} \right) + \left(\boldsymbol{\Phi}_{T_2F}^{\mathrm{visc}}(\underline{\boldsymbol{u}}_{T_2}) + p_{T_2} \boldsymbol{n}_{T_2F} \right) = \boldsymbol{0}, \tag{8.54a}$$

$$\boldsymbol{u}_F \cdot \boldsymbol{n}_{T_1F} + \boldsymbol{u}_F \cdot \boldsymbol{n}_{T_2F} = 0. \tag{8.54b}$$

Proof The proof adapts that of Lemma 2.21 accounting for the following differences: first, the velocity is a vector-valued unknown; second, the continuity of the mass flux is not enforced by the scheme, but rather built into the single-valuedness of face velocity unknowns.

The following equivalent expression for the viscous bilinear form a_h defined by (8.27) is inferred as in Lemma 2.25, where the scalar case is considered: For all $\underline{\boldsymbol{v}}_h \in \underline{\boldsymbol{U}}_{h,0}^k$,

$$\nu \mathrm{a}_h(\underline{\boldsymbol{u}}_h, \underline{\boldsymbol{v}}_h)$$
$$= \sum_{T \in \mathcal{T}_h} \left[\nu \int_T \nabla \mathbf{r}_T^{k+1} \underline{\boldsymbol{u}}_T : \nabla \boldsymbol{v}_T - \sum_{F \in \mathcal{F}_T} \int_F \boldsymbol{\Phi}_{TF}^{\mathrm{visc}}(\underline{\boldsymbol{u}}_T) \cdot (\boldsymbol{v}_F - \boldsymbol{v}_T) \right]. \tag{8.55}$$

Moreover, writing the definition (8.34) of b_h for $q_h = p_h$ and expanding, for all $T \in \mathcal{T}_h$, $\mathrm{D}_T^k \underline{\boldsymbol{v}}_T$ according to (8.20) with $q = p_T$, we obtain

$$\mathrm{b}_h(\underline{\boldsymbol{v}}_h, p_h) = - \sum_{T \in \mathcal{T}_h} \left[\int_T p_T \, (\nabla \cdot \boldsymbol{v}_T) + \sum_{F \in \mathcal{F}_T} \int_F p_T \boldsymbol{n}_{TF} \cdot (\boldsymbol{v}_F - \boldsymbol{v}_T) \right]. \tag{8.56}$$

Plugging (8.55) and (8.56) into the discrete momentum equation (8.42a), and expanding D_T^k in (8.42b) (see also (8.34)) according to its definition (8.19), we see that $(\underline{u}_h, p_h)$ solves (8.42) if and only if, for all $\underline{v}_h \in \underline{U}_{h,0}^k$ and all $q_h \in \mathbb{P}^k(\mathcal{T}_h)$,

$$\sum_{T\in\mathcal{T}_h}\left[\int_T \nu\nabla \mathbf{r}_T^{k+1}\underline{u}_T:\nabla v_T - \int_T p_T\,(\nabla\cdot v_T)\right.$$
$$\left. + \sum_{F\in\mathcal{F}_T}\int_F \left(\Phi_{TF}^{\rm visc}(\underline{u}_T) + p_T n_{TF}\right)\cdot(v_T - v_F)\right] = \sum_{T\in\mathcal{T}_h}\int_T f\cdot v_T \tag{8.57}$$

and

$$\sum_{T\in\mathcal{T}_h}\left(\int_T u_T\cdot\nabla q_T - \sum_{F\in\mathcal{F}_T}\int_F (u_F\cdot n_{TF})q_T\right) = 0. \tag{8.58}$$

The momentum flux balance (8.53a) and conservation (8.54a) follow from (8.57) as in the proof of Lemma 2.25, by selecting as test vectors $\underline{v}_h$ elements of the canonical basis of $\underline{U}_h^k$: (8.53a) is obtained by taking $\underline{v}_h$ such that v_T spans $\mathbb{P}^k(T)^d$ for a selected mesh element $T \in \mathcal{T}_h$ while $v_{T'} = \mathbf{0}$ for all $T' \in \mathcal{T}_h\setminus\{T\}$ and $v_F = \mathbf{0}$ for all $F \in \mathcal{F}_h$; (8.54a) corresponds to $\underline{v}_h$ such that $v_T = \mathbf{0}$ for all $T \in \mathcal{T}_h$, v_F spans $\mathbb{P}^k(F)^d$ for a selected interface $F \in \mathcal{F}_h^{\rm i}$ such that $F \subset \partial T_1 \cap \partial T_2$ for distinct mesh elements $T_1, T_2 \in \mathcal{T}_h$, and $v_{F'} = \mathbf{0}$ for all $F' \in \mathcal{F}_h \setminus \{F\}$.

The mass balance equation (8.53b) clearly corresponds to testing (8.58) against the canonical basis of $\mathbb{P}^k(\mathcal{T}_h)$, obtained by selecting, for each element $T \in \mathcal{T}_h$, q_h such that q_T spans $\mathbb{P}^k(T)$ while $q_{T'} = 0$ for all $T' \in \mathcal{T}_h \setminus \{T\}$. Being able to take $\mathbb{P}^k(T)$, and not just P_h^k, as space for the test functions in (8.42b) is crucial to do so; see Remark 8.13 on this subject.

Finally, the continuity of the mass fluxes expressed by (8.54b) is an immediate consequence of the single-valuedness of face unknowns, and the fact that $n_{T_1F} + n_{T_2F} = \mathbf{0}$ whenever F is an interface between the two cells T_1, T_2. □

8.5 Error Analysis

We carry out in this section the convergence analysis for the HHO scheme (8.42).

8.5.1 Energy Error Estimate

We start, as usual, by a convergence estimate for the discretisation error measured in discrete norm of X_h^k which we take such that, for all $(\underline{v}_h, q_h) \in X_h^k$, in accordance with (A.18) and the choice of norms made in the proof of Lemma 8.15,

$$\|(\underline{v}_h, q_h)\|_{X,h} := \left(\nu\|\underline{v}_h\|_{1,h}^2 + \nu^{-1}\|q_h\|_{L^2(\Omega)}^2\right)^{\frac{1}{2}}. \tag{8.59}$$

Theorem 8.18 (Discrete Energy Error Estimate) *Let $(\mathcal{M}_h)_{h\in\mathcal{H}}$ denote a regular mesh sequence in the sense of Definition 1.9. Let a polynomial degree $k \geq 0$ be fixed. Let $(\boldsymbol{u}, p) \in \boldsymbol{U} \times P$ denote the unique solution to the continuous problem (8.3) (or, equivalently, (8.7)), for which we assume the additional regularity $\boldsymbol{u} \in H^{r+2}(\mathcal{T}_h)^d$ and $p \in H^1(\Omega) \cap H^{r+1}(\mathcal{T}_h)$ for some $r \in \{0, \ldots, k\}$. For all $h \in \mathcal{H}$, let $(\underline{\boldsymbol{u}}_h, p_h) \in \underline{\boldsymbol{U}}_{h,0}^k \times P_h^k$ denote the unique solution to the discrete problem (8.42) (or, equivalently, (8.46)) with stabilisation bilinear forms s_T, $T \in \mathcal{T}_h$, in (8.28) satisfying Assumption 8.10. Then,*

$$\|(\underline{\boldsymbol{u}}_h - \underline{\boldsymbol{I}}_h^k\boldsymbol{u}, p_h - \pi_h^{0,k}p)\|_{X,h}$$
$$\lesssim C_{\mathcal{A}}^{-1} h^{r+1}\left(\nu^{\frac{1}{2}}|\boldsymbol{u}|_{H^{r+2}(\mathcal{T}_h)^d} + \nu^{-\frac{1}{2}}|p|_{H^{r+1}(\mathcal{T}_h)}\right), \tag{8.60}$$

where the hidden constant is independent of $\boldsymbol{u}$, p, h and ν, while $C_{\mathcal{A}}$ depends on the stability constants of a_h (see (8.31)) and b_h (see (8.36)) according to

$$C_{\mathcal{A}} := \left[C_{\mathrm{a}}^2\left(1 + 2C_{\mathrm{b}}^{-2}C_{\mathrm{a}}^2\right)^2 + 4C_{\mathrm{b}}^{-2}\right]^{-\frac{1}{2}}. \tag{8.61}$$

Proof We invoke Corollary A.13 with $\mathrm{U} = \boldsymbol{U}$, $\mathrm{P} = P$, $\mathrm{U}_h = \underline{\boldsymbol{U}}_{h,0}^k$ equipped with the norm $\nu^{\frac{1}{2}}\|\cdot\|_{1,h}$ and the velocity interpolator $\mathrm{I}_h = \underline{\boldsymbol{I}}_h^k$, $\mathrm{P}_h = P_h^k$ equipped with the norm $\nu^{-\frac{1}{2}}\|\cdot\|_{L^2(\Omega)}$ and the pressure interpolator $\mathrm{J}_h = \pi_T^{0,k}$, $\mathtt{a}_h = \nu\mathrm{a}_h$ (so that, by (8.31), $\mathtt{a}_h$ is coercive with constant $\alpha = C_{\mathrm{a}}^{-1}$ and $\|\mathtt{a}_h\|_{\mathrm{U}_h\times\mathrm{U}_h} \leq C_{\mathrm{a}}$), $\mathtt{b}_h = \mathrm{b}_h$ (so that, by (8.36), $\mathtt{b}_h$ satisfies the inf–sup condition with constant $\beta = C_{\mathrm{b}}$). With these choices, it follows from Lemma A.11 that $\mathcal{A}_h$ satisfies an inf–sup condition on X_h^k with constant given by (8.61). Moreover, by the assumed regularity on $\boldsymbol{u}$ and p, (8.1) holds almost everywhere in Ω; hence, $\boldsymbol{f} = -\nu\Delta\boldsymbol{u} + \nabla p$ and the consistency error defined by (A.24) can be written

$$\mathcal{E}_h((\boldsymbol{u}, p); (\underline{\boldsymbol{v}}_h, q_h)) = \int_\Omega \boldsymbol{f}\cdot\boldsymbol{v}_h - \nu\mathrm{a}_h(\underline{\boldsymbol{I}}_h^k\boldsymbol{u}, \underline{\boldsymbol{v}}_h) - \mathrm{b}_h(\underline{\boldsymbol{v}}_h, \pi_h^{0,k}p)$$
$$+ \mathrm{b}_h(\underline{\boldsymbol{I}}_h^k\boldsymbol{u}, q_h) \tag{8.62}$$

$$= \nu\left(-\int_\Omega \Delta \boldsymbol{u}\cdot\boldsymbol{v}_h - \mathrm{a}_h(\underline{\boldsymbol{I}}_h^k\boldsymbol{u}, \underline{\boldsymbol{v}}_h)\right) + \left(\int_\Omega \nabla p\cdot\boldsymbol{v}_h - \mathrm{b}_h(\underline{\boldsymbol{v}}_h, \pi_h^{0,k} p)\right) + \mathrm{b}_h(\underline{\boldsymbol{I}}_h^k\boldsymbol{u}, q_h). \tag{8.63}$$

Using the definitions (8.33) of the viscous consistency error and (8.38) of the pressure–velocity coupling consistency error for the first two terms in parentheses, along with the consistency property (8.35) of b_h together with the mass balance equation (8.3b) to write $\mathrm{b}_h(\underline{\boldsymbol{I}}_h^k\boldsymbol{u}, q_h) = b(\boldsymbol{u}, q_h) = 0$, we obtain

$$\mathcal{E}_h((\boldsymbol{u}, p); (\underline{\boldsymbol{v}}_h, q_h)) = \nu\mathcal{E}_{\mathrm{a},h}(\boldsymbol{u}; \underline{\boldsymbol{v}}_h) + \mathcal{E}_{\mathrm{b},h}(p; \underline{\boldsymbol{v}}_h).$$

Hence, by the consistency properties (8.32) and (8.37) of the bilinear forms a_h and b_h, respectively, it is inferred, with $\|\cdot\|_{X,h,\star}$ denoting the norm dual to $\|\cdot\|_{X,h}$,

$$\|\mathcal{E}_h((\boldsymbol{u}, p); (\underline{\boldsymbol{v}}_h, q_h))\|_{X,h,\star} \lesssim h^{r+1}\left(\nu^{\frac{1}{2}}|\boldsymbol{u}|_{H^{r+2}(\mathcal{T}_h)^d} + \nu^{-\frac{1}{2}}|p|_{H^{r+1}(\mathcal{T}_h)}\right), \tag{8.64}$$

and the conclusion follows. □

8.5.2 Improved L^2-Error Estimates for the Velocity

We derive here improved L^2-error estimates for the velocity. Recalling Remark 8.1, since the topics of this section will not be further developed for the Navier–Stokes problem, we work under the assumption

$$\nu = 1. \tag{8.65}$$

As for the Poisson problem, we need further regularity for the continuous operator. Specifically, we assume that, for all $\boldsymbol{g} \in L^2(\Omega)^d$, the unique solution of the problem: Find $(\boldsymbol{w}_{\boldsymbol{g}}, r_{\boldsymbol{g}}) \in X$ such that

$$\mathcal{A}((\boldsymbol{w}_{\boldsymbol{g}}, r_{\boldsymbol{g}}), (\boldsymbol{v}, q)) = \int_\Omega \boldsymbol{g}\cdot\boldsymbol{v} \qquad \forall(\boldsymbol{v}, q) \in X \tag{8.66}$$

satisfies the a priori estimate

$$\|\boldsymbol{w}_{\boldsymbol{g}}\|_{H^2(\Omega)^d} + \|r_{\boldsymbol{g}}\|_{H^1(\Omega)} \le C\|\boldsymbol{g}\|_{L^2(\Omega)^d}, \tag{8.67}$$

with real number C depending only on Ω. This elliptic regularity is known, for example, if Ω is a convex polygon [217] or polyhedron [138].

Remark 8.19 (Elliptic Regularity for the Dual Problem) Elliptic regularity of the primal problem implies elliptic regularity also for the dual problem. Given $\boldsymbol{g} \in L^2(\Omega)^d$, the dual problem reads: Find $(\boldsymbol{z}_{\boldsymbol{g}}, s_{\boldsymbol{g}}) \in X$ such that

$$\mathcal{A}((\boldsymbol{v}, q), (\boldsymbol{z}_{\boldsymbol{g}}, s_{\boldsymbol{g}})) = \int_\Omega \boldsymbol{g}\cdot\boldsymbol{v} \qquad \forall(\boldsymbol{v}, q) \in X. \tag{8.68}$$

The definition (8.6) of $\mathcal{A}$ shows that $(\boldsymbol{z}_{\boldsymbol{g}}, -s_{\boldsymbol{g}})$ solves the primal problem with the same right-hand side so that, by (8.67), we have the a priori estimate

$$\|\boldsymbol{z}_{\boldsymbol{g}}\|_{H^2(\Omega)^d} + \|s_{\boldsymbol{g}}\|_{H^1(\Omega)} \leq C\|\boldsymbol{g}\|_{L^2(\Omega)^d}. \tag{8.69}$$

The L^2-error estimate is expressed in terms of the global velocity reconstruction $\mathbf{r}_h^{k+1} : \underline{\boldsymbol{U}}_h^k \to \mathbb{P}^{k+1}(\mathcal{T}_h)^d$ such that, for all $\underline{\boldsymbol{v}}_h \in \underline{\boldsymbol{U}}_h^k$,

$$(\mathbf{r}_h^{k+1}\underline{\boldsymbol{v}}_h)_{|T} := \mathbf{r}_T^{k+1}\underline{\boldsymbol{v}}_T \qquad \forall T \in \mathcal{T}_h. \tag{8.70}$$

Theorem 8.20 (L^2-Error Estimate for the Velocity) *Let $(\mathcal{M}_h)_{h\in\mathcal{H}}$ denote a regular mesh sequence in the sense of Definition 1.9. Let a polynomial degree $k \geq 0$ be fixed and assume (8.65). Let $(\boldsymbol{u}, p) \in X$ denote the unique solution of the continuous problem (8.3), for which we assume the additional regularity $\boldsymbol{u} \in H^{r+2}(\mathcal{T}_h)^d$ and $p \in H^1(\Omega) \cap H^{r+1}(\mathcal{T}_h)$ for some $r \in \{0, \ldots, k\}$. For all $h \in \mathcal{H}$, let $(\underline{\boldsymbol{u}}_h, p_h) \in X_h^k$ denote the unique solution to the discrete problem (8.42) with stabilisation bilinear forms s_T, $T \in \mathcal{T}_h$, in (8.28) satisfying Assumption 8.10. Under the elliptic regularity assumption, and further assuming that $\boldsymbol{f} \in H^1(\mathcal{T}_h)^d$ if $k = 0$, it holds*

$$\|\mathbf{r}_h^{k+1}\underline{\boldsymbol{u}}_h - \boldsymbol{u}\|_{L^2(\Omega)^d} \lesssim \begin{cases} h^2\|\boldsymbol{f}\|_{H^1(\mathcal{T}_h)^d} & \text{if } k = 0, \\ h^{r+2}\left(|\boldsymbol{u}|_{H^{r+2}(\mathcal{T}_h)^d} + |p|_{H^{r+1}(\mathcal{T}_h)}\right) & \text{if } k \geq 1, \end{cases} \tag{8.71}$$

with hidden constant independent of h, but possibly depending on Ω, d, ϱ, and k.

The proof of Theorem 8.20 follows the arguments of the proof of Theorem 2.32 (not repeated here for the sake of brevity) with the following lemma playing the role of Lemma 2.33.

Lemma 8.21 (Superconvergence of Velocity Element Unknowns) *Under the assumptions and notations of Theorem 8.20, it holds that*

$$\|\boldsymbol{u}_h - \boldsymbol{\pi}_h^{0,k}\boldsymbol{u}\|_{L^2(\Omega)^d} \lesssim \begin{cases} h^2\|\boldsymbol{f}\|_{H^1(\mathcal{T}_h)^d} & \text{if } k = 0, \\ h^{r+2}\left(|\boldsymbol{u}|_{H^{r+2}(\mathcal{T}_h)^d} + |p|_{H^{r+1}(\mathcal{T}_h)}\right) & \text{if } k \geq 1, \end{cases} \tag{8.72}$$

where, for any $\boldsymbol{v} \in L^1(\Omega)^d$, $\boldsymbol{\pi}_h^{0,k}\boldsymbol{v}$ *is the* L^2*-orthogonal projection of* $\boldsymbol{v}$ *on the broken polynomial space* $\mathbb{P}^k(\mathcal{T}_h)^d$ *such that* $(\boldsymbol{\pi}_h^{0,k}\boldsymbol{v})_{|T} := \boldsymbol{\pi}_T^{0,k}\boldsymbol{v}_{|T}$ *for all* $T \in \mathcal{T}_h$.

Proof The result follows from Lemma A.14 with the same setting as in Theorem 8.18 after accounting for (8.65), that is $\mathtt{U} = \boldsymbol{U}$, $\mathtt{P} = P$, $\mathtt{U}_h = \underline{\boldsymbol{U}}_{h,0}^k$ equipped with the $\|\cdot\|_{1,h}$-norm and the velocity interpolator $\mathtt{I}_h = \underline{\boldsymbol{I}}_h^k$, $\mathtt{P}_h = P_h^k$ equipped with the L^2-norm and the pressure interpolator $\mathtt{J}_h = \pi_h^{0,k}$, $\mathtt{a}_h = \mathrm{a}_h$, and $\mathtt{b}_h = \mathrm{b}_h$. We additionally let $\mathtt{L} = L^2(\Omega)^d$ with velocity reconstruction $\mathtt{r}_h$ equal to the mapping $\underline{\boldsymbol{U}}_{h,0}^k \ni \underline{\boldsymbol{v}}_h \mapsto \boldsymbol{v}_h \in \mathbb{P}^k(\mathcal{T}_h)^d$.

With this setting, taking $(\underline{\boldsymbol{z}}_{\boldsymbol{g}}, s_{\boldsymbol{g}})$ the solution to the dual problem (8.68) with given body force $\boldsymbol{g} \in L^2(\Omega)^d$ and letting, for the sake of brevity,

$$\underline{\hat{\boldsymbol{u}}}_h := \underline{\boldsymbol{I}}_h^k\boldsymbol{u}, \qquad \hat{p}_h := \pi_h^{0,k}p, \qquad \underline{\hat{\boldsymbol{z}}}_{\boldsymbol{g},h} := \underline{\boldsymbol{I}}_h^k\boldsymbol{z}_{\boldsymbol{g}}, \qquad \hat{s}_{\boldsymbol{g},h} := \pi_h^{0,k}s_{\boldsymbol{g}},$$

the error estimate (A.30) translates into

$$\begin{aligned} &\|\boldsymbol{u}_h - \hat{\boldsymbol{u}}_h\|_{L^2(\Omega)^d} \\ &\lesssim \underbrace{\|(\underline{\boldsymbol{u}}_h - \underline{\hat{\boldsymbol{u}}}_h, p_h - \hat{p}_h)\|_{X,h} \sup_{\boldsymbol{g}\in L^2(\Omega)^d, \|\boldsymbol{g}\|_{L^2(\Omega)^d}\leq 1} \|\mathcal{E}_h^{\mathrm{d}}((\boldsymbol{z}_{\boldsymbol{g}}, s_{\boldsymbol{g}}); \cdot)\|_{X,h,\star}}_{\mathfrak{E}_1} \\ &\qquad + \underbrace{\sup_{\boldsymbol{g}\in L^2(\Omega)^d, \|\boldsymbol{g}\|_{L^2(\Omega)^d}\leq 1} \mathcal{E}_h((\boldsymbol{u}, p); (\underline{\hat{\boldsymbol{z}}}_{\boldsymbol{g},h}, \hat{s}_{\boldsymbol{g},h}))}_{\mathfrak{E}_2}, \end{aligned} \tag{8.73}$$

where the primal consistency error $\mathcal{E}_h((\boldsymbol{u}, p); \cdot)$ is given by (8.63), while the dual consistency error is such that, for any $(\underline{\boldsymbol{v}}_h, q_h) \in X_h^k$,

$$\mathcal{E}_h^{\mathrm{d}}((\boldsymbol{z}_{\boldsymbol{g}}, s_{\boldsymbol{g}}); (\underline{\boldsymbol{v}}_h, q_h)) = \int_\Omega \boldsymbol{g}\cdot\boldsymbol{v}_h - \mathcal{A}_h((\underline{\boldsymbol{v}}_h, q_h), (\underline{\hat{\boldsymbol{z}}}_{\boldsymbol{g},h}, \hat{s}_{\boldsymbol{g},h})). \tag{8.74}$$

The error estimate (8.72) follows from (8.73) after bounding the terms in the right-hand side.

(i) *Estimate of* $\mathcal{E}_1$. The first factor in $\mathcal{E}_1$ is readily estimated using (8.60) as follows:

$$\|(\underline{\boldsymbol{u}}_h - \hat{\underline{\boldsymbol{u}}}_h, p_h - \hat{p}_h)\|_{X,h} \lesssim h^{r+1} \left(|\boldsymbol{u}|_{H^{r+2}(\mathcal{T}_h)^d} + |p|_{H^{r+1}(\mathcal{T}_h)}\right). \tag{8.75}$$

Noticing that the dual consistency error (8.74) is identical to the primal consistency error (8.62) with $(\boldsymbol{u}, p, \boldsymbol{f})$ replaced by $(\boldsymbol{z_g}, -s_{\boldsymbol{g}}, \boldsymbol{g})$ and applied to $(\underline{\boldsymbol{v}}_h, -q_h)$ instead of $(\underline{\boldsymbol{v}}_h, q_h)$, the second factor in $\mathcal{E}_1$ is estimated by using Remark 8.19 and (8.64) with $r = 1$, and by invoking the regularity property (8.69):

$$\sup_{\boldsymbol{g} \in L^2(\Omega)^d, \|\boldsymbol{g}\|_{L^2(\Omega)^d} \le 1} \|\mathcal{E}_h^{\mathrm{d}}((\boldsymbol{z_g}, s_{\boldsymbol{g}}); \cdot)\|_{X,h,\star} \lesssim h. \tag{8.76}$$

Combining (8.75) and (8.76), we get

$$\mathcal{E}_1 \lesssim h^{r+2} \left(|\boldsymbol{u}|_{H^{r+2}(\mathcal{T}_h)^d} + |p|_{H^{r+1}(\mathcal{T}_h)}\right). \tag{8.77}$$

(ii) *Estimate of* $\mathcal{E}_2$. To estimate the primal-dual consistency error, we distinguish two different cases: $k \ge 1$ and $k = 0$.

(ii.A) *The case* $k \ge 1$. Recalling that (8.1a) is satisfied almost everywhere in Ω and that $\nu = 1$, we can replace $\boldsymbol{f}$ by $-\Delta \boldsymbol{u} + \nabla p$ in the expression (8.63) of the consistency error evaluated at $(\underline{\boldsymbol{v}}_h, q_h) = (\hat{\underline{\boldsymbol{z}}}_{\boldsymbol{g},h}, \hat{s}_{\boldsymbol{g},h})$ to write

$$\mathcal{E}_h((\boldsymbol{u}, p); (\hat{\underline{\boldsymbol{z}}}_{\boldsymbol{g},h}, \hat{s}_{\boldsymbol{g},h})) = \underbrace{-\int_\Omega \Delta \boldsymbol{u} \cdot \hat{\boldsymbol{z}}_{\boldsymbol{g},h} - \mathrm{a}_h(\hat{\underline{\boldsymbol{u}}}_h, \hat{\underline{\boldsymbol{z}}}_{\boldsymbol{g},h})}_{\mathfrak{T}_1}$$
$$+ \underbrace{\int_\Omega \nabla p \cdot \hat{\boldsymbol{z}}_{\boldsymbol{g},h} - \mathrm{b}_h(\hat{\underline{\boldsymbol{z}}}_{\boldsymbol{g},h}, \hat{p}_h)}_{\mathfrak{T}_2} + \underbrace{\mathrm{b}_h(\hat{\underline{\boldsymbol{z}}}_{\boldsymbol{g},h}, \hat{s}_{\boldsymbol{g},h})}_{\mathfrak{T}_3}.$$

Proceeding as in Point (ii.A) of the proof of Lemma 2.33 and using (8.69), we get for the first term

$$|\mathfrak{T}_1| \lesssim h^{r+2} |\boldsymbol{u}|_{H^{r+2}(\mathcal{T}_h)^d} \|\boldsymbol{g}\|.$$

For the second term, using (8.41) with $q = p$ and $\underline{\boldsymbol{v}}_h = \hat{\underline{\boldsymbol{z}}}_{\boldsymbol{g},h}$ together with $h_T \le h$ for all $T \in \mathcal{T}_h$ and a discrete Cauchy–Schwarz inequality on the sum over $T \in \mathcal{T}_h$, we obtain

$$|\mathfrak{T}_2| \lesssim h^{r+1} |p|_{H^{r+1}(\mathcal{T}_h)} \left(\sum_{T \in \mathcal{T}_h} |\hat{\underline{\boldsymbol{z}}}_{\boldsymbol{g},T}|_{1,\partial T}^2 \right)^{\frac{1}{2}} \lesssim h^{r+2} |p|_{H^{r+1}(\mathcal{T}_h)} |\boldsymbol{z_g}|_{H^2(\Omega)^d},$$

where the conclusion follows estimating the discrete boundary seminorm as in (2.78). Finally, by the consistency property (8.35) of b_h applied to $\boldsymbol{v} = \boldsymbol{z_g}$ and $q_h = \hat{s}_{\boldsymbol{g},h}$, and since $\nabla\cdot\boldsymbol{z_g} = 0$, we have

$$\mathfrak{T}_3 = 0.$$

Using the above bounds for $\mathfrak{T}_1$, $\mathfrak{T}_2$ and $\mathfrak{T}_3$, we conclude that

$$\mathcal{E}_2 \lesssim h^{r+2}\left(|\boldsymbol{u}|_{H^{r+2}(\mathcal{T}_h)^d} + |p|_{H^{r+1}(\mathcal{T}_h)}\right) \quad \text{if } k \geq 1. \tag{8.78}$$

(ii.B) *The case* $k = 0$. We start from a different decomposition of the primal-dual consistency error. Proceeding as in Point (ii.B) in the proof of Lemma 2.33, we notice that

$$\begin{aligned}
\int_\Omega \boldsymbol{f}\cdot\boldsymbol{\pi}_h^{0,0}\boldsymbol{z_g} &= \int_\Omega \boldsymbol{\pi}_h^{0,0}\boldsymbol{f}\cdot\boldsymbol{z_g} \\
&= \int_\Omega (\boldsymbol{\pi}_h^{0,0}\boldsymbol{f} - \boldsymbol{f})\cdot\boldsymbol{z_g} + \int_\Omega \boldsymbol{f}\cdot\boldsymbol{z_g} \\
&= \int_\Omega (\boldsymbol{\pi}_h^{0,0}\boldsymbol{f} - \boldsymbol{f})\cdot(\boldsymbol{z_g} - \boldsymbol{\pi}_h^{0,0}\boldsymbol{z_g}) + \int_\Omega (-\Delta\boldsymbol{u} + \nabla p)\cdot\boldsymbol{z_g} \\
&= \int_\Omega (\boldsymbol{\pi}_h^{0,0}\boldsymbol{f} - \boldsymbol{f})\cdot(\boldsymbol{z_g} - \boldsymbol{\pi}_h^{0,0}\boldsymbol{z_g}) + \int_\Omega \nabla\boldsymbol{u}:\nabla\boldsymbol{z_g} - \cancel{\int_\Omega p(\nabla\cdot\boldsymbol{z_g})},
\end{aligned}$$

where we have used $\boldsymbol{f} = -\Delta\boldsymbol{u} + \nabla p$ and the definition of the L^2-orthogonal projector to insert $\boldsymbol{\pi}_h^{0,0}\boldsymbol{z_g}$ in the third line, an integration by parts to pass to the fourth line, and recalled the fact that $\nabla\cdot\boldsymbol{z_g} = 0$ to cancel the last term in the right-hand side. Plugging this expression into the definition (8.62) of the consistency error with $\underline{\boldsymbol{v}}_h = \underline{\hat{\boldsymbol{z}}}_{\boldsymbol{g}}$, we can write

$$\begin{aligned}
\mathcal{E}_h((\boldsymbol{u}, p); (\underline{\hat{\boldsymbol{z}}}_{\boldsymbol{g},h}, \hat{s}_{\boldsymbol{g},h})) = &\underbrace{\int_\Omega (\boldsymbol{\pi}_h^{0,0}\boldsymbol{f} - \boldsymbol{f})\cdot(\boldsymbol{z_g} - \boldsymbol{\pi}_h^{0,0}\boldsymbol{z_g})}_{\mathfrak{T}_1} \\
&+ \underbrace{\int_\Omega \nabla\boldsymbol{u}:\nabla\boldsymbol{z_g} - \mathrm{a}_h(\underline{\hat{\boldsymbol{u}}}_h, \underline{\hat{\boldsymbol{z}}}_{\boldsymbol{g},h})}_{\mathfrak{T}_2} \underbrace{- \mathrm{b}_h(\underline{\hat{\boldsymbol{z}}}_{\boldsymbol{g},h}, \hat{p}_h) + \mathrm{b}_h(\underline{\hat{\boldsymbol{u}}}_h, \hat{s}_{\boldsymbol{g},h})}_{\mathfrak{T}_3}.
\end{aligned}$$

For the first term, using the approximation properties (1.74) of the L^2-orthogonal projector with X successively equal to the elements in $\mathcal{T}_h$, $l = 0$, $p = 2$, $s = 1$ and $m = 0$, it is readily inferred

$$|\mathfrak{T}_1| \lesssim h^2 |\boldsymbol{f}|_{H^1(\mathcal{T}_h)^d} |\boldsymbol{z_g}|_{H^1(\Omega)^d}.$$

For the second term, proceeding as in Point (ii.B) of the proof of Lemma 2.33, we get

$$|\mathfrak{T}_2| \lesssim h^2 |\boldsymbol{u}|_{H^2(\Omega)^d} |\boldsymbol{z}_{\boldsymbol{g}}|_{H^2(\Omega)^d}.$$

Finally, recalling the consistency property (8.35) of b_h and the fact that $\nabla\cdot\boldsymbol{u} = \nabla\cdot\boldsymbol{z}_{\boldsymbol{g}} = 0$, we have for the third term

$$\mathfrak{T}_3 = 0.$$

Using the above bounds for $\mathfrak{T}_1$, $\mathfrak{T}_2$ and $\mathfrak{T}_3$ to estimate the primal-dual consistency error, and recalling that, by the elliptic regularity estimate (8.67), $|\boldsymbol{u}|_{H^2(\Omega)^d} + |p|_{H^1(\Omega)} \lesssim \|\boldsymbol{f}\|_{L^2(\Omega)^d} \le \|\boldsymbol{f}\|_{H^1(\mathcal{T}_h)^d}$, we conclude that

$$\mathcal{E}_2 \lesssim h^2 \|\boldsymbol{f}\|_{H^1(\mathcal{T}_h)^d} \quad \text{if } k = 0. \tag{8.79}$$

(iii) *Conclusion.* Plug the estimates (8.77), (8.78), and (8.79) into (8.73). □

8.5.3 Other Hybrid Methods

Several hybrid methods have been developed for the Stokes problem on standard meshes. In the Hybridisable Discontinuous Galerkin method of [245], polynomials of degree k are used for the flux, velocity, and pressure variables, and convergence in h^{k+1} is experimentally observed for the L^2-norm of the error in each variable (recall that, in our case, the L^2-norm of the velocity converges as h^{k+2}, see Theorem 8.20). Similar considerations apply to the methods considered in [128]. In [223], the authors propose a Hybridisable Discontinuous Galerkin method where the velocity unknowns are polynomials of degree k at mesh elements and faces. Also in this case, the L^2-norm of the errors on both the velocity and the pressure converges as h^{k+1}. In [182], on the other hand, a method based on polynomials of degree k for the velocity and $(k-1)$ for the pressure is proposed, and its hp-convergence analysis is carried out. In this case, both the L^2-norms of the strain rate and of the pressure converge as h^k. Moving to general polyhedral meshes, we can cite: the original HHO method of [9], which hinges on the hybridised version of the Mixed High-Order method for the viscous terms (see Sect. 5.4) and an equal-order, fully discontinuous approximation of the pressure; the nonconforming Virtual Element Method of [93], which takes element-based velocity unknowns one degree less than face-based velocity unknowns and pressure unknowns; the two-dimensional $\boldsymbol{H}(\mathrm{div};\Omega)$-conforming Virtual Element Method of [52], where the velocity degrees of freedom include, for any $T \in \mathcal{T}_h$ and $k \ge 2$, nodal and edge values, moments with respect to the L^2-orthogonal complement of $\nabla\mathbb{P}^{k-1}(T)$ in $\mathbb{P}^{k-2}(T)^2$, and polynomial moments of the divergence up to degree $(k-1)$.

8.5.4 Numerical Example

We close this section with a numerical example that corroborates the theoretical results. We let $\Omega = (0,1)^2$ and consider the exact solution with $\nu = 1$, velocity components

$$u_1(\boldsymbol{x}) = -e^{x_1}(x_2 \cos x_2 + \sin x_2), \qquad u_2(\boldsymbol{x}) = e^{x_1} x_2 \sin x_2,$$

and pressure

$$p(\boldsymbol{x}) = 2e^{x_1} \sin x_2 - 2(e-1)(1-\cos 1).$$

The domain is discretised by means of a refined sequence of unstructured triangular meshes, the first four refinements of which are depicted in Fig. 3.1a. We consider polynomial degrees $k \in \{0, \dots, 3\}$. We report in Table 8.1 the following quantities:

Table 8.1 Two-dimensional test case

$N_{\mathrm{dof},h}$	$\|\underline{\boldsymbol{u}}_h - \underline{\boldsymbol{I}}_h^k \boldsymbol{u}\|_{\mathrm{a},h}$	EOC	$\|\boldsymbol{u}_h - \boldsymbol{\pi}_h^{0,k} \boldsymbol{u}\|_{L^2(\Omega)^d}$	EOC	$\|p_h - \pi_h^{0,k} p\|_{L^2(\Omega)}$	EOC
$k = 0$						
134	4.54e−01	–	2.10e−02	–	2.25e−01	–
574	2.60e−01	0.81	5.98e−03	1.82	9.95e−02	1.18
2366	1.42e−01	0.87	1.67e−03	1.84	4.58e−02	1.12
9274	7.19e−02	0.98	4.29e−04	1.96	2.27e−02	1.01
38,048	3.56e−02	1.01	1.05e−04	2.03	1.11e−02	1.03
$k = 1$						
268	2.46e−02	–	4.06e−04	–	1.32e−02	–
1148	6.68e−03	1.88	5.79e−05	2.81	4.09e−03	1.69
4732	1.81e−03	1.88	8.00e−06	2.86	1.06e−03	1.95
18,548	4.58e−04	1.98	9.98e−07	3.00	2.62e−04	2.02
76,096	1.15e−04	2.00	1.24e−07	3.01	6.50e−05	2.01
$k = 2$						
402	1.11e−03	–	1.50e−05	–	7.07e−04	–
1722	1.35e−04	3.04	8.81e−07	4.09	8.51e−05	3.05
7098	1.88e−05	2.85	6.13e−08	3.84	1.11e−05	2.93
27,822	2.39e−06	2.97	3.90e−09	3.97	1.40e−06	2.99
114,144	2.91e−07	3.04	2.38e−10	4.03	1.72e−07	3.03
$k = 3$						
536	2.51e−05	–	3.10e−07	–	1.77e−05	–
2296	1.70e−06	3.88	9.58e−09	5.02	1.09e−06	4.03
9464	1.22e−07	3.80	3.65e−10	4.72	7.67e−08	3.82
37,096	3.48e−08	1.81*	4.75e−11	2.94*	4.84e−09	3.99*
152,192	8.49e−08	−1.29*	5.70e−11	−0.26*	4.64e−10	3.38*

Starred orders of convergence are affected by machine precision

(1) $\|\underline{\boldsymbol{u}}_h - \underline{\boldsymbol{I}}_h^k \boldsymbol{u}\|_{\mathrm{a},h}$, the error on the velocity measured in the norm associated with the viscous bilinear form a_h—notice that, recalling the norm equivalence (8.31), an estimate analogous to (8.60) holds for this quantity; (2) $\|\boldsymbol{u}_h - \boldsymbol{\pi}_h^{0,k} \boldsymbol{u}\|_{L^2(\Omega)^d}$, the L^2-error on the velocity; (3) $\|p_h - \pi_h^{0,k} p\|_{L^2(\Omega)}$, the L^2-error for the pressure. All errors are relative to the corresponding norms of the discrete solution. The number of degrees of freedom $N_{\mathrm{dof},h}$ corresponds to the number of unknowns after static condensation and is defined by (8.43). In each case, we display the Estimated Order of Convergence (EOC) which, denoting by e_i an error on the ith mesh refinement, is computed as

$$\mathrm{EOC} = \frac{\log e_i - \log e_{i+1}}{\log h_i - \log h_{i+1}}. \tag{8.80}$$

As expected, the energy error estimate on the velocity and the L^2-error estimate on the pressure converge as h^{k+1}, whereas the L^2-error estimate on the velocity converges as h^{k+2}. For the last two mesh refinements with $k = 3$, a saturation of the error is observed, with a decreased convergence rate for the velocity (and on the pressure for the last refinement).

8.6 A Pressure-Robust Variation

In this section, we consider a variation of the HHO method (8.42) which delivers an estimate of the velocity independent of both the pressure and the viscosity, and which is therefore referred to as *pressure-robust*.

8.6.1 A Key Remark

We start by highlighting a key property of the continuous problem, namely that modifying the irrotational part of the body force only affects the pressure, not the velocity.

Proposition 8.22 (Independence of the Velocity from Irrotational Body Forces) *For any $\psi \in H^1(\Omega)$ such that $\int_\Omega \psi = 0$, if $(\boldsymbol{u}, p) \in \boldsymbol{U} \times P$ solves the weak problem* (8.3) *with body force $\boldsymbol{f}$ then, denoting by $(\tilde{\boldsymbol{u}}, \tilde{p}) \in \boldsymbol{U} \times P$ the solution of the weak problem with body force $\boldsymbol{f} + \nabla\psi$, it holds $(\tilde{\boldsymbol{u}}, \tilde{p}) := (\boldsymbol{u}, p + \psi)$.*

Proof By definition, $(\tilde{\boldsymbol{u}}, \tilde{p})$ satisfies

$$\begin{aligned} \nu a(\tilde{\boldsymbol{u}}, \boldsymbol{v}) + b(\boldsymbol{v}, \tilde{p}) &= \int_\Omega (\boldsymbol{f} + \nabla\psi)\cdot\boldsymbol{v} && \forall \boldsymbol{v} \in \boldsymbol{U}, \\ -b(\tilde{\boldsymbol{u}}, q) &= 0 && \forall q \in L^2(\Omega). \end{aligned}$$

Integrating by parts the second contribution in the right-hand side of the momentum equation and recalling the definition (8.4) of the bilinear form b, we can write

$$\int_\Omega \nabla\psi\cdot\boldsymbol{v} = -\int_\Omega \psi\,(\nabla\cdot\boldsymbol{v}) + \cancel{\int_{\partial\Omega} \psi\,(\boldsymbol{v}\cdot\boldsymbol{n})} = b(\boldsymbol{v},\psi), \tag{8.81}$$

where $\boldsymbol{n}$ denotes the unit normal vector field on $\partial\Omega$ with exterior orientation, and we have used the fact that $\boldsymbol{v}$ has zero trace on $\partial\Omega$ to cancel the boundary term. Hence, rearranging, we have that

$$\begin{aligned} \nu a(\tilde{\boldsymbol{u}},\boldsymbol{v}) + b(\boldsymbol{v},\tilde{p}-\psi) &= \int_\Omega \boldsymbol{f}\cdot\boldsymbol{v} &\qquad& \forall \boldsymbol{v}\in\boldsymbol{U},\\ -b(\tilde{\boldsymbol{u}},q) &= 0 && \forall q\in L^2(\Omega). \end{aligned}$$

From the well-posedness of problem (8.3) we deduce that $(\boldsymbol{u},p) = (\tilde{\boldsymbol{u}},\tilde{p}-\psi)$, so that $(\tilde{\boldsymbol{u}},\tilde{p}) = (\boldsymbol{u},p+\psi)$. □

8.6.2 An Abstract Modification of the Right-Hand Side

It can be checked that a discrete counterpart of the property highlighted in the previous section does not hold for the scheme (8.42). The reason is that relation (8.81) fails at the discrete level. This remark prompts us to consider the following variation of the scheme: Find $(\underline{\boldsymbol{u}}_h, p_h) \in \underline{\boldsymbol{U}}_{h,0}^k \times P_h^k$ such that

$$\nu \mathrm{a}_h(\underline{\boldsymbol{u}}_h,\underline{\boldsymbol{v}}_h) + \mathrm{b}_h(\underline{\boldsymbol{v}}_h,p_h) = \ell_h(\boldsymbol{f},\underline{\boldsymbol{v}}_h) \qquad \forall \underline{\boldsymbol{v}}_h\in\underline{\boldsymbol{U}}_{h,0}^k, \tag{8.82a}$$

$$-\mathrm{b}_h(\underline{\boldsymbol{u}}_h,q_h) = 0 \qquad \forall q_h\in\mathbb{P}^k(\mathcal{T}_h), \tag{8.82b}$$

where the discretisation ℓ_h of the body force satisfies the following assumption.

Assumption 8.23 (Pressure-Robust Discretisation of Body Forces) *The bilinear form $\ell_h : L^2(\Omega)^d \times \underline{\boldsymbol{U}}_{h,0}^k \to \mathbb{R}$ satisfies the following properties:*

(L1) Velocity invariance. *For all $\psi\in H^1(\Omega)$,*

$$\ell_h(\nabla\psi,\underline{\boldsymbol{v}}_h) = \mathrm{b}_h(\underline{\boldsymbol{v}}_h,\pi_h^{0,k}\psi) \qquad \forall \underline{\boldsymbol{v}}_h\in\underline{\boldsymbol{U}}_{h,0}^k. \tag{8.83}$$

(L2) Consistency. *It holds for all $r\in\{0,\ldots,k\}$ and all $\boldsymbol{w}\in H^r(\mathcal{T}_h)^d$,*

(continued)

Assumption 8.23 (continued)

$$\sup_{\underline{v}_h \in \underline{U}_{h,0}^k, \|\underline{v}_h\|_{1,h}=1} \left|\mathcal{E}_{\ell,h}(w; \underline{v}_h)\right| \lesssim h^{r+1} |w|_{H^r(\mathcal{T}_h)^d}, \tag{8.84}$$

where the hidden constant is independent of w *and* h, *and the linear form* $\mathcal{E}_{\ell,h}(w; \cdot) : \underline{U}_{h,0}^k \to \mathbb{R}$ *representing the consistency error is such that*

$$\mathcal{E}_{\ell,h}(w; \underline{v}_h) := \ell_h(w, \underline{v}_h) - \int_\Omega w \cdot v_h. \tag{8.85}$$

Condition (L1) restores (8.81) at the discrete level, while condition (L2) guarantees that the new discretisation of body forces preserves the original accuracy of the scheme.

8.6.3 *Pressure-Robust Error Estimate*

In the following theorem, we investigate the effect of the novel formulation of the right-hand side in (8.82a) on the error estimate.

Theorem 8.24 (Pressure-Robust Energy Error Estimate) *Let* $(\mathcal{M}_h)_{h\in\mathcal{H}}$ *denote a regular mesh sequence in the sense of Definition 1.9. Let a polynomial degree* $k \geq 0$ *be fixed. Let* $(u, p) \in U \times P$ *denote the unique solution to the continuous problem* (8.3), *for which we assume the additional regularity* $u \in H^{r+2}(\mathcal{T}_h)^d$ *and* $p \in H^1(\Omega) \cap H^{r+1}(\mathcal{T}_h)$ *for some* $r \in \{0, \ldots, k\}$. *For all* $h \in \mathcal{H}$, *let* $(\underline{u}_h, p_h) \in \underline{U}_{h,0}^k \times P_h^k$ *denote the unique solution to the discrete problem* (8.82) *with stabilisation bilinear forms* s_T, $T \in \mathcal{T}_h$, *in* (8.28) *satisfying Assumptions 8.10 and bilinear form* ℓ_h *satisfying Assumption 8.23. Then,*

$$\|(\underline{u}_h - \underline{I}_h^k u, p_h - \pi_h^{0,k} p)\|_{X,h} \lesssim h^{r+1} \nu^{\frac{1}{2}} |u|_{H^{r+2}(\mathcal{T}_h)^d}, \tag{8.86}$$

where the hidden constant is independent of u, p, h, *and* ν.

Remark 8.25 (Robustness of the Error Estimate (8.86)*)* The error estimate (8.86) reveals a crucial difference with respect to (8.60), namely that the multiplicative constant in the right-hand side is independent of the pressure. Accounting for Proposition 8.22, this shows that the approximation error

$\|(\underline{\boldsymbol{u}}_h - \underline{\boldsymbol{I}}_h^k \boldsymbol{u}, p_h - \pi_h^{0,k} p)\|_{X,h}$ is unaffected by the presence of irrotational body forces, which leave the continuous velocity (hence the factor multiplying h^{r+1} in the right-hand side of (8.86)) unchanged. As remarked in [232], this is a practically relevant feature for, e.g., buoyancy-driven flows such as the one considered in [167], or when the Coriolis force is added to the incompressible Navier–Stokes equations as in [135, 136].

A related property is viscosity-robustness of the velocity estimate. Expanding the $\|\cdot\|_{X,h}$-norm according to its definition (8.59), we can write the following separate estimates for the velocity and the pressure:

$$\|\underline{\boldsymbol{u}}_h - \underline{\boldsymbol{I}}_h^k \boldsymbol{u}\|_{1,h} \lesssim h^{r+1} |\boldsymbol{u}|_{H^{r+2}(\mathcal{T}_h)^d}, \qquad \|p_h - \pi_h^{0,k} p\|_{L^2(\Omega)} \lesssim \nu h^{r+1} |\boldsymbol{u}|_{H^{r+2}(\mathcal{T}_h)^d}.$$

Crucially, the multiplicative constant in the first bound is independent of the viscosity.

Proof Proceeding as in the proof of Theorem 8.18, we infer the following expression for the consistency error:

$$\mathcal{E}_h((\boldsymbol{u}, p); (\underline{\boldsymbol{v}}_h, q_h)) = \ell_h(\boldsymbol{f}, \underline{\boldsymbol{v}}_h) - \nu \mathrm{a}_h(\underline{\boldsymbol{I}}_h^k \boldsymbol{u}, \underline{\boldsymbol{v}}_h) - \mathrm{b}_h(\underline{\boldsymbol{v}}_h, \pi_h^{0,k} p) + \mathrm{b}_h(\underline{\boldsymbol{I}}_h^k \boldsymbol{u}, q_h).$$

Using the fact that $\boldsymbol{f} = -\nu\Delta\boldsymbol{u} + \boldsymbol{\nabla} p$ along with the linearity of ℓ_h in its first argument, inserting $\pm \int_\Omega \nu\Delta\boldsymbol{u}\cdot\boldsymbol{v}_h$, and recalling the definitions (8.33) and (8.85) of the consistency error linear forms, we can go on writing

$$\begin{aligned}\mathcal{E}_h((\boldsymbol{u}, p); (\underline{\boldsymbol{v}}_h, q_h)) &= \underbrace{\ell_h(-\nu\Delta\boldsymbol{u}, \underline{\boldsymbol{v}}_h) + \int_\Omega \nu\Delta\boldsymbol{u}\cdot\boldsymbol{v}_h}_{\mathcal{E}_{\ell,h}(-\nu\Delta\boldsymbol{u};\underline{\boldsymbol{v}}_h)} \underbrace{- \int_\Omega \nu\Delta\boldsymbol{u}\cdot\boldsymbol{v}_h - \nu \mathrm{a}_h(\underline{\boldsymbol{I}}_h^k \boldsymbol{u}, \underline{\boldsymbol{v}}_h)}_{\nu\mathcal{E}_{\mathrm{a},h}(\boldsymbol{u};\underline{\boldsymbol{v}}_h)} \\ &\quad + \cancel{\ell_h(\boldsymbol{\nabla} p, \underline{\boldsymbol{v}}_h) - \mathrm{b}_h(\underline{\boldsymbol{v}}_h, \pi_h^{0,k} p)} + \cancel{\mathrm{b}_h(\underline{\boldsymbol{I}}_h^k \boldsymbol{u}, \underline{\boldsymbol{v}}_h)},\end{aligned}$$

we have used, respectively, (L1) and the consistency property (8.35) of b_h together with (8.3b) to cancel the terms in the last line. Passing to the $\|\cdot\|_{X,h,\star}$-norm and using the consistency properties (8.32) of a_h and (L2) of ℓ_h, (8.86) follows. □

8.6.4 A Discretisation of Body Forces Based on a Raviart–Thomas–Nédélec Velocity Reconstruction

In this section, following the ideas of [69, 73, 160], we build a discrete bilinear form ℓ_h satisfying Assumption 8.23 when $\mathcal{T}_h$ is a matching simplicial mesh in the sense of Definition 1.7.

8.6.4.1 The Raviart–Thomas–Nédélec Space

The formulation of the discrete bilinear form ℓ_h hinges on a local velocity reconstruction in the Raviart–Thomas–Nédélec [244, 255] space

$$\mathbb{RTN}^k(T) := \mathbb{P}^k(T)^d + \boldsymbol{x}\mathbb{P}^k(T).$$

Functions in $\mathbb{RTN}^k(T)$ have divergence in $\mathbb{P}^k(T)$ and normal traces in $\mathbb{P}^k(F)$ for all $F \in \mathcal{F}_T$; see, e.g., [196, Lemma 3.6]. For future use, we note the following estimate, which results from a scaling argument:

$$\|\boldsymbol{w}_T\|_T^2 \simeq \|\pi_T^{0,k-1}\boldsymbol{w}_T\|_T^2 + \sum_{F\in\mathcal{F}_T} h_F\|\boldsymbol{w}_T\cdot\boldsymbol{n}_{TF}\|_F^2 \qquad \forall \boldsymbol{w}_T \in \mathbb{RTN}^k(T), \tag{8.87}$$

with hidden constant independent of h, T, and $\boldsymbol{w}_T$, but possibly depending on ϱ and k.

The global Raviart–Thomas–Nédélec space is defined as

$$\mathbb{RTN}^k(\mathcal{T}_h) := \left\{\boldsymbol{v}_h \in \boldsymbol{H}(\mathrm{div};\Omega) \ :\ (\boldsymbol{v}_h)_{|T} \in \mathbb{RTN}^k(T) \quad \forall T \in \mathcal{T}_h\right\}.$$

8.6.4.2 An $H(\mathrm{div};\Omega)$-Conforming Velocity Reconstruction

Let $T \in \mathcal{T}_h$. The velocity reconstruction $\mathfrak{r}_T^k : \underline{\boldsymbol{U}}_T^k \to \mathbb{RTN}^k(T)$ is defined such that, for all $\underline{\boldsymbol{v}}_T \in \underline{\boldsymbol{U}}_T^k$,

$$\int_T (\mathfrak{r}_T^k\underline{\boldsymbol{v}}_T - \boldsymbol{v}_T)\cdot\boldsymbol{w} = 0 \qquad \forall \boldsymbol{w} \in \mathbb{P}^{k-1}(T)^d, \tag{8.88a}$$

$$\int_F (\mathfrak{r}_T^k\underline{\boldsymbol{v}}_T - \boldsymbol{v}_F)\cdot\boldsymbol{n}_{TF}\,q = 0 \qquad \forall F \in \mathcal{F}_T,\quad \forall q \in \mathbb{P}^k(F). \tag{8.88b}$$

The fact that these relations define $\mathfrak{r}_T^k\underline{\boldsymbol{v}}_T$ uniquely is a consequence of [57, Proposition 2.3.4]; see also [196, Theorem 3.3]. Recalling the definition (1.57) of the L^2-orthogonal projector along with the properties of the local Raviart–Thomas–Nédélec space, it is also immediate to see that (8.88) can be equivalently reformulated as follows:

$$\boldsymbol{\pi}_T^{0,k-1}(\mathfrak{r}_T^k\underline{\boldsymbol{v}}_T) = \boldsymbol{\pi}_T^{0,k-1}\boldsymbol{v}_T \quad \text{and} \quad (\mathfrak{r}_T^k\underline{\boldsymbol{v}}_T)\cdot\boldsymbol{n}_{TF} = \boldsymbol{v}_F\cdot\boldsymbol{n}_{TF} \quad \forall F \in \mathcal{F}_T. \tag{8.89}$$

From these relations, it follows that, for any $\underline{\boldsymbol{v}}_T \in \underline{\boldsymbol{U}}_T^k$ and any $q \in \mathbb{P}^k(T)$,

$$\begin{aligned}\int_T (\nabla\cdot\mathfrak{r}_T^k\underline{\boldsymbol{v}}_T)\, q &= -\int_T \mathfrak{r}_T^k\underline{\boldsymbol{v}}_T\cdot\nabla q + \sum_{F\in\mathcal{F}_T}\int_F (\mathfrak{r}_T^k\underline{\boldsymbol{v}}_T\cdot\boldsymbol{n}_{TF})\, q \\ &= -\int_T \boldsymbol{v}_T\cdot\nabla q + \sum_{F\in\mathcal{F}_T}\int_F (\boldsymbol{v}_F\cdot\boldsymbol{n}_{TF})\, q = \int_T \mathrm{D}_T^k\underline{\boldsymbol{v}}_T\, q,\end{aligned}$$

where we have used an integration by parts in the first line followed by (8.89) together with $\nabla q \in \mathbb{P}^{k-1}(T)^d$ and $q_{|F} \in \mathbb{P}^k(F)$ to obtain the second equality. The conclusion is a consequence of the definition (8.19) of D_T^k. Hence, since both $\nabla\cdot\mathfrak{r}_T^k\underline{\boldsymbol{v}}_T$ and $\mathrm{D}_T^k\underline{\boldsymbol{v}}_T$ belong to $\mathbb{P}^k(T)$,

$$\nabla\cdot\mathfrak{r}_T^k\underline{\boldsymbol{v}}_T = \mathrm{D}_T^k\underline{\boldsymbol{v}}_T \qquad \forall\underline{\boldsymbol{v}}_T \in \underline{\boldsymbol{U}}_T^k. \tag{8.90}$$

A global velocity reconstruction $\mathfrak{r}_h^k : \underline{\boldsymbol{U}}_h^k \to \mathbb{RTN}^k(\mathcal{T}_h)$ is obtained patching the local contributions: For all $\underline{\boldsymbol{v}}_h \in \underline{\boldsymbol{U}}_h^k$,

$$(\mathfrak{r}_h^k\underline{\boldsymbol{v}}_h)_{|T} := \mathfrak{r}_T^k\underline{\boldsymbol{v}}_T \qquad \forall T \in \mathcal{T}_h.$$

Notice that, by Lemma 1.17, $\mathfrak{r}_h^k\underline{\boldsymbol{v}}_h$ indeed belongs to $\boldsymbol{H}(\mathrm{div};\Omega)$ since its normal component across each mesh interface is single-valued as a consequence of (8.89).

8.6.4.3 Pressure-Robust Bilinear Form ℓ_h

We define the bilinear form $\ell_h : L^2(\Omega)^d \times \underline{\boldsymbol{U}}_{h,0}^k \to \mathbb{R}$ such that, for any $\boldsymbol{f} \in L^2(\Omega)^d$ and any $\underline{\boldsymbol{v}}_h \in \underline{\boldsymbol{U}}_{h,0}^k$,

$$\ell_h(\boldsymbol{f}, \underline{\boldsymbol{v}}_h) := \int_\Omega \boldsymbol{f}\cdot\mathfrak{r}_h^k\underline{\boldsymbol{v}}_h. \tag{8.91}$$

Lemma 8.26 (Pressure-Robust Bilinear from ℓ_h) *The bilinear form ℓ_h defined by* (8.91) *satisfies Assumption 8.23.*

Proof (i) *Proof of (L1).* Let $\psi \in H^1(\Omega)$. It holds, for all $\underline{\boldsymbol{v}}_h \in \underline{\boldsymbol{U}}_{h,0}^k$,

$$\begin{aligned}\ell_h(\nabla\psi, \underline{\boldsymbol{v}}_h) &= \int_\Omega \nabla\psi\cdot\mathfrak{r}_h^k\underline{\boldsymbol{v}}_h \\ &= -\int_\Omega \psi\,(\nabla\cdot\mathfrak{r}_h^k\underline{\boldsymbol{v}}_h) + \cancel{\int_{\partial\Omega} \psi\,(\mathfrak{r}_h^k\underline{\boldsymbol{v}}_h\cdot\boldsymbol{n})}\end{aligned}$$

$$
\begin{aligned}
&= -\sum_{T\in\mathcal{T}_h}\int_T \psi\,(\nabla\cdot \mathbf{r}_T^k \underline{\boldsymbol{v}}_T)\\
&= -\sum_{T\in\mathcal{T}_h}\int_T \psi\,\mathrm{D}_T^k \underline{\boldsymbol{v}}_T\\
&= -\sum_{T\in\mathcal{T}_h}\int_T \pi_T^{0,k}\psi\,\mathrm{D}_T^k \underline{\boldsymbol{v}}_T = \mathrm{b}_h(\underline{\boldsymbol{v}}_h, \pi_h^{0,k}\psi),
\end{aligned}
$$

where we have used an integration by parts to pass to the second line along with the fact that, by (8.89), the normal traces of the velocity reconstruction vanish on $\partial\Omega$, (8.90) to pass to the fourth line, the definition (1.57) of the local L^2-orthogonal projector to pass to the fifth line, and the definitions (1.59) and (8.34) of the global L^2-orthogonal projector and of the bilinear form b_h to conclude.

(ii) *Proof of (L2).* With $\boldsymbol{w}$ as in the statement of property (L2), and assuming first $k \geq 1$, it holds, for all $\underline{\boldsymbol{v}}_h \in \underline{\boldsymbol{U}}_{h,0}^k$,

$$
\left|\mathcal{E}_{\ell,h}(\boldsymbol{w};\underline{\boldsymbol{v}}_h)\right| = \left|\sum_{T\in\mathcal{T}_h}\int_T \boldsymbol{w}\cdot(\mathbf{r}_T^k\underline{\boldsymbol{v}}_T - \boldsymbol{v}_T)\right| \tag{8.92}
$$

$$
\begin{aligned}
&= \left|\sum_{T\in\mathcal{T}_h}\int_T \left(\boldsymbol{w} - \boldsymbol{\pi}_T^{0,k-1}\boldsymbol{w}\right)\cdot(\mathbf{r}_T^k\underline{\boldsymbol{v}}_T - \boldsymbol{v}_T)\right|\\
&\leq \left(\sum_{T\in\mathcal{T}_h}\|\boldsymbol{w} - \boldsymbol{\pi}_T^{0,k-1}\boldsymbol{w}\|_T^2\right)^{\frac12}\left(\sum_{T\in\mathcal{T}_h}\|\mathbf{r}_T^k\underline{\boldsymbol{v}}_T - \boldsymbol{v}_T\|_T^2\right)^{\frac12}
\end{aligned}
$$

$$
\lesssim h^r|\boldsymbol{w}|_{H^r(\mathcal{T}_h)^d}\left(\sum_{T\in\mathcal{T}_h}\|\mathbf{r}_T^k\underline{\boldsymbol{v}}_T - \boldsymbol{v}_T\|_T^2\right)^{\frac12}, \tag{8.93}
$$

where we have used the definition (8.88a) of $\mathbf{r}_T^k\underline{\boldsymbol{v}}_T$ to insert $\boldsymbol{\pi}_T^{0,k-1}\boldsymbol{w}$ in the second line, a Cauchy–Schwarz inequality in the third line, and the approximation properties (1.74) of the local L^2-orthogonal projector with $X = T$, $l = k-1$, $p = 2$, $s = r$, and $m = 0$ to conclude. If $k = 0$, then $r = 0$ and $|\boldsymbol{w}|_{H^r(\mathcal{T}_h)^d} = \|\boldsymbol{w}\|_{L^2(\Omega)^d}$; hence, straightforward Cauchy–Schwarz inequalities on (8.92) yield (8.93). Let now $T \in \mathcal{T}_h$, and observe that, applying (8.87) to $\boldsymbol{w}_T = \mathbf{r}_T^k\underline{\boldsymbol{v}}_T - \boldsymbol{v}_T \in \mathbb{RTN}^k(T)$ and recalling (8.89), we have

$$\begin{aligned}\|\mathfrak{r}_T^k\underline{\boldsymbol{v}}_T - \boldsymbol{v}_T\|_T^2 &\lesssim \sum_{F\in\mathcal{F}_T} h_F \|(\mathfrak{r}_T^k\underline{\boldsymbol{v}}_T - \boldsymbol{v}_T)\cdot\boldsymbol{n}_{TF}\|_F^2 \\ &= \sum_{F\in\mathcal{F}_T} h_F \|(\boldsymbol{v}_F - \boldsymbol{v}_T)\cdot\boldsymbol{n}_{TF}\|_F^2 \le h_T^2 |\underline{\boldsymbol{v}}_T|_{1,\partial T}^2,\end{aligned} \tag{8.94}$$

where the conclusion follows from $h_F \le h_T$, a Hölder inequality together with $\|\boldsymbol{n}_{TF}\|_{L^\infty(F)^d} = 1$, and the definition (8.15) of $|\underline{\boldsymbol{v}}_T|_{1,\partial T}$. Plugging (8.94) into (8.93) and recalling the definition (8.24) of the global discrete H^1-like norm, we arrive at

$$\left|\mathcal{E}_{\ell,h}(\boldsymbol{w};\underline{\boldsymbol{v}}_h)\right| \lesssim h^{r+1}|\boldsymbol{w}|_{H^r(\mathcal{T}_h)^d}\|\underline{\boldsymbol{v}}_h\|_{1,h}.$$

Passing to the supremum over $\underline{\boldsymbol{v}}_h \in \underline{\boldsymbol{U}}_{h,0}^k$ such that $\|\underline{\boldsymbol{v}}_h\|_{1,h} = 1$ yields (L2). □

8.6.5 Numerical Examples

We illustrate with numerical examples, taken from [160], the difference between the original HHO scheme (8.42) and the modified version (8.82). For further numerical tests we refer the reader to [160].

8.6.5.1 Viscosity-Independence

To illustrate the viscosity-independence of the velocity approximation discussed in Remark 8.25, we consider on the unit cube domain $\Omega = (0, 1)^3$ the exact solution such that $\boldsymbol{u} = \nabla\psi$ with harmonic function ψ such that, for all $\boldsymbol{x} \in \Omega$,

$$\begin{aligned}\psi(\boldsymbol{x}) = 5x_1^6 - 90x_1^4x_2^2 + 120x_1^2x_2^4 + 15x_1^4x_3^2 \\ + 5x_3^6 - 90x_2^2x_3^4 + 120x_2^4x_3^2 + 15x_1^2x_3^4 - 16x_2^6 - 180x_1^2x_2^2x_3^2\end{aligned}$$

and $p(\boldsymbol{x}) = x_1^5 + x_2^5 + x_3^5 - \frac{1}{2}$. We show in Fig. 8.3 the velocity and pressure errors corresponding to $\nu \in \{1, 10^{-1}, 10^{-2}, 10^{-3}\}$ and polynomial degrees $k \in \{0, 1, 2\}$. All computations are realised on a fixed unstructured grid with 360 tetrahedra. The pressure-robust variation (8.82) yields significantly better results compared to the original version (8.42). The independence of the velocity approximation from the viscosity highlighted in Remark 8.25 is confirmed while, as expected, the pressure approximation does depend on the viscosity.

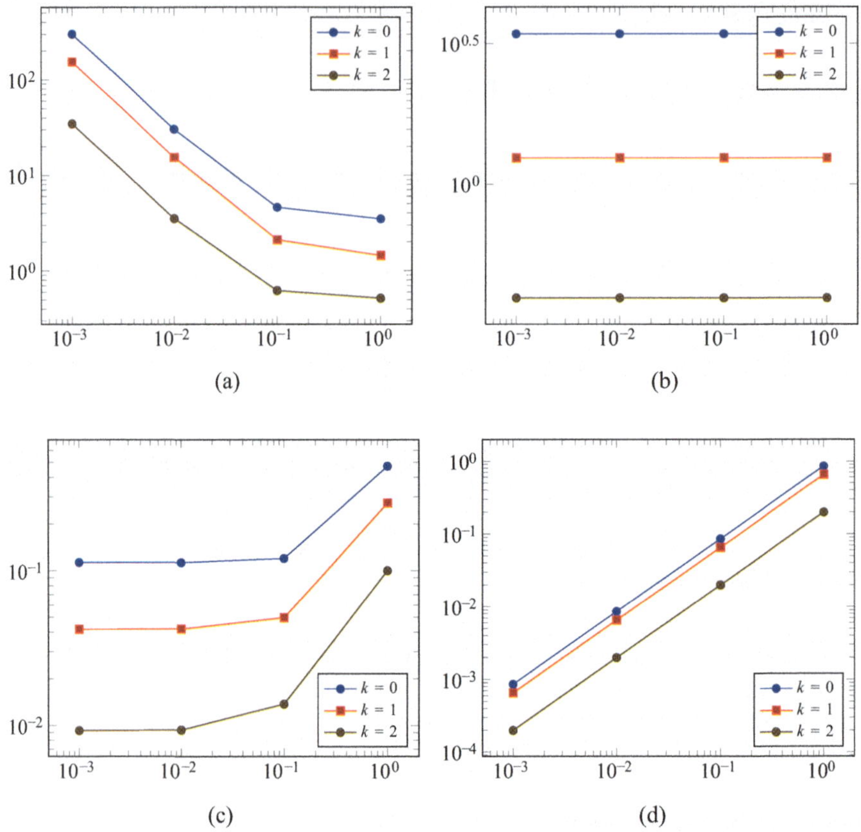

Fig. 8.3 Results for the numerical example of Sect. 8.6.5.1. **(a)** $\|\underline{u}_h - \underline{I}_h^k u\|_{a,h}$ vs. ν, classical. **(b)** $\|\underline{u}_h - \underline{I}_h^k u\|_{a,h}$ vs. ν, pressure-robust. **(c)** $\|p_h - \pi_h^{0,k} p\|_{L^2(\Omega)}$ vs. ν, classical. **(d)** $\|p_h - \pi_h^{0,k} p\|_{L^2(\Omega)}$ vs. ν, pressure-robust

8.6.5.2 Convergence

We next confirm the theoretically predicted convergence rates by considering the following solution on the unit cube domain $\Omega = (0, 1)^3$:

$$u = \begin{pmatrix} \frac{1}{2}\sin(2\pi x_1)\cos(2\pi x_2)\cos(2\pi x_3) \\ \frac{1}{2}\cos(2\pi x_1)\sin(2\pi x_2)\cos(2\pi x_3) \\ -\cos(2\pi x_1)\cos(2\pi x_2)\cos(2\pi x_3) \end{pmatrix}, \qquad p(x) = \sin(2\pi x_1)\sin(2\pi x_2)\sin(2\pi x_3).$$

The viscosity is taken equal to 1, while the value of the inhomogeneous Dirichlet boundary condition as well as that of the body force are inferred from the expressions of u and p. As for the numerical example of Sect. 8.5.4, for polynomial values $k \in \{0, 1\}$, we display in Table 8.2: (1) $\|\underline{\boldsymbol{u}}_h - \underline{\boldsymbol{I}}_h^k \boldsymbol{u}\|_{\mathrm{a},h}$, the error on the velocity measured in the norm associated with the viscous bilinear form a_h—notice that, recalling the norm equivalence (8.31), an estimate analogous to (8.60) holds for this quantity; (2) $\|p_h - \pi_{\mathcal{T}}^{0,k} p\|_{L^2(\Omega)}$, the L^2-error for the pressure. Each quantity is accompanied by the corresponding EOC defined according to (8.80). The convergence rates are coherent with those predicted by the error estimates (8.60) and (8.86).

Table 8.2 Results for the numerical example of Sect. 8.6.5.2

	Classical HHO scheme (8.42)				Pressure–robust HHO scheme (8.82)			
$\mathrm{card}(\mathcal{T}_h)$	$\Vert\underline{\boldsymbol{u}}_h - \underline{\boldsymbol{I}}_h^k\boldsymbol{u}\Vert_{\mathrm{a},h}$	EOC	$\Vert p_h - \pi_h^{0,k} p\Vert_{L^2(\Omega)}$	EOC	$\Vert\underline{\boldsymbol{u}}_h - \underline{\boldsymbol{I}}_h^k\boldsymbol{u}\Vert_{\mathrm{a},h}$	EOC	$\Vert p_h - \pi_h^{0,k} p\Vert$	EOC
$k = 0$								
44	4.36	–	0.68	–	3.05	–	1.58	–
360	3.52	0.31	0.48	0.51	3.4	−0.15	0.87	0.87
2883	1.97	0.84	0.27	0.81	1.97	0.78	0.41	1.09
23,077	1.02	0.95	0.15	0.90	1.05	0.92	0.2	1.00
$k = 1$								
44	4.34	–	0.82	–	3.52	–	2.18	–
360	1.46	1.57	0.28	1.57	1.24	1.50	0.66	1.72
2883	0.45	1.71	$8.71\cdot 10^{-2}$	1.66	0.4	1.62	0.18	1.92
23,077	0.12	1.95	$2.33\cdot 10^{-2}$	1.90	0.11	1.91	$4.36\cdot 10^{-2}$	2.01

Chapter 9
Navier–Stokes

In this chapter, we discuss HHO discretisations of the steady incompressible Navier–Stokes equations. These equations, which model the motion of fluids, were originally derived by Navier [242] and Poisson [251] using a molecular approach, while a more specific derivation is due to Saint–Venant [34] and Stokes [262] based on a linear relation between the stress tensor and the strain rate tensor. The main difference with respect to the Stokes equations treated in Chap. 8 is the presence of a nonlinear contribution in the momentum balance to model convective inertial forces. Our focus is therefore on the design and analysis of HHO trilinear forms to discretise this term. From a mathematical point of view, a relevant property of the convective term is that it does not contribute to the kinetic energy balance, obtained taking the velocity as a test function in the momentum equation. This property, referred to as "non-dissipativity" in what follows, is reproduced at the discrete level, as it plays an important role in the analysis.

The presence of the nonlinear term also entails relevant differences in the analysis with respect to the Stokes problem. Specifically, uniqueness of the discrete solution and error estimates require a data smallness condition. Convergence for general data, on the other hand, can be proved resorting to the compactness techniques introduced in Chap. 6, which do not deliver an estimate on the convergence rate.

The material is organised as follows. In Sect. 9.1 we establish the continuous setting for the model, state the weak formulation of the incompressible Navier–Stokes equations, discuss the non-dissipativity of the continuous convective trilinear form, and derive two equivalent reformulations to be used as inspiration for its discrete counterpart.

In Sect. 9.2 we formulate an HHO discretisation based on an abstract discrete convective trilinear form. Under the proposed design conditions, we prove the existence of a discrete solution using a topological degree argument, then show that uniqueness holds under a data smallness condition. In Sect. 9.3 we prove an energy error estimate under a data smallness assumption. Specifically, for sufficiently regular exact solutions, we prove convergence in h^{k+1} (with h and $k \geq 0$ denoting,

D. A. Di Pietro, J. Droniou, *The Hybrid High-Order Method for Polytopal Meshes*,
MS&A 19, https://doi.org/10.1007/978-3-030-37203-3_9

as usual, the meshsize and the polynomial degree) for the H^1-like norm of the error on the velocity and the L^2-norm of the error on the pressure. We close this section by briefly describing how convective stabilisation can be incorporated into the scheme.

In Sect. 9.5 we discuss two examples of discrete convective trilinear forms that match the design conditions of Sect. 9.2. The first example, inspired by [151], is obtained from a skew-symmetric reformulation of the continuous trilinear form by replacing the continuous gradient operator with a discrete gradient reconstructed in the space of polynomials of degree $2k$. The second discrete trilinear form is inspired by [268], with reconstructions of the advective derivative of degree k and of the divergence of degree $2k$ replacing the corresponding continuous operators.

In Sect. 9.6 we prove convergence for general data using a compactness argument. Specifically, after proving preliminary results concerning the compactness of sequences of HHO functions bounded uniformly in the discrete H^1-norm and the strong convergence of the interpolates of smooth functions, we prove: strong convergence of the discrete velocity in $L^q(\Omega)^d$ with $q \in [1, \infty)$ if $d = 2$, $q \in [1, 6)$ if $d = 3$; strong convergence of the gradient of the reconstructed velocity in $L^2(\Omega)^{d\times d}$; strong convergence of the pressure in $L^2(\Omega)$; convergence to zero of the stabilisation seminorm.

Finally, in Sect. 9.7 we numerically demonstrate the performance of the method on classical benchmark problems.

9.1 Model

We start by discussing the continuous setting for the model.

9.1.1 The Navier–Stokes Problem

Let $d \in \{2, 3\}$, and take $\Omega \subset \mathbb{R}^d$ that satisfies Assumption 1.3. As in Chap. 8, this domain is assumed to have a Lipschitz-continuous boundary. Let $\nu > 0$ denote a real number representing the kinematic viscosity, and let $\boldsymbol{f} : \Omega \to \mathbb{R}^d$ denote a body force. The steady incompressible Navier–Stokes problem for a uniform density, Newtonian fluid consists in finding the velocity $\boldsymbol{u} : \Omega \to \mathbb{R}^d$ and the pressure $p : \Omega \to \mathbb{R}$ such that

$$-\nu\Delta\boldsymbol{u} + (\boldsymbol{u}\cdot\nabla)\boldsymbol{u} + \nabla p = \boldsymbol{f} \qquad \text{in } \Omega, \tag{9.1a}$$

$$\nabla\cdot\boldsymbol{u} = 0 \qquad \text{in } \Omega, \tag{9.1b}$$

$$\boldsymbol{u} = \boldsymbol{0} \qquad \text{on } \partial\Omega, \tag{9.1c}$$

$$\int_\Omega p = 0. \tag{9.1d}$$

In (9.1a), we have introduced the convective derivative such that, if $\boldsymbol{u} = (u_j)_{1\le j\le d}$, then $(\boldsymbol{u}\cdot\nabla)\boldsymbol{u} = \sum_{j=1}^{d} u_j\partial_j\boldsymbol{u}$. As for the Stokes problem, Eq. (9.1a) expresses the momentum balance where, with respect to (8.1a), an additional nonlinear term appears accounting for convective effects. This term is the source of nonlinearity in the Navier–Stokes equations, and is at the root of physically relevant phenomena such as turbulence. We will focus, for the sake of simplicity, on the homogeneous Dirichlet (wall) boundary condition (9.1c); other standard boundary conditions can be treated without difficulties. Finally, condition (9.1d) is introduced to uniquely identify the pressure, which would otherwise be defined only up to an additive constant.

Remark 9.1 (Conservative Reformulation of the Momentum Equation) Recalling the definitions (7.1) of the tensor product of two vectors and (7.3) of the divergence of a tensor, the momentum equation (9.1a) admits the following reformulation:

$$\nabla\cdot(-\nu\nabla\boldsymbol{u} + \boldsymbol{u}\otimes\boldsymbol{u} + p\boldsymbol{I}_d) = \boldsymbol{f} \qquad \text{in } \Omega, \tag{9.2}$$

which highlights the expression of the conserved momentum flux under the divergence operator. To derive (9.1a) from (9.2), it suffices to observe that

$$\nabla\cdot(\boldsymbol{u}\otimes\boldsymbol{u}) = (\boldsymbol{u}\cdot\nabla)\boldsymbol{u} + \cancel{\boldsymbol{u}(\nabla\cdot\boldsymbol{u})},$$

where we have used the mass balance equation (9.1b) to cancel the second term.

9.1.2 Weak Formulation

Recalling the velocity and pressure spaces defined in (8.2), that is,

$$\boldsymbol{U} := H_0^1(\Omega)^d, \qquad P := \left\{ q \in L^2(\Omega) \,:\, \int_\Omega q = 0 \right\},$$

a classical weak formulation of problem (9.1) reads: Find $(\boldsymbol{u}, p) \in \boldsymbol{U}\times P$ such that

$$\nu a(\boldsymbol{u},\boldsymbol{v}) + t(\boldsymbol{u},\boldsymbol{u},\boldsymbol{v}) + b(\boldsymbol{v},p) = \int_\Omega \boldsymbol{f}\cdot\boldsymbol{v}, \qquad \forall \boldsymbol{v}\in\boldsymbol{U}, \tag{9.3a}$$

$$-b(\boldsymbol{u},q) = 0 \qquad \forall q \in L^2(\Omega), \tag{9.3b}$$

with bilinear forms $a : \boldsymbol{U}\times\boldsymbol{U}\to\mathbb{R}$ and $b : \boldsymbol{U}\times L^2(\Omega)\to\mathbb{R}$ defined by (8.4), that is,

$$a(\boldsymbol{w},\boldsymbol{v}) := \int_\Omega \nabla\boldsymbol{w}:\nabla\boldsymbol{v}, \quad b(\boldsymbol{v},q) := -\int_\Omega (\nabla\cdot\boldsymbol{v})\, q,$$

and trilinear form $t : \boldsymbol{U} \times \boldsymbol{U} \times \boldsymbol{U} \to \mathbb{R}$ such that

$$t(\boldsymbol{w}, \boldsymbol{v}, \boldsymbol{z}) := \int_{\Omega} (\boldsymbol{w} \cdot \nabla) \boldsymbol{v} \cdot \boldsymbol{z}, \tag{9.4}$$

where we remind the reader that, if $\boldsymbol{w} = (w_i)_{1 \le i \le d}$, $\boldsymbol{v} = (v_i)_{1 \le i \le d}$, and $\boldsymbol{z} = (z_i)_{1 \le i \le d}$, then $(\boldsymbol{w} \cdot \nabla) \boldsymbol{v} \cdot \boldsymbol{z} = \sum_{i=1}^{d} \sum_{j=1}^{d} (w_j \partial_j v_i) z_i$.

The existence of a solution to problem (9.3) will be a side result of the convergence analysis in Sect. 9.6; see Remark 9.33. Uniqueness, on the other hand, can be proved assuming that the L^2-norm of the body force $\boldsymbol{f}$ is small enough, the so-called *data smallness* condition; see, e.g., [199, Eq. (2.12), Chapter IV]. In the analysis, besides the properties of the bilinear forms a and b discussed in Sect. 8.1.2, a key role is played by the non-dissipativity of the trilinear form t. This property deserves a more in-depth discussion, which makes the object of the following section.

9.1.3 Non-dissipativity of the Convective Term

Let us examine the non-dissipativity property of t in order to illustrate the strategy adopted to design its discrete counterpart. We start by noting the following integration by parts formula: For all $\boldsymbol{w}, \boldsymbol{v}, \boldsymbol{z} \in H^1(\Omega)^d$,

$$\int_{\Omega} (\boldsymbol{w} \cdot \nabla) \boldsymbol{v} \cdot \boldsymbol{z} + \int_{\Omega} (\boldsymbol{w} \cdot \nabla) \boldsymbol{z} \cdot \boldsymbol{v} + \int_{\Omega} (\nabla \cdot \boldsymbol{w})(\boldsymbol{v} \cdot \boldsymbol{z}) = \int_{\partial\Omega} (\boldsymbol{w} \cdot \boldsymbol{n}_{\Omega})(\boldsymbol{v} \cdot \boldsymbol{z}), \tag{9.5}$$

where $\boldsymbol{n}_{\Omega}$ denotes the outward unit vector normal to $\partial\Omega$. Writing (9.5) for $\boldsymbol{w} = \boldsymbol{v} = \boldsymbol{z} = \boldsymbol{u}$ (with $\boldsymbol{u}$ velocity solution to (9.3)), we get

$$t(\boldsymbol{u}, \boldsymbol{u}, \boldsymbol{u}) = \int_{\Omega} (\boldsymbol{u} \cdot \nabla) \boldsymbol{u} \cdot \boldsymbol{u} = -\frac{1}{2} \int_{\Omega} (\nabla \cdot \boldsymbol{u})(\boldsymbol{u} \cdot \boldsymbol{u}) + \frac{1}{2} \int_{\partial\Omega} (\boldsymbol{u} \cdot \boldsymbol{n}_{\Omega})(\boldsymbol{u} \cdot \boldsymbol{u}) = 0, \tag{9.6}$$

where we have used (9.3b) to infer $\nabla \cdot \boldsymbol{u} = 0$ and cancel the first term, and the fact that $\boldsymbol{u}$ vanishes on $\partial\Omega$ to cancel the second. This relation expresses the fact that the convective term does not contribute to the kinetic energy balance, obtained taking $\boldsymbol{v} = \boldsymbol{u}$ in (9.3a).

When attempting to reproduce property (9.6) at the discrete level, a difficulty arises: the discrete counterparts of the terms in the right-hand side of (9.6) may not vanish, since the discrete solution may not be "sufficiently" divergence-free (see Remark 9.21) and/or it may not be zero on $\partial\Omega$. To overcome this difficulty, the following modified expression for t can be used as a starting point, an idea which can be traced back to Temam [268]:

$$\tilde{t}(\boldsymbol{w}, \boldsymbol{v}, \boldsymbol{z}) = \int_\Omega (\boldsymbol{w}\cdot\nabla)\boldsymbol{v}\cdot\boldsymbol{z} + \frac{1}{2}\int_\Omega (\nabla\cdot\boldsymbol{w})(\boldsymbol{v}\cdot\boldsymbol{z}) - \frac{1}{2}\int_{\partial\Omega} (\boldsymbol{w}\cdot\boldsymbol{n}_\Omega)(\boldsymbol{v}\cdot\boldsymbol{z}). \tag{9.7}$$

With this choice, the skew-symmetric nature of convective terms becomes apparent, as we can write

$$\begin{aligned}\tilde{t}(\boldsymbol{w}, \boldsymbol{v}, \boldsymbol{z}) &= \frac{1}{2}\int_\Omega (\boldsymbol{w}\cdot\nabla)\boldsymbol{v}\cdot\boldsymbol{z} \\ &\quad - \frac{1}{2}\left(-\int_\Omega (\boldsymbol{w}\cdot\nabla)\boldsymbol{v}\cdot\boldsymbol{z} - \int_\Omega (\nabla\cdot\boldsymbol{w})(\boldsymbol{v}\cdot\boldsymbol{z}) + \int_{\partial\Omega} (\boldsymbol{w}\cdot\boldsymbol{n}_\Omega)(\boldsymbol{v}\cdot\boldsymbol{z})\right)\end{aligned}$$

and, using (9.5) to reformulate the term in parentheses,

$$\tilde{t}(\boldsymbol{w}, \boldsymbol{v}, \boldsymbol{z}) = \frac{1}{2}\int_\Omega (\boldsymbol{w}\cdot\nabla)\boldsymbol{v}\cdot\boldsymbol{z} - \frac{1}{2}\int_\Omega (\boldsymbol{w}\cdot\nabla)\boldsymbol{z}\cdot\boldsymbol{v}. \tag{9.8}$$

It is a simple matter to check that the expressions (9.7) and (9.8) for the convective trilinear form can be used in place of (9.4) in (9.3) without modifying this weak formulation. From a numerical standpoint, they are more appropriate as a starting point to derive a discretisation of the convective term, as they satisfy the following generalised version of property (9.6): For all $\boldsymbol{w}, \boldsymbol{v} \in H^1(\Omega)^d$,

$$\tilde{t}(\boldsymbol{w}, \boldsymbol{v}, \boldsymbol{v}) = 0.$$

This means, in particular, that $\tilde{t}$ is non-dissipative even if $\boldsymbol{w}$ is not divergence free and $\boldsymbol{v}$ does not vanish on $\partial\Omega$ (as may be the case for the discrete velocity).

9.2 Discrete Problem

In this section we formulate an HHO scheme based on a set of design properties for the discrete convective trilinear form, and we discuss existence and uniqueness of the discrete solution.

9.2.1 Discrete Problem and Design Properties for the Discrete Trilinear Form

Let the discrete velocity and pressure spaces be defined by (8.26), that is

$$\underline{\boldsymbol{U}}_{h,0}^k := \left\{\underline{\boldsymbol{v}}_h \in \underline{\boldsymbol{U}}_h^k \;:\; \boldsymbol{v}_F = \boldsymbol{0} \quad \forall F \in \mathcal{F}_h^{\mathrm{b}}\right\}, \qquad P_h^k := \mathbb{P}^k(\mathcal{T}_h) \cap P,$$

with $\underline{U}_h^k$ defined by (8.22). Let the viscous bilinear form a_h be given by (8.27) with local stabilisation bilinear forms in (8.28) matching Assumption 8.10, and let the pressure–velocity coupling bilinear form b_h be given by (8.34). We consider the following HHO approximation to (9.3): Find $(\underline{\boldsymbol{u}}_h, p_h) \in \underline{\boldsymbol{U}}_{h,0}^k \times P_h^k$ such that

$$\nu\mathrm{a}_h(\underline{\boldsymbol{u}}_h, \underline{\boldsymbol{v}}_h) + \mathrm{t}_h(\underline{\boldsymbol{u}}_h, \underline{\boldsymbol{u}}_h, \underline{\boldsymbol{v}}_h) + \mathrm{b}_h(\underline{\boldsymbol{v}}_h, p_h) = \int_\Omega \boldsymbol{f}\cdot\boldsymbol{v}_h \qquad \forall \underline{\boldsymbol{v}}_h \in \underline{\boldsymbol{U}}_{h,0}^k, \tag{9.9a}$$

$$-\mathrm{b}_h(\underline{\boldsymbol{u}}_h, q_h) = 0 \qquad \forall q_h \in \mathbb{P}^k(\mathcal{T}_h). \tag{9.9b}$$

The assumptions on the trilinear form t_h are as follows.

Assumption 9.2 (Trilinear Form t_h) *The trilinear form $\mathrm{t}_h : \underline{\boldsymbol{U}}_h^k \times \underline{\boldsymbol{U}}_h^k \times \underline{\boldsymbol{U}}_h^k \to \mathbb{R}$ satisfies the following properties:*

(T1) Non-dissipativity. *For all $\underline{\boldsymbol{w}}_h, \underline{\boldsymbol{v}}_h \in \underline{\boldsymbol{U}}_{h,0}^k$, it holds that*

$$\mathrm{t}_h(\underline{\boldsymbol{w}}_h, \underline{\boldsymbol{v}}_h, \underline{\boldsymbol{v}}_h) = 0. \tag{9.10}$$

(T2) Boundedness. *There is $C_\mathrm{t} \ge 0$ independent of h such that, for all $\underline{\boldsymbol{w}}_h, \underline{\boldsymbol{v}}_h, \underline{\boldsymbol{z}}_h \in \underline{\boldsymbol{U}}_{h,0}^k$,*

$$|\mathrm{t}_h(\underline{\boldsymbol{w}}_h, \underline{\boldsymbol{v}}_h, \underline{\boldsymbol{z}}_h)| \le C_\mathrm{t} \|\underline{\boldsymbol{w}}_h\|_{1,h} \|\underline{\boldsymbol{v}}_h\|_{1,h} \|\underline{\boldsymbol{z}}_h\|_{1,h}, \tag{9.11}$$

with $\|\cdot\|_{1,h}$-norm defined by (8.24) *and* (8.15).

(T3) Consistency. *For all $r \in \{0, \ldots, k\}$ and all $\boldsymbol{w} \in \boldsymbol{U} \cap W^{r+1,4}(\mathcal{T}_h)^d$ such that $\nabla\cdot\boldsymbol{w} = 0$,*

$$\sup_{\underline{\boldsymbol{z}}_h \in \underline{\boldsymbol{U}}_{h,0}^k, \|\underline{\boldsymbol{z}}_h\|_{1,h}=1} \left| \int_\Omega (\boldsymbol{w}\cdot\nabla)\boldsymbol{w}\cdot\boldsymbol{z}_h - \mathrm{t}_h(\underline{\boldsymbol{I}}_h^k \boldsymbol{w}, \underline{\boldsymbol{I}}_h^k \boldsymbol{w}, \underline{\boldsymbol{z}}_h) \right| \lesssim h^{r+1} \|\boldsymbol{w}\|_{W^{1,4}(\Omega)^d} |\boldsymbol{w}|_{W^{r+1,4}(\mathcal{T}_h)^d}, \tag{9.12}$$

with hidden constant independent of both $\boldsymbol{w}$ and h.

Some remarks are of order.

Remark 9.3 (Efficient Implementation) Assume that the trilinear form is designed so that its stencil is the same as that of the diffusive bilinear form, that is, the coupling between neighbouring elements is only established through the unknowns attached to a common face (this is the case for the two examples of trilinear forms provided in Sect. 9.5). Then, when solving the system of nonlinear algebraic equations corresponding to (9.9) by a first-order (e.g., Newton) algorithm, all element-based velocity unknowns and all but one pressure unknowns per element can be locally eliminated at each iteration by computing the corresponding Schur complement element-wise. As all the computations are local, this static condensation procedure is a trivially parallel task which can fully benefit from multi-thread and multi-processor architectures. This procedure has been described in detail for the Stokes problem in [160, Section 6.2]. The only variation here is that also the linearised convective term appears in the matrices therein denoted by A_T. After further eliminating the boundary unknowns by strongly enforcing the wall condition (9.1c), one ends up solving at each iteration a linear system of size

$$d\,\mathrm{card}(\mathcal{F}_h^{\mathrm{i}})\binom{k+d-1}{d-1} + \mathrm{card}(\mathcal{T}_h).$$

Remark 9.4 (Weak Enforcement of Boundary Conditions) An interesting variation of the HHO scheme (9.9) is obtained by weakly enforcing the wall boundary condition adapting Nitsche's techniques [247]. The weak enforcement of boundary conditions can improve the resolution of boundary layers, since the boundary unknowns are not constrained to a fixed value, and it can simplify the parallel implementation of the method. We do not develop further this subject here, and refer the interested reader to [70].

Remark 9.5 (Pressure-Robust Variations) As we did in Sect. 8.6 for the Stokes problem, it is possible to devise pressure-robust variations of the HHO scheme (9.9) for the Navier–Stokes problem. In addition to modifying the right-hand side in order to comply with Assumption 8.23, a corresponding modification of the convective trilinear form is required in this case. We refer the reader to [99] for further details.

9.2.2 Existence and Uniqueness of a Discrete Solution

The existence of a solution to problem (9.9) can be proved using the following topological degree lemma (cf., e.g., [139]), as originally proposed in [187] in the context of Finite Volumes for nonlinear hyperbolic problems; see also [144, 189] concerning the incompressible Navier–Stokes equations.

Lemma 9.6 (Topological Degree) *Let W be a finite-dimensional vector space equipped with a norm denoted by $\|\cdot\|_W$, and let the function $\Psi : W \times [0,1] \to W$ satisfy the following assumptions:*

(i) Ψ is continuous;
(ii) There exists $\mu > 0$ such that, for any $(w,\rho) \in W \times [0,1]$, $\Psi(w,\rho) = 0$ implies $\|w\|_W \neq \mu$;
(iii) $\Psi(\cdot,0)$ is an affine function and the equation $\Psi(w,0) = 0$ has a solution $w \in W$ such that $\|w\|_W < \mu$.

Then, there exists $w \in W$ such that $\Psi(w,1) = 0$ and $\|w\|_W < \mu$.

Theorem 9.7 (Existence and a Priori Bounds) *There exists a solution to (9.9). Moreover, any solution* $(\underline{\boldsymbol{u}}_h, p_h) \in \underline{\boldsymbol{U}}_{h,0}^k \times P_h^k$ *to this problem satisfies the a priori bounds*

$$\begin{aligned} \nu \|\underline{\boldsymbol{u}}_h\|_{1,h} &\leq C_{\mathrm{a}} C_{\mathrm{P}} \|\boldsymbol{f}\|_{L^2(\Omega)^d}, \\ \|p_h\|_{L^2(\Omega)} &\leq C \left(\|\boldsymbol{f}\|_{L^2(\Omega)^d} + \nu^{-2} \|\boldsymbol{f}\|_{L^2(\Omega)^d}^2 \right), \end{aligned} \tag{9.13}$$

with C_{a} *as in* (8.31)*,* C_{P} *as in* (8.47)*, and* $C > 0$ *real number independent of* ν*,* h*,* $\underline{\boldsymbol{u}}_h$ *and* p_h*.*

Proof We consider the finite-dimensional space $X_h^k = \underline{\boldsymbol{U}}_{h,0}^k \times P_h^k$ (see (8.44)) equipped with following the norm (notice that this definition is slightly different from the one considered for the Stokes problem, see (8.59), hence the triple-bar notation):

$$|||(\underline{\boldsymbol{v}}_h, q_h)|||_{X,h} := \nu \|\underline{\boldsymbol{v}}_h\|_{1,h} + \|q_h\|_{L^2(\Omega)} \qquad \forall (\underline{\boldsymbol{v}}_h, q_h) \in X_h^k,$$

and the function $\Psi : X_h^k \times [0, 1] \to X_h^k$ such that, for given $(\underline{\boldsymbol{w}}_h, r_h) \in X_h^k$ and $\rho \in [0, 1]$, $(\underline{\boldsymbol{\xi}}_h, \zeta_h) = \Psi((\underline{\boldsymbol{w}}_h, r_h), \rho)$ is defined as the unique element of X_h^k that satisfies: For all $\underline{\boldsymbol{v}}_h \in \underline{\boldsymbol{U}}_{h,0}^k$ and all $q_h \in P_h^k$,

$$(\underline{\boldsymbol{\xi}}_h, \underline{\boldsymbol{v}}_h)_{0,h} = \nu \mathrm{a}_h(\underline{\boldsymbol{w}}_h, \underline{\boldsymbol{v}}_h) + \rho \mathrm{t}_h(\underline{\boldsymbol{w}}_h, \underline{\boldsymbol{w}}_h, \underline{\boldsymbol{v}}_h) + \mathrm{b}_h(\underline{\boldsymbol{v}}_h, r_h) - \int_\Omega \boldsymbol{f} \cdot \boldsymbol{v}_h \tag{9.14a}$$

$$(\zeta_h, q_h)_{L^2(\Omega)} = -\mathrm{b}_h(\underline{\boldsymbol{w}}_h, q_h), \tag{9.14b}$$

where $(\cdot, \cdot)_{0,h}$ is the L^2-like scalar product on $\underline{\boldsymbol{U}}_h^k$ such that, for all $\underline{\boldsymbol{w}}_h, \underline{\boldsymbol{v}}_h \in \underline{\boldsymbol{U}}_h^k$,

$$(\underline{\boldsymbol{w}}_h, \underline{\boldsymbol{v}}_h)_{0,h} := \int_\Omega \boldsymbol{w}_h \cdot \boldsymbol{v}_h + \sum_{T \in \mathcal{T}_h} \sum_{F \in \mathcal{F}_T} h_F \int_F (\boldsymbol{w}_F - \boldsymbol{w}_T) \cdot (\boldsymbol{v}_F - \boldsymbol{v}_T).$$

We next check the assumptions of the topological degree lemma.

(i) Since X_h^k is a finite-dimensional space, the bilinear forms a_h and b_h, the trilinear form t_h, and the scalar products are continuous, and so is the case for the function Ψ.

(ii) Let $(\underline{w}_h, r_h) \in X_h^k$ be such that $\Psi((\underline{w}_h, r_h), \rho) = (\underline{0}, 0)$ for some $\rho \in [0, 1]$. We next show that

$$\nu\|\underline{w}_h\|_{1,h} \le C_{\mathrm{a}} C_{\mathrm{P}} \|f\|_{L^2(\Omega)^d} \text{ and } \|r_h\|_{L^2(\Omega)} \le C\left(\|f\|_{L^2(\Omega)^d} + \nu^{-2}\|f\|_{L^2(\Omega)^d}^2\right),$$

with constant C in the second estimate independent of ν, h, ρ, $\underline{w}_h$ and r_h. This proves Point (ii) in Lemma 9.6 for

$$\mu = (C_{\mathrm{a}} C_{\mathrm{P}} + C)\|f\|_{L^2(\Omega)^d} + C\nu^{-2}\|f\|_{L^2(\Omega)^d}^2 + \epsilon$$

with $\epsilon > 0$. Make $\underline{v}_h = \underline{w}_h$ in (9.14a). Recalling the coercivity of a_h expressed by the first inequality in (8.31), that $\mathrm{t}_h(\underline{w}_h, \underline{w}_h, \underline{w}_h) = 0$ by the non-dissipativity property (9.10), and that $\mathrm{b}_h(\underline{w}_h, r_h) = 0$ owing to (9.14b) with $q_h = r_h$, we have

$$\begin{aligned}\nu C_{\mathrm{a}}^{-1}\|\underline{w}_h\|_{1,h}^2 \le \nu \mathrm{a}_h(\underline{w}_h, \underline{w}_h) &= \int_\Omega f \cdot w_h \\ &\le \|f\|_{L^2(\Omega)^d}\|w_h\|_{L^2(\Omega)^d} \le C_{\mathrm{P}}\|f\|_{L^2(\Omega)^d}\|\underline{w}_h\|_{1,h},\end{aligned}$$

where we have used the discrete Poincaré inequality (8.47) to conclude. The bound on $\underline{w}_h$ follows. To prove the bound on r_h, we proceed as follows:

$$\begin{aligned}\|r_h\|_{L^2(\Omega)} &\lesssim \sup_{\underline{v}_h \in \underline{U}_{h,0}^k, \|\underline{v}_h\|_{1,h}=1} \mathrm{b}_h(\underline{v}_h, r_h) \\ &= \sup_{\underline{v}_h \in \underline{U}_{h,0}^k, \|\underline{v}_h\|_{1,h}=1} \left(\int_\Omega f \cdot v_h - \nu \mathrm{a}_h(\underline{w}_h, \underline{v}_h) - \rho \mathrm{t}_h(\underline{w}_h, \underline{w}_h, \underline{v}_h)\right) \\ &\lesssim \|f\|_{L^2(\Omega)^d} + \nu\|\underline{w}_h\|_{1,h} + \rho\|\underline{w}_h\|_{1,h}^2 \\ &\lesssim \|f\|_{L^2(\Omega)^d} + \nu^{-2}\|f\|_{L^2(\Omega)^d}^2,\end{aligned}$$

where we have used the inf–sup condition (8.36) on b_h in the first line and (9.14a) to pass to the second line; the Cauchy–Schwarz and the discrete Poincaré inequalities together with the boundedness of a_h and t_h, respectively expressed by the second inequality in (8.31) and by property (T2), are used to pass to the third line; finally, the bound on $\|\underline{w}_h\|_{1,h}$ and the fact that $\rho \le 1$ allowed us to conclude.

(iii) $\Psi(\cdot, 0)$ is an affine function from X_h^k to X_h^k. The fact that $\Psi(\cdot, 0)$ is invertible corresponds to the well-posedness of the HHO scheme for the Stokes problem. Additionally, the unique solution $(\underline{w}_h, r_h) \in X_h^k$ to the equation $\Psi((\underline{w}_h, r_h), 0) = (\underline{0}, 0)$ satisfies $|||(\underline{w}_h, r_h)|||_{X,h} < \mu$ as a consequence of Point (ii).

The existence of a solution to (9.9) is then an immediate consequence of Lemma 9.6 after observing that, if $(\underline{u}_h, p_h) \in X_h^k$ is such that $\Psi((\underline{u}_h, p_h), 1) = (\underline{0}, 0)$, then $(\underline{u}_h, p_h)$ solves (9.9). The bounds (9.13) follow from Point (ii) above. □

We next consider uniqueness, which can be classically proved under a data smallness condition.

Theorem 9.8 (Uniqueness of the Discrete Solution) *Assume that the body force verifies, for some* $\chi \in [0, 1)$,

$$\|f\|_{L^2(\Omega)^d} \leq \chi \frac{\nu^2}{C_a^2 C_t C_P} \tag{9.15}$$

where C_a, C_t, *and* C_P *are as in* (8.31), (9.11), *and* (8.47), *respectively. Then, the solution* $(\underline{u}_h, p_h) \in \underline{U}_{h,0}^k \times P_h^k$ *of* (9.9) *is unique.*

Remark 9.9 (Data Smallness Condition) The scaling in ν^2 of the smallness assumption on the source term f already appears in the proof of uniqueness for the solution to the continuous problem (9.3); see, e.g., [268, Chapter 2, Theorem 1.6].

Proof Let $(\underline{u}_{1,h}, p_{1,h}), (\underline{u}_{2,h}, p_{2,h}) \in \underline{U}_{h,0}^k \times P_h^k$ solve (9.9), and set

$$\underline{w}_h := \underline{u}_{1,h} - \underline{u}_{2,h} \text{ and } r_h := p_{1,h} - p_{2,h}.$$

Uniqueness if proved if we show that $(\underline{w}_h, r_h) = (\underline{0}, 0)$.

We start by proving that $\underline{w}_h = \underline{0}$, which expresses uniqueness for the velocity. Taking the difference of the discrete momentum balance equation (9.9a) written first for $(\underline{u}_h, p_h) = (\underline{u}_{1,h}, p_{1,h})$ then for $(\underline{u}_h, p_h) = (\underline{u}_{2,h}, p_{2,h})$, inserting $\pm \mathrm{t}_h(\underline{u}_{1,h}, \underline{u}_{2,h}, \underline{v}_h)$, and using the linearity of t_h in its first and second arguments, we infer that it holds, for all $\underline{v}_h \in \underline{U}_{h,0}^k$,

$$\nu \mathrm{a}_h(\underline{w}_h, \underline{v}_h) + \mathrm{t}_h(\underline{u}_{1,h}, \underline{w}_h, \underline{v}_h) + \mathrm{t}_h(\underline{w}_h, \underline{u}_{2,h}, \underline{v}_h) + \mathrm{b}_h(\underline{v}_h, r_h) = 0. \tag{9.16}$$

Making $\underline{v}_h = \underline{w}_h$ in the above equation, observing that $\mathrm{t}_h(\underline{u}_{1,h}, \underline{w}_h, \underline{w}_h) = 0$ owing to the non-dissipativity property (9.10), that $\mathrm{b}_h(\underline{w}_h, r_h) = 0$ as a consequence of the discrete mass balance equation (9.9b) written for $\underline{u}_{1,h}$ and $\underline{u}_{2,h}$ with $q_h = r_h$, and using the coercivity of a_h expressed by the first inequality in (8.31) and the boundedness of t_h expressed by (9.11), we obtain

$$\left(\nu C_a^{-1} - C_t \|\underline{u}_{2,h}\|_{1,h}\right) \|\underline{w}_h\|_{1,h}^2 \leq 0.$$

By the first a priori bound in (9.13) and the assumption (9.15) on $\boldsymbol{f}$, the first factor in the left-hand side is > 0. As a result, $\underline{\boldsymbol{w}}_h = \underline{\boldsymbol{0}}$.

Using $\underline{\boldsymbol{w}}_h = \underline{\boldsymbol{0}}$ in (9.16), it is inferred that it holds, for all $\underline{\boldsymbol{v}}_h \in \underline{\boldsymbol{U}}_{h,0}^k$, $\mathrm{b}_h(\underline{\boldsymbol{v}}_h, r_h) = 0$. The inf–sup stability (8.36) of b_h then gives

$$\|r_h\|_{L^2(\Omega)} \lesssim \sup_{\underline{\boldsymbol{v}}_h \in \underline{\boldsymbol{U}}_{h,0}^k, \|\underline{\boldsymbol{v}}_h\|_{1,h}=1} b_h(\underline{\boldsymbol{v}}_h, r_h) = 0,$$

which proves uniqueness for the pressure and concludes the proof. □

9.3 Energy Error Estimate for Small Data

We prove in this section an energy-norm error estimate valid under a small data assumption. To state this assumption, we introduce the continuous Poincaré constant C_Ω, which depends only on Ω and is such that, for all $\boldsymbol{v} \in \boldsymbol{U}$,

$$\|\boldsymbol{v}\|_{L^2(\Omega)^d} \le C_\Omega \|\nabla \boldsymbol{v}\|_{L^2(\Omega)^{d\times d}}. \tag{9.17}$$

Theorem 9.10 (Discrete Energy Error Estimate for Small Data) *Let $(\mathcal{M}_h)_{h\in\mathcal{H}}$ denote a regular mesh sequence in the sense of Definition 1.9. Let a polynomial degree $k \ge 0$ be fixed. Assume that the forcing term $\boldsymbol{f}$ satisfies, for some $\chi \in [0, 1)$,*

$$\|\boldsymbol{f}\|_{L^2(\Omega)^d} \le \chi \frac{\nu^2}{d^{\frac{1}{2}} C_I C_{\mathrm{a}} C_{\mathrm{t}} C_\Omega}, \tag{9.18}$$

with C_I, C_{a}, C_{t} and C_Ω as in (8.25), (8.31), (9.11) and (9.17), respectively. Let $(\boldsymbol{u}, p) \in \boldsymbol{U} \times P$ and $(\underline{\boldsymbol{u}}_h, p_h) \in \underline{\boldsymbol{U}}_{h,0}^k \times P_h^k$ solve problems (9.3) and (9.9), respectively. Assume, moreover, the additional regularity $\boldsymbol{u} \in H^{r+2}(\mathcal{T}_h)^d$ and $p \in H^1(\Omega) \cap H^{r+1}(\mathcal{T}_h)$ for some $r \in \{0, \ldots, k\}$. Then,

$$\begin{aligned}
&\nu\|\underline{\boldsymbol{u}}_h - \underline{\boldsymbol{I}}_h^k \boldsymbol{u}\|_{1,h} + \|p_h - \pi_h^{0,k} p\|_{L^2(\Omega)} \\
&\quad \lesssim h^{r+1} \left(\nu |\boldsymbol{u}|_{H^{r+2}(\mathcal{T}_h)^d} + \|\boldsymbol{u}\|_{W^{1,4}(\Omega)^d} |\boldsymbol{u}|_{W^{r+1,4}(\mathcal{T}_h)^d} + |p|_{H^{r+1}(\mathcal{T}_h)}\right),
\end{aligned} \tag{9.19}$$

where the norm $\|{\cdot}\|_{1,h}$ is defined by (8.24) and (8.15), and the hidden constant is independent of h and ν, but possibly depends on d, Ω, k, ϱ, χ, C_I, C_{a}, C_{t} and C_Ω.

Remark 9.11 (Regularity for the Velocity) Since $d \leq 3$ and each $T \in \mathcal{T}_h$ is polytopal, the Sobolev embeddings used in each element show that $H^{r+2}(\mathcal{T}_h) \subset W^{r+1,4}(\mathcal{T}_h)$. The embedding constant of this inclusion is however not independent of h, which is why the $W^{r+1,4}(\mathcal{T}_h)$-norm is explicitly used in (9.19).

Notice also that, since the exact velocity is in $\boldsymbol{U} = H_0^1(\Omega)^d$, both its jumps across interfaces and its trace on boundary faces vanish. Hence, the regularity $\boldsymbol{u} \in W^{1,4}(\mathcal{T}_h)^d$ implies $\boldsymbol{u} \in W^{1,4}(\Omega)^d$ owing to Lemma 1.21 with $p = 4$.

Remark 9.12 (Convergence Rate for High Reynolds Number) The error estimate (9.19) is valid only for small data, which correspond to small Reynolds numbers. In passing, the scaling in ν^2 in the smallness condition on $\boldsymbol{f}$ also appears when establishing error estimates for Finite Element approximations of the Navier–Stokes equations, see [268, Chapter 2, Eq. (3.72)].

In order to get an idea of the convergence rate for high Reynolds numbers, one can consider the linearised version corresponding to the Oseen problem. It has been shown in [8] that, in this case, a similar behaviour as the one outlined in Theorem 3.32 is to be expected: mesh elements $T \in \mathcal{T}_h$ for which diffusion dominates contribute with a term in h_T^{k+1}, whereas mesh elements for which convection dominates contribute with a term in $h_T^{k+\frac{1}{2}}$.

Proof (Theorem 9.10) Let, for the sake of brevity,

$$\underline{\hat{\boldsymbol{u}}}_h := \underline{\boldsymbol{I}}_h^k \boldsymbol{u}, \qquad \hat{p}_h := \pi_h^{0,k} p, \qquad \underline{\boldsymbol{e}}_h := \underline{\boldsymbol{u}}_h - \underline{\hat{\boldsymbol{u}}}_h, \qquad \epsilon_h := p_h - \hat{p}_h. \tag{9.20}$$

The proof proceeds in three steps: in the first step, we identify the consistency error and derive a lower bound in terms of $\|\underline{\boldsymbol{e}}_h\|_{1,h}$ using the data smallness assumption; in the second step, we estimate the error on the velocity; in the third step, we estimate the error on the pressure.

(i) *Consistency error and lower bound.* Even though the problem is nonlinear and the results of Appendix A cannot be directly applied, the general principles can be adapted, as we did in Chap. 6 for the p-Laplace equation. For starter, the consistency error is designed to be the right-hand side of an error equation on the difference between the approximate solution and the interpolate of the exact solution (see (A.7) in the linear setting). Here, a first error equation is readily inferred from the discrete momentum equation (9.9a): For all $\underline{\boldsymbol{v}}_h \in \underline{\boldsymbol{U}}_{h,0}^k$,

$$\nu \mathrm{a}_h(\underline{\boldsymbol{e}}_h, \underline{\boldsymbol{v}}_h) + \mathrm{t}_h(\underline{\boldsymbol{u}}_h, \underline{\boldsymbol{u}}_h, \underline{\boldsymbol{v}}_h) - \mathrm{t}_h(\underline{\hat{\boldsymbol{u}}}_h, \underline{\hat{\boldsymbol{u}}}_h, \underline{\boldsymbol{v}}_h) + \mathrm{b}_h(\underline{\boldsymbol{v}}_h, \epsilon_h) = \mathcal{E}_h((\boldsymbol{u}, p); \underline{\boldsymbol{v}}_h), \tag{9.21}$$

with consistency error $\mathcal{E}_h((\boldsymbol{u}, p); \cdot) : \underline{\boldsymbol{U}}_{h,0}^k \to \mathbb{R}$ such that, for all $\underline{\boldsymbol{v}}_h \in \underline{\boldsymbol{U}}_{h,0}^k$,

$$\mathcal{E}_h((\boldsymbol{u}, p); \underline{\boldsymbol{v}}_h) := \int_\Omega \boldsymbol{f} \cdot \boldsymbol{v}_h - \nu \mathrm{a}_h(\underline{\hat{\boldsymbol{u}}}_h, \underline{\boldsymbol{v}}_h) - \mathrm{t}_h(\underline{\hat{\boldsymbol{u}}}_h, \underline{\hat{\boldsymbol{u}}}_h, \underline{\boldsymbol{v}}_h) - \mathrm{b}_h(\underline{\boldsymbol{v}}_h, \hat{p}_h).$$

The second error equation is: For all $q_h \in \mathbb{P}^k(\mathcal{T}_h)$,

$$\mathrm{b}_h(\underline{\boldsymbol{e}}_h, q_h) = \mathrm{b}_h(\underline{\boldsymbol{u}}_h, q_h) - b_h(\underline{\hat{\boldsymbol{u}}}_h, q_h) = 0, \tag{9.22}$$

which follows from the discrete mass conservation (9.9b) together with the consistency property (8.35) of the pressure–velocity coupling bilinear form and the continuous mass balance equation (9.3b), that allow us to write $\mathrm{b}_h(\underline{\hat{\boldsymbol{u}}}_h, q_h) = b(\boldsymbol{u}, q_h) = 0$. Make $\underline{\boldsymbol{v}}_h = \underline{\boldsymbol{e}}_h$ in (9.21) and $q_h = \epsilon_h$ in (9.22). Observing that $\mathrm{t}_h(\underline{\boldsymbol{u}}_h, \underline{\boldsymbol{u}}_h, \underline{\boldsymbol{e}}_h) = \mathrm{t}_h(\underline{\boldsymbol{u}}_h, \underline{\hat{\boldsymbol{u}}}_h, \underline{\boldsymbol{e}}_h)$, owing to the linearity of t_h in its second argument along with the non-dissipativity property (9.10), we infer

$$\begin{aligned}
\mathcal{E}_h((\boldsymbol{u}, p); \underline{\boldsymbol{e}}_h) &= \nu \|\underline{\boldsymbol{e}}_h\|_{\mathrm{a},h}^2 + \mathrm{t}_h(\underline{\boldsymbol{e}}_h, \underline{\hat{\boldsymbol{u}}}_h, \underline{\boldsymbol{e}}_h) \\
&\geq \nu C_{\mathrm{a}}^{-1} \|\underline{\boldsymbol{e}}_h\|_{1,h}^2 - C_{\mathrm{t}} \|\underline{\hat{\boldsymbol{u}}}_h\|_{1,h} \|\underline{\boldsymbol{e}}_h\|_{1,h}^2 \\
&\geq \left(\nu C_{\mathrm{a}}^{-1} - d^{\frac{1}{2}} C_{\mathrm{t}} C_I C_\Omega \nu^{-1} \|\boldsymbol{f}\|_{L^2(\Omega)^d}\right) \|\underline{\boldsymbol{e}}_h\|_{1,h}^2 \\
&\geq (1-\chi) C_{\mathrm{a}}^{-1} \nu \|\underline{\boldsymbol{e}}_h\|_{1,h}^2,
\end{aligned} \tag{9.23}$$

where we have used the coercivity of a_h expressed by the first inequality in (8.31) together with the boundedness (9.11) of t_h to pass to the second line, and invoked the boundedness (8.25) of $\underline{\boldsymbol{I}}_h^k$ together with the definition of the H^1-seminorm (corresponding to (1.16) with $X = \Omega$, $s = 1$, and $p = 2$) and the standard a priori estimate $\|\nabla \boldsymbol{u}\|_{L^2(\Omega)^{d\times d}} \leq C_\Omega \nu^{-1} \|\boldsymbol{f}\|_{L^2(\Omega)^d}$ on the exact velocity to infer

$$\|\underline{\hat{\boldsymbol{u}}}_h\|_{1,h} \leq C_I |\boldsymbol{u}|_{H^1(\Omega)^d} \leq d^{\frac{1}{2}} C_I \|\nabla \boldsymbol{u}\|_{L^2(\Omega)^{d\times d}} \leq d^{\frac{1}{2}} C_I C_\Omega \nu^{-1} \|\boldsymbol{f}\|_{L^2(\Omega)^d} \tag{9.24}$$

and pass to the third line; the proof of (9.23) was concluded using the data smallness assumption (9.18).

(ii) *Estimate on the velocity.* Observing that $\boldsymbol{f} = -\nu \Delta \boldsymbol{u} + (\boldsymbol{u}\cdot\nabla)\boldsymbol{u} + \nabla p$ almost everywhere in Ω (cf. (9.1a)), it holds, for all $\underline{\boldsymbol{v}}_h \in \underline{\boldsymbol{U}}_{h,0}^k$,

$$\begin{aligned}
\mathcal{E}_h((\boldsymbol{u}, p); \underline{\boldsymbol{v}}_h) = &\underbrace{-\nu \left(\int_\Omega (\Delta \boldsymbol{u})\cdot \boldsymbol{v}_h + \mathrm{a}_h(\underline{\hat{\boldsymbol{u}}}_h, \underline{\boldsymbol{v}}_h)\right)}_{\mathfrak{T}_1} \\
&+ \underbrace{\int_\Omega (\boldsymbol{u}\cdot\nabla)\boldsymbol{u}\cdot\boldsymbol{v}_h - \mathrm{t}_h(\underline{\hat{\boldsymbol{u}}}_h, \underline{\hat{\boldsymbol{u}}}_h, \underline{\boldsymbol{v}}_h)}_{\mathfrak{T}_2} + \underbrace{\int_\Omega \nabla p\cdot\boldsymbol{v}_h - \mathrm{b}_h(\underline{\boldsymbol{v}}_h, \hat{p}_h)}_{\mathfrak{T}_3}.
\end{aligned}$$

Using the consistency property (8.32) of the viscous bilinear form a_h, we infer for the first term

$$|\mathfrak{T}_1| \lesssim h^{r+1} \nu |\boldsymbol{u}|_{H^{r+2}(\mathcal{T}_h)^d} \|\underline{\boldsymbol{v}}_h\|_{1,h}.$$

Assumption (9.12) on the discrete convective trilinear form gives for the second term

$$|\mathfrak{T}_2| \lesssim h^{r+1} \|\boldsymbol{u}\|_{W^{1,4}(\Omega)^d} |\boldsymbol{u}|_{W^{r+1,4}(\mathcal{T}_h)^d} \|\underline{\boldsymbol{v}}_h\|_{1,h}.$$

Finally, the consistency property (8.37) of the pressure–velocity coupling bilinear form b_h yields

$$|\mathfrak{T}_3| \lesssim h^{r+1} |p|_{H^{r+1}(\mathcal{T}_h)} \|\underline{\boldsymbol{v}}_h\|_{1,h}.$$

Collecting the above bounds, we get

$$\begin{aligned} \mathcal{S} &:= \sup_{\underline{\boldsymbol{v}}_h \in \underline{\boldsymbol{U}}_{h,0}^k, \|\underline{\boldsymbol{v}}_h\|_{1,h}=1} |\mathcal{E}_h((\boldsymbol{u}, p); \underline{\boldsymbol{v}}_h)| \\ &\lesssim h^{r+1} \left(\nu |\boldsymbol{u}|_{H^{r+2}(\mathcal{T}_h)^d} + \|\boldsymbol{u}\|_{W^{1,4}(\Omega)^d} |\boldsymbol{u}|_{W^{r+1,4}(\mathcal{T}_h)^d} + |p|_{H^{r+1}(\mathcal{T}_h)} \right), \end{aligned} \tag{9.25}$$

so that, in particular,

$$\begin{aligned} \mathcal{E}_h((\boldsymbol{u}, p); \underline{\boldsymbol{e}}_h) &\le \mathcal{S} \|\underline{\boldsymbol{e}}_h\|_{1,h} \\ &\lesssim h^{r+1} \left(\nu |\boldsymbol{u}|_{H^{r+2}(\mathcal{T}_h)^d} + \|\boldsymbol{u}\|_{W^{1,4}(\Omega)^d} |\boldsymbol{u}|_{W^{r+1,4}(\mathcal{T}_h)^d} + |p|_{H^{r+1}(\mathcal{T}_h)} \right) \|\underline{\boldsymbol{e}}_h\|_{1,h}. \end{aligned} \tag{9.26}$$

Combining (9.23) with (9.26), the estimate on the velocity in (9.19) follows.

(iii) *Estimate on the pressure.* Let us now estimate the error on the pressure. We have

$$\begin{aligned} \|\epsilon_h\|_{L^2(\Omega)} &\lesssim \sup_{\underline{\boldsymbol{v}}_h \in \underline{\boldsymbol{U}}_{h,0}^k, \|\underline{\boldsymbol{v}}_h\|_{1,h}=1} \mathrm{b}_h(\underline{\boldsymbol{v}}_h, \epsilon_h) \\ &= \sup_{\underline{\boldsymbol{v}}_h \in \underline{\boldsymbol{U}}_{h,0}^k, \|\underline{\boldsymbol{v}}_h\|_{1,h}=1} \Big(\mathcal{E}_h((\boldsymbol{u}, p); \underline{\boldsymbol{v}}_h) - \nu \mathrm{a}_h(\underline{\boldsymbol{e}}_h, \underline{\boldsymbol{v}}_h) - \mathrm{t}_h(\underline{\boldsymbol{u}}_h, \underline{\boldsymbol{u}}_h, \underline{\boldsymbol{v}}_h) \\ &\qquad + \mathrm{t}_h(\underline{\hat{\boldsymbol{u}}}_h, \underline{\hat{\boldsymbol{u}}}_h, \underline{\boldsymbol{v}}_h) \Big) \\ &= \sup_{\underline{\boldsymbol{v}}_h \in \underline{\boldsymbol{U}}_{h,0}^k, \|\underline{\boldsymbol{v}}_h\|_{1,h}=1} \Big(\mathcal{E}_h((\boldsymbol{u}, p); \underline{\boldsymbol{v}}_h) - \nu \mathrm{a}_h(\underline{\boldsymbol{e}}_h, \underline{\boldsymbol{v}}_h) - \mathrm{t}_h(\underline{\boldsymbol{e}}_h, \underline{\boldsymbol{u}}_h, \underline{\boldsymbol{v}}_h) \\ &\qquad - \mathrm{t}_h(\underline{\hat{\boldsymbol{u}}}_h, \underline{\boldsymbol{e}}_h, \underline{\boldsymbol{v}}_h) \Big) \\ &\lesssim \mathcal{S} + \left(\nu + \|\underline{\boldsymbol{u}}_h\|_{1,h} + \|\underline{\hat{\boldsymbol{u}}}_h\|_{1,h} \right) \|\underline{\boldsymbol{e}}_h\|_{1,h} \\ &\lesssim \mathcal{S} + \left(\nu + \nu^{-1} \|\boldsymbol{f}\|_{L^2(\Omega)^d} \right) \|\underline{\boldsymbol{e}}_h\|_{1,h} \\ &\lesssim \mathcal{S} + \nu \|\underline{\boldsymbol{e}}_h\|_{1,h}. \end{aligned} \tag{9.27}$$

In (9.27), we have used the inf–sup inequality (8.36) on b_h in the first line and the error equation (9.21) to pass to the second line; to pass to the third line, we have inserted $\pm\mathrm{t}_h(\hat{\underline{\boldsymbol{u}}}_h, \underline{\boldsymbol{u}}_h, \underline{\boldsymbol{v}}_h)$, used the linearity of t_h in its first and second arguments, and recalled the definition (9.20) of $\underline{\boldsymbol{e}}_h$; to pass to the fourth line, we have used the boundedness properties (8.31) of a_h and (9.11) of t_h; to pass to the fifth line, we have used the a priori bounds (9.13) on $\|\underline{\boldsymbol{u}}_h\|_{1,h}$ and (9.24) on $\|\hat{\underline{\boldsymbol{u}}}_h\|_{1,h}$; the data smallness assumption (9.18) gives the conclusion. The estimate on the pressure then follows using (9.25) and the estimate on the velocity established in Point (ii) to further bound the addends in the right-hand side of (9.27). □

As usual, from the discrete error estimate (9.19) we can derive an error estimate based on the global velocity reconstruction $\mathbf{r}_h^{k+1} : \underline{\boldsymbol{U}}_h^k \to \mathbb{P}^{k+1}(\mathcal{T}_h)^d$ defined by (8.70), that is, for all $\underline{\boldsymbol{v}}_h \in \underline{\boldsymbol{U}}_h^k$,

$$(\mathbf{r}_h^{k+1}\underline{\boldsymbol{v}}_h)_{|T} := \mathbf{r}_T^{k+1}\underline{\boldsymbol{v}}_T \qquad \forall T \in \mathcal{T}_h. \tag{9.28}$$

Corollary 9.13 (Energy Error Estimate for Small Data) *Under the assumptions and notations of Theorem 9.10, and denoting by $\boldsymbol{\nabla}_h$ the broken gradient operator acting on vector fields and defined as in* (1.21)*, it holds*

$$\begin{aligned}\nu\left(\|\boldsymbol{\nabla}_h\mathbf{r}_h^{k+1}\underline{\boldsymbol{u}}_h - \boldsymbol{\nabla}\boldsymbol{u}\|_{L^2(\Omega)^{d\times d}} + |\underline{\boldsymbol{u}}_h|_{\mathrm{s},h}\right) + \|p_h - p\|_{L^2(\Omega)} \lesssim \\ h^{r+1}\left(\nu|\boldsymbol{u}|_{H^{r+2}(\mathcal{T}_h)^d} + \|\boldsymbol{u}\|_{W^{1,4}(\Omega)^d}|\boldsymbol{u}|_{W^{r+1,4}(\mathcal{T}_h)^d} + |p|_{H^{r+1}(\mathcal{T}_h)}\right),\end{aligned} \tag{9.29}$$

where we have defined the stabilisation seminorm such that, for all $\underline{\boldsymbol{v}}_h \in \underline{\boldsymbol{U}}_h^k$,

$$|\underline{\boldsymbol{v}}_h|_{\mathrm{s},h} := \left(\sum_{T\in\mathcal{T}_h} \mathrm{s}_T(\underline{\boldsymbol{v}}_T, \underline{\boldsymbol{v}}_T)\right)^{\frac{1}{2}}. \tag{9.30}$$

Proof The estimate on the bracketed term in the left-hand side of (9.29) is done exactly as in the proof of Theorem 2.28: insert $\pm\boldsymbol{\nabla}_h\mathbf{r}_h^{k+1}\hat{\underline{\boldsymbol{u}}}_h$ into $\|\boldsymbol{\nabla}_h\mathbf{r}_h^{k+1}\underline{\boldsymbol{u}}_h - \boldsymbol{\nabla}\boldsymbol{u}\|_{L^2(\Omega)^{d\times d}}$ and $\pm\hat{\underline{\boldsymbol{u}}}_h$ into $|\underline{\boldsymbol{u}}_h|_{\mathrm{s},h}$, apply a triangle inequality, then use (9.19) together with the norm equivalence (8.31) to estimate $\|\boldsymbol{\nabla}_h\mathbf{r}_h^{k+1}(\underline{\boldsymbol{u}}_h - \hat{\underline{\boldsymbol{u}}}_h)\|_{L^2(\Omega)^{d\times d}} + |\underline{\boldsymbol{u}}_h - \hat{\underline{\boldsymbol{u}}}_h|_{\mathrm{s},h}$, and bound the remaining term $\|\boldsymbol{\nabla}_h\mathbf{r}_h^{k+1}\hat{\underline{\boldsymbol{u}}}_h - \boldsymbol{\nabla}\boldsymbol{u}\|_{L^2(\Omega)^{d\times d}} + |\hat{\underline{\boldsymbol{u}}}_h|_{\mathrm{s},h}$ as $\mathfrak{T}_2$ in the proof of Theorem 2.28.

To estimate $\|p_h - p\|_{L^2(\Omega)}$, insert $\pm\pi_h^{0,k}p$ into the norm, use a triangle inequality, and recall (9.19) and the approximation property (1.74) with Lebesgue exponent $p = 2$, $l = k$, $s = r + 1$, and $m = 0$:

$$\begin{aligned}\|p_h - p\|_{L^2(\Omega)} &\le \|p_h - \pi_h^{0,k}p\|_{L^2(\Omega)} + \|\pi_h^{0,k}p - p\|_{L^2(\Omega)} \\ &\lesssim h^{r+1}\left(\nu|\boldsymbol{u}|_{H^{r+2}(\mathcal{T}_h)^d} + \|\boldsymbol{u}\|_{W^{1,4}(\Omega)^d}|\boldsymbol{u}|_{W^{r+1,4}(\mathcal{T}_h)^d} + |p|_{H^{r+1}(\mathcal{T}_h)}\right) + h^{r+1}|p|_{H^{r+1}(\Omega)}.\end{aligned}$$

□

9.4 Convective Stabilisation

When dealing with high-Reynolds flows, it is sometimes desirable to strengthen stability by penalising the difference between face and element unknowns. Fix $\rho : \mathbb{R} \to [0, \infty)$ a Lipschitz-continuous function and $\underline{\boldsymbol{w}}_h \in \underline{\boldsymbol{U}}_h^k$ a vector of discrete unknowns, and define the convective stabilisation bilinear form $\mathrm{j}_h(\underline{\boldsymbol{w}}_h; \cdot, \cdot) : \underline{\boldsymbol{U}}_h^k \times \underline{\boldsymbol{U}}_h^k \to \mathbb{R}$ such that, for all $\underline{\boldsymbol{v}}_h, \underline{\boldsymbol{z}}_h \in \underline{\boldsymbol{U}}_h^k$,

$$\mathrm{j}_h(\underline{\boldsymbol{w}}_h; \underline{\boldsymbol{v}}_h, \underline{\boldsymbol{z}}_h) := \sum_{T \in \mathcal{T}_h} \sum_{F \in \mathcal{F}_T} \int_F \frac{\nu}{h_F} \rho(\mathrm{Pe}_{TF}(\boldsymbol{w}_F))(\boldsymbol{v}_F - \boldsymbol{v}_T)\cdot(\boldsymbol{z}_F - \boldsymbol{z}_T). \tag{9.31}$$

Here, for all $T \in \mathcal{T}_h$ and all $F \in \mathcal{F}_T$, the local (oriented) Péclet number $\mathrm{Pe}_{TF} : \mathbb{P}^k(F)^d \to \mathbb{R}$ is such that, for all $\boldsymbol{w} \in \mathbb{P}^k(F)^d$,

$$\mathrm{Pe}_{TF}(\boldsymbol{w}) := h_F \frac{\boldsymbol{w}\cdot\boldsymbol{n}_{TF}}{\nu}.$$

As already pointed out in [44, 103, 159], using the generic function ρ in the definition of the convective stabilisation terms enables a unified treatment of several classical discretisations (in the notations of [44], $A(s) = \rho(s) + \frac{1}{2}s$ and $B(s) = -\rho(s) + \frac{1}{2}s$; in the notations of [159], $\rho = \frac{1}{2}|A|$). Specifically, the HHO version of classical convective stabilisations is obtained with the following choices of ρ:

- *Centred scheme*: $\rho = 0$.
- *Upwind scheme*: $\rho(s) = \frac{1}{2}|s|$. In this case, the definition (9.31) of $\mathrm{j}_h(\underline{\boldsymbol{w}}_h; \cdot, \cdot)$ simplifies to (compare with (3.73))

$$\mathrm{j}_h(\underline{\boldsymbol{w}}_h; \underline{\boldsymbol{v}}_h, \underline{\boldsymbol{z}}_h) := \sum_{T \in \mathcal{T}_h} \sum_{F \in \mathcal{F}_T} \int_F \frac{|\boldsymbol{w}_F\cdot\boldsymbol{n}_{TF}|}{2}(\boldsymbol{v}_F - \boldsymbol{v}_T)\cdot(\boldsymbol{z}_F - \boldsymbol{z}_T).$$

- *Locally upwinded θ-scheme*: $\rho(s) = \frac{1}{2}(1 - \theta(s))|s|$, where $\theta \in C_c^1(-1, 1)$, $0 \le \theta \le 1$ and $\theta \equiv 1$ on $[-\frac{1}{2}, \frac{1}{2}]$. This choice in (9.31) corresponds to the centred scheme if $|\mathrm{Pe}(\boldsymbol{w}_F)| \le \frac{1}{2}$ (dominating viscosity) and to the upwind scheme if $|\mathrm{Pe}(\boldsymbol{w}_F)| \ge 1$ (dominating advection).
- *Scharfetter–Gummel scheme [257]*: $\rho(s) = \frac{s}{2}\coth(\frac{s}{2}) - 1$.

The advantage of the locally upwinded θ-scheme and the Scharfetter–Gummel scheme over the upwind scheme is that they behave as the centred scheme, and thus introduce less numerical diffusion, when $|\mathrm{Pe}(\boldsymbol{w}_F)|$ is not too large (dominating viscosity). See, e.g., the discussion in [170, Section 4.1] for the Scharfetter–Gummel scheme.

The HHO scheme with convective stabilisation reads: Find $(\underline{\boldsymbol{u}}_h, p_h) \in \underline{\boldsymbol{U}}_{h,0}^k \times P_h^k$ such that, for all $(\underline{\boldsymbol{v}}_h, q_h) \in \underline{\boldsymbol{U}}_{h,0}^k \times \mathbb{P}^k(\mathcal{T}_h)$,

$$\nu \mathrm{a}_h(\underline{\boldsymbol{u}}_h, \underline{\boldsymbol{v}}_h) + \mathrm{t}_h(\underline{\boldsymbol{u}}_h, \underline{\boldsymbol{u}}_h, \underline{\boldsymbol{v}}_h) + \mathrm{j}_h(\underline{\boldsymbol{u}}_h; \underline{\boldsymbol{u}}_h, \underline{\boldsymbol{v}}_h) + \mathrm{b}_h(\underline{\boldsymbol{v}}_h, p_h) = \int_\Omega \boldsymbol{f} \cdot \boldsymbol{v}_h, \tag{9.32a}$$

$$-\mathrm{b}_h(\underline{\boldsymbol{u}}_h, q_h) = 0. \tag{9.32b}$$

Both the error estimate of Theorem 9.10 and the convergence analysis of Theorem 9.32 below can be adapted to incorporate the convective stabilisation terms. These developments are not further pursued here for the sake of conciseness; the interested reader can consult [70].

9.5 Examples of Discrete Convective Trilinear Forms

We present here two examples of discrete convective trilinear forms that match the design properties in Assumption 9.2. A gradient reconstruction $\mathbf{G}_T^l$, which generalises (4.37) to polynomial degrees different from k, is first introduced and analysed. It is then used to construct a discrete version of (9.8), which leads to the first example of a convective trilinear form. The second example consists in discretising the form (9.7) based on Temam's device, using for the divergence the trace of $\mathbf{G}_T^{2k}$ and for $\boldsymbol{w} \cdot \boldsymbol{\nabla}$ a directional derivative inspired by the one introduced for scalar advection–diffusion–reaction models (see (3.64)). The major difference between the two examples of t_h we construct here is that, contrary to the first one, the second one enables us to write a flux formulation of the scheme (9.9); see Remark 9.28.

9.5.1 A Local Gradient Reconstruction

Let $l \geq 0$ be an integer. We define here a generalisation of the gradient used in Sect. 4.2 and Chap. 6, consisting in a tensorial gradient reconstruction in a local polynomial space of degree l instead of k. Precisely, given a mesh element $T \in \mathcal{T}_h$ and following (4.37), we define the gradient operator $\mathbf{G}_T^l : \underline{\boldsymbol{U}}_T^k \to \mathbb{P}^l(T)^{d \times d}$ such that, for all $\underline{\boldsymbol{v}}_T \in \underline{\boldsymbol{U}}_T^k$,

$$\int_T \mathbf{G}_T^l \underline{\boldsymbol{v}}_T : \boldsymbol{\tau} = -\int_T \boldsymbol{v}_T \cdot (\boldsymbol{\nabla} \cdot \boldsymbol{\tau}) + \sum_{F \in \mathcal{F}_T} \int_F \boldsymbol{v}_F \cdot (\boldsymbol{\tau} \boldsymbol{n}_{TF}) \qquad \forall \boldsymbol{\tau} \in \mathbb{P}^l(T)^{d \times d}. \tag{9.33}$$

By the Riesz representation theorem in $\mathbb{P}^l(T)^{d\times d}$ endowed with the $L^2(T)^{d\times d}$-inner product, $\mathbf{G}_T^l\underline{\boldsymbol{v}}_T$ is uniquely defined. Integrating by parts the first term in the right-hand side of (9.33), we obtain the following characterisation of $\mathbf{G}_T^l$: For all $\underline{\boldsymbol{v}}_T \in \underline{\boldsymbol{U}}_T^k$ and all $\boldsymbol{\tau} \in \mathbb{P}^l(T)^{d\times d}$,

$$\int_T \mathbf{G}_T^l\underline{\boldsymbol{v}}_T{:}\boldsymbol{\tau} = \int_T \nabla\boldsymbol{v}_T{:}\boldsymbol{\tau} + \sum_{F\in\mathcal{F}_T}\int_F (\boldsymbol{v}_F - \boldsymbol{v}_T)\cdot(\boldsymbol{\tau}\boldsymbol{n}_{TF}). \tag{9.34}$$

Taking two polynomial degrees $l, m \geq 0$ and applying the definitions (9.33) of $\mathbf{G}_T^l$ and $\mathbf{G}_T^m$ to $\boldsymbol{\tau} \in \mathbb{P}^{\min(l,m)}(T)^{d\times d}$, the two right-hand sides are identical and thus, for all $\underline{\boldsymbol{v}}_T \in \underline{\boldsymbol{U}}_T^k$,

$$\int_T \mathbf{G}_T^l\underline{\boldsymbol{v}}_T : \boldsymbol{\tau} = \int_T \mathbf{G}_T^m\underline{\boldsymbol{v}}_T : \boldsymbol{\tau} \qquad \forall\boldsymbol{\tau} \in \mathbb{P}^{\min(l,m)}(T)^{d\times d}. \tag{9.35}$$

In other words,

$$\boldsymbol{\pi}_T^{0,\min(l,m)}(\mathbf{G}_T^l\underline{\boldsymbol{v}}_T) = \boldsymbol{\pi}_T^{0,\min(l,m)}(\mathbf{G}_T^m\underline{\boldsymbol{v}}_T). \tag{9.36}$$

The other properties of this gradient reconstruction relevant to our analysis are summarised in the following proposition.

Proposition 9.14 (Properties of the Local Gradient Reconstruction) *For any $T \in \mathcal{T}_h$ and any $l \geq 0$, the gradient reconstruction defined by* (9.33) *satisfies the following properties:*

(i) Boundedness. *For all $\underline{\boldsymbol{v}}_T \in \underline{\boldsymbol{U}}_T^k$, it holds with local seminorm $\|\cdot\|_{1,T}$ defined by* (8.15)*:*

$$\|\mathbf{G}_T^l\underline{\boldsymbol{v}}_T\|_{L^2(T)^{d\times d}} \lesssim \|\underline{\boldsymbol{v}}_T\|_{1,T}. \tag{9.37}$$

(ii) Consistency. *For any $r \in \{0,\ldots,l+1\}$ if $l \leq k$, $r \in \{0,\ldots,k\}$ if $l > k$, it holds that*

$$\|\mathbf{G}_T^l\underline{\boldsymbol{I}}_T^k\boldsymbol{v} - \nabla\boldsymbol{v}\|_{L^2(T)^{d\times d}} \lesssim h_T^r|\boldsymbol{v}|_{H^{r+1}(T)^d} \qquad \forall\boldsymbol{v} \in H^{r+1}(T)^d. \tag{9.38a}$$

Moreover, if $r \geq 1$,

$$\|\mathbf{G}_T^l\underline{\boldsymbol{I}}_T^k\boldsymbol{v} - \nabla\boldsymbol{v}\|_{L^2(\partial T)^{d\times d}} \lesssim h_T^{r-\frac{1}{2}}|\boldsymbol{v}|_{H^{r+1}(T)^d} \qquad \forall\boldsymbol{v} \in H^{r+1}(T)^d. \tag{9.38b}$$

Above, the hidden constants are independent of both h and T, but possibly depend on d, ϱ, k, and l.

Proof (i) *Boundedness.* Making $\boldsymbol{\tau} = \mathrm{G}^l_T \underline{\boldsymbol{v}}_T$ in (9.34), using a Cauchy–Schwarz inequality for the volumetric term, a generalised Hölder inequality with exponents $(2, 2, \infty)$ together with $\|\boldsymbol{n}_{TF}\|_{L^\infty(F)^d} = 1$, and the discrete trace inequality (1.55) with $p = 2$ for the boundary terms yields the conclusion, in a similar way as in (6.23).

(ii) *Consistency.* Let $\boldsymbol{v} \in H^{r+1}(T)^d$. If $l \le k$, applying (9.36) to $m = k$ and $\underline{\boldsymbol{v}}_T = \underline{\boldsymbol{I}}^k_T \boldsymbol{v}$, and recalling that $\mathrm{G}^k_T \underline{\boldsymbol{I}}^k_T \boldsymbol{v} = \boldsymbol{\pi}^{0,k}_T(\nabla \boldsymbol{v})$ (see (4.40)), we have $\mathrm{G}^l_T \underline{\boldsymbol{I}}^k_T \boldsymbol{v} = \boldsymbol{\pi}^{0,l}_T(\nabla \boldsymbol{v})$. The estimates (9.38) then immediately follow from the approximation properties of the L^2-orthogonal projector; see Theorem 1.45.

Consider now $l > k$. We start by noticing the following estimate, which is deduced from a triangle inequality and (3.92) with $w =$ components of $\boldsymbol{v}$:

$$
\begin{aligned}
\|\boldsymbol{\pi}^{0,k}_F \boldsymbol{v} - \boldsymbol{v}\|_{L^2(F)^d} &\le \|\boldsymbol{\pi}^{0,k}_F \boldsymbol{v} - \boldsymbol{\pi}^{0,k}_T \boldsymbol{v}\|_{L^2(F)^d} + \|\boldsymbol{\pi}^{0,k}_T \boldsymbol{v} - \boldsymbol{v}\|_{L^2(F)^d} \\
&\lesssim h_T^{r+\frac{1}{2}} |\boldsymbol{v}|_{H^{r+1}(T)^d}.
\end{aligned}
\tag{9.39}
$$

For all $\boldsymbol{\tau} \in \mathbb{P}^l(T)^{d\times d}$, plugging the definition (8.14) of $\underline{\boldsymbol{I}}^k_T \boldsymbol{v}$ into (9.33) and subtracting $\int_T \nabla \boldsymbol{v}:\boldsymbol{\tau} = -\int_T \boldsymbol{v}\cdot(\nabla\cdot\boldsymbol{\tau}) + \sum_{F\in\mathcal{F}_T} \int_F \boldsymbol{v}\cdot(\boldsymbol{\tau}\boldsymbol{n}_{TF})$, we get

$$
\int_T (\mathrm{G}^l_T \underline{\boldsymbol{I}}^k_T \boldsymbol{v} - \nabla \boldsymbol{v}):\boldsymbol{\tau} = -\int_T (\boldsymbol{\pi}^{0,k}_T \boldsymbol{v} - \boldsymbol{v})\cdot(\nabla\cdot\boldsymbol{\tau}) + \sum_{F\in\mathcal{F}_T} \int_F (\boldsymbol{\pi}^{0,k}_F \boldsymbol{v} - \boldsymbol{v})\cdot(\boldsymbol{\tau}\boldsymbol{n}_{TF}).
$$

Make $\boldsymbol{\tau} = \mathrm{G}^l_T \underline{\boldsymbol{I}}^k_T \boldsymbol{v} - \boldsymbol{\pi}^{0,l}_T(\nabla \boldsymbol{v})$ and notice that, by orthogonality property of $\boldsymbol{\pi}^{0,l}_T$, the term $\nabla \boldsymbol{v}$ in the left-hand side can be replaced with $\boldsymbol{\pi}^{0,l}_T(\nabla \boldsymbol{v})$. Using a Cauchy–Schwarz inequality for the volumetric term and generalised Hölder inequalities with exponents $(2, 2, \infty)$ together with $\|\boldsymbol{n}_{TF}\|_{L^\infty(F)^d} = 1$ for the boundary terms, we obtain

$$
\begin{aligned}
\|\mathrm{G}^l_T \underline{\boldsymbol{I}}^k_T \boldsymbol{v} - \boldsymbol{\pi}^{0,l}_T(\nabla \boldsymbol{v})\|^2_{L^2(T)^{d\times d}} &\le \|\boldsymbol{\pi}^{0,k}_T \boldsymbol{v} - \boldsymbol{v}\|_{L^2(T)^d} \|\nabla\cdot \left(\mathrm{G}^l_T \underline{\boldsymbol{I}}^k_T \boldsymbol{v} - \boldsymbol{\pi}^{0,l}_T(\nabla \boldsymbol{v})\right)\|_{L^2(T)^d} \\
&\quad + \sum_{F\in\mathcal{F}_T} \|\boldsymbol{\pi}^{0,k}_F \boldsymbol{v} - \boldsymbol{v}\|_{L^2(F)^d} \|\mathrm{G}^l_T \underline{\boldsymbol{I}}^k_T \boldsymbol{v} - \boldsymbol{\pi}^{0,l}_T(\nabla \boldsymbol{v})\|_{L^2(F)^{d\times d}}.
\end{aligned}
$$

Invoking the discrete inverse (1.46) and trace (1.55) inequalities (both with $p = 2$ and $v =$ components of $\mathrm{G}^l_T \underline{\boldsymbol{I}}^k_T \boldsymbol{v} - \boldsymbol{\pi}^{0,l}_T(\nabla \boldsymbol{v})$), together with the approximation property (1.74) of $\boldsymbol{\pi}^{0,k}_T$ (with $s = r + 1$) and (9.39), we infer

$$
\|\mathrm{G}^l_T \underline{\boldsymbol{I}}^k_T \boldsymbol{v} - \boldsymbol{\pi}^{0,l}_T(\nabla \boldsymbol{v})\|_{L^2(T)^{d\times d}} \lesssim h^r_T |\boldsymbol{v}|_{H^{r+1}(T)^d}. \tag{9.40}
$$

Hence, using the triangle inequality and the approximation properties (1.74) of $\boldsymbol{\pi}_T^{0,l}$, we obtain

$$\|\mathbf{G}_T^l \underline{\boldsymbol{I}}_T^k \boldsymbol{v} - \boldsymbol{\nabla}\boldsymbol{v}\|_{L^2(T)^{d\times d}} \le \|\mathbf{G}_T^l \underline{\boldsymbol{I}}_T^k \boldsymbol{v} - \boldsymbol{\pi}_T^{0,l}(\boldsymbol{\nabla}\boldsymbol{v})\|_{L^2(T)^{d\times d}} + \|\boldsymbol{\pi}_T^{0,l}(\boldsymbol{\nabla}\boldsymbol{v}) - \boldsymbol{\nabla}\boldsymbol{v}\|_{L^2(T)^{d\times d}}$$
$$\lesssim h_T^r |\boldsymbol{v}|_{H^{r+1}(T)^d},$$

and (9.38a) follows. To prove (9.38b), we observe that it holds

$$\begin{aligned}
&\|\mathbf{G}_T^l \underline{\boldsymbol{I}}_T^k \boldsymbol{v} - \boldsymbol{\nabla}\boldsymbol{v}\|_{L^2(\partial T)^{d\times d}} \\
&\quad\le \|\mathbf{G}_T^l \underline{\boldsymbol{I}}_T^k \boldsymbol{v} - \boldsymbol{\pi}_T^{0,l}(\boldsymbol{\nabla}\boldsymbol{v})\|_{L^2(\partial T)^{d\times d}} + \|\boldsymbol{\pi}_T^{0,l}(\boldsymbol{\nabla}\boldsymbol{v}) - \boldsymbol{\nabla}\boldsymbol{v}\|_{L^2(\partial T)^{d\times d}} \\
&\quad\lesssim h_T^{-\frac{1}{2}} \|\mathbf{G}_T^l \underline{\boldsymbol{I}}_T^k \boldsymbol{v} - \boldsymbol{\pi}_T^{0,l}(\boldsymbol{\nabla}\boldsymbol{v})\|_{L^2(T)^{d\times d}} + h_T^{r-\frac{1}{2}} |\boldsymbol{v}|_{H^{r+1}(T)^d},
\end{aligned}$$

where we have inserted $\pm\boldsymbol{\pi}_T^{0,l}(\boldsymbol{\nabla}\boldsymbol{v})$ inside the norm and used the triangle inequality in the first bound, followed by the discrete trace inequality (1.55) with $p = 2$ together with the approximation properties (1.75) of $\boldsymbol{\pi}_T^{0,l}$ in the second. Using (9.40) to estimate the first term in the right-hand side of the above inequality, (9.38b) is proved. □

9.5.2 A Skew-Symmetric Trilinear Form Using a Gradient-Based Approximation of the Convective Derivative

The first discrete trilinear form that we consider, originally introduced in [151], is inspired by the skew-symmetric formulation (9.8) of the continuous trilinear form. The key idea consists in replacing the gradient operator by the discrete counterpart $\mathbf{G}_T^{2k}$, which amounts to using a gradient-based approximation of the convective derivative.

9.5.2.1 A Gradient-Based Discrete Directional Derivative

We preliminarily study a gradient-based discrete directional derivative. Specifically, for $\boldsymbol{w} = (w_i)_{1\le i\le d} \in L^2(T)^d$, we define $\boldsymbol{w}\cdot\mathbf{G}_T^{2k} : \underline{\boldsymbol{U}}_T^k \to L^2(T)^d$ by

$$(\boldsymbol{w}\cdot\mathbf{G}_T^{2k})\underline{\boldsymbol{v}}_T := \left(\sum_{j=1}^d w_j (\mathbf{G}_T^{2k}\underline{\boldsymbol{v}}_T)_{ij}\right)_{1\le i\le d} \qquad \forall \underline{\boldsymbol{v}}_T \in \underline{\boldsymbol{U}}_T^k. \tag{9.41}$$

Recalling that $(\mathbf{G}_T^{2k}\underline{\boldsymbol{v}}_T)_{ij}$ approximates the partial derivative with respect to the jth space variable of the ith component of the function represented by $\underline{\boldsymbol{v}}_T$, $\boldsymbol{w}\cdot\mathbf{G}_T^{2k}$ can be regarded as a discrete version of $\boldsymbol{w}\cdot\boldsymbol{\nabla}$.

Lemma 9.15 below states the properties of the operator (9.41) that are relevant to our analysis. The proof of this lemma hinges on the following discrete Sobolev embedding, obtained by applying Theorem 6.40, with $p = 2$ and $q = 4$, to $\underline{v}_h =$ components of $\underline{\boldsymbol{v}}_h$:

$$\|\boldsymbol{v}_h\|_{L^4(\Omega)^d} \lesssim \|\underline{\boldsymbol{v}}_h\|_{1,h} \qquad \forall \underline{\boldsymbol{v}}_h \in \underline{\boldsymbol{U}}_{h,0}^k, \tag{9.42}$$

where the hidden multiplicative constant depends on Ω, d, ϱ, and k. We note, in passing, that the continuous Sobolev embeddings are essential, in the theoretical analysis of Navier–Stokes equations, to deal with the nonlinear term. It is therefore no surprise if a discrete version thereof is required for the numerical analysis of these equations.

Lemma 9.15 (Properties of the Gradient-Based Discrete Directional Derivative) *The operator defined by* (9.41) *satisfies the following properties:*

(i) Boundedness. *For all* $\underline{\boldsymbol{w}}_h, \underline{\boldsymbol{v}}_h, \underline{\boldsymbol{z}}_h \in \underline{\boldsymbol{U}}_{h,0}^k$, *it holds*

$$\left|\sum_{T\in\mathcal{T}_h}\int_T (\boldsymbol{w}_T\cdot\mathbf{G}_T^{2k})\underline{\boldsymbol{v}}_T\cdot\boldsymbol{z}_T\right| \lesssim \|\underline{\boldsymbol{w}}_h\|_{1,h}\|\underline{\boldsymbol{v}}_h\|_{1,h}\|\underline{\boldsymbol{z}}_h\|_{1,h}, \tag{9.43}$$

with hidden constant independent of h, $\underline{\boldsymbol{w}}_h$, $\underline{\boldsymbol{v}}_h$ *and* $\underline{\boldsymbol{z}}_h$.

(ii) Consistency. *If* $r \in \{0, \ldots, k\}$ *and* $\boldsymbol{w} \in H_0^1(\Omega)^d \cap W^{r+1,4}(\mathcal{T}_h)^d$ *then, setting* $\underline{\hat{\boldsymbol{w}}}_h := \underline{\boldsymbol{I}}_h^k \boldsymbol{w}$, *it holds*

$$\left|\int_\Omega (\boldsymbol{w}\cdot\nabla)\boldsymbol{w}\cdot\boldsymbol{z}_h - \sum_{T\in\mathcal{T}_h}\int_T (\hat{\boldsymbol{w}}_T\cdot\mathbf{G}_T^{2k})\underline{\hat{\boldsymbol{w}}}_T\cdot\boldsymbol{z}_T\right|$$
$$\lesssim h^{r+1}|\boldsymbol{w}|_{W^{r+1,4}(\mathcal{T}_h)^d}|\boldsymbol{w}|_{W^{1,4}(\Omega)^d}\|\boldsymbol{z}_h\|_{L^4(\Omega)^d} \qquad \forall \boldsymbol{z}_h \in \mathbb{P}^k(\mathcal{T}_h)^d, \tag{9.44}$$

where $\boldsymbol{z}_T = (\boldsymbol{z}_h)_{|T}$ *for all* $T \in \mathcal{T}_h$, *and the hidden constant is independent of* h, $\boldsymbol{w}$, *and* $\boldsymbol{z}_h$.

Proof (i) *Boundedness.* We have

$$\begin{aligned}
\left|\sum_{T\in\mathcal{T}_h}\int_T (\boldsymbol{w}_T\cdot\mathbf{G}_T^{2k})\underline{\boldsymbol{v}}_T\cdot\boldsymbol{z}_T\right| &\le \sum_{T\in\mathcal{T}_h}\|\boldsymbol{w}_T\|_{L^4(T)^d}\|\mathbf{G}_T^{2k}\underline{\boldsymbol{v}}_T\|_{L^2(T)^{d\times d}}\|\boldsymbol{z}_T\|_{L^4(T)^d}\\
&\lesssim \sum_{T\in\mathcal{T}_h}\|\boldsymbol{w}_T\|_{L^4(T)^d}\|\underline{\boldsymbol{v}}_T\|_{1,T}\|\boldsymbol{z}_T\|_{L^4(T)^d}\\
&\le \|\boldsymbol{w}_h\|_{L^4(\Omega)}\|\underline{\boldsymbol{v}}_h\|_{1,h}\|\boldsymbol{z}_h\|_{L^4(\Omega)^d}\\
&\lesssim \|\underline{\boldsymbol{w}}_h\|_{1,h}\|\underline{\boldsymbol{v}}_h\|_{1,h}\|\underline{\boldsymbol{z}}_h\|_{1,h},
\end{aligned}$$

where we have used generalised Hölder inequalities with exponents $(4, 2, 4)$ on the integrals in the first inequality, the boundedness property (9.37) of the local gradient reconstruction in the second inequality, generalised Hölder inequalities with exponents $(4, 2, 4)$ on the sum over $T \in \mathcal{T}_h$ in the third inequality, and the discrete Sobolev embedding (9.42) to conclude.

(ii) *Consistency.* Inserting $\pm(\boldsymbol{w}\cdot\mathbf{G}_T^{2k})\underline{\hat{\boldsymbol{w}}}_T$ into the integrals, we have

$$\begin{aligned}
&\int_\Omega (\boldsymbol{w}\cdot\nabla)\boldsymbol{w}\cdot\boldsymbol{z}_h - \sum_{T\in\mathcal{T}_h}\int_T (\hat{\boldsymbol{w}}_T\cdot\mathbf{G}_T^{2k})\underline{\hat{\boldsymbol{w}}}_T\cdot\boldsymbol{z}_T \\
&= \sum_{T\in\mathcal{T}_h}\int_T \left[(\boldsymbol{w}\cdot\nabla)\boldsymbol{w} - (\boldsymbol{w}\cdot\mathbf{G}_T^{2k})\underline{\hat{\boldsymbol{w}}}_T\right]\cdot\boldsymbol{z}_T + \sum_{T\in\mathcal{T}_h}\int_T \left[(\boldsymbol{w}-\hat{\boldsymbol{w}}_T)\cdot\mathbf{G}_T^{2k}\right]\underline{\hat{\boldsymbol{w}}}_T\cdot\boldsymbol{z}_T \\
&=: \mathfrak{T}_1 + \mathfrak{T}_2.
\end{aligned} \tag{9.45}$$

For the first term, we start by noticing the following relation, obtained applying (9.35) to $(l, m) = (2k, k)$ and $\underline{\boldsymbol{v}}_T = \underline{\hat{\boldsymbol{w}}}_T$, and recalling that $\mathbf{G}_T^k\underline{\hat{\boldsymbol{w}}}_T = \boldsymbol{\pi}_T^{0,k}(\nabla\boldsymbol{w})$ (see (4.40)):

$$\int_T (\mathbf{G}_T^{2k}\underline{\hat{\boldsymbol{w}}}_T - \nabla\boldsymbol{w}):\boldsymbol{\tau} = 0 \qquad \forall\boldsymbol{\tau}\in\mathbb{P}^k(T)^{d\times d}. \tag{9.46}$$

Recalling the definition (7.2) of the Frobenius product and using (9.46) with $\boldsymbol{\tau} = \boldsymbol{z}_T\otimes\boldsymbol{\pi}_T^{0,0}\boldsymbol{w}\in\mathbb{P}^k(T)^{d\times d}$, we infer

$$\begin{aligned}
\mathfrak{T}_1 &= \sum_{T\in\mathcal{T}_h}\int_T (\nabla\boldsymbol{w} - \mathbf{G}_T^{2k}\underline{\hat{\boldsymbol{w}}}_T):(\boldsymbol{z}_T\otimes\boldsymbol{w}) \\
&= \sum_{T\in\mathcal{T}_h}\int_T (\nabla\boldsymbol{w} - \mathbf{G}_T^{2k}\underline{\hat{\boldsymbol{w}}}_T):\big(\boldsymbol{z}_T\otimes(\boldsymbol{w}-\boldsymbol{\pi}_T^{0,0}\boldsymbol{w})\big).
\end{aligned}$$

Hence, using generalised Hölder inequalities with exponents $(2, 4, 4)$, we obtain

$$\begin{aligned}
|\mathfrak{T}_1| &\le \sum_{T\in\mathcal{T}_h}\|\nabla\boldsymbol{w} - \mathbf{G}_T^{2k}\underline{\hat{\boldsymbol{w}}}_T\|_{L^2(T)^{d\times d}}\|\boldsymbol{z}_T\|_{L^4(T)^d}\|\boldsymbol{w}-\boldsymbol{\pi}_T^{0,0}\boldsymbol{w}\|_{L^4(T)^d} \\
&\lesssim \sum_{T\in\mathcal{T}_h} h_T^{r+1}|\boldsymbol{w}|_{H^{r+1}(T)^d}\|\boldsymbol{z}_T\|_{L^4(T)^d}|\boldsymbol{w}|_{W^{1,4}(T)^d} \\
&\lesssim h^{r+1}|\boldsymbol{w}|_{H^{r+1}(\mathcal{T}_h)^d}\|\boldsymbol{z}_h\|_{L^4(\Omega)^d}|\boldsymbol{w}|_{W^{1,4}(\Omega)^d},
\end{aligned}$$

where we have used the consistency property (9.38a) of the discrete gradient reconstruction with $l = 2k$ together with the approximation properties (1.74) of the L^2-orthogonal projector $\boldsymbol{\pi}_T^{0,0}$ (with $X = T$, $p = 4$, $s = 1$ and $m = 0$) to pass to the second line, and a generalised Hölder inequality with exponents (2, 4, 4) on the sum over $T \in \mathcal{T}_h$ to pass to the third line.

For the second term, we can write

$$
\begin{aligned}
|\mathfrak{T}_2| &\leq \sum_{T\in\mathcal{T}_h} \|\boldsymbol{w} - \hat{\boldsymbol{w}}_T\|_{L^4(T)^d} \|\mathbf{G}_T^{2k}\underline{\hat{\boldsymbol{w}}}_T\|_{L^2(T)^{d\times d}} \|\boldsymbol{z}_T\|_{L^4(T)^d} \\
&\lesssim \sum_{T\in\mathcal{T}_h} h_T^{r+1} |\boldsymbol{w}|_{W^{r+1,4}(T)^d} \|\underline{\hat{\boldsymbol{w}}}_T\|_{1,T} \|\boldsymbol{z}_T\|_{L^4(T)^d} \\
&\lesssim h^{r+1} |\boldsymbol{w}|_{W^{r+1,4}(\mathcal{T}_h)^d} |\boldsymbol{w}|_{H^1(\Omega)^d} \|\boldsymbol{z}_h\|_{L^4(\Omega)^d},
\end{aligned}
$$

where we have used generalised Hölder inequalities with exponents (4, 2, 4) in the first line, the approximation properties (1.74) of the L^2-orthogonal projector with $X = T$, $l = k$, $p = 4$, $s = r + 1$, and $m = 0$ together with the boundedness (9.37) of the local gradient reconstruction to pass to the second line, and a generalised Hölder inequality with exponents (4, 2, 4) on the sum over $T \in \mathcal{T}_h$ together with the boundedness property (8.25) of the global interpolator to conclude.

Estimate (9.44) follows by gathering the above bounds on $\mathfrak{T}_1$ and $\mathfrak{T}_2$ in (9.45), and by noticing that $|\boldsymbol{w}|_{H^1(\Omega)^d} \lesssim |\boldsymbol{w}|_{W^{1,4}(\Omega)^d}$ and $|\boldsymbol{w}|_{H^{r+1}(\mathcal{T}_h)^d} \lesssim |\boldsymbol{w}|_{W^{r+1,4}(\mathcal{T}_h)^d}$. □

9.5.2.2 Discrete Trilinear Form

We define $\mathrm{t}_h^{\mathrm{ss}} : \underline{\boldsymbol{U}}_h^k \times \underline{\boldsymbol{U}}_h^k \times \underline{\boldsymbol{U}}_h^k \to \mathbb{R}$ such that, for all $\underline{\boldsymbol{w}}_h, \underline{\boldsymbol{v}}_h, \underline{\boldsymbol{z}}_h \in \underline{\boldsymbol{U}}_h^k$,

$$
\mathrm{t}_h^{\mathrm{ss}}(\underline{\boldsymbol{w}}_h, \underline{\boldsymbol{v}}_h, \underline{\boldsymbol{z}}_h) := \frac{1}{2} \sum_{T\in\mathcal{T}_h} \left[\int_T (\boldsymbol{w}_T \cdot \mathbf{G}_T^{2k}) \underline{\boldsymbol{v}}_T \cdot \boldsymbol{z}_T - \int_T (\boldsymbol{w}_T \cdot \mathbf{G}_T^{2k}) \underline{\boldsymbol{z}}_T \cdot \boldsymbol{v}_T \right]. \tag{9.47}
$$

Remark 9.16 (Implementation of the Trilinear Form (9.47)*)* In the practical implementation, one does not need to actually compute $\mathbf{G}_T^{2k}$ to evaluate $\mathrm{t}_h^{\mathrm{ss}}$. Instead, the following expression can be used, obtained applying (9.33) twice to expand the terms involving $\mathbf{G}_T^{2k}$:

$$
\begin{aligned}
\mathrm{t}_h^{\mathrm{ss}}(\underline{\boldsymbol{w}}_h, \underline{\boldsymbol{v}}_h, \underline{\boldsymbol{z}}_h) &= \frac{1}{2} \sum_{T\in\mathcal{T}_h} \int_T [(\boldsymbol{w}_T \cdot \nabla)\boldsymbol{v}_T \cdot \boldsymbol{z}_T - \boldsymbol{v}_T \cdot (\boldsymbol{w}_T \cdot \nabla)\boldsymbol{z}_T] \\
&\quad + \frac{1}{2} \sum_{T\in\mathcal{T}_h} \sum_{F\in\mathcal{F}_T} \int_F (\boldsymbol{w}_T \cdot \boldsymbol{n}_{TF})(\boldsymbol{v}_F \cdot \boldsymbol{z}_T - \boldsymbol{v}_T \cdot \boldsymbol{z}_F).
\end{aligned} \tag{9.48}
$$

Proposition 9.17 (Properties of the Convective Trilinear Form (9.47)**)** *The convective trilinear form* $\mathrm{t}_h^{\mathrm{ss}}$ *defined by* (9.47) *satisfies Assumption 9.2.*

Proof (T1) *Non-dissipativity.* This property is straightforward from the inherently skew-symmetric definition of $\mathrm{t}_h^{\mathrm{ss}}$.

(T2) *Boundedness.* Apply (9.43) twice, swapping $\underline{\boldsymbol{v}}_h$ and $\underline{\boldsymbol{z}}_h$ the second time.

(T3) *Consistency.* For the sake of brevity, set $\underline{\hat{\boldsymbol{w}}}_h := \underline{\boldsymbol{I}}_h^k \boldsymbol{w}$. Integrating by parts element by element, recalling that $\nabla{\cdot}\boldsymbol{w} = 0$, and using the single-valuedness of $(\boldsymbol{w}{\cdot}\boldsymbol{n}_F)\boldsymbol{w}$ at interfaces together with the fact that $\boldsymbol{z}_F = \mathbf{0}$ on boundary faces to insert $\boldsymbol{z}_F$ into the third term, we have

$$
\begin{aligned}
\int_\Omega (\boldsymbol{w}{\cdot}\nabla)\boldsymbol{w}{\cdot}\boldsymbol{z}_h &= \frac{1}{2}\sum_{T\in\mathcal{T}_h}\left(\int_T (\boldsymbol{w}{\cdot}\nabla)\boldsymbol{w}{\cdot}\boldsymbol{z}_T - \int_T (\boldsymbol{w}{\cdot}\nabla)\boldsymbol{z}_T{\cdot}\boldsymbol{w}\right)\\
&\quad - \frac{1}{2}\sum_{T\in\mathcal{T}_h}\sum_{F\in\mathcal{F}_T}\int_F (\boldsymbol{w}{\cdot}\boldsymbol{n}_{TF})(\boldsymbol{z}_F - \boldsymbol{z}_T){\cdot}\boldsymbol{w}.
\end{aligned}
\tag{9.49}
$$

On the other hand, starting from the definition (9.47) of $\mathrm{t}_h^{\mathrm{ss}}$ and using inside each element $T \in \mathcal{T}_h$ the characterisation (9.34) of $\mathbf{G}_T^{2k}$ with $\underline{\boldsymbol{v}}_T = \underline{\boldsymbol{z}}_T$ and $\boldsymbol{\tau} = \hat{\boldsymbol{w}}_T \otimes \hat{\boldsymbol{w}}_T$, we have

$$
\begin{aligned}
\mathrm{t}_h^{\mathrm{ss}}(\underline{\hat{\boldsymbol{w}}}_h, \underline{\hat{\boldsymbol{w}}}_h, \underline{\boldsymbol{z}}_h) &= \frac{1}{2}\sum_{T\in\mathcal{T}_h}\left(\int_T (\hat{\boldsymbol{w}}_T{\cdot}\mathbf{G}_T^{2k})\underline{\hat{\boldsymbol{w}}}_T{\cdot}\boldsymbol{z}_T - \int_T (\hat{\boldsymbol{w}}_T{\cdot}\nabla)\boldsymbol{z}_T{\cdot}\hat{\boldsymbol{w}}_T\right)\\
&\quad - \frac{1}{2}\sum_{T\in\mathcal{T}_h}\sum_{F\in\mathcal{F}_T}\int_F (\hat{\boldsymbol{w}}_T{\cdot}\boldsymbol{n}_{TF})(\boldsymbol{z}_F - \boldsymbol{z}_T){\cdot}\hat{\boldsymbol{w}}_T.
\end{aligned}
\tag{9.50}
$$

Subtracting (9.50) from (9.49) and inserting into the right-hand side of the resulting expression the quantity

$$
\pm\frac{1}{2}\sum_{T\in\mathcal{T}_h}\left(\int_T (\hat{\boldsymbol{w}}_T{\cdot}\nabla)\boldsymbol{z}_T{\cdot}\boldsymbol{w} + \sum_{F\in\mathcal{F}_T}\int_F (\boldsymbol{w}{\cdot}\boldsymbol{n}_{TF})(\boldsymbol{z}_F - \boldsymbol{z}_T){\cdot}\hat{\boldsymbol{w}}_T\right),
$$

we arrive at

$$\int_\Omega (\boldsymbol{w}\cdot\nabla)\boldsymbol{w}\cdot\boldsymbol{z}_h - \mathrm{t}_h^{\mathrm{ss}}(\underline{\hat{\boldsymbol{w}}}_h, \underline{\hat{\boldsymbol{w}}}_h, \underline{\boldsymbol{z}}_h)$$

$$= \underbrace{\frac{1}{2}\left[\sum_{T\in\mathcal{T}_h}\int_T (\boldsymbol{w}\cdot\nabla)\boldsymbol{w}\cdot\boldsymbol{z}_T - \int_T (\hat{\boldsymbol{w}}_T\cdot\mathbf{G}_T^{2k})\underline{\hat{\boldsymbol{w}}}_T\cdot\boldsymbol{z}_T\right]}_{\mathfrak{T}_1}$$

$$+ \underbrace{\frac{1}{2}\sum_{T\in\mathcal{T}_h}\int_T (\hat{\boldsymbol{w}}_T\cdot\nabla)\boldsymbol{z}_T\cdot(\hat{\boldsymbol{w}}_T - \boldsymbol{w})}_{\mathfrak{T}_2} + \underbrace{\frac{1}{2}\sum_{T\in\mathcal{T}_h}\int_T \big((\hat{\boldsymbol{w}}_T - \boldsymbol{w})\cdot\nabla\big)\boldsymbol{z}_T\cdot\boldsymbol{w}}_{\mathfrak{T}_3}$$

$$+ \underbrace{\frac{1}{2}\sum_{T\in\mathcal{T}_h}\sum_{F\in\mathcal{F}_T}\int_F ((\hat{\boldsymbol{w}}_T - \boldsymbol{w})\cdot\boldsymbol{n}_{TF})(\boldsymbol{z}_F - \boldsymbol{z}_T)\cdot\hat{\boldsymbol{w}}_T}_{\mathfrak{T}_4}$$

$$+ \underbrace{\frac{1}{2}\sum_{T\in\mathcal{T}_h}\sum_{F\in\mathcal{F}_T}\int_F (\boldsymbol{w}\cdot\boldsymbol{n}_{TF})(\boldsymbol{z}_F - \boldsymbol{z}_T)\cdot(\hat{\boldsymbol{w}}_T - \boldsymbol{w})}_{\mathfrak{T}_5}. \tag{9.51}$$

The first term is estimated using (9.44) together with the discrete Sobolev embedding (9.42) to bound $\|\boldsymbol{z}_h\|_{L^4(\Omega)}$:

$$|\mathfrak{T}_1| \lesssim h^{r+1}|\boldsymbol{w}|_{W^{r+1,4}(\mathcal{T}_h)^d}|\boldsymbol{w}|_{W^{1,4}(\Omega)^d}\|\underline{\boldsymbol{z}}_h\|_{1,h}. \tag{9.52}$$

Proceeding similarly as for the estimate of $\mathfrak{T}_2$ in the proof of Lemma 9.15, after applying Hölder inequalities with exponents $(4, 2, 4)$ and invoking the approximation properties of the L^2-orthogonal projector, we have for the second and third terms

$$\begin{aligned}
|\mathfrak{T}_2| + |\mathfrak{T}_3| &\lesssim h^{r+1}\left(\|\hat{\boldsymbol{w}}_h\|_{L^4(\Omega)^d} + \|\boldsymbol{w}\|_{L^4(\Omega)^d}\right)|\boldsymbol{w}|_{W^{r+1,4}(\mathcal{T}_h)^d}\|\nabla_h\boldsymbol{z}_h\|_{L^2(\Omega)^{d\times d}} \\
&\lesssim h^{r+1}\left(\|\underline{\hat{\boldsymbol{w}}}_h\|_{1,h} + \|\boldsymbol{w}\|_{H^1(\Omega)^d}\right)|\boldsymbol{w}|_{W^{r+1,4}(\mathcal{T}_h)^d}\|\underline{\boldsymbol{z}}_h\|_{1,h} \\
&\lesssim h^{r+1}\|\boldsymbol{w}\|_{H^1(\Omega)^d}|\boldsymbol{w}|_{W^{r+1,4}(\mathcal{T}_h)^d}\|\underline{\boldsymbol{z}}_h\|_{1,h},
\end{aligned} \tag{9.53}$$

where we have used discrete (9.42) and continuous Sobolev embeddings to bound the terms in parentheses together with the definitions (8.24) and (8.15) of the $\|\cdot\|_{1,h}$- and $\|\cdot\|_{1,T}$-seminorms to bound the third factor when passing to the second line, and the boundedness (8.25) of $\underline{\boldsymbol{I}}_h^k$ to further write $\|\underline{\hat{\boldsymbol{w}}}_h\|_{1,h} \lesssim |\boldsymbol{w}|_{H^1(\Omega)^d}$ and conclude.

Finally, for the fourth and fifth terms in (9.51), using generalised Hölder inequalities with exponents $(4, \infty, 2, 4)$ together with $\|\boldsymbol{n}_{TF}\|_{L^\infty(F)^d} = 1$ and the trace approximation properties (1.75) of the L^2-orthogonal projector (with $l = k$,

$p = 4$, $s = r + 1$, and $m = 0$), we obtain

$$
\begin{aligned}
&|\mathfrak{T}_4| + |\mathfrak{T}_5| \\
&\quad \lesssim \sum_{T\in\mathcal{T}_h} \sum_{F\in\mathcal{F}_T} h^{r+1-\frac{1}{4}} |\boldsymbol{w}|_{W^{r+1,4}(T)^d} \left(\|\hat{\boldsymbol{w}}_T\|_{L^4(F)^d} + \|\boldsymbol{w}\|_{L^4(F)^d} \right) \|\boldsymbol{z}_F - \boldsymbol{z}_T\|_{L^2(F)^d} \\
&\quad \lesssim h^{r+1} |\boldsymbol{w}|_{W^{r+1,4}(\mathcal{T}_h)^d} \left[\sum_{T\in\mathcal{T}_h} h_T \left(\|\hat{\boldsymbol{w}}_T\|_{L^4(\partial T)^d}^4 + \|\boldsymbol{w}\|_{L^4(\partial T)^d}^4 \right) \right]^{\frac{1}{4}} \left(\sum_{T\in\mathcal{T}_h} |\underline{\boldsymbol{z}}_T|_{1,\partial T}^2 \right)^{\frac{1}{2}} \\
&\quad \lesssim h^{r+1} |\boldsymbol{w}|_{W^{r+1,4}(\mathcal{T}_h)^d} \|\boldsymbol{w}\|_{W^{1,4}(\Omega)^d} \|\underline{\boldsymbol{z}}_h\|_{1,h},
\end{aligned}
\tag{9.54}
$$

where, to pass to the second line, we have used a generalised Hölder inequality on the sums with exponents $(4, 4, 2)$ and distributed $h_T^{-\frac{1}{4}}$ as $h_T^{\frac{1}{4}}$ on $\left(\|\hat{\boldsymbol{w}}_T\|_{L^4(F)^d} + \|\boldsymbol{w}\|_{L^4(F)^d} \right)$ and $h_T^{-\frac{1}{2}}$ on $\|\boldsymbol{z}_F - \boldsymbol{z}_T\|_{L^2(F)^d}$; to pass to the last line, we have used the continuous (1.51) and discrete (1.55) trace inequalities with $p = 4$ and the L^4-boundedness of $\boldsymbol{\pi}_T^{0,k}$ (see (1.77)) for the second factor, whilst the definitions (8.24) and (8.15) of the $\|\cdot\|_{1,h}$- and $\|\cdot\|_{1,T}$-seminorms have been used to bound the third factor. Taking absolute values in (9.51), and using (9.52)–(9.54) to bound the right-hand side, (9.12) follows after observing that $\|\boldsymbol{w}\|_{H^1(\Omega)^d} \lesssim \|\boldsymbol{w}\|_{W^{1,4}(\Omega)^d}$. □

9.5.3 *A Trilinear Form Incorporating Temam's Device for Stability*

The second trilinear form discussed here, originally introduced in [70], is inspired by Temam's formulation (9.7), where the operator $(\boldsymbol{w}\cdot\nabla)$ and the divergence are replaced, respectively, by a discrete reconstruction of the directional derivative and by the trace of the gradient $\mathbf{G}_T^{2k}$ defined by (9.33).

9.5.3.1 Discrete Directional Derivative

Given $\underline{\boldsymbol{w}}_T \in \underline{\boldsymbol{U}}_T^k$, the directional derivative reconstruction $\boldsymbol{G}_T^k(\underline{\boldsymbol{w}}_T;\cdot) : \underline{\boldsymbol{U}}_T^k \to \mathbb{P}^k(T)^d$ is such that, for all $\underline{\boldsymbol{v}}_T \in \underline{\boldsymbol{U}}_T^k$,

$$
\begin{aligned}
&\int_T \boldsymbol{G}_T^k(\underline{\boldsymbol{w}}_T;\underline{\boldsymbol{v}}_T)\cdot\boldsymbol{z} \\
&\qquad = \int_T (\boldsymbol{w}_T\cdot\nabla)\boldsymbol{v}_T\cdot\boldsymbol{z} + \sum_{F\in\mathcal{F}_T} \int_F (\boldsymbol{w}_F\cdot\boldsymbol{n}_{TF})(\boldsymbol{v}_F - \boldsymbol{v}_T)\cdot\boldsymbol{z} \qquad \forall \boldsymbol{z} \in \mathbb{P}^k(T)^d.
\end{aligned}
\tag{9.55}
$$

This expression mimics (3.65) with the role of the advective velocity inside the element and on its faces played by $\boldsymbol{w}_T$ and $(\boldsymbol{w}_F)_{F\in\mathcal{F}_T}$, respectively. For all $\boldsymbol{z} \in \mathbb{P}^k(T)^d$, writing (9.34) for $l = 2k$ and $\boldsymbol{\tau} = \boldsymbol{z} \otimes \boldsymbol{w}_T$, recalling the notation (9.41) for $(\boldsymbol{w}_T{\cdot}\mathbf{G}_T^{2k})\underline{\boldsymbol{v}}_T$ and comparing with (9.55), one can see that it holds

$$\begin{aligned}\int_T \boldsymbol{\mathcal{G}}_T^k(\underline{\boldsymbol{w}}_T;\underline{\boldsymbol{v}}_T){\cdot}\boldsymbol{z} &= \int_T (\boldsymbol{w}_T{\cdot}\mathbf{G}_T^{2k})\underline{\boldsymbol{v}}_T{\cdot}\boldsymbol{z}\\ &+ \sum_{F\in\mathcal{F}_T}\int_F (\boldsymbol{w}_F-\boldsymbol{w}_T){\cdot}\boldsymbol{n}_{TF}(\boldsymbol{v}_F-\boldsymbol{v}_T){\cdot}\boldsymbol{z} \qquad \forall \boldsymbol{z}\in\mathbb{P}^k(T)^d.\end{aligned} \tag{9.56}$$

This shows that $\boldsymbol{\mathcal{G}}_T^k(\underline{\boldsymbol{w}}_T;\underline{\boldsymbol{v}}_T)$ differs from $(\boldsymbol{w}_T{\cdot}\mathbf{G}_T^{2k})\underline{\boldsymbol{v}}_T$ in that $\boldsymbol{w}_F$ replaces $\boldsymbol{w}_T$ in the boundary term. The properties of the discrete directional derivative relevant to the analysis are summarised in the following proposition.

Proposition 9.18 (Properties of the Discrete Directional Derivative) *The discrete directional derivative defined by* (9.55) *satisfies the following properties:*

(i) Boundedness. *For all* $\underline{\boldsymbol{w}}_h, \underline{\boldsymbol{v}}_h, \underline{\boldsymbol{z}}_h \in \underline{\boldsymbol{U}}_{h,0}^k$, *it holds*

$$\left|\sum_{T\in\mathcal{T}_h}\int_T \boldsymbol{\mathcal{G}}_T^k(\underline{\boldsymbol{w}}_T;\underline{\boldsymbol{v}}_T){\cdot}\boldsymbol{z}_T\right| \lesssim \|\underline{\boldsymbol{w}}_h\|_{1,h}\|\underline{\boldsymbol{v}}_h\|_{1,h}\|\underline{\boldsymbol{z}}_h\|_{1,h}, \tag{9.57}$$

with hidden constant independent of h, $\underline{\boldsymbol{w}}_h$, $\underline{\boldsymbol{v}}_h$, *and* $\underline{\boldsymbol{z}}_h$.

(ii) Consistency. *If* $r \in \{0,\ldots,k\}$ *and* $\boldsymbol{w} \in H_0^1(\Omega)^d \cap W^{r+1,4}(\mathcal{T}_h)^d$, *then, setting* $\underline{\hat{\boldsymbol{w}}}_h := \underline{\boldsymbol{I}}_h^k \boldsymbol{w}$, *it holds*

$$\begin{aligned}&\left|\sum_{T\in\mathcal{T}_h}\int_T \left[(\boldsymbol{w}{\cdot}\boldsymbol{\nabla})\boldsymbol{w} - \boldsymbol{\mathcal{G}}_T^k(\underline{\hat{\boldsymbol{w}}}_T;\underline{\hat{\boldsymbol{w}}}_T)\right]{\cdot}\boldsymbol{z}_T\right|\\ &\qquad\lesssim h^{r+1}|\boldsymbol{w}|_{W^{r+1,4}(\mathcal{T}_h)^d}|\boldsymbol{w}|_{W^{1,4}(\Omega)^d}\|\boldsymbol{z}_h\|_{L^4(\Omega)^d} \qquad \forall \boldsymbol{z}_h\in\mathbb{P}^k(\mathcal{T}_h)^d,\end{aligned} \tag{9.58}$$

where $\boldsymbol{z}_T = (\boldsymbol{z}_h)_{|T}$ *for all* $T \in \mathcal{T}_h$, *and the hidden constant is independent of* h, $\boldsymbol{w}$ *and* $\boldsymbol{z}_h$.

Proof (i) *Boundedness.* Using (9.56), we can write

$$\begin{aligned}
&\left|\sum_{T\in\mathcal{T}_h}\int_T \boldsymbol{G}_T^k(\underline{\boldsymbol{w}}_T;\underline{\boldsymbol{v}}_T)\cdot\boldsymbol{z}_T\right| \\
&\quad\le \left|\sum_{T\in\mathcal{T}_h}\int_T(\boldsymbol{w}_T\cdot\mathbf{G}_T^{2k})\underline{\boldsymbol{v}}_T\cdot\boldsymbol{z}_T\right| + \underbrace{\left|\sum_{T\in\mathcal{T}_h}\sum_{F\in\mathcal{F}_T}\int_F(\boldsymbol{w}_F-\boldsymbol{w}_T)\cdot\boldsymbol{n}_{TF}(\boldsymbol{v}_F-\boldsymbol{v}_T)\cdot\boldsymbol{z}_T\right|}_{\mathfrak{T}_1}. \\
&\quad\lesssim \|\underline{\boldsymbol{w}}_h\|_{1,h}\|\underline{\boldsymbol{v}}_h\|_{1,h}\|\underline{\boldsymbol{z}}_h\|_{1,h}+\mathfrak{T}_1, \qquad (9.59)
\end{aligned}$$

where we have used (9.43) to pass to the second line. To estimate $\mathfrak{T}_1$, we write

$$\begin{aligned}
\mathfrak{T}_1 &\le \sum_{T\in\mathcal{T}_h}\sum_{F\in\mathcal{F}_T}\|\boldsymbol{w}_F-\boldsymbol{w}_T\|_{L^4(F)^d}\|\boldsymbol{v}_F-\boldsymbol{v}_T\|_{L^2(F)^d}\|\boldsymbol{z}_T\|_{L^4(F)^d} \\
&\lesssim \|\boldsymbol{z}_h\|_{L^4(\Omega)^d}\sum_{T\in\mathcal{T}_h}\sum_{F\in\mathcal{F}_T} h_F^{-\frac14}\|\boldsymbol{w}_F-\boldsymbol{w}_T\|_{L^4(F)^d}\|\boldsymbol{v}_F-\boldsymbol{v}_T\|_{L^2(F)^d} \\
&\lesssim \|\boldsymbol{z}_h\|_{L^4(\Omega)^d}\sum_{T\in\mathcal{T}_h}\sum_{F\in\mathcal{F}_T} h_F^{-\frac14}|F|_{d-1}^{-\frac14}\|\boldsymbol{w}_F-\boldsymbol{w}_T\|_{L^2(F)^d}\|\boldsymbol{v}_F-\boldsymbol{v}_T\|_{L^2(F)^d} \\
&\lesssim \|\boldsymbol{z}_h\|_{L^4(\Omega)^d}\sum_{T\in\mathcal{T}_h}\sum_{F\in\mathcal{F}_T} h_F^{-\frac12}\|\boldsymbol{w}_F-\boldsymbol{w}_T\|_{L^2(F)^d}\, h_F^{-\frac12}\|\boldsymbol{v}_F-\boldsymbol{v}_T\|_{L^2(F)^d} \\
&\lesssim \|\underline{\boldsymbol{z}}_h\|_{1,h}\|\underline{\boldsymbol{w}}_h\|_{1,h}\|\underline{\boldsymbol{v}}_h\|_{1,h}, \qquad (9.60)
\end{aligned}$$

where we have used a generalised Hölder inequality with exponents $(4,\infty,2,4)$ together with $\|\boldsymbol{n}_{TF}\|_{L^\infty(F)^d}=1$ in the first line, the discrete trace inequality (1.55) with $p=4$ followed by $\|\boldsymbol{z}_T\|_{L^4(T)^d}\le\|\boldsymbol{z}_h\|_{L^4(\Omega)^d}$ for all $T\in\mathcal{T}_h$ in the second line, the inverse Lebesgue embedding (1.35) with $X=F$ (this choice is possible in view of Remark 1.27), $q=4$, and $m=2$ in the third line, the bound $h_F^{-\frac14}|F|_{d-1}^{-\frac14}\lesssim h_F^{-\frac14-\frac{d-1}{4}}\lesssim h_F^{-1}=h_F^{-\frac12}h_F^{-\frac12}$ (valid since $d\le 3$) in the fourth line, and the discrete Sobolev embedding (9.42) together with a discrete Cauchy–Schwarz inequality on the sums over $T\in\mathcal{T}_h$ and $F\in\mathcal{F}_T$, and the definitions (8.24) and (8.15) of $\|{\cdot}\|_{1,h}$ and $\|{\cdot}\|_{1,T}$ to conclude. Plugging the bound (9.60) into (9.59) yields the estimate (9.57).

(ii) *Consistency.* Writing (9.56) for $\underline{w}_T = \underline{v}_T = \underline{\hat{w}}_T$, we get

$$\sum_{T\in\mathcal{T}_h}\int_T \left[(w\cdot\nabla)w - \mathcal{G}_T^k(\underline{\hat{w}}_T;\underline{\hat{w}}_T)\right]\cdot z_T = \underbrace{\sum_{T\in\mathcal{T}_h}\int_T \left[(w\cdot\nabla)w - (\hat{w}_T\cdot \mathbf{G}_T^{2k})\underline{\hat{w}}_T\right]\cdot z_T}_{\mathfrak{T}_1}$$
$$- \underbrace{\sum_{T\in\mathcal{T}_h}\sum_{F\in\mathcal{F}_T}\int_F (\hat{w}_F - \hat{w}_T)\cdot n_{TF}(\hat{w}_F - \hat{w}_T)\cdot z_T}_{\mathfrak{T}_2}. \tag{9.61}$$

The first term is estimated using (9.44):

$$|\mathfrak{T}_1| \lesssim h^{r+1}|w|_{W^{r+1,4}(\mathcal{T}_h)^d}|w|_{W^{1,4}(\Omega)^d}\|z_h\|_{L^4(\Omega)^d}. \tag{9.62}$$

For the second term, we first observe that, owing to the linearity, idempotency, and boundedness of $\pi_F^{0,k}$ (see Lemma 1.44), it holds, for $\alpha \in \{2, 4\}$,

$$\|\hat{w}_F - \hat{w}_T\|_{L^\alpha(F)^d} = \|\pi_F^{0,k}(w - \pi_T^{0,k}w)\|_{L^\alpha(F)^d} \lesssim \|w - \pi_T^{0,k}w\|_{L^\alpha(F)^d}. \tag{9.63}$$

Hence, generalised Hölder inequalities with exponents $(2, \infty, 4, 4)$ along with $\|n_{TF}\|_{L^\infty(F)^d} = 1$ give

$$\begin{aligned}|\mathfrak{T}_2| &\lesssim \sum_{T\in\mathcal{T}_h}\sum_{F\in\mathcal{F}_T}\|w - \pi_T^{0,k}w\|_{L^2(F)^d}\|w - \pi_T^{0,k}w\|_{L^4(F)^d}\|z_T\|_{L^4(F)^d}\\ &\lesssim h^{r+1}|w|_{H^{r+1}(\mathcal{T}_h)^d}|w|_{W^{1,4}(\Omega)^d}\|z_h\|_{L^4(\Omega)^d},\end{aligned} \tag{9.64}$$

where we have used the trace approximation properties (1.75) of the L^2-orthogonal projector with $l = k$ and, respectively, $(p, s, m) = (2, r + 1, 0)$ and $(p, s, m) = (4, 1, 0)$ to bound the first two factors inside the summation, the discrete trace inequality (1.55) with $p = 4$ to bound the third one, and another generalised Hölder inequality with exponents $(2, 4, 4)$ on the sums to conclude.

Combining (9.61), (9.62) and (9.64) yields (9.58). □

9.5.3.2 Discrete Divergence and Integration by Parts Formula

Given a mesh element $T \in \mathcal{T}_h$ and a polynomial degree $l \geq 0$, we next define the generalised discrete divergence such that, for all $\underline{v}_T \in \underline{U}_T^k$,

$$\mathrm{D}_T^l\underline{v}_T := \mathrm{tr}(\mathbf{G}_T^l\underline{v}_T) = \sum_{i=1}^d(\mathbf{G}_T^l\underline{v}_T)_{ii} = \mathbf{G}_T^l\underline{v}_T : I_d, \tag{9.65}$$

where I_d denotes the identity matrix of $\mathbb{R}^{d\times d}$. For future use, we record the following characterisation of D_T^l, obtained from (9.34) with $\tau = qI_d$: For all $\underline{v}_T \in \underline{U}_T^k$,

$$\int_T \mathrm{D}_T^l \underline{v}_T\, q = \int_T (\nabla\cdot v_T)\, q + \sum_{F\in\mathcal{F}_T} \int_F (v_F - v_T)\cdot n_{TF}\, q \qquad \forall q \in \mathbb{P}^l(T). \tag{9.66}$$

This formula shows that, for $l = k$, we indeed recover the discrete divergence used in Chap. 8 (see (8.20)). With this definition, we can prove a discrete integration by parts formula which plays the role of (9.5) at the discrete level.

Proposition 9.19 (Discrete Integration by Parts Formula) *For all $\underline{w}_h, \underline{v}_h, \underline{z}_h \in \underline{U}_h^k$ it holds*

$$\begin{aligned}\sum_{T\in\mathcal{T}_h} \int_T \Big(\mathcal{G}_T^k(\underline{w}_T;\underline{v}_T)\cdot z_T + v_T\cdot\mathcal{G}_T^k(\underline{w}_T;\underline{z}_T) + \mathrm{D}_T^{2k}\underline{w}_T(v_T\cdot z_T)\Big) \\ = -\sum_{T\in\mathcal{T}_h}\sum_{F\in\mathcal{F}_T}\int_F (w_F\cdot n_{TF})(v_F - v_T)\cdot(z_F - z_T) \\ + \sum_{F\in\mathcal{F}_h^{\mathrm{b}}}\int_F (w_F\cdot n_F)(v_F\cdot z_F).\end{aligned} \tag{9.67}$$

Remark 9.20 (Comparison with (9.5)*)* Compared with its continuous counterpart (9.5), formula (9.67) contains one additional term in the right-hand side where the differences between face and element unknowns in $\underline{v}_h$ and $\underline{z}_h$ appear. This term reflects the non-conformity of the HHO space.

Proof (Proposition 9.19) Let an element $T \in \mathcal{T}_h$ be fixed. Expanding first $\mathcal{G}_T^k(\underline{w}_T;\underline{v}_T)$ according to its definition (9.55) with $z = z_T$, then integrating by parts the volumetric term, we obtain

$$\begin{aligned}&\int_T \mathcal{G}_T^k(\underline{w}_T;\underline{v}_T)\cdot z_T \\ &\quad = \int_T (w_T\cdot\nabla)v_T\cdot z_T + \sum_{F\in\mathcal{F}_T}\int_F (w_F\cdot n_{TF})(v_F - v_T)\cdot z_T \\ &\quad = -\int_T v_T\cdot(w_T\cdot\nabla)z_T - \int_T (\nabla\cdot w_T)(v_T\cdot z_T) \\ &\qquad + \sum_{F\in\mathcal{F}_T}\int_F [(w_F\cdot n_{TF})(v_F\cdot z_T) - (w_F\cdot n_{TF})(v_T\cdot z_T) + (w_T\cdot n_{TF})(v_T\cdot z_T)] \\ &\quad =: \mathfrak{T}_1 + \mathfrak{T}_2 + \mathfrak{T}_3.\end{aligned} \tag{9.68}$$

Using again (9.55), this time with $\underline{v}_T = \underline{z}_T$ and $z = v_T$, we obtain for the first term

$$\mathfrak{T}_1 = -\int_T v_T \cdot \boldsymbol{G}_T^k(\underline{w}_T; \underline{z}_T) + \sum_{F \in \mathcal{F}_T} \int_F (w_F \cdot n_{TF})\, v_T \cdot (z_F - z_T). \tag{9.69}$$

Invoking the characterisation (9.66) of the discrete divergence reconstruction with $l = 2k$ and $q = v_T \cdot z_T$, we get for the second term

$$\mathfrak{T}_2 = -\int_T \mathrm{D}_T^{2k} \underline{w}_T (v_T \cdot z_T) + \sum_{F \in \mathcal{F}_T} \int_F (w_F - w_T) \cdot n_{TF} (v_T \cdot z_T). \tag{9.70}$$

Plugging (9.69) and (9.70) into (9.68) and rearranging, we obtain

$$\int_T \left(\boldsymbol{G}_T^k(\underline{w}_T; \underline{v}_T) \cdot z_T + v_T \cdot \boldsymbol{G}_T^k(\underline{w}_T; \underline{z}_T) + \mathrm{D}_T^{2k} \underline{w}_T (v_T \cdot z_T) \right) \\ = \sum_{F \in \mathcal{F}_T} \int_F (w_F \cdot n_{TF})(v_T \cdot z_F - v_T \cdot z_T + v_F \cdot z_T).$$

Summing the above equality over $T \in \mathcal{T}_h$ and adding the quantity

$$-\sum_{T \in \mathcal{T}_h} \sum_{F \in \mathcal{F}_T} \int_F (w_F \cdot n_{TF})(v_F \cdot z_F) + \sum_{F \in \mathcal{F}_h^{\mathrm{b}}} \int_F (w_F \cdot n_F)(v_F \cdot z_F) = 0, \tag{9.71}$$

the conclusion follows after observing that $v_T \cdot z_F - v_T \cdot z_T + v_F \cdot z_T - v_F \cdot z_F = -(v_F - v_T) \cdot (z_F - z_T)$. Formula (9.71) is justified observing that, for any internal face $F \in \mathcal{F}_h^{\mathrm{i}}$ such that $F \in \mathcal{F}_{T_1} \cap \mathcal{F}_{T_2}$ for distinct mesh elements $T_1, T_2 \in \mathcal{T}_h$, it holds that $(w_F \cdot n_{T_1 F})(v_F \cdot z_F) + (w_F \cdot n_{T_2 F})(v_F \cdot z_F) = 0$ owing to the single-valuedness of w_F, v_F and z_F. □

9.5.3.3 Discrete Trilinear Form

We can now define the discrete convective trilinear form $\mathrm{t}_h^{\mathrm{tm}} : \underline{U}_h^k \times \underline{U}_h^k \times \underline{U}_h^k \to \mathbb{R}$ inspired by (9.7): For all $\underline{w}_h, \underline{v}_h, \underline{z}_h \in \underline{U}_h^k$,

$$\mathrm{t}_h^{\mathrm{tm}}(\underline{w}_h, \underline{v}_h, \underline{z}_h) := \sum_{T \in \mathcal{T}_h} \int_T \boldsymbol{G}_T^k(\underline{w}_T; \underline{v}_T) \cdot z_T \\ + \frac{1}{2} \sum_{T \in \mathcal{T}_h} \left(\int_T \mathrm{D}_T^{2k} \underline{w}_T (v_T \cdot z_T) + \sum_{F \in \mathcal{F}_T} \int_F (w_F \cdot n_{TF})(v_F - v_T) \cdot (z_F - z_T) \right), \tag{9.72}$$

where the terms in the second line embody Temam's device for stability.

Remark 9.21 (Discrete Incompressibility Constraint and Temam's Device) Equation (9.9b) is equivalent to $\mathrm{D}_T^k \underline{\boldsymbol{u}}_T = 0$ for all $T \in \mathcal{T}_h$, and expresses at the discrete level the fact that the HHO velocity field solution to (9.9) is incompressible. Notice, however, that the fact that $\mathrm{D}_T^k \underline{\boldsymbol{u}}_T = 0$ for all $T \in \mathcal{T}_h$ does not imply, in general, that $\mathrm{D}_T^{2k} \underline{\boldsymbol{u}}_T = 0$, which justifies the introduction of the second term in (9.72).

Remark 9.22 (Implementation of the Trilinear Form (9.72) *and Link with* (9.47)*)* Expanding the discrete directional derivatives appearing in (9.75) below according to their definition (9.55), we arrive at the following reformulation of $\mathrm{t}_h^{\mathrm{tm}}$, which shows that, in the computer implementation, one does not actually need to compute $\boldsymbol{G}_T^k(\cdot\,;\cdot)$ nor D_T^{2k}: For any $\underline{\boldsymbol{w}}_h, \underline{\boldsymbol{v}}_h, \underline{\boldsymbol{z}}_h \in \underline{\boldsymbol{U}}_h^k$,

$$
\begin{aligned}
\mathrm{t}_h^{\mathrm{tm}}(\underline{\boldsymbol{w}}_h, \underline{\boldsymbol{v}}_h, \underline{\boldsymbol{z}}_h) &= \frac{1}{2} \sum_{T\in\mathcal{T}_h} \int_T [(\boldsymbol{w}_T\cdot\nabla)\boldsymbol{v}_T\cdot\boldsymbol{z}_T - \boldsymbol{v}_T\cdot(\boldsymbol{w}_T\cdot\nabla)\boldsymbol{z}_T] \\
&\quad + \frac{1}{2} \sum_{T\in\mathcal{T}_h} \sum_{F\in\mathcal{F}_T} \int_F (\boldsymbol{w}_F\cdot\boldsymbol{n}_{TF})(\boldsymbol{v}_F\cdot\boldsymbol{z}_T - \boldsymbol{v}_T\cdot\boldsymbol{z}_F).
\end{aligned}
\tag{9.73}
$$

Comparing with (9.48), it can be seen that the only difference is that $\boldsymbol{w}_F$ replaces $\boldsymbol{w}_T$ in the boundary term, and that the following relation holds:

$$
\begin{aligned}
\mathrm{t}_h^{\mathrm{tm}}(\underline{\boldsymbol{w}}_h, \underline{\boldsymbol{v}}_h, \underline{\boldsymbol{z}}_h) &= \mathrm{t}_h^{\mathrm{ss}}(\underline{\boldsymbol{w}}_h, \underline{\boldsymbol{v}}_h, \underline{\boldsymbol{z}}_h) \\
&\quad + \frac{1}{2} \sum_{T\in\mathcal{T}_h} \sum_{F\in\mathcal{F}_T} \int_F (\boldsymbol{w}_F - \boldsymbol{w}_T)\cdot\boldsymbol{n}_{TF}(\boldsymbol{v}_F\cdot\boldsymbol{z}_T - \boldsymbol{v}_T\cdot\boldsymbol{z}_F).
\end{aligned}
\tag{9.74}
$$

Proposition 9.23 (Properties of the Convective Trilinear Form (9.72)**)** *The convective trilinear form* $\mathrm{t}_h^{\mathrm{tm}}$ *defined by* (9.72) *satisfies Assumption 9.2.*

Proof (T1) *Non-dissipativity.* Using the discrete integration by parts formula (9.67), we can write, for any $\underline{\boldsymbol{w}}_h, \underline{\boldsymbol{v}}_h \in \underline{\boldsymbol{U}}_{h,0}^k$,

$$
\begin{aligned}
&\frac{1}{2} \sum_{T\in\mathcal{T}_h} \left(\int_T \mathrm{D}_T^{2k} \underline{\boldsymbol{w}}_T (\boldsymbol{v}_T\cdot\boldsymbol{z}_T) + \sum_{F\in\mathcal{F}_T} \int_F (\boldsymbol{w}_F\cdot\boldsymbol{n}_{TF})(\boldsymbol{v}_F - \boldsymbol{v}_T)\cdot(\boldsymbol{z}_F - \boldsymbol{z}_T) \right) \\
&\qquad = -\frac{1}{2} \sum_{T\in\mathcal{T}_h} \int_T \left(\boldsymbol{G}_T^k(\cdot\,; \underline{\boldsymbol{v}}_T)\cdot\boldsymbol{z}_T + \boldsymbol{v}_T\cdot\boldsymbol{G}_T^k(\cdot\,; \underline{\boldsymbol{z}}_T) \right).
\end{aligned}
$$

Plugging this equation into the definition (9.72) of $\mathrm{t}_h^{\mathrm{tm}}$, we arrive at the following reformulation, which makes the skew-symmetry of $\mathrm{t}_h^{\mathrm{tm}}$ evident:

$$\mathrm{t}_h^{\mathrm{tm}}(\underline{\boldsymbol{w}}_h, \underline{\boldsymbol{v}}_h, \underline{\boldsymbol{z}}_h) = \frac{1}{2} \sum_{T\in\mathcal{T}_h} \int_T \left(\boldsymbol{G}_T^k(\underline{\boldsymbol{w}}_T\,;\underline{\boldsymbol{v}}_T)\cdot\boldsymbol{z}_T - \boldsymbol{v}_T\cdot\boldsymbol{G}_T^k(\underline{\boldsymbol{w}}_T\,;\underline{\boldsymbol{z}}_T)\right). \tag{9.75}$$

The conclusion then follows letting $\underline{\boldsymbol{z}}_h = \underline{\boldsymbol{v}}_h$.

(T2) *Boundedness.* Accounting for (9.75), the boundedness follows applying (9.57) twice, with $\underline{\boldsymbol{v}}_h$ and $\underline{\boldsymbol{z}}_h$ swapped the second time.

(T3) *Consistency.* Set, for the sake of brevity, $\underline{\hat{\boldsymbol{w}}}_h := \underline{\boldsymbol{I}}_h^k \boldsymbol{w}$. Recalling the definition (9.72) of $\mathrm{t}_h^{\mathrm{tm}}$, we decompose the argument of the supremum in (9.12) into the sum of the following terms:

$$\mathfrak{T}_1 := \sum_{T\in\mathcal{T}_h} \int_T \left[(\boldsymbol{w}\cdot\nabla)\boldsymbol{w} - \boldsymbol{G}_T^k(\underline{\hat{\boldsymbol{w}}}_T;\underline{\hat{\boldsymbol{w}}}_T)\right]\cdot\boldsymbol{z}_T,$$

$$\mathfrak{T}_2 := -\frac{1}{2} \sum_{T\in\mathcal{T}_h} \int_T \mathrm{D}_T^{2k}\underline{\hat{\boldsymbol{w}}}_T(\hat{\boldsymbol{w}}_T\cdot\boldsymbol{z}_T)$$

$$\mathfrak{T}_3 := -\frac{1}{2} \sum_{T\in\mathcal{T}_h} \sum_{F\in\mathcal{F}_T} \int_F (\hat{\boldsymbol{w}}_F\cdot\boldsymbol{n}_{TF})(\hat{\boldsymbol{w}}_F - \hat{\boldsymbol{w}}_T)\cdot(\boldsymbol{z}_F - \boldsymbol{z}_T).$$

Using the approximation properties (9.58) of the discrete directional derivative followed by the discrete Sobolev embedding (9.42), it is inferred for the first term:

$$|\mathfrak{T}_1| \lesssim h^{r+1} |\boldsymbol{w}|_{W^{r+1,4}(\mathcal{T}_h)^d} |\boldsymbol{w}|_{W^{1,4}(\Omega)^d} \|\underline{\boldsymbol{z}}_h\|_{1,h}. \tag{9.76}$$

After observing that $\mathrm{D}_T^{2k}\underline{\hat{\boldsymbol{w}}}_T$ is L^2-orthogonal to functions in $\mathbb{P}^k(T)$ as a consequence of $\nabla\cdot\boldsymbol{w} = 0$ together with (9.46) written for $\boldsymbol{\tau} = q\boldsymbol{I}_d$ with q spanning $\mathbb{P}^k(T)$, and that $\boldsymbol{\pi}_T^{0,0}\boldsymbol{w}\cdot\boldsymbol{z}_T \in \mathbb{P}^k(T)$, we can write

$$\begin{aligned}
|\mathfrak{T}_2| &= \frac{1}{2} \left| \sum_{T\in\mathcal{T}_h} \int_T (\nabla\cdot\boldsymbol{w} - \mathrm{D}_T^{2k}\underline{\hat{\boldsymbol{w}}}_T)(\hat{\boldsymbol{w}}_T - \boldsymbol{\pi}_T^{0,0}\boldsymbol{w})\cdot\boldsymbol{z}_T \right| \\
&\lesssim \sum_{T\in\mathcal{T}_h} \|\nabla\cdot\boldsymbol{w} - \mathrm{D}_T^{2k}\underline{\hat{\boldsymbol{w}}}_T\|_{L^2(T)} \|\hat{\boldsymbol{w}}_T - \boldsymbol{\pi}_T^{0,0}\boldsymbol{w}\|_{L^4(T)^d} \|\boldsymbol{z}_T\|_{L^4(T)^d} \\
&\lesssim h^{r+1} |\boldsymbol{w}|_{H^{r+1}(\mathcal{T}_h)^d} |\boldsymbol{w}|_{W^{1,4}(\Omega)^d} \|\underline{\boldsymbol{z}}_h\|_{1,h}.
\end{aligned} \tag{9.77}$$

To pass from the second to the third line, we have used: the approximation properties of the divergence reconstruction resulting from (9.38a) with $l = 2k$ to bound the first factor; the linearity, idempotency, and L^4-boundedness of $\boldsymbol{\pi}_T^{0,k}$ followed by the approximation properties (1.74) of the L^2-orthogonal projector with $l = 0$, $p = 4$, $m = 0$, and $s = 1$ to estimate the second factor as follows:

$$\|\hat{\boldsymbol{w}}_T - \boldsymbol{\pi}_T^{0,0}\boldsymbol{w}\|_{L^4(T)^d} = \|\boldsymbol{\pi}_T^{0,k}(\boldsymbol{w} - \boldsymbol{\pi}_T^{0,0}\boldsymbol{w})\|_{L^4(T)^d}$$
$$\lesssim \|\boldsymbol{w} - \boldsymbol{\pi}_T^{0,0}\boldsymbol{w}\|_{L^4(T)^d} \lesssim h_T|\boldsymbol{w}|_{W^{1,4}(T)^d};$$

a generalised Hölder inequality on the sum over $T \in \mathcal{T}_h$ with exponents $(2, 4, 4)$, and the discrete Sobolev embedding (9.42) on $\underline{z}_h$ to conclude.

To estimate the third term, using a generalised Hölder inequality with exponents $(4, \infty, 4, 2)$ we obtain, after accounting for $\|\boldsymbol{n}_{TF}\|_{L^\infty(F)^d} = 1$,

$$|\mathfrak{T}_3| \le \frac{1}{2}\sum_{T\in\mathcal{T}_h}\sum_{F\in\mathcal{F}_T} \|\hat{\boldsymbol{w}}_F\|_{L^4(F)^d}\|\hat{\boldsymbol{w}}_F - \hat{\boldsymbol{w}}_T\|_{L^4(F)^d}\|\boldsymbol{z}_F - \boldsymbol{z}_T\|_{L^2(F)^d}.$$

For the first factor inside the summations, we use the L^4-boundedness of $\boldsymbol{\pi}_F^{0,k}$ followed by the local trace inequality (1.51) with $p = 4$ and the fact that $h_T \le \operatorname{diam}(\Omega) \lesssim 1$ to write

$$\|\hat{\boldsymbol{w}}_F\|_{L^4(F)^d} \lesssim \|\boldsymbol{w}\|_{L^4(F)^d} \lesssim h_T^{-\frac{1}{4}}\left(\|\boldsymbol{w}\|_{L^4(T)^d} + h_T\|\nabla\boldsymbol{w}\|_{L^4(T)^{d\times d}}\right) \lesssim h_T^{-\frac{1}{4}}\|\boldsymbol{w}\|_{W^{1,4}(T)^d}.$$

For the second factor, using (9.63) with $\alpha = 4$ followed by the optimal approximation properties of $\boldsymbol{\pi}_T^{0,k}$ we obtain

$$\|\hat{\boldsymbol{w}}_F - \hat{\boldsymbol{w}}_T\|_{L^4(F)^d} \lesssim \|\boldsymbol{w} - \boldsymbol{\pi}_T^{0,k}\boldsymbol{w}\|_{L^4(F)^d} \lesssim h_T^{r+\frac{3}{4}}|\boldsymbol{w}|_{W^{r+1,4}(T)^d}.$$

Collecting the above estimates, we can go on writing

$$\begin{aligned}
|\mathfrak{T}_3| &\lesssim \sum_{T\in\mathcal{T}_h}\sum_{F\in\mathcal{F}_T} h_T^{-\frac{1}{4}}\|\boldsymbol{w}\|_{W^{1,4}(T)^d}\, h_T^{r+\frac{3}{4}}|\boldsymbol{w}|_{W^{r+1,4}(T)^d}\,\|\boldsymbol{z}_F - \boldsymbol{z}_T\|_{L^2(F)^d} \\
&\lesssim h^{r+1}\|\boldsymbol{w}\|_{W^{1,4}(\Omega)^d}|\boldsymbol{w}|_{W^{r+1,4}(\mathcal{T}_h)^d}\left(\sum_{T\in\mathcal{T}_h}|\underline{\boldsymbol{z}}_T|_{1,\partial T}^2\right)^{\frac{1}{2}} \\
&\lesssim h^{r+1}\|\boldsymbol{w}\|_{W^{1,4}(\Omega)^d}|\boldsymbol{w}|_{W^{r+1,4}(\mathcal{T}_h)^d}\|\underline{\boldsymbol{z}}_h\|_{1,h},
\end{aligned} \tag{9.78}$$

where we have used generalised Hölder inequalities with exponents $(4, 4, 2)$ on the sums over $T \in \mathcal{T}_h$ and $F \in \mathcal{F}_T$ together with $h_F \le h_T \le h$ to pass to the second line, and the definitions (8.24) of $\|\cdot\|_{1,h}$ and (8.15) of $\|\cdot\|_{1,T}$ to conclude.

Collecting the bounds (9.76), (9.77), and (9.78), and observing that $|\boldsymbol{w}|_{H^{r+1}(\mathcal{T}_h)^d} \lesssim |\boldsymbol{w}|_{W^{r+1,4}(\mathcal{T}_h)^d}$, the conclusion follows. □

Remark 9.24 (Comparison with Hybridisable Discontinuous Galerkin Methods) Let $k \geq 0$ and define the enriched space of unknowns

$$\underline{\boldsymbol{U}}_h^{k+1,k} := \Big\{ \underline{\boldsymbol{v}}_h = ((\boldsymbol{v}_T)_{T\in\mathcal{T}_h}, (\boldsymbol{v}_F)_{F\in\mathcal{F}_h}) \ : \\ \boldsymbol{v}_T \in \mathbb{P}^{k+1}(T)^d \quad \forall T \in \mathcal{T}_h \text{ and } \boldsymbol{v}_F \in \mathbb{P}^k(F)^d \quad \forall F \in \mathcal{F}_h \Big\},$$

where the difference with respect to (8.22) is that element-based unknowns are polynomials of degree $(k+1)$ instead of k. The reformulation (9.73) enables a comparison with the Hybridisable Discontinuous Galerkin (HDG) trilinear form $\mathrm{t}_h^{\mathrm{HDG}} : \underline{\boldsymbol{U}}_h^{k+1,k} \times \underline{\boldsymbol{U}}_h^{k+1,k} \times \underline{\boldsymbol{U}}_h^{k+1,k} \to \mathbb{R}$, originally proposed in [253] (cf., in particular, Definition 3.3 therein and also [101]):

$$\begin{aligned}
\mathrm{t}_h^{\mathrm{HDG}}(\underline{\boldsymbol{w}}_h, \underline{\boldsymbol{v}}_h, \underline{\boldsymbol{z}}_h) := \frac{1}{2}\sum_{T\in\mathcal{T}_h}\int_T [(\boldsymbol{w}_T\cdot\nabla)\boldsymbol{v}_T\cdot\boldsymbol{z}_T - \boldsymbol{v}_T\cdot(\boldsymbol{w}_T\cdot\nabla)\boldsymbol{z}_T] \\
+ \frac{1}{2}\sum_{T\in\mathcal{T}_h}\sum_{F\in\mathcal{F}_T}\int_F (\boldsymbol{w}_F\cdot\boldsymbol{n}_{TF})(\boldsymbol{v}_F\cdot\boldsymbol{z}_T - \boldsymbol{z}_F\cdot\boldsymbol{v}_T) \\
+ \frac{\eta}{2}\sum_{T\in\mathcal{T}_h}\sum_{F\in\mathcal{F}_T}\int_F |\boldsymbol{w}_F\cdot\boldsymbol{n}_{TF}|(\boldsymbol{v}_F - \boldsymbol{v}_T)\cdot(\boldsymbol{z}_F - \boldsymbol{z}_T).
\end{aligned}$$

There are two main differences with respect to the HHO method discussed in this section. The first one is that, in the HDG method, the element-based unknowns are polynomials of degree $(k+1)$ instead of k. Correspondingly, the viscous term is discretised as in [228, 249] in order to improve the convergence rates to match the ones of HHO methods; see [129] for further details, in particular Remark 2.2 therein. This variation corresponds to the HHO method with enriched element unknowns discussed in Sect. 5.1. The second difference is the presence of an upwind convective stabilisation term; see the discussion in Sect. 9.4 on this point.

9.5.3.4 Flux Formulation

Recalling Remark 9.1, local balance relations with continuous fluxes can be identified for the Navier–Stokes problem in a similar way as we did for the Stokes problem in Sect. 8.4. Specifically, denoting by $(\boldsymbol{u}, p) \in \boldsymbol{U} \times P$ a solution to (9.3) and assuming sufficient regularity for the boundary integrals to be well-defined, it holds: For all $T \in \mathcal{T}_h$ and all $(\boldsymbol{v}_T, q_T) \in \mathbb{P}^k(T)^d \times \mathbb{P}^k(T)$,

$$\int_T \nu \nabla \boldsymbol{u}{:}\nabla \boldsymbol{v}_T - \int_T (\boldsymbol{u}\cdot\nabla)\boldsymbol{v}_T\cdot\boldsymbol{u} - \int_T p\,(\nabla\cdot\boldsymbol{v}_T)$$
$$+ \sum_{F\in\mathcal{F}_T} \int_F (-\nu\nabla\boldsymbol{u} + \boldsymbol{u}\otimes\boldsymbol{u} + p\boldsymbol{I}_d)_{|T}\,\boldsymbol{n}_{TF}\cdot\boldsymbol{v}_T = \int_T \boldsymbol{f}\cdot\boldsymbol{v}_T, \tag{9.79a}$$

$$\int_T \boldsymbol{u}\cdot\nabla q_T - \int_T (\boldsymbol{u}_{|T}\cdot\boldsymbol{n}_{TF})\,q_T = 0. \tag{9.79b}$$

Compared to (8.50a), the balance relation (9.79a) contains two additional terms accounting for the nonlinear convection: the volumetric term $-\int_T(\boldsymbol{u}\cdot\nabla)\boldsymbol{v}_T\cdot\boldsymbol{u}$, and the contribution $(\boldsymbol{u}\otimes\boldsymbol{u})_{|T}\boldsymbol{n}_{TF}$ to the momentum flux $(-\nu\nabla\boldsymbol{u}+\boldsymbol{u}\otimes\boldsymbol{u}+p\boldsymbol{I}_d)_{|T}\,\boldsymbol{n}_{TF}$. Also in this case, the normal traces of the fluxes are continuous across interfaces, i.e., for all $F \in \mathcal{F}_h^{\rm i}$ such that $F \subset \partial T_1 \cap \partial T_2$ with distinct mesh elements $T_1, T_2 \in \mathcal{T}_h$, it holds that

$$(-\nu\nabla\boldsymbol{u} + \boldsymbol{u}\otimes\boldsymbol{u} + p\boldsymbol{I}_d)_{|T_1}\,\boldsymbol{n}_{T_1F} + (-\nu\nabla\boldsymbol{u} + \boldsymbol{u}\otimes\boldsymbol{u} + p\boldsymbol{I}_d)_{|T_2}\,\boldsymbol{n}_{T_2F} = \boldsymbol{0}, \tag{9.80a}$$

$$\boldsymbol{u}_{|T_1}\cdot\boldsymbol{n}_{T_1F} + \boldsymbol{u}_{|T_2}\cdot\boldsymbol{n}_{T_2F} = 0. \tag{9.80b}$$

The interest of the trilinear form (9.72) built on Temam's device over the trilinear form defined by (9.47) is that the former enables a flux formulation that reproduces the relations (9.79) and (9.80) at the discrete level. This can be essential for discretisations of coupled systems involving the Navier–Stokes equations and advection processes.

Lemma 9.25 (Flux Formulation) *Let $\mathcal{M}_h$ denote a polytopal mesh in the sense of Definition 1.4. Let the viscous bilinear form a_h be given by (8.27), with local stabilisation bilinear forms in (8.28) matching Assumption 8.10. Let the pressure–velocity coupling bilinear form b_h be given by (8.34), and the trilinear form $\mathrm{t}_h^{\rm tm}$ be given by (9.72).*

Let $(\underline{\boldsymbol{u}}_h, p_h) \in \underline{\boldsymbol{U}}_{h,0}^k \times P_h^k$ and, for all $T \in \mathcal{T}_h$ and all $F \in \mathcal{F}_T$, define the numerical normal traces $\boldsymbol{\Phi}_{TF}^{\rm visc}(\underline{\boldsymbol{u}}_T) \in \mathbb{P}^k(F)^d$ and $\boldsymbol{\Phi}_{TF}^{\rm conv}(\underline{\boldsymbol{u}}_T) \in \mathbb{P}^k(F)^d$ of the viscous and convective momentum fluxes as follows:

$$\boldsymbol{\Phi}_{TF}^{\rm visc}(\underline{\boldsymbol{u}}_T) := \nu\left(-\nabla\mathbf{r}_T^{k+1}\underline{\boldsymbol{u}}_T\boldsymbol{n}_{TF} + \boldsymbol{R}_{TF}^k\underline{\boldsymbol{u}}_T\right),$$

$$\boldsymbol{\Phi}_{TF}^{\rm conv}(\underline{\boldsymbol{u}}_T) := \boldsymbol{\pi}_F^{0,k}\left(\frac{\boldsymbol{u}_F+\boldsymbol{u}_T}{2}\otimes\boldsymbol{u}_F\right)\boldsymbol{n}_{TF} = \boldsymbol{\pi}_F^{0,k}\left[(\boldsymbol{u}_F\cdot\boldsymbol{n}_{TF})\frac{\boldsymbol{u}_F+\boldsymbol{u}_T}{2}\right],$$

(continued)

Lemma 9.25 (continued)

with $\boldsymbol{R}_{TF}^k$ defined by (8.52).

Then, $(\underline{\boldsymbol{u}}_h, p_h)$ solves (9.9) if and only if the following two properties hold:

(i) Local momentum and mass balance. *For all $T \in \mathcal{T}_h$ and all $(\boldsymbol{v}_T, q_T) \in \mathbb{P}^k(T)^d \times \mathbb{P}^k(T)$,*

$$\begin{aligned}
&\int_T \nu \nabla \mathbf{r}_T^{k+1} \underline{\boldsymbol{u}}_T : \nabla \boldsymbol{v}_T - \int_T (\boldsymbol{u}_T \cdot \nabla) \boldsymbol{v}_T \cdot \boldsymbol{u}_T \\
&\quad - \int_T p_T \, (\nabla \cdot \boldsymbol{v}_T) - \frac{1}{2} \int_T \mathrm{D}_T^{2k} \underline{\boldsymbol{u}}_T (\boldsymbol{u}_T \cdot \boldsymbol{v}_T) \\
&\quad + \sum_{F \in \mathcal{F}_T} \int_F \left(\boldsymbol{\Phi}_{TF}^{\mathrm{visc}}(\underline{\boldsymbol{u}}_T) + \boldsymbol{\Phi}_{TF}^{\mathrm{conv}}(\underline{\boldsymbol{u}}_T) + p_T \boldsymbol{n}_{TF} \right) \cdot \boldsymbol{v}_T = \int_T \boldsymbol{f} \cdot \boldsymbol{v}_T,
\end{aligned} \tag{9.81a}$$

$$\int_T \boldsymbol{u}_T \cdot \nabla q_T - \sum_{F \in \mathcal{F}_T} \int_F (\boldsymbol{u}_F \cdot \boldsymbol{n}_{TF}) q_T = 0. \tag{9.81b}$$

(ii) Continuity of the numerical normal traces of the momentum and mass fluxes. *For any interface $F \in \mathcal{F}_h^{\mathrm{i}}$ such that $F \subset \partial T_1 \cap \partial T_2$ for distinct mesh elements $T_1, T_2 \in \mathcal{T}_h$, it holds*

$$\begin{aligned}
&\left(\boldsymbol{\Phi}_{T_1 F}^{\mathrm{visc}}(\underline{\boldsymbol{u}}_{T_1}) + \boldsymbol{\Phi}_{T_1 F}^{\mathrm{conv}}(\underline{\boldsymbol{u}}_{T_1}) + p_{T_1} \boldsymbol{n}_{T_1 F} \right) \\
&\qquad + \left(\boldsymbol{\Phi}_{T_2 F}^{\mathrm{visc}}(\underline{\boldsymbol{u}}_{T_2}) + \boldsymbol{\Phi}_{T_2 F}^{\mathrm{conv}}(\underline{\boldsymbol{u}}_{T_2}) + p_{T_2} \boldsymbol{n}_{T_2 F} \right) = \boldsymbol{0},
\end{aligned} \tag{9.82a}$$

$$\boldsymbol{u}_F \cdot \boldsymbol{n}_{T_1 F} + \boldsymbol{u}_F \cdot \boldsymbol{n}_{T_2 F} = 0. \tag{9.82b}$$

Before proving this lemma, some remarks are in order.

Remark 9.26 (Extension to Convective Stabilisation) This lemma is also valid if a convective stabilisation term is added to the scheme, that is, for the HHO scheme (9.32). In this case, an additional term $-\boldsymbol{\pi}_F^{0,k}\left[\frac{\nu}{h_F} \rho(\mathrm{Pe}(\boldsymbol{u}_F))(\boldsymbol{u}_F - \boldsymbol{u}_T)\right]$ must be added to $\boldsymbol{\Phi}_{TF}^{\mathrm{conv}}(\underline{\boldsymbol{u}}_T)$, see [70, Proposition 16].

Remark 9.27 (Finite Volume Momentum Balance) In (9.81a), an additional term $\int_T \mathrm{D}_T^{2k} \underline{\boldsymbol{u}}_T (\boldsymbol{u}_T \cdot \boldsymbol{v}_T)$ is present with respect to the continuous momentum balance (9.79a). This term, however, does not appear in the lowest-order version of the discrete momentum balance obtained taking $\boldsymbol{v}_T \in \mathbb{P}^0(T)^d$ since, using (9.35) with $(l, m) = (2k, k)$ and recalling that $\mathrm{D}_T^l = \mathrm{tr}(\mathbf{G}_T^l)$, we have in this case $\int_T \mathrm{D}_T^{2k} \underline{\boldsymbol{u}}_T (\boldsymbol{u}_T \cdot \boldsymbol{v}_T) = \int_T \mathrm{D}_T^k \underline{\boldsymbol{u}}_T (\boldsymbol{u}_T \cdot \boldsymbol{v}_T) = 0$, where the conclusion follows using (9.9b). Hence, the HHO scheme (9.9) with convective trilinear form $\mathrm{t}_h^{\mathrm{tm}}$ given by (9.72) satisfies the following Finite Volume-like local momentum and mass balances: For all $T \in \mathcal{T}_h$,

$$\sum_{F \in \mathcal{F}_T} \int_F \left(\boldsymbol{\Phi}_{TF}^{\text{visc}}(\underline{\boldsymbol{u}}_T) + \boldsymbol{\Phi}_{TF}^{\text{conv}}(\underline{\boldsymbol{u}}_T) + p_T \boldsymbol{n}_{TF} \right) = \int_T \boldsymbol{f}, \tag{9.83a}$$

$$\sum_{F \in \mathcal{F}_T} \int_F (\boldsymbol{u}_F \cdot \boldsymbol{n}_{TF}) = 0, \tag{9.83b}$$

as can be checked taking $\boldsymbol{v}_T$ in (9.81a) successively equal to the vectors of the canonical basis of $\mathbb{R}^d$ and q_T in (9.81b) equal to 1.

Proof (Lemma 9.25) The proof follows that of Lemma 8.17, after assessing what additional terms the trilinear form $\mathrm{t}_h^{\text{tm}}(\underline{\boldsymbol{u}}_h, \underline{\boldsymbol{u}}_h, \underline{\boldsymbol{v}}_h)$ brings to (8.53a) and (8.54a).

Using the discrete integration by parts formula (9.67), for any $\underline{\boldsymbol{v}}_h \in \underline{\boldsymbol{U}}_{h,0}^k$ we have

$$\mathrm{t}_h^{\text{tm}}(\underline{\boldsymbol{u}}_h, \underline{\boldsymbol{u}}_h, \underline{\boldsymbol{v}}_h) = -\sum_{T \in \mathcal{T}_h} \left[\int_T \boldsymbol{u}_T \cdot \boldsymbol{G}_T^k(\underline{\boldsymbol{u}}_T; \underline{\boldsymbol{v}}_T) + \frac{1}{2} \int_T \mathrm{D}_T^{2k} \underline{\boldsymbol{u}}_T (\boldsymbol{u}_T \cdot \boldsymbol{v}_T) \right.$$
$$\left. + \frac{1}{2} \sum_{F \in \mathcal{F}_T} \int_F (\boldsymbol{u}_F \cdot \boldsymbol{n}_{TF})(\boldsymbol{u}_F - \boldsymbol{u}_T) \cdot (\boldsymbol{v}_F - \boldsymbol{v}_T) \right].$$

Hence, expanding each $\boldsymbol{G}_T^k(\underline{\boldsymbol{u}}_T; \underline{\boldsymbol{v}}_T)$ according to its definition (9.55) with $\underline{\boldsymbol{w}}_T = \underline{\boldsymbol{u}}_T$ and $\boldsymbol{z} = \boldsymbol{u}_T$ and rearranging the terms, we obtain

$$\begin{aligned} \mathrm{t}_h^{\text{tm}}(\underline{\boldsymbol{u}}_h, \underline{\boldsymbol{u}}_h, \underline{\boldsymbol{v}}_h) = -\sum_{T \in \mathcal{T}_h} \left[\int_T (\boldsymbol{u}_T \cdot \nabla) \boldsymbol{v}_T \cdot \boldsymbol{u}_T + \frac{1}{2} \int_T \mathrm{D}_T^{2k} \underline{\boldsymbol{u}}_T (\boldsymbol{u}_T \cdot \boldsymbol{v}_T) \right. \\ \left. + \sum_{F \in \mathcal{F}_T} \int_F \boldsymbol{\Phi}_{TF}^{\text{conv}}(\underline{\boldsymbol{u}}_T) \cdot (\boldsymbol{v}_F - \boldsymbol{v}_T) \right], \end{aligned} \tag{9.84}$$

where we have further observed that $(\boldsymbol{v}_F - \boldsymbol{v}_{T|F}) \in \mathbb{P}^k(F)^d$ to insert $\boldsymbol{\pi}_F^{0,k}$ into the expression of the convective flux. We then easily realise that, using element basis functions (resp. face basis functions) for $\underline{\boldsymbol{v}}_T$, $\mathrm{t}_h^{\text{tm}}(\underline{\boldsymbol{u}}_h, \underline{\boldsymbol{u}}_h, \underline{\boldsymbol{v}}_h)$ is responsible for the terms $-\int_T (\boldsymbol{u}_T \cdot \nabla) \boldsymbol{v}_T \cdot \boldsymbol{u}_T - \frac{1}{2} \int_T \mathrm{D}_T^{2k} \boldsymbol{u}_T (\boldsymbol{u}_T \cdot \boldsymbol{v}_T)$ and $\boldsymbol{\Phi}_{TF}^{\text{conv}}(\underline{\boldsymbol{u}}_T) \cdot \boldsymbol{v}_T$ in (9.81a) (resp. the terms $\boldsymbol{\Phi}_{T_1F}^{\text{conv}}(\underline{\boldsymbol{u}}_{T_1})$ and $\boldsymbol{\Phi}_{T_2F}^{\text{conv}}(\underline{\boldsymbol{u}}_{T_2})$ in (9.82a)). □

Remark 9.28 (Lack of a Flux Formulation for the Skew-Symmetric Trilinear Form (9.47)) Recalling (9.74), the relation (9.84) gives the following expression for $\mathrm{t}_h^{\mathrm{ss}}$:

$$\begin{aligned}\mathrm{t}_h^{\mathrm{ss}}(\underline{\boldsymbol{u}}_h, \underline{\boldsymbol{u}}_h, \underline{\boldsymbol{v}}_h) = -\sum_{T\in\mathcal{T}_h}\Bigg[\int_T (\boldsymbol{u}_T\cdot\boldsymbol{\nabla})\boldsymbol{v}_T\cdot\boldsymbol{u}_T + \frac{1}{2}\int_T \mathrm{D}_T^{2k}\underline{\boldsymbol{u}}_T(\boldsymbol{u}_T\cdot\boldsymbol{v}_T)\\ + \sum_{F\in\mathcal{F}_T}\int_F \Phi_{TF}^{\mathrm{conv}}(\underline{\boldsymbol{u}}_T)\cdot(\boldsymbol{v}_F - \boldsymbol{v}_T)\\ - \frac{1}{2}\sum_{F\in\mathcal{F}_T}\int_F (\boldsymbol{u}_F - \boldsymbol{u}_T)\cdot\boldsymbol{n}_{TF}(\boldsymbol{u}_F\cdot\boldsymbol{v}_T - \boldsymbol{u}_T\cdot\boldsymbol{v}_F)\Bigg].\end{aligned}$$

To identify continuous fluxes (that is, fluxes satisfying (9.82a)) for the gradient-based scheme using $\mathrm{t}_h^{\mathrm{ss}}$, one has to gather the terms in the scheme that multiply the face test function $\boldsymbol{v}_F$. This would lead here to setting, as fluxes for the gradient-based scheme,

$$\boldsymbol{\Phi}_{TF}^{\mathrm{conv,ss}}(\underline{\boldsymbol{u}}_T) = \boldsymbol{\Phi}_{TF}^{\mathrm{conv}}(\underline{\boldsymbol{u}}_T) + \frac{1}{2}(\boldsymbol{u}_F - \boldsymbol{u}_T)\cdot\boldsymbol{n}_{TF}\boldsymbol{u}_T.$$

The balance relation for these fluxes would then be obtained by considering the terms in the scheme involving the element test function $\boldsymbol{v}_T$. Here, these terms can be written

$$\begin{aligned}\int_T \nu\boldsymbol{\nabla}\mathbf{r}_T^{k+1}\underline{\boldsymbol{u}}_T:\boldsymbol{\nabla}\boldsymbol{v}_T - \int_T (\boldsymbol{u}_T\cdot\boldsymbol{\nabla})\boldsymbol{v}_T\cdot\boldsymbol{u}_T - \int_T p_T\,(\boldsymbol{\nabla}\cdot\boldsymbol{v}_T)\\ - \frac{1}{2}\int_T \mathrm{D}_T^{2k}\underline{\boldsymbol{u}}_T(\boldsymbol{u}_T\cdot\boldsymbol{v}_T) - \boxed{\frac{1}{2}\sum_{F\in\mathcal{F}_T}\int_F (\boldsymbol{u}_F - \boldsymbol{u}_T)\cdot\boldsymbol{n}_{TF}(\boldsymbol{u}_F - \boldsymbol{u}_T)\cdot\boldsymbol{v}_T}\\ + \sum_{F\in\mathcal{F}_T}\int_F \left(\boldsymbol{\Phi}_{TF}^{\mathrm{visc}}(\underline{\boldsymbol{u}}_T) + \boldsymbol{\Phi}_{TF}^{\mathrm{conv,ss}}(\underline{\boldsymbol{u}}_T) + p_T\boldsymbol{n}_{TF}\right)\cdot\boldsymbol{v}_T = \int_T \boldsymbol{f}\cdot\boldsymbol{v}_T.\end{aligned} \tag{9.85}$$

As seen for example in Lemmas 2.25, 3.17 and 3.30, the balance equations for flux formulations of HHO methods are expected to contain a volumetric contribution that vanishes when $\boldsymbol{v}_T$ is constant inside T, and face contributions solely involving the fluxes. The boxed term in (9.85) does not fall into any of these two categories. On the one hand, it cannot be incorporated as a volumetric contribution into the local balance equation as it does not necessarily vanish for $\boldsymbol{v}_T \in \mathbb{P}^0(T)^d$. On the other hand, if we incorporated it into the fluxes, the latter would no longer be continuous. This shows that the gradient-based scheme using $\mathrm{t}_h^{\mathrm{ss}}$ does not admit, contrary to the scheme using $\mathrm{t}_h^{\mathrm{tm}}$, a flux formulation.

9.6 Convergence for General Data

The error estimate of Theorem 9.10 is valid only under the data smallness assumption (9.18). In this section, we prove convergence for general data using compactness techniques which, as discussed in Chap. 6, do not require additional regularity on the exact solution or the data, at the expense of not delivering estimates on the convergence rate. For the sake of simplicity, we focus on the skew-symmetric discrete trilinear form $\mathrm{t}_h^{\mathrm{ss}}$ defined by (9.47), and leave as an exercise to the reader the adaptation of the proofs to the trilinear form $\mathrm{t}_h^{\mathrm{tm}}$ defined by (9.72).

9.6.1 *Discrete Compactness and Strong Convergence of the Interpolates*

In this section we establish two important preliminary results, namely a compactness property for sequences of vectors of discrete unknowns uniformly bounded in the $\|\cdot\|_{1,h}$-seminorm, and the convergence of gradient reconstructions applied to interpolates of sufficiently smooth functions.

To state these results, we need the global velocity reconstruction $\mathbf{r}_h^{k+1} : \underline{\boldsymbol{U}}_h^k \to \mathbb{P}^{k+1}(\mathcal{T}_h)^d$ defined by (9.28) and, for a given integer $l \geq 0$, the global gradient reconstruction $\mathbf{G}_h^l : \underline{\boldsymbol{U}}_h^k \to \mathbb{P}^l(\mathcal{T}_h)^{d\times d}$ such that, for all $\underline{\boldsymbol{v}}_h \in \underline{\boldsymbol{U}}_h^k$,

$$(\mathbf{G}_h^l \underline{\boldsymbol{v}}_h)_{|T} := \mathbf{G}_T^l \underline{\boldsymbol{v}}_T \qquad \forall T \in \mathcal{T}_h.$$

The following lemma is the equivalent of Theorem 6.41 with $p = 2$ and $\mathbf{G}_T^l$ instead of $\mathbf{G}_T^k$.

Theorem 9.29 (Discrete Compactness) *Let $k \geq 0$ be a polynomial degree and $(\mathcal{M}_h)_{h\in\mathcal{H}}$ be a regular mesh sequence in the sense of Definition 1.9. Let $(\underline{\boldsymbol{v}}_h)_{h\in\mathcal{H}} \in (\underline{\boldsymbol{U}}_{h,0}^k)_{h\in\mathcal{H}}$, and assume the existence of a real number $C > 0$ such that $\|\underline{\boldsymbol{v}}_h\|_{1,h} \leq C$ for all $h \in \mathcal{H}$. Then, there exists $\boldsymbol{v} \in H_0^1(\Omega)^d$ such that, up to a subsequence as $h \to 0$,*

(i) $\boldsymbol{v}_h \to \boldsymbol{v}$ *strongly in* $L^q(\Omega)^d$ *for all* $q \in [1, \infty)$ *if* $d = 2$, *and all* $q \in [1, 6)$ *if* $d = 3$;
(ii) $\mathbf{G}_h^l \underline{\boldsymbol{v}}_h \rightharpoonup \boldsymbol{\nabla}\boldsymbol{v}$ *weakly in* $L^2(\Omega)^{d\times d}$ *for any integer* $l \geq 0$;
(iii) $\boldsymbol{\nabla}_h \mathbf{r}_h^{k+1} \underline{\boldsymbol{v}}_h \rightharpoonup \boldsymbol{\nabla}\boldsymbol{v}$ *weakly in* $L^2(\Omega)^{d\times d}$.

Remark 9.30 (Extension to the Non-Hilbertian Setting) It is possible to extend this compactness result to the non-Hilbertian setting considered in Theorem 6.41. The details are left to the reader as this extension will not be needed in the following discussion.

Proof *(Theorem 9.29)* Applying Theorem 6.41 with $p = 2$ to the components of $\underline{v}_h$, we obtain the existence of $v \in H_0^1(\Omega)^d$ such that, up to a subsequence, for q as in Point (i) of the statement it holds $v_h \to v$ and $\mathbf{r}_h^{k+1}\underline{v}_h \to v$ in $L^q(\Omega)^d$, and $\mathbf{G}_h^k\underline{v}_h \rightharpoonup \nabla v$ weakly in $L^2(\Omega)^{d\times d}$.

It remains to prove the convergence of $\mathbf{G}_h^l\underline{v}_h$, for a generic $l \geq 0$, and of $\nabla_h\mathbf{r}_h^{k+1}\underline{v}_h$. The norm equivalence (8.31), the boundedness of $\|\underline{v}_h\|_{1,h}$, and the estimate (9.37) show that both $(\nabla_h\mathbf{r}_h^{k+1}\underline{v}_h)_{h\in\mathcal{H}}$ and $(\mathbf{G}_h^l\underline{v}_h)_{h\in\mathcal{H}}$ are bounded in $L^2(\Omega)^{d\times d}$, and thus converge up to a subsequence weakly in this space to τ and ξ, respectively. The proof is complete if we show that $\tau = \xi = \nabla v$.

Let us first identify ξ. For all $\Phi \in C_c^\infty(\Omega)^{d\times d}$, setting $n := \min(l, k)$, we have

$$\begin{aligned}\int_\Omega \mathbf{G}_h^l\underline{v}_h{:}\Phi &= \int_\Omega \mathbf{G}_h^l\underline{v}_h{:}\pi_h^{0,n}\Phi + \underbrace{\int_\Omega \mathbf{G}_h^l\underline{v}_h{:}\left(\Phi - \pi_h^{0,n}\Phi\right)}_{\mathfrak{T}_{1,h}} \\ &= \int_\Omega \mathbf{G}_h^k\underline{v}_h{:}\pi_h^{0,n}\Phi + \mathfrak{T}_{1,h},\end{aligned} \tag{9.86}$$

where we have inserted $\pm\pi_h^{0,n}\Phi$ in the first line, and used the property (9.35) with $m = k$ in the second line. By regularity of Φ and Theorem 1.45, it holds $\pi_h^{0,n}\Phi \to \Phi$ in $L^2(\Omega)^{d\times d}$ as $h \to 0$. The boundedness of $(\mathbf{G}_h^l\underline{v}_h)_{h\in\mathcal{H}}$ in $L^2(\Omega)^{d\times d}$ and a Cauchy–Schwarz inequality then show that $\mathfrak{T}_{1,h} \to 0$ as $h \to 0$. Taking the limit in h of (9.86) and recalling that $\mathbf{G}_h^k\underline{v}_h \rightharpoonup \nabla v$ weakly in $L^2(\Omega)^{d\times d}$ thus yields $\int_\Omega \xi{:}\Phi = \int_\Omega \nabla v{:}\Phi$, which proves that $\xi = \nabla v$ as required.

We now turn to τ. Let $T \in \mathcal{T}_h$. Recalling Remark 4.9, we see that $\nabla\mathbf{r}_T^{k+1}\underline{v}_T$ is the L^2-orthogonal projection of $\mathbf{G}_T^k\underline{v}_T$ on $\nabla\mathbb{P}^{k+1}(T)^d$. Projecting further on $\mathbb{P}^0(T)^{d\times d} = \nabla\mathbb{P}^1(T)^d \subset \nabla\mathbb{P}^{k+1}(T)^d$, we obtain $\pi_T^{0,0}(\nabla\mathbf{r}_T^{k+1}\underline{v}_T) = \pi_T^{0,0}(\mathbf{G}_T^k\underline{v}_T)$. Patching these relations yields $\pi_h^{0,0}(\nabla\mathbf{r}_h^{k+1}\underline{v}_h) = \pi_h^{0,0}(\mathbf{G}_h^k\underline{v}_h)$. We can then follow the reasoning above, applying (9.86) with $n = 0$ and $\nabla\mathbf{r}_h^{k+1}\underline{v}_h$ instead of $\mathbf{G}_h^l\underline{v}_h$, to deduce that $\tau = \nabla v$. □

The second preliminary result, the strong convergence of the discrete gradients for interpolates of smooth functions, is stated in the following proposition.

Proposition 9.31 (Strong Convergence of the Interpolates) *Let $(\mathcal{M}_h)_{h\in\mathcal{H}}$ be a regular mesh sequence in the sense of Definition 1.9 and let the global interpolator $\underline{I}_h^k$ be defined by* (8.23)*. Let $l \in \mathbb{N}$ be such that*

$$l = 0 \text{ if } k = 0, \quad l \geq 0 \text{ if } k \geq 1. \tag{9.87}$$

Then, for all $v \in H^1(\Omega)^d$,

$$\mathbf{G}_h^l\underline{I}_h^k v \to \nabla v \text{ strongly in } L^2(\Omega)^{d\times d} \text{ as } h \to 0 \tag{9.88}$$

and

$$\nabla_h \mathbf{r}_h^{k+1} \underline{\boldsymbol{I}}_h^k \boldsymbol{v} \to \nabla \boldsymbol{v} \text{ strongly in } L^2(\Omega)^{d\times d} \text{ as } h \to 0. \tag{9.89}$$

Moreover, denoting by $(\mathrm{s}_T)_{T\in\mathcal{T}_h}$ *a family of stabilisation bilinear forms matching Assumption 8.10, for all* $\boldsymbol{v} \in H^2(\Omega)^d$ *it holds that*

$$|\underline{\boldsymbol{I}}_h^k \boldsymbol{v}|_{\mathrm{s},h} \to 0 \text{ as } h \to 0, \tag{9.90}$$

with seminorm $|\cdot|_{\mathrm{s},h}$ *defined by* (9.30).

Proof Throughout the proof, hidden constants are independent of both h and $\boldsymbol{v}$.

(i) *Proof of* (9.88) *and* (9.89). We reason by density. Specifically, we let $(\boldsymbol{v}_\epsilon)_{\epsilon>0}$ denote a sequence in $H^2(\Omega)^d$ that converges to $\boldsymbol{v}$ in $H^1(\Omega)^d$ as $\epsilon \to 0$ and we write, inserting $\pm(\mathbf{G}_h^l \underline{\boldsymbol{I}}_h^k \boldsymbol{v}_\epsilon - \nabla \boldsymbol{v}_\epsilon)$ into the norm and using the triangle inequality,

$$\begin{aligned}
&\|\mathbf{G}_h^l \underline{\boldsymbol{I}}_h^k \boldsymbol{v} - \nabla \boldsymbol{v}\|_{L^2(\Omega)^{d\times d}} \\
&\quad\le \|\mathbf{G}_h^l \underline{\boldsymbol{I}}_h^k (\boldsymbol{v} - \boldsymbol{v}_\epsilon)\|_{L^2(\Omega)^{d\times d}} + \|\mathbf{G}_h^l \underline{\boldsymbol{I}}_h^k \boldsymbol{v}_\epsilon - \nabla \boldsymbol{v}_\epsilon\|_{L^2(\Omega)^{d\times d}} + \|\nabla(\boldsymbol{v}_\epsilon - \boldsymbol{v})\|_{L^2(\Omega)^{d\times d}} \\
&\quad\lesssim |\boldsymbol{v} - \boldsymbol{v}_\epsilon|_{H^1(\Omega)^d} + \|\mathbf{G}_h^l \underline{\boldsymbol{I}}_h^k \boldsymbol{v}_\epsilon - \nabla \boldsymbol{v}_\epsilon\|_{L^2(\Omega)^{d\times d}},
\end{aligned} \tag{9.91}$$

where we have used the boundedness (9.37) of the gradient reconstruction and (8.25) of the global interpolator to write $\|\mathbf{G}_h^l \underline{\boldsymbol{I}}_h^k (\boldsymbol{v} - \boldsymbol{v}_\epsilon)\|_{L^2(\Omega)^{d\times d}} \lesssim \|\underline{\boldsymbol{I}}_h^k (\boldsymbol{v} - \boldsymbol{v}_\epsilon)\|_{1,h} \lesssim |\boldsymbol{v} - \boldsymbol{v}_\epsilon|_{H^1(\Omega)^d}$. Using, for all $T \in \mathcal{T}_h$, the consistency properties (9.38a) of the gradient reconstruction with $r = 1$ (this choice is possible under assumption (9.87)), we get for the second term

$$\|\mathbf{G}_h^l \underline{\boldsymbol{I}}_h^k \boldsymbol{v}_\epsilon - \nabla \boldsymbol{v}_\epsilon\|_{L^2(\Omega)^{d\times d}} \lesssim h|\boldsymbol{v}_\epsilon|_{H^2(\Omega)^d},$$

which shows that this term tends to zero as $h \to 0$. Taking, in this order, the supremum limit of (9.91) as $h \to 0$, then the limit of the resulting inequality as $\epsilon \to 0$ concludes the proof of (9.88).

The proof of (9.89) is obtained in a similar way, with the norm equivalence (8.31) replacing (9.37) in the estimate of the first term in the second line of (9.91) and the approximation properties (1.78) of the elliptic projector (together with $\mathbf{r}_h^{k+1} \circ \underline{\boldsymbol{I}}_h^k = \boldsymbol{\pi}_h^{1,k+1}$, see (2.14) for the scalar case) replacing (9.38a) in the estimate of the second term in the third line of (9.91).

(ii) *Proof of* (9.90). Proceeding as in the proof of Proposition 2.14 (where the scalar case is considered), one gets, for all $T \in \mathcal{T}_h$, $\mathrm{s}_T(\underline{\boldsymbol{I}}_h^k \boldsymbol{v}, \underline{\boldsymbol{I}}_h^k \boldsymbol{v}) \lesssim h_T^2 |\boldsymbol{v}|_{H^2(T)^d}^2$, where the hidden constant is additionally independent of T. Summing these bounds over $T \in \mathcal{T}_h$, taking the square root, and letting $h \to 0$ yields $|\underline{\boldsymbol{I}}_h^k \boldsymbol{v}|_{\mathrm{s},h} \to 0$, thus proving (9.90). □

9.6.2 Convergence by Compactness

We are now ready to prove the convergence of the HHO scheme for the Navier–Stokes equations with general data.

Theorem 9.32 (Convergence for General Data) *Let $(\mathcal{M}_h)_{h\in\mathcal{H}}$ denote a regular mesh sequence in the sense of Definition 1.9, let $k \geq 0$ be a polynomial degree, and let $((\underline{\boldsymbol{u}}_h, p_h))_{h\in\mathcal{H}}$ be such that, for all $h \in \mathcal{H}$, $(\underline{\boldsymbol{u}}_h, p_h) \in \underline{\boldsymbol{U}}_{h,0}^k \times P_h^k$ solves (9.9). Then, up to a subsequence as $h \to 0$, it holds*

(i) $\boldsymbol{u}_h \to \boldsymbol{u}$ *strongly in* $L^q(\Omega)^d$ *for all* $q \in [1,\infty)$ *if* $d = 2$ *and* $q \in [1, 6)$ *if* $d = 3$,
(ii) $\nabla_h \mathbf{r}_h^{k+1}\underline{\boldsymbol{u}}_h \to \nabla\boldsymbol{u}$ *strongly in* $L^2(\Omega)^{d\times d}$,
(iii) $|\underline{\boldsymbol{u}}_h|_{\mathrm{s},h} \to 0$,
(iv) $p_h \to p$ *strongly in* $L^2(\Omega)$,

where $(\boldsymbol{u}, p) \in \boldsymbol{U} \times P$ is a solution to the continuous problem (9.3). If, in addition, the solution to (9.3) is unique (which is the case, e.g., if the smallness condition detailed in [199, Eq. (2.12), Chapter IV] holds for $\boldsymbol{f}$), convergence extends to the whole sequence.

Proof The proof proceeds in four steps: (1) we start by proving the existence of a limit for the sequence of discrete solutions; (2) we next show that this limit is indeed a solution of the continuous problem (9.3); (3) we then prove the strong convergence of the velocity gradient and of the jumps; (4) we conclude by proving the strong convergence of the pressure.

Step 1: Existence of a Limit Since, for all $h \in \mathcal{H}$, $(\underline{\boldsymbol{u}}_h, p_h) \in \underline{\boldsymbol{U}}_{h,0}^k \times P_h^k$ solves (9.9), combining the a priori bounds (9.13) and Theorem 9.29 we infer that there exists $(\boldsymbol{u}, p) \in \boldsymbol{U} \times P$ such that, up to a subsequence as $h \to 0$:

(a) $\boldsymbol{u}_h \to \boldsymbol{u}$ strongly in $L^q(\Omega)^d$ for all $q \in [1,\infty)$ if $d = 2$ and $q \in [1, 6)$ if $d = 3$;
(b) $\mathbf{G}_h^l\underline{\boldsymbol{u}}_h \rightharpoonup \nabla\boldsymbol{u}$ weakly in $L^2(\Omega)^{d\times d}$ for all $l \geq 0$;
(c) $\nabla_h \mathbf{r}_h^{k+1}\underline{\boldsymbol{u}}_h \rightharpoonup \nabla\boldsymbol{u}$ weakly in $L^2(\Omega)^{d\times d}$;
(d) $p_h \rightharpoonup p$ weakly in $L^2(\Omega)$.

Step 2: Identification of the Limit We next prove that $(\boldsymbol{u}, p) \in \boldsymbol{U} \times P$ is a solution to (9.3). To do so, we examine the convergence of each term in the discrete problem (9.9) when the test function is the interpolate of a smooth function, then conclude by a density argument.

Let $\boldsymbol{\phi} \in C_c^\infty(\Omega)^d$ and consider the discrete momentum equation (9.9a) with $\underline{\boldsymbol{v}}_h = \underline{\boldsymbol{I}}_h^k\boldsymbol{\phi}$. For the viscous term, we have

$$\mathrm{a}_h(\underline{\boldsymbol{u}}_h, \underline{\boldsymbol{I}}_h^k\boldsymbol{\phi}) = \int_\Omega \nabla_h \mathbf{r}_h^{k+1}\underline{\boldsymbol{u}}_h : \nabla_h \mathbf{r}_h^{k+1}\underline{\boldsymbol{I}}_h^k\boldsymbol{\phi} + \sum_{T\in\mathcal{T}_h} \mathrm{s}_T(\underline{\boldsymbol{u}}_T, \underline{\boldsymbol{I}}_T^k\boldsymbol{\phi}_{|T}) =: \mathfrak{T}_1 + \mathfrak{T}_2.$$

For the first term, a weak-strong convergence argument (recall Point (c) in **Step 1** for the first factor and (9.89) for the second) readily gives $\mathfrak{T}_1 \to \int_\Omega \nabla\boldsymbol{u}:\nabla\boldsymbol{\phi}$. For the second term, we can write, using a Cauchy–Schwarz inequality together with the definition (9.30) of the $|\cdot|_{\mathrm{s},h}$-seminorm and the norm equivalence (8.31), $|\mathfrak{T}_2| \le |\underline{\boldsymbol{u}}_h|_{\mathrm{s},h}|\underline{\boldsymbol{I}}_h^k\boldsymbol{\phi}|_{\mathrm{s},h} \lesssim \|\underline{\boldsymbol{u}}_h\|_{1,h}|\underline{\boldsymbol{I}}_h^k\boldsymbol{\phi}|_{\mathrm{s},h}$. Combined with the uniform a priori bound (9.13) on the first factor and the strong convergence (9.90) of the second factor, this shows that $|\mathfrak{T}_2| \to 0$ as $h \to 0$. In conclusion, we have

$$\mathrm{a}_h(\underline{\boldsymbol{u}}_h, \underline{\boldsymbol{I}}_h^k\boldsymbol{\phi}) \to \int_\Omega \nabla\boldsymbol{u}:\nabla\boldsymbol{\phi} = a(\boldsymbol{u}, \boldsymbol{\phi}). \tag{9.92}$$

For the convective term, recalling the definition (9.47) of $\mathrm{t}_h^{\mathrm{ss}}$ and (7.1) of the tensor product, we have

$$\mathrm{t}_h^{\mathrm{ss}}(\underline{\boldsymbol{u}}_h, \underline{\boldsymbol{u}}_h, \underline{\boldsymbol{I}}_h^k\boldsymbol{\phi}) = \frac{1}{2}\int_\Omega \mathbf{G}_h^{2k}\underline{\boldsymbol{u}}_h : \boldsymbol{\pi}_h^{0,k}\boldsymbol{\phi} \otimes \boldsymbol{u}_h - \frac{1}{2}\int_\Omega \mathbf{G}_h^{2k}\underline{\boldsymbol{I}}_h^k\boldsymbol{\phi} : \boldsymbol{u}_h \otimes \boldsymbol{u}_h =: \mathfrak{T}_3 + \mathfrak{T}_4.$$

Since $\boldsymbol{u}_h \to \boldsymbol{u}$ and $\boldsymbol{\pi}_h^{0,k}\boldsymbol{\phi} \to \boldsymbol{\phi}$ strongly in $L^4(\Omega)^d$, $\boldsymbol{\pi}_h^{0,k}\boldsymbol{\phi} \otimes \boldsymbol{u}_h \to \boldsymbol{\phi} \otimes \boldsymbol{u}$ strongly in $L^2(\Omega)^{d\times d}$. Hence, recalling that $\mathbf{G}_h^{2k}\underline{\boldsymbol{u}}_h \rightharpoonup \nabla\boldsymbol{u}$ weakly in $L^2(\Omega)^{d\times d}$ owing to Point (b) in **Step 1**, we infer that $\mathfrak{T}_3 \to \frac{1}{2}\int_\Omega \nabla\boldsymbol{u}:\boldsymbol{\phi}\otimes\boldsymbol{u} = \frac{1}{2}\int_\Omega(\boldsymbol{u}\cdot\nabla)\boldsymbol{u}\cdot\boldsymbol{\phi}$. For the second term, observing that $\boldsymbol{u}_h \otimes \boldsymbol{u}_h \to \boldsymbol{u}\otimes\boldsymbol{u}$ strongly in $L^2(\Omega)^{d\times d}$ (since $\boldsymbol{u}_h \to \boldsymbol{u}$ strongly in $L^4(\Omega)^d$) and $\mathbf{G}_h^{2k}\underline{\boldsymbol{I}}_h^k\boldsymbol{\phi} \to \nabla\boldsymbol{\phi}$ strongly in $L^2(\Omega)^{d\times d}$ (see (9.88)), we get $\mathfrak{T}_4 \to -\frac{1}{2}\int_\Omega \nabla\boldsymbol{\phi}:\boldsymbol{u}\otimes\boldsymbol{u} = -\frac{1}{2}\int_\Omega(\boldsymbol{u}\cdot\nabla)\boldsymbol{\phi}\cdot\boldsymbol{u}$. In conclusion, recalling the definition (9.8) of the continuous skew-symmetric trilinear form $\tilde{t}$, we have

$$\mathrm{t}_h^{\mathrm{ss}}(\underline{\boldsymbol{u}}_h, \underline{\boldsymbol{u}}_h, \underline{\boldsymbol{I}}_h^k\boldsymbol{\phi}) \to \frac{1}{2}\int_\Omega (\boldsymbol{u}\cdot\nabla)\boldsymbol{u}\cdot\boldsymbol{\phi} - \frac{1}{2}\int_\Omega (\boldsymbol{u}\cdot\nabla)\boldsymbol{\phi}\cdot\boldsymbol{u} = \tilde{t}(\boldsymbol{u}, \boldsymbol{u}, \boldsymbol{\phi}). \tag{9.93}$$

For the pressure–velocity coupling term in the momentum equation, recalling the definition (8.34) of b_h and using a weak-strong convergence argument, it is readily inferred that

$$\mathrm{b}_h(\underline{\boldsymbol{I}}_h^k\boldsymbol{\phi}, p_h) = -\int_\Omega \mathbf{G}_h^k\underline{\boldsymbol{I}}_h^k\boldsymbol{\phi}:p_h\boldsymbol{I}_d \to -\int_\Omega \nabla\boldsymbol{\phi}:p\boldsymbol{I}_d = -\int_\Omega (\nabla\cdot\boldsymbol{\phi})\, p = b(\boldsymbol{\phi}, p), \tag{9.94}$$

where we have additionally used the fact that, by definition (9.65), $\mathrm{D}_T^k\underline{\boldsymbol{v}}_T = \mathbf{G}_T^k\underline{\boldsymbol{v}}_T:\boldsymbol{I}_d$ for all $T \in \mathcal{T}_h$ and all $\underline{\boldsymbol{v}}_T \in \underline{\boldsymbol{U}}_T^k$.

Collecting (9.92), (9.93) and (9.94), and observing that $\int_\Omega \boldsymbol{f}\cdot\boldsymbol{\pi}_h^{0,k}\boldsymbol{\phi} \to \int_\Omega \boldsymbol{f}\cdot\boldsymbol{\phi}$ since $\boldsymbol{\pi}_h^{0,k}\boldsymbol{\phi} \to \boldsymbol{\phi}$ strongly in $L^2(\Omega)^d$, we conclude that $(\boldsymbol{u}, p)$ satisfies

$$\nu a(\boldsymbol{u}, \boldsymbol{\phi}) + \tilde{t}(\boldsymbol{u}, \boldsymbol{u}, \boldsymbol{\phi}) + b(\boldsymbol{\phi}, p) = \int_\Omega \boldsymbol{f}\cdot\boldsymbol{\phi} \qquad \forall \boldsymbol{\phi} \in C_{\rm c}^\infty(\Omega)^d. \tag{9.95}$$

Moving now to the mass balance equation, we observe that it holds, for all $\psi \in C_{\rm c}^\infty(\Omega)$,

$$-\mathrm{b}_h(\underline{\boldsymbol{u}}_h, \pi_h^{0,k}\psi) = \int_\Omega \mathbf{G}_h^k\underline{\boldsymbol{u}}_h{:}\pi_h^{0,k}\psi \boldsymbol{I}_d \to \int_\Omega \nabla\boldsymbol{u}{:}\psi\boldsymbol{I}_d = \int_\Omega (\nabla\cdot\boldsymbol{u})\,\psi = -b(\boldsymbol{u}, \psi),$$

where we have used the weak convergence in $L^2(\Omega)^{d\times d}$ of the first factor to $\nabla\boldsymbol{u}$, resulting from Point (b) in **Step 1**, together with the strong convergence $\pi_h^{0,k}\psi \to \psi$ in $L^2(\Omega)$. Hence, the limit $\boldsymbol{u}$ satisfies

$$-\,b(\boldsymbol{u}, \psi) = 0 \qquad \forall \psi \in C_{\rm c}^\infty(\Omega). \tag{9.96}$$

Combining (9.95) and (9.96), and using the density of $C_{\rm c}^\infty(\Omega)^d$ in $\boldsymbol{U}$ and of $C_{\rm c}^\infty(\Omega)$ in $L^2(\Omega)$, we conclude that $(\boldsymbol{u}, p) \in \boldsymbol{U} \times P$ is a solution to (9.3).

Step 3: Strong Convergence of the Velocity Gradient and of the Jumps Making $\underline{\boldsymbol{v}}_h = \underline{\boldsymbol{u}}_h$ in (9.9a) and observing that $\mathrm{t}_h^{\rm ss}(\underline{\boldsymbol{u}}_h, \underline{\boldsymbol{u}}_h, \underline{\boldsymbol{u}}_h) = 0$ owing to the non-dissipativity property (9.10), and that $\mathrm{b}_h(\underline{\boldsymbol{u}}_h, p_h) = 0$ owing to (9.9b) with $q_h = p_h$, we have

$$\nu\|\nabla_h\mathbf{r}_h^{k+1}\underline{\boldsymbol{u}}_h\|_{L^2(\Omega)^{d\times d}}^2 + \nu|\underline{\boldsymbol{u}}_h|_{{\rm s},h}^2 = \nu\mathrm{a}_h(\underline{\boldsymbol{u}}_h, \underline{\boldsymbol{u}}_h) = \int_\Omega \boldsymbol{f}\cdot\boldsymbol{u}_h.$$

Since $\boldsymbol{u}_h$ converges to $\boldsymbol{u}$ strongly in $L^2(\Omega)^d$ and $\boldsymbol{u}$ is a solution to (9.3), we have

$$\begin{aligned}\nu \limsup_{h\to 0}\left(\|\nabla_h\mathbf{r}_h^{k+1}\underline{\boldsymbol{u}}_h\|_{L^2(\Omega)^{d\times d}}^2 + |\underline{\boldsymbol{u}}_h|_{{\rm s},h}^2\right) &= \limsup_{h\to 0}\int_\Omega \boldsymbol{f}\cdot\boldsymbol{u}_h \\ &= \int_\Omega \boldsymbol{f}\cdot\boldsymbol{u} = \nu\|\nabla\boldsymbol{u}\|_{L^2(\Omega)^{d\times d}}^2.\end{aligned} \tag{9.97}$$

This implies, in particular,

$$\limsup_{h\to 0}\|\nabla_h\mathbf{r}_h^{k+1}\underline{\boldsymbol{u}}_h\|_{L^2(\Omega)^{d\times d}}^2 \le \|\nabla\boldsymbol{u}\|_{L^2(\Omega)^{d\times d}}^2. \tag{9.98}$$

Combined with the weak convergence $\nabla_h\mathbf{r}_h^{k+1}\underline{\boldsymbol{u}}_h \rightharpoonup \nabla\boldsymbol{u}$ in $L^2(\Omega)^{d\times d}$, this inequality establishes the strong convergence $\nabla_h\mathbf{r}_h^{k+1}\underline{\boldsymbol{u}}_h \to \nabla\boldsymbol{u}$ of the velocity gradient in $L^2(\Omega)^{d\times d}$. Hence, the inequality in (9.98) is an equality which, plugged back into (9.97), gives

$$|\underline{\boldsymbol{u}}_h|^2_{\mathrm{s},h} \to 0. \tag{9.99}$$

Step 4: Strong Convergence of the Pressure Observing that $p_h \in P$, from Lemma 8.3 we infer the existence of $\boldsymbol{v}_{p_h} \in \boldsymbol{U}$ such that

$$\nabla\cdot\boldsymbol{v}_{p_h} = p_h \text{ and } \|\boldsymbol{v}_{p_h}\|_{H^1(\Omega)^d} \lesssim \|p_h\|_{L^2(\Omega)}, \tag{9.100}$$

with hidden constant depending only on Ω. Let us study the properties of the sequence $(\underline{\boldsymbol{I}}_h^k \boldsymbol{v}_{p_h})_{h\in\mathcal{H}}$. For all $h \in \mathcal{H}$, it holds

$$\|\underline{\boldsymbol{I}}_h^k \boldsymbol{v}_{p_h}\|_{1,h} \lesssim |\boldsymbol{v}_{p_h}|_{H^1(\Omega)^d} \lesssim \|p_h\|_{L^2(\Omega)} \lesssim \|\boldsymbol{f}\|_{L^2(\Omega)^d} + \nu^{-2}\|\boldsymbol{f}\|^2_{L^2(\Omega)^d}, \tag{9.101}$$

where we have used the boundedness (8.25) of $\underline{\boldsymbol{I}}_h^k$ in the first inequality, (9.100) in the second, and the a priori bound (9.13) on the pressure to conclude. Then, by Theorem 9.29 applied to $\underline{\boldsymbol{v}}_h = \underline{\boldsymbol{I}}_h^k \boldsymbol{v}_{p_h}$, there exists $\boldsymbol{v}_p \in \boldsymbol{U}$ such that, up to a subsequence, $\boldsymbol{\pi}_h^{0,k}\boldsymbol{v}_{p_h} \to \boldsymbol{v}_p$ strongly in $L^q(\Omega)^d$ for all $q \in [1,4]$, $\mathbf{G}_h^{2k}\underline{\boldsymbol{I}}_h^k \boldsymbol{v}_{p_h} \rightharpoonup \nabla\boldsymbol{v}_p$ weakly in $L^2(\Omega)^{d\times d}$, and $\nabla_h \mathbf{r}_h^{k+1}\underline{\boldsymbol{I}}_h^k \boldsymbol{v}_{p_h} \rightharpoonup \nabla\boldsymbol{v}_p$ weakly in $L^2(\Omega)^{d\times d}$. Moreover, by uniqueness of the limit in the distributional sense, it holds that

$$\nabla\cdot\boldsymbol{v}_p = p. \tag{9.102}$$

We can now write

$$\begin{aligned}
\|p_h\|^2_{L^2(\Omega)} &= -b(\boldsymbol{v}_{p_h}, p_h)\\
&= -\mathrm{b}_h(\underline{\boldsymbol{I}}_h^k \boldsymbol{v}_{p_h}, p_h)\\
&= \nu\mathrm{a}_h(\underline{\boldsymbol{u}}_h, \underline{\boldsymbol{I}}_h^k \boldsymbol{v}_{p_h}) + \mathrm{t}_h^{\mathrm{ss}}(\underline{\boldsymbol{u}}_h, \underline{\boldsymbol{u}}_h, \underline{\boldsymbol{I}}_h^k \boldsymbol{v}_{p_h}) - \int_\Omega \boldsymbol{f}\cdot\boldsymbol{\pi}_h^{0,k}\boldsymbol{v}_{p_h},
\end{aligned} \tag{9.103}$$

where we have used the consistency property (8.35) of b_h to pass to the second line and the discrete momentum balance equation (9.9a) with $\underline{\boldsymbol{v}}_h = \underline{\boldsymbol{I}}_h^k \boldsymbol{v}_{p_h}$ to conclude. We study the limit of the three terms on the right of (9.103) using the convergence properties for the discrete solution proved in the previous steps. Combining the strong converge of $\nabla_h \mathbf{r}_h^{k+1}\underline{\boldsymbol{u}}_h$ with the weak convergence of $\nabla_h \mathbf{r}_h^{k+1}\underline{\boldsymbol{I}}_h^k \boldsymbol{v}_{p_h}$ gives $\int_\Omega \nabla\mathbf{r}_h^{k+1}\underline{\boldsymbol{u}}_h{:}\nabla\mathbf{r}_h^{k+1}\underline{\boldsymbol{I}}_h^k \boldsymbol{v}_{p_h} \to \int_\Omega \nabla\boldsymbol{u}{:}\nabla\boldsymbol{v}_p$. Moreover, the convergence (9.99) of the jumps of $\underline{\boldsymbol{u}}_h$ and the uniform bound (9.101) imply $|\mathrm{s}_h(\underline{\boldsymbol{u}}_h, \underline{\boldsymbol{I}}_h^k \boldsymbol{v}_{p_h})| \le |\underline{\boldsymbol{u}}_h|_{\mathrm{s},h}|\underline{\boldsymbol{I}}_h^k \boldsymbol{v}_{p_h}|_{\mathrm{s},h} \to 0$, so that, in conclusion, we have for the viscous term

$$\mathrm{a}_h(\underline{\boldsymbol{u}}_h, \underline{\boldsymbol{I}}_h^k \boldsymbol{v}_{p_h}) \to a(\boldsymbol{u}, \boldsymbol{v}_p).$$

Observing that the convergence properties of the sequences $(\underline{\boldsymbol{u}}_h)_{h\in\mathcal{H}}$ and $(\underline{\boldsymbol{I}}_h^k \boldsymbol{v}_{p_h})_{h\in\mathcal{H}}$ are sufficient to mimic the reasoning for the convective term in **Step 2** of this proof, we deduce that

$$\mathrm{t}_h^{\mathrm{ss}}(\underline{\boldsymbol{u}}_h, \underline{\boldsymbol{u}}_h, \underline{\boldsymbol{I}}_h^k \boldsymbol{v}_{p_h}) \to t(\boldsymbol{u}, \boldsymbol{u}, \boldsymbol{v}_p).$$

Finally, by the strong convergence $\boldsymbol{\pi}_h^{0,k} \boldsymbol{v}_{p_h} \to \boldsymbol{v}_p$ in $L^2(\Omega)^d$, we readily infer for the source term

$$\int_\Omega \boldsymbol{f}\cdot\boldsymbol{\pi}_h^{0,k} \boldsymbol{v}_{p_h} \to \int_\Omega \boldsymbol{f}\cdot\boldsymbol{v}_p.$$

Collecting the above convergence results in (9.103) and using the momentum balance equation (9.3a) together with (9.102) leads to

$$\limsup_{h\to 0} \|p_h\|_{L^2(\Omega)}^2 \le \nu a(\boldsymbol{u}, \boldsymbol{v}_p) + t(\boldsymbol{u}, \boldsymbol{u}, \boldsymbol{v}_p) - \int_\Omega \boldsymbol{f}\cdot\boldsymbol{v}_p = -b(\boldsymbol{v}_p, p) = \|p\|_{L^2(\Omega)}^2,$$

and the strong convergence $p_h \to p$ in $L^2(\Omega)$ classically follows. □

Remark 9.33 (Existence of a Solution to the Continuous Problem) A by-product of Theorem 9.32 is the existence of a solution to the continuous problem (9.3).

9.7 Numerical Examples

In this section we showcase the performance of the method on classical benchmark problems. The numerical results are taken from [70] and are obtained using the discrete trilinear form (9.72) incorporating Temam's device for stability. Numerical examples for the skew-symmetric trilinear form (9.47) can be found in [150, 151].

9.7.1 Kovasznay Flow

We start by assessing the convergence properties of the method using Kovasznay's analytical solution; see [220]. Specifically, in dimension $d = 2$ we solve on the square domain $\Omega := (-0.5, 1.5) \times (0, 2)$ the Dirichlet problem corresponding to the exact solution $(\boldsymbol{u}, p)$ such that, introducing the Reynolds number $\mathrm{Re} := \frac{1}{2\nu}$ and letting $\lambda := \mathrm{Re} - \left(\mathrm{Re}^2 + 4\pi^2\right)^{\frac{1}{2}}$, the velocity components are given by

$$u_1(\boldsymbol{x}) = 1 - \exp(\lambda x_1)\cos(2\pi x_2), \qquad u_2(\boldsymbol{x}) = \frac{\lambda}{2\pi}\exp(\lambda x_1)\sin(2\pi x_2),$$

while the pressure is given by

$$p(\boldsymbol{x}) = -\frac{1}{2}\exp(2\lambda x_1) + \frac{\lambda}{2}\left(\exp(4\lambda) - 1\right).$$

We take here $\nu = 0.025$, corresponding to $\mathrm{Re} = 20$, and consider polynomial degrees $k \in \{0, \ldots, 5\}$ over a sequence of uniformly h-refined Cartesian grids with 2^i, $i \in \{2, 3, \ldots, 7\}$, elements in each direction. We report in Table 9.1 the results for the method (9.32) with upwind convective stabilisation; see Sect. 9.4. The following quantities are monitored: $N_{\mathrm{dof},h}$, the number of degrees of freedom after static condensation (see Remark 9.3); $N_{\mathrm{nz},h}$, the number of non-zero entries of the matrix to be inverted at each nonlinear iteration; $\|\underline{\boldsymbol{u}}_h - \underline{\boldsymbol{I}}_h^k \boldsymbol{u}\|_{\nu,h} := \nu^{\frac{1}{2}} \|\underline{\boldsymbol{u}}_h - \underline{\boldsymbol{I}}_h^k \boldsymbol{u}\|_{\mathrm{a},h}$, the energy norm of the error on the velocity (combining the norm equivalence (8.31) with the error estimate (9.19), we readily infer an estimate in h^{k+1} for this norm under the assumptions of Theorem 9.10); $\|\boldsymbol{u}_h - \boldsymbol{\pi}_h^{0,k} \boldsymbol{u}\|_{L^2(\Omega)^d}$, the L^2-error on the velocity; and $\|p_h - \pi_h^{0,k} p\|_{L^2(\Omega)}$, the L^2-error on the pressure. Denoting by e_i and h_i, respectively, the error in a given norm and the meshsize corresponding to a refinement iteration i, the estimated order of convergence (EOC) is obtained according to the following formula:

$$\mathrm{EOC} = \frac{\log e_i - \log e_{i+1}}{\log h_i - \log h_{i+1}}.$$

The numerical results essentially confirm the theoretical estimate of Theorem 9.10, and a convergence in h^{k+2} is additionally observed for the L^2-norm of the velocity. It can be numerically checked that the slightly suboptimal order of convergence observed for the velocity, in particular for $k = 0$, is to be ascribed to the upwind stabilisation (which, in turn, facilitates convergence on the coarser meshes). Indeed, the highest jumps between element and faces unknowns are observed for under-resolved low degree computations.

9.7.2 *Lid-Driven Cavity Flow*

We next consider the lid-driven cavity flow, one of the most extensively studied problems in fluid mechanics. The computational domain is the unit square $\Omega = (0, 1)^2$. Homogeneous (wall) boundary conditions are enforced at all but the top horizontal wall (at $x_2 = 1$), where we enforce a unit tangential velocity, that is, $\boldsymbol{u} = (1, 0)$. We note that this boundary condition is incompatible with the formulation (9.3), even modified to account for non-homogeneous boundary conditions, since the corresponding exact solution does not belong to $H^1(\Omega)^2$. This is however, as mentioned, a very classical and well-understood test that informs on the quality of the numerical scheme. The velocity magnitude contours for two different values of the Reynolds number are plotted in Fig. 9.1.

Table 9.1 Convergence results for the Kovasznay problem at Re = 20 with upwind stabilisation

$N_{\mathrm{dof},h}$	$N_{\mathrm{nz},h}$	$\Vert\underline{\boldsymbol{u}}_h - \underline{\boldsymbol{I}}_h^k\boldsymbol{u}\Vert_{\nu,h}$	EOC	$\Vert\boldsymbol{u}_h - \boldsymbol{\pi}_h^{0,k}\boldsymbol{u}\Vert_{L^2(\Omega)^d}$	EOC	$\Vert p_h - \pi_h^{0,k}p\Vert_{L^2(\Omega)}$	EOC
$k = 0$							
65	736	9.37e−01	–	1.40e−01	–	6.84e−01	–
289	3808	1.13e+00	−0.27	5.50e−01	−1.98	1.96e−01	1.80
1217	17,056	9.14e−01	0.31	2.26e−01	1.28	1.02e−01	0.94
4993	71,968	6.26e−01	0.55	7.89e−02	1.52	3.52e−02	1.54
20, 225	295,456	3.87e−01	0.70	2.47e−02	1.68	9.78e−03	1.85
81, 409	1,197,088	2.47e−01	0.65	8.06e−03	1.61	3.09e−03	1.66
$k = 1$							
113	2464	7.31e−01	–	5.37e−01	–	2.49e−01	–
513	13,056	3.83e−01	0.93	1.54e−01	1.80	4.29e−02	2.54
2177	59,008	1.02e−01	1.90	2.13e−02	2.85	3.98e−03	3.43
8961	249,984	2.93e−02	1.80	2.97e−03	2.84	6.54e−04	2.61
36, 353	1,028,224	8.23e−03	1.83	3.99e−04	2.90	1.28e−04	2.35
146, 433	4,169,856	2.26e−03	1.86	5.21e−05	2.94	2.65e−05	2.27
$k = 2$							
161	5216	3.50e−01	–	2.09e−01	–	6.42e−02	–
737	27,872	3.76e−02	3.22	1.34e−02	3.96	2.07e−03	4.95
3137	126,368	6.96e−03	2.43	1.31e−03	3.36	1.48e−04	3.80
12, 929	536,096	1.06e−03	2.72	9.48e−05	3.79	1.77e−05	3.07
52, 481	2,206,496	1.55e−04	2.77	6.36e−06	3.90	2.27e−06	2.96
211, 457	8,951,072	2.21e−05	2.81	4.13e−07	3.95	2.72e−07	3.06
$k = 3$							
209	8992	7.93e−02	–	4.41e−02	–	7.58e−03	–
961	48,256	6.23e−03	3.67	1.98e−03	4.48	2.97e−04	4.67
4097	219,136	4.16e−04	3.90	6.43e−05	4.95	1.32e−05	4.49
16, 897	930,304	3.09e−05	3.75	2.20e−06	4.87	8.19e−07	4.01
68, 609	3,830,272	2.28e−06	3.76	7.40e−08	4.89	5.12e−08	4.00
276, 481	15,540,736	1.63e−07	3.81	2.42e−09	4.93	3.14e−09	4.03
$k = 4$							
257	13,792	1.42e−02	–	7.89e−03	–	1.83e−03	–
1185	74,208	4.24e−04	5.07	1.14e−04	6.11	2.05e−05	6.48
5057	337,312	1.81e−05	4.55	2.57e−06	5.48	6.39e−07	5.00
20, 865	1,432,608	6.90e−07	4.71	4.55e−08	5.82	2.28e−08	4.81
84, 737	5,899,552	2.59e−08	4.74	7.59e−10	5.91	7.64e−10	4.90
341, 505	23,938,848	9.53e−10	4.76	1.23e−11	5.95	2.42e−11	4.98
$k = 5$							
305	19,616	2.28e−03	–	1.05e−03	–	1.70e−04	–
1409	105,728	4.01e−05	5.83	1.05e−05	6.65	2.05e−06	6.37
6017	480,896	7.21e−07	5.80	8.98e−08	6.87	3.21e−08	6.00
24, 833	2,043,008	1.37e−08	5.72	7.89e−10	6.83	5.43e−10	5.88
100, 865	8,414,336	2.56e−10	5.74	6.72e−12	6.88	9.14e−12	5.89

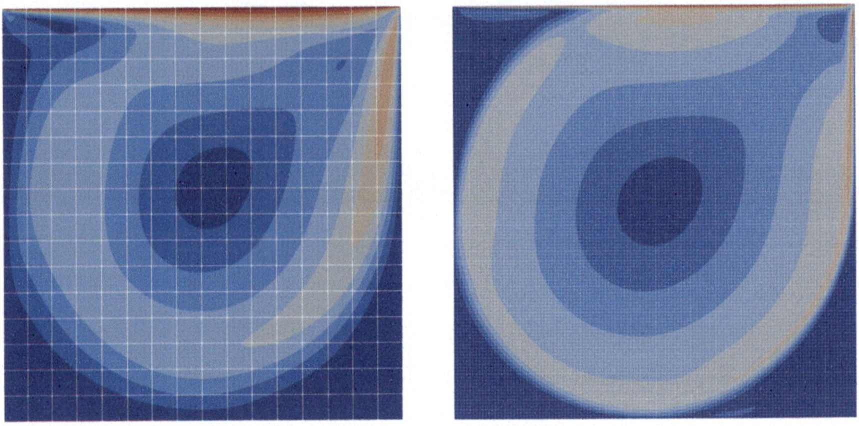

Fig. 9.1 Lid-driven cavity flow, velocity magnitude contours (10 equispaced values in the range [0, 1]) for $k = 7$. Computations at $\mathrm{Re} = 1000$ (left: 16×16 grid) and $\mathrm{Re} = 20{,}000$ (right: 128×128 grid)

In Figs. 9.2, 9.3, 9.4, and 9.5 we report the horizontal component u_1 of the velocity along the vertical centreline $x_1 = \frac{1}{2}$ and the vertical component u_2 of the velocity along the horizontal centreline $x_2 = \frac{1}{2}$ for the two dimensional flow at Reynolds numbers $\mathrm{Re} := \frac{1}{\nu}$ respectively equal to 1000, 5000, 10,000, and 20,000. The reference computation is carried out on a 128×128 Cartesian mesh with $k = 1$. For the sake of comparison, we also include very high-order computations with $k = 7$ on progressively finer Cartesian grids: 16×16 for $\mathrm{Re} = 1000$, 32×32 for $\mathrm{Re} = 5000$, 64×64 for $\mathrm{Re} = 10{,}000$, and 128×128 for $\mathrm{Re} = 20{,}000$. The high-order solutions corresponding to $\mathrm{Re} = 1000$ and $\mathrm{Re} = 20{,}000$ are displayed in Fig. 9.1. When available, reference solutions from the literature [185, 197] are also plotted for the sake of comparison.

We remark that the solid blue and red lines outlining, respectively, the behavior of low-order ($k = 1$) and high-order ($k = 7$) velocity approximations are perfectly superimposed at low Reynolds numbers, while significant differences are present starting from $\mathrm{Re} = 10{,}000$. In particular, at $\mathrm{Re} = 20{,}000$, computations with $k = 1$ are in better agreement with reference solutions by Erturk et al. [185]. Nevertheless, since high-polynomial degrees over coarse meshes provide accurate results at low Reynolds numbers, we are led to think that the HHO computations with $k = 1$ are over-dissipative at high Reynolds numbers. Indeed, strong velocity gradients observed close to cavity walls and multiple counter-rotating vortices developing at the bottom corners are known to be very demanding, both from the stability and the accuracy viewpoints. Note that the thin jet originating at the top-right corner is contained in exactly one mesh element, both on the 16×16 grid for $\mathrm{Re} = 1000$ and on the 128×128 grid for $\mathrm{Re} = 20{,}000$, see Fig. 9.1.

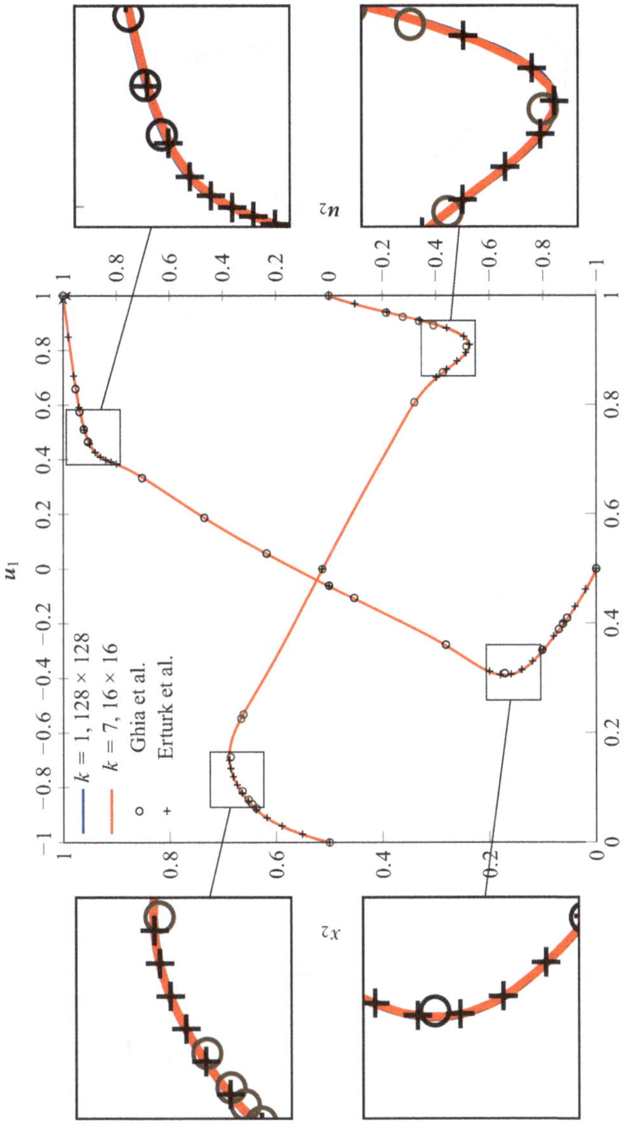

Fig. 9.2 Lid-driven cavity flow, horizontal component u_1 of the velocity along the vertical centreline $x_1 = \frac{1}{2}$ and vertical component u_2 of the velocity along the horizontal centreline $x_2 = \frac{1}{2}$ for Re $= 1000$

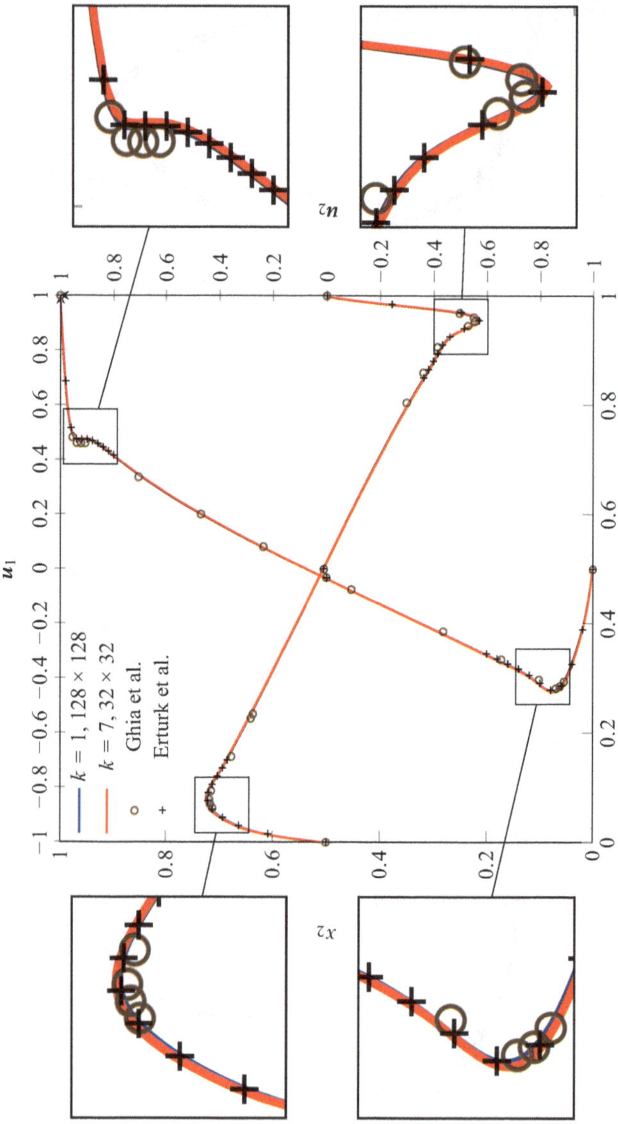

Fig. 9.3 Lid-driven cavity flow, horizontal component u_1 of the velocity along the vertical centreline $x_1 = \frac{1}{2}$ and vertical component u_2 of the velocity along the horizontal centreline $x_2 = \frac{1}{2}$ for Re $= 5000$

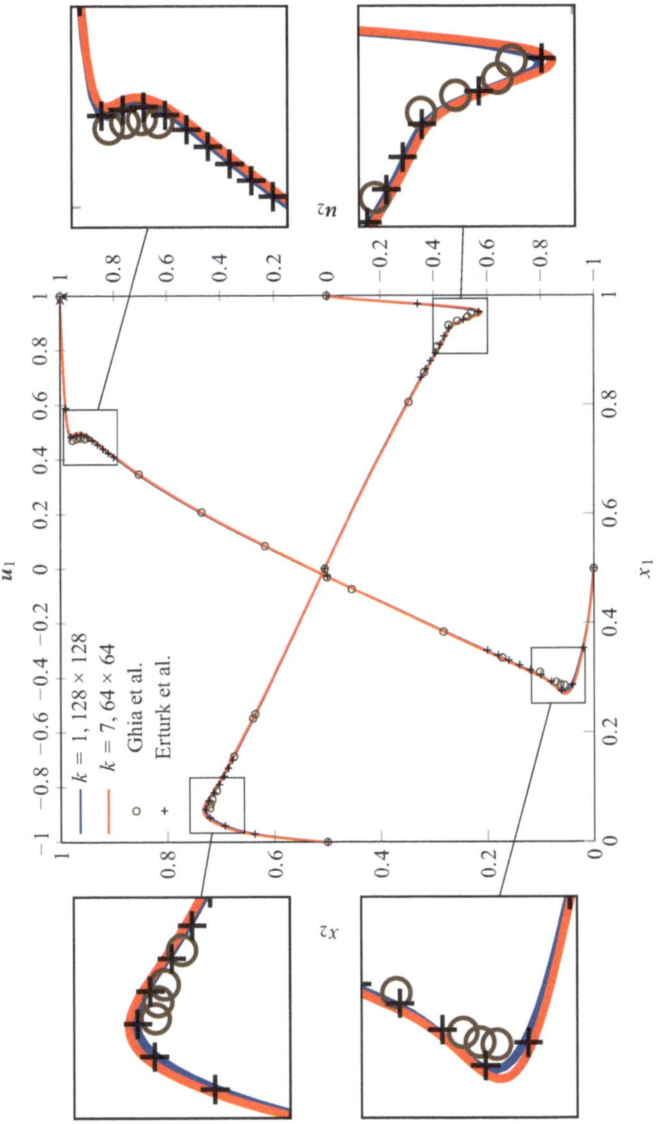

Fig. 9.4 Lid-driven cavity flow, horizontal component u_1 of the velocity along the vertical centreline $x_1 = \frac{1}{2}$ and vertical component u_2 of the velocity along the horizontal centreline $x_2 = \frac{1}{2}$ for Re $= 10{,}000$

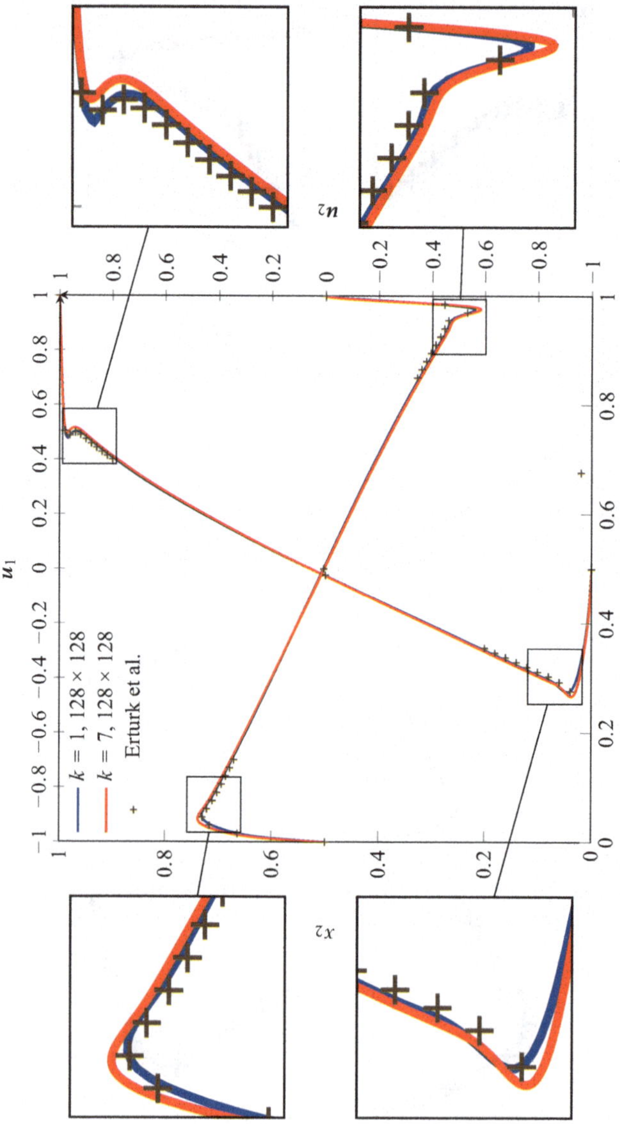

Fig. 9.5 Lid-driven cavity flow, horizontal component u_1 of the velocity along the vertical centreline $x_1 = \frac{1}{2}$ and vertical component u_2 of the velocity along the horizontal centreline $x_2 = \frac{1}{2}$ for Re = 20,000

Correction to: The Hybrid High-Order Method for Polytopal Meshes

Correction to:
D. A. Di Pietro, J. Droniou,
***The Hybrid High-Order Method for Polytopal Meshes*, MS&A 19,**
https://doi.org/10.1007/978-3-030-37203-3

Inadvertently, the author name was captured in the metadata of the chapters as **Daniele Di Antonio Pietro**, whereas it should be **Daniele Antonio Di Pietro**. This has been corrected now in the metadata.

The updated online version of the book can be found at
https://doi.org/10.1007/978-3-030-37203-3

D. A. Di Pietro, J. Droniou, *The Hybrid High-Order Method for Polytopal Meshes*,
MS&A 19, https://doi.org/10.1007/978-3-030-37203-3_10

Appendix A
Error Analysis Setting for Schemes in Fully Discrete Formulation

We recall here the basic setting and results from the theory presented in [143].

A.1 General Case

A.1.1 Setting

We consider a setting where the continuous and discrete problems are both written in classical variational formulations. For the continuous problem, we choose:

- a Hilbert space $\mathtt{U}$;
- a continuous bilinear form $\mathtt{a} : \mathtt{U} \times \mathtt{U} \to \mathbb{R}$;
- a continuous linear form $\mathtt{l} : \mathtt{U} \to \mathbb{R}$.

The problem we aim at approximating, and whose unique solution is assumed to exist, is

$$\text{Find } u \in \mathtt{U} \text{ such that } \quad \mathtt{a}(u, v) = \mathtt{l}(v) \quad \forall v \in \mathtt{U}. \tag{A.1}$$

This problem is referred to as *continuous problem*, since $\mathtt{U}$ is usually infinite-dimensional.

The approximation is written in fully discrete formulation, using an approximation space that is not necessarily a space of functions—it is not necessarily embedded in any natural space in which $\mathtt{U}$ is also embedded. We consider thus:

- a space $\mathtt{U}_h$, with norm $\|\cdot\|_{\mathtt{U}_h}$;
- an interpolator $\mathtt{I}_h : \mathtt{U} \to \mathtt{U}_h$;
- a bilinear form $\mathtt{a}_h : \mathtt{U}_h \times \mathtt{U}_h \to \mathbb{R}$;
- a linear form $\mathtt{l}_h : \mathtt{U}_h \to \mathbb{R}$.

D. A. Di Pietro, J. Droniou, *The Hybrid High-Order Method for Polytopal Meshes*, MS&A 19, https://doi.org/10.1007/978-3-030-37203-3

Note that $\mathtt{I}_h$ is not required to be linear or even continuous. Likewise, the continuity of $\mathtt{a}_h$ or $\mathtt{l}_h$ is not directly used, but is always verified in practice, and of course usually required to ensure the existence of a solution. The index h represents a discretisation parameter (e.g., the meshsize) which characterises the space $\mathtt{U}_h$, and such that convergence of the method (in a sense to be made precise) is expected when $h \to 0$.

The approximation of (A.1) is

$$\text{Find } u_h \in \mathtt{U}_h \text{ such that } \quad \mathtt{a}_h(u_h, v_h) = \mathtt{l}_h(v_h) \quad \forall v_h \in \mathtt{U}_h. \tag{A.2}$$

In what follows, (A.2) will be referred to as the *discrete problem*, since the space $\mathtt{U}_h$ is usually finite dimensional.

We intend to compare the solutions to (A.1) and (A.2) by estimating $u_h - \mathtt{I}_h u$, where u is the solution to (A.1).

A.1.2 Third Strang Lemma

We now describe a notion of stability of $\mathtt{a}_h$ that yields a bound on the solutions to (A.2)

Definition A.1 (Inf–sup Stability) The bilinear form $\mathtt{a}_h$ is inf–sup stable for $\|\cdot\|_{\mathtt{U}_h}$ if

$$\exists \gamma > 0 \text{ such that } \sup_{v_h \in \mathtt{U}_h \backslash \{0\}} \frac{\mathtt{a}_h(u_h, v_h)}{\|v_h\|_{\mathtt{U}_h}} \geq \gamma \|u_h\|_{\mathtt{U}_h} \quad \forall u_h \in \mathtt{U}_h. \tag{A.3}$$

Remark A.2 (Uniform inf–sup Stability) In practice, one typically requires the real number γ independent of discretisation parameters such as the meshsize, so that the condition (A.3) is verified uniformly. This is needed to have optimal error estimates.

Remark A.3 (Coercivity) The inf–sup stability is of course satisfied if $\mathtt{a}_h$ is coercive in the sense that $\mathtt{a}_h(v_h, v_h) \geq \gamma \|v_h\|_{\mathtt{U}_h}^2$ for all $v_h \in \mathtt{U}_h$, where γ does not depend on v_h. To check it, it suffices to write, for any $u_h \in \mathtt{U}_h$,

$$\gamma \|u_h\|_{\mathtt{U}_h}^2 \leq \mathtt{a}_h(u_h, u_h) \leq \left(\sup_{v_h \in \mathtt{U}_h \backslash \{0\}} \frac{\mathtt{a}_h(u_h, v_h)}{\|v_h\|_{\mathtt{U}_h}} \right) \|u_h\|_{\mathtt{U}_h}.$$

If Z is a Banach space with norm $\|\cdot\|_Z$, the dual norm of a linear form $\mu : Z \to \mathbb{R}$ is classically defined by

$$\|\mu\|_{Z^\star} = \sup_{z \in Z \backslash \{0\}} \frac{|\mu(z)|}{\|z\|_Z}. \tag{A.4}$$

Proposition A.4 (Stability of (A.2)) *If* $\mathtt{a}_h$ *is inf–sup stable in the sense of Definition A.1,* $m_h : \mathtt{U}_h \to \mathbb{R}$ *is linear and* w_h *satisfies*

$$\mathtt{a}_h(w_h, v_h) = m_h(v_h) \qquad \forall v_h \in \mathtt{U}_h,$$

then,

$$\|w_h\|_{\mathtt{U}_h} \leq \gamma^{-1} \|m_h\|_{\mathtt{U}_h^\star}.$$

Proof Take $v_h \in \mathtt{U}_h \backslash \{0\}$ and write, by definition of $\|\cdot\|_{\mathtt{U}_h^\star}$,

$$\frac{\mathtt{a}_h(w_h, v_h)}{\|v_h\|_{\mathtt{U}_h}} = \frac{m_h(v_h)}{\|v_h\|_{\mathtt{U}_h}} \leq \|m_h\|_{\mathtt{U}_h^\star}.$$

The proof is completed by taking the supremum over such v_h and using (A.3). □

The next notion of consistency enables us, in combination with the inf–sup stability, to prove an estimate on the error $u_h - \mathtt{I}_h u$ in the $\mathtt{U}_h$ norm.

Definition A.5 (Consistency Error and Consistency) Let u be the solution to the continuous problem (A.1). The *consistency error* is the linear form $\mathcal{E}_h(u; \cdot) : \mathtt{U}_h \to \mathbb{R}$ such that, for any $v_h \in \mathtt{U}_h$,

$$\mathcal{E}_h(u; v_h) := \mathtt{l}_h(v_h) - \mathtt{a}_h(\mathtt{I}_h u, v_h). \tag{A.5}$$

Let now a family $(\mathtt{U}_h, \mathtt{a}_h, \mathtt{l}_h)_{h \to 0}$ of spaces and forms be given, and consider the corresponding family of discrete problems (A.2). We say that *consistency* holds if

$$\|\mathcal{E}_h(u; \cdot)\|_{\mathtt{U}_h^\star} \to 0 \text{ as } h \to 0.$$

Remark A.6 (Choice of $\mathtt{I}_h$*)* No particular property is required here on $\mathtt{I}_h u$; it could actually be any element of $\mathtt{U}_h$. However, for the estimates that follow to be meaningful, it is expected that $\mathtt{I}_h u$ is computed from u, not necessarily in a linear way, so that it encodes meaningful properties of u itself. In particular, $\mathtt{I}_h u$ should draw some information from the fact that u solves (A.1), to ensure a certain smallness of the consistency error. See, for example, the proof of Point (ii) in Lemma 2.18 in Chap. 2.

The following lemma establishes estimates on $u_h - \mathtt{I}_h u$, and is at the core of the proofs of Theorems 2.27, 3.18, 3.32, 3.38, 4.16, 7.33 and 7.45. The name "Strang 3" refers to the fact that, while this result is obtained in a similar spirit of the first two Strang lemmas [263, 264], it covers the more general case of schemes written in fully discrete formulation.

Lemma A.7 (Strang 3) *Assume that* $\mathtt{a}_h$ *is inf–sup stable in the sense of Definition A.1. Let* u *be a solution to* (A.1), *and recall the definition* (A.5) *of the consistency error* $\mathcal{E}_h(u;\cdot)$. *If* u_h *is a solution to* (A.2), *then*

$$\|u_h - \mathtt{I}_h u\|_{\mathtt{U}_h} \le \gamma^{-1}\|\mathcal{E}_h(u;\cdot)\|_{\mathtt{U}_h^\star}. \tag{A.6}$$

As a consequence, letting a family $(\mathtt{U}_h, \mathtt{a}_h, \mathtt{l}_h)_{h\to 0}$ *of spaces and forms be given, if consistency holds, then we have* convergence *in the following sense:*

$$\|u_h - \mathtt{I}_h u\|_{\mathtt{U}_h} \to 0 \text{ as } h \to 0.$$

Proof For any $v_h \in \mathtt{U}_h$, the scheme (A.2) yields

$$\mathtt{a}_h(u_h - \mathtt{I}_h u, v_h) = \mathtt{a}_h(u_h, v_h) - \mathtt{a}_h(\mathtt{I}_h u, v_h) = \mathtt{l}_h(v_h) - \mathtt{a}_h(\mathtt{I}_h u, v_h).$$

Recalling the definition of the consistency error, we infer that the error $u_h - \mathtt{I}_h u$ is a solution to the following problem

$$\mathtt{a}_h(u_h - \mathtt{I}_h u, v_h) = \mathcal{E}_h(u; v_h) \qquad \forall v_h \in \mathtt{U}_h, \tag{A.7}$$

which is therefore referred to as the *error equation*. The proof is completed by applying Proposition A.4 to $m_h = \mathcal{E}_h(u;\cdot)$ and $w_h = u_h - \mathtt{I}_h u$. □

Remark A.8 (Quasi-Optimality of the Error Estimate) Let

$$\|\mathtt{a}_h\|_{\mathtt{U}_h\times\mathtt{U}_h} := \sup_{w_h, v_h\in\mathtt{U}_h\setminus\{0\}} \frac{|\mathtt{a}_h(w_h, v_h)|}{\|w_h\|_{\mathtt{U}_h}\|v_h\|_{\mathtt{U}_h}} \tag{A.8}$$

be the standard norm of the bilinear form $\mathtt{a}_h$. The error equation (A.7) shows that

$$\|\mathcal{E}_h(u;\cdot)\|_{\mathtt{U}_h^\star} \le \|\mathtt{a}_h\|_{\mathtt{U}_h\times\mathtt{U}_h}\|u_h - \mathtt{I}_h u\|_{\mathtt{U}_h}.$$

Hence, if $\|\mathtt{a}_h\|_{\mathtt{U}_h\times\mathtt{U}_h}$ (and γ, see Remark A.2) remains bounded with respect to h as $h \to 0$, which is always the case in practice, the estimate (A.6) is quasi-optimal in the sense that, for some $C > 0$ not depending on h,

$$C^{-1}\|\mathcal{E}_h(u;\cdot)\|_{\mathtt{U}_h^\star} \le \|u_h - \mathtt{I}_h u\|_{\mathtt{U}_h} \le C\|\mathcal{E}_h(u;\cdot)\|_{\mathtt{U}_h^\star}.$$

A.1.3 Aubin–Nitsche Trick

Assume now that $\mathtt{U}$ is continuously embedded in a Banach space $\mathtt{L}$, with norm denoted by $\|\cdot\|_{\mathtt{L}}$, and that there exists a linear reconstruction operator

$$\mathtt{r}_h : \mathtt{U}_h \to \mathtt{L}. \tag{A.9}$$

If $\mathtt{r}_h$ is continuous with continuity constant C, then (A.6) readily gives

$$\|\mathtt{r}_h(u_h - \mathtt{I}_h u)\|_{\mathtt{L}} \le C\gamma^{-1}\|\mathcal{E}_h(u;\cdot)\|_{\mathtt{U}_h^\star}.$$

Our aim here is to improve this estimate by using an Aubin–Nitsche trick. To this purpose, we assume that, for all $\mathtt{g} \in \mathtt{L}^\star$ (space of linear forms $\mathtt{L} \to \mathbb{R}$), there exists a solution to the continuous dual problem: Find $z_{\mathtt{g}} \in \mathtt{U}$ such that

$$\mathtt{a}(w, z_{\mathtt{g}}) = \mathtt{g}(w) \qquad \forall w \in \mathtt{U}. \tag{A.10}$$

Definition A.9 (Dual Consistency Error) Under Assumption (A.9), let $\mathtt{g} \in \mathtt{L}^\star$ and $z_{\mathtt{g}}$ be a solution to the dual problem (A.10). The dual consistency error of $z_{\mathtt{g}}$ is the linear form $\mathcal{E}_h^{\mathrm{d}}(z_{\mathtt{g}};\cdot) : \mathtt{U}_h \to \mathbb{R}$ such that, for all $v_h \in \mathtt{U}_h$,

$$\mathcal{E}_h^{\mathrm{d}}(z_{\mathtt{g}}; v_h) := \mathtt{g}(\mathtt{r}_h(v_h)) - \mathtt{a}_h(v_h, \mathtt{I}_h z_{\mathtt{g}}).$$

The improved estimate in the weaker norm is stated in the following lemma. Examples of usage of this lemma can be found in the proofs of Lemma 2.33 and Theorems 3.42, 4.21, and 7.37.

Lemma A.10 (Aubin–Nitsche) *Assume* (A.9) *and that the dual problem* (A.10) *has a solution* $z_{\mathtt{g}}$ *for any* $\mathtt{g} \in \mathtt{L}^\star$. *Let* $B_{\mathtt{L}^\star} = \{\mathtt{g} \in \mathtt{L}^\star : \|\mathtt{g}\|_{\mathtt{L}^\star} \le 1\}$ *be the unit ball in* $\mathtt{L}^\star$. *Let* u *and* u_h *be solutions to* (A.1) *and* (A.2), *respectively. Then,*

$$\|\mathtt{r}_h(u_h - \mathtt{I}_h u)\|_{\mathtt{L}} \le \|u_h - \mathtt{I}_h u\|_{\mathtt{U}_h} \sup_{\mathtt{g}\in B_{\mathtt{L}^\star}} \|\mathcal{E}_h^{\mathrm{d}}(z_{\mathtt{g}};\cdot)\|_{\mathtt{U}_h^\star} + \sup_{\mathtt{g}\in B_{\mathtt{L}^\star}} \mathcal{E}_h(u; \mathtt{I}_h z_{\mathtt{g}}). \tag{A.11}$$

Proof Let $\mathtt{g} \in B_{\mathtt{L}^\star}$. We have, by definition of $\mathcal{E}_h^{\mathrm{d}}(z_{\mathtt{g}};\cdot)$ and for $w_h \in \mathtt{U}_h$,

$$\mathtt{g}(\mathtt{r}_h(w_h)) = \mathcal{E}_h^{\mathrm{d}}(z_{\mathtt{g}}; w_h) + \mathtt{a}_h(w_h, \mathtt{I}_h z_{\mathtt{g}}).$$

Applied to $w_h = u_h - \mathtt{I}_h u$ and recalling the error equation (A.7), this gives

$$\mathtt{g}(\mathtt{r}_h(u_h - \mathtt{I}_h u)) = \mathcal{E}_h^{\mathrm{d}}(z_{\mathtt{g}}; u_h - \mathtt{I}_h u) + \mathcal{E}_h(u; \mathtt{I}_h z_{\mathtt{g}}).$$

Taking the supremum over $\mathtt{g} \in B_{\mathtt{L}^\star}$ and recalling that $\sup_{\mathtt{g}\in B_{\mathtt{L}^\star}} \mathtt{g}(w) = \|w\|_{\mathtt{L}}$ for all $w \in \mathtt{L}$, we infer

$$\|\mathtt{r}_h(u_h - \mathtt{I}_h u)\|_{\mathtt{L}} \le \sup_{\mathtt{g}\in B_{\mathtt{L}^\star}} \mathcal{E}_h^{\mathrm{d}}(z_{\mathtt{g}}; u_h - \mathtt{I}_h u) + \sup_{\mathtt{g}\in B_{\mathtt{L}^\star}} \mathcal{E}_h(u; \mathtt{I}_h z_{\mathtt{g}}). \tag{A.12}$$

The proof is completed recalling the definition (A.4) of the dual norm to write

$$\mathcal{E}_h^{\mathrm{d}}(z_{\mathtt{g}}; u_h - \mathtt{I}_h u) \le \|u_h - \mathtt{I}_h u\|_{\mathtt{U}_h} \|\mathcal{E}_h^{\mathrm{d}}(z_{\mathtt{g}};\cdot)\|_{\mathtt{U}_h^\star}.$$

□

A.2 Saddle-Point Problems

A.2.1 Setting

We now consider a special situation where the continuous and discrete problems are saddle-point equations, typically resulting from the formulation of constrained minimisation problems using Lagrange multipliers. For the continuous problem, consider:

- two Hilbert spaces $\mathtt{U}$ and $\mathtt{P}$, respectively continuously embedded in Banach spaces $\mathtt{L}$ and $\mathtt{L}'$;
- two continuous bilinear forms $\mathtt{a} : \mathtt{U} \times \mathtt{U} \to \mathbb{R}$ and $\mathtt{b} : \mathtt{U} \times \mathtt{P} \to \mathbb{R}$;
- two continuous linear forms $\mathtt{l} : \mathtt{L} \to \mathbb{R}$ and $\mathtt{m} : \mathtt{L}' \to \mathbb{R}$.

We then write: Find $(u, p) \in \mathtt{U} \times \mathtt{P}$ such that

$$\mathtt{a}(u, v) + \mathtt{b}(v, p) = \mathtt{l}(v) \qquad \forall v \in \mathtt{U}, \tag{A.13a}$$

$$-\mathtt{b}(u, q) = \mathtt{m}(q) \qquad \forall q \in \mathtt{P}. \tag{A.13b}$$

Reminiscent of the application to the Stokes problem treated in Chap. 8, we will refer to u and p as *velocity* and *pressure*. Problem (A.13) can be recast into the variational form (A.1) defining the Cartesian product space $\mathtt{X} := \mathtt{U} \times \mathtt{P}$ and the global bilinear form $\mathtt{A} : \mathtt{X} \times \mathtt{X} \to \mathbb{R}$ such that, for all $(u, p), (v, q) \in \mathtt{X}$,

$$\mathtt{A}((u, p), (v, q)) := \mathtt{a}(u, v) + \mathtt{b}(v, p) - \mathtt{b}(u, q),$$

and writing: Find $(u, p) \in \mathtt{X}$ such that

$$\mathtt{A}((u, p), (v, q)) = \mathtt{l}(v) + \mathtt{m}(q) \qquad \forall (v, q) \in \mathtt{X}. \tag{A.14}$$

The approximation is based on the following spaces and forms:

- two spaces $\mathtt{U}_h$ and $\mathtt{P}_h$, with respective norms $\|\cdot\|_{\mathtt{U}_h}$ and $\|\cdot\|_{\mathtt{P}_h}$;
- two interpolation operators $\mathtt{I}_h : \mathtt{U} \to \mathtt{U}_h$ and $\mathtt{J}_h : \mathtt{P} \to \mathtt{P}_h$;
- two continuous bilinear forms $\mathtt{a}_h : \mathtt{U}_h \times \mathtt{U}_h \to \mathbb{R}$ and $\mathtt{b}_h : \mathtt{U}_h \times \mathtt{P}_h \to \mathbb{R}$;
- two linear forms $\mathtt{l}_h : \mathtt{U}_h \to \mathbb{R}$ and $\mathtt{m}_h : \mathtt{P}_h \to \mathbb{R}$.

The dual norms $\|\cdot\|_{\mathtt{U}_h,\star}$ and $\|\cdot\|_{\mathtt{P}_h,\star}$ are defined on the dual spaces of $\mathtt{U}_h$ and $\mathtt{P}_h$, respectively, in a similar way as (A.4).

The approximation of (A.13) reads: Find $(u_h, p_h) \in \mathtt{U}_h \times \mathtt{P}_h$ such that

$$\mathtt{a}_h(u_h, v_h) + \mathtt{b}_h(v_h, p_h) = \mathtt{l}_h(v_h) \qquad \forall v_h \in \mathtt{U}_h, \tag{A.15a}$$

$$-\mathtt{b}_h(u_h, q_h) = \mathtt{m}_h(q_h) \qquad \forall q_h \in \mathtt{P}_h. \tag{A.15b}$$

As for the continuous problem, we can recast (A.15) into the variational form (A.2) introducing the global space $\mathtt{X}_h := \mathtt{U}_h \times \mathtt{P}_h$ and the global bilinear form $\mathtt{A}_h : \mathtt{X}_h \times \mathtt{X}_h \to \mathbb{R}$ such that, for all $(u_h, p_h), (v_h, q_h) \in \mathtt{X}_h$,

$$\mathtt{A}_h((u_h, p_h), (v_h, q_h)) := \mathtt{a}_h(u_h, v_h) + b_h(v_h, p_h) - b_h(u_h, q_h), \tag{A.16}$$

and writing: Find $(u_h, p_h) \in \mathtt{X}_h$ such that

$$\mathtt{A}_h((u_h, p_h), (v_h, q_h)) = \mathtt{l}_h(v_h) + \mathtt{m}_h(q_h) \qquad \forall (v_h, q_h) \in \mathtt{X}_h. \tag{A.17}$$

A.2.2 *Stability and Energy Error Estimate*

For saddle-point problems, it is usually easier to check the stability of the discrete bilinear forms $\mathtt{a}_h$ and $\mathtt{b}_h$ separately. Equip $\mathtt{X}_h$ with the norm such that, for all $(v_h, q_h) \in \mathtt{X}_h$,

$$\|(v_h, q_h)\|_{\mathtt{X}_h} := \left(\|v_h\|_{\mathtt{U}_h}^2 + \|q_h\|_{\mathtt{P}_h}^2\right)^{\frac{1}{2}}. \tag{A.18}$$

The following result identifies sufficient conditions on $\mathtt{a}_h$ and $\mathtt{b}_h$ under which the inf–sup stability in the sense of Definition A.1 holds for $\mathtt{A}_h$.

Lemma A.11 (Saddle-Point Stability) *Assume that $\mathtt{a}_h$ is coercive and $\mathtt{b}_h$ is inf–sup stable, that is,*

$$\exists \alpha > 0 \text{ such that } \mathtt{a}_h(v_h, v_h) \geq \alpha \|v_h\|_{\mathtt{U}_h}^2 \quad \forall v_h \in \mathtt{U}_h\,, \tag{A.19}$$

$$\exists \beta > 0 \text{ such that } \sup_{v_h \in \mathtt{U}_h \setminus \{0\}} \frac{\mathtt{b}_h(v_h, q_h)}{\|v_h\|_{\mathtt{U}_h}} \geq \beta \|q_h\|_{\mathtt{P}_h} \quad \forall q_h \in \mathtt{P}_h. \tag{A.20}$$

Then, recalling the definition (A.8) *of the norm of $\mathtt{a}_h$ and setting*

$$\gamma := \left[\alpha^{-2}\left(1 + 2\beta^{-2}\|\mathtt{a}_h\|_{\mathtt{U}_h \times \mathtt{U}_h}^2\right)^2 + 4\beta^{-2}\right]^{-\frac{1}{2}}, \tag{A.21}$$

it holds that

$$\sup_{(v_h, q_h) \in \mathtt{X}_h \setminus \{0\}} \frac{\mathtt{A}_h((u_h, p_h), (v_h, q_h))}{\|(v_h, q_h)\|_{\mathtt{X}_h}} \geq \gamma \|(u_h, p_h)\|_{\mathtt{X}_h} \quad \forall (u_h, p_h) \in \mathtt{X}_h. \tag{A.22}$$

Remark A.12 (Inf-sup Stable $\mathtt{a}_h$) A more general set of conditions is obtained replacing (A.19) with the inf–sup stability of $\mathtt{a}_h$ in the kernel of the operator on $\mathtt{U}_h$ associated to $\mathtt{b}_h$; see, e.g., [57].

Proof Denote by $\mathcal{S}$ the supremum in the left-hand side of (A.22). Using the coercivity (A.19) of a_h together with the definition (A.16) of A_h followed by a passage to the supremum, it is inferred that

$$\|u_h\|_{\mathrm{U}_h}^2 \le \alpha^{-1}\mathrm{a}_h(u_h, u_h) = \alpha^{-1}\mathrm{A}_h((u_h, p_h), (u_h, p_h)) \le \alpha^{-1}\mathcal{S}\|(u_h, p_h)\|_{\mathrm{X}_h}. \tag{A.23}$$

On the other hand, using the inf–sup condition (A.20) on b_h followed by the definition (A.16) of A_h and the continuity of a_h, we can write

$$\begin{aligned}
\|p_h\|_{\mathrm{P}_h} &\le \beta^{-1} \sup_{v_h\in\mathrm{U}_h\setminus\{0\}} \frac{\mathrm{b}_h(v_h, p_h)}{\|v_h\|_{\mathrm{U}_h}} \\
&= \beta^{-1} \sup_{v_h\in\mathrm{U}_h\setminus\{0\}} \frac{\mathrm{A}_h((u_h, p_h), (v_h, 0)) - \mathrm{a}_h(u_h, v_h)}{\|v_h\|_{\mathrm{U}_h}} \\
&\le \beta^{-1}\left(\mathcal{S} + \|\mathrm{a}_h\|_{\mathrm{U}_h\times\mathrm{U}_h}\|u_h\|_{\mathrm{U}_h}\right).
\end{aligned}$$

Squaring the above inequality, adding it to (A.23), using the inequality $(a+b)^2 \le 2(a^2+b^2)$ with $a = \mathcal{S}$ and $b = \|\mathrm{a}_h\|_{\mathrm{U}_h\times\mathrm{U}_h}\|u_h\|_{\mathrm{U}_h}$, and recalling (A.18), we have that

$$\begin{aligned}
\|(u_h, p_h)\|_{\mathrm{X}_h}^2 &\le \alpha^{-1}\mathcal{S}\|(u_h, p_h)\|_{\mathrm{X}_h} + 2\beta^{-2}\mathcal{S}^2 + 2\beta^{-2}\|\mathrm{a}_h\|_{\mathrm{U}_h\times\mathrm{U}_h}^2\|u_h\|_{\mathrm{U}_h}^2 \\
&\le \alpha^{-1}\mathcal{S}\|(u_h, p_h)\|_{\mathrm{X}_h} + 2\beta^{-2}\mathcal{S}^2 + 2\beta^{-2}\alpha^{-1}\|\mathrm{a}_h\|_{\mathrm{U}_h\times\mathrm{U}_h}^2\mathcal{S}\|(u_h, p_h)\|_{\mathrm{X}_h} \\
&= \alpha^{-1}\left(1 + 2\beta^{-2}\|\mathrm{a}_h\|_{\mathrm{U}_h\times\mathrm{U}_h}^2\right)\mathcal{S}\|(u_h, p_h)\|_{\mathrm{X}_h} + 2\beta^{-2}\mathcal{S}^2,
\end{aligned}$$

where we have used (A.23) to pass to the second line and rearranged the terms to pass to the third. Using the Young inequality on the first term in the right-hand side and rearranging, we finally arrive at

$$\|(u_h, p_h)\|_{\mathrm{X}_h}^2 \le \underbrace{\left[\alpha^{-2}\left(1 + 2\beta^{-2}\|\mathrm{a}_h\|_{\mathrm{U}_h\times\mathrm{U}_h}^2\right)^2 + 4\beta^{-2}\right]}_{\gamma^{-2}}\mathcal{S}^2.$$

Taking the square root yields (A.22). □

The energy error estimate for the approximation of the saddle point problem is stated in the following corollary, used in the proof of Theorem 8.18.

Corollary A.13 (Abstract Error Estimate and Convergence in Energy Norm for Saddle Point Problems) *Let the assumptions and notations of Lemma A.11 hold. Let (u, p) be a solution to* (A.13) *(or, equivalently,* (A.14)*), and define the consistency error linear form $\mathcal{E}_h((u,p);\cdot) : \mathtt{X}_h \to \mathbb{R}$ such that, for all $(v_h, q_h) \in \mathtt{X}_h$,*

$$\mathcal{E}_h((u,p);(v_h,q_h)) = \mathtt{l}_h(v_h) - \mathtt{a}_h(\mathtt{I}_h u, v_h) - \mathtt{b}_h(v_h, \mathtt{J}_h p) \\ + \mathtt{m}_h(q_h) + \mathtt{b}_h(\mathtt{I}_h u, q_h) \tag{A.24}$$

or, equivalently,

$$\mathcal{E}_h((u,p);(v_h,q_h)) = \mathtt{l}_h(v_h) + \mathtt{m}_h(q_h) - \mathtt{A}_h((\mathtt{I}_h u, \mathtt{J}_h p),(v_h,q_h)). \tag{A.25}$$

Then, if $(u_h, p_h) \in \mathtt{X}_h$ is a solution to problem (A.15) *(or, equivalently,* (A.17)*), we have that*

$$\|(u_h - \mathtt{I}_h u, p - \mathtt{J}_h p)\|_{\mathtt{X}_h} \le \gamma^{-1} \|\mathcal{E}_h((u,p);\cdot)\|_{\mathtt{X}_h^\star} \tag{A.26}$$

with γ given by (A.21)*. As a consequence, letting a family $(\mathtt{U}_h, \mathtt{P}_h, \mathtt{a}_h, \mathtt{b}_h, \mathtt{l}_h, \mathtt{m}_h)_{h\to 0}$ of spaces and forms be given, if consistency holds, then we have* convergence *in the following sense:*

$$\|(u_h - \mathtt{I}_h u, p - \mathtt{J}_h p)\|_{\mathtt{X}_h} \to 0 \text{ as } h \to 0.$$

Proof Combine Lemmas A.11 and A.7. □

A.2.3 Improved Error Estimate in a Weaker Norm

We derive in this section an improved $\mathtt{L}$-norm error estimate for the velocity. Assume that there exists a linear velocity reconstruction operator

$$\mathtt{r}_h : \mathtt{U}_h \to \mathtt{L}. \tag{A.27}$$

We moreover assume that, for all $\mathtt{g} \in \mathtt{L}^\star$, there exists a solution to the continuous dual problem: Find $(z_\mathtt{g}, s_\mathtt{g}) \in \mathtt{X}$ such that

$$\mathtt{A}((w,r),(z_\mathtt{g}, s_\mathtt{g})) = \mathtt{g}(w) \qquad \forall (w,r) \in \mathtt{X}. \tag{A.28}$$

We define the dual consistency error linear form $\mathcal{E}_h^{\rm d}((z_\mathtt{g}, s_\mathtt{g});\cdot) : \mathtt{X}_h \to \mathbb{R}$ such that, for all $(v_h, q_h) \in \mathtt{X}_h$,

$$\mathcal{E}_h^{\rm d}((z_\mathtt{g}, s_\mathtt{g});(v_h,q_h)) := \mathtt{g}(\mathtt{r}_h(v_h)) - \mathtt{A}_h((v_h,q_h),(\mathtt{I}_h z_\mathtt{g}, \mathtt{J}_h s_\mathtt{g})). \tag{A.29}$$

The improved estimate in a weaker norm for the velocity unknown is stated in the following lemma. Lemma 8.21 gives an example of an application of this result.

Lemma A.14 (Improved Velocity Estimate in the $\mathtt{L}$-Norm) *Assume* (A.27) *and that the dual problem* (A.28) *has a solution* $(z_{\mathtt{g}}, s_{\mathtt{g}}) \in \mathtt{X}$ *for any* $\mathtt{g} \in \mathtt{L}^\star$. *Let* $B_{\mathtt{L}^\star} = \{\mathtt{g} \in \mathtt{L}^\star : \|\mathtt{g}\|_{\mathtt{L}^\star} \leq 1\}$ *be the unit ball in* $\mathtt{L}^\star$. *Let* $(u, p) \in \mathtt{X}$ *and* $(u_h, p_h) \in \mathtt{X}_h$ *be solutions to* (A.14) *and* (A.17), *respectively. Then,*

$$\|\mathtt{r}_h(u_h - \mathtt{I}_h u)\|_{\mathtt{L}} \leq \|(u_h - \mathtt{I}_h u, p_h - \mathtt{J}_h p)\|_{\mathtt{X}_h} \sup_{\mathtt{g} \in B_{\mathtt{L}^\star}} \|\mathcal{E}_h^{\mathrm{d}}((z_{\mathtt{g}}, s_{\mathtt{g}}); \cdot)\|_{\mathtt{X}_h^\star} + \sup_{\mathtt{g} \in B_{\mathtt{L}^\star}} \mathcal{E}_h((u, p); (\mathtt{I}_h z_{\mathtt{g}}, \mathtt{J}_h s_{\mathtt{g}})). \tag{A.30}$$

Proof The proof closely resembles that of Lemma A.10. Using the definition (A.29) of the dual error with $(v_h, q_h) = (u_h - \mathtt{I}_h u, p_h - \mathtt{J}_h p)$, we have that

$$\begin{aligned}
&\mathtt{g}(\mathtt{r}_h(u_h - \mathtt{I}_h u)) \\
&\quad = \mathcal{E}_h^{\mathrm{d}}((z_{\mathtt{g}}, s_{\mathtt{g}}); (v_h, q_h)) + \mathtt{A}_h((u_h - \mathtt{I}_h u, p_h - \mathtt{J}_h p), (\mathtt{I}_h z_{\mathtt{g}}, \mathtt{J}_h s_{\mathtt{g}})) \\
&\quad = \mathcal{E}_h^{\mathrm{d}}((z_{\mathtt{g}}, s_{\mathtt{g}}); (v_h, q_h)) + \mathtt{A}_h((u_h, p_h), (\mathtt{I}_h z_{\mathtt{g}}, \mathtt{J}_h s_{\mathtt{g}})) \\
&\qquad - \mathtt{A}_h((\mathtt{I}_h u, \mathtt{J}_h p), (\mathtt{I}_h z_{\mathtt{g}}, \mathtt{J}_h s_{\mathtt{g}})) \\
&\quad = \mathcal{E}_h^{\mathrm{d}}((z_{\mathtt{g}}, s_{\mathtt{g}}); (v_h, q_h)) + \mathtt{l}(\mathtt{I}_h z_{\mathtt{g}}) + \mathtt{m}(\mathtt{J}_h s_{\mathtt{g}}) \\
&\qquad - \mathtt{A}_h((\mathtt{I}_h u, \mathtt{J}_h p), (\mathtt{I}_h z_{\mathtt{g}}, \mathtt{J}_h s_{\mathtt{g}})) \\
&\quad = \mathcal{E}_h^{\mathrm{d}}((z_{\mathtt{g}}, s_{\mathtt{g}}); (v_h, q_h)) + \mathcal{E}_h((u, p); (\mathtt{I}_h z_{\mathtt{g}}, \mathtt{J}_h s_{\mathtt{g}})),
\end{aligned}$$

where we have used the linearity of $\mathtt{A}_h$ in its first argument to pass to the second line, the problem (A.17) to pass to the third line, and the definition (A.25) of the consistency error with $(v_h, q_h) = (\mathtt{I}_h z_{\mathtt{g}}, \mathtt{J}_h s_{\mathtt{g}})$ in the last line. The conclusion follows proceeding as in the proof of Lemma A.10. □

Appendix B
Implementation

In this chapter we discuss some practical aspects concerning the implementation of the Hybrid High-Order scheme (2.88) for the Poisson problem (2.85) with mixed boundary conditions; see Sect. 2.4 for further details. The material is organised as follows: in Sect. B.1 we introduce polynomial bases and degrees of freedom; Sect. B.2 addresses the local construction; in Sect. B.3 we discuss the assembly and efficient numerical resolution of the discrete problem.

B.1 Polynomial Bases and Degrees of Freedom

For any mesh element $T \in \mathcal{T}_h$ and any integer $l \geq 0$, we fix a basis for $\mathbb{P}^l(T)$ denoted by

$$\Phi_T^l := \{\varphi_i^T\}_{1 \leq i \leq N_d^l},$$

where we recall that N_d^l defined by (1.30) is such that

$$N_d^l = \binom{l+d}{d}.$$

We assume, for the sake of simplicity, that the basis is hierarchical so that, in particular

$$\Phi_T^{m-1} \subset \Phi_T^m \qquad \forall m \in \{1, \ldots, l\}.$$

This assumption could be removed, but we keep it here on the one hand because it simplifies the discussion, on the other hand because it potentially leads to more efficient implementations: since the basis Φ_T^k for the element-based discrete

D. A. Di Pietro, J. Droniou, *The Hybrid High-Order Method for Polytopal Meshes*, MS&A 19, https://doi.org/10.1007/978-3-030-37203-3

unknowns (that is, v_T for $\underline{v}_T = (v_T, (v_F)_{F\in\mathcal{F}_T}) \in \underline{U}_T^k$) is a subset of the basis Φ_T^{k+1} used to express the potential reconstruction $\mathrm{p}_T^{k+1}\underline{v}_T$, it suffices to construct and evaluate the latter basis at quadrature nodes to perform the local construction.

For any face $F \in \mathcal{F}_h$, we fix a basis for $\mathbb{P}^k(F)$ denoted by

$$\Phi_F^k := \{\varphi_i^F\}_{1\le i\le N_{d-1}^k}.$$

In this case, we do not require that Φ_F^k is hierarchical.

A basis for the global HHO space $\underline{U}_h^k$ is obtained taking the Cartesian product of the bases for the local polynomial spaces:

$$\Phi_h^k := \left(\bigtimes_{T\in\mathcal{T}_h} \Phi_T^k\right) \times \left(\bigtimes_{F\in\mathcal{F}_h} \Phi_F^k\right).$$

We next fix the degrees of freedom, i.e., a set of linear functionals that form a basis for the dual space of $\underline{U}_h^k$. While other choices are possible, in what follows we take them equal to the functionals that map a given vector of discrete unknowns in $\underline{U}_h^k$ to the coefficients of its expansion in the selected basis. Specifically, the degrees of freedom applied to a given

$$\underline{v}_h = ((v_T)_{T\in\mathcal{T}_h}, (v_F)_{F\in\mathcal{F}_h}) \in \underline{U}_h^k$$

return the real numbers

$$\begin{aligned} &\mathsf{V}_i^T \text{ with } 1 \le i \le N_d^k \quad \text{and } T \in \mathcal{T}_h,\\ &\mathsf{V}_i^F \text{ with } 1 \le i \le N_{d-1}^k \text{ and } F \in \mathcal{F}_h \end{aligned} \tag{B.1}$$

such that

$$v_T = \sum_{i=1}^{N_d^k} \mathsf{V}_i^T \varphi_i^T \text{ for all } T \in \mathcal{T}_h \text{ and } v_F = \sum_{i=1}^{N_{d-1}^k} \mathsf{V}_i^F \varphi_i^F \text{ for all } F \in \mathcal{F}_h. \tag{B.2}$$

For the sake of brevity, with a little abuse in terminology, we henceforth refer to the real numbers (B.1) as the *degrees of freedom of* $\underline{v}_h$.

B.1.1 Choice of Basis Functions

HHO methods are geared towards meshes containing polytopal elements of arbitrary shape so that, in general, the notion of reference element cannot be used to generate the basis Φ_T^l. We next discuss a few options to generate hierarchical bases directly on the physical element. The design of face basis functions Φ_F^k is not explicitly discussed here, as these functions can be obtained exactly as the element basis functions, provided that we consider a local set of coordinates on the hyperplane in which the face is contained.

Let a mesh element $T \in \mathcal{T}_h$ be fixed, as well as a point $\boldsymbol{x}_T \in \overline{T}$. A basis for the polynomial space $\mathbb{P}^l(T)$ can be obtained setting

$$\Phi_T^l := \left\{ \prod_{i=1}^{d} \xi_{T,i}^{\alpha_i} \, : \boldsymbol{\alpha} \in \boldsymbol{A}_d^l \text{ and } \xi_{T,i}(\boldsymbol{x}) := \frac{x_i - x_{T,i}}{h_T} \right. \\ \left. \text{for all } \boldsymbol{x} \in \overline{T} \text{ and all } 1 \le i \le d \right\}, \tag{B.3}$$

where we recall that the set of multi-indices $\boldsymbol{A}_d^l$ defined by (1.15) is such that

$$\boldsymbol{A}_d^l = \left\{ \boldsymbol{\alpha} \in \mathbb{N}^d \; : \; \|\boldsymbol{\alpha}\|_1 \le l \right\}.$$

The basis Φ_T^l constructed above is composed of monomial functions in the locally translated and scaled coordinates $(\xi_{T,i})_{1\le i\le d}$. This choice ensures that the basis is invariant with respect to translations of the element T, but not with respect to rotations. This is perfectly acceptable when working with isotropic meshes or low degrees, but can lead to badly conditioned systems when stretched elements are present or higher degrees are required.

The situation can be improved in this case by rotating the local reference frame so that it is aligned with the principal axes of inertia of the element, and scaling with different local length scales in each direction. A further improvement consists in orthonormalising the resulting basis by a Gram–Schmidt algorithm, as originally proposed in [38] in the context of Discontinuous Galerkin methods. It should be noted that an implementation using orthonormal basis functions comes with a higher computational cost, as each evaluation of a basis function requires more floating point operations; the usage of an orthonormal basis is, however, sometimes necessary to preserve numerical convergence.

The issue of badly conditioned systems when using monomial basis functions is illustrated in Fig. B.1. Therein, the results obtained with two implementations of the HHO scheme (2.48) with $k = 3$ for the Poisson problem are presented; the first implementation is based on the monomial basis functions (B.3), whilst the second relies on the basis functions obtained by a Gram–Schmidt orthonormalisation of these monomial basis functions (we did not apply here any rotation of the reference frame). The meshes used for this test are the distorted "Kershaw" meshes from [207] (see Fig. 4.6, right). As can be seen in Fig. B.1, when using monomial basis functions, the severe anisotropy of the mesh leads to round-off errors propagation that quickly results in stagnant errors. On the contrary, the usage of orthonormalised basis functions preserves the convergence rates.

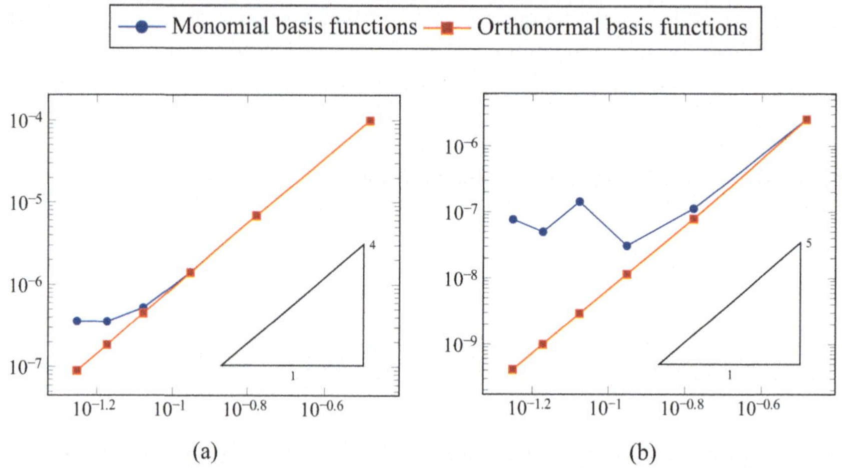

Fig. B.1 Error vs. h for the HHO scheme (2.48), comparison between monomial basis functions (B.3) and orthonormal basis functions. The reference slope indicates the expected order of convergence. (**a**) $\|\underline{I}_h^k u - \underline{u}_h\|_{a,h}$ vs. h, Kershaw meshes, $k = 3$. (**b**) $\|\pi_h^{0,k} u - u_h\|$ vs. h, Kershaw meshes, $k = 3$

B.2 Local Construction

We describe here the practical implementation of the local construction discussed in Sect. 2.1. We assume that the integrals that appear in what follows can be computed numerically up to rounding errors if the integrand is a polynomial, or suitably approximated if this is not the case. Numerical integration on polyhedra can be performed, e.g., by considering a submesh composed of standard elements for which quadrature rules are available. For homogeneous polynomial functions, one can also resort to the techniques of [110] based on a repeated use of Stokes formula to compute the integrals as combinations of vertex values. Throughout this section, we work on a fixed mesh element $T \in \mathcal{T}_h$.

B.2.1 Local Potential Reconstruction Operator

In what follows, we adopt the convention that vectors in $\mathbb{R}^m$ ($m \geq 1$) are denoted in sans-serif font and matrix in boldface sans-serif font. The starting point is the computation of the potential reconstruction operator p_T^{k+1}. Let $\underline{v}_T \in \underline{U}_T^k$ be given, and denote by $\underline{\mathsf{V}}_T$ the corresponding vector of degrees of freedom partitioned as follows:

$$\underline{\mathsf{V}}_T = \begin{bmatrix} \mathsf{V}_T \\ \mathsf{V}_{F_1} \\ \vdots \\ \mathsf{V}_{F_{N_{\partial,T}}} \end{bmatrix} \in \mathbb{R}^{N_{\mathrm{dof},T}}$$

with subvectors

$$\mathsf{V}_T = [V_i^T]_{1 \le i \le N_d^k} \in \mathbb{R}^{N_d^k}, \qquad \mathsf{V}_F = [V_i^F]_{1 \le i \le N_{d-1}^k} \in \mathbb{R}^{N_{d-1}^k} \quad \forall F \in \mathcal{F}_T,$$

where, setting $N_{\partial,T} := \mathrm{card}(\mathcal{F}_T)$, we have defined the integer

$$N_{\mathrm{dof},T} := \dim(\underline{U}_T^k) = N_d^k + N_{\partial,T} N_{d-1}^k$$

representing the number of local degrees of freedom associated with T and its faces, and we have introduced a numbering of the faces of T from 1 to $N_{\partial,T}$.

We collect the coefficients of the expansion of $\mathrm{p}_T^{k+1}\underline{v}_T$ on the basis Φ_T^{k+1} in the vector $\mathsf{P}_T = (P_i^T)_{1 \le i \le N_d^{k+1}}$, so that

$$\mathrm{p}_T^{k+1}\underline{v}_T = \sum_{i=1}^{N_d^{k+1}} P_i^T \varphi_i^T. \tag{B.4}$$

Recall the characterisation (2.13) of $\mathrm{p}_T^{k+1}\underline{v}_T$, for a fixed non-zero number λ_T. Setting $\mu_T := \lambda_T/|T|_d$, we have $\lambda_T \pi_T^{0,0} w = \mu_T (w, 1)_T$, and (2.13) therefore reads: For all $w \in \mathbb{P}^{k+1}(T)$,

$$\begin{aligned}(\nabla \mathrm{p}_T^{k+1}\underline{v}_T, \nabla w)_T &+ \mu_T(\mathrm{p}_T^{k+1}\underline{v}_T, 1)_T (w, 1)_T \\ &= (\nabla v_T, \nabla w)_T + \sum_{F \in \mathcal{F}_T} (v_F - v_T, \nabla w{\cdot}\boldsymbol{n}_{TF})_F + \mu_T (v_T, 1)_T (w, 1)_T.\end{aligned} \tag{B.5}$$

Plugging the decompositions (B.2) and (B.4) of v_T, $(v_F)_{F \in \mathcal{F}_T}$, and $\mathrm{p}_T^{k+1}\underline{v}_T$ into (B.5) applied to all $w \in \Phi_T^{k+1}$, we obtain the algebraic realisation

$$\left(\mathbf{S}_T + \mu_T \mathsf{L}_T^{k+1} (\mathsf{L}_T^{k+1})^\intercal\right) \mathsf{P}_T = \left(\mathbf{B}_{P,T} + \mu_T \mathsf{L}_T^{k+1} (\mathsf{L}_T^k)^\intercal\right) \mathsf{V}_T + \sum_{F \in \mathcal{F}_T} \mathbf{B}_{P,F} \mathsf{V}_F, \tag{B.6}$$

where

$$\mathsf{S}_T := \left[(\nabla\varphi_i^T, \nabla\varphi_j^T)_T\right]_{1\le i,j\le N_d^{k+1}},$$

$$\mathsf{B}_{P,T} := \left[(\nabla\varphi_i^T, \nabla\varphi_j^T)_T - \sum_{F\in\mathcal{F}_T}(\nabla\varphi_i^T\cdot \boldsymbol{n}_{TF}, \varphi_j^T)_F\right]_{1\le i\le N_d^{k+1}, 1\le j\le N_d^k},$$

$$\mathsf{B}_{P,F} := \left[(\nabla\varphi_i^T\cdot \boldsymbol{n}_{TF}, \varphi_j^F)_F\right]_{1\le i\le N_d^{k+1}, 1\le j\le N_{d-1}^k}$$

and, for $l \ge 0$ a polynomial degree, L_T^l is the column vector

$$\mathsf{L}_T^l := \left[(\varphi_i^T, 1)_T\right]_{1\le i\le N_d^l}$$

and $(\mathsf{L}_T^l)^\mathsf{T}$ is its transposed vector.

Being a linear operator, p_T^{k+1} can be represented by a matrix P_T of size $N_d^{k+1} \times N_{\mathrm{dof},T}$ once we have fixed a basis for $\underline{U}_T^k$ and one for $\mathbb{P}^{k+1}(T)$. This matrix can be computed repeatedly solving (B.6) for $\underline{\mathsf{V}}_T = \mathsf{e}_j$, $1 \le j \le N_{\mathrm{dof},T}$, with e_j denoting the jth vector of the canonical basis of $\mathbb{R}^{N_{\mathrm{dof},T}}$. In other words, the matrix P_T is the solution of the linear system

$$\left(\mathsf{S}_T + \mu_T \mathsf{L}_T^{k+1}(\mathsf{L}_T^{k+1})^\mathsf{T}\right)\mathsf{P}_T = \left[\mathsf{B}_{P,T} + \mu_T \mathsf{L}_T^{k+1}(\mathsf{L}_T^k)^\mathsf{T} \;\; \mathsf{B}_{P,F_1} \;\; \cdots \;\; \mathsf{B}_{P,F_{N_{\partial,T}}}\right]. \tag{B.7}$$

Remark B.1 (Choice of μ_T) The matrix S_T is singular, and has as kernel the vectors corresponding to constant polynomials in the cell. The component $\mu_T \mathsf{L}_T^{k+1}(\mathsf{L}_T^{k+1})^\mathsf{T}$ is therefore essential for (B.7) to be well-posed. To ensure a proper conditioning of this system, μ_T should be chosen such that this component has a similar magnitude as S_T. For example,

$$\mu_T = \frac{\|\mathsf{S}_T\|}{\|\mathsf{L}_T^{k+1}(\mathsf{L}_T^{k+1})^\mathsf{T}\|},$$

or, since S_T is symmetric positive semidefinite,

$$\mu_T = \frac{\mathrm{tr}(\mathsf{S}_T)}{\|\mathsf{L}_T^{k+1}\|^2}.$$

Remark B.2 (Alternative Approach) If Φ_T^{k+1} is hierarchical, then φ_1^T is a constant function, and $\{\nabla\varphi_i\}_{2\le i\le N_d^{k+1}}$ is a basis of $\nabla\mathbb{P}^{k+1}(T)$. In this case,

$$\nabla \mathrm{p}_T^{k+1}\underline{v}_T = \sum_{i=2}^{N_d^{k+1}} P_i^T \nabla\varphi_i^T$$

is entirely determined by the Eq. (2.12), whose algebraic realisation is

$$\widetilde{\mathbf{S}}_T \widehat{\mathbf{P}}_T = \widehat{\mathbf{B}}_{P,T} \mathsf{V}_T + \sum_{F \in \mathcal{F}_T} \widehat{\mathbf{B}}_{P,F} \mathsf{V}_F,$$

with $\widetilde{\mathbf{S}}_T$ denoting the $(N_d^{k+1} - 1) \times (N_d^{k+1} - 1)$ matrix obtained removing the first row and column of $\mathbf{S}_T$, $\widehat{\mathbf{P}}_T$ is the $(N_d^{k+1} - 1) \times 1$ vector obtained removing the first component (row) of $\mathbf{P}_T$, and $\widehat{\mathbf{B}}_{P,*}$ (for $* = T$ or F) is the matrix obtained removing the first row of $\mathbf{B}_{P,*}$. Making $\underline{\mathsf{V}}_T$ a generic vector then leads, in a similar way as (B.7), to

$$\widetilde{\mathbf{S}}_T \widehat{\mathbf{P}}_T = \left[\widehat{\mathbf{B}}_{P,T} \ \widehat{\mathbf{B}}_{P,F_1} \ \cdots \ \widehat{\mathbf{B}}_{P,F_{N_{\partial,T}}} \right].$$

The matrix $\widehat{\mathbf{P}}_T$ corresponds to $\mathbf{P}_T$ with the first row removed, and the vector $\widehat{\mathbf{P}}_T \underline{\mathsf{V}}_T$ gives the coefficients $\{P_i^T\}_{2 \le i \le N_d^{k+1}}$ of $\mathrm{p}_T^{k+1} \underline{v}_T$ on $\{\varphi_i^T\}_{2 \le i \le N_d^{k+1}}$. The coefficient P_1^T on φ_1^T can then be inferred by imposing the closure condition (2.11b):

$$P_1^T (\varphi_1^T, 1)_T = \sum_{i=1}^{N_d^k} V_i^T (\varphi_i^T, 1)_T - \sum_{i=2}^{N_d^{k+1}} P_i^T (\varphi_i^T, 1)_T.$$

Note however that, when implementing the HHO scheme with original stabilisation (2.22), the potential reconstruction $\mathrm{p}_T^{k+1} \underline{v}_T$ only needs to be known up to an additive constant (such an additive constant disappears in the consistent component $(\nabla \mathrm{p}_T^{k+1} \underline{u}_T, \nabla \mathrm{p}_T^{k+1} \underline{v}_T)_T$ in (2.15), and also in the contributions $(\delta_{TF}^k - \delta_T^k) \underline{v}_T$ appearing in the stabilisation term). In this case, the knowledge of $\widehat{\mathbf{P}}_T$ alone is sufficient to implement the scheme.

B.2.2 Difference Operators

We next discuss the computation of the difference operators defined by (2.19), that is: For all $\underline{v}_T \in \underline{U}_T^k$,

$$\delta_T^k \underline{v}_T := \pi_T^{0,k} (\mathrm{p}_T^{k+1} \underline{v}_T - v_T), \qquad \delta_{TF}^k \underline{v}_T := \pi_F^{0,k} (\mathrm{p}_T^{k+1} \underline{v}_T - v_F) \quad \forall F \in \mathcal{F}_T. \tag{B.8}$$

These operators are a key ingredient to devising high-order stabilisation terms. For integers $l, m \ge 0$, we define the local element mass matrix

$$\mathbf{M}_{TT}^{l,m} := \left[(\varphi_i^T, \varphi_j^T)_T \right]_{1 \le i \le N_d^l, 1 \le j \le N_d^m}.$$

The element difference operator $\delta_T^k : \underline{U}_T^k \to \mathbb{P}^k(T)$ is represented by the matrix $\mathbf{D}_T \in \mathbb{R}^{N_d^k \times N_{\mathrm{dof},T}}$ such that

$$\mathbf{D}_T := (\mathbf{M}_{TT}^{k,k})^{-1}\mathbf{M}_{TT}^{k,k+1}\mathbf{P}_T - \begin{bmatrix} \mathbf{I}_{N_d^k} & 0 \cdots 0 \end{bmatrix},$$

where the zero blocks in the rightmost term fill the columns of face unknowns.

Remark B.3 (Hierarchical Matrices) Having chosen hierarchical basis functions, we notice that the element mass matrices are also hierarchical: $\mathbf{M}_{TT}^{k,k}$ is a sub-matrix of $\mathbf{M}_{TT}^{k,k+1}$, and thus only the latter needs to be computed.

Let a face $F \in \mathcal{F}_T$ be fixed and, for given integers $l, m \geq 0$, define the face-element and face-face mass matrices

$$\mathbf{M}_{FT}^{l,m} := \Big[(\varphi_i^F, \varphi_j^T)_F\Big]_{1\le i\le N_{d-1}^l, 1\le j\le N_d^m}, \qquad \mathbf{M}_{FF}^{l,m} := \Big[(\varphi_i^F, \varphi_j^F)_F\Big]_{1\le i\le N_{d-1}^l, 1\le j\le N_{d-1}^m}.$$

The face difference operator $\delta_{TF}^k : \underline{U}_T^k \to \mathbb{P}^k(F)$ is represented by the matrix $\mathbf{D}_{TF} \in \mathbb{R}^{N_{d-1}^k \times N_{\mathrm{dof},T}}$ such that

$$\mathbf{D}_{TF} := (\mathbf{M}_{FF}^{k,k})^{-1}\mathbf{M}_{FT}^{k,k+1}\mathbf{P}_T - \begin{bmatrix} 0 & 0 \cdots \mathbf{I}_{N_{d-1}^k} \cdots 0 \end{bmatrix},$$

where the identity matrix fills the column block corresponding to the unknowns attached to the considered face F.

Remark B.4 (Interpretation of the Mass Matrices) The combinations of mass matrices that appear in the formulas above have the following interpretations:

(i) $(\mathbf{M}_{TT}^{k,k})^{-1}\mathbf{M}_{TT}^{k,k+1}$ is the matrix, in the bases Φ_T^{k+1} and Φ_T^k, of the linear operator $(\pi_T^{0,k})_{|\mathbb{P}^{k+1}(T)} : \mathbb{P}^{k+1}(T) \to \mathbb{P}^k(T)$.

(ii) $(\mathbf{M}_{FF}^{k,k})^{-1}\mathbf{M}_{FT}^{k,k+1}$ is the matrix, in the bases Φ_T^{k+1} and Φ_F^k, of the linear operator $\pi_F^{0,k} \circ \gamma_{TF}^{k+1} : \mathbb{P}^{k+1}(T) \to \mathbb{P}^k(F)$, where $\gamma_{TF}^{k+1} : \mathbb{P}^{k+1}(T) \to \mathbb{P}^{k+1}(F)$ is the restriction operator.

B.2.3 Local Contribution

Recall the definition (2.15) of the local bilinear form:

$$\mathrm{a}_T(\underline{u}_T, \underline{v}_T) := (\nabla \mathrm{p}_T^{k+1}\underline{u}_T, \nabla \mathrm{p}_T^{k+1}\underline{v}_T)_T + \mathrm{s}_T(\underline{u}_T, \underline{v}_T).$$

This form is associated with a symmetric positive semidefinite local matrix $\mathbf{A}_T$, which represents the contribution of element T to the system matrix (see the definition (2.39) of a_h). This local contribution can be decomposed into its consistency and stability terms as

$$\mathbf{A}_T = \mathbf{A}_T^{\text{cons}} + \mathbf{A}_T^{\text{stab}} \in \mathbb{R}^{N_{\text{dof},T} \times N_{\text{dof},T}}. \tag{B.9}$$

The consistency contribution reads

$$\mathbf{A}_T^{\text{cons}} = \mathbf{P}_T^{\mathsf{T}} \mathbf{S}_T \mathbf{P}_T.$$

Several choices are possible for stabilisations that satisfy Assumption 2.4. The following expression for $\mathbf{A}_T^{\text{stab}}$ corresponds to the one discussed in Example 2.7:

$$\mathbf{A}_T^{\text{stab}} = \sum_{F \in \mathcal{F}_T} h_F^{-1} (\mathbf{D}_{TF} - (\mathbf{M}_{FF}^{k,k})^{-1} \mathbf{M}_{FT}^{k,k} \mathbf{D}_T)^{\mathsf{T}} \mathbf{M}_{FF}^{k,k} (\mathbf{D}_{TF} - (\mathbf{M}_{FF}^{k,k})^{-1} \mathbf{M}_{FT}^{k,k} \mathbf{D}_T), \tag{B.10}$$

whereas the one of Example 2.8 reads

$$\mathbf{A}_T^{\text{stab}} = h_T^{-2} \mathbf{D}_T^{\mathsf{T}} \mathbf{M}_{TT}^{k,k} \mathbf{D}_T + \sum_{F \in \mathcal{F}_T} h_F^{-1} \mathbf{D}_{TF}^{\mathsf{T}} \mathbf{M}_{FF}^{k,k} \mathbf{D}_{TF}.$$

Remark B.5 In a similar way as in Remark B.4, we notice that the matrix $(\mathbf{M}_{FF}^{k,k})^{-1} \mathbf{M}_{FT}^{k,k}$ represents, in the bases Φ_T^k and Φ_F^k, the restriction map $\gamma_{TF}^k : \mathbb{P}^k(T) \to \mathbb{P}^k(F)$.

The local contribution, from the element T, to the source term is

$$\mathsf{B}_T = \begin{bmatrix} \mathsf{B}_T^T \\ 0 \end{bmatrix} \in \mathbb{R}^{N_{\text{dof},T}}, \qquad \mathsf{B}_T^T := \left[(f, \varphi_i^T)_T \right]_{1 \le i \le N_d^k},$$

where the 0 block fills the rows corresponding to all face unknowns.

B.3 Discrete Problem

In this section, we discuss the assembly and efficient numerical resolution of the discrete problem.

B.3.1 Assembly and Enforcement of Boundary Conditions

The following global matrix and vector are assembled element-wise, relying on the usual technique based on a global table of degrees of freedom to ensure that interface unknowns match from one element to the adjacent one:

$$\widetilde{\mathsf{A}}_h = \sum_{T\in\mathcal{T}_h} \mathsf{A}_T \qquad \widetilde{\mathsf{B}}_h = \sum_{T\in\mathcal{T}_h} \mathsf{B}_T. \tag{B.11}$$

We assume the following ordering for the degrees of freedom: first those attached to mesh elements, then those attached to interfaces and Neumann boundary faces, finally those attached to Dirichlet boundary faces. Recalling the definition (2.87) of the set $\mathcal{F}_h^{\not D}$ collecting interfaces and non-Dirichlet boundary faces, this ordering induces the following block structure on $\widetilde{\mathsf{A}}_h$ and $\widetilde{\mathsf{B}}_h$:

$$\widetilde{\mathsf{A}}_h = \begin{bmatrix} \mathsf{A}_{\mathcal{T}_h\mathcal{T}_h} & \mathsf{A}_{\mathcal{T}_h\mathcal{F}_h^{\not D}} & \mathsf{A}_{\mathcal{T}_h\mathcal{F}_h^{D}} \\ \mathsf{A}_{\mathcal{F}_h^{\not D}\mathcal{T}_h} & \mathsf{A}_{\mathcal{F}_h^{\not D}\mathcal{F}_h^{\not D}} & \mathsf{A}_{\mathcal{F}_h^{\not D}\mathcal{F}_h^{D}} \\ \mathsf{A}_{\mathcal{F}_h^{D}\mathcal{T}_h} & \mathsf{A}_{\mathcal{F}_h^{D}\mathcal{F}_h^{\not D}} & \mathsf{A}_{\mathcal{F}_h^{D}\mathcal{F}_h^{D}} \end{bmatrix}, \qquad \widetilde{\mathsf{B}}_h = \begin{bmatrix} \mathsf{B}_{\mathcal{T}_h} \\ 0 \\ 0 \end{bmatrix}.$$

Denote by $\underline{\mathsf{U}}_{h,\mathrm{D}}$ the vector of degrees of freedom corresponding to $\underline{u}_{h,\mathrm{D}}$ defined by (2.86), partitioned as follows:

$$\underline{\mathsf{U}}_{h,\mathrm{D}} = \begin{bmatrix} 0 \\ 0 \\ \mathsf{U}_{\mathcal{F}_h^{D}} \end{bmatrix},$$

where the zero subvectors correspond to the degrees of freedom attached to element and non-Dirichlet faces. Define the following vector accounting for the non-homogeneous Neumann boundary condition:

$$\mathsf{B}_{h,\mathrm{N}} = \left[\mathsf{B}_{F,\mathrm{N}}\right]_{F\in\mathcal{F}_h^{\not D}} \text{ with } \mathsf{B}_{F,\mathrm{N}} := \begin{cases} 0 & \text{if } F\in\mathcal{F}_h^{\mathrm{i}}, \\ \left[(g_{\mathrm{N}},\varphi_i^F)_F\right]_{1\le i\le N_{d-1}^k} & \text{if } F\in\mathcal{F}_h^{\mathrm{N}}. \end{cases}$$

Denoting by

$$N_{\mathrm{dof},h} := \mathrm{card}(\mathcal{T}_h)N_d^k + \mathrm{card}(\mathcal{F}_h^{\not D})N_{d-1}^k$$

the global number of degrees of freedom, the algebraic realisation of the discrete problem (2.88) reads: Find $\underline{\mathsf{U}}_{h,0}\in\mathbb{R}^{N_{\mathrm{dof},h}}$ such that

$$\begin{bmatrix} \mathsf{A}_{\mathcal{T}_h\mathcal{T}_h} & \mathsf{A}_{\mathcal{T}_h\mathcal{F}_h^{\not D}} \\ \mathsf{A}_{\mathcal{F}_h^{\not D}\mathcal{T}_h} & \mathsf{A}_{\mathcal{F}_h^{\not D}\mathcal{F}_h^{\not D}} \end{bmatrix} \begin{bmatrix} \mathsf{U}_{\mathcal{T}_h,0} \\ \mathsf{U}_{\mathcal{F}_h^{\not D},0} \end{bmatrix} = \begin{bmatrix} \mathsf{B}_{\mathcal{T}_h} \\ \mathsf{B}_{h,\mathrm{N}} \end{bmatrix} - \begin{bmatrix} \mathsf{A}_{\mathcal{T}_h\mathcal{F}_h^{D}} \\ \mathsf{A}_{\mathcal{F}_h^{\not D}\mathcal{F}_h^{D}} \end{bmatrix} \mathsf{U}_{\mathcal{F}_h^{D}} =: \begin{bmatrix} \mathsf{C}_{\mathcal{T}_h} \\ \mathsf{C}_{\mathcal{F}_h^{\not D}} \end{bmatrix}. \tag{B.12}$$

B.3.2 Static Condensation

The submatrix $\mathsf{A}_{\mathcal{T}_h\mathcal{T}_h}$ is block-diagonal (with each block corresponding to one mesh element) and symmetric positive definite, and is therefore inexpensive to invert. The block-diagonal structure is a consequence of the fact that, for a fixed mesh element $T \in \mathcal{T}_h$, the discrete unknown u_T attached to T interacts with the other discrete unknowns only through the face unknowns u_F, $F \in \mathcal{F}_T$. The fact that $\mathsf{A}_{\mathcal{T}_h\mathcal{T}_h}$ is positive definite, on the other hand, follows from (2.16) after observing that, for any $v_T \in \mathbb{P}^k(T)$,

$$\mathrm{a}_T((v_T,\underline{0}),(v_T,\underline{0})) \gtrsim \|(v_T,\underline{0})\|_{1,T}^2 = \|\nabla v_T\|_T^2 + \sum_{F\in\mathcal{F}_T} h_F^{-1}\|v_T\|_F^2,$$

and the quantity in the right-hand side is a norm on $\mathbb{P}^k(T)$. Here, the hidden constant is independent of both h and T.

This remark suggests to solve the linear system (B.12) in two steps:

(i) First, element-based degrees of freedom in $\mathsf{U}_{\mathcal{T}_h,0}$ are expressed in terms of $\mathsf{C}_{\mathcal{T}_h}$ and $\mathsf{U}_{\mathcal{F}_h^{\mathrm{D}},0}$ by the inexpensive solution of the first block equation:

$$\mathsf{U}_{\mathcal{T}_h,0} = \mathsf{A}_{\mathcal{T}_h\mathcal{T}_h}^{-1}\left(\mathsf{C}_{\mathcal{T}_h} - \mathsf{A}_{\mathcal{T}_h\mathcal{F}_h^{\mathrm{D}}}\mathsf{U}_{\mathcal{F}_h^{\mathrm{D}},0}\right). \tag{B.13a}$$

This step is referred to as *static condensation* in the Finite Element literature;

(ii) Second, face-based coefficients in $\mathsf{U}_{\mathcal{F}_h^{\mathrm{D}},0}$ are obtained solving the following global problem involving quantities attached to the mesh skeleton:

$$\underbrace{\left(\mathsf{A}_{\mathcal{F}_h^{\mathrm{D}}\mathcal{F}_h^{\mathrm{D}}} - \mathsf{A}_{\mathcal{F}_h^{\mathrm{D}}\mathcal{T}_h}\mathsf{A}_{\mathcal{T}_h\mathcal{T}_h}^{-1}\mathsf{A}_{\mathcal{T}_h\mathcal{F}_h^{\mathrm{D}}}\right)}_{=:\mathsf{A}_h^{\mathrm{sc}}}\mathsf{U}_{\mathcal{F}_h^{\mathrm{D}},0} = \mathsf{C}_{\mathcal{F}_h^{\mathrm{D}}} - \mathsf{A}_{\mathcal{F}_h^{\mathrm{D}}\mathcal{T}_h}\mathsf{A}_{\mathcal{T}_h\mathcal{T}_h}^{-1}\mathsf{C}_{\mathcal{T}_h}. \tag{B.13b}$$

This computationally more intensive step requires to invert the symmetric positive definite matrix $\mathsf{A}_h^{\mathrm{sc}}$ which has size

$$N_{\mathrm{dof},h}^{\mathrm{sc}} := \mathrm{card}(\mathcal{F}_h^{\mathrm{D}})N_{d-1}^k. \tag{B.13c}$$

The fact that $\mathsf{A}_h^{\mathrm{sc}}$ is positive definite can be deduced observing that it is in fact the Schur complement of $\mathsf{A}_{\mathcal{T}_h\mathcal{T}_h}$ in

$$\mathsf{A}_h := \begin{bmatrix} \mathsf{A}_{\mathcal{T}_h\mathcal{T}_h} & \mathsf{A}_{\mathcal{T}_h\mathcal{F}_h^{\mathrm{D}}} \\ \mathsf{A}_{\mathcal{F}_h^{\mathrm{D}}\mathcal{T}_h} & \mathsf{A}_{\mathcal{F}_h^{\mathrm{D}}\mathcal{F}_h^{\mathrm{D}}} \end{bmatrix}.$$

Since A_h is symmetric and both A_h and $\mathsf{A}_{\mathcal{T}_h\mathcal{T}_h}$ are positive definite, a classical result in linear algebra yields that $\mathsf{A}_h^{\mathrm{sc}}$ is also positive definite (see, e.g., [212]).

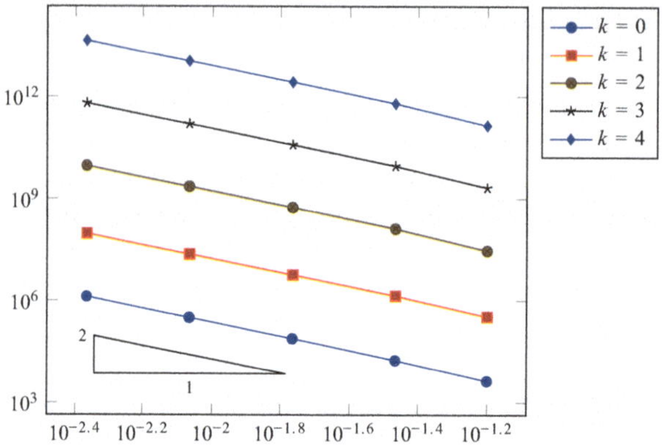

Fig. B.2 Condition number in the 1-norm vs. h for the matrix in the left-hand side of (B.13b) corresponding to the polygonal mesh of Fig. 1.1c

Remark B.6 (Stencil of the System) The local matrices A_T couple the unknowns in the element T and on its faces $F \in \mathcal{F}_T$. After element-wise assembly (B.11) and accounting for the boundary conditions, the stencil of each unknown on a face $F \in \mathcal{F}_h$ in the matrix of (B.12) is made of the unknowns associated with the elements on each side of F, as well as all their non-Dirichlet faces. This stencil is not increased after static condensation, since this condensation is performed element-wise. Hence, in the final system (B.13b), each unknown on a face F is only coupled with the unknowns on the faces that share an element with F.

Remark B.7 (Solution of the Linear System) As for standard Finite Element Methods, the condition number of the matrix in the left-hand side of (B.13b) grows with both h and k; see Fig. B.2 for an example. This typically requires the development of ad hoc solution strategies when large problems are considered. For the numerical tests in this book, direct solvers were used whenever possible in two space dimensions while, for the three-dimensional tests, a variety of (direct or iterative) solution strategies were used depending on the features of the problem at hand.

Generally speaking, the development of efficient algorithms for the resolution of the linear systems resulting from high-order skeletal polytopal methods (i.e., methods such as HHO or Virtual Elements, for which the globally coupled unknowns are attached to the mesh skeleton) remains an open field of research. Very recent works on p-multilevel solution strategies for HHO, albeit in their infancy, seem extremely promising; see, e.g., [41], where the Poisson and Stokes problems in two and three space dimensions are considered. On the other hand, geometric h-multigrid resolution strategies seem less obvious than, say, for Discontinuous Galerkin methods, owing to the need to coarsen the space of skeletal unknowns. Important efforts are currently being undertaken to tackle this problem; we cite, in particular, the `fast4hho` project (*Agence Nationale de la Recherche* grant ANR-17-CE23-0019), involving both academic and industrial partners.

References

1. I. Aavatsmark, T. Barkve, Ø. Bøe, T. Mannseth, Discretization on unstructured grids for inhomogeneous, anisotropic media. I. Derivation of the methods. SIAM J. Sci. Comput. **19**(5), 1700–1716 (1998). https://doi.org/10.1137/S1064827595293582
2. I. Aavatsmark, T. Barkve, Ø. Bøe, T. Mannseth, Discretization on unstructured grids for inhomogeneous, anisotropic media. II. Discussion and numerical results. SIAM J. Sci. Comput. **19**(5), 1717–1736 (1998). https://doi.org/10.1137/S1064827595293594
3. I. Aavatsmark, G.T. Eigestad, B.T. Mallison, J.M. Nordbotten, A compact multipoint flux approximation method with improved robustness. Numer. Methods Partial Differential Equations **24**(5), 1329–1360 (2008). https://doi.org/10.1002/num.20320
4. M. Abbas, A. Ern, N. Pignet, Hybrid High-Order methods for finite deformations of hyperelastic materials. Comput. Mech. **62**(4), 909–928 (2018). https://doi.org/10.1007/s00466-018-1538-0
5. Y. Achdou, C. Bernardi, F. Coquel, A priori and a posteriori analysis of finite volume discretizations of Darcy's equations. Numer. Math. **96**(1), 17–42 (2003). https://doi.org/10.1007/s00211-002-0436-7
6. R.A. Adams, J.J.F. Fournier, *Sobolev Spaces*. Pure and Applied Mathematics (Amsterdam), vol. 140, 2nd edn. (Elsevier/Academic, Amsterdam, 2003), pp. xiv+305. ISBN: 0-12-044143-8
7. L. Agélas, D.A. Di Pietro, J. Droniou, The G method for heterogeneous anisotropic diffusion on general meshes. ESAIM: Math. Model Numer. Anal. **44**(4), 597–625 (2010). https://doi.org/10.1051/m2an/2010021
8. J. Aghili, D.A. Di Pietro, An advection-robust Hybrid High-Order method for the Oseen problem. J. Sci. Comput. **77**(3), 1310–1338 (2018). https://doi.org/10.1007/s10915-018-0681-2
9. J. Aghili, S. Boyaval, D.A. Di Pietro, Hybridization of mixed high-order methods on general meshes and application to the Stokes equations. Comput. Meth. Appl. Math. **15**(2), 111–134 (2015). https://doi.org/10.1515/cmam-2015-0004
10. J. Aghili, D.A. Di Pietro, B. Ruffini, An *hp*-Hybrid High-Order method for variable diffusion on general meshes. Comput. Meth. Appl. Math. **17**(3), 359–376 (2017). https://doi.org/10.1515/cmam-2017-0009
11. M. Ainsworth, B. Senior, Aspects of an adaptive hp-finite element method: adaptive strategy, conforming approximation and efficient solvers. Comput. Meth. Appl. Mech. Eng. **150**(1), 65–87 (1997). https://doi.org/S0045782597001011
12. G. Allaire, *Analyse Numérique et Optimisation* (Les éditions de l'École Polytechnique, Palaiseau, 2009)

D. A. Di Pietro, J. Droniou, *The Hybrid High-Order Method for Polytopal Meshes*,
MS&A 19, https://doi.org/10.1007/978-3-030-37203-3

13. C. Amrouche, V. Girault, On the existence and regularity of the solution of Stokes problem in arbitrary dimension. Proc. Jpn. Acad. **67**, 171–175 (1991)
14. D. Anderson, J. Droniou, An arbitrary order scheme on generic meshes for miscible displacements in porous media. SIAM J. Sci. Comput. **40**(4), B1020–B1054 (2018). https://doi.org/10.1137/17M1138807
15. B. Andreianov, F. Boyer, F. Hubert, Discrete duality finite volume schemes for Leray-Lions-type elliptic problems on general 2D meshes. Numer. Methods Partial Differential Equations **23**(1), 145–195 (2007)
16. P.F. Antonietti, F. Brezzi, L.D. Marini, Bubble stabilization of discontinuous Galerkin methods. Comput. Methods Appl. Mech. Eng. **198**(21–26), 1651–1659 (2009). https://doi.org/10.1016/j.cma.2008.12.033
17. P.F. Antonietti, S. Giani, P. Houston, *hp*-version composite discontinuous Galerkin methods for elliptic problems on complicated domains. SIAM J. Sci. Comput. **35**(3), A1417–A1439 (2013). https://doi.org/10.1137/120877246
18. P.F. Antonietti, A. Cangiani, J. Collis, Z. Dong, E.H. Georgoulis, S. Giani, P. Houston, Review of discontinuous Galerkin finite element methods for partial differential equations on complicated domains, in *Building Bridges: Connections and Challenges in Modern Approaches to Numerical Partial Differential Equations*. Lecture Notes in Computational Science and Engineering, vol. 114 (Springer, Cham, 2016), pp. 279–308
19. P.F. Antonietti, G. Manzini, M. Verani, The fully nonconforming virtual element method for biharmonic problems. Math. Models Methods Appl. Sci. **28**(2), 387–407 (2018). https://doi.org/10.1142/S0218202518500100
20. D.N. Arnold, An interior penalty finite element method with discontinuous elements. SIAM J. Numer. Anal. **19**(4), 742–760 (1982). https://doi.org/10.1137/0719052
21. D. Arnold, *Finite Element Exterior Calculus* (SIAM, Philadelphia, 2018). ISBN: 978-1-611975-53-6
22. D.N. Arnold, D. Boffi, R.S. Falk, Approximation by quadrilateral finite elements. Math. Comp. **71**(239), 909–922 (2002). https://doi.org/10.1090/S0025-5718-02-01439-4
23. D.N. Arnold, F. Brezzi, B. Cockburn, L.D. Marini, Unified analysis of discontinuous Galerkin methods for elliptic problems. SIAM J. Numer. Anal. **39**(5), 1749–1779 (2002)
24. D.N. Arnold, R.S. Falk, R. Winther, Finite element exterior calculus, homological techniques, and applications. Acta Numer. **15**, 1–155 (2006). https://doi.org/10.1017/S0962492906210018
25. D.N. Arnold, R.S. Falk, R. Winther, Finite element exterior calculus: from Hodge theory to numerical stability. Bull. Am. Math. Soc. (N.S.) **47**(2), 281–354 (2010). https://doi.org/10.1090/S0273-0979-10-01278-4
26. B. Ayuso de Dios, L.D. Marini, Discontinuous Galerkin methods for advection-diffusion-reaction problems. SIAM J. Numer. Anal. **47**(2), 1391–1420 (2009). https://doi.org/10.1137/080719583
27. B. Ayuso de Dios, K. Lipnikov, G. Manzini, The nonconforming virtual element method. ESAIM: Math. Model Numer. Anal. **50**(3), 879–904 (2016). https://doi.org/10.1051/m2an/2015090
28. I. Babuška, The finite element method with penalty. Math. Comp. **27**, 221–228 (1973). https://doi.org/10.2307/2005611
29. I. Babuška, M. Suri, The optimal convergence rate of the *p*-version of the finite element method. SIAM J. Numer. Anal. **24**(4), 750–776 (1987)
30. I. Babuška, M. Suri, Locking effects in the finite element approximation of elasticity problems. Numer. Math. **62**(4), 439–463 (1992). https://doi.org/10.1007/BF01396238
31. I. Babuška, B. Szabo, On the rates of convergence of the finite element method. Int. J. Numer. Methods Eng. **18**(3), 323–341 (1982). https://doi.org/10.1002/nme.1620180302
32. I. Babuška, M. Zlámal, Nonconforming elements in the finite element method with penalty. SIAM J. Numer. Anal. **10**, 863–875 (1973). https://doi.org/10.1137/0710071

33. C. Bacuta, J.H. Bramble, Regularity estimates for solutions of the equations of linear elasticity in convex plane polygonal domains. Z. Angew. Math. Phys. **54**(5), 874–878 (2003). [Special issue dedicated to Lawrence E. Payne]. https://doi.org/10.1007/s00033-003-3211-4
34. A.J.-C. Barré de Saint Venant, Note à joindre au Mémoire sur la dynamique des fluides, présenté le 14 avril 1834. Compt. Rend. Acad. Sci. Paris **17**, 1240–1243 (1843)
35. J.W. Barrett, W.B. Liu, Quasi-norm error bounds for the finite element approximation of a non-Newtonian flow. Numer. Math. **68**(4), 437–456 (1994)
36. F. Bassi, S. Rebay, A high-order accurate discontinuous finite element method for the numerical solution of the compressible Navier–Stokes equations. J. Comput. Phys. **131**(2), 267–279 (1997). https://doi.org/10.1006/jcph.1996.5572
37. F. Bassi, S. Rebay, G. Mariotti, S. Pedinotti, M. Savini, A high-order accurate discontinuous finite element method for inviscid and viscous turbomachinery flows, in *Proceedings of the 2nd European Conference on Turbomachinery Fluid Dynamics and Thermodynamics*, ed. by R. Decuypere, G. Dibelius (1997), pp. 99–109
38. F. Bassi, L. Botti, A. Colombo, D.A. Di Pietro, P. Tesini, On the flexibility of agglomeration based physical space discontinuous Galerkin discretizations. J. Comput. Phys. **231**(1), 45–65 (2012). https://doi.org/10.1016/j.jcp.2011.08.018
39. F. Bassi, L. Botti, A. Colombo, S. Rebay, Agglomeration based discontinuous Galerkin discretization of the Euler and Navier–Stokes equations. Comput. Fluids **61**, 77–85 (2012). https://doi.org/10.1016/j.compfluid.2011.11.002
40. F. Bassi, L. Botti, A. Colombo, Agglomeration-based physical frame dG discretizations: an attempt to be mesh free. Math. Models Methods Appl. Sci. **24**(8), 1495–1539 (2014). https://doi.org/10.1142/S0218202514400028
41. F. Bassi, L. Botti, A. Colombo, F. Massa. *p-Multilevel Solution Strategies for HHO Methods* (2019). https://imag.umontpellier.fr/~di-pietro/poems2019/lorenzo_botti.pdf
42. M. Bebendorf, A note on the Poincaré inequality for convex domains. Z. Anal. Anwendungen **22**(4), 751–756 (2003). https://doi.org/10.4171/ZAA/1170
43. L. Beirão da Veiga, G. Manzini, A virtual element method with arbitrary regularity. IMA J. Numer. Anal. **34**(2), 759–781 (2014). https://doi.org/10.1093/imanum/drt018
44. L. Beirão da Veiga, J. Droniou, M. Manzini, A unified approach for handling convection terms in finite volumes and mimetic discretization methods for elliptic problems. IMA J. Numer. Anal. **31**(4), 1357–1401 (2011). https://doi.org/10.1093/imanum/drq018
45. L. Beirão da Veiga, F. Brezzi, A. Cangiani, G. Manzini, L.D. Marini, A. Russo, Basic principles of virtual element methods. Math. Models Methods Appl. Sci. (M3AS) **199**(23), 199–214 (2013). https://doi.org/10.1142/S0218202512500492
46. L. Beirão da Veiga, F. Brezzi, L.D. Marini, Virtual elements for linear elasticity problems. SIAM J. Numer. Anal. **2**(51), 794–812 (2013). https://doi.org/10.1142/S0218202512500492
47. L. Beirão da Veiga, K. Lipnikov, G. Manzini. *The Mimetic Finite Difference Method for Elliptic Problems*. MS&A Modeling, Simulation and Applications, vol. 11 (Springer, Cham, 2014), pp. xvi+392. ISBN: 978-3-319-02662-6; 978-3-319-02663-3. https://doi.org/10.1007/978-3-319-02663-3
48. L. Beirão da Veiga, F. Brezzi, L.D. Marini, A. Russo, $H(\mathrm{div})$ and $H(\mathrm{curl})$-conforming VEM. Numer. Math. **133**, 303–332 (2016). https://doi.org/10.1007/s00211-015-0746-1
49. L. Beirão da Veiga, F. Brezzi, L.D. Marini, A. Russo, Mixed virtual element methods for general second order elliptic problems on polygonal meshes. ESAIM: Math. Model. Numer. Anal. **50**(3), 727–747 (2016). https://doi.org/10.1051/m2an/2015067
50. L. Beirão da Veiga, F. Brezzi, L.D. Marini, A. Russo, Virtual element method for general second-order elliptic problems on polygonal meshes. Math. Models Methods Appl. Sci. **26**(4), 729–750 (2016). https://doi.org/10.1142/S0218202516500160
51. L. Beirão da Veiga, A. Chernov, L. Mascotto, A. Russo, Basic principles of *hp* virtual elements on quasiuniform meshes. Math. Models Methods Appl. Sci. **26**(8), 1567–1598 (2016). https://doi.org/10.1142/S021820251650038X

52. L. Beirão da Veiga, C. Lovadina, G. Vacca, Divergence free virtual elements for the Stokes problem on polygonal meshes. ESAIM: Math. Model. Numer. Anal. (M2AN) **51**(2), 509–535 (2017). https://doi.org/10.1051/m2an/2016032
53. L. Beirão da Veiga, A. Russo, G. Vacca. *The Virtual Element Method with Curved Edges* (2017). Preprint arXiv. https://arxiv.org/abs/1711.04306
54. L. Beirão da Veiga, C. Lovadina, A. Russo, Stability analysis for the virtual element method. Math. Models Methods Appl. Sci. **27**(13), 2557–2594 (2017). https://doi.org/10.1142/S021820251750052X
55. M.F. Benedetto, S. Berrone, S. Pieraccini, S. Scialò, The virtual element method for discrete fracture network simulations. Comput. Methods Appl. Mech. Eng. **280**, 135–156 (2014). https://doi.org/10.1016/j.cma.2014.07.016
56. D. Boffi, D.A. Di Pietro, Unified formulation and analysis of mixed and primal discontinuous skeletal methods on polytopal meshes. ESAIM: Math. Model Numer. Anal. **52**(1), 1–28 (2018). https://doi.org/10.1051/m2an/2017036
57. D. Boffi, F. Brezzi, M. Fortin, *Mixed Finite Element Methods and Applications*. Springer Series in Computational Mathematics, vol. 44 (Springer, Heidelberg, 2013), pp. xiv+685. https://doi.org/10.1007/978-3-642-36519-5
58. D. Boffi, M. Botti, D.A. Di Pietro, A nonconforming high-order method for the Biot problem on general meshes. SIAM J. Sci. Comput. **38**(3), A1508–A1537 (2016). https://doi.org/10.1137/15M1025505
59. M. Bogovskiĭ, Theory of cubature formulas and the application of functional analysis to problems of mathematical physics. Novosibirsk, Russia: Akad. Nauk SSSR Sibirsk. Otdel. Inst. Mat. Chap. Solutions of some problems of vector analysis associated with the operators div and grad. Trudy Sem. S. L. Soboleva. **149**(1), 5–40 (1980)
60. F. Bonaldi, D.A. Di Pietro, G. Geymonat, F. Krasucki, A hybrid high-order method for Kirchhoff–Love plate bending problems. ESAIM: Math. Model Numer. Anal. **52**(2), 393–421 (2018). https://doi.org/10.1051/m2an/2017065
61. J. Bonelle, Compatible discrete operator schemes on polyhedral meshes for elliptic and Stokes equations. PhD Thesis, University of Paris-Est, 2014
62. J. Bonelle, A. Ern, Analysis of compatible discrete operator schemes for elliptic problems on polyhedral meshes. ESAIM: Math. Model. Numer. Anal. **48**, 553–581 (2014). https://doi.org/10.1051/m2an/2013104
63. J. Bonelle, A. Ern, Analysis of compatible discrete operator schemes for the Stokes equations on polyhedral meshes. IMA J. Numer. Anal. **35**, 1672–1697 (2015). https://doi.org/10.1093/imanum/dru051
64. J. Bonelle, D.A. Di Pietro, A. Ern, Low-order reconstruction operators on polyhedral meshes: application to compatible discrete operator schemes. Comput. Aided Geom. Des. **35–36**, 27–41 (2015). https://doi.org/10.1016/j.cagd.2015.03.015
65. L. Botti, Influence of reference-to-physical frame mappings on approximation properties of discontinuous piecewise polynomial spaces. J. Sci. Comput. **52**(3), 675–703 (2012). https://doi.org/10.1007/s10915-011-9566-3
66. L. Botti, D.A. Di Pietro, Numerical assessment of hybrid high-order methods on curved meshes and comparison with discontinuous Galerkin methods. J. Comput. Phys. **370**, 58–84 (2018). https://doi.org/10.1016/j.jcp.2018.05.017
67. L. Botti, A. Colombo, F. Bassi, *h*-multigrid agglomeration based solution strategies for discontinuous Galerkin discretizations of incompressible flow problems. J. Comput. Phys. **347**, 382–415 (2017). https://doi.org/10.1016/j.jcp.2017.07.002
68. M. Botti, D.A. Di Pietro, P. Sochala, A Hybrid High-Order method for nonlinear elasticity. SIAM J. Numer. Anal. **55**(6), 2687–2717 (2017). https://doi.org/10.1137/16M1105943
69. L. Botti, D.A. Di Pietro, J. Droniou, A Hybrid High-Order discretisation of the Brinkman problem robust in the Darcy and Stokes limits. Comput. Methods Appl. Mech. Eng. **341**, 278–310 (2018). https://doi.org/10.1016/j.cma.2018.07.004

70. L. Botti, D.A. Di Pietro, J. Droniou, A Hybrid High-Order method for the incompressible Navier–Stokes equations based on Temam's device. J. Comput. Phys. **376**, 786–816 (2019). https://doi.org/10.1016/j.jcp.2018.10.014
71. M. Botti, D.A. Di Pietro, A. Guglielmana, A low-order nonconforming method for linear elasticity on general meshes. Comput. Meth. Appl. Mech. Eng. **354**, 96–118 (2019). https://doi.org/10.1016/j.cma.2019.05.031
72. M. Botti, D.A. Di Pietro, O. Le Maître, P. Sochala. *Numerical Approximation of Poroelasticity with Random Coefficients Using Polynomial Chaos and High-Order Polyhedral Methods*. Comput. Meth. Appl. Mech. Eng. **361** (2020). https://dx.doi.org/10.1016/j.cma.2019.112736
73. M. Botti, D.A. Di Pietro, P. Sochala, A Hybrid High-Order discretization method for nonlinear poroelasticity. Comput. Meth. Appl. Math. (2019). https://doi.org/10.1515/cmam-2018-0142
74. F. Boyer, P. Fabrie. *Mathematical Tools for the Study of the Incompressible Navier–Stokes Equations and Related Models*. Applied Mathematical Sciences, vol. 183. (Springer, New York, 2013), pp. xiv+525. ISBN: 978-1-4614-5974-3; 978-1-4614-5975-0. https://doi.org/10.1007/978-1-4614-5975-0
75. S.C. Brenner, A nonconforming mixed multigrid method for the pure displacement problem in planar linear elasticity. SIAM J. Numer. Anal. **30**(1), 116–135 (1993). https://doi.org/10.1137/0730006
76. S.C. Brenner, Poincaré-Friedrichs inequalities for piecewise H^1 functions. SIAM J. Numer. Anal. **41**(1), 306–324 (2003). https://doi.org/10.1137/S0036142902401311
77. S.C. Brenner, R. Scott. *The Mathematical Theory of Finite Element Methods*. Texts in Applied Mathematics, vol. 15, 3rd edn. (Springer, New York, 2008), pp. xviii+397. ISBN: 978-0-387-75933-3. https://doi.org/10.1007/978-0-387-75934-0
78. S.C. Brenner, L.-Y. Sung, Linear finite element methods for planar linear elasticity. Math. Comp. **59**(200), 321–338 (1992). https://doi.org/10.2307/2153060
79. S.C. Brenner, L.-Y. Sung, Virtual element methods on meshes with small edges or faces. Math. Models Methods Appl. Sci. **28**(7), 1291–1336 (2018). https://doi.org/10.1142/S0218202518500355
80. S.C. Brenner, Q. Guan, L.-Y. Sung, Some estimates for virtual element methods. Comput. Methods Appl. Math. **17**(4), 553–574 (2017). https://doi.org/10.1515/cmam-2017-0008
81. H. Brézis. *Functional Analysis, Sobolev Spaces, and Partial Differential Equations*. Universitext (Springer, New York, 2011). ISBN: 978-0-387-70913-0
82. F. Brezzi, K. Lipnikov, M. Shashkov, Convergence of the mimetic finite difference method for diffusion problems on polyhedral meshes. SIAM J. Numer. Anal. **43**(5), 1872–1896 (2005). https://doi.org/10.1137/040613950
83. F. Brezzi, K. Lipnikov, V. Simoncini, A family of mimetic finite difference methods on polygonal and polyhedral meshes. Math. Models Methods Appl. Sci. **15**(10), 1533–1551 (2005)
84. F. Brezzi, B. Cockburn, L.D. Marini, E. Süli, Stabilization mechanisms in discontinuous Galerkin finite element methods. Comput. Methods Appl. Mech. Eng. **195**(25–28), 3293–3310 (2006). https://doi.org/10.1016/j.cma.2005.06.015
85. F. Brezzi, K. Lipnikov, M. Shashkov, Convergence of mimetic finite difference method for diffusion problems on polyhedral meshes with curved faces. Math. Models Methods Appl. Sci. **16**(2), 275–297 (2006). https://doi.org/10.1142/S0218202506001157
86. F. Brezzi, A. Buffa, K. Lipnikov, Mimetic finite differences for elliptic problems. M2AN Math. Model. Numer. Anal. **43**(2), 277–295 (2009). https://doi.org/10.1051/m2an:2008046
87. F. Brezzi, R.S. Falk, L.D. Marini, Basic principles of mixed virtual element methods. ESAIM Math. Model. Numer. Anal. **48**(4), 1227–1240 (2014). https://doi.org/10.1051/m2an/2013138
88. A. Buffa, C. Ortner, Compact embeddings of broken Sobolev spaces and applications. IMA J. Numer. Anal. **29**(4), 827–855 (2009). https://doi.org/10.1093/imanum/drn038
89. E. Burman, A. Ern, An unfitted hybrid high-order method for elliptic interface problems. SIAM J. Numer. Anal. **56**(3), 1525–1546 (2018). https://doi.org/10.1137/17M1154266

90. E. Burman, S. Claus, P. Hansbo, M.G. Larson, A. Massing, CutFEM: discretizing geometry and partial differential equations. Int. J. Numer. Methods Eng. **104**(7), 472–501 (2015). https://doi.org/10.1002/nme.4823
91. A. Cangiani, E.H. Georgoulis, P. Houston, *hp*-version discontinuous Galerkin methods on polygonal and polyhedral meshes. Math. Models Methods Appl. Sci. **24**(10), 2009–2041 (2014). https://doi.org/10.1142/S0218202514500146
92. A. Cangiani, Z. Dong, E.H. Georgoulis, P. Houston, *hp*version discontinuous Galerkin methods for advection-diffusion-reaction problems on polytopic meshes. ESAIM Math. Model. Numer. Anal. **50**(3), 699–725 (2016). https://doi.org/10.1051/m2an/2015059
93. A. Cangiani, V. Gyrya, G. Manzini, The nonconforming virtual element method for the Stokes equations. SIAM J. Numer. Anal. **54**(6), 3411–3435 (2016). https://doi.org/10.1137/15M1049531
94. A. Cangiani, Z. Dong, E.H. Georgoulis, P. Houston. *hp-Version Discontinuous Galerkin Methods on Polygonal and Polyhedral Meshes*. Springer Briefs in Mathematics (Springer, Cham, 2017), pp. viii+131. ISBN: 978-3-319-67671-5; 978-3-319-67673-9
95. A. Cangiani, G. Manzini, O.J. Sutton, Conforming and nonconforming virtual element methods for elliptic problems. IMA J. Numer. Anal. **37**(3), 1317–1354 (2017). https://doi.org/10.1093/imanum/drw036
96. P. Cantin, A. Ern, Vertex-based compatible discrete operator schemes on polyhedral meshes for advection-diffusion equations. Comput. Methods Appl. Math. **16**(2), 187–212 (2016). https://doi.org/10.1515/cmam-2016-0007
97. C. Carstensen, M. Feischl, M. Page, D. Praetorius, Axioms of adaptivity. Comput. Math. Appl. Int. J.**67**(6), 1195–1253 (2014). https://doi.org/10.1016/j.camwa.2013.12.003
98. K.L. Cascavita, J. Bleyer, X. Chateau, A. Ern, Hybrid discretization methods with adaptive yield surface detection for Bingham pipe flows. J. Sci. Comput. **77**(3), 1424–1443 (2018). https://doi.org/10.1007/s10915-018-0745-3
99. D. Castanon Quiroz, D.A. Di Pietro. *A Hybrid High-Order Method for the Incompressible Navier–Stokes Problem Robust for Large Irrotational Body Forces* (2020). https://dx.doi.org/10.1016/j.camwa.2019.12.005
100. P. Castillo, B. Cockburn, I. Perugia, D. Schötzau, An a priori error analysis of the local discontinuous Galerkin method for elliptic problems. SIAM J. Numer. Anal. **38**, 1676–1706 (2000). https://doi.org/10.1137/S0036142900371003
101. A. Cesmelioglu, B. Cockburn, W. Qiu, Analysis of a hybridizable discontinuous Galerkin method for the steady-state incompressible Navier–Stokes equations. Math. Comp. **86**(306), 1643–1670 (2017). https://doi.org/10.1090/mcom/3195
102. C. Chainais-Hillairet, J. Droniou, Convergence analysis of a mixed finite volume scheme for an elliptic-parabolic system modeling miscible fluid flows in porous media. SIAM J. Numer. Anal. **45**(5), 2228–2258 (2007). https://doi.org/10.1137/060657236
103. C. Chainais-Hillairet, J. Droniou, Finite-volume schemes for noncoercive elliptic problems with Neumann boundary conditions. IMA J. Numer. Anal. **31**(1), 61–85 (2011). https://doi.org/10.1093/imanum/drp009
104. C. Chainais-Hillairet, S. Krell, A. Mouton, Convergence analysis of a DDFV scheme for a system describing miscible fluid flows in porous media. Numer. Methods Partial Differential Equations **31**(3), 723–760 (2015). https://doi.org/10.1002/num.21913
105. F. Chave, D.A. Di Pietro, F. Marche, F. Pigeonneau, A Hybrid High-Order method for the Cahn–Hilliard problem in mixed form. SIAM J. Numer. Anal. **54**(3), 1873–1898 (2016). https://doi.org/10.1137/15M1041055
106. F. Chave, D.A. Di Pietro, L. Formaggia, A Hybrid High-Order method for Darcy flows in fractured porous media. SIAM J. Sci. Comput. **40**(2), A1063–A1094 (2018). https://doi.org/10.1137/17M1119500
107. F. Chave, D.A. Di Pietro, L. Formaggia, A Hybrid High-Order method for passive transport in fractured porous media. Int. J. Geomath. **10**, 12 (2019). https://doi.org/10.1007/s13137-019-0114-x

108. Y. Chen, B. Cockburn, Analysis of variable-degree HDG methods for convection-diffusion equations. Part II: semimatching nonconforming meshes. Math. Comp. **83**(285), 87–111 (2014). https://doi.org/10.1090/S0025-5718-2013-02711-1
109. W. Chen, Y. Wang, Minimal degree $H(\mathrm{curl})$ and $H(\mathrm{div})$ conforming finite elements on polytopal meshes. Math. Comp. **86**(307), 2053–2087 (2017). https://doi.org/10.1090/mcom/3152
110. E.B. Chin, J.B. Lasserre, N. Sukumar, Numerical integration of homogeneous functions on convex and nonconvex polygons and polyhedra. Comput. Mech. **56**(6), 967–981 (2015). https://doi.org/10.1007/s00466-015-1213-7
111. N.H. Christ, R. Friedberg, T.D. Lee, Weights of links and plaquettes in a random lattice. Nuclear Phys. B **210**(3), FS 6, 337–346 (1982). https://doi.org/10.1016/0550-3213(82)90124-9
112. E.T. Chung, B. Engquist, Optimal discontinuous Galerkin methods for the acoustic wave equation in higher dimensions. SIAM J. Numer. Anal. **47**(5), 3820–3848 (2009). https://doi.org/10.1137/080729062
113. P.G. Ciarlet, *The Finite Element Method for Elliptic Problems*. Classics in Applied Mathematics, vol. 40 . Reprint of the 1978 original [North-Holland, Amsterdam; MR0520174 (58 #25001)] (Society for Industrial and Applied Mathematics (SIAM), Philadelphia, 2002), pp. xxviii+530. ISBN: 0-89871-514-8
114. M. Cicuttin, D.A. Di Pietro, A. Ern, Implementation of discontinuous Skeletal methods on arbitrary-dimensional, polytopal meshes using generic programming. J. Comput. Appl. Math. **344**, 852–874 (2018). https://doi.org/10.1016/j.cam.2017.09.017
115. M. Cicuttin, A. Ern, S. Lemaire, A Hybrid High-Order method for highly oscillatory elliptic problems. Comput. Meth. Appl. Math. **19**, 723–748 (2018). https://doi.org/10.1515/cmam-2018-0013
116. B. Cockburn, Static condensation, hybridization, and the devising of the HDG methods, in *Building Bridges: Connections and Challenges in Modern Approaches to Numerical Partial Differential Equations*. Lecture Notes in Computational Science and Engineering, vol. 114 (Springer, Cham, 2016), pp. 129–177
117. B. Cockburn. *The Weak Galerkin Methods are Rewritings of the Hybridizable Discontinuous Galerkin Methods* (2018). https://arxiv.org/abs/1812.08146
118. B. Cockburn, G. Fu, Superconvergence by M-decompositions. Part II: construction of two-dimensional finite elements. ESAIM: Math. Model. Numer. Anal. **51**, 165–186 (2017). https://doi.org/10.1051/m2an/2016016
119. B. Cockburn, G. Fu, Superconvergence by M-decompositions. Part III: construction of three-dimensional finite elements. ESAIM: Math. Model. Numer. Anal. **51**, 365–398 (2017). https://doi.org/10.1051/m2an/2016023
120. B. Cockburn, C.-W. Shu, TVB Runge-Kutta local projection discontinuous Galerkin finite element method for conservation laws. II. General framework. Math. Comp. **52**(186), 411–435 (1989). https://doi.org/10.2307/2008474
121. B. Cockburn, C.-W. Shu, The Runge-Kutta local projection P^1-discontinuous-Galerkin finite element method for scalar conservation laws. RAIRO Modél. Math. Anal. Numér. **25**(3), 337–361 (1991). https://doi.org/10.1051/m2an/1991250303371
122. B. Cockburn, C.-W. Shu, The Runge-Kutta discontinuous Galerkin method for conservation laws. V. Multidimensional systems. J. Comput. Phys. **141**(2), 199–224 (1998). https://doi.org/10.1006/jcph.1998.5892
123. B. Cockburn, C.-W. Shu, The local discontinuous Galerkin method for time-dependent convection-diffusion systems. SIAM J. Numer. Anal. **35**(6), 2440–2463 (1998). https://doi.org/10.1137/S0036142997316712
124. B. Cockburn, S.Y. Lin, C.-W. Shu, TVB Runge-Kutta local projection discontinuous Galerkin finite element method for conservation laws. III. One-dimensional systems. J. Comput. Phys. **84**(1), 90–113 (1989). https://doi.org/10.1016/0021-9991(89)90183-6

125. B. Cockburn, S. Hou, C.-W. Shu, The Runge-Kutta local projection discontinuous Galerkin finite element method for conservation laws. IV. The multidimensional case. Math. Comp. **54**(190), 545–581 (1990). https://doi.org/10.2307/2008501
126. B. Cockburn, B. Dong, J. Guzmán, M. Restelli, R. Sacco, A hybridizable discontinuous Galerkin method for steady-state convection-diffusion-reaction problems. SIAM J. Sci. Comput. **31**(5), 3827–3846 (2009). https://doi.org/10.1137/080728810
127. B. Cockburn, J. Gopalakrishnan, R. Lazarov, Unified hybridization of discontinuous Galerkin, mixed, and continuous Galerkin methods for second order elliptic problems. SIAM J. Numer. Anal. **47**(2), 1319–1365 (2009). https://doi.org/10.1137/070706616
128. B. Cockburn, N.C. Nguyen, J. Peraire, A comparison of HDG methods for Stokes flow. J. Sci. Comput. **45**(1–3), 215–237 (2010). https://doi.org/10.1007/s10915-010-9359-0
129. B. Cockburn, D.A. Di Pietro, A. Ern, Bridging the Hybrid High-Order and hybridizable discontinuous Galerkin methods. ESAIM: Math. Model. Numer. Anal. **50**(3), 635–650 (2016). https://doi.org/10.1051/m2an/2015051
130. B. Cockburn, G. Fu, F.J. Sayas, Superconvergence by M-decompositions. Part I: general theory for HDG methods for diffusion. Math. Comp. **86**(306), 1609–1641 (2017). https://doi.org/10.1090/mcom/3140
131. L. Codecasa, R. Specogna, F. Trevisan, Symmetric positive-definite constitutive matrices for discrete eddy-current problems. IEEE Trans. Magn. **43**(2), 510–515 (2007)
132. L. Codecasa, R. Specogna, F. Trevisan, Base functions and discrete constitutive relations for staggered polyhedral grids. Comput. Methods Appl. Mech. Eng. **198**(9–12), 1117–1123 (2009). https://doi.org/10.1016/j.cma.2008.11.021
133. L. Codecasa, R. Specogna, F. Trevisan, A new set of basis functions for the discrete geometric approach. J. Comput. Phys. **19**(299), 7401–7410 (2010). https://doi.org/10.1016/j.jcp.2010.06.023
134. L. Codecasa, B. Kapidani, R. Specogna, F. Trevisan, Novel FDTD technique over tetrahedral grids for conductive media. IEEE Trans. Antennas Propag. **66**(10), 5387–5396 (2018)
135. R. Codina, Numerical solution of the incompressible Navier–Stokes equations with Coriolis forces based on the discretization of the total time derivative. J. Comput. Phys. **148**(2), 467–496 (1999). https://doi.org/10.1006/jcph.1998.6126
136. R. Codina, O. Soto, Finite element solution of the Stokes problem with dominating Coriolis force. Comput. Methods Appl. Mech. Eng. **142**(3–4), 215–234 (1997). https://doi.org/10.1016/S0045-7825(96)01141-3
137. Y. Coudière, F. Hubert, A 3D discrete duality finite volume method for nonlinear elliptic equations. SIAM J. Sci. Comput. **33**(4), 1739–1764 (2011)
138. M. Dauge, Stationary Stokes and Navier–Stokes systems on two- or three-dimensional domains with corners. I. Linearized equations. SIAM J. Math. Anal. **20**(1), 74–97 (1989). https://doi.org/10.1137/0520006
139. K. Deimling. *Nonlinear Functional Analysis*. (Springer, Berlin, 1985), pp. xiv+450. ISBN: 3-540-13928-1
140. D.A. Di Pietro, Cell centered Galerkin methods for diffusive problems. ESAIM: Math. Model. Numer. Anal. **46**(1), 111–144 (2012). https://doi.org/10.1051/m2an/2011016
141. D.A. Di Pietro, J. Droniou, $W^{s,p}$-approximation properties of elliptic projectors on polynomial spaces, with application to the error analysis of a Hybrid High-Order discretisation of Leray–Lions problems. Math. Models Methods Appl. Sci. **27**(5), 879–908 (2017). https://doi.org/10.1142/S0218202517500191
142. D.A. Di Pietro, J. Droniou, A Hybrid High-Order method for Leray–Lions elliptic equations on general meshes. Math. Comp. **86**(307), 2159–2191 (2017). https://doi.org/10.1090/mcom/3180
143. D.A. Di Pietro, J. Droniou, A third Strang lemma for schemes in fully discrete formulation. Calcolo **55**(40) (2018). https://doi.org/10.1007/s10092-018-0282-3
144. D.A. Di Pietro, A. Ern, Discrete functional analysis tools for discontinuous Galerkin methods with application to the incompressible Navier–Stokes equations. Math. Comp. **79**(271), 1303–1330 (2010). https://doi.org/10.1090/S0025-5718-10-02333-1

145. D.A. Di Pietro, A. Ern, *Mathematical Aspects of Discontinuous Galerkin Methods*. Mathématiques & Applications [Mathematics & Applications], vol. 69 (Springer, Berlin, 2012), pp. xviii+384. ISBN: 978-3-642-22979-4. https://doi.org/10.1007/978-3-642-22980-0
146. D.A. Di Pietro, A. Ern, A hybrid high-order locking-free method for linear elasticity on general meshes. Comput. Meth. Appl. Mech. Eng. **283**, 1–21 (2015). https://doi.org/10.1016/j.cma.2014.09.009
147. D.A. Di Pietro, A. Ern, Equilibrated tractions for the Hybrid High-Order method. C. R. Acad. Sci. Paris, Ser. I **353**, 279–282 (2015). https://doi.org/10.1016/j.crma.2014.12.009
148. D.A. Di Pietro, A. Ern, Hybrid high-order methods for variable-diffusion problems on general meshes. C. R. Acad. Sci. Paris, Ser. I **353**, 31–34 (2015). https://doi.org/10.1016/j.crma.2014.10.013
149. D.A. Di Pietro, A. Ern, Arbitrary-order mixed methods for heterogeneous anisotropic diffusion on general meshes. IMA J. Numer. Anal. **37**(1), 40–63 (2017). https://doi.org/10.1093/imanum/drw003
150. D.A. Di Pietro, S. Krell, Benchmark session: the 2D Hybrid High-Order method, in *Finite Volumes for Complex Applications VIII–Methods and Theoretical Aspects*, ed. by C. Cancès, P. Omnes (Springer, Cham, 2017), pp. 91–106
151. D.A. Di Pietro, S. Krell, A Hybrid High-Order method for the steady incompressible Navier–Stokes problem. J. Sci. Comput. **74**(3), 1677–1705 (2018). https://doi.org/10.1007/s10915-017-0512-x
152. D.A. Di Pietro, S. Lemaire, An extension of the Crouzeix–Raviart space to general meshes with application to quasi-incompressible linear elasticity and Stokes flow. Math. Comp. **84**(291), 1–31 (2015). https://doi.org/10.1090/S0025-5718-2014-02861-5
153. D.A. Di Pietro, S. Nicaise, A locking-free discontinuous Galerkin method for linear elasticity in locally nearly incompressible heterogeneous media. Appl. Num. Math. **63**, 105–116 (2013). https://doi.org/10.1016/j.apnum.2012.09.009
154. D.A. Di Pietro, R. Specogna, An a posteriori-driven adaptive Mixed High-Order method with application to electrostatics. J. Comput. Phys. **326**(1), 35–55 (2016). https://doi.org/10.1016/j.jcp.2016.08.041
155. D.A. Di Pietro, R. Tittarelli, An introduction to hybrid high-order methods, in *Numerical Methods for PDEs. State of the Art Techniques*, ed. by L. Formaggia, D.A. Di Pietro, A. Ern. SEMA-SIMAI, vol. 15 (Springer, Berlin, 2018). ISBN: 978-3-319-94675-7 (Print) 978-3-319-94676-4 (eBook). https://doi.org/10.1007/978-3-319-94676-4_4
156. D.A. Di Pietro, M. Vohralík, A review of recent advances in discretization methods, a posteriori error analysis, and adaptive algorithms for numerical modeling in geosciences. Oil Gas Sci. Technol. **69**(4), 701–730 (2014). https://doi.org/10.2516/ogst/2013158
157. D.A. Di Pietro, A. Ern, J.-L. Guermond, Discontinuous Galerkin methods for anisotropic semidefinite diffusion with advection. SIAM J. Numer. Anal. **46**(2), 805–831 (2008). https://doi.org/10.1137/060676106
158. D.A. Di Pietro, A. Ern, S. Lemaire, An arbitrary-order and compact-stencil discretization of diffusion on general meshes based on local reconstruction operators. Comput. Meth. Appl. Math. **14**(4), 461–472 (2014). https://doi.org/10.1515/cmam-2014-0018
159. D.A. Di Pietro, J. Droniou, A. Ern, A discontinuous-skeletal method for advection-diffusion-reaction on general meshes. SIAM J. Numer. Anal. **53**(5), 2135–2157 (2015). https://doi.org/10.1137/140993971
160. D.A. Di Pietro, A. Ern, S. Lemaire, A review of hybrid high-order methods: formulations, computational aspects, comparison with other methods, in *Building Bridges: Connections and Challenges in Modern Approaches to Numerical Partial Differential Equations*. Lecture Notes in Computational Science and Engineering, vol. 114 (Springer, Berlin, 2016). ISBN: 978-3-319-41638-0 (Print) 978-3-319-41640-3 (eBook). https://doi.org/10.1007/978-3-319-41640-3
161. D.A. Di Pietro, A. Ern, A. Linke, F. Schieweck, A discontinuous skeletal method for the viscosity-dependent Stokes problem. Comput. Meth. Appl. Mech. Eng. **306**, 175–195 (2016). https://doi.org/10.1016/j.cma.2016.03.033

162. D.A. Di Pietro, B. Kapidani, R. Specogna, F. Trevisan, An arbitrary order discontinuous skeletal method for solving electrostatics on general polyhedral meshes. IEEE Trans. Magn. **53**(6), 1–4 (2017). https://doi.org/10.1109/TMAG.2017.2666546
163. D.A. Di Pietro, J. Droniou, G. Manzini, Discontinuous Skeletal Gradient Discretisation methods on polytopal meshes. J. Comput. Phys. **355**, 397–425 (2018). https://doi.org/10.1016/j.jcp.2017.11.018
164. J.I. Diaz, F. de Thélin, On a nonlinear parabolic problem arising in some models related to turbulent flows. SIAM J. Math. Anal. **25**(4), 1085–1111 (1994). https://doi.org/10.1137/S0036141091217731
165. P. Dłotko, R. Specogna, Cohomology in 3D magneto-quasistatics modeling. Commun. Comput. Phys. **14**(1), 48–76 (2013)
166. K. Domelevo, P. Omnes, A finite volume method for the Laplace equation on almost arbitrary two-dimensional grids. M2AN Math. Model. Numer. Anal. **39**(6), 1203–1249 (2005)
167. O. Dorok, W. Grambow, L. Tobiska, Aspects of finite element discretizations for solving the Boussinesq approximation of the Navier–Stokes Equations, in *Notes on Numerical Fluid Mechanics: Numerical Methods for the Navier–Stokes Equations. Proceedings of the International Workshop held at Heidelberg, October 1993*, vol. 47 (1994), pp. 50–61
168. J. Douglas Jr., T. Dupont, Interior penalty procedures for elliptic and parabolic Galerkin methods, *Computing Methods in Applied Sciences*. Lecture Notes in Physics, vol. 58 (Springer, Berlin, 1976), pp. 207–216
169. J. Droniou, Finite volume schemes for fully non-linear elliptic equations in divergence form. M2AN Math. Model. Numer. Anal. **40**(6) (2006), 1069–1100 (2007). https://doi.org/10.1051/m2an:2007001
170. J. Droniou, Remarks on discretizations of convection terms in hybrid mimetic mixed methods. Netw. Heterog. Media **5**(3), 545–563 (2010). https://doi.org/10.3934/nhm.2010.5.545
171. J. Droniou, Finite volume schemes for diffusion equations: introduction to and review of modern methods. Math. Models Methods Appl. Sci. **24**(8), 1575–1619 (2014). https://doi.org/10.1142/S0218202514400041
172. J. Droniou, R. Eymard, A mixed finite volume scheme for anisotropic diffusion problems on any grid. Numer. Math. **105**, 35–71 (2006). https://doi.org/10.1007/s00211-006-0034-1
173. J. Droniou, R. Eymard, Study of the mixed finite volume method for Stokes and Navier–Stokes equations. Numer. Methods Partial Differential Equations **25**(1), 137–171 (2009). https://doi.org/10.1002/num.20333
174. J. Droniou, N. Nataraj, Improved L^2 estimate for gradient schemes and super-convergence of the TPFA finite volume scheme. IMA J. Numer. Anal. **38**(3), 1254–1293 (2018). https://doi.org/10.1093/imanum/drx028
175. J. Droniou, K.S. Talbot, On a miscible displacement model in porous media flow with measure data. SIAM J. Math. Anal. **46**(5), 3158–3175 (2014). https://doi.org/10.1137/130949294
176. J. Droniou, R. Eymard, T. Gallouët, R. Herbin, A unified approach to mimetic finite difference, hybrid finite volume and mixed finite volume methods. Math. Models Methods Appl. Sci. (M3AS) **20**(2), 1–31 (2010). https://doi.org/10.1142/S0218202510004222
177. J. Droniou, R. Eymard, T. Gallouët, R. Herbin, Gradient schemes: a generic framework for the discretisation of linear, nonlinear and nonlocal elliptic and parabolic equations. Math. Models Methods Appl. Sci. (M3AS) **23**(13), 2395–2432 (2013). https://doi.org/10.1142/S0218202513500358
178. J. Droniou, R. Eymard, T. Gallouët, C. Guichard, R. Herbin. *The Gradient Discretisation Method*. Mathematics & Applications, vol. 82 (Springer, Berlin, 2018), p. 511. ISBN: 978-3-319-79041-1 (Softcover) 978-3-319-79042-8 (eBook). https://doi.org/10.1007/978-3-319-79042-8
179. T. Dupont, R. Scott, Polynomial approximation of functions in Sobolev spaces. Math. Comp. **34**(150), 441–463 (1980). https://doi.org/10.2307/2006095
180. R.G. Durán, M.A. Muschietti, An explicit right inverse of the divergence operator which is continuous in weighted norms. Studia Math. **148**(3), 207–219 (2001). https://doi.org/10.4064/sm148-3-2

181. M.G. Edwards, C.F. Rogers, Finite volume discretization with imposed flux continuity for the general tensor pressure equation. Comput. Geosci. **2**(4), 259–290 (1998/1999). https://doi.org/10.1023/A:1011510505406
182. H. Egger, C. Waluga, *hp* analysis of a hybrid DG method for Stokes flow. IMA J. Numer. Anal. **33**(2), 687–721 (2013). https://doi.org/10.1093/imanum/drs018
183. A. Ern, J.-L. Guermond, *Theory and Practice of Finite Elements*. Applied Mathematical Sciences, vol. 159 (Springer, New York, 2004). https://doi.org/10.1007/978-1-4757-4355-5
184. A. Ern, A.F. Stephansen, M. Vohralík, Guaranteed and robust discontinuous Galerkin a posteriori error estimates for convection-diffusion-reaction problems. J. Comput. Appl. Math. **234**(1), 114–130 (2010). https://doi.org/10.1016/j.cam.2009.12.009
185. E. Erturk, T.C. Corke, C. Gökçöl, Numerical solutions of 2-D steady incompressible driven cavity flow at high Reynolds. Int. J. Numer. Meth. Fluids **48**, 747–774 (2005). https://doi.org/10.1002/fld.953
186. L.C. Evans, *Partial Differential Equations*. Graduate Studies in Mathematics, vol. 19 (American Mathematical Society, Providence, 1998), pp. xviii+662. ISBN: 0-8218-0772-2
187. R. Eymard, T. Gallouët, M. Ghilani, R. Herbin, Error estimates for the approximate solutions of a nonlinear hyperbolic equation given by finite volume schemes. IMA J. Numer. Anal. **18**(4), 563–594 (1998)
188. R. Eymard, T. Gallouët, R. Herbin, Finite volume methods, in *Hand-book of Numerical Analysis, Vol. VII*. Handbook of Numerical Analysis, vol. VII (North-Holland, Amsterdam, 2000), pp. 713–1020
189. R. Eymard, R. Herbin, J.-C. Latché, Convergence analysis of a colocated finite volume scheme for the incompressible Navier–Stokes equations on general 2D or 3D meshes. SIAM J. Numer. Anal. **45**(1), 1–36 (2007)
190. R. Eymard, T. Gallouët, R. Herbin, Discretization of heterogeneous and anisotropic diffusion problems on general nonconforming meshes. SUSHI: a scheme using stabilization and hybrid interfaces. IMA J. Numer. Anal. **30**(4), 1009–1043 (2010). https://doi.org/10.1093/imanum/drn084
191. R. Eymard, C. Guichard, R. Herbin, Small-stencil 3D schemes for diffusive flows in porous media. ESAIM Math. Model. Numer. Anal. **46**(2), 265–290 (2012). https://doi.org/10.1051/m2an/2011040
192. I. Faille, M. Thibaut, M.-C. Cacas, P. Havé, F. Willien, S. Wolf, L. Agélas, S. Pegaz-Fiornet, Modeling fluid flow in faulted basins. Oil Gas Sci. Technol. **69**(4), 529–553 (2014). https://doi.org/10.2516/ogst/2013204
193. G. Fichera, Asymptotic behaviour of the electric field and density of the electric charge in the neighbourhood of singular points of a conducting surface. Russ. Math. Surv. **30**(3), 107 (1975). http://stacks.iop.org/0036-0279/30/i=3/a=R03
194. M. Fortin, An analysis of the convergence of mixed finite element methods. RAIRO Anal. Numér. **11**(4), 341–354, iii (1977). https://doi.org/10.1051/m2an/1977110403411
195. F. Gastaldi, A. Quarteroni, On the coupling of hyperbolic and parabolic systems: analytical and numerical approach. Appl. Numer. Math. **6**(1–2), 3–31 (1989/90). Spectral multi-domain methods (Paris, 1988). https://doi.org/10.1016/0168-9274(89)90052-4
196. G.N. Gatica. *A Simple Introduction to the Mixed Finite Element Method: Theory and Applications*. SpringerBriefs in Mathematics (Springer, Cham, 2014), pp. xii+132. ISBN: 978-3-319-03694-6; 978-3-319-03695-3. https://doi.org/10.1007/978-3-319-03695-3
197. U. Ghia, K.N. Ghia, C.T. Shin, High-Re solutions for incompressible flow using the Navier–Stokes equations and a multigrid method. J. Comput. Phys. **48**, 387–411 (1982). https://doi.org/10.1016/0021-9991(82)90058-4
198. A. Gillette, A. Rand, C. Bajaj, Construction of scalar and vector finite element families on polygonal and polyhedral meshes. Comput. Methods Appl. Math. **16**(4), 667–683 (2016). https://doi.org/10.1515/cmam-2016-0019

199. V. Girault, P.-A. Raviart, *Finite Element Methods for Navier–Stokes Equations: Theory and Algorithms*. Springer Series in Computational Mathematics, vol. 5 (Springer, Berlin, 1986), pp. x+374. ISBN: 3-540-15796-4
200. R. Glowinski, *Numerical Methods for Nonlinear Variational Problems*. Springer Series in Computational Physics (Springer, New York, 1984), pp. xv+493. ISBN: 0-387-12434-9. https://doi.org/10.1007/978-3-662-12613-4
201. R. Glowinski, J. Rappaz, Approximation of a nonlinear elliptic problem arising in a non-Newtonian fluid flow model in glaciology. M2AN Math. Model. Numer. Anal. **37**(1), 175–186 (2003). https://doi.org/10.1051/m2an:2003012
202. T. Goudon, S. Krell, A DDFV scheme for incompressible Navier–Stokes equations with variable density, in *Finite Volumes for Complex Applications VII. Elliptic, Parabolic and Hyperbolic Problems*. Springer Proceedings in Mathematics & Statistics, vol. 78 (Springer, Cham, 2014), pp. 627–635. https://doi.org/10.1007/978-3-319-05591-6_62. https://doi.org.ezproxy.lib.monash.edu.au/10.1007/978-3-319-05591-6_62
203. T. Goudon, S. Krell, G. Lissoni, DDFV method for Navier–Stokes problem with outflow boundary conditions. Numer. Math. **142**(1), 55–102 (2019). https://doi.org/10.1007/s00211-018-1014-y. https://doi-org.ezproxy.lib.monash.edu.au/10.1007/s00211-018-1014-y
204. P.L. Gould, *Introduction to Linear Elasticity* (Springer, Berlin, 2013). ISBN: 978-1-4614-4832-7 (Print) 978-1-4614-4833-4 (eBook). https://doi.org/10.1007/978-1-4614-4833-4
205. P. Grisvard, *Singularities in Boundary Value Problems* (Masson, Paris, 1992)
206. R.J. Guyan, Reduction of stiffness and mass matrices. AIAA J. **3**(2), 380 (1965)
207. R. Herbin, F. Hubert, Benchmark on discretization schemes for anisotropic diffusion problems on general grids, in *Finite Volumes for Complex Applications V*, ed. by R. Eymard, J.-M. Hérard (Wiley, Hoboken, 2008), pp. 659–692
208. F. Hermeline, Une méthode de volumes finis pour les équations elliptiques du second ordre. C. R. Acad. Sci. Paris Sér. I Math. **326**(12), 1433–1436 (1998). https://doi.org/10.1016/S076-44442(98)80406-0
209. J.S. Hesthaven, T. Warburton, *Nodal Discontinuous Galerkin Methods*. Texts in Applied Mathematics. Algorithms, Analysis, and Applications, vol. 54 (Springer, New York, 2008), pp. xiv+500. ISBN: 978-0-387-72065-4. https://doi.org/10.1007/978-0-387-72067-8
210. H. Hiyoshi, K. Sugihara, Two generalizations of an interpolant based on Voronoi diagrams. Int. J. Shape Model. **5**(2), 219–231 (1999)
211. C.O. Horgan, Korn's inequalities and their applications in continuum mechanics. SIAM Rev. **37**(4), 491–511 (1995). https://doi.org/10.1137/1037123
212. R.A. Horn, F. Zhang. *The Schur Complement and Its Applications*, Chap. Basic Properties of the Schur Complement (Springer, Berlin, 2006), pp. 17–46
213. P. Houston, C. Schwab, E. Süli, Discontinuous *hp*-finite element methods for advection-diffusion-reaction problems. SIAM J. Numer. Anal. **39**(6), 2133–2163 (2002). https://doi.org/10.1137/S0036142900374111
214. B.M. Irons, Structural eigenvalue problems–elimination of unwanted variables. AIAA J. **3**(5), 961–962 (1965)
215. C. Johnson, J. Pitkäranta, An analysis of the discontinuous Galerkin method for a scalar hyperbolic equation. Math. Comp. **46**(173), 1–26 (1986) https://doi.org/10.2307/2008211
216. O.A. Karakashian, F. Pascal, A posteriori error estimates for a discontinuous Galerkin approximation of second-order elliptic problems. SIAM J. Numer. Anal. **41**(6), 2374–2399 (2003). https://doi.org/10.1137/S0036142902405217
217. R.B. Kellogg, J.E. Osborn, A regularity result for the Stokes problem in a convex polygon. J. Funct. Anal. **21**(4), 397–431 (1976)
218. K.Y. Kim, A posteriori error estimators for locally conservative methods of nonlinear elliptic problems. Appl. Numer. Math. **57**(9), 1065–1080 (2007). https://doi.org/10.1016/j.apnum.2006.09.010
219. R.A. Klausen, A.F. Stephansen, Convergence of multi-point flux approximations on general grids and media. Int. J. Numer. Anal. Model. **9**(3), 584–606 (2012)

220. L.I.G. Kovasznay, Laminar flow behind a two-dimensional grid. Math. Proc. Camb. Philos. Soc. **44**(1), 58–62 (1948). https://doi.org/10.1017/S0305004100023999
221. Y. Kuznetsov, S. Repin, Mixed finite element method on polygonal and polyhedral meshes, in *Numerical Mathematics and Advanced Applications* (Springer, Berlin, 2004), pp. 615–622
222. Y. Kuznetsov, K. Lipnikov, M. Shashkov, Mimetic finite difference method on polygonal meshes for diffusion-type problems. Comput. Geosci. **8**, 301–324 (2004)
223. R.J. Labeur, G.N. Wells, Energy stable and momentum conserving hybrid finite element method for the incompressible Navier–Stokes equations. SIAM J. Sci. Comput. **34**(2), A889–A913 (2012). https://doi.org/10.1137/100818583
224. P. Ladevèze, Comparaison de modèles de milieux continus. PhD Thesis, Université Pierre et Marie Curie (Paris 6), 1975
225. A. Lasis, E. Süli, *Poincaré-Type Inequalities for Broken Sobolev Spaces*. Technical Report NI03067-CPD (Isaac Newton Institute for Mathematical Sciences, Oxford, 2003). http://www.newton.ac.uk/preprints2003.html
226. P.D. Lax, A.N. Milgram, Parabolic equations, in *Contributions to the Theory of Partial Differential Equations*. Annals of Mathematics Studies, vol. 33 (Princeton University Press, Princeton, 1954), pp. 167–190
227. C. Le Potier, A finite volume method for the approximation of highly anisotropic diffusion operators on unstructured meshes, in *Finite Volumes for Complex Applications IV* (ISTE, London, 2005), pp. 401–412
228. C. Lehrenfeld, Hybrid discontinuous Galerkin methods for solving incompressible flow problems. PhD Thesis, Rheinisch-Westfälischen Technischen Hochschule Aachen, 2010
229. S. Lemaire, *Bridging the Hybrid High-Order and Virtual Element Methods*. (Oxford University Press, Oxford, 2018). https://hal.archives-ouvertes.fr/hal-01902962
230. J. Leray, J.-L. Lions, Quelques résulatats de Višik sur les problèmes elliptiques nonlinéaires par les méthodes de Minty-Browder. Bull. Soc. Math. France **93**, 97–107 (1965). http://www.numdam.org/item?id=BSMF_1965__93__97_0
231. Y. Li, I. Babuška, A convergence analysis of an h-version finite element method with high-order elements for two-dimensional elasto-plasticity problems. SIAM J. Numer. Anal. **34**(3), 998–1036 (1997). https://doi.org/10.1137/S0036142994263578
232. A. Linke, Collision in a cross-shaped domain—a steady 2d Navier–Stokes example demonstrating the importance of mass conservation in CFD. Comput. Methods Appl. Mech. Eng. **198**(41–44), 3278–3286 (2009). https://doi.org/10.1016/j.cma.2009.06.016
233. J.-L. Lions, E. Magenes, *Non-homogeneous Boundary Value Problems and Applications. Vol. I* (translated from the French by P. Kenneth). Die Grundlehren der mathematischen Wissenschaften, Band, vol. 181 (Springer, New York, 1972), pp. xvi+357
234. K. Lipnikov, G. Manzini, A high-order mimetic method on unstructured polyhedral meshes for the diffusion equation. J. Comput. Phys. **272**, 360–385 (2014). https://doi.org/10.1016/j.jcp.2014.04.021
235. K. Lipnikov, M. Shashkov, D. Svyatskiy, The mimetic finite difference discretization of diffusion problem on unstructured polyhedral meshes. J. Comput. Phys. **211**(2), 473–491 (2006). https://doi.org/10.1016/j.jcp.2005.05.028
236. A. Lozinski. *A Primal Discontinuous Galerkin Method with Static Condensation on Very General Meshes* (2018). Preprint. arXiv. https://arxiv.org/abs/1803.06846
237. S.G. Mikhlin, *Variational Methods in Mathematical Physics* (trans: T. Boddington; editorial introduction by L.I.G. Chambers). A Pergamon Press Book (The Macmillan, New York, 1964), pp. xxxii+582
238. G.J. Minty, On a "monotonicity" method for the solution of non-linear equations in Banach spaces. Proc. Natl. Acad. Sci. U. S. A. **50**, 1038–1041 (1963)
239. I. Moulitsas, G. Karypis, *MGridGen/ParmGridGen, Serial/Parallel Library for Generating Coase Meshes for Multigrid Methods*. Technical Report Version 1.0, University of Minnesota, Department of Computer Science/Army HPC Research Center (2001)
240. L. Mu, J. Wang, X. Ye, A weak Galerkin finite element method with polynomial reduction. J. Comput. Appl. Math. **285**, 45–58 (2015). https://doi.org/10.1016/j.cam.2015.02.001

241. L. Mu, J. Wang, X. Ye, Weak Galerkin finite element methods on polytopal meshes. Int. J. Numer. Anal. Model. **12**(1), 31–53 (2015)
242. C.-L.M.H. Navier, Mémoire sur les lois de l'équilibre et du mouvement des corps solides élastiques [read on 14 May, 1821]. Mém. Acad. R. Sci. **7**, 375–393 (1827)
243. J. Nečas. *Equations Aux Dérivées Partielles* (Presses de l'Université de Montréal, Montréal, 1965)
244. J.-C. Nédélec, Mixed finite elements in $\mathbf{R}^3$. Numer. Math. **35**(3), 315–341 (1980). https://doi.org/10.1007/BF01396415
245. N.C. Nguyen, J. Peraire, B. Cockburn, A hybridizable discontinuous Galerkin method for Stokes flow. Comput. Methods Appl. Mech. Eng. **199**(9–12), 582–597 (2010). https://doi.org/10.1016/j.cma.2009.10.007
246. J. Nitsche, Über ein Variationsprinzip zur Lösung von Dirichlet-Problemen bei Verwendung von Teilräumen, die keinen Randbedingungen unterworfen sind. Abh. Math. Sem. Univ. Hamburg **36**, 9–15 (1971). [Collection of articles dedicated to Lothar Collatz on his sixtieth birthday]. https://doi.org/10.1007/BF02995904
247. J. Nitsche, On Dirichlet problems using subspaces with nearly zero boundary conditions, in *The Mathematical Foundations of the Finite Element Method with Applications to Partial Differential Equations (Proceedings of Symposium, University of Maryland, Baltimore, 1972)* (Academic, New York, 1972), pp. 603–627
248. Y. Notay, An aggregation-based algebraic multigrid method. Electron. Trans. Numer. Anal. **37**(6), 123–146 (2010)
249. I. Oikawa, A hybridized discontinuous Galerkin method with reduced stabilization. J. Sci. Comput. **65**(1), 327–340 (2015). https://doi.org/10.1007/s10915-014-9962-6
250. L.E. Payne, H.F. Weinberger, An optimal Poincaré inequality for convex domains. Arch. Rational Mech. Anal. **5**, 286–292 (1960). https://doi.org/10.1007/BF00252910
251. S.D. Poisson, Mémoire sur les équations générales de l'équilibre et du mouvement des corps solides élastiques et des fluides. J. École Polytech. **13**, 1–174 (1831)
252. W. Prager, J.L. Synge, Approximations in elasticity based on the concept of function space. Quart. Appl. Math. **5**, 241–269 (1947). https://doi.org/10.1090/qam/25902
253. W. Qiu, K. Shi, A superconvergent HDG method for the incompressible Navier–Stokes equations on general polyhedral meshes. IMA J. Numer. Anal. **36**(4), 1943–1967 (2016). https://doi.org/10.1093/imanum/drv067
254. A. Quarteroni, A. Valli. *Numerical Approximation of Partial Differential Equations*. Springer Series in Computational Mathematics, vol. 23 (Springer, Berlin, 1994), pp. xvi+543. ISBN: 3-540-57111-6
255. P.A. Raviart, J.M. Thomas, A mixed finite element method for 2nd order elliptic problems, in *Mathematical Aspects of the Finite Element Method*, ed. by I. Galligani, E. Magenes (Springer, New York, 1977)
256. W.H. Reed, T.R. Hill, *Triangular Mesh Methods for the Neutron Transport Equation*. Technical Report LA-UR-73-0479 (Los Alamos Scientific Laboratory, Los Alamos, 1973). http://lib-www.lanl.gov/cgi-bin/getfile%7B?%7D00354107.pdf
257. D.L. Scharfetter, H.K. Gummel, Large signal analysis of a silicon Read diode. IEEE Trans. Elec. Dev. **16**, 64–77 (1969)
258. J. Schöberl, NETGEN An advancing front 2D/3D-mesh generator based on abstract rules. Comput. Vis. Sci. **1**(1), 41–52 (1997). https://doi.org/10.1007/s007910050004
259. V.A. Solonnikov, L^p-estimates for solutions of the heat equation in a dihedral angle. Rend. Mat. Appl. **21**, 1–15 (2001)
260. R. Specogna, Complementary geometric formulations for electrostatics. Int. J. Numer. Methods Eng. **86**(8), 1041–1068 (2011). https://doi.org/10.1002/nme.3089
261. R. Specogna, F. Trevisan, A discrete geometric approach to solving time independent Schrödinger equation. J. Comput. Phys. **230**(4), 1370–1381 (2011). https://doi.org/10.1016/j.jcp.2010.11.007
262. G.G. Stokes, On the theories of the internal friction of fluids in motion, and of the equilibrium and motion of elastic solids. Trans. Camb. Philos. Soc. **8**, 287–305 (1845)

263. G. Strang, Variational crimes in the finite element method, in *The Mathematical Foundations of the Finite Element Method with Applications to Partial Differential Equations (Proceedings of Symposium, University of Maryland, Baltimore, 1972)* (Academic, New York, 1972), pp. 689–710
264. G. Strang, G. Fix, *An Analysis of the Finite Element Method*, 2nd edn. (Wellesley-Cambridge Press, Wellesley, 2008), pp. x+402
265. N. Sukumar, A. Tabarraei, Conforming polygonal finite elements. Int. J. Numer. Methods Eng. **61**(12), 2045–2066 (2004). https://doi.org/10.1002/nme.1141
266. A. Tabarraei, N. Sukumar, Application of polygonal finite elements in linear elasticity. Int. J. Comput. Methods **3**(4), 503–520 (2006). https://doi.org/10.1142/S021987620600117X
267. A. Tabarraei, N. Sukumar, Extended finite element method on polygonal and quadtree meshes. Comput. Methods Appl. Mech. Eng. **197**(5), 425–438 (2007). https://doi.org/10.1016/j.cma.2007.08.013
268. R. Temam, *Navier–Stokes Equations: Theory and Numerical Analysis*. Studies in Mathematics and its Applications, With an appendix by F. Thomasset, Revised. vol. 2 (North-Holland, Amsterdam, 1979), pp. x+519. ISBN: 0-444-85307-3; 0-444-85308-1
269. P. Tesini, An h-multigrid approach for high-order discontinuous Galerkin methods. PhD Thesis, Università di Bergamo, 2008
270. A. Tiero, On Korn's inequality in the second case. J. Elast. **54**(3), 187–191 (1999). https://doi.org/10.1023/A:1007549427722
271. A.N. Tihonov, A.A. Samarskii, Homogeneous difference schemes. Ž. Vyčisl. Mat. i Mat. Fiz. **1**, 5–63 (1961)
272. E. Tonti, *On the Formal Structure of Physical Theories* (Istituto di Matematica del Politecnico di Milano, Milano, 1975)
273. E. Tonti, *The Mathematical Structure of Classical and Relativistic Physics*. Modeling and Simulation in Science, Engineering and Technology. A general classification diagram (Birkhäuser/Springer, New York, 2013), pp. xxxvi+514. ISBN: 978-1-4614-7421-0; 978-1-4614-7422-7. https://doi.org/10.1007/978-1-4614-7422-7
274. R. Verfürth, *A Review of a Posteriori Error Estimation and Adaptive Mesh-Refinement Techniques* (Teubner-Wiley, Stuttgart, 1996), p. 127. ISBN: 3-519-02605-8
275. M. Vohralík, B.I. Wohlmuth, Mixed finite element methods: implementation with one unknown per element, local flux expressions, positivity, polygonal meshes, and relations to other methods. Math. Models Methods Appl. Sci. (M3AS) **23**(5), 803–838 (2013). https://doi.org/10.1142/S0218202512500613
276. E.L. Wachspress, *A Rational Finite Element Basis*. Mathematics in Science and Engineering, vol. 114 (Academic [A subsidiary of Harcourt Brace Jovanovich, Publishers], New York, 1975), pp. xiii+331
277. J. Wang, X. Ye, A weak Galerkin mixed finite element method for second order elliptic problems. Math. Comp. **83**(289), 2101–2126 (2014)
278. M.F. Wheeler, An elliptic collocation-finite element method with interior penalties. SIAM J. Numer. Anal. **15**(1), 152–161 (1978). https://doi.org/10.1137/0715010
279. H. Whitney, *Geometric Integration Theory* (Princeton University Press, Princeton, 1957)

Model Index

D. A. Di Pietro, J. Droniou, *The Hybrid High-Order Method for Polytopal Meshes*, MS&A 19, https://doi.org/10.1007/978-3-030-37203-3

V

General Index

D. A. Di Pietro, J. Droniou, *The Hybrid High-Order Method for Polytopal Meshes*, MS&A 19, https://doi.org/10.1007/978-3-030-37203-3

R

S

Author Index

D. A. Di Pietro, J. Droniou, *The Hybrid High-Order Method for Polytopal Meshes*, MS&A 19, https://doi.org/10.1007/978-3-030-37203-3

O

P

Q

R

S

T

V

W

Y

Z

The manufacturer's authorised representative in the EU is Springer
Nature Customer Service Centre GmbH, Europaplatz 3, 69115 Heidelberg,
Germany. If you have any concerns regarding our products, please
contact ProductSafety@springernature.com

Printed and bound by CPI Group (UK) Ltd, Croydon, CR0 4YY
30/09/2026
02225646-0018